MW00760831

High Performance Computing in Science and Engineering, Garching/Munich 2009

Siegfried Wagner • Matthias Steinmetz •
Arndt Bode • Markus Michael Müller
Editors

High Performance Computing in Science and Engineering, Garching/Munich 2009

Transactions of the Fourth Joint HLRB and KONWIHR Review and Results Workshop, Dec. 8–9, 2009, Leibniz Supercomputing Centre, Garching/Munich, Germany

Springer

Editors
Siegfried Wagner
Universität Stuttgart
Institut für Aerodynamik
und Gasdynamik
Pfaffenwaldring 21
70550 Stuttgart
Germany
wagner@iag.uni-stuttgart.de

Arndt Bode
Markus Michael Müller
Leibniz-Rechenzentrum
Boltzmannstr. 1
85748 Garching b. München
Germany
arndt.bode@lrz.de
markus.michael.mueller@lrz.de

Matthias Steinmetz
Astrophysikalisches Institut Potsdam
An der Sternwarte 16
14482 Potsdam
Germany
msteinmetz@aip.de

Front cover figure: Visualization of the SGS turbulence energy density K_{sgs} in a 512^3 LES with 2/3 solenoidal and 1/3 compressive forcing modes

ISBN 978-3-642-13871-3 e-ISBN 978-3-642-13872-0
DOI 10.1007/978-3-642-13872-0
Springer Heidelberg Dordrecht London New York

Library of Congress Control Number: 2010932776

Mathematics Subject Classification (2010): 65-06, 65-XX, 68-04, 76-04, 81T25, 83-04, 85-04, 86-04, 92-04, 92Exx

Cover design: WMXDesign GmbH, Heidelberg

Printed on acid-free paper

Springer is part of Springer Science+Business Media (www.springer.com)

Preface

The Leibniz Supercomputing Centre (LRZ) and the Bavarian Competence Network for Technical and Scientific High Performance Computing (KONWIHR) publish in the present book results of numerical simulations facilitated by the High Performance Computer System in Bavaria (HLRB II) within the last two years. The papers were presented at the Fourth Joint HLRB and KONWIHR Review and Result Workshop in Garching on 8th and 9th December 2009, and were selected from all progress reports of projects that use the HLRB II. Similar to the workshop two years ago, the majority of the contributed papers belong to the area of computational fluid dynamics (CFD), condensed matter physics, astrophysics, chemistry, computer sciences and high-energy physics. We note a considerable increase of the user community in some areas: Compared to 2007, the number of papers increased from 6 to 12 in condensed matter physics and from 2 to 5 in high-energy physics. Bio sciences contributed only one paper in 2007, but four papers in 2009. This indicates that the area of application of supercomputers is continuously growing and entering new fields of research.

The year 2007 saw two major events of particular importance for the LRZ. First, after a substantial upgrade with dual-core processors the SGI Altix 4700 supercomputer reached a peak performance of more than 62 Teraflop/s. And second, the non-profit organization *Gauss Centre for Supercomputing e. V. (GCS)* was founded on April 13th. Its founding members were four representatives of the Jülich Research Centre, the Bavarian Academy of Sciences, and the University of Stuttgart, the heads of the three national supercomputing centres in Germany, namely the John von Neumann Institute for Computing/Jülich Supercomputing Centre (NIC/JSC), the LRZ (Munich), and the High Performance Computing Centre Stuttgart (HLRS), and the chairpersons of their respective steering committees. The foundation of the GCS marks an important strategic step towards creating a powerful HPC infrastructure at the Tier-0 level in Germany. Furthermore, on 17th April 2007 a European Tier-0 High Performance Computing Service was established at a meeting chaired by the German Federal Ministry of Education and Research in Berlin involving 14 partner nations in Europe. Today, 21 European nations form the Partnership for Advanced Computing in Europe (PRACE) with the goal to implement a Tier-0 HPC system infrastructure. Six nations plan to offer HPC cycles with a value of EUR 100 million each over a period of 5 years. These nations are called 'hosting members' of PRACE, one of which is Germany. Via the GCS the LRZ will thus be one of the PRACE Tier-0 centres in the future.

It is of utmost importance that users of the HLRB II gain further experience in the development of sophisticated numerical methods and advanced algorithms. The workshop in December 2009 has demonstrated remarkable progress in this respect over the past two years: The number of cores that were employed in several applications has increased remarkably, though it is still fairly below the approximately 10,000 cores available at the HLRB II. The next generation of supercomputers will have a number of cores that is an order of magnitude larger in order to allow for computations in the Petaflop/s range. Scaling applications to such a high number of cores will be a formidable challenge for the near future and will require major revisions in the simulation software, with a special emphasize on new parallelization techniques.

Besides the aforementioned founding role in PRACE, the LRZ is an active partner in the European HPC infrastructure project DEISA (Distributed European Infrastructure for Supercomputing Applications). Customers of the HLRB II thus also have access to other European High Performance Computers and can choose whatever architecture is best suited for their specific project. Analogously, HLRB II serves customers from other European countries for the same reason.

Acknowledgments

We gratefully acknowledge the continued support of the State of Bavaria, the German Research Foundation (DFG), the German Federal Ministry of Education and Research (BMBF), and other institutions in promoting high performance computing. We thank the referees and the HLRB Steering Committee for the review of the contributions and for their useful hints. Without their efforts it would not be possible to sustain the high scientific quality.

Finally, we would like to thank the members of LRZ's HPC Support Group — in particular Mrs. Cornelia Wendler — for their help during the workshop and its preparation.

Garching bei München,
April 2010

Siegfried Wagner
Matthias Steinmetz
Arndt Bode
Markus Michael Müller

Contents

Part I
Computer Science

Complexities of Performance Prediction for Bandwidth-Limited Loop Kernels on Multi-Core Architectures

Jan Treibig, Georg Hager, and Gerhard Wellein

Abstract The balance metric is a simple approach to estimate the performance of bandwidth-limited loop kernels. However, applying the method to modern multi-core architectures yields unsatisfactory results. This paper analyzes the influence of cache hierarchy design on performance predictions for bandwidth-limited loop kernels on current mainstream processors. We present a diagnostic model with improved predictive power, correcting the limitations of the simple balance metric. The importance of code execution overhead even in bandwidth-bound situations is emphasized.

1 Introduction

Many algorithms are limited by bandwidth, meaning that the memory subsystem cannot provide the data as fast as the arithmetic core could process it. One solution to this problem is to introduce multi-level memory hierarchies with low-latency and high-bandwidth caches, which exploit spatial and (hopefully) temporal locality in an application's data access pattern. In many scientific algorithms the bandwidth bottleneck is still severe, however. A popular way to estimate the performance in such situations is the memory bandwidth balance metric [1]. This metric can estimate loop kernel performance very well on vector systems and previous generations of cache-based processors. We will show why the balance model fails on recent processors (Intel Nehalem) To overcome these limitations we introduce a "diagnostic"

J. Treibig · G. Hager · G. Wellein
Regionales Rechenzentrum Erlangen, Friedrich-Alexander Universität Erlangen-Nürnberg, Martensstr. 1, D-91058 Erlangen, Germany
e-mail: jan.treibig@rrze.uni-erlangen.de
e-mail: georg.hager@rrze.uni-erlangen.de
e-mail: gerhard.wellein@rrze.uni-erlangen.de

S. Wagner et al. (eds.), *High Performance Computing in Science and Engineering, Garching/Munich 2009*, DOI 10.1007/978-3-642-13872-0_1,

performance model based on the real cache architectures and data transfer paths. The model is "diagnostic" in the sense that it abstracts many complexities in processor design and makes no claim to provide accurate predictions from a complete low-level coverage of architectural features. The application of the model is demonstrated on elementary data transfer operations (load, store and copy operations) and benchmarked on two x86-type test machines. In addition, as a prototype for many streaming algorithms we use the STREAM triad $\mathbf{A} = \mathbf{B} + \alpha * \mathbf{C}$, which matches the performance characteristics of many real algorithms [4].

This paper is organized as follows. Section 2 gives an overview on the microarchitectures and technical specifications of the test machines. In Section 3 we first present the original balance model as introduced in [1] and demonstrate its limitations, using a Jacobi relaxation solver. We then use a thorough analysis of cache hierarchies to develop our diagnostic model.

2 Experimental Test Bed

An overview of the machines used for benchmarking can be found in Table 1. As representatives of current x86 architectures we have chosen Intel's "Core 2 Quad" and "Core i7" processors. The cache group structure, i.e., which cores share caches of what size, is illustrated in Figure 1. For detailed information about microarchitecture and cache organization, see the Intel Optimization Handbook [3].

Table 1 Test machine specifications. The cacheline size is 64 bytes for all processors and cache levels

Microarchitecture	Intel Core 2	Intel Nehalem
Model	Core2 Q9550	Core i7 920
Clock [GHz]	2.83	2.67
Instr. throughput	4 μops/cycle	4 μops/cycle
Peak FP rate Mult+Add	4 flops/cycle	4 flops/cycle
L1 Cache	32 kB	32 kB
Parallelism	4 banks, dual ported	4 banks, dual ported
L2 Cache	2x6 MB (inclusive)	4x256 kB
L3 Cache (shared)	-	8 MB (inclusive)
Main Memory	DDR2-800	DDR3-1066
Channels	2	3
Memory clock [MHz]	800	1066
Theoretical bandwidth [GB/s]	12.8	25.6
STREAM triad 1 thread [GB/s]	6.8	13.9
STREAM triad node [GB/s]	7.1	22.2

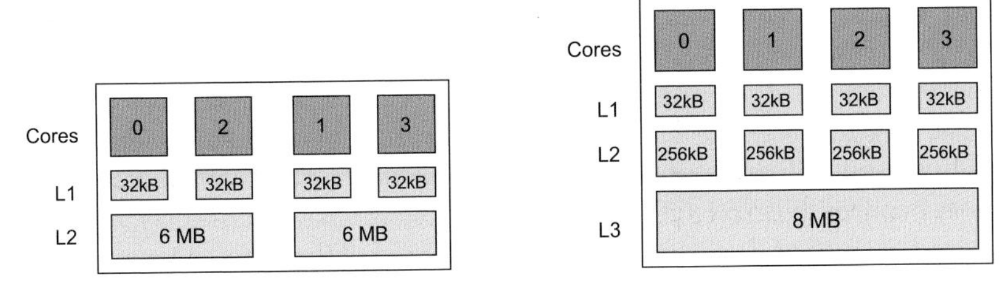

Fig. 1 Cache group structure of the multi-core architectures in the test-bed for Core 2 Q9550 (left) and Core i7 920 (right)

3 Bandwidth

3.1 Memory Bandwidth Balance Model

The balance metric [1] sets into relation the number of data words a processor can transfer from memory per time unit to the number of arithmetic operations it can execute per time unit. This relation is also referred to as "machine balance", B_M. The "algorithmic balance" B_A is the ratio between the number of words a given loop code needs per iteration to the number of arithmetic operations it performs with this data. The expected efficiency (fraction of peak performance) of the loop on a certain machine is then determined by the relationship $\ell = \frac{B_M}{B_A}$. If $\ell \geq 1$, performance is not limited by bandwidth any more and the balance model is not applicable.

3.2 Limitations of the Memory Balance Model

Using the example of a single-threaded 3D Jacobi smoother on the Nehalem architecture, this section will explain why performance predictions based on the balance metric can fail even if the algorithm is bandwidth-limited. The stencil scheme for the Jacobi smoother uses an eight-point update operation [2].

```
for(int i=1; i<n-1; i++) {
  for(int j=1; j<n-1; j++) {
    for(int k=1; k<n-1; k++) {
      tn[i][j][k] = frac * t[i][j][k] + frac
                  * (t[i-1][j][k] + t[i+1][j][k]
                  + t[i][j-1][k] + t[i][j+1][k]
                  + t[i][j][k-1] + t[i][j][k+1] );
    }
  }
}
```

This variant of the Jacobi smoother performs eight flops per update (six additions and two multiplication). The Nehalem node described in Section 2 was used for the benchmarks. It is important to note that peak performance on this architecture can only be achieved with an equal distribution between additions and multiplications and full usage of packed SSE instructions. The peak theoretical main memory bandwidth is 25.6 GBytes/s or 3.2 double precision GWords/s, resulting in a machine balance of $B_M = \frac{3.2\text{GWords/s}}{10.64\text{GFlops/s}} \approx 0.30$ Words/Flop. This value is often considered as an upper limit for memory-bound performance. At an algorithmic balance of $B_A = \frac{8\text{words}}{8\text{flops}} \approx 1.0$ Words/Flop a performance of 3192 MFlops/s can be expected. In certain cases this can be within reasonable range of real measurements. However, this is pure coincidence and caused by a cancellation of two effects: The large memory bandwidth of the Nehalem architecture as compared to L3 performance, and our ignorance towards the real runtime contributions and data streams that have to be sustained from memory. As will be shown in the following, care must be taken that the model is applied in a sensible way.

The machine balance based on peak properties considers upper limits which cannot be reached even in the theoretical case by the Jacobi algorithm. It is necessary to adapt the machine balance to reflect unavoidable architectural inefficiencies and the limitations of the algorithm under consideration. Hence, a more realistic estimate is to consider the peak performance for the present arithmetic instruction mix and the sustained single-threaded main memory performance of the STREAM triad benchmark (as listed in Table 1). We choose the STREAM benchmark because it achieves an upper limit for memory bandwidth that cannot be surpassed by any application code. This results in a new machine balance of $B_M = \frac{1.74\text{GWord/s}}{6.65\text{GFlops/s}} \approx 0.262$ Words/Flop. The initial algorithmic balance is based on the properties of the Jacobi stencil update, but can be significantly wrong on cache-based processors. Here the size of the grid determines how many streams have to be loaded from main memory. As we consider the balance model in the main memory domain, only the streams to memory must be taken into account.

Depending on cache capacity spatial blocking may be necessary to ensure the minimum of two streams to main memory [2]. The resulting algorithmic balance is 0.375 Words/Flop and the performance prediction based on the balance metric results in 4648 MFlops/s, while the measured performance is 3024 MFlops/s. This indicates that the simple balance model, while accurate for situations with high pressure on the memory subsystem, fails when many loads come from the outer-level cache.

A more realistic analysis of the Jacobi algorithm will be performed in Section 3.3.3, revealing the exact reasons for the balance metric failing in certain situations. The main assumption of the balance metric is that the contribution of in-cache data transfers and the execution of the instructions can be neglected against the time required to transfer the data from memory. The STREAM triad results (see Table 1) are used as the memory bandwidth component in the machine balance. The triad has a certain relation between runtime spent on-chip and runtime used to transfer data from main memory. In case of the Jacobi algorithm this relation is different and also depends on the ratio between the number of data streams toward main memory and cache.

Using the STREAM triad as the absolute limit for memory performance is only justified for kernels which have similar on-chip contributions to overall runtime, or on systems with very bandwidth-starved memory access. The latter is not the case on the Nehalem architecture as will be shown in Section 3.3.3. Note also that the model in the form presented here strictly applies only to single-threaded execution; multiple data streams to separate cores require substantial refinements.

3.3 Diagnostic Performance Model for Bandwidth-Limited Loop Kernels

This model proposes an approach to analytically predict the performance of bandwidth-limited algorithms in all memory hierarchy levels. The basic building block of a streaming algorithm is its computational kernel in the inner loop body. Since all loads and stores come from and go to L1 cache, the kernel's execution time on a cacheline basis is governed by the maximum number of L1 load and store accesses per cycle and the capability of the pipelined, superscalar core to execute instructions. All lower levels of the memory hierarchy are reduced to their bandwidth properties, with data paths and transfer volumes based on the real cache architecture. The minimum transfer size between memory levels is one cacheline.

Based on the transfer volumes and bandwidths, the total execution time per cacheline is obtained by adding all contributions from data transfers between caches and kernel execution times in L1. To lowest order, we assume that there is no access latency (i.e., all latencies can be effectively hidden by software pipelining and prefetching) and that the different components of overall execution time do not overlap.

It must be stressed that a correct application of this model requires an intimate knowledge of cache architecture and data paths. This information is available from processor manufacturers [3], but sometimes the level of detail is insufficient for fixing all parameters, and relevant information must be inferred from measurements.

3.3.1 Theoretical Analysis

In this section we substantiate the model outlined above by providing the necessary architectural details for current Intel processors. Using simple kernel loops, we derive performance predictions which will be compared to actual measurements in the following section. All results are given in CPU cycles per cacheline; if n streams are processed in the kernel, the number of cycles denotes the time required to process one cacheline per stream.

Basic data operations in L1 cache are limited by cache bandwidth, which is determined by the load and store instructions that can execute per cycle. The Intel

Table 2 Theoretical prediction of execution time contributions from different cache levels and memory for eight loop iterations (one cacheline per stream) on Core 2 Quad (A) and Core i7 (B) processors

	L1		L2		L3	Memory	
	A	B	A	B	B	A	B
Load	4	4	6	6	8	20	15
Store	4	4	8	8	12	36	26
Copy	4	4	10	10	16	52	36
Triad	8	8	16	16	24	72	51

cores can retire one 128-bit load and one 128-bit store in every cycle. L1 bandwidth is thus limited to 16 bytes per cycle if only loads (or stores) are used, and reaches its peak of 32 bytes per cycle only for a copy operation.

For load-only and store-only kernels, there is only one data stream, i.e., exactly one cacheline is processed at any time. With copy and stream triad kernels, this number increases to two and three, respectively. Together with the execution limits described above it is possible to predict the number of cycles needed to execute the instructions necessary to process one cacheline per stream (see the "L1" columns in Table 2).

L2 cache bandwidth as seen from the execution core is influenced by three factors: (i) the finite bus width between L1 and L2 cache for refills and evictions, (ii) the fact that *either* ALU access *or* L1 cache refill can occur at any one time, and (iii) the L2 cache access latency. Both architectures have a 256-bit bus connection between L1 and L2 and use a write back and write allocate strategy for stores. In case of an L1 store miss, the cacheline is first moved from L2 to L1 before it can be updated (write allocate). Together with its later eviction to L2, this results in an effective data transfer of 128 bytes per cacheline write miss update.

On Intel processors, a load miss incurs only a single cacheline transfer from L2 to L1, because the cache hierarchy is inclusive. The L2 cache of Core i7 is not strictly inclusive, but for the benchmarks covered here (no cacheline sharing and no reuse) an inclusive behavior was assumed due to the lack of detailed documentation about the L2 cache.

Up to the L2 level the model is the same for both processors. Overall execution time of the loop kernel on one cacheline per stream is the sum of (i) the time needed to transfer the cacheline(s) between L2 and L1 and (ii) the runtime of the loop kernel in L1 cache. Table 2 shows the different contributions for pure load, pure store, copy and triad operations. Looking at, e.g., the copy operation, the model predicts that only 6 cycles out of 10 can be used to transfer data from L2 to L1 cache. The remaining 4 cycles are spent with the execution of the loop kernel in L1. This explains the well-known performance breakdown for streaming kernels when data does not fit into L1 any more, although the nominal L1 and L2 bandwidths are identical (taking loads and stores into account; for loads alone, L2 can theoretically provide enough bandwidth for two cores). All results are included in the "L2" columns of Table 2.

Not much is known about the L3 cache architecture on Intel Core i7. It can be assumed that the bus width between the caches is 256 bits, which was confirmed by our measurements. Our model assumes a strictly inclusive cache hierarchy, in

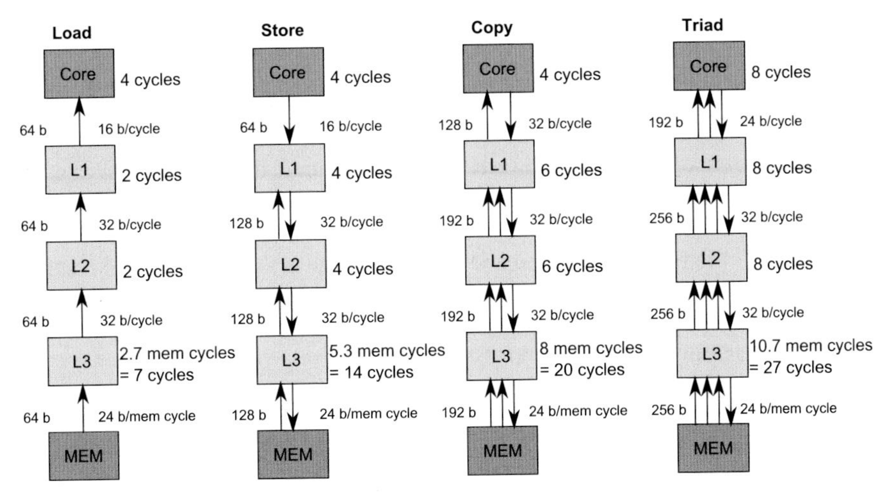

Fig. 2 Main memory performance model for Intel Core i7. There are separate buses connecting the different cache levels

which L3 cache is "just another level". Under these assumptions, the model predicts the required number of cycles in the same way as for the L2 case above. The "L3" column in Table 2 show the results.

If data resides in main memory, we again assume a strictly hierarchical (inclusive) data load. The cycles for main memory transfers are computed using the effective memory clock and bus width and are converted into CPU cycles. For consistency reasons, non-temporal ("streaming") stores were not used for the main memory regime (non-temporal stores bypass the cache hierarchy, reducing the pressure on the memory interface because no write allocate is required on store misses). Data transfer volumes and rates, and predicted cycles for a cacheline update are illustrated in Figure 2. They are also included in the "Memory" columns of Table 2.

3.3.2 Measurements

Measured cycles for a cacheline update, the ratio of predicted versus measured cycles, and the real and effective bandwidths are listed in Table 3 for Load, Store, Copy and Triad benchmarks. Here, "effective bandwidth" means the bandwidth available to the application, whereas "real bandwidth" refers to the actual data transfer taking place. For every layer in the hierarchy the working set size was chosen to fit into the appropriate level, but not into higher ones. The measurements confirm the predictions of the model well in the L1 regime.

Also the L2 results confirm the predictions. One exception is the store performance of the Intel Core i7, which is significantly better than the prediction. This indicates that the model does not describe the store behavior correctly. At the moment we have no additional information about the L2 behavior on Core i7 to solve this problem. The overhead for accessing the L2 cache with a streaming data access pattern scales with the number of involved cachelines, as can be derived from a comparison of the measured cacheline update cycles in Table 3 and the predictions in Table 2. The highest cost occurs on the Core 2 with 2 cycles per cacheline for the triad. Core i7 has a very low L2 access overhead of 0.5 cycles per cache line. Still, all Core i7 results must be interpreted with caution until the L2 behavior can be predicted correctly by a revised model. Both architectures are good at hiding cache latencies for streaming patterns.

On Core i7 the behavior with regard to the L3 cache is similar to the L2 results: The store result is better than the prediction, which influences all other test cases involving a store. A possible explanation could be an overlap between L3-L2 refill and execution from L1 (see below).

As for main memory access, one must distinguish between the classic frontside bus concept as used with all Core 2 designs, and the newer architectures with on-chip memory controller. The former has much larger overhead, which is why Core 2 shows mediocre efficiencies of around 60 %. The Core i7 shows results better than the theoretical prediction on all memory levels except L1. This might be caused either by a potential overlap between the different contributions (which is ignored by our model), or by deficiencies in the model caused by insufficient knowledge about the details of data paths inside the cache hierarchy.

Table 3 Benchmark results and comparisons to the predictions of the diagnostic model for Load, Store, Copy and Triad kernels

	L1				L2			
	Load	Store	Copy	Triad	Load	Store	Copy	Triad
Core 2 [%]	96.0	93.8	92.7	99.5	83.1	94.1	74.9	70.4
CL update	4.17	4.26	4.31	8.04	7.21	8.49	13.34	22.72
GB/s	43.5	42.5	84.1	67.7	25.1	42.7	40.7	31.9
eff. GB/s	-	-	-	-	-	21.3	27.2	23.9
Nehalem [%]	97.1	95.3	94.1	96.0	83.5	120.9	91.4	91.7
CL update	4.12	4.20	4.26	8.34	7.18	6.61	10.94	17.45
GB/s	41.3	40.5	79.8	61.2	23.7	51.5	46.7	39.0
eff. GB/s	-	-	-	-	-	25.7	31.1	29.3
	L3				Memory			
	Load	Store	Copy	Triad	Load	Store	Copy	Triad
Core 2 [%]					67.6	49,9	58.7	66.6
CL update					29.60	72.04	88.61	108.15
GB/s					6.1	5.0	6.1	6.7
eff. GB/s					-	2.5	4.1	5.0
Nehalem [%]	95.3	121.4	103.9	96.3	106.8	142.2	123	119.4
CL update	8.39	9.88	15.4	24.91	14.02	18.27	29.25	42.72
GB/s	20.3	34.4	33.2	27.3	12.1	18.6	17.4	15.9
eff. GB/s	-	17.2	22.1	20.5	-	9.3	11.6	11.9

3.3.3 Application to the Jacobi Smoother on Nehalem

An analysis of the Jacobi algorithm is shown in Figure 3. The prediction based on this analysis for the case with three data streams to memory is 2745 MFlops/s, while we measure 3024 MFlops/s. It must be noted that two important details of the Nehalem processor are not documented: First, measurements show that runtime is overestimated if an RFO transfer from L2 to L1 is assumed for each store miss. This issue was already taken into account in the model analysis in Figure 3, and indicates that the L1/L2 hierarchy is not accurately described by a simple inclusive structure. Second, the possibility to overlap data transfers in different hierarchy levels is neglected in our model. The measured performance indicates that our pre-

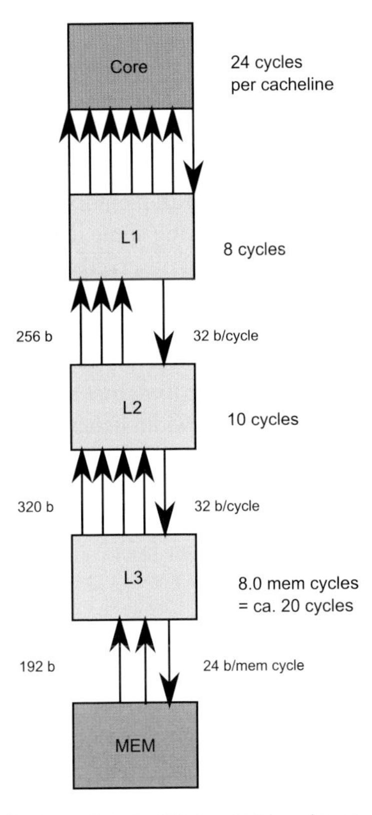

Fig. 3 An instruction analysis shows that for 3D Jacobi 24 cycles are needed to update one cacheline per data stream. Here it is assumed that four planes fit into the L3 cache and seven lines fit into the L1 cache. This results in the cacheline transfers shown. Every arrow is one 64-byte cacheline transfer. Bus width between cache levels is 256 bit, resulting in a data transfer rate of 32 bytes/cycle. For the data rate to memory, the memory clock and memory bus width is taken into account. This results in 8 cycles to transfer the necessary cachelines from L2 to L1 cache, 10 cycles to transfer the cachelines from L3 and L2 cache and 20 cycles to load the cachelines from main memory into the L3 cache. Total runtime for one cache line (per stream) update according to the model is 62 cycles

diction overestimates the time required by a cacheline update by six cycles. Since these predictions were based on bandwidth capabilities we must conclude that indeed the cache hierarchy allows an overlap between data transfers and execution of the instructions from L1 cache.

Finally, the comparison between the model predictions and the measured performance show that on the Nehalem architecture it is a good approximation for stream-oriented algorithms to neglect data access latencies.

4 Conclusion and Outlook

Using single-threaded, stream-based benchmarks and a Jacobi solver, we have demonstrated that performance modeling by a simple balance metric fails for current multi-core architectures (Intel Core 2, Nehalem), because it neglects the runtime contributions from different cache levels. The popular assumption that those contributions perfectly overlap with memory transfers is false. Based on these results we have introduced a diagnostic performance model, which led to a deeper understanding of runtime contributions in all memory hierarchy levels and finally to more accurate predictions. However, lacking some important details about the covered microarchitectures, up to now not all observed effects can yet be explained by the model.

Future work will include a thorough analysis of multi-threaded interleaving effects for shared caches and memory. We will substantiate our findings with performance counter measurements and develop tools that allow end users to gain a deeper understanding of their applications' bandwidth behavior.

Acknowledgements We thank Darren Kerbyson (LANL), Herbert Cornelius (Intel Germany), Michael Meier (RRZE), and Matthias Müller (ZIH) for fruitful discussions. This work was financially supported by the KONWIHR-II project "Omi4papps".

References

1. W. Schönauer: Scientific Supercomputing: Architecture and Use of Shared and Distributed Memory Parallel Computers. Self-edition, Karlsruhe (2000).
2. K. Datta, M. Murphy, V. Volkov, S. Williams, J. Carter, L. Oliker, D. Patterson, J. Shalf, K. Yelick: Stencil Computation Optimization and Auto-tuning on State-of-the-Art Multicore Architectures. In: ACM/IEEE (Ed.): Proceedings of the ACM/IEEE SC 2008 Conference (Supercomputing Conference '08, Austin, TX, Nov 15–21, 2008).
3. Intel Corporation: Intel 64 and IA-32 Architectures Optimization Reference Manual. (2008) Document Number: 248966–17.
4. W. Jalby, C. Lemuet and X. Le Pasteur: WBTK: a New Set of Microbenchmarks to Explore Memory System Performance for Scientific Computing. International Journal of High Performance Computing Applications, Vol. 18, 211–224 (2004).

Performance Limitations for Sparse Matrix-Vector Multiplications on Current Multi-Core Environments

Gerald Schubert, Georg Hager, and Holger Fehske

Abstract The increasing importance of multi-core processors calls for a reevaluation of established numerical algorithms in view of their ability to profit from this new hardware concept. In order to optimize the existent algorithms, a detailed knowledge of the different performance-limiting factors is mandatory. In this contribution we investigate sparse matrix-vector multiplications, which are the dominant operation in many sparse eigenvalue solvers. Two conceptually different storage schemes and computational kernels have been conceived in the past to target cache-based and vector architectures, respectively: compressed row and jagged diagonal storage. Starting from a series of microbenchmarks to single out performance limitations, we apply the gained insight to optimize sparse MVM implementations, reviewing serial and OpenMP-parallel performance on state-of-the-art multi-core systems.

1 Introduction

Large sparse eigensystems or systems of linear equations emerge from the mathematical description of many problems in physics, chemistry, and continuum dynamics. Solving those systems often requires multiplication of a sparse matrix with a vector as the dominant operation. Depending on the matrix dimension, its spar-

G. Schubert · G. Hager
Regionales Rechenzentrum Erlangen, Friedrich-Alexander Universität Erlangen-Nürnberg, Martensstr. 1, D-91058 Erlangen, Germany
e-mail: gerald.schubert@rrze.uni-erlangen.de
e-mail: georg.hager@rrze.uni-erlangen.de

H. Fehske
Ernst-Moritz-Arndt-Universität Greifswald, Institut für Physik, Felix-Hausdorff-Str. 6, D-17489 Greifswald, Germany
e-mail: fehske@physik.uni-greifswald.de

S. Wagner et al. (eds.), *High Performance Computing in Science and Engineering, Garching/Munich 2009*, DOI 10.1007/978-3-642-13872-0_2,

sity and the complexity of the remaining algorithm, thc fraction spent in the sparse matrix-vector multiplication (SpMVM) may easily constitute over 99% of total run time. An efficient implementation of the SpMVM on current hardware is thus fundamental, and complements the development of sophisticated high-level algorithms like, e.g., Krylov-subspace techniques [13], preconditioners [7, 10], and multi-grid methods [5, 6].

The exponential growth in available computational power enables larger problems, finer grids and more complex physics to be tackled on present-day supercomputers. Distributed-memory parallelization has always been instrumental in exploiting the full potential of sparse solvers on massively parallel architectures, but with the advent of multi-core processors the per-core performance has begun to decline, and parallel computers have developed a strongly hierarchical structure. Efficient implementations of the SpMVM must hence be adapted to the new shared-memory and cache topology complexities as well. These optimizations will be the focus of this contribution. We target various current systems, including the recent Intel "Nehalem" and AMD "Shanghai" processors.

In recent years, the performance of various SpMVM algorithms has been evaluated by several groups [4, 9, 17]. Covering different matrix storage formats and implementations on various types of hardware, they reviewed a more or less large number of publicly available matrices and reported on the obtained performance. Their results account for the crucial influence of the sparsity pattern on the obtained performance: Similar to a fingerprint, the positions of the non-zero entries in the matrix determine which storage and multiplication scheme is best suited and how fast this operation can be performed on a given platform.

We will take a different approach by isolating the individual contributions to the run time of the SpMVM on a generic level, without resorting (at first) to specific matrices. Based on the assumption that erratic, indirect memory access patterns are decisive for SpMVM performance, an in-depth investigation of latency and bandwidth effects, hardware prefetching, and the influence of non-local memory access on cache-coherent non uniform memory access (ccNUMA) architectures (as implemented in almost all current HPC platforms) will clearly reveal the performance-limiting aspects on the multi-core and multi-socket (node) level. A successful performance model will be predictive for the expected performance of various SpMVM implementations for a given matrix on the basis of its sparsity pattern, and give a hint to the respective optimal storage scheme.

This paper is organized as follows. In Section 2 we describe the dominant sparse matrix storage schemes, CRS and JDS, together with some promising refinements to JDS. Section 4 presents low-level benchmarks for SpMVM-like access patterns and performance results for standard and optimized storage layouts. In Section 5 we elaborate on multi-threaded SpMVM performance, both in multi-core and ccNUMA contexts. Finally, Section 6 gives a summary and outlook to future work.

2 Common Storage Schemes

There is a variety of options for storing sparse matrices [3]. Despite similar memory requirements, the chosen format may substantially impact the SpMVM performance. Prior knowledge about the sparsity pattern (symmetry features, dense subblocks, off-diagonal structures etc.) can be exploited to generate a specialized format, suitable only for one type of physical problem but optimal with respect to data access properties. In absence of such additional information one is limited to general storage schemes. Among these, the most popular one is the compressed row storage (CRS) format, which is also the basis of many sparse solver packages like, e.g., SuperLU [8] and PETSc [1]. In a nutshell, the CRS format stores the matrix in three (one-dimensional) arrays. For each non-zero element, its value (`val`) and column index (`col_idx`) are stored. The third array (`row_ptr`) holds the offset into `val` where the entries belonging to a new matrix row start. If the number of matrix rows is `N_r`, the SpMVM kernel code for CRS looks like this:

```
do i = 1, N_r
  do j = row_ptr(i), row_ptr(i+1) - 1
    resvec(i) = resvec(i) + val(j) * invec(col_idx(j))
  enddo
enddo
```

The success of the CRS format lies in its simplicity and the high computational performance on most cache-based architectures, which results from the inner loop having the characteristics of a sparse scalar product with an algorithmic balance of 10 bytes/floating point operation (Flop) [12, 14]. On vector systems, however, this format has the serious drawback of a short inner loop if the number of non-zeros per row is not at least a couple of hundreds.

To make SpMVM run well on vector processors, alternative formats have been conceived, the most prominent being "jagged diagonals storage" (JDS). In JDS, large vector lengths are given priority over minimizing data transfer: In a first step the rows and columns of the matrix are permuted such that the number of elements per row decreases with increasing row index. All further calculations are then performed in this permuted basis. In each row the permuted non-zero elements are shifted to the left. The resulting columns of decreasing length are called *jagged diagonals* and are stored consecutively in memory. There is one array for non-zero values (`val`) and one for column indices (`col_idx`), just as with CRS. A third array (`jd_ptr`) holds the starting offsets of the jagged diagonals in `val` and `col_idx`. If `N_j` is the number of diagonals, the SpMVM kernel code for JDS looks like this:

```
do diag = 1, N_j
  diagLen = jd_ptr(diag+1) - jd_ptr(diag)
  offset = jd_ptr(diag) - 1
  do i = 1, diagLen
    resvec(i) = resvec(i) + val(offset+i) * invec(col_idx(offset+i))
  enddo
enddo
```

Since the inner loop has the characteristics of a sparse vector triad with an algorithmic balance of 18 bytes/Flop [12], its computational intensity is smaller than with CRS, but due to the large loop length it is much better suited for vector processors and similar machines.

In view of the changing supercomputing landscape, where traditional cache-based and vector architectures are being replaced by hybrid, hierarchical systems comprising multi-core chips and accelerator hardware, it is worthwhile pursuing both CRS and JDS as potentially promising approaches to storing general sparse matrices. In order to adapt the JDS format to the bandwidth-starved situation on current multi-core architectures, we have to balance the vectorization aspect against the potential of cache and register reuse. Blocking ("NBJDS") and outer loop unrolling ("NUJDS") techniques seem most promising in this respect, and can reduce the algorithmic balance of JDS considerably, so that it eventually becomes equal to CRS balance [12]. Standard blocking cuts all jagged diagonals into blocks of a given size. Instead of updating the complete result vector for each jagged diagonal as a whole, only the elements of the current block are processed for all jagged diagonals that have entries in this block, to the effect that the corresponding part of the result vector remains in cache. During a block update, accessing a new jagged diagonal requires skipping entries in the `val` and `col_idx` arrays. In the "RBJDS" scheme we avoid this non-contiguous access by storing all elements of a block consecutively. In the outer-loop-unrolled "NUJDS" scheme each element of the result vector is updated by two (or more) jagged diagonals simultaneously. If the unrolling factor equals the number of jagged diagonals, this variant is identical to the CRS scheme, aside from working in the permuted basis.

While all optimizations on the plain JDS format reduce memory traffic for writing the result, the discontinuous access to the input vector remains a performance bottleneck. Depending on the sparsity of the matrix and the anticipated number of required SpMVMs it may be beneficial to use a more sophisticated initial ordering of the elements. In the "SOJDS" scheme the elements in a row are sorted such that within each column of a block the input vector is accessed with stride one (or as close as possible). Figure 1 summarizes the ordering of the `val` and `col_idx` arrays for the different storage schemes.

3 Test Bed

All benchmark tests were performed on three test systems:

Woodcrest: A two-socket UMA-type node based on 3 GHz dual-core Intel Xeon 5160 processors (Core 2 architecture) with a 1333 MHz frontside bus. The two cores in a socket share a 4 MB L2 cache. Measured STREAM Triad bandwidth is about 6.5 GB/s.

Shanghai: A two-socket ccNUMA-type node based on quad-core AMD Opteron 2378 processors at 2.4 GHz with two-channel DDR2-800 memory. All four cores

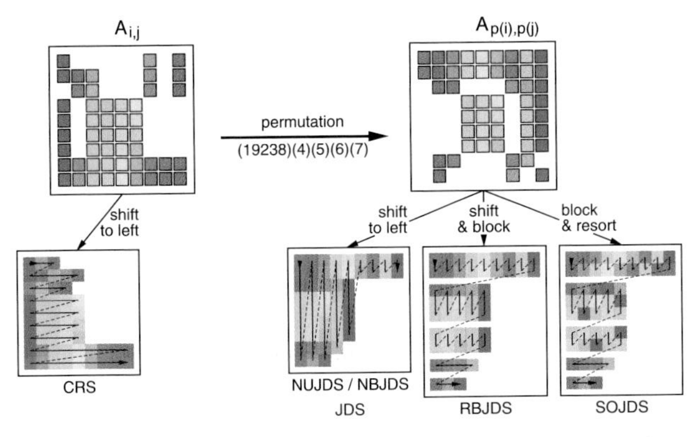

Fig. 1 Construction of the `val` and `col_idx` arrays for different storage schemes for a given matrix $A_{i,j}$. All JDS-flavored storage formats are based on the permuted matrix $A_{p(i),p(j)}$, with an additional sorting operation for SOJDS. The arrays within each sub-panel indicate the storage ordering of the elements. While the storage pattern for JDS, NUJDS and NBJDS is identical, these methods differ in the corresponding access pattern in the multiplication

on a chip share a 6 MB L3 cache. Measured STREAM Triad bandwidth is about 20 GB/s.

Nehalem: A two-socket ccNUMA-type node based on quad-core Intel "Core i7" (Nehalem) X5550 processors at 2.66 GHz with 3-channel DDR3-1333 memory. The four cores on a chip share an 8 MB L3 cache. Measured STREAM Triad bandwidth is about 35 GB/s.

For comparison we also obtained performance data on one node of HLRB-II.

4 Limitations of Serial Performance

As compared to the peak performance of a processor, the performance for a SpMVM is governed by memory access, and will usually be far less than 10% of peak. To single out different aspects of this performance reduction, in a first step we focus on basic operations that are the building blocks for the SpMVM in the following sections.

4.1 Basic Sparse Vector Operations

The inner loops of CRS and JDS differ in essence by the amount of data which is written in each iteration. While for CRS the result may be kept in a register and written to memory once after completing the inner loop, in the JDS case the whole result vector is written to memory `N_j` times. Concerning data to be read, the two building blocks are identical. Besides one (large) consecutive array, `val`, the elements of the

Table 1 Basic sparse vector operations, addition (ADD) and scalar product (SCP). Implementations are packed dense (PD), direct constant stride k (CS) and indirect addressing (IS/IR). For the latter we distinguish two cases: either constant stride in the index array for IS, that is `ind(i)=k*i` or random stride for IR. Then k is the mean stride

	ADD	SCP
PD	`s = s + B(i)`	`s = s + A(i) * B(i)`
CS	`s = s + B(k*i)`	`s = s + A(i) * B(k*i)`
IS / IR	`s = s + B(ind(i))`	`s = s + A(i) * B(ind(i))`

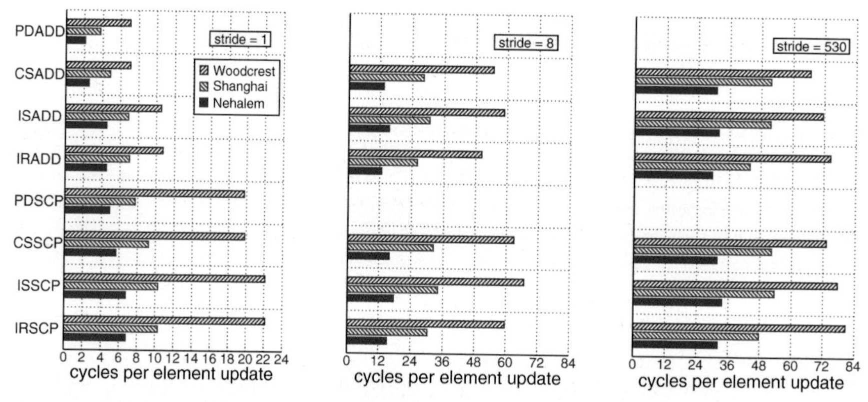

Fig. 2 Performance of the basic sparse operations given in Tab. 1 on different hardware architectures. Abstracting from the used hardware, performance is given in required cycles per non-zero element update. The considered strides correspond to dense packing, one entry per cache line and one entry per memory page. In the latter case we used stride $k = 530$ in order to circumvent the performance penalties by cache trashing effects for $k = 512$ [see Fig. 3(a)]

second array, `invec`, are determined indirectly by the indexing array `col_idx`. The access to `invec` comprises three performance penalties, which we will discuss by means of microbenchmarks. First, the indirect addressing of `invec` costs an additional 4 bytes per iteration for reading the (dense) indexing vector. Second, even if we replace the indirect addressing, `invec(col_idx(i))`, by a direct access with non-unit stride, `invec(k*i)`, unnecessary data is transfered since for each entry a whole cache line is read. Third, the indirect addressing may make efficient hardware prefetching almost impossible. Therefore, the relevant restriction for the performance might be latency, not memory bandwidth.

In an attempt to factor out the influence of loading the dense vector `val`, we abstract even more from the inner loop body by considering (indirect) sparse vector additions and scalar products. In Fig. 2 we give the required cycles per element update for the sparse scalar products and additions summarized in Tab. 1. Hereby the dense operations for stride $k = 1$ serve as a baseline. Note that the array lengths have been chosen such that the problem does not fit in any cache-level. Therefore the performance is limited by memory bandwidth, and the benefit of using packed SSE loads (`addpd`) instead of scalar ones (`addsd`) in the assembler code is marginal.[1]

[1] We do observe the expected factor of two for problem sizes which fit into cache.

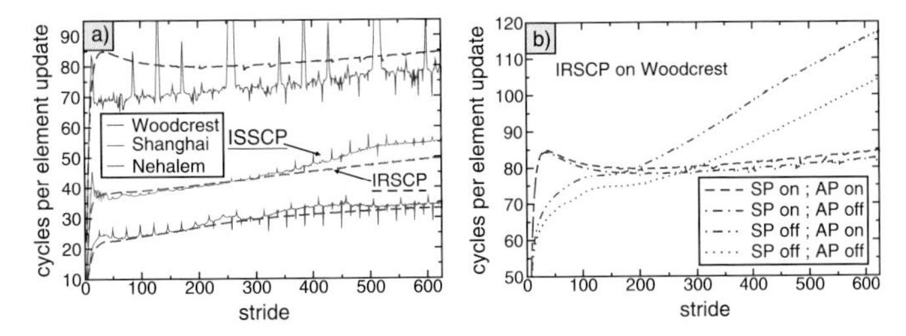

Fig. 3 Left panel: Performance of indirect scalar products for constant (ISSCP – thin solid lines) and random (IRSCP – thick dashed lines) strided access. Right panel: Performance for deactivated strided (SP) and adjacent cache line (AP) prefetcher for IRSCP on Woodcrest

The obtained performance using scalar loads is identical to the CSADD case which means that the additional multiplication by the stride k does not cause a performance penalty as long as the innermost loop is sufficiently unrolled. Indirect addressing causes an overhead of around 50% for ISADD, which is in accordance with the excess transferred data for the indexing array. Increasing the stride to $k = 8$ we observe the anticipated performance drop since for each element a whole cache line is read and seven eighths of it are useless in this case. This also holds for $k = 530$, but the number of translation lookaside buffer (TLB) misses is drastically larger for this stride, leading to a further penalty. A detailed comparison of the performance as a function of constant and random stride is given in Fig. 3(a) for ISSCP and IRSCP. For the IRSCP benchmark we generated a non-zero element for each entry of `invec` for which a drawn random number is smaller than a threshold given by the inverse mean stride $p = 1/k$.

Especially on the Woodcrest system, we encounter drastic performance drops for ISSCP at strides which are multiples of powers of two. On the other architectures these spikes also exist but are less severe. They can be attributed to an effective reduction of cache-size due to cache trashing. Randomizing the strides in IRSCP the distinct peaks disappear. A persisting feature is the bulge of reduced performance for $k < 25$, again most pronounced for the Woodcrest system. To pin down its origins, we deactivated the hardware prefetching mechanisms of the processor: (i) The adjacent cache line prefetch (AP), which loads two instead of one cache line on each miss, and (ii) the strided prefetcher (SP), which detects regular access patterns and tries to maintain continuous data streams, hiding latency. As shown in Fig. 3(b), turning off the strided prefetcher, the bulge vanishes and up to strides around $k \approx 200$ the performance stays better than for active prefetchers. However, this is not a general option since for larger strides SP is crucial for the performance and disabling it results in massive penalties. If we disable the adjacent cache line prefetcher, we observe an additional performance gain, either for activated or disabled SP, the latter being more pronounced. This can be attributed to the reduced memory traffic since now for each element only one cache line is fetched from memory instead of two. In general it is surprising how efficiently hardware prefetching works even with irregular strides.

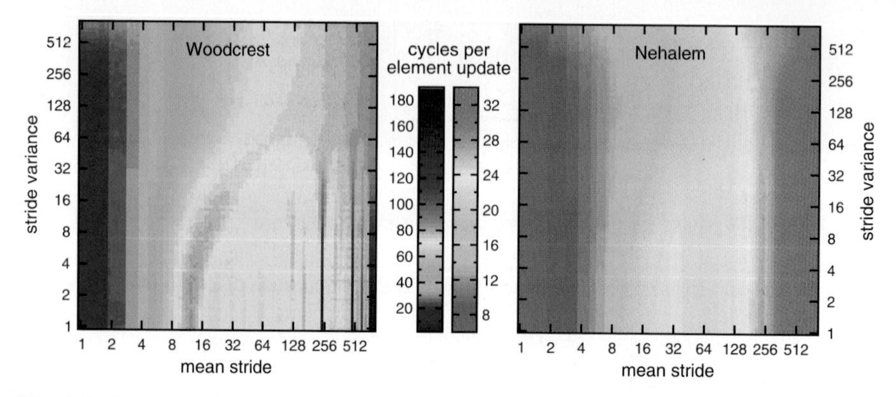

Fig. 4 Performance of IRSCP for an index vector with strides drawn from a Gaussian probability distribution with given mean value and variance of the stride

Up to now we have focused on inner loop kernels, for which `B` is accessed in a strictly monotonic order with random positive strides only. Clearly, the complete access pattern of the SpMVM kernels is not monotonic since negative strides arise when the first element of a row refers to an entry of `invec` that has a lower index than the last entry of the previous row. In order to investigate the influence of such backward jumps, we extend our IRSCP studies to Gaussian-distributed strides (Fig. 4). Fixing mean stride and variance of the distribution independently, we obtain a refined control over distribution characteristics allowing for negative strides provided the variance is large enough. The obtained results for the Woodcrest system underline the findings shown in Fig. 3(a). The peak structure in the ISSCP data is reproduced for small variances with only a minor effect on performance due to the stride jitter. For large stride variance the smooth performance variation agrees with the IRSCP data in Fig. 3(a) but the bulge observed there is missing. We therefore attribute the bulge to the peculiarities of the above stride distribution for which the grows with the mean stride as $k(k-1)$. Hence, the curves in Fig. 3(a) correspond to a cut along a tilted axis in Fig. 4, combining properties for small (large) variances at small (large) mean strides. On the Nehalem system the fine performance structures are missing and we only observe an overall decrease of the performance with increasing mean stride.

For Nehalem we also investigated the possibility of using non-temporal loads, which bypass the cache hierarchy when moving data from memory to registers. An instruction with the corresponding hint to the architecture (`movntdqa`) was introduced in SSE4.1 that is supported by the Nehalem processor. The non-temporal load may reduce the penalty connected with loading full cache lines of which only a fraction is actually used. However we could not see any effect, positive or negative, perhaps due to the hint-status of `movntdqa`. On current processors it behaves like a standard load.

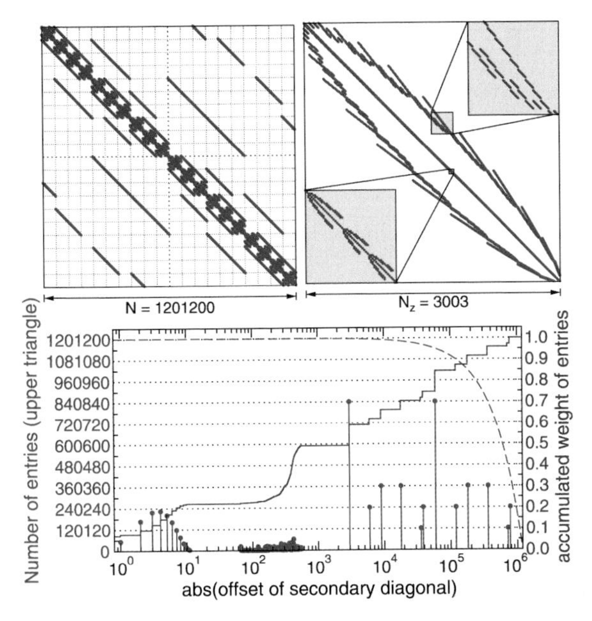

Fig. 5 Top left panel: Sparsity pattern of the Holstein-Hubbard Hamiltonian with dimension $N = 1201200$. Top right panel: detailed look at the matrix structure near the diagonal. This panel magnifies the matrix by a factor of 400, i.e. its linear size corresponds to $1/20$ of a dotted square in the left panel. In the further magnifying insets ($N_{zz} = 54, 250$) individual matrix entries can be distinguished. Bottom panel: Compressed information on the matrix structure in terms of offset and occupation of secondary diagonals. The dashed red line gives the total (zero and non-zero) number of elements for each secondary diagonal

4.2 Resulting Performance for SpMVM

The low-level results from the previous section can be applied to a wide variety of sparse numerical problems. Since the sparsity pattern of a matrix influences the data access characteristics in an essential way, this investigation will focus on a special physical problem, which is characterized by a Holstein-Hubbard Hamiltonian [16]. Two basic distributions of non-zero elements are present in the corresponding sparse matrix: From the sparsity pattern in Fig. 5 we see that a considerable fraction of the matrix entries is concentrated in (rather dense) secondary diagonals. The remaining elements are scattered evenly over a band containing several hundred secondary diagonals, impeding the use of multi-diagonal storage schemes. Such a split structure is typical for electron-phonon systems. In addition, since the eigenvalues of a Hamiltonian are real-valued physical observables the corresponding matrix is Hermitian (symmetric in the real case). From the point of view of data transfer in the SpMVM this potentially allows for a further optimization which we do not investigate here. Tracing back the matrix properties to the characteristics of our previous benchmarks, the information contained in the sparsity pattern is too detailed. In an attempt to compress this information we show in the lower panel of Fig. 5 the num-

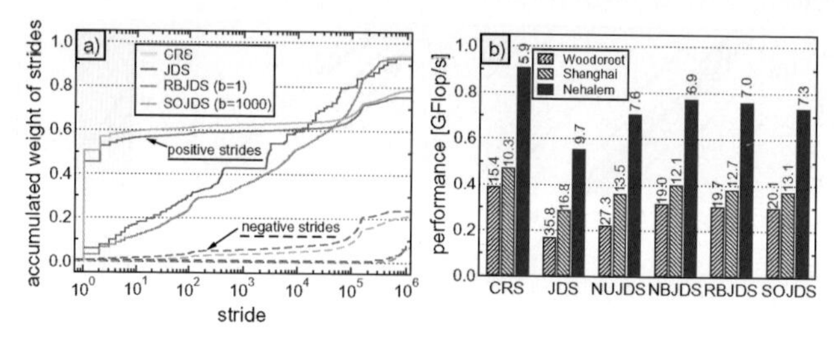

Fig. 6 Left panel: Distribution function of strides for the different storage schemes. The used block size for RBJDS (SOJDS) is 1 (1000). Solid (dashed) lines correspond to forward (backward) jumps in `invec`. Right panel: Serial performance for the complete SpMVM of the test matrix using the different kernels. For the blocked versions the block sizes have been chosen to optimize the performance (see Fig. 7). In addition to the performance in Flop/s the required cycles per element update are indicated for comparison with previous results

ber of non-zero elements as a function of their distance to the main diagonal together with the corresponding distribution function. From the latter we see that about 50% of the non-zero elements are contained in the twelve outermost secondary diagonals. Each of those is a potential candidate for special treatment by a dense storage scheme. Such hybrid implementations have been proposed in the literature [4], but we will not go into more detail on this aspect here.

The stride distribution in the SpMVM kernel is influenced by more aspects than just the secondary diagonal structure. A further important factor is the storage scheme [see Fig. 6(a)]. The stride distribution for CRS directly reflects the secondary diagonal structure, with negative offsets only once at the beginning of a new row. Since the matrix has on average 14 non-zero elements per row, the accumulated weight of backward jumps is around 7%. The positive strides are almost evenly distributed. The required permutation of the matrix for setting up the JDS format does not change this distribution significantly if the elements are still accessed row by row. This is the case for the RBJDS with block size $b = 1$. Increasing the block size the distribution gradually approaches the limiting case of JDS, corresponding to a block size of the matrix dimension. For the JDS almost 60% of the strides are smaller than 8 entries (64 bytes), enhancing cache line reuse. The drawback is a tripled amount of backward jumps. The structure of our test matrix leaves only limited potential for SOJDS and the optimized stride distribution does not differ significantly from JDS. Comparing the resulting performance of all schemes [Fig. 6(b)], the CRS outperforms all JDS-flavored variants. Interestingly, despite the improved memory access pattern, RBJDS and SOJDS cannot outperform NBJDS if an optimal block size is chosen in all cases. As expected, the benefit of the advanced blocked JDS formats is a wider range of suitable block sizes (see Fig. 7). This may be important with parallel execution if load balancing becomes an issue, which is however not the case for the considered test matrix.

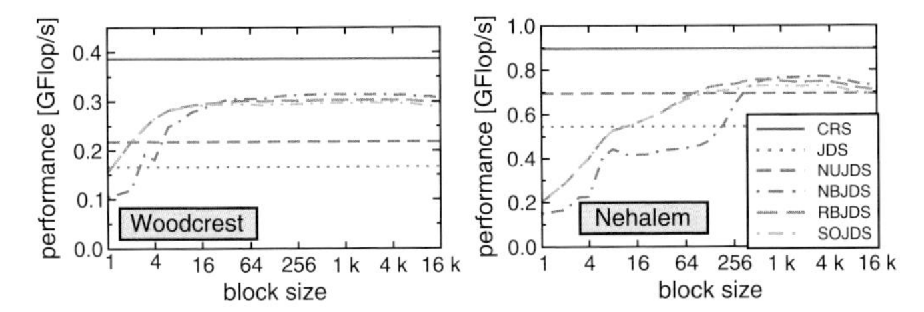

Fig. 7 Block size dependence of the serial performance for the SpMVM with the test matrix. For comparison also the performance for the schemes without blocking (CRS, JDS, NUJDS) is given. Data for Shanghai is not shown because the characteristics are very similar to Nehalem, although on a lower level

5 Shared-Memory Parallel SpMVM

In order to fully exploit the potential of multi-processor architectures, parallel execution is mandatory. In this work we consider OpenMP parallelization of our SpMVM code on the systems described in Sec. 3. Since all current commodity HPC systems are of the ccNUMA type, we distinguish clearly between intra-socket and inter-socket scaling behavior. Of course, pinning all threads to the physical cores is crucial for obtaining reliable performance data, because NUMA placement and bus contention effects are important factors for the performance of all bandwidth bound algorithms. While for a serial process this can be accomplished easily using the `taskset` command, pinning OpenMP threads is more involved. We implemented thread affinity by overloading the pthread library to ensure correct pinning of each thread created during program execution [2].

5.1 Intra-Socket Performance

In current multi-core designs like Nehalem and Shanghai, a single thread is not able to saturate the memory bandwidth of a socket. The reasons for this are complex [15] and beyond the scope of this contribution. For memory bound applications like the SpMVM it is therefore obligatory to use several threads per socket. In order to minimize parallelization overhead, it is however desirable to use the smallest number of threads that saturate the memory bandwidth. In Fig. 8 we compare the performance of the SpMVM for our test matrix versus the number of OpenMP threads on different architectures. While for Nehalem and Shanghai the performance scales up to three threads per socket, using a second thread per socket on Woodcrest results in no performance gain. As expected from the STREAM bandwidth numbers (see Sect. 3) Nehalem outperforms Shanghai by a factor of roughly two. The more recent AMD Istanbul processor has the same memory subsystem as Shanghai and is thus not expected to achieve better performance per socket.

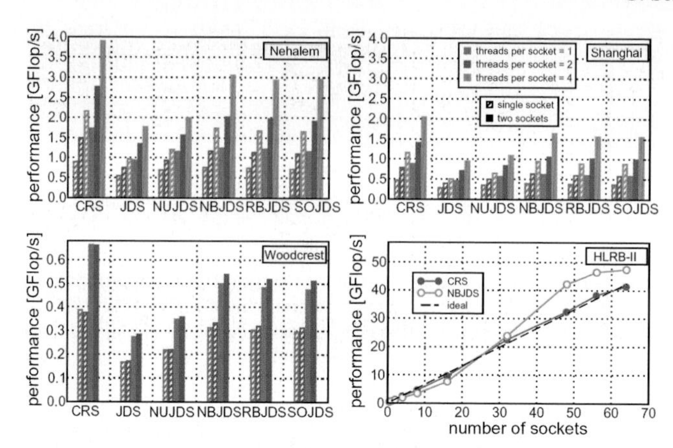

Fig. 8 Performance of OpenMP parallel SpMVM kernels for one and two sockets versus number of threads per socket. Data is based on static scheduling and a block size of 1000. In the lower right panel the measured (ideal) speedup on a HLRB II node is given by solid (dashed) lines. Two threads per node were used on this machine, and the CRS and NBJDS baselines were identical

5.2 Inter-Socket Performance

Using two sockets, performance on Woodcrest increases by about 60%, which is expected because this UMA-type FSB-based node architecture is known to show scalability problems [11]. The ccNUMA systems scale distinctly better, provided that proper page placement is implemented by employing first touch initialization. Apart from minor deficiencies which can be attributed to the access to `invec` our data reflects this feature (see Fig. 8). Placement of the input vector is imperfect by design as non-local accesses from other NUMA domains cannot be avoided. This is a property that strongly depends on the matrix structure, of course.

Choosing the correct loop scheduling is vital on ccNUMA nodes. Because of first touch placement, dynamic and guided scheduling are usually ruled out, except for strongly load imbalanced problems. In Fig. 9 we investigate the impact of different OpenMP schedulings and chunk sizes using eight threads on the Nehalem system. As expected, best overall performance is achieved with static scheduling and the CRS format. Small chunk sizes that lead to data blocks smaller than a memory page are hazardous for performance since page placement becomes random. For all JDS flavors, using large block sizes together with large OpenMP chunks leads to underutilized threads because the number of chunks becomes too small (top right sector in all blocked JDS panels).

In summary, the superior performance of the CRS is maintained even for dual socket ccNUMA systems. For the considered Hamiltonian the possible benefits of load balancing by guided or dynamic loop scheduling does not outweigh the penalties from disrupted NUMA locality.

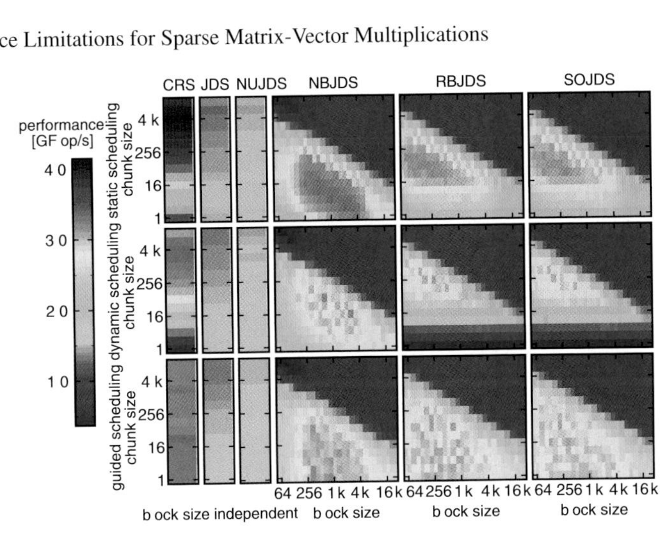

Fig. 9 Performance of the SpMVM kernels as a function of block size, scheduling policy and scheduling chunk size for 2×4 threads on Nehalem

5.3 HLRB-II Scalability

For comparison we performed scaling runs on a single node in the HLRB-II bandwidth partition (two cores per locality domain). In order to put the results shown in Fig. 8 into perspective, we must note that the whole matrix fits into the aggregated L3 caches of only 12 sockets. Thus we expect two effects: (i) superlinear speedup might be observed and (ii) JDS should have a significant performance advantage on many threads, because of the negative performance impact of short loops (as in the CRS kernel) on the Itanium2. Although CRS is slightly faster on 32 sockets or less, NBJDS dominates for large thread counts, which confirms our predictions.

6 Conclusion and Outlook

We analyzed the achievable performance of different SpMVM implementations on current multi-core multi-socket architectures. By a series of microbenchmarks, the basic operations of the SpMVM have been investigated in detail. The hardware prefetching mechanism of current x86 processors have been shown to work unexpectedly well, even for moderately random data access patterns.

Several SpMVM storage schemes have been benchmarked with respect to their serial and OpenMP-parallel performance, using a Hamiltonian matrix from solid state physics. In summary, the CRS format outperforms the best cache-blocked JDS schemes by at least 20%. Nevertheless we believe that JDS deserves further attention due to its long inner loop lengths, which might be advantageous on future processor and accelerator designs.

Future work will encompass a hardware counter analysis of SpMVM in order to get even more detailed information on its data access requirements. In view of massively parallel systems distributed memory and hybrid implementations will be thoroughly investigated.

Acknowledgements We thank J. Treibig and G. Wellein for valuable discussion and acknowledge financial support from KONWIHR II (project HQS@HPC II). We are also indebted to LRZ München for providing access to HLRB-II.

References

1. URL `http://www.mcs.anl.gov/petsc/petsc-as/`
2. URL `http://code.google.com/p/likwid`
3. Bai, Z., Demmel, J., Dongarra, J., Ruhe, A., van der Vorst, H.: Templates for the Solution of Algebraic Eigenvalue Problems: A Practical Guide. SIAM, Philadelphia (2000)
4. Bell, N., Garland, M.: Implementing sparse matrix-vector multiplication on throughput-oriented processors. In: Proceedings of Supercomputing Conference 2009 (2009). To be published
5. Brandt, A.: Guide to multigrid development. Lect. Notes Math. **960**, 220 (1981)
6. Briggs, W.L., Henson, V.E., McCormick, S.F.: A Multigrid Tutorial. SIAM, Philadelphia (2000)
7. Demmel, J.: Applied Numerical Linear Algebra. SIAM, Philadelphia (1997)
8. Demmel, J.W., Gilbert, J.R., Li, X.S.: An asynchronous parallel supernodal algorithm for sparse gaussian elimination. SIAM J. Matrix Analysis and Applications **20**(4), 915–952 (1999)
9. Goumas, G., Kourtis, K., Anastopoulos, N., Karakasis, V., Koziris, N.: Performance evaluation of the sparse matrix-vector multiplication on modern architectures. J. Supercomputing (2008). DOI 10.1007/s11227-008-0251-8
10. Greenbaum, A.: Iterative Methods for Solving Linear Systems. SIAM, Philadelphia (1997)
11. Hager, G., Stengel, H., Zeiser, T., Wellein, G.: Rzbench: Performance evaluation of current hpc architectures using low-level and application benchmarks. In: S. Wagner, M. Steinmetz, A. Bode, M. Brehm (eds.) High Performance Computing in Science and Engineering, Garching/Munich 2007. Transactions of the Third Joint HLRB and KONWIHR Status and Result Workshop, Dec 3-4, 2007, LRZ Garching, pp. 485–501 (2009). arXiv:0712.3389
12. Hager, G., Wellein, G.: Optimization techniques for modern high performance computers. Lect. Notes Phys. **739**, 731 (2008)
13. Saad, Y.: Iterative Methods for Sparse Linear Systems, 2 edn. SIAM, Philadelphia (2003). URL `http://www-users.cs.umn.edu/~saad/books.html`
14. Schönauer, W.: Scientific Supercomputing: Architecture and Use of Shared and Distributed Memory Parallel Computers. Self-edition (2000). URL `http://www.rz.uni-karlsruhe.de/~rx03/book`
15. Treibig, J., Hager, G., Wellein, G.: Complexities of performance prediction for bandwidth-limited loop kernels on multi-core architectures. In: High Performance Computing in Science and Engineering, Garching/Munich 2009, p. 3–12 (2010)
16. Wellein, G., Röder, H., Fehske, H.: Polarons and bipolarons in strongly interacting electron-phonon systems. Phys. Rev. B **53**, 9666 (1996)
17. Williams, S., Oliker, L., Vuduc, R., Shalf, J., Yelick, K., Demmel, J.: Optimization of sparse matrix-vector multiplications on emerging multicore platforms. Parallel Comput. **35**, 178 (2009)

waLBerla: Optimization for Itanium-based Systems with Thousands of Processors

S. Donath, J. Götz, C. Feichtinger, K. Iglberger, and U. Rüde

Abstract Performance optimization is an issue at different levels, in particular for computing and communication intensive codes like free surface lattice Boltzmann. This method is used to simulate liquid-gas flow phenomena such as bubbly flows and foams. Due to a special treatment of the gas phase, an aggregation of bubble volume data is necessary in every time step. In order to accomplish efficient parallel scaling, the all-to-all communication schemes used up to now had to be replaced with more sophisticated patterns that work in a local vicinity. With this approach, scaling could be improved such that simulation runs on up to 9 152 processor cores are possible with more than 90% efficiency. Due to the computation of surface tension effects, this method is also computational intensive. Therefore, also optimization of single core performance plays a tremendous role. The characteristics of the Itanium processor require programming techniques that assist the compiler in efficient code vectorization, especially for complex C++ codes like the waLBerla framework. An approach using variable length arrays shows promising results.

1 Introduction

Applications from computational fluid dynamics are often very demanding in terms of computing resources. The lattice Boltzmann method (LBM) is known as an alternative to Navier-Stokes, promising high flexibility for code implementation as well as extension to complex application schemes. A free surface extension to the LBM [8, 13] enables simulation of bubbly flows and even foams. A challenging industrial application is the simulation of the liquid water transport in the porous medium of a fuel cell (see Sect. 5), because a fine resolution and thus a large amount of memory is required, as it is only available on a large-scale system like the

S. Donath · J. Götz · C. Feichtinger · K. Iglberger · U. Rüde
Computer Science Department 10 (System Simulation), University of Erlangen-Nuremberg, Erlangen, Germany
e-mail: Stefan.Donath@informatik.uni-erlangen.de

S. Wagner et al. (eds.), *High Performance Computing in Science and Engineering, Garching/Munich 2009*, DOI 10.1007/978-3-642-13872-0_3,

HLRB 2 [6]. In order to achieve reasonable parallel efficiency on several thousand processes consuming tera-bytes of memory, the parallel implementation of the free surface extension has been optimized. While the former version [11] is based on all-to-all communication schemes, the new algorithm uses a more local approach (see Sect. 4). It has been implemented in the *waLBerla* framework which facilitates implementation of parallel applications based on the LBM.

Since HLRB 2 is an Itanium 2-based machine, the C++ code of waLBerla had to be optimized for the EPIC architecture. By rewriting the kernel routines to assist the compiler in code vectorization, a tremendous improvement of single core performance has been obtained (see Sect. 6).

2 The waLBerla Framework

The *waLBerla* framework [2] serves as a platform for sophisticated extensions to the LBM. Besides blood flow and Brownian motion, most prominent examples are multi-component flows, multi-phase or free surface flows, and flows with moving objects. Besides, waLBerla features basic tools for data management and parallelization. Most recently, large-scale simulation of fluid-structure interaction (FSI) has been accomplished (Fig. 1, [3]). With the successful integration of large-scale free surface method, simulation in fuel cells (see Sect. 5) is feasible, and possibly the combination with FSI will be achieved in future.

In order to achieve high flexibility, waLBerla provides concepts for easy integration of user-defined extensions: The computational domain is split into so-called *patches* that may have different features implemented by the user. Since several patches can be assigned to a single process, waLBerla internally realizes data exchange by either process-local or MPI communication. Each data exchange consists strictly of a pair of send and blocking receive, in exactly this order. For each iteration of the time loop, the sequence control executes a user-defined order of so-called *sweep* routines that for example traverse the grid and manipulate data. Before each

Fig. 1 Parallel simulation of fluid-structure interaction. Visualization of many objects in a funnel, with liquid flow from the bottom upwards (left). Right image illustrates the boundaries of the 128 processes involved in fluid simulation by transparent planes

sweep, communication of required data takes place. The sweep function can also request a reiteration before the next sweep is executed.

3 Lattice Boltzmann Method

The basic advantage of the LBM is the locality of the method: For each operation on a single discretization point, only data from the local neighborhood is needed, which is the reason why the LBM is said to be easy to parallelize. While this is true for the basic single phase LBM, any more sophisticated extension raises more challenges. Especially for the free surface method, which is explained in this section, achieving locality is complex and results in a considerable communication to computation ratio [11].

3.1 Single Phase Lattice Boltzmann Method

The LBM [12] is a mesoscopic CFD simulation scheme that is based on kinetic fluid theory and uses a stencil-based approach on a Cartesian grid, to solve time-dependent quasi-incompressible flows in continuum mechanics. In tensor notation, the time and space discretization of this model is given by

$$f_\alpha(x_i + e_{\alpha,i}\delta t, t+\delta t) - f_\alpha(x_i,t) = -\frac{\delta t}{\tau}\left[f_\alpha(x_i,t) - f_\alpha^{(eq)}(\rho(x_i,t), u_i(x_i,t))\right], \quad (1)$$

where $f_\alpha(x_i,t)$ is the discrete particle distribution function (PDF) which is defined as the expected fraction of particles in the volume δx^3 located at the lattice position x_i with the lattice velocity $e_{\alpha,i}$. For the sake of simplicity, quantities depending on x_i and t will be written without their dependencies, e.g. $f_\alpha = f_\alpha(x_i,t)$. The D3Q19 model [9], illustrated in Fig. 2, uses $\alpha \in [0,18]$ lattice directions, having dimension i within $[0,2]$. For the isothermal case, the equilibrium distribution

$$f_\alpha^{(eq)}(\rho, u_i) = \rho \cdot w_\alpha \cdot \left[1 + \frac{1}{c_s^2}(e_{\alpha,i} \cdot u_i) + \frac{1}{2c_s^4}(e_{\alpha,i} \cdot u_i)^2 - \frac{1}{2c_s^2}u_i^2\right] \quad (2)$$

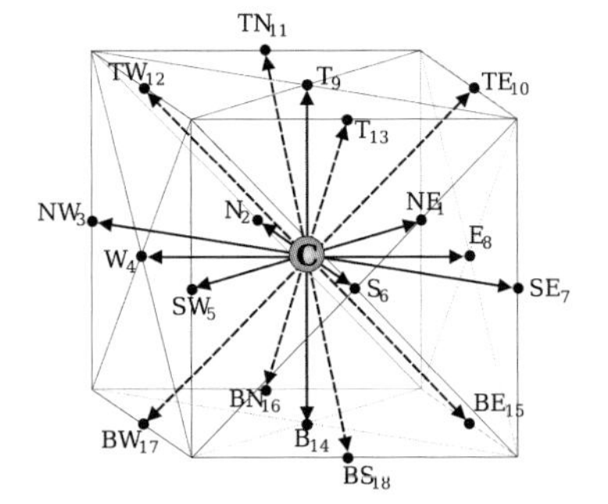

Fig. 2 Discretization model D3Q19 in three dimensions with 18 discrete velocity directions and the center representing the fraction of non-moving particles

depends on the macroscopic velocity u_i (3) and the macroscopic density ρ (4). It is a Maxwell-Boltzmann distribution function discretized for low mach numbers. In the D3Q19 model $c_s = \frac{1}{\sqrt{3}}$, which is the thermodynamic speed of sound. Additionally, it has the same order of magnitude as the quadratic mean of the fluid particles' molecular velocities. The macroscopic quantities of interest (ρ, u_i) can be determined from the first two moments of the distribution functions:

$$\rho u_i = \sum_{\alpha=0}^{18} e_{\alpha,i} \cdot f_\alpha = \sum_{\alpha=0}^{18} e_{\alpha,i} \cdot f_\alpha^{(eq)}(\rho, u_i), \tag{3}$$

$$\rho = \sum_{\alpha=0}^{18} f_\alpha = \sum_{\alpha=0}^{18} f_\alpha^{(eq)}(\rho, u_i). \tag{4}$$

For the LBM, the macroscopic pressure is given by $p = c_s^2 \rho$. The lattice velocities $e_{\alpha,i}$ and the lattice weights w_α for the D3Q19 model are:

$$e_{\alpha,i} = \begin{cases} (0,0,0), \\ (\pm1,0,0),(0,\pm1,0),(0,0,\pm1), \\ (\pm1,\pm1,0),(0,\pm1,\pm1),(\pm1,0,\pm1) \end{cases} \quad w_\alpha = \begin{cases} 1/3, & \alpha = \mathrm{C} \\ 1/18, & \alpha = \mathrm{W, E, N, S, T, B} \\ 1/36, & \alpha = \mathrm{NW, NE, SW, SE, TN, TS,} \\ & \quad\mathrm{BN, BS, TW, TE, BW, BE} \end{cases}$$

Algorithmically, a lattice Boltzmann time step is split into two parts:

- The *stream step* represents the propagation of the fluid particles. It is modeled by the left hand side of (1). Here, the particle distribution functions are moved to their neighboring cells, which corresponds to an advection of the fluid particles.
- The *collision step* describes the collisions of the fluid particles and is modeled by the right hand side of (1). Here, τ determines the relaxation of the distribution functions towards equilibrium and is therefore named the relaxation time.

For a detailed description of the LBM see [4, 5, 15].

3.2 Free Surface Extension

The free surface extension introduces different cell types for gas, liquid and the interface in between. For liquid cells, standard LBM is performed, while in gas cells no collision takes place. For all cells, a fill value is introduced which specifies the normalized quantity of liquid in a cell. Gas and interface cells store an identifier (ID) of the region of gas (*bubble*) they belong to (see Fig. 3). Conversion rules ensure closure of interface during movement, prohibiting direct adjacency of gas and liquid cells. In interface cells, PDFs are missing from the gas phase and thus have to be reconstructed such that the velocities of both phases are equal, and the force performed by the gas is balanced by the fluid's force. To realize this, all PDFs with $\mathbf{e}_i \cdot \mathbf{n} < 0$ are computed like

$$f_i(\mathbf{x} - \mathbf{e}_i \Delta t, t + \Delta t) = f_i^{eq}(\rho_G, \mathbf{u}) + f_{\bar{i}}^{eq}(\rho_G, \mathbf{u}) - f_{\bar{i}}(\mathbf{x}, t), \tag{5}$$

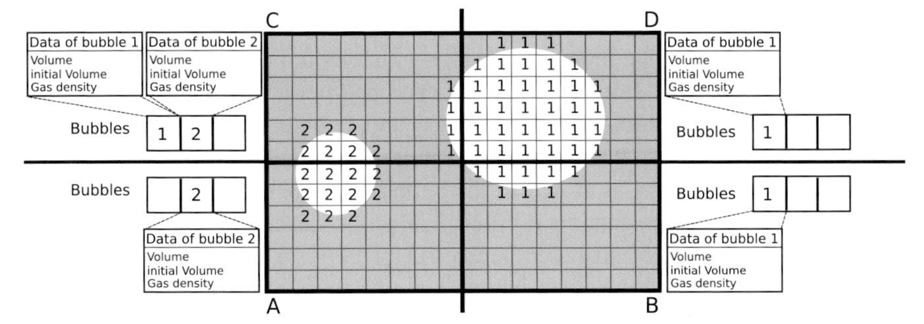

Fig. 3 Data representation of a bubble in free-surface LBM: Gas and interface cells store an ID of the bubble. Additionally, each process stores a list of bubble entries with volume data

where $f_{\bar{i}}$ points into the opposite direction of f_i, i.e. $\mathbf{e}_{\bar{i}} = -\mathbf{e}_i$. The gas density is calculated by $\rho_G = 3p_G + 6\sigma\kappa(\mathbf{x},t)$, which arises from the energy balance $p_G \cdot dV = \sigma \cdot dA$ with surface tension σ on the surface dA and gas pressure p_G in volume dV. Gas pressure is calculated from the volume of the bubble, implying that any volume change resulting from change of fill values or conversion of cells has to be tracked throughout the simulation. The surface normal $\mathbf{n}$ is computed as the gradient of the fill values using a Parker-Youngs approximation [10]. It is also used for computing surface points for each cell, needed for determination of curvature κ. This is the most computing-intensive part of the method.

Details on the free surface extension can be found in [7] and [11].

4 Localized Bubble Merge Algorithm

In previous implementations, handling of bubble data was not localized. Every process stored volume data of every bubble, irrespective where the bubble was located, and volume changes were communicated in each time step among all processes. Pohl [11] implemented two techniques to achieve this information exchange: Using all-to-all communication primitives of MPI library turned out to be less efficient than arranging all processes in a chain, where the two ends start to send packets upwards and downwards, respectively, and every participant in between merges the own information to the message which is passed on. The latter communication scheme results in $N-1$ iteration steps for N processes, but involves only direct neighbor communication, minimizing messages over slow connections if processes are properly placed in the network.

In order to enable massively parallel free surface simulations, the novel algorithm avoids global communication by storing bubble data only on processes that require it. As a consequence, bubbles crossing the boundary of a process have to be sent to the neighbor, and leaving of a bubble has to be recognized to allow deletion of volume data. Moreover, bubble coalescence has to be handled locally among the involved processes of the bubbles, which may become complicated if several bubble volumes have to be merged, possibly with data unknown to a process. The key of the concept is to store a handle for each process a bubble resides on, and communicate this data among all processes of this bubble. Additionally, information on coalescence with other bubbles, including the name of the process that is responsible for the fusion, is stored in the bubble data. With this information distributed, the requirements can be fulfilled: For proper detection whether a bubble crosses the boundary, the cell-based bubble ID (see Fig. 3) is used. If an unknown ID appears in the halo, data for a bubble has to be received. Likewise, if an ID touches the border to a neighbor that is not contained in the process list of this bubble, it has to be sent. At the same time, all other processes knowing this bubble have to be informed on the change. Handling merges is more complicated. The novel algorithm works with variable number of iterations, depending on the complexity of situation. Higher bubble IDs are merged to the lower, and a process responsible to perform a certain merge pair has to wait until all higher merges are done. If a merge is performed, all processes that knew one of the involved bubbles will be informed on the change. Conflicts can arise if bubble merge information is invalidated by another merge. In this case, the algorithm ensures restoration of consistency by rippling messages through the corresponding processes. Since MPI library requires an explicit receive issued in order to establish exchange of information, processes only passively involved in the merge action (i.e. know bubbles but not perform the merge themselves) have to communicate in each iteration until the expected data arrives. More details on this algorithm, including a description of its implementation in waLBerla, can be found in [1].

Consequently, the novel algorithm possibly means more communication in case of merges. However, this communication occurs on a more local vicinity, i.e. the neighborhood of processes harboring the bubbles involved. Since merges of bubbles occur rarely (a bubble usually covers the distance of one cell in approx. 1 000 time steps), the more complicated merge algorithm is paid off by the performance gain due to the saved all-to-all messages. Thus, parallel efficiency benefits by the locality of volume data exchange compared to the all-to-all communication in each time step. In Fig. 4 a test scenario with two bubbles far away of each other scales similar to a simulation with only one bubble. This shows that the local volume update exchanges occur independent from each other and do not affect performance

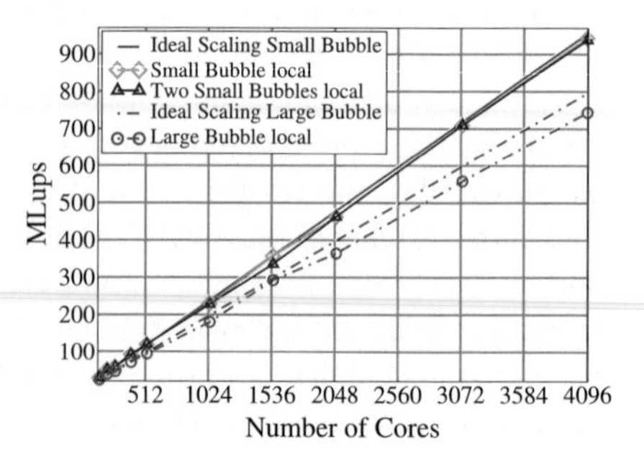

Fig. 4 Weak scaling on HLRB 2. The domain scales from 950^3 ($\approx$ 439 GB) on 128 cores to 3010^3 ($\approx$ 14 TB) on 4080 cores

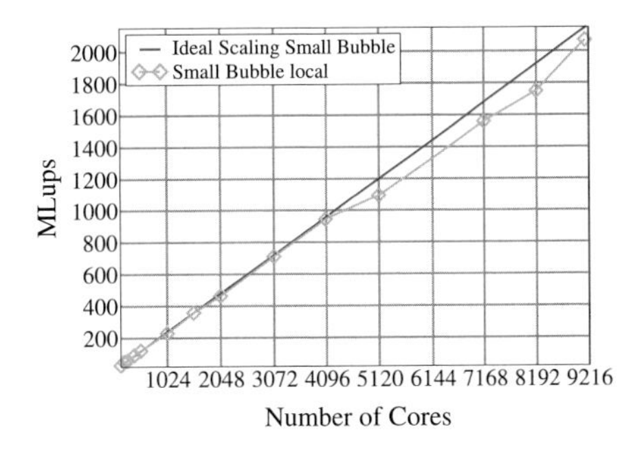

Fig. 5 Weak scaling on HLRB 2. The domain scales from 950^3 (≈ 439 GB) on 128 cores to 3940^3 (≈ 31 TB) on 9 152 cores

scaling. For a larger bubble, the performance is worse, of course, due to the higher number of interface cells. Since more processes are involved in the volume update exchange, scaling is not as ideal as for small bubbles. Figure 5 proves that this algorithm is suitable for large-scale parallel runs on up to 9 152 cores.

5 Large-Scale Free Surface Applications

A recent project with other universities, research institutes, and industry involves the simulation of liquid water in a polymer-electrolyte fuel cell (PEFC) with a proton exchange membrane. On the cathode side of the membrane, reaction of protons, electrons, and oxygen results in liquid water, which is to be evacuated from the reaction zone to sustain electrical performance. Hence, optimization of the structure and properties of the porous membrane is of particular interest. Since experiments cannot accomplish reliable quantification of water throughput in relation to material parameters due to the micron scales, simulation will assist improvement of process. The gas-diffusion layer (GDL) of a PEFC is characterized by a porous structure consisting of many thin fibers. Water generated in the layer below this structure has to evacuate towards a flow channel above. Figure 6 shows a flooding scenario in a similar geometry, indicating the suitability of the novel algorithm for large-scale simulations. The simulation shown in Fig. 6 was carried out on a domain of $2.7 \cdot 10^7$ lattice cells. Final evaluations will be extended to domain sizes of $2.5 \cdot 10^9$ lattice cells.

Figure 7 shows that the novel algorithm enables finely resolved simulations of many bubbles: A domain of $7.7 \cdot 10^8$ lattice cells contains 3 000 rising bubbles of different diameters, consuming approx. 400 GB of memory. Thanks to the improved parallel efficiency, 256 processes compute 5 000 time steps within 7 hours only. Since visualization of such huge data amounts is computationally expensive, also a smaller simulation with only 1 000 bubbles in a domain of $1.4 \cdot 10^8$ lattice cells has been performed in order to be able to visualize the movement of the bubbles (see Fig. 8). Here, 64 processes simulated 140 000 time steps in only six days. This scenario can be downloaded as video from the author's home page. The fluid in both simulations is a viscous mineral oil with a high surface tension.

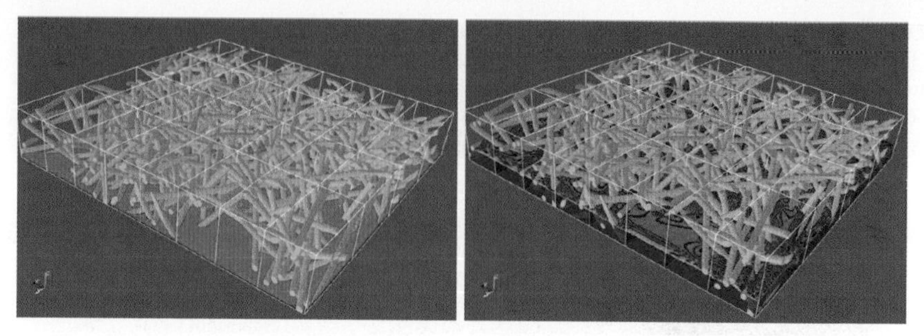

Fig. 6 Small extract of simulation in fiber geometry similar to gas diffusion layer of a fuel cell. Colors on liquid surface depict fill values

Fig. 7 Simulation of 3 000 bubbles rising and coalescing in mineral oil

6 Itanium-Specific Optimizations

The Itanium 2 architecture is known to be prone to short loops and high performance can only be achieved if the code is vectorized and access to data is optimized. Already in a previous work, the tremendous influence of memory layout has been proven for LBM on Itanium 2 architectures [14]. While this work focused on a highly optimized Fortran kernel and used the Intel compiler in version 8, Fig. 9 shows, that things change a bit for sophisticated C++ software and the newer compiler version 11. In general, of course, the variant with fused stream and collide step is still faster, irrespective of the compiler and memory layout used. The GNU compiler still shows the expected behavior of a drop in the performance curve for the array-of-structures (AoS) layout (called "collision optimized" in [14]) when running

Fig. 8 Simulation of 1 000 bubbles rising and coalescing in mineral oil

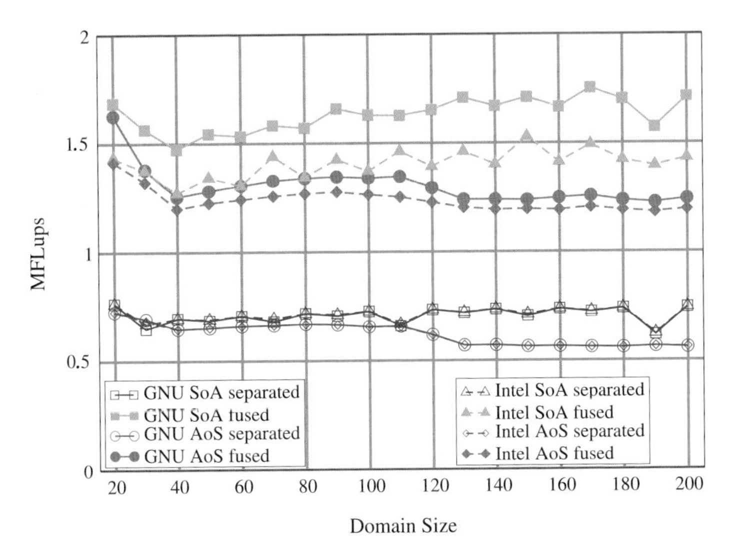

Fig. 9 Performance comparison of memory layouts "structure-of-arrays" (SoA) and "array-of-structures" (AoS) for separated and fused stream-collide with GNU and Intel compiler

out of cache, which is evaded by the structure-of-arrays (SoA) layout (the "propagation optimized" layout). Due to a more optimal memory pattern for vectorization, the SoA layout results in an overall better performance. A well-known problem of Itanium is still the performance quality of the compiler, which can be experienced

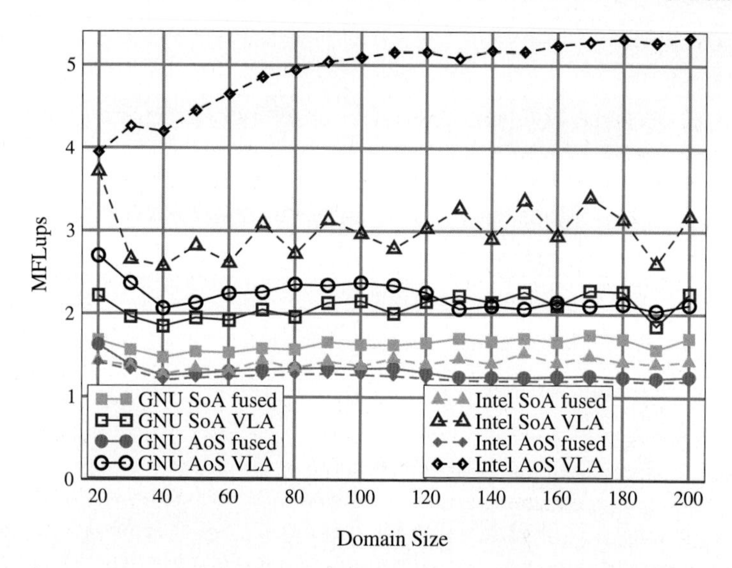

Fig. 10 Performance comparison of optimized fused stream-collide routines with variable length arrays (VLA) for the "structure-of-arrays" (SoA) and "array-of-structures" (AoS) memory layout, as well as GNU and Intel compiler

by each version change of Intel compiler. The new version 11 now is able to optimize the AoS layout to achieve the same performance as the SoA, even without the effect of the memory wall when running out of cache. However, in the end the GNU compiler still reaches better performance.

Comparing the absolute performance values with the highly optimized Fortran Kernel [14], the C++ code of waLBerla falls far behind. Here the reason is probably the inability of the compiler to optimize the address calculation, which is shown in a high number of integer loads and also leads to an inefficient access of the data. However, in the special case of the waLBerla implementation, the compiler is unable to efficiently do that because the dimensions and lengths of the data arrays are not visible to the kernel routine, but abstracted by data container classes. By the use of special C-based kernels that get access to data by variable length arrays, the compiler is able to optimize the address calculation based on the knowledge about the structure of the array dimensions. Figure 10 shows that this leads to a tremendous performance improvement for both compilers, independent of the memory layout. Interestingly, with this technique the AoS layout can again benefit more. Most notably is that by this simple change the Intel compiler now outperforms the GNU compiler by a factor of two.

This short study shows that architecture-dependent optimization is still important for the Itanium 2 architecture. Especially for C++ codes this can be a challenge because modular design and high performance due to simple kernels, which easily can be optimized by the compiler, here are often clear contradictions.

7 Conclusion

This contribution presents two important considerations concerning the performance optimization of a complex C++ code for massive-parallel systems based on Itanium processors like used in the HLRB 2. The waLBerla framework combines many different lattice Boltzmann related applications in one software and aims for large-scale parallelism in order to be able to simulate engineering-relevant scenarios from computational fluid dynamics. The free surface extension that can be used to simulate bubbly flows or the liquid water in a fuel cell, has to exchange aggregated data of gas volumes among the processes. For parallel runs with more than a few hundreds or even thousands of processes, the formerly used all-to-all communication schemes resulted in a poor performance scaling. A localized communication algorithm manages to exchange data only among processes that really require it and thus saves a lot of communication which improves the scaling.

For Itanium architectures, resulting performance is still depending on the abilites of the compiler. Especially in C++ care has to be taken that the compiler is able to do proper vectorization, because a modular and flexible design often hides important information from the compiler that would be required to optimize the code. Replacing the kernel routines by C-based code and using variable length arrays without abstraction resulted in a tremendous performance improve for single core performance.

Acknowledgements Special thanks to the HLRB 2 administrators for their valued support and their willingness to carry out the large-scale performance runs. Part of this work is carried out within the framework of *walberlaMC*, a project funded by the "Kompetenznetzwerk für Technisch-Wissenschaftliches Hoch- und Höchstleistungsrechnen in Bayern" (*KONWIHR*). It is also partially funded by the European Commission in the project *DECODE* (CORDIS project 213295) and by the Bundesministerium für Bildung und Forschung under *SKALB* (grant 01IH08003A).

References

1. Donath, S., Feichtinger, C., Götz, J., Deserno, F., Iglberger, K., Rüde, U.: waLBerla: On implementation details of a localized parallel algorithm for bubble coalescence. Tech. Rep. 09-3, Computer Science Department 10 (System Simulation), University of Erlangen-Nuremberg (2009)
2. Feichtinger, C., Götz, J., Donath, S., Iglberger, K., Rüde, U.: Concepts of waLBerla Prototype 0.1. Tech. Rep. 07–10, Computer Science Department 10 (System Simulation), University of Erlangen-Nuremberg (2007)
3. Götz, J., Feichtinger, C., Iglberger, K., Donath, S., Rüde, U.: Large scale simulation of fluid structure interaction using lattice Boltzmann methods and the 'physics engine'. In: G.N. Mercer, A.J. Roberts (eds.) Proceedings of CTAC 2008, *ANZIAM J.*, vol. 50, pp. C166–C188 (2008)
4. Hänel, D.: Molekulare Gasdynamik. Springer (2004)
5. He, X., Luo, L.S.: Theory of the lattice Boltzmann method: From the Boltzmann equation to the lattice Boltzmann equation. Phys. Rev. E **56**(6), 6811–6817 (1997). DOI 10.1103/PhysRevE.56.6811

6. Information on the HLRB. http://www.lrz-muenchen.de/services/compute/hlrb/ (2007)
7. Körner, C., Thies, M., Hofmann, T., Thürey, N., Rüde, U.: Lattice Boltzmann Model for Free Surface Flow for Modeling Foaming. Journal of Statistical Physics **121**(1-2), 179–196 (2005).
8. Körner, C., Thies, M., Singer, R.F.: Modeling of metal foaming with lattice Boltzmann automata. Adv. Eng. Mat. **4**, 765–769 (2002)
9. Mei, R., Shyy, W., Yu, D., Luo, L.S.: Lattice Boltzmann Method for 3-D Flows with Curved Boundary. J. Comp. Phys. **161**, 680–699 (2000)
10. Parker, B.J., Youngs, D.L.: Two and three dimensional Eulerian simulation of fluid flow with material interfaces. Tech. Rep. 01/92, UK Atomic Weapons Establishment, Berkshire (1992)
11. Pohl, T.: High Performance Simulation of Free Surface Flows Using the Lattice Boltzmann Method. Ph.D. thesis, Univ. of Erlangen (2008)
12. Succi, S.: The Lattice Boltzmann Equation – For Fluid Dynamics and Beyond. Clarendon Press (2001)
13. Thürey, N., Rüde, U.: Free surface lattice-Boltzmann fluid simulations with and without level sets. pp. 199–208. VMV, IOS Press (2004)
14. Wellein, G., Zeiser, T., Hager, G., Donath, S.: On the single processor performance of simple lattice boltzmann kernels. Computers & Fluids **35**(8–9), 910–919 (2006)
15. Yu, D., Mei, R., Luo, L.S., Shyy, W.: Viscous flow computation with method of lattice Boltzmann equation. Progress in Aerospace Sciences **39**(5), 329–367 (2003)

Fast 3D Block Parallelisation for the Matrix Multiplication Prefix Problem
Application in Quantum Control

K. Waldherr, T. Huckle, T. Auckenthaler, U. Sander, and T. Schulte-Herbrüggen

Abstract For exploiting the power of supercomputers like the HLRB II cluster, developing parallel algorithms becomes increasingly important. The matrix prefix problem belongs to a class of issues lending themselves for parallelisation. We compare the tree-based parallel prefix scheme, which is adapted from a recursive approach, with a sequential multiplication scheme where only the individual matrix multiplications are parallelised. We show that this fine-grain approach outperforms the parallel prefix scheme by a factor of $2-3$ and also leads to less memory requirements. Unlike the tree-based scheme, the fine-grain approach enables many options in the choice of the number of parallel processors and shows a better speedup performance when increasing the matrix sizes. The usage of the fine-grain approach in a quantum control algorithm instead of the coarse-grain approach allows us both to deal with systems of higher dimensions and to choose a finer discretisation.

1 Introduction: The Prefix Problem

In general, the *prefix problem* is given as follows: Let $\circ$ be a binary operator and $\mathscr{A}$ a set, which is closed under the operator $\circ$. Furthermore, let $\circ$ be associative, i.e. the identity $(a \circ b) \circ c = a \circ (b \circ c)$ holds for all $a, b, c \in \mathscr{A}$. Then, for given elements $x_1, \dots, x_M \in \mathscr{A}$, the prefix problem means the computation of all the products

$$y_i = x_1 \circ \cdots \circ x_i. \tag{1}$$

K. Waldherr · T. Huckle · T. Auckenthaler
Dept. of Computer Science, TU Munich, D-85748 Garching, Germany
e-mail: huckle@in.tum.de

U. Sander · T. Schulte-Herbrüggen
Dept. of Chemistry, TU Munich, D-85747 Garching, Germany
e-mail: tosh@ch.tum.de

S. Wagner et al. (eds.), *High Performance Computing in Science and Engineering, Garching/Munich 2009*, DOI 10.1007/978-3-642-13872-0_4,

Analogously, the *suffix problem* amounts to computing of all the products

$$z_i = x_{i+1} \circ \cdots \circ x_M. \tag{2}$$

In our applications, the elements x_i denote quadratic matrices and the operator '$\circ$' describes the multiplication of two matrices.

1.1 Scope and Organisation of the Paper

This account comprises two sections. First, we present two approaches for the parallelisation of the prefix problem, the coarse-grain tree-based and the fine-grain 3D block approach. The coarse-grain approach reorganises the numbering of the matrix multiplication in such a way that the products may be computed in parallel, whereas the fine-grain approach parallelises the individual matrix multiplications. In Sect. 2.1 we describe the two methods, demonstrate their pros and cons and underpin these statements with numerical results, i.e. speedup measurements taken on the ALTIX and computation time measurements on different architectures.

Second, we present applications showing how matrix methods improve the computation time of solutions to cutting-edge optimal quantum control problems. In turn, these control methods pave the way to finding optimised experimental steerings of quantum devices in realistic settings as they occur in a broad array of applications in quantum electronics, nanotechnology, spectroscopy, and quantum computation.

1.2 Hardware and Software Setup

The improvements in matrix multiplication methods directly translate into faster algorithms for the parallel prefix problem. As a main test bed, here we use high-dimensional matrices as occurring in large quantum systems. To this end, we extended our parallelised C++ code of the GRAPE package described in [5].

Computations were performed on the HLRB II cluster currently providing an SGI Altix 4700 platform equipped with 9728 Intel Itanium2 Montecito Dual Core processors with a clock rate of 1.6 GHz, which give a total LINPACK performance of 63.3 TFlops/s. The sequential linear algebra tasks were executed by usage of the MATH KERNEL LIBRARY(MKL).

For comparing the performance of our program on different architectures, computations were also performed on an *Infiniband cluster* with 32 Opteron nodes; each node contains four AMD Opteron 850 processors (2.4 GHz) connected to 8 GB or 16 GB of shared memory. Each node is equipped with one MT23108 InfiniBand Host Channel Adapter card, which is thus shared by 4 processors for communication. The sequential linear algebra tasks were executed by the implementation of AMD'S CORE MATH LIBRARY (ACML).

2 Parallelising the Prefix Problem

With regard to parallelisation, the matrix prefix problem defined in Eq. 1 offers both a fine-grain and a coarse-grain approach.

2.1 Coarse-Grain versus Fine-Grain Approach

The idea of the fine-grain approach is to simply compute the matrix products $\Pi_{j=1}^{k} U_j$ sequentially for $k = 1,...,M$ and to parallelise the individual matrix multiplications. In contrast, the coarse-grain approach applies the following divide-and-conquer approach to compute a product $U_{k_1:k_2}$:

1. compute the products $U_{k_1:\kappa}$ for all $\kappa = k_1,...,\hat{k}-1$, with $\hat{k} = \left\lceil \frac{1}{2}(k_1+k_2) \right\rceil$.
2. compute the products $U_{\hat{k}:\kappa}$ for all $\kappa = \hat{k},\ldots,k_2$; steps 1 and 2 can be executed in parallel;
3. compute the products $U_{k_1:\kappa} := U_{k_1:(\hat{k}-1)} U_{\hat{k}:\kappa}$ for all $\kappa = \hat{k},\ldots,k_2$; all (k_2-k_1+1) products of this step can be computed in parallel.

This coarse-grain approach can then be extended recursively to a tree-like multiplication scheme as given in Fig. 1, as was first presented in [9]. In order to compute all the interior products $U_{1:k}$ for $k = 1,...,M$, we may at most use $\frac{M}{2}$ processors in parallel. If a larger number of processors is available, the individual matrix multiplication itself would have to be parallelised as well. The tree-like coarse-grain approach offers some advantages: the communication pattern is simpler than in the fine-grain approach and we can expect good parallel speedups of the tree algorithm, because the individual matrix multiplications are strictly sequential. However, the tree-like approach increases the total computational work by a logarithmic

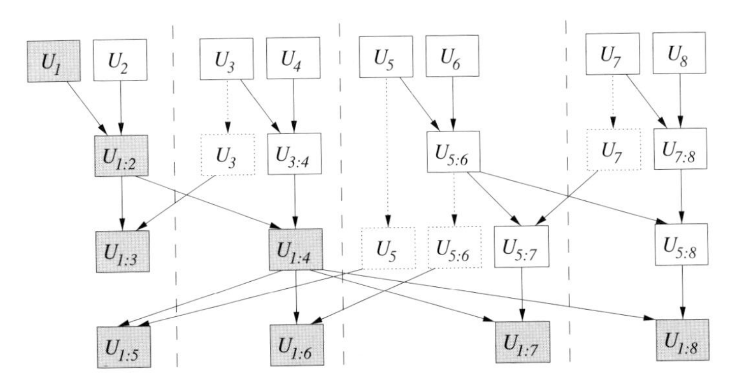

Fig. 1 Divide-and-conquer scheme for the parallel prefix computation. The dashed lines define the processor scopes; solid lines denote communication between processors, and dotted lines and boxes indicate where matrices have to be retained for next-level computations. Note that the result matrices (grey boxes) are not balanced among the processes

factor, because the parallel computation requires $\mathcal{O}(\log M)$ subsequent steps, which are given by the horizontal levels in Fig. 1. This additional overhead is saved in the fine-grain approach, where the individual matrix multiplications themselves are parallelised. Another disadvantage of the tree scheme is the fact, that the resulting matrices $U_{1:k}$ are not balanced among the processes: In the special case of $p = \frac{M}{2}$ parallel processes, the first process holds $\frac{M}{2}$ result matrices, whereas the last process holds only one of them. However, the balance missing here is recovered in the fine-grain approach. For a detailed analysis of the tree scheme, we refer the reader to our last report [11].

2.2 3D Block-Oriented Parallel Matrix Multiplication

In its most general form, the computation of the matrix product $C = AB$ can be formulated via the algorithm

1: **for all** $(i,j,k) \in \{1,\dots,n\} \times \{1,\dots,n\} \times \{1,\dots,n\}$ **do**
2: $\quad C_{ik} \leftarrow C_{ik} + A_{ij}B_{jk}$
3: **end for**

(the matrix C is assumed to be initialised with zeros), where the C_{ik}, A_{ij}, and B_{jk} may be simple matrix elements or even matrix blocks. Note, that the body of the for-loop, i.e. all block operations of step 2, may be executed entirely in parallel. In Fig. 2, these block operations are visualised as a cube of size $n \times n \times n$, where the projections in the k-, i- and j- direction indicate which matrix blocks A_{ij}, B_{jk} and C_{ik} are to be accessed, respectively.

Depending on the data distribution and also on the distribution of the block operations to the available processors, we may classify the algorithms for parallel matrix multiplication into the following types:

1D algorithms parallelise the block operations into planes in the cube, which corresponds to column-wise computation of the result matrix.

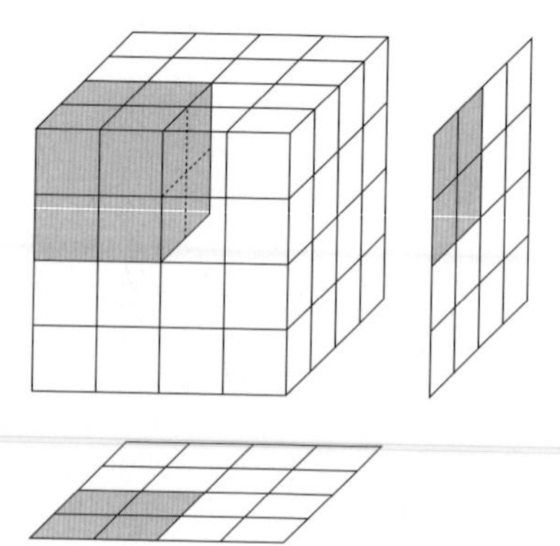

Fig. 2 Block operations in the 3D block multiplication algorithm. The projections onto the planes below and right illustrate the accessed matrix blocks of the result matrix C and of one of the operand matrices, respectively

2D algorithms distribute the block operations into columns; usually these columns work on individual blocks in the result matrix, such that all computations on that result block are executed by a single processor ("owner computes").

3D algorithms define a blocking on the three nested main loops of the algorithm. Hence, the parallelisation is then purely *work-oriented* in the sense that a 3D subblock of block operations is assigned to each processor. Both result and operand matrices may be distributed over several processors.

Naturally, these three classes of algorithms lead to different performance properties. Most importantly, 3D algorithms feature the lowest communication effort, namely $\mathcal{O}(n^2p^{1/3})$ instead of $\mathcal{O}(n^2p^{1/2})$ for 2D, and $\mathcal{O}(n^2p)$ for 1D algorithms (compare [6, 7]).

Under the influence of the performance study [2], where 3D algorithms showed the best runtime performance for our problem setting, we implemented a 3D block algorithm. The improved performance comes at the cost of a slightly higher memory requirement: $\mathcal{O}(n^2p^{1/3})$ additional matrix elements have to be distributed on p processors, which is less than one additional matrix per processor. As we typically store several matrices (up to 256 in our examples in Sect. 2.3) on each CPU, this additional requirement is easily affordable. Details may be found in [1].

In the 3D block algorithm, the number D of blocks per index dimension is chosen such that D is the smallest power of 2, which fulfills $D^3 \geq p$ (p is the number of processes). Hence, each of the D^2 blocks of the result matrix C is computed by p/D^2 processes, and each processor performs D^3/p block operations, all of which work on a single matrix block C_{ik}. Altogether, each process will perform the following two steps:

for all local block operations (i, j, k) **do**
  fetch A_{ij} and B_{jk} from remote processes
  $C_{ik} \leftarrow C_{ik} + A_{ij}B_{jk}$
end for
accumulate all results for block C_{ik} (group-collective operation).

The accumulation of the results in C_{ik} is a group-collective procedure, which is performed in parallel by all p/D^2 processes that compute the same block C_{ik}. The accumulation is organised as a pairwise accumulation of data, as illustrated in Fig. 3.

After $\log_2(p/D^2)$ subsequent steps, each process has computed one part of the block C_{ik}—these parts are then broadcasted to the other processes. Note that, within the for-loop to compute the block operations, communication and computation may be overlapped such as to hide communication costs.

During a computation $U_{1:k} = U_{1:k-1}U_k$, only blocks of the matrices $U_{1:k-1}$ and U_k are accessed. Hence, if $U_{1:k-1}$ and U_k were stored on only one process each, we would obtain a communication bottleneck. We therefore distribute each matrix onto all available processes before starting the prefix computation. The granularity of this distribution is given by the subblocks used in the accumulation step (compare Fig. 3). Subblock l of a matrix U_k will be stored on process $(l+k) \mod p$ (p the number of processes). After each accumulation process, the computed matrix $U_{1:k}$

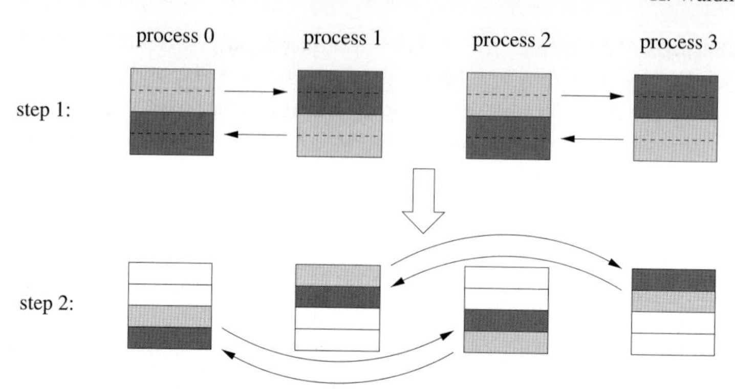

Fig. 3 Communication pattern for the 3D block approach during the accumulation of results within a single matrix block C_{ik}. In each successive step, pairs of processes exchange and accumulate their partial results: the number of exchanged rows is halved and the distance between two communicating processes is doubled. Gray boxes represent rows that are sent to the corresponding process, black boxes represent rows that are received and accumulated to local data

will be correctly distributed to all processes. After the entire prefix loop, the result matrices $U_{1:k}$ will be assembled in a final global communication step, such that again all matrices are stored on a separate processor.

2.3 Numerical Results

In this section, we want to present some performance results of the 3D block approach. As already mentioned in the hardware setup at the beginning of our report, we evaluated these results both on the HLRB II cluster and on an Infiniband cluster.

Figure 4 shows the speedups achieved when using 8, 16, 32, 64, 128 or 256 processors for $M = 2048$ matrices of size 512×512 and $M = 256$ matrices of size 1024×1024 respectively and when using 32, 64, 128, 256 or 512 processors for $M = 512$ matrices of size 2048×2048 and accordingly $M = 64$ matrices of size 4096×4096. We see that the speedups become increasingly better by augmenting the matrix size: for matrices of size 1024×1024, the achieved parallel efficiency on the HLRB II is only about 30% for 256 processors, whereas for matrices of size 4096×4096, we get an efficiency of about 65% for 256 processors. This is both due to communication overhead and due to decreased performance of the sequential matrix multiplication for very small matrix blocks. The illustrated results also show that due to the better communication interconnect, the HLRB II enables larger speedups than the Infiniband cluster.

For comparing the fine-grain to the coarse-grain approach in terms of runtime performance, we measured the computation times of both methods on the HLRB II for the same problem settings and numbers of processes as in the previous speedup measurements. Despite the less-than-optimal speedups, the 3D block multiplica-

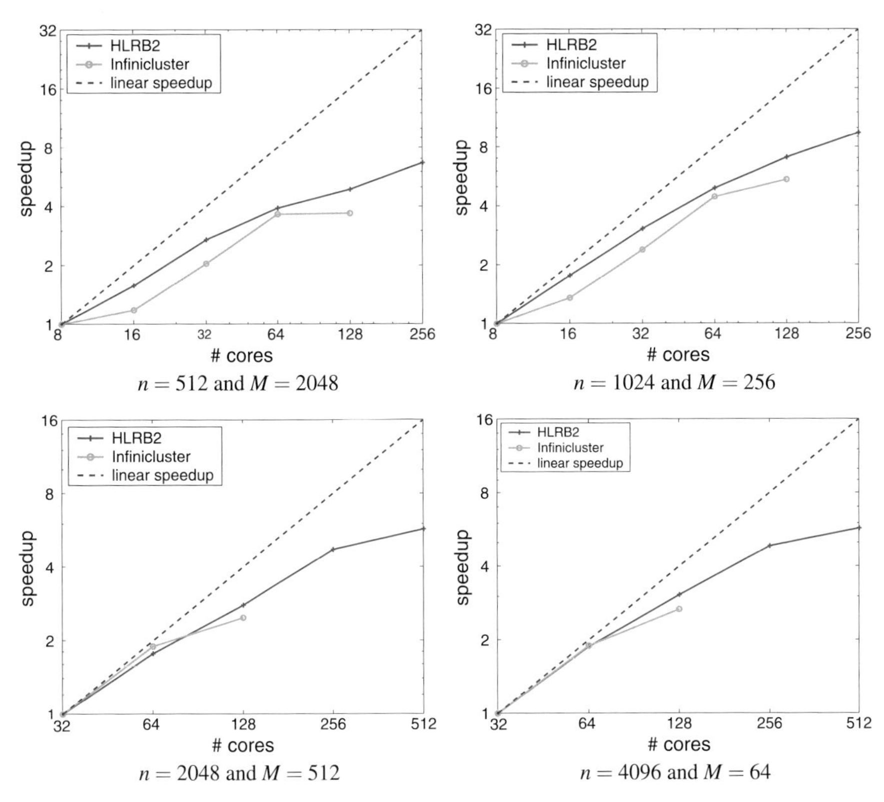

Fig. 4 The speedups achieved on the HLRB II cluster compared to the speedups on the Infiniband cluster. These speedups were measured for different configurations of the matrix size n and number M of matrices being multiplied. Please note, that the Infiniband cluster only offers 128 cores

tion outperforms the parallel prefix scheme, which can be seen from Fig. 5. As expected, the performance advantage of the 3D block method grows for larger matrix sizes: for matrices of size 512×512, we get about the same computation times for 256 parallel processes, whereas for matrices of size 2048×2048, the 3D approach is more than 2 times faster than the tree-wise method. Recall that for the coarse-grain tree-like method, at most $\frac{M}{2}$ parallel processors may be used. This is why no further speedups are to be expected in the problem configurations of Fig. 5 c) and d).

3 Application: Optimal Quantum Control

In view of novel quantum- and nano-technology, quantum control plays a key role for steering quantum hardware systems [4]. Moreover for exploiting the power of such quantum devices, one has to manipulate them by classical controls with the

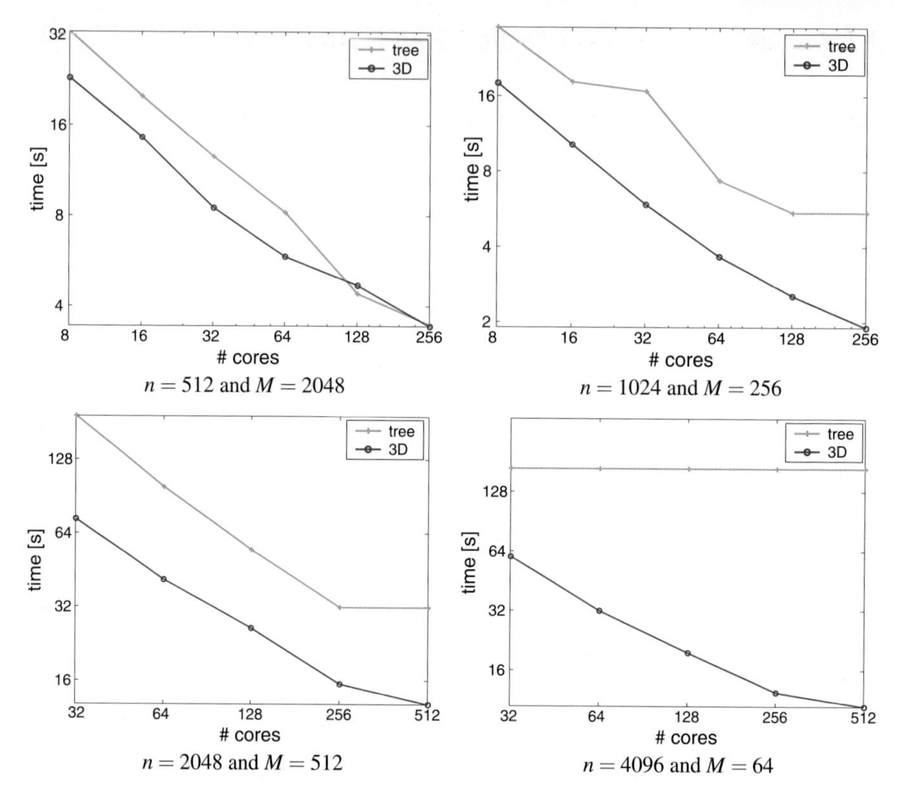

Fig. 5 Comparison of the tree-like and the 3D block approach due to computation time. These results were achieved on the HLRB II cluster

shapes of these controls critically determining the performance of the quantum system. Providing computational infrastructure for devising optimal shapes by using high-performance computer clusters is therefore tantamount to using the power of present and future quantum resources. To this end, numerical schemes like the GRAPE (GRadient Ascent Pulse Engineering) algorithm [8] have been introduced. — However, the task of finding optimised quantum controls is computationally highly demanding in the sense that the classical resources needed grow exponentially with the size of the quantum system.

The examples presented here refer to a particularly interesting class of quantum states known as *cluster states*. Their power lies in the fact that they are highly correlated (i.e. *entangled*). While preparing those states as initial states is challenging, it makes the actual quantum information processing step very efficient [3, 10], because the latter then reduces to local operations on each qubit. So minimising the time cost for this preparatory step is relegated to our numerical algorithm GRAPE, which is significantly improved by the fast matrix multiplication schemes presented.

3.1 Linear Algebra Tasks in the GRAPE Algorithm

The GRAPE algorithm provides a method for optimal quantum control based on gradient flows. In order to get a first impression about the complexity of the GRAPE

0: define as quality function $f\big(U(t_M)\big) := \Re\,\mathrm{Tr}\{U_G^{\dagger}U(t_M)\}$;
1: set initial control amplitudes $u_j^{(0)}(t_k)$ for all times t_k with $k = 1,\ldots,M$;
2: starting from $U_0 = I$, calculate the forward-propagation for all $t_1,\ldots,t_M$
for simplicity, take uniform $\Delta t := t_{k+1} - t_k$:

$$U(t_k) = e^{-i\Delta t H_k} e^{-i\Delta t H_{k-1}} \cdots e^{-i\Delta t H_1},$$

where $H_k := H_0 + \sum_j u_j(t_k)H_j$ comprises a non-switchable drift term H_0, the control amplitudes $u_j(t_k)$ and control Hamiltonians H_j.
3: likewise, starting with $T = t_M$ and $W(T) = e^{i\phi} \cdot U_G$ compute the back-propagation for all $t_M, t_{M-1},\ldots,t_{k+1}$ (which we naturally label as $W^{\dagger}(t_k)$, because ideally $W^{\dagger}(t_k)U(t_k) = e^{-i\phi}\,\mathbb{1}$)

$$W^{\dagger}(t_k) := W^{\dagger}(T)\, e^{+i\Delta t H_M} \cdots e^{+i\Delta t H_{k+1}};$$

4: calculate $\frac{\partial f(U(t_k))}{\partial u_j} = \Re\,\mathrm{Tr}\big\{W^{\dagger}(t_k)(-i\Delta t H_j)U(t_k)\big\}$
5: with $u_j^{(1)}(t_k) = u_j^{(0)}(t_k) + \varepsilon \frac{\partial f}{\partial u_j}\big|_{t=t_k}$ update all the piece-wise constant Hamiltonians H_k and continue with step 2.

Algorithm 1: Gradient Flow Algorithm GRAPE for Optimal Quantum Control [8]

algorithm, we consider the pseudo code given in algorithm 1. It describes a step-wise optimisation in conjugate gradients: In each iteration step, we compute the Hamiltonian quantum evolution given by the matrix exponentials at every time step k. Then forward (step 2) and backward propagation (step 3) are given by a sequence of evolutions of the quantum system under M piecewise constant Hamiltonians H_k. The gradient $\frac{\partial f(U(t_k))}{\partial u_j}$ of the performance function f (i.e. the projection of the time evolution $U(T)$ onto the desired target $W(T)$) is given by the trace (calculated in step 4) and needed for the update of the control amplitudes in step 5. The control amplitudes may be initialised by a suitable guess for the initial values $u_j^{(0)}(t_k)$ in step 1 or by setting them to some constant.

From a computational point of view, the GRAPE algorithm makes heavy use of the following linear algebra tasks:

1. the computation of the matrix exponentials $U_k := e^{-i\Delta t H_k}$, where H_k denotes a large and sparse Hermitian matrix,
2. the matrix multiplications in the prefix and postfix problems (steps 2 and 3).

The trace evaluations in step 4 are about in the same complexity class as the matrix exponentials, but they offer no possibility for algorithmic improvement.

The computation of the matrix exponentials was already an issue in the report [11] of 2008. By using a Chebyshev series approach instead of dealing with the eigendecomposition of the Hamiltonians, we could save a factor of 30 % in computation time without loss of accuracy. For details, see [1, 11].

Therefore, only the computation of the forward and backward propagation, which is an instance of the prefix problem Eq. 1 or the suffix problem Eq. 2, respectively, is left for computational improvement.

3.2 Numerical Results

For the computation of the forward and backward propagation, we may now use and compare the methods considered in Sect. 2. A 1D algorithm (see Sect. 2.2) had already been considered in the original work by Gradl et al. [5]. However, it was found to be inferior to the coarse-grain parallel prefix scheme, both with respect to runtime as well as due to increased memory and communication requirements. Therefore, the additional computation costs of $\log(M)$ matrix multiplications involved by the tree-like propagation were accepted in order to be able to deal with systems of larger size (see [11]). With the 3D block approach presented in Sect. 2.2, we now have a second algorithm beside the tree-like propagation for the computation of the forward- and backward-propagation (steps 2 and 3 of algorithm 1).

Fig. 6 gives the runtimes for one complete iteration of the GRAPE algorithm on the HLRB II cluster using the tree-like ('old') or the 3D block ('new') approach. For either setting, the total runtime is split up into three parts: computing the matrix

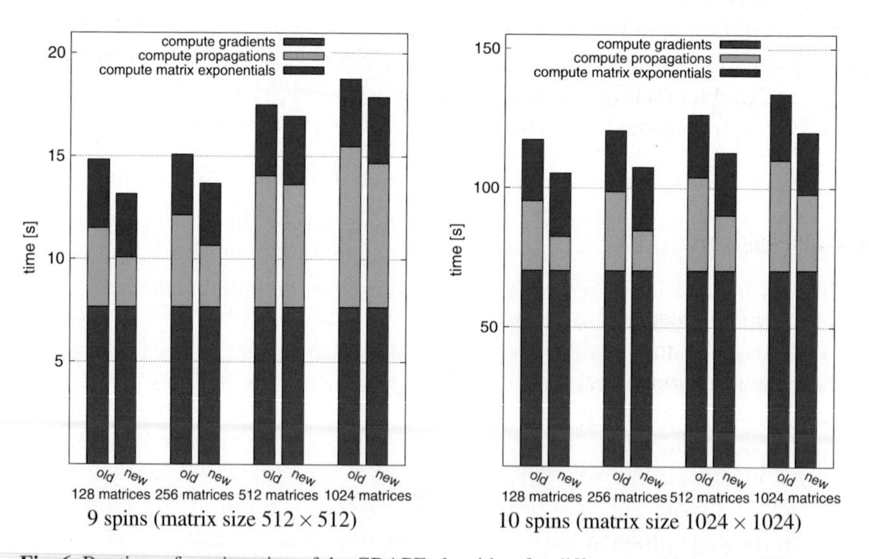

Fig. 6 Runtime of one iteration of the GRAPE algorithm for different problem sizes. We compare the runtimes for using the parallel prefix scheme ('old') vs. using the 3D parallel matrix multiplication ('new') for systems with 9 and 10 spins, where the number M of matrices varies between 128 and 1024. The number of parallel processes is chosen as $p = \frac{M}{2}$

Table 1 Parallel runtimes for the computation of the first 100 GRAPE iterations on p processors for different matrix sizes $2^n \times 2^n$ and number M of matrices. Note that n directly corresponds to the number of qubits (spins)

Problem Size			Parallel Runtime $[s]$	
matrix dimension $2^n \times 2^n$	# matrices M	# processes	tree	3D block
$n = 8$	64	32	324	362
$n = 8$	64	128	324	191
$n = 8$	128	64	388	391
$n = 8$	128	128	388	240
$n = 8$	256	128	836	427
$n = 10$	128	64	36842	21260
$n = 10$	128	128	36842	10609

exponentials, forward- and backward-propagation, and computation of the gradients (steps 4 and 5 of algorithm 1). The number p of processors was chosen as $p = \frac{M}{2}$, which means best performance for the tree-like method (cp. Sect. 2.1).

Using the 3D matrix multiplication gives a substantially better performance even for the 9-spin problem, where the matrix sizes are comparably small (512×512). For the 10-spin problem (matrix size 1024×1024), the 3D matrix multiplication is already more than 2 times faster than the parallel prefix scheme. Note that compared to the tree-like approach any other number of parallel processes would lead to even better results for the block approach and that due to Sect. 2.3, the speedup behaviour of the 3D method will become increasingly better when raising the number of spins (i.e. the matrix size).

After demonstrating benchmark results, we finally present a practical application. Table 1 summarises the computation times for the calculation of 100 iterations for two real problem settings, namely an 8-spin and a-10 spin instance of finding optimal controls for a quantum dynamic state-to-state transfer preparing cluster states.

4 Conclusions

Summing up, in order to fully exploit the power of the HLRB II cluster, we devised a 3D approach for the parallel prefix multiplication of matrices. It efficiently parallelises the three loops occurring in matrix multiplication while keeping memory demands as well as communication costs low. Compared to the tree-like 'divide and conquer' matrix multiplication scheme (reported previously [11]), the 3D approach presented here saves the additional logarithmic factor in the total computational work, provides more opportunities for parallelisation and promises good parallel speedups for large problem sizes. With matrix multiplication being in the core of many algorithms, we anticipate that the general scheme shown here awaits broad application for improving parallel prefix algorithms on high-speed clusters.

As an instance, we demonstrated the gain in CPU time in an algorithm optimising the preparation of quantum states in an important class of problems.

Acknowledgements This work was supported in part by the integrated EU project QAP, by the Bavarian excellence initiative ENB in the PhD programme QCCC, and by *Deutsche Forschungsgemeinschaft*, DFG, within the collaborative research centre SFB-631.

References

1. Auckenthaler, T., Bader, M., Huckle, T., Spörl, A., Waldherr, K.: Matrix Exponentials and Parallel Prefix Computation in a Quantum Control Problem. Parallel Computing: Architectures, Algorithms and Applications, Special issue of the PMAA 2008 (submitted 2008)
2. Bader, M., Hanigk, S., Huckle, T.: Parallelisation of block recursive matrix multiplication in prefix computations. In: C. Bischof, et al. (eds.) Parallel Computing: Architectures, Algorithms and Applications, Proceedings of the ParCo, Parallel Computing 2007, *NIC Series*, vol. 38, pp. 175–184 (2008)
3. Briegel, H.J., Raussendorf, R.: Persistent Entanglement in Arrays of Interacting Particles. Phys. Rev. Lett. **86**, 910–913 (2001)
4. Dowling, J., Milburn, G.: Quantum Technology: The Second Quantum Revolution. Phil. Trans. R. Soc. Lond. A **361**, 1655–1674 (2003)
5. Gradl, T., Spörl, A.K., Huckle, T., Glaser, S.J., Schulte-Herbrüggen, T.: Parallelising Matrix Operations on Clusters for an Optimal-Control-Based Quantum Compiler. Lect. Notes Comput. Sci. **4128**, 751–762 (2006). Proceedings of the EURO-PAR 2006
6. Gupta, A., Kumar, V.: Scalability of Parallel Algorithms for Matrix Multiplication. In: International Conference on Parallel Processing – ICPP'93, vol. 3, pp. 115–123 (1993)
7. Irony, D., Toledo, S., Tiskin, A.: Communication Lower Bounds for Distributed-Memory Matrix Multiplication. J. Parallel Distrib. Comput. **64**, 1017–1026 (2004)
8. Khaneja, N., Reiss, T., Kehlet, C., Schulte-Herbrüggen, T., Glaser, S.J.: Optimal Control of Coupled Spin Dynamics: Design of NMR Pulse Sequences by Gradient Ascent Algorithms. J. Magn. Reson. **172**, 296–305 (2005)
9. Ladner, R.E., Fischer, M.J.: Parallel Prefix Computation. J. ACM **27**, 831–838 (1980)
10. Raussendorf, R., Briegel, H.J.: A One-Way Quantum Computer. Phys. Rev. Lett. **86**, 5188–5191 (2001)
11. Schulte-Herbrüggen, T., Spörl, A.K., Waldherr, K., Gradl, T., Glaser, S.J., Huckle, T.: in: High-Performance Computing in Science and Engineering, Garching 2007, chap. Using the HLRB Cluster as Quantum CISC Compiler: Matrix Methods and Applications for Advanced Quantum Control by Gradient-Flow Algorithms on Parallel Clusters, pp. 517–533. Springer, Berlin (2008)

OMI4papps: Optimisation, Modelling and Implementation for Highly Parallel Applications

Volker Weinberg, Matthias Brehm, and Iris Christadler

Abstract This article reports on first results of the KONWIHR-II project OMI4papps at the Leibniz Supercomputing Centre (LRZ). The first part describes Apex-MAP, a tunable synthetic benchmark designed to simulate the performance of typical scientific applications. Apex-MAP mimics common memory access patterns and different computational intensity of scientific codes. An approach for modelling LRZ's application mix is given which makes use of performance counter measurements of real applications running on "HLRB II", an SGI Altix system based on 9728 Intel Montecito dual-cores. The second part will show how the Apex-MAP benchmark could be used to simulate the performance of two mathematical kernels frequently used in scientific applications: a dense matrix-matrix multiplication and a sparse matrix-vector multiplication. The performance of both kernels has been intensively studied on x86 cores and hardware accelerators. We will compare the predicted performance with measured data to validate our Apex-MAP approach.

1 Performance Modelling Using the Apex-MAP Benchmark

A simple synthetic benchmark with tunable hardware independent parameters that mimics the behaviour of typical scientific applications is very useful for the evaluation of new hardware platforms for a certain job mix. Mapping application performance data measured on a production system to specific parameter combinations of the synthetic benchmark allows to model the performance of a wide spectrum of applications with a simple approach.

Volker Weinberg · Matthias Brehm · Iris Christadler
Leibniz-Rechenzentrum der Bayerischen Akademie der Wissenschaften, Boltzmannstr. 1, 85748 Garching bei München, Germany
e-mail: volker.weinberg@lrz.de
e-mail: matthias.brehm@lrz.de
e-mail: iris.christadler@lrz.de

S. Wagner et al. (eds.), *High Performance Computing in Science and Engineering, Garching/Munich 2009*, DOI 10.1007/978-3-642-13872-0_5,

To get insight in the performance patterns of the applications running on HLRB II, samples of the most important hardware counters (currently 25 counters) are taken from all processors in 10 minute intervals and are stored in a huge database at LRZ. Though the measurements do not only include production runs of optimised user codes, but also badly optimised programs and test runs etc., the results give a deep insight into LRZ's job mix and the typical performance of the system. Details about the measurement process, the sampling method, the database scheme and the data analysis can be found in the LRZ technical report 2006-06 [1].

1.1 The Apex-MAP Benchmark

To synthetically model the performance behaviour of LRZ's application mix we extended the Apex-MAP benchmark (*A*pplication *pe*rformance *ch*aracterisation project – *M*emory *A*ccess *P*robe) originally developed by E. Strohmeier & H. Shang from the Future Technology Group at the Lawrence Berkeley National Lab (LBNL), California [2, 3].

The initial idea of the Apex project is the assumption that the performance behaviour of any scientific application can be characterised by a small set of application-specific and architecture independent performance factors. Combining these performance factors, synthetic benchmarks that avoid any hardware specific model can be designed to simulate typical application performance. Assuming that the combination of memory accesses and computational intensity is the dominant performance factor, the Apex-MAP benchmark simulates typical memory access patterns of scientific applications.

Concerning the regularity of the memory access, the original Apex-MAP benchmark focused on random access patterns inside an allocated memory block. Our implementation also considers strided access patterns, which are common in many scientific applications. The benchmark written in the style of Apex-MAP has the following 6 parameters:

- M — The total size of the allocated memory block `data` in which data accesses are simulated,
- L — the vector length of data access, (sub-blocks of length $L < M$ starting at `ind[i]` are accessed in succession), describes the *Spatial Locality*,
- α — the shape parameter of power distribution function ($0 \leq \alpha \leq 1$) determines the random starting addresses `ind[i]`, describes the *Temporal Locality*,
- S — the stride width,
- C — a parameter used to increase the *Computational Intensity* by calling the subroutine `compute(C)`,
- I — the length of the index buffer `ind[]`.

In the case of strided access only the parameters M, S and C are relevant.

The kernel routine for strided access sums up every S-th element of the allocated memory block `data[M]`.

```
for (int k = 0; k < M/S; k+=1) {
    W0 += c0*data[k*S];
    W0 += compute(C);
}
```

To increase the computational intensity, i.e. the ratio of the number of floating point operations and memory accesses, we added calls to the subroutine `compute(C)`:

```
double compute(int C){
  double s0,s1,s2,s3,s4,s5,s6,s7;
  s0=s1=s2=s3=s4=s5=s6=s7=0.;
  for(int i=1;i<=C;i++){
    dummy(&s0,&s1,&s2,&s3,&s4,&s5,&s6,&s7);
    s0+=(x[0]*y[0])+(x[0]*y[1])+(x[0]*y[2])+(x[0]*y[3])+
        (x[0]*y[4])+(x[0]*y[5])+(x[0]*y[6])+(x[0]*y[7]);
    s1+=(x[1]*y[0])+(x[1]*y[1])+(x[1]*y[2])+(x[1]*y[3])+
        (x[1]*y[4])+(x[1]*y[5])+(x[1]*y[6])+(x[1]*y[7]);
    ...
    s7+=(x[7]*y[0])+(x[7]*y[1])+(x[7]*y[2])+(x[7]*y[3])+
        (x[7]*y[4])+(x[7]*y[5])+(x[7]*y[6])+(x[7]*y[7]);
  }
  return s0+s1+s2+s3+s4+s5+s6+s7;
}
```

Performance is usually a mixture of hardware and compiler properties. Braces and calls to a `dummy` routine have been inserted into the `compute` routine to assure that the 128 floating point operations in the loop body are really executed and not cancelled by optimisations of the compiler. On Itanium the generated assembler code contains 64 consecutive `fma` (fused multiply-add) instructions which make optimal use of the floating point registers. One 128 Byte cacheline is sufficient to hold the two data arrays `x[8]` and `y[8]`. The `compute` routine is thus able to run with nearly peak performance on Itanium.

In the case of random access patterns M, L, α, C and I are the relevant parameters. The kernel routine for random memory access is:

```
for (i = 0; i < I; i++) {
    for (k = 0; k < L; k++) {
      W0 += c0*data[ind[i]+k];
      W0 += compute(C);
    }
}
```

In this mode I subblocks of length L are accessed. The vector length L is the number of contiguous memory locations accessed in succession starting at `ind[i]`. L characterises the spatial locality of the data access. The starting addresses of the subblocks are kept in the index buffer `ind[]`. This access pattern is illustrated in Fig. 1 (a).

The starting addresses are random numbers drawn from a power distribution function and are defined as follows:

`ind[j] = (L*pow(drand48(), 1/`α`) * (M/L -1))` $\in$ `[0;M-L[`

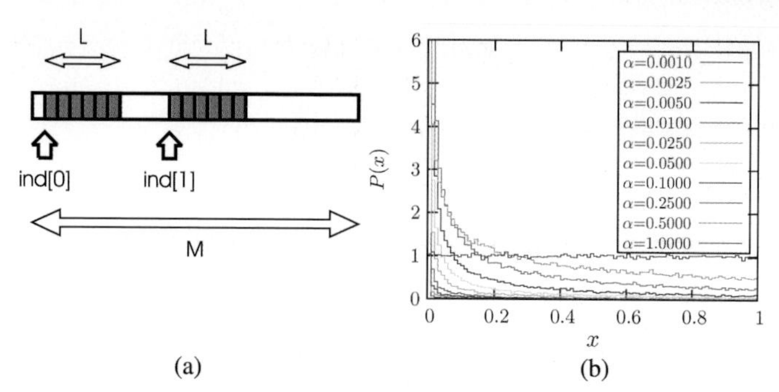

Fig. 1 Random access pattern of Apex-MAP: the left figure illustrates the indexed random access using the index buffer `ind[]`. The starting addresses kept in this array are random numbers drawn from a power distribution function with a probability distribution shown on the right for various values of α

The parameter $\alpha \in [0;1]$ of this distribution function defines the temporal reuse of data. Figure 1 (b) shows the probability distribution of the power function `pow(drand48(), 1/α)`. For $\alpha = 1$ the random numbers are just deviates with a uniform probability distribution, while the smaller α is, the more the distribution function is peaked near 0 and the higher the temporal reuse of data is. For $\alpha = 0$ always the same starting address is used.

1.2 Comparison of Apex-MAP with Real Application Performance

To use Apex-MAP for comparing the average memory bandwidth and the floating point performance of real applications the Itanium performance counters *FP_OPS_RETIRED*, *CPU_OP_CYCLES_ALL* and *L3_MISSES* have been measured and aggregated for various combinations of the Apex-MAP input parameters. The L3 cacheline size of the Itanium is 128 Bytes and can hold 16 64-bit (double-precision) values. The consumed bandwidth between memory and L3 cache is given by *L3_MISSES* $\times$ 128 Bytes.

Figure 2 shows the number of floating point operations per cycle (*FP_OPS_RETIRED / CPU_OP_CYCLES_ALL*) versus the memory bandwidth, expressed by L3 misses in Bytes/cycle (*L3_MISSES/CPU_OP_CYCLES_ALL* $\times$ 128 Bytes). Figure 2 (a) on the left shows this data for real applications running on HLRB II, while Fig. 2 (b) on the right shows data from simulations using the Apex-MAP benchmark with various input parameters.

For the left picture the hardware counters were sampled every 10 minutes on all processors of HLRB II for approximately 3 days with a sampling time of 10 seconds. More than 3.2 Mio. samples are taken into account. The average floating point operations per cycle for this 3-day interval is 0.48, which is equal to 770

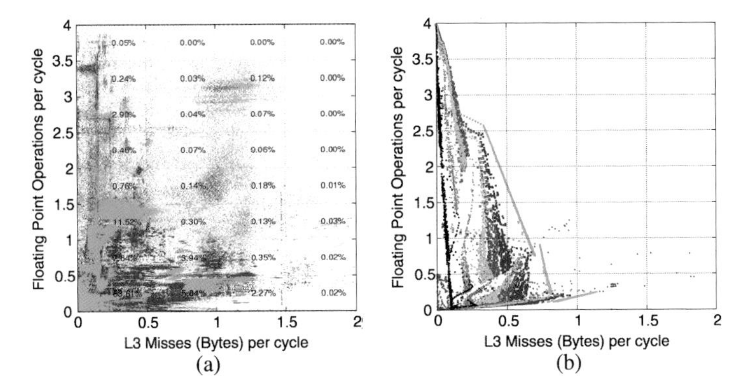

Fig. 2 Comparison of floating point operations per cycle vs. L3 Misses (in Bytes) per cycle for real applications running on HLRB II (a) and for simulations using the Apex-MAP benchmark with various simulation parameters (b)

MFlops per core and 12% of the Itanium's peak performance. The mean for the L3 misses is 0.2 Bytes per cycle.

The parameter space is divided into 32 rectangles of size 0.5 L3 Misses/cycle × 0.5 Flops/cycle. The percentage of data points falling into each rectangle is given.

For the right picture around 23000 different combinations of the Apex-MAP benchmark are used. This picture includes various runs, using both random and strided memory access as well as serial and parallel runs using OpenMP to cover the same areas as the measured data on the left. The range of the simulation parameters for Fig. 2 (b) is $L < M = 1$ GB, $0 \leq \alpha \leq 1$, $2 \leq S \leq 400$, $0 \leq C \leq 1000$, $\mathrm{I} = 50$.

The two pictures demonstrate that it is possible to model the performance of real applications by using suitable combinations of input parameters for the Apex-MAP benchmark. Comparing the two pictures shows that every region in the left picture with significant percentage (i.e. above 0.5%) of data points can be covered by a specific combination of Apex-MAP input parameters. In total, the Apex-MAP runs are able to cover 98.9% of the measured real-application performance data.

The measurements in the left picture are partly based on MPI-parallelised programs. Performance counters are only implicitly able to measure the impact of additional communication overheads, e.g. waiting time for external data on remote processors. Although Apex-MAP focuses on single processor performance it is able to mimic the behaviour of parallel applications as long as the network characteristics stay roughly the same.

1.3 Modelling LRZ's Application Mix

It has been shown that Apex-MAP is able to cover the parameter space that is attained by real applications. It is assumed that Fig. 2 (a) gives a general overview of the application mix running on HLRB II. A good indication for this is given by the

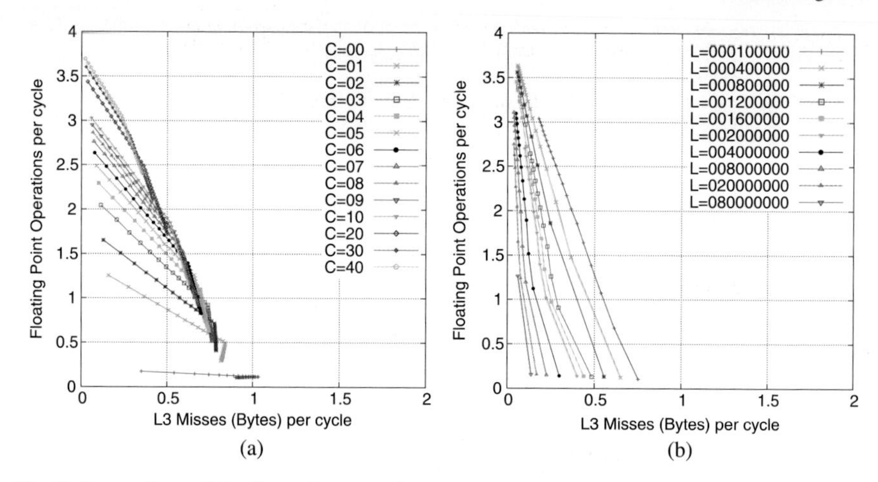

Fig. 3 Comparison of floating point operations per cycle vs. L3 Misses (in Bytes) per cycle for increasing computational intensity C and variations of S for the strided access memory pattern (a) and variations of L for random access memory patterns (b)

fact that the mean MFlops-rate for this 3-day interval is 770 MFlops per core or 12% of peak performance, which is a good approximation of the overall mean application performance of HLRB II (see also [4]). Therefore the weights associated with each rectangle (percentages in Fig. 2 (a)) are used to model the general application mix.

Besides the weights for each rectangle, the most suitable combinations of input parameters for Apex-MAP needs to be found. Figure 3 shows the achievable combinations of Flops versus L3 Misses for different versions of the strided access (a) and random access (b) memory patterns using a serial version of the code. Figure 3 (a) visualises the common understanding of the influence of a stride memory access to performance. Every line corresponds to an increase in computational intensity C. As long as the computational intensity is low (e.g., C stays small), only 1 Flop per clock cycle is possible (which is equal to 25% of peak performance). As the computational intensity grows larger, the codes are able to run at maximum speed with nearly 4 Flops in every cycle. As said before, the L3 cacheline size of the Itanium is 128 Bytes; 16 doubles fit in one cacheline. Therefore with an increase in stride along each line from 1 (contiguous access) to 16 (access only one item per cache line), the performance drops and stays at a minimum for figures above 16 which always need a new cache line.

Figure 4 shows the data points that have been chosen to model the application mix of LRZ. The corresponding Apex-MAP input parameters multiplied with the derived weights are being used to compute an overall performance of the application mix in MFlops. Running the adapted Apex-MAP benchmark on HLRB II yields a performance estimate of 898 MFlops per core. This is quite close to the actual application performance on HLRB II: 14% deviation from the measured 3-day interval.

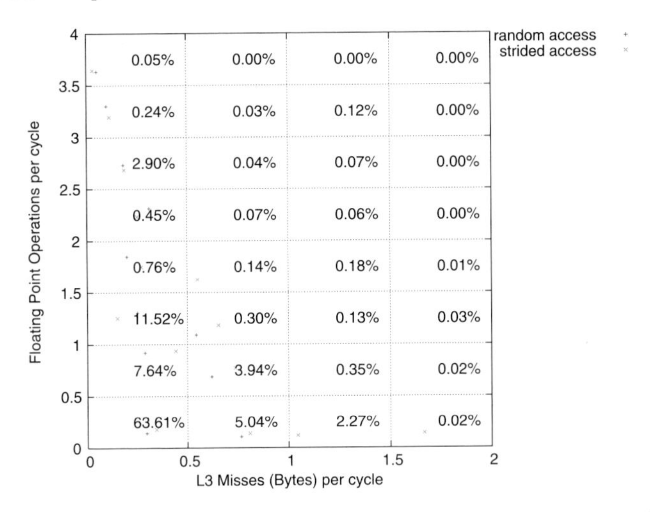

Fig. 4 Chosen data points to model the application mix. These data points represent input parameters of the Apex-MAP benchmark

2 Validation Using the EuroBen Mathematical Kernels

The validation of our Apex-MAP version will be done by using two mathematical kernels typical for many scientific applications. Within the EC FP7 funded project "Partnership for Advanced Computing in Europe" (PRACE, [5]), several mathematical kernels from the EuroBen benchmark suite [8] have been chosen as templates for commonly used scientific applications. To validate Apex-MAP two very distinct codes have been chosen:

- mod2am, a dense matrix-matrix multiplication,
- mod2as, a sparse CSR (compressed sparse row) matrix-vector multiplication.

Within PRACE these codes have been ported to several new languages and architectures; results will be published on the PRACE website in deliverable D6.6 and D8.3.2 [6]. The PRACE surveys analysed the current standards for parallel programming and their evolution, PGAS languages, the languages introduced as a consequence of the DARPA HPCS project and the languages, paradigms and environments for hardware accelerators. Performance data has been gathered for various architectures.

The performance of these benchmarks is well known; many different performance runs have been measured, suitable reference input data sets exists and LRZ was responsible for the MKL, CUDA and RapidMind ports. The first benchmark mod2am has a high computational intensity and is well suited for the use of highly multi-threaded devices. The second benchmark has a low computational intensity and is a template for codes which will benefit from a higher memory bandwidth. Using these benchmarks will ensure that Apex-MAP is able to model the two extremes in terms of computational intensity versus memory access and will allow to validate Apex-MAP for the use on hardware accelerators in the future.

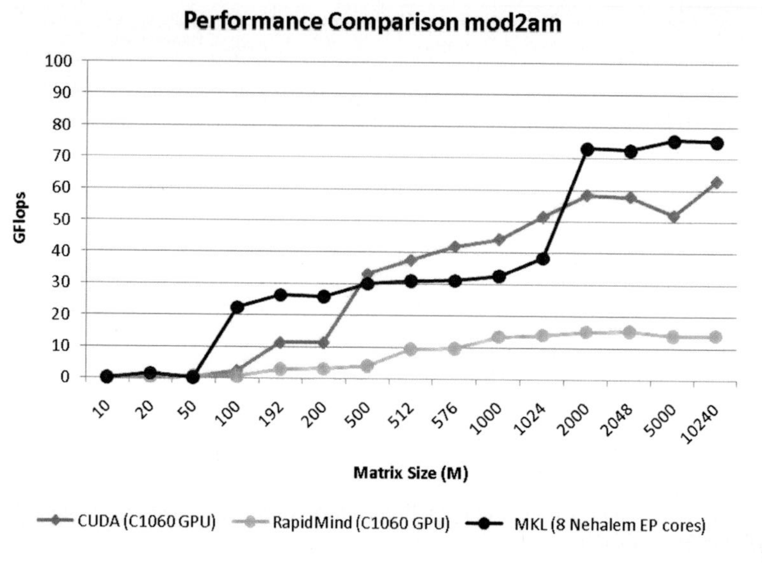

Fig. 5 Comparison of the performance of the dense matrix-matrix multiplications (mod2am, double-precision) for various matrix sizes using RapidMind's CUDA backend, Nvidia's cuBLAS and Intel's Math Kernel Library. The peak performance of one C1060 GPU is comparable to 8 Nehalem EP cores (78 vs. 80 GFlops)

2.1 mod2am: Dense Matrix-Matrix Multiplication

Several PRACE implementations of the matrix-matrix multiplication are based on the BLAS Level 3 routine dgemm. For the x86 implementation the `cblas_dgemm` routine from Intel's MKL (Math Kernel Library) has been used; the CUDA implementation is based on cuBLAS. The RapidMind implementation uses a code-example from the RapidMind developer portal [9] for a general matrix-matrix multiplication code which was slightly adapted. This code is optimised for the use on GPUs.

Figure 5 shows performance measurements from the PRACE project. It compares the performance of the CUDA and RapidMind implementations on an Nvidia C1060 GPU, which is used in Nvidia's Tesla boxes, with the performance of an MKL version on 8 Intel Nehalem EP cores. The reference input data sets are those from PRACE which operate on quadratic matrices. They are described in Deliverable D6.6 available from [6] and have been chosen to firstly, represent frequently used problem sizes and secondly, show the dependency between problem size and performance, especially on hardware accelerators. The double-precision peak performance of one C1060 (78 GFlops) is comparable to 8 Nehalem cores (80 GFlops). The diagram shows that the RapidMind implementation is a factor of 4 slower than the highly optimised cuBLAS library. However, the RapidMind implementation follows roughly the same trend as the CUDA version.

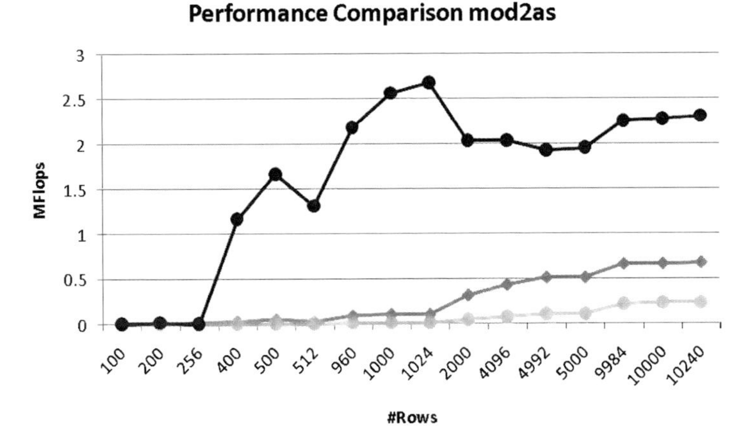

Fig. 6 Comparison of the performance of the sparse matrix-vector multiplications (mod2as, double-precision) for various numbers of rows using RapidMind's CUDA backend, Nvidia CUDA and Intel's MKL. The peak performance of one C1060 GPU is comparable to 8 Nehalem EP cores (78 vs. 80 GFlops)

2.2 mod2as: Sparse Matrix-Vector Multiplication

In the case of the mod2as benchmark the input matrix is stored in the 3-array variation of the CSR (compressed sparse row) format. Using this format only the nonzero elements of the input matrix are stored in one array, and the other two arrays contain information to compute the row and the column of the nonzero elements. The entries of the input matrix are computed using a random number generator but could be reproduced for several runs by using the same seed.

Figure 6 compares the performance of the RapidMind implementation with the CUDA and MKL version. The MKL version makes use of a library call to `mkl_dcsrmv`. The CUDA implementation is based on the paper "Efficient Sparse Matrix-Vector Multiplication on CUDA" [10]. A description of the RapidMind implementation can be found in Deliverable D8.3.2 at [6]. Again, the reference input data sets from PRACE have been used; all data sets contain quadratic matrices of different sizes and fill ratios. The diagram shows that the RapidMind version is a factor of 3 slower than the optimised CUDA version. The trend of both is very similar.

2.3 Validation of Apex-MAP

Validating Apex-MAP by using the two mathematical kernels needs several steps:

1. Measure the performance of mod2am/as on the original hardware (HLRB II).
2. Measure the hardware counters for mod2am/as on HLRB II.

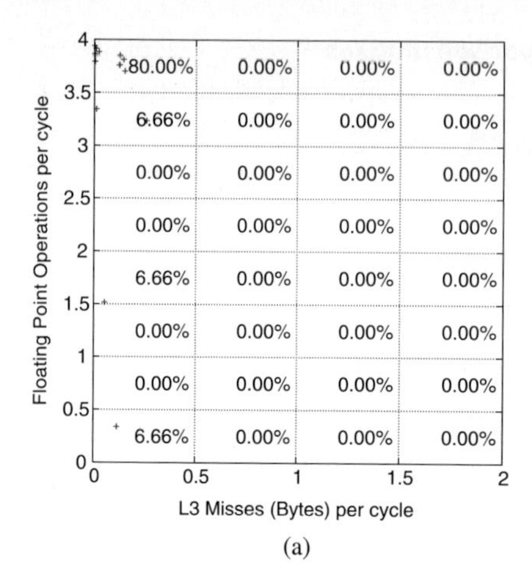

(a)

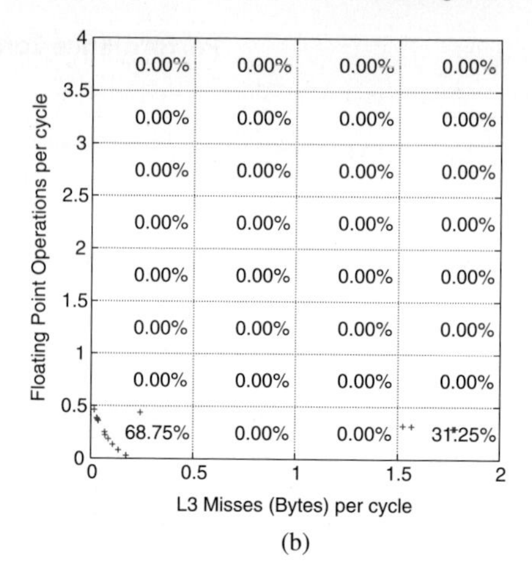

(b)

Fig. 7 Floating point operations per cycle versus L3 Misses (in Bytes) as measured by the hardware counters on HLRB II for mod2am (a) and mod2as (b) (Step 2). The numbers indicate the resulting weights associated with each rectangle (Step 3)

3. Generate weights for each rectangle and each kernel.
4. Measure the performance of mod2am/as on the target hardware (Nehalem EP).
5. Run Apex-MAP with the weights for mod2am/as on Nehalem and HLRB II.
6. Compare the predicted results (Step 5) with the actual results (Steps 1 and 4).

Step 1 yields a mean performance on HLRB II for all reference input data sets of 5.4 GFlops per core for mod2am (84% of peak) and 0.5 GFlops for mod2as (8% peak). Figure 7 shows the hardware counter measurements done in Step 2 and the derived weights for the Apex-MAP runs (Step 3): It can be clearly seen, that the dense matrix-matrix multiplication (a) is compute bound while the sparse matrix-vector multiplication (b) is memory bound.

Step 4 shows an actual performance on the target architecture Nehalem EP of 8.0 GFlops (80%) for mod2am and 0.9 GFlops (9%) for mod2as. The results of Step 5 can be seen in Fig. 8; the mean performance predicted by Apex-MAP deviates only slightly from the measured data on both architectures. The measurements on Nehalem are slightly worse, since the `compute` routine, which has been optimised for Itanium is not able to reach peak performance on Nehalem.

3 Conclusion and Outlook

It has been shown that an adaptation of the Apex-MAP benchmark can be used to model the application mix in order to use it for benchmarking the suitability of new architectures for a computing centre. The adapted benchmark has been validated by using it to predict the performance of two mathematical kernels on two architectures.

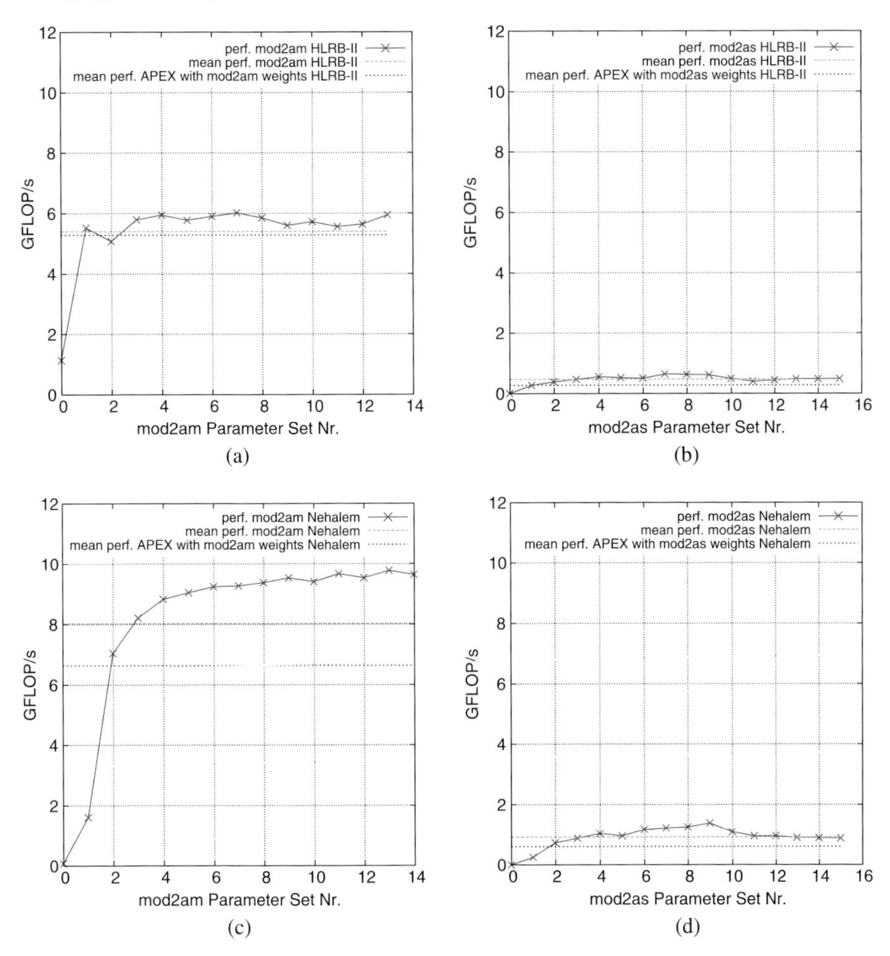

Fig. 8 All figures show the actual performance measured for each reference input data set (red curve) together with the mean performance measured by the mathematical kernel (green line) and predicted by Apex-MAP (blue line). The first line shows results on HLRB II (a,b), the second line results on Nehalem EP (c,d). The left diagrams are based on mod2am (a,c) and the diagrams on the right on mod2as (b,d)

Future work will go mainly into two directions. Firstly we want to investigate in more detail the quality of the predictions to refine the benchmark and ensure that it adapts easily to new environments. Secondly we want to use Apex-MAP to investigate if hardware accelerators are advantageous for our application mix. Hardware accelerators like GPUs and the CELL processor with an enormous peak performance have recently gained much interest in the community. A programming model that was evaluated at LRZ is the multi-core development platform RapidMind, which is a tool that allows generating code for GPUs, the CELL processor and multi-core CPUs with the same source file. Using a RapidMind version of the Apex-MAP benchmark could offer an easy way to simulate typical application performance patterns on a broad range of architectures. RapidMind ports of the two mathematical kernels are already available and could be used to validate a RapidMind Apex-MAP version.

Acknowledgements This work was financially supported by the KONWIHR-II project "OMI4papps" and by the PRACE project funded in part by the EU's 7th Framework Programme (FP7/2007-2013) under grant agreement no. RI-211528.

References

1. R. Patra, M. Brehm, R. Bader, R. Ebner, S. Haupt, Performance Monitoring – A Generic Approach (LRZ-Bericht 2006-06) `http://www.lrz-muenchen.de/wir/berichte/TB/LRZ-Bericht-2006-06.pdf`
2. `https://ftg.lbl.gov/ApeX/ApeX.shtml`
3. E. Strohmaier, H. Shan, Architecture Independent Performance Characterisation and Benchmarking for Scientific Applications, `https://ftg.lbl.gov/ApeX/mascots.pdf`
4. `http://www.lrz-muenchen.de/services/compute/hlrb/betriebszustand/hlrb2_4weeks.html`
5. `http://www.prace-project.eu/`
6. `http://www.prace-project.eu/documents/public-deliverables-1/`
7. `http://www.rapidmind.com/`
8. `http://www.euroben.nl/`
9. `https://developer.rapidmind.com/sample-code/matrix-multiplication-samples/rm-sgemm-gpu-5938.zip`
10. N. Bell, M. Garland, Efficient Sparse Matrix-Vector Multiplication on CUDA, `http://www.nvidia.com/object/nvidia_research_pub_001.html`

Computational Steering of Complex Flow Simulations

Atanas Atanasov, Hans-Joachim Bungartz, Jérôme Frisch, Miriam Mehl, Ralf-Peter Mundani, Ernst Rank, and Christoph van Treeck

Abstract Computational Steering, the combination of a simulation back-end with a visualisation front-end, offers great possibilities to exploit and optimise scenarios in engineering applications. Due to its interactivity, it requires fast grid generation, simulation, and visualisation and, therefore, mostly has to rely on coarse and inaccurate simulations typically performed on rather small interactive computing facilities and not on much more powerful high-performance computing architectures operated in batch-mode. This paper presents a steering environment that intends to bring these two worlds – the interactive and the classical HPC world – together in an integrated way. The environment consists of efficient fluid dynamics simulation codes and a steering and visualisation framework providing a user interface, communication methods for distributed steering, and parallel visualisation tools. The gap between steering and HPC is bridged by a hierarchical approach that performs fast interactive simulations for many scenario variants increasing the accuracy via hierarchical refinements in dependence of the time the user wants to wait. Finally, the user can trigger large simulations for selected setups on an HPC architecture exploiting the pre-computations already done on the interactive system.

Atanas Atanasov · Hans-Joachim Bungartz · Miriam Mehl
Department of Informatics, Technische Universität München, München, Germany
e-mail: atanasoa@in.tum.de
e-mail: bungartz@in.tum.de
e-mail: mehl@in.tum.de

Jérôme Frisch · Ralf-Peter Mundani · Ernst Rank
Chair for Computation in Engineering, Technische Universität München, München, Germany
e-mail: frisch@bv.tum.de
e-mail: mundani@bv.tum.de
e-mail: rank@bv.tum.de

Christoph van Treeck
Fraunhofer-Institut für Bauphysik, Holzkirchen, Germany
e-mail: treeck@ibp.fraunhofer.de

S. Wagner et al. (eds.), *High Performance Computing in Science and Engineering, Garching/Munich 2009*, DOI 10.1007/978-3-642-13872-0_6,

1 Introduction

Computational Steering is the coupling of a simulation back-end with a visualisation front-end in order to interactively exploit design alternatives and/or optimise (material) parameters and shape. Therefore, different aspects such as grid generation, efficient algorithms and data structures, code optimisation, and parallel computing play a dominant role to provide quick results (i. e. several simulation and visualisation updates per second in case of modifications of the underlying data) to keep up the principle of cause and effect, which is necessary to gain better insight and a deeper understanding of problems from the field of engineering applications. Nevertheless, even nowadays interactivity and high-performance computing (HPC) are still a contradiction, as most HPC systems do not provide interactive access to the hardware.

As a remedy for the latter one, a two-stage approach (i. e. interactive preprocessing of "low level" problems and parallel processing of "high level" problems) helps to bridge the gap between small – and typically interactive – systems for a quick quantitative analysis and large – and typically batch – HPC systems for a complex qualitative analysis. Such an approach also provides the advantage of reducing the amount of long and, thus, expensive simulation runs to those necessary only without waisting additional computing time for redundant computations. To ensure a *seamless* transition from "low level" problems on coarse grids with few thousands of unknowns to "high level" problems on fine grids with many millions of unknowns, hierarchical approaches are indispensable.

This also has a significant relevance for the practical usage of computational steering and HPC in industrial applications, as most approaches there suffer from a insufficient integration of HPC into the workflow of industrial processes. Hence, from the very beginning one of our main objectives was to provide a framework for engineering applications that not only addresses challenging mathematical and computer science related questions, but also combines and consolidates the two conflicting aspects of interactivity and high-performance computing. Therefore, we will show the benefits of our framework for the interactive control of different engineering applications running on parallel architectures.

The remainder of this paper is as follows. Section 2 presents the ingredients of the steering environment. As this environment does not cover the whole range of applicability of the underlying approaches, Sect. 3 describes two further applications that have been or will be coupled to the steering and visualisation framework. Finally, we draw a short conclusion and give an outlook on the future work in Sect. 4.

2 Computational Steering Environment

In order to increase the performance, i. e. decrease simulation and visualisation response time of our steering environment as well as to prepare it for a later HPC usage, several measures have been taken. This was done with a straight focus on the two-way approach as described above, where small systems are used for an interactive data exploration before a high-quality analysis (based on the parameters explored) is launched as (massively) parallel job on large HPC systems.

2.1 Hierarchical Approach

The main idea in joining the interactively computed small systems with the large parallel systems computed on HPC architectures is to exploit hierarchies of grid levels or discretisation orders. As a response on each user input, a simulation on a very coarse grid or with lowest discretisation order is triggered such that first visualised results are available very fast. Depending on the time given – that is the time the user wants to wait for more accurate results – the simulation is refined in a recursive manner. Each of these refinement steps adds a new layer of grid points to decrease the mesh width or additional degrees of freedom at existing grid points to enhance the approximation order. This allows to quickly check results for numerous input configurations, to examine those that seem to be relevant more accurately and, finally, to start large HPC simulations only for a few scenarios of particular interest. Hereby the refined simulations already profit from the coarser ones in a full multigrid manner. Codes such as iFluids, Peano, and the p-FEM structural mechanics codes described below naturally fit this approach as they inherently already provide the required hierarchy.

2.2 iFluids

The kernel of our steering framework is a Lattice-Boltzmann fluid solver which has been developed by our group and ported to the former HPC system – the pseudo-vector computer Hitachi SR8000-F1 – installed at Leibniz-Rechenzentrum (LRZ). This fluid solver – called iFluids [15] – was running interactively on the Hitachi while coupled with the interactive visualisation nodes also available at LRZ for computational steering applications. Due to the replacement of the old HPC system with the SGI Altix 4700 severe changes of iFluids became necessary in order to run it successfully on the new system. These changes comprise to switch from a pure MPI-based implementation to a cache-efficient hybrid approach (MPI/OpenMP) to benefit also from the Itanium CPUs' local shared memory as well as to modifiy the communication and data distribution pattern, such that it optimally suits the underlying network topology (2D tori connected via a fat tree) in order to minimise latency.

As the porting of iFluids is still work in progress, current performance measurements (up to 1024 processes) on the Altix do not yet reveal the full potential of the parallel code, nevertheless already sound very promising. For a problem size with 7.5 million degrees of freedom a nearly linear speedup (strong scaling) up to $p = 64$

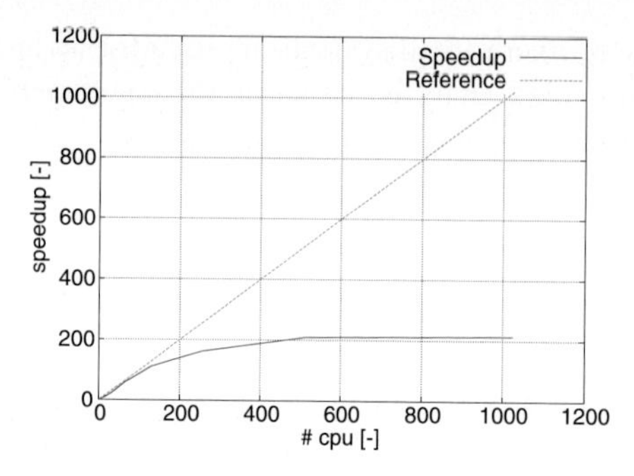

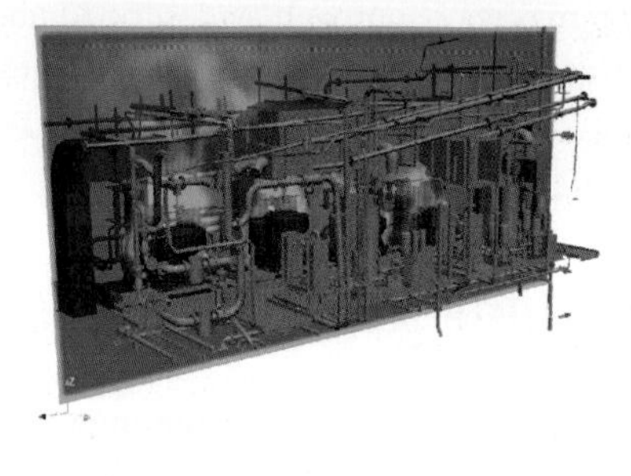

Fig. 1 Performance results running iFluids on the SGI Altix 4700 – left-hand side shows strong speedup values, right-hand side shows simulation results for a complex geometry

processes could be observed which strongly drops for growing numbers of p (see Fig. 1).

Further investigations on this behaviour showed that the major drawback of the current parallelisation is the regular block decomposition of the domain that leads to partitions consisting of mostly or entirely obstacle cells only, for which no computation have to be performed. This leads to an unbalanced load situation. Therefore and due to the frequent geometry and refinement depth changes in a steering environment, a more enhanced adaptive and dynamical load balancing strategy is inevitable. A modified master-slave concept which has been developed by our group (see next section) is being incorporated into iFluids at the moment.

2.3 Adaptive Load Balancing

Within a related project for structural analysis using the p-version finite element method (p-FEM, [12]) – i. e. increasing the polynomial degree p of the shape functions for better accuracy without changing the discretisation – a similar behaviour regarding unbalanced load situations has been observed when using a hierarchical approach (octrees) for domain decomposition [8]. Therefore, we have implemented an adaptive load balancing strategy based on the idea of task stealing—a modified master-slave concept, that takes into account varying workload on the grid nodes.

Here, a master process first analyses the tree and estimates the total amount of work (measured in floating-point operations) per node. In the next step, those nodes are assigned to processes called *traders* – an intermediate layer between master and slaves – to prevent communication bottlenecks in the master and, thus, making this approach also scaleable for large amounts of processes. The traders define tasks (i. e. systems of linear equations for domain partitions), "advertise" them via the master to the slaves, and take care about the corresponding data transfer. They also keep track about dependencies between the tasks and update those dependencies with each result sent back from a slave. Benchmark computations with different ratios of traders and slaves have shown good results with respect to the average percentage a single slave is busy during the entire runtime. This is important to obtain high update rates in case of frequent re-computations which are necessary for interactive computational steering applications.

Hence, iFluids can also benefit from this approach. By applying a hierarchical organisation of the computational domain, the master process could easily identify regions mostly consisting of obstacle cells when doing its work load estimation. Such a region could then be combined with neighbouring regions to a larger task which is processed by a single slave to achieve a better computation-communication-ratio. As this is still work in progress, there are no current results so far.

2.4 Remote Visualisation and Steering Framework

For fast visualisation and user interaction, a remote and parallel visualisation and steering framework has been developed in [2]. It is based on the idea of a distributed application. That is, the steering and visualisation application, the underlying simulation, and the user interface run on separate computing facilties. The interaction between these components is realised via remote procedure calls (RCP) and TCP sockets. As our task is to bring together interactive simulations and visualisations with HPC applications, i. e. large systems of equations to be solved and large data sets to be visualised, the visualisation and simulation are parallel processes themselves as displayed in Fig. 2.

The visualisation is based on the Visualization Toolkit (VTK, [7, 10]). For scalar data sets, it provides a colour mapping as well as iso-lines or iso-surfaces enhanced by cutting planes that can be displaced and rotated interactively. Vector data such as flow velocities are visualised using streamlines, dashed streamlines with glyphs, or streambands. Geometries are represented by surface triangulations and a bounding box widget that allows to scale, displace, or rotate the geometry.

The user interface consists of a 3D-viewer, a geometry catalogue, a geometry browser, and a control panel. It allows the user to change geometries (add, delete, move, or scale geometrical objects), choose data to be visualised (velocities or pressure, e. g.), select visualisation techniques (streamlines or streambands, e. g.), and to examine simulation results from different views and with different techniques. Figure 3 shows a screenshot of the user interface with a visualisation of a fluid dynamics scenario.

The visualisation is parallelised following a data parallel approach. Visualisations are performed in parallel for subdomains of the entire scenario. The bottleneck of this approach is the composition of all subdomain pictures to a picture of the entire scenario at the end of the visualisation process. A binary space partition (BSP) tree approach avoids the accumulation of the whole composition work in one master process. It recursively joins pictures associated to the same father in a bottom-up traversal of the BSP tree.

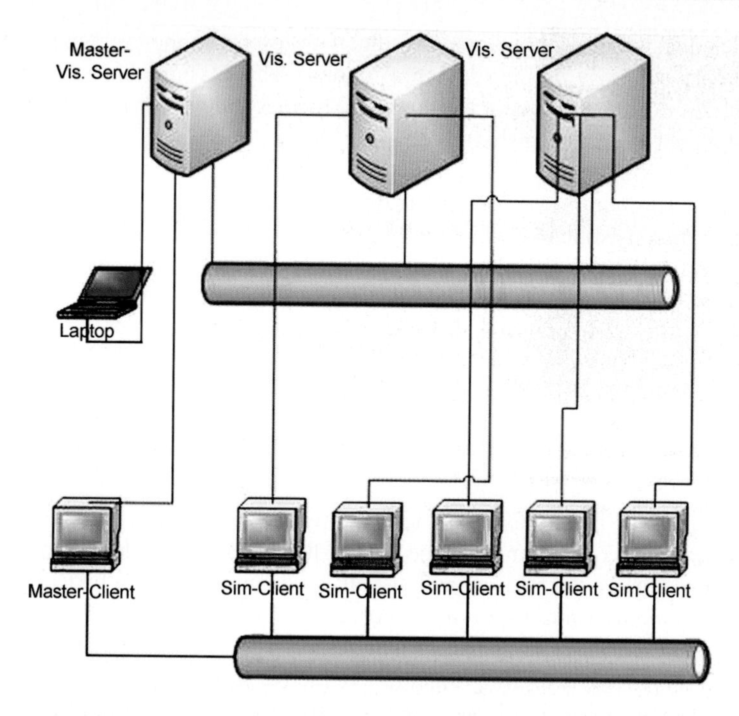

Fig. 2 Steering environment with parallel remote visualisation and parallel simulation (taken from [2])

Figure 4 shows an example of a domain splitting using a BSP tree. In this example, the subdomains D and E would be joined first to a larger domain DE. In a second step, A and B would be joined to AB. In parallel, DE would be joined with C to CDE, and, finally, AB and CDE would be joined to the entire scenario. In our applications such as Peano, we use a particular form of BSP trees – octree-like space-partitioning trees.

In case of Peano (see Sect. 3.1) as simulation code, it is not neccessary to define a new BSP-tree decomposition of the domain for visualisation purposes as Peano already provides it for its own domain decomposition. As this decomposition is already done in a load balanced way and, in case of a non-p-adaptive code such as Peano, simulation costs as well as visualisation costs per inner domain node are approximately constant, it can be efficiently used also for parallel visualisation. Test runs with the steering framework and the CFD solver Peano have been performed at the Linux Cluster (eight-way AMD Opteron, 2.6 GHz, 32 GByte RAM per node) at Leibniz Supercomputing Center (LRZ) in Garching. The visualisation has been done on a Sun X4600 Server with eight quad-core Opterons with 256 GByte RAM per processor and four Nvidia Quadro FX5800 graphic cards. Figure 5 shows the resulting speedup and the costs for picture composition. These results are preliminary and still offer a wide range of optimisation properties both in terms of the number of processors used and in terms of the speedup.

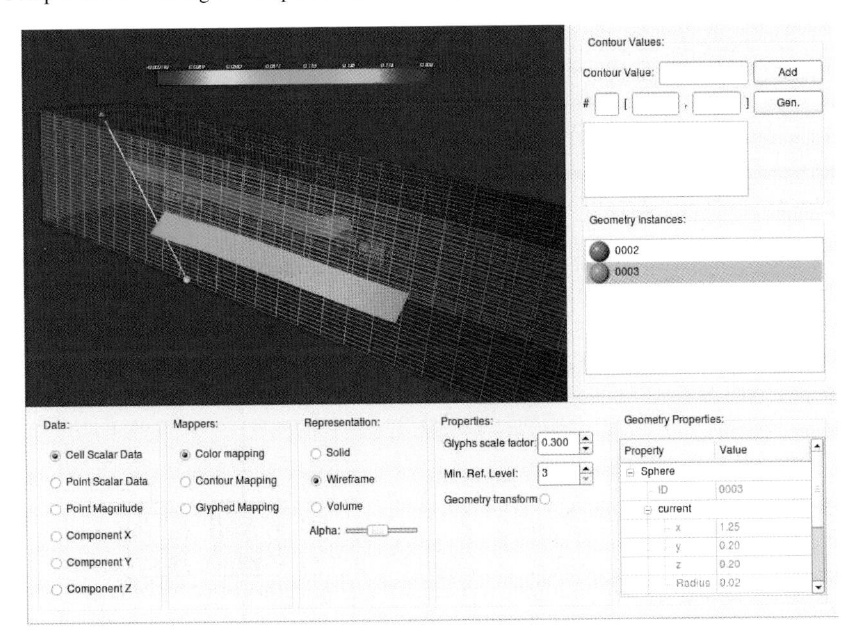

Fig. 3 Steering User Interface with a streamband visualisation for a flow computed with the Peano CFD solver. The scenario is a channel flow with a spherical obstacle. The second sphere has been added at runtime. At the right bottom, properties of the geometry are displayed (taken from [2])

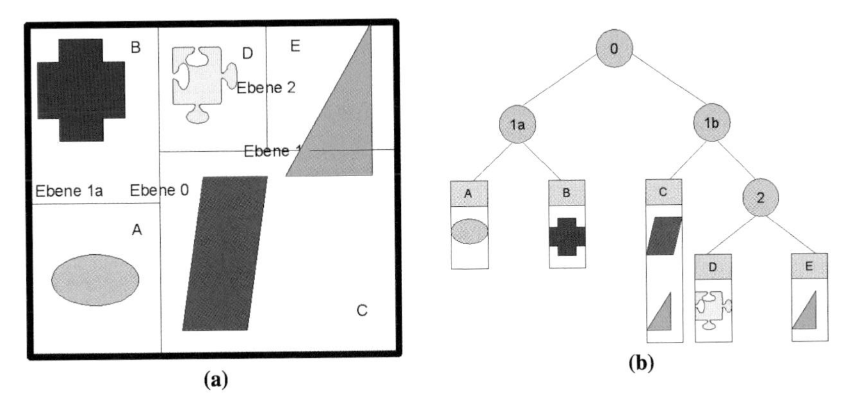

Fig. 4 Example for a BSP-tree domain partitioning for parallel visualisation (taken from [2]). (a) Spatial decomposition according to the BSP-tree; (b) BSP-tree and data structure for this example

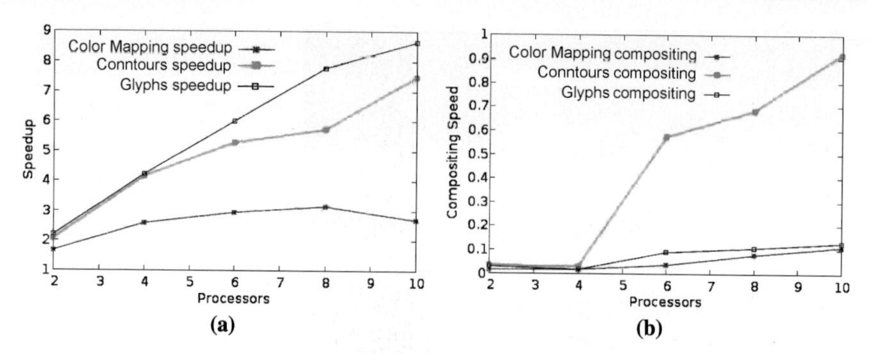

Fig. 5 Speedup evaluation of the parallel visulisation with a domain partitioning defined by the Peano solver and costs for the visualisation composition (taken from [2]). (a) Speedup of the colour mapping, contour generation, and glyphing. (b) Runtime for the composition of pictures

3 Related Applications

In the following, we will highlight some related applications that have been developed independent from iFluids. The first one, the Navier-Stokes solver of the framework Peano, has been the test application during the development of the steering framework. The second one, a thermal comfort assessment application, is a steering application not yet directly related to high-performance computing. However, to refine the underlying model – which will be neccessary in the future – also fluid dynamics will have to be included in the model which will than strongly be related to the main focus of this paper.

3.1 Peano

Peano is a solver framework for partial differential equations (PDE) that works on adaptively refined Cartesian grids corresponding to octree-like tree structures, so-called space-partitioning grids [16]. Within this framework, a Navier-Stokes solver with dynamical grid refinement is implemented [9]. This code fits perfectly with the steering concept described above as it naturally provides the grid hierarchy required for the hierarchical integration of interactive simulations with large HPC batch jobs for selected scenarios. Figure 6 (a) shows the grid hierarchy for a simple two-dimensional example.

The unique selling points of Peano are low memory requirements in combination with high cache hit-rates, efficient multiscale solvers, and efficient and parallel tree-based domain decomposition. Peano has been run on the HLRB II at the Leibniz Supercomputing Center in Garching on up to 900 processors with a speedup of 700 [4]. It can handle moving objects leading to arbitrarily large geometry or even topology changes as it is based on a fixed (Eulerian) grid. Only the adaptive grid refinement is adjusted according a deforming, moving, deleted, or added object (see

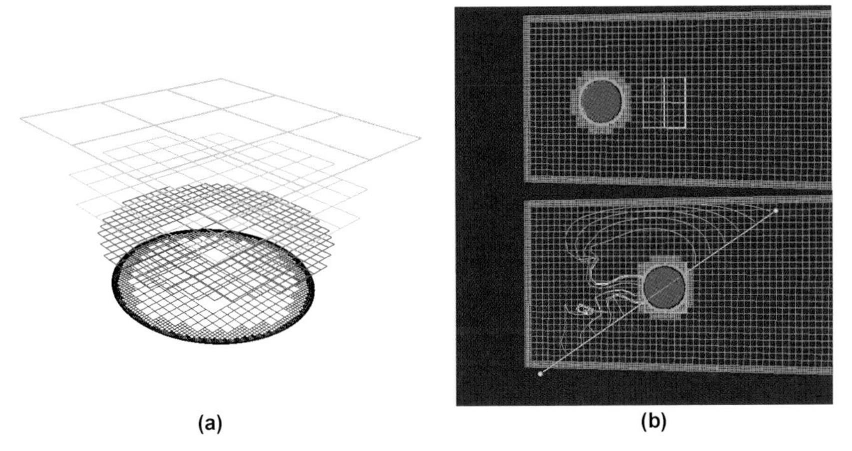

Fig. 6 Peano grids for two-dimensional examples. (a) Grid hierarchy for a spherical domain; (b) adaptive grid refinement following a moving shpere (taken from [2])

Fig. 6 (b)). Such, also particles advected in a flow field can be simulated in a very efficient way [3].

Due to its suitability for both the hierarchical integration approach and the parallel tree-based domain decomposition that can also be used for parallel visualisation, the test runs for the steering framework described in the previous section, have been performed with Peano as a simulation code.

3.2 Thermal Comfort Assessment

3.2.1 Motivation

Indoor climate predictions in office buildings gained increasing importance in the past. The aim of reducing the energy consumption of buildings, and maintaining reasonable indoor temperatures for the occupants at the same time, can be accomplished using simulation tools in the early design stages of the design phase.

In the broader context of the underlying research project COMFSIM [15] three modules were defined. In a first study, a *virtual climate chamber* [14] was designed, which makes use of a human thermoregulation model according to Fiala [5]. Occupants can be situated in a rectangular enclosure with well-defined boundary conditions, such as room and surface temperatures, relative humidity, air velocity and metabolic rate. The latter quantities can be changed during an ongoing simulation using the computational steering concept.

The numerical thermal manikin can be coupled with iFluids [15]. After a series of iterations of the CFD solver, the current boundary conditions at the surface of the manikin shall be delivered to the thermoregulation interface. The existing interface provides the thermal state of the manikin in terms of the resultant surface temperatures and heat fluxes, which may act as new boundary conditions of the manikin in the next CFD step. Using these resulting surface temperatures, a local comfort vote can be calculated using a 7 point ASHRAE scale [1], for example, indicating the comfort state of the manikin. The developed local assessment method of our postprocessing tool has already been published by the authors in [14]. Coupling CFD with the numerical manikin offers the possibility to predict the indoor thermal comfort situation in detail, such as assessing the draught risk, asymmetric radiation, etc. [13]

3.2.2 Thermoregulation Modeling

Thermoregulatory reactions of the central nervous system are an answer of multiple functions of signals from core and peripherals. Local changes in skin temperature additionally cause local reactions such as modifying the sweating rate or the local vasodilatation. Significant indicators are the mean skin temperature and its variation over time and the hypothalamus temperature. The indicators can be correlated with the autonomic responses in order to form a detailed thermoregulation model [5, 11].

Detailed manikin models usually consist of a passive system dealing with physical and physiological properties, including the blood circulation and an active thermoregulation system for the afferent signals analysis [11]. Local clothing parameters are taken into account and the response of the metabolism can be simulated over a wide range of ambient conditions. Besides two-node models (Gagge) [6], multi-segment models are known which are founded on the early work of Stolwijk [11]. Most models use a decomposition of the human body into layers and segments for the passive system which are in thermodynamic contact with each other and with the ambient environment.

As mentioned in section above, the numerical approach for the evaluation of the human thermoregulation for this application was chosen to be the Fiala model. Detailed information can be found in [5].

3.2.3 Computational Steering Approach

The above mentioned procedure can be embedded in a computational steering context. Figure 7 shows the coupling of the virtual climate chamber (VCC) with the thermoregulation interface. The user loads the geometry in to the virtual climate chamber for visualization. There global boundary conditions can be set, governing the chamber climate. The data is transfered to the thermoregulation interface which is coupled to a numerical solver. The aim of the interface is to provide standard interface functions in a way that the numerical model could be exchanged easily. The numerical model computes a small timestep and delivers the results to the interface which sends them to the virtual climate chamber for visualisation purposes. Depending on the just shown results, the user might want to alter some of the boundary

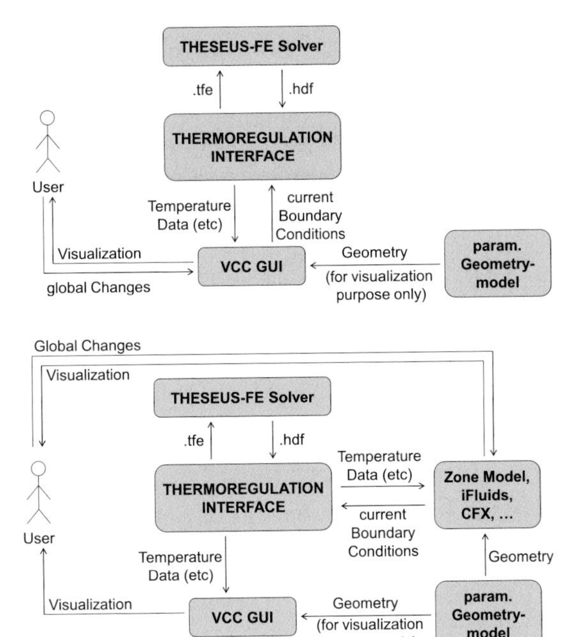

Fig. 7 Coupling concept: *virtual climate chamber* in computational steering mode

Fig. 8 Coupling concept: external CFD solvers or zonal models coupled with the thermoregulation simulation

conditions which will be again transfered to the interface for further treatment and so on.

This procedure is nice for test cases, but is hardly applicable in real applications. Therefore a more realistic coupling is depicted in Fig. 8. The user starts a CFD computation which loads the geometry and scene information. Manikins are now embedded in the geometry and classified as *thermal active components*. The CFD code computes a fixed amount of timestep and delivers the local velocities and temperatures at the manikin's surface which will be transfered to the thermoregulation interface who will pass them on to the solver and deliver the results back to the interface. The resultant surface temperatures are given to the CFD computation which will act as new boundary conditions in the next CFD step. The virtual climate chamber is connected to the thermoregulation interface in *view only* mode in order to observe further detailed information about the numerical thermoregulation simulation like mean values for the whole body as mean skin temperature etc.

4 Summary and Outlook

We proposed tools that combine efficient HPC flow solvers with a steering environment in order to allow both fast interactive simulations for many different scenarios and large HPC simulations for selected scenarios in a hierarchical manner. First tests measuring the performance of the parallel visualisation tools and the simulation codes on high-performance graphics hardware and HPC architectures, resp., show promising results.

In the future, the combination of the presented tools shall be applied to further scenarios and, accordingly, enhanced with more functionality. In particular, the domain decomposition approach of iFluids will be improved and particle simulation methods will be implemented in iFluids and enhanced in Peano.

Acknowledgements Parts of this work have been carried out with the financial support of KONWIHR – the Kompetenznetzwerk für Technisch-Wissenschaftliches Hoch- und Höchstleistungsrechnen in Bayern.

References

1. ASHRAE: Standard 55: Thermal Environmental Conditions for Human Occupancy. American Society of Heating, Refrigerating and Air-Conditioning Engineers, Atlanta (2004)
2. Atanasov, A.: Design and implementation of a computational steering framework for CFD simulations. Diploma Thesis, Institut für Informatik, Technische Universität München (2009)
3. Brenk, M., Bungartz, H.J., Mehl, M., Muntean, I.L., Neckel, T., Weinzierl, T.: Numerical simulation of particle transport in a drift ratchet. SIAM Journal of Scientific Computing **30**(6), 2777–2798 (2008). URL `http://dx.doi.org/10.1137/070692212`
4. Bungartz, H.J., Mehl, M., Neckel, T., Weinzierl, T.: The pde framewirk peano applied to computational fluid dynamics. Computational Mechanics (2009). Accepted
5. Fiala, D.: Dynamic simulation of human heat transfer and thermal comfort. Band 41, De Monfort University Leicester, HFT Stuttgart (1998)
6. Gagge, A.: Rational temp. indices of man's thermal env. and their use with a 2-node model of his temp. reg. Fed. Proc. **32**, 1572–1582 (1973)
7. Laboratories, S.N.: Vtk – visualization toolkit. URL `www.vtk.org`
8. Mundani, R.P., Düster, A., Knežević, J., Niggl, A., Rank, E.: Dynamic load balancing strategies for hierarchical *p*-FEM solvers. In: 16th EuroPVM/MPI Conf., pp. 305–312 (2009)
9. Neckel, T.: The PDE Framework Peano: An Environment for Efficient Flow Simulations. Verlag Dr. Hut (2009)
10. Schroeder, W., Nartin, K., Lorenson, B.: Visualisation Toolkit: An Object-Oriented Approach to 3D Graphics. Kitware (2006)
11. Stolwijk, J.: A mathematical model of physiological temperature regulation in man. Contractor report NASA CR-1855, National Aeronautics and Space Administration, Washington D.C. (1971)
12. Szabó, B., Düster, A., Rank, E.: Encyclopedia of Computational Mechanics, chap. The *p*-version of the Finite Element Method, pp. 119–139. John Wiley & Sons (2004)
13. van Treeck, C., Frisch, J., Egger, M., Rank, E.: Model-adaptive analysis of indoor thermal comfort. In: Building Simulation 2009. Glasgow, Scotland (2009)
14. van Treeck, C., Frisch, J., Pfaffinger, M., Rank, E., Paulke, S., Schweinfurth, I., Schwab, R., Hellwig, R., Holm, A.: Integrated thermal comfort analysis using a parametric manikin model for interactive real-time simulation. J Building Performance Simulation **in press** (2009)
15. van Treeck, C., Wenisch, P., Borrmann, A., Pfaffinger, M., Wenisch, O., Rank, E.: ComfSim - Interaktive Simulation des thermischen Komforts in Innenräumen auf Höchstleistungsrechnern. Bauphysik **29**(1), 2–7 (2007). DOI: 10.1002/bapi.200710000
16. Weinzierl, T.: A Framework for Parallel PDE Solvers on Multiscale Adaptive Cartesian Grids. Verlag Dr. Hut (2009)

Part II
Computational Fluid Dynamics

Numerical Experiments for Quantification of Small-Scale Effects in Particle-Laden Turbulent Flow

Ch. Gobert and M. Manhart

Abstract The present work contains results from numerical simulations of particle-laden isotropic turbulence at high Reynolds and Stokes numbers (up to $Re_\lambda = 265$ and $St = 100$ based on the Taylor length scale and Kolmogorov time scale, respectively). The focus is on the effect of small-scale turbulence on the particles, a modelling issue for LES of particle-laden flow. The results show that in dependence of Stokes number, particles tend to cluster in regions where kinetic energy of the unresolved scales is lower than average. This effect was so far neglected in most LES models for particle-laden flow. Furthermore, the results show that locations for particle clustering and mechanisms leading to clustering are dominated by large-scale dynamics, a promising result for LES.

1 Introduction

Particle-laden turbulent flows can be found in many applications, ranging from medical applications (e.g. deposition in the respiratory tract) to geophysical problems. In the last decades, several methods for the numerical simulation of such flows were developed. Recently, Guha [11] as well as Balachandar and Eaton [4] reviewed the state of the art of computational methods for this field.

In so called Eulerian-Lagrangian methods, the carrier fluid is computed by solving the Navier-Stokes equations and the particle dynamics are computed by tracing single particles in the flow field. If the fluid flow can be computed by direct numerical simulation (DNS), then state of the art Eulerian-Lagrangian methods produce reliable results [4]. On the other hand, in typical applications the Reynolds number is too high to make DNS possible in the near future. With nowadays computers, often Large Eddy Simulation (LES) is the method of choice.

Ch. Gobert · M. Manhart
TU München, FG Hydromechanik, Arcisstr. 21, 80333 München, Germany
e-mail: ch.gobert@bv.tum.de

S. Wagner et al. (eds.), *High Performance Computing in Science and Engineering, Garching/Munich 2009*, DOI 10.1007/978-3-642-13872-0_7,

In LES, only the largest scales are resolved. The effect of the unresolved scales must be modelled. For single phase flows, modelling in LES means modelling the effect of the unresolved scales (small scales or subgrid scales, SGS) on the resolved scales (large scales). Such models are refered to hereinafter as 'fluid-LES models'. For simple configurations, reliable fluid-LES models are readily available [20].

For particle-laden flow, another modelling issue comes into play, namely modelling the effect of the subgrid scales on the particles. A large number of works (see e.g. [2, 9, 15, 26]) showed that in general these effects cannot be neglected. Thus, the need for a corresponding model arises. Hereinafter these models are referred to as 'particle-LES models'.

So far, there is still no satisfactory particle-LES model available. One issue is that, in order to construct such a model, reliable reference data is needed. This data can be computed by DNS of turbulent flow at high Reynolds number but such computations are very expensive in terms of computational costs because high Reynolds number means high computational costs for the carrier fluid. Additionally, the particles must be computed, increasing the computational costs again. Such computations can only be conducted on a supercomputer such as HLRB2.

The present study presents results from such computations. First, the numerical methods for computing flow and particles are presented, sections 2 and 3. In section 4, results for code validation are shown. Sections 5 and 6 discuss the effect of small-scale turbulence on particles. The computational requirements for all simulations are summarized in section 7.

2 Numerical Simulation of the Carrier Flow

In the present work we analyze particle dynamics in forced isotropic turbulence by DNS and LES. For the simulation of the carrier fluid, we use a second-order Finite-Volume method together with a third-order Runge-Kutta scheme proposed by Williamsson [25] for advancement in time. The conservation of mass is satisfied by solving the Poisson equation for the pressure using an iterative solver proposed by Stone [21]. The flow is driven using the deterministic forcing scheme proposed by Sullivan *et al.* [22]. More details on the flow solver can be found in [16].

The flow was computed at three Reynolds numbers, namely $Re_\lambda = 52$, $Re_\lambda = 99$ and $Re_\lambda = 265$. The Reynolds number $Re_\lambda = \frac{\lambda u_{\rm rms}}{\nu}$ is based on the transverse Taylor microscale λ and the rms value of one (arbitrary) component of the fluctuations $u_{\rm rms}$.

In all computations the flow was solved in a cube on a staggered Cartesian equidistant grid. The size of the computational box and the cell width was chosen in dependence of the Reynolds number as recommended by Pope [19]. The testcase $Re_\lambda = 265$ serves for validation of the code. Here, the computational box was chosen smaller than the requirements stated by Pope [19] due to computational limitations. The testcases $Re_\lambda = 52$ and $Re_\lambda = 99$ serve for analysis of small-scale turbulence. Here it is important to choose the computational box such that all scales

Table 1 Simulation parameters and Eulerian statistics from DNS of forced isotropic turbulence. Reynolds number Re_λ, Number of grid points N, length of computational box L, cell width Δx, rate of dissipation ε, Kolmogorov length scale η_K, Kolmogorov time scale τ_K, integral length scale L_f and time scale of energy containing eddies k_f/ε

	DNS		
Re_λ	52	99	265
N	256^3	512^3	1030^3
L/λ	23.8	39.9	48.3
L/L_f	11.9	12.2	14.76
$\Delta x/\lambda$	0.093	0.078	0.047
$\Delta x/\eta_K$	1.34	1.54	1.54
$\varepsilon\lambda^2/u_{rms}^2/\nu$	14.99	15.8	14.8
η_K/λ	0.070	0.050	0.030
$\tau_K u_{rms}/\lambda$	0.248	0.252	0.28
L_f/λ	2.00	3.27	3.27
$k_f/\varepsilon\, u_{rms}/\lambda$	5.15	9.38	28.5

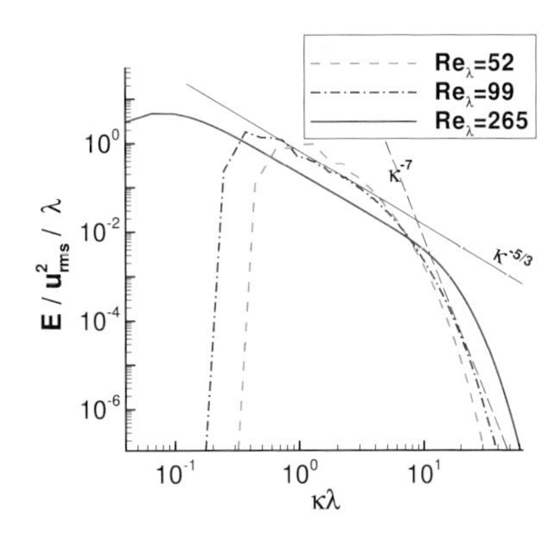

Fig. 1 Instantaneous energy spectrum functions from DNS together with a line proportional to $\kappa^{-5/3}$ (thin continuous line) and a line proportional to κ^{-7} (thin long dashed line)

are resolved. Therefore in these cases all criteria stated by Pope [19] were fulfilled, cf. table 1.

Instantaneous energy spectra $E(\kappa)$ are plotted in figure 1. The simulation at $Re_\lambda = 99$ shows a well established inertial subrange with $E(\kappa) \sim \kappa^{-5/3}$. This is a prerequisite for investigating effects at high Reynolds number turbulence.

3 Discrete Particle Simulation

In this study we consider dilute suspensions of small particles. Thus, effects of the particles on the fluid and particle-particle interactions are neglected (one way coupling).

In all simulations the density of the particles was set to $\rho_p = 1800\rho$ where ρ is the density of the fluid. In each simulation the particles were divided in several fractions with different diameter d. The maximum diameter was always chosen such that the diameter of the largest particles is smaller than the Kolmogorov length scale. Then, particles can be treated as point particles. The particle relaxation time

$$\tau_p = \frac{\rho_p}{\rho} \frac{d^2}{18\nu} \tag{1}$$

ranges from $\tau_p = 0.1\tau_K$ to $\tau_p = 100\tau_K$. Corresponding Stokes numbers $St = \frac{\tau_p}{\tau_K}$ based on the Kolmogorov time scale τ_K range from $St = 0.1$ to $St = 100$.

Based on the works of Armenio and Fiorotto [1] and Kubik and Kleiser [14], we assumed that in the given configurations the acceleration of a particle $\frac{d\mathbf{v}}{dt}$ is given by Stokes drag only,

$$\frac{d\mathbf{v}}{dt} = -\frac{c_D Re_p}{24\tau_p}(\mathbf{v} - \mathbf{u}_{f@p}). \tag{2}$$

Here, $\mathbf{v}(t)$ denotes the particle velocity and $\mathbf{u}_{f@p}$ the fluid velocity at the particle position. The particle Reynolds number Re_p is based on particle diameter and particle slip velocity $\|\mathbf{u}_{f@p} - \mathbf{v}\|$ which leads to a nonlinear term for the Stokes drag. The drag coefficient c_D was computed in dependence of Re_p according to the scheme proposed by Clift *et al.* [7]. In the present simulations, time averaged Re_p range from 0.005 to 7.2. At these particle Reynolds numbers, the correction term $c_D Re_p/24$ ranges from 1 to 1.6.

The fluid velocity $\mathbf{u}_{f@p}$ must be evaluated at the particle position $\mathbf{x}_p(t)$, i.e. $\mathbf{u}_{f@p} = \mathbf{u}(\mathbf{x}_p(t),t)$. Hence, these values must be interpolated. In the present work, a standard fourth-order interpolation scheme was implemented, following the recommendations of Yeung & Pope [27] and Balachandar & Maxey [5]. For the analysis of preferential concentration (section 6), Meyer & Jenny [18] pointed out that conservativity is important. Thus, we implemented the conservative second-order scheme presented in [10] for these simulations.

Equation (2) is a stiff ordinary differential equation. The numerical scheme for integrating equation (2) must be capable to handle this. Therefore, equation (2) was solved by a Rosenbrock-Wanner method [12]. This method is a fourth-order method with adaptive time stepping. The stiff term in equation (2) is linearized in each time step and discretized by an implicit Runge-Kutta scheme.

In all simulations the particles were initialized at random positions (homogeneous distribution) inside the computational box and traced until a statistical steady state was obtained. Then, statistics were recorded. The time span for recording statistics was set in dependence of the highest Stokes number such that statistics were sampled over at least 10 times the particle relaxation time.

4 Validation of the Code

The code was validated via comparison against experimental results and results from DNS found in literature. One very sensitive and thus very well suited quantity for

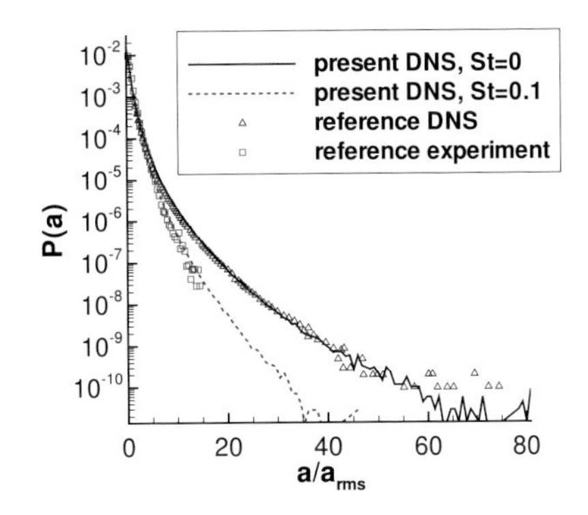

Fig. 2 Probability density function of particle acceleration. X-axis is normalized with respect to the rms value of **a**. Triangles: reference DNS of $St = 0$ particles conducted by Biferale *et al.* [6]. Squares: reference experiment of $St = 0.09 \pm 0.03$ particles conducted by Ayyalasomayajula *et al.* [3] (renormalized)

validation is the probability density function (PDF) for the particle acceleration. To this end, a DNS of forced isotropic turbulence at $Re_\lambda = 265$ on 1030^3 grid points was conducted. This data was then compared to data from a DNS conducted by Biferale *et al.* [6] and an experiment conducted by Ayyalasomayajula *et al.* [3]. Biferale *et al.* conducted a DNS at $Re_\lambda = 280$ and traced inertia free particles (i.e. $St = 0$). Ayyalasomayajula *et al.'s* experiment was at $Re_\lambda = 250$ with particle Stokes numbers $St = 0.09 \pm 0.03$. Correspondingly, in the present simulation two particle fractions were traced, one at $St = 0$ and another at $St = 0.1$. Each fraction consists of 960000 particles. Figure 2 shows that the results from the present simulations agree very well with the reference data.

5 Effect of the SGS Turbulence on the Kinetic Energy Seen by the Particles

Most particle-SGS models proposed so far are based on the reconstruction of the fluid velocity seen by the particles. Accordingly, we present in the present section statistics of the kinetic seen by the particles. It will be shown that these do not monotonously depend on Stokes number because of particle clustering. This is a challenge for models reconstructing the fluid velocity seen by the particles.

In the present section small-scale effects on the particles are quantified by a priori analysis. A priori analysis means comparison of DNS data against filtered DNS data. In our study, we extracted large-scale fluctuations using a box filter, i.e. we computed

$$\hat{\mathbf{u}}(\mathbf{x},t) = \frac{1}{\Delta^3} \iiint\limits_{[-\Delta/2,\Delta/2]^3} \mathbf{u}(\mathbf{x}+\mathbf{r},t)d\mathbf{r}. \tag{3}$$

$\mathbf{u}$ is the fluid velocity computed from DNS and Δ is the filter width. We set Δ such that the energy of the filtered field $\hat{k}_f = \langle \hat{u}_i^2 \rangle /2$ is 87 - 88 % of the energy of the unfiltered field $k_f = \langle u_i^2 \rangle /2$, cf. table 2. $\langle \cdot \rangle$ denotes spatial and temporal averaging.

Table 2 Simulation parameters for filtered DNS. Reynolds number Re_λ, filter width Δ and ratio of kinetic energy of the filtered field $\hat{k}_f$ to kinetic energy of the unfiltered field k_f

	filtered DNS	
Re_λ	52	99
$\Delta/\Delta x$	7	9
$\hat{k}_f/k_f$	87%	88%

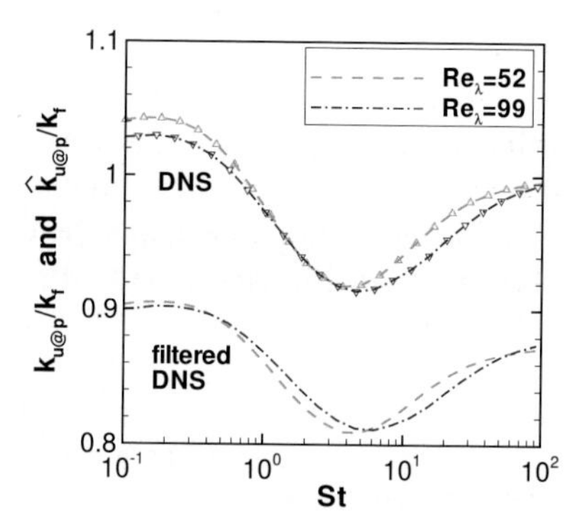

Fig. 3 A priori analysis: Unfiltered and filtered kinetic energy of the fluid seen by the particles computed from DNS

LES with a perfect fluid- and particle-SGS model would give, in a statistical sense, the same particle trajectories as DNS. Therefore we analyze small-scale turbulence along exact particle trajectories in a priori analysis, i.e. we solve the particle transport equation (2) using the (unfiltered) fluid velocity $\mathbf{u}_{f@p}$ and not the filtered velocity $\hat{\mathbf{u}}_{f@p}$. We record $\hat{\mathbf{u}}_{f@p}$ along the particle paths to differentiate between large-scale and small-scale turbulence but $\hat{\mathbf{u}}_{f@p}$ does not affect the particle trajectories.

In figure 3, the unfiltered and filtered kinetic energies of the fluid seen by the particles

$$k_{u@p} = \frac{1}{2}\left\langle u_{f@p,i}^2 \right\rangle = \frac{1}{2}\left\langle u_i^2(\mathbf{x}_p(t),t) \right\rangle \text{ and} \tag{4}$$

$$\hat{k}_{u@p} = \frac{1}{2}\left\langle \hat{u}_{f@p,i}^2 \right\rangle = \frac{1}{2}\left\langle \hat{u}_i^2(\mathbf{x}_p(t),t) \right\rangle \tag{5}$$

computed from DNS are depicted. $\langle\cdot\rangle$ denotes averaging over particles with same diameter and additonally averaging over time.

Figure 3 shows that the general shape of the function $k_{u@p}(St)$ is similar in all configurations. For $St = 0$ (not shown due to logarithmic scale), the particles follow the fluid exactly and $k_{u@p}$ equals the kinetic energy of the fluid k_f. For $St \to \infty$, particles are at rest and $k_{u@p}$ also converges with respect to k_f. In the intermediate range, $k_{u@p}$ attains different values due to particle clustering.

The Reynolds number effect on the shape of $k_{u@p}$ is evident. At the higher Reynolds number, $k_{u@p}$ approaches the limit value k_f more slowly indicating a wider spread of the length scales involved in the particles' dynamics.

For $St \lesssim 1$, filtering obscures the Reynolds number effects on kinetic energy seen by the particles. This indicates that the clustering of small particles is mainly dependent on small scales.

It is well understood that particles tend to cluster in low-vorticity and high-strain-rate regions [23]. The results presented additionally show that, dependent on Stokes number, particles may cluster in regions where the kinetic energy of the carrier fluid is either higher or lower than average. Particles at $St \lesssim 1$ cluster in regions with higher turbulent kinetic energy and those at $1 < St \lesssim 10$ in regions with lower turbulent kinetic energy.

Figure 3 shows that filtering leads to an essentially Stokes number independent shift of $k_{u@p}$, $\hat{k}_{u@p}(St)k_{u@p}(0) \approx k_{u@p}(St)\hat{k}_{u@p}(0)$. This means that the large eddies determine the locations for particle clustering.

In order to gain more insight into small-scale effects, we decomposed $\mathbf{u}_{f@p}$ into large-scale velocity $\hat{\mathbf{u}}_{f@p}$ and small-scale (SGS) velocity $\mathbf{u}'_{f@p} := \mathbf{u}_{f@p} - \hat{\mathbf{u}}_{f@p}$. The kinetic energy can be decomposed by

$$k_{u@p} = \frac{1}{2}\left\langle u^2_{f@p,i}\right\rangle = \hat{k}_{u@p} + \left\langle \hat{u}_{f@p,i}u'_{f@p,i}\right\rangle + \underbrace{\frac{1}{2}\left\langle u'_{f@p,i}u'_{f@p,i}\right\rangle}_{=k'_{u@p}}. \tag{6}$$

$\left\langle \hat{u}_{f@p,i}u'_{f@p,i}\right\rangle$ is the covariance of large and small scales. It should be noted that for sharp spectral filters, $\left\langle \hat{u}_{f@p,i}u'_{f@p,i}\right\rangle = 0$ but for most other filters, including Smagorinsky and top hat filter, this does not hold [8].

In figure 4, $\left\langle \hat{u}_{f@p,i}u'_{f@p,i}\right\rangle$ and $k'_{u@p}$ are depicted. At the lower Reynolds number most of the residual turbulent kinetic energy is in the covariance between large and small scales. Due to the limited spectral range, the filter cannot clearly separate the small scales from the large scales. Towards the higher Reynolds number, the magnitude of the covariance $\left\langle \hat{u}_{f@p,i}u'_{f@p,i}\right\rangle$ decreases whereas the magnitude

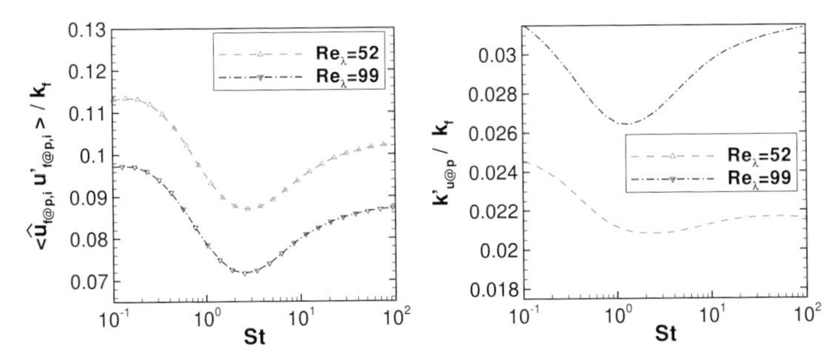

Fig. 4 A priori analysis: Covariance of filtered velocity and SGS fluctuations seen by the particles (left) and kinetic energy of the SGS fluctuations seen by the particles (right)

of $k'_{u@p}$ increases. This is due to the widening of the energy spectrum and increase of filter width with respect to the Kolmogorov length scale. For LES at very high Reynolds numbers, the filter width will be very large in comparison to the Kolmogorov length scale and it can be expected that $k'_{u@p}$ becomes dominant in comparison to $\langle \hat{u}_{f@p,i} u'_{f@p,i} \rangle$ in this case.

Furthermore, figure 4 shows that $k'_{u@p}$ depends clearly on Stokes number, in particular at the higher Reynolds number. For typical LES it can be expected that $k'_{u@p}$ becomes strongly dependent on Stokes number. Particles with $St \approx 1$ tend to cluster in regions with lower SGS activity than the ones with very small or very large Stokes numbers. They even cluster in regions with lower than average SGS activity. A model for the small-scale effects on particles would need to mimic this.

6 Effect of the SGS Turbulence on Preferential Concentration

In most turbulent flows one can observe particle clustering due to the interaction of centrifugal forces and Stokes drag, called *preferential concentration*.

One method to detect preferential concentration is to divide the test section into boxes and to compute a histogram of the number of particles per box. This method was proposed by Fessler *et al.*. The histogram leads to the distribution P_{pc}. For homogeneously distributed particles, P_{pc} is the Poisson distribution $P_{Poisson}$. If preferential concentration occurs then P_{pc} and $P_{Poisson}$ differ. The first moment of these distributions is the average number of particles per box and therefore the first moments of both distributions are equal. Therefore one compares second moments and defines as measure for preferential concentration

$$\Sigma = \frac{\sigma_{pc} - \sigma_{Poisson}}{\sigma^2_{Poisson}}. \tag{7}$$

Here, σ denotes the standard deviation of P_{pc} and $\sigma_{Poisson}$ the standard deviation of $P_{Poisson}$, respectively.

In the present section the effect of the unresolved scales on preferential concentration is analyzed. To this end, an a posteriori analysis at $Re_\lambda = 52$ was conducted, i.e. we compared DNS against LES data. As LES model, the Lagrangian dynamic Smagorinsky model proposed by Meneveau *et al.* [17] was implemented. The resolution of the LES was chosen such that the kinetic energy resolved by LES $\bar{k}_f$ and the kinetic energy of the filtered DNS field $\hat{k}_f$ are equal. This leads to a resolution of 42^3 points. The LES velocity is denoted by $\bar{\mathbf{u}}$.

In order to analyze preferential concentration, the number of particles must be large enough to resolve the scale on which clustering occurs. In isotropic turbulence, these scales are known to be in the range of $2-6\eta_K$ (cf. e.g. [13]). Therefore in the present work Σ was computed by using boxes of dimensions $(3\eta_K)^3$. Computational requirements for the particles limit the box size and therefore the Reynolds number. In the present work, all results on preferential concentration were obtained from

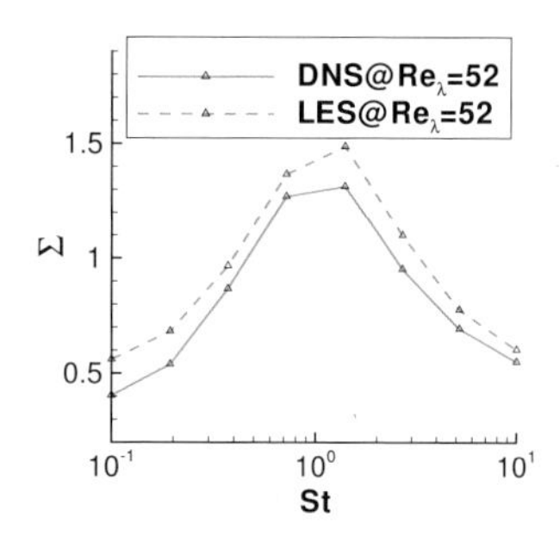

Fig. 5 A posteriori analysis of preferential concentration: Continuous lines: DNS, dashed lines: LES

the $Re_\lambda = 52$ testcase. We traced 5 Mio. particles per fraction. Thus, in a box of dimensions $(3\eta_K)^3$ one will find in average 3.4 particles.

In these simulations, the particles' Stokes numbers range from 0.1 to 10. In each simulation, 10 time samples were taken. The time lag between the samples is approximately 10 Kolmogorov time scales.

In figure 5 the accumulation Σ is depicted. For interpolation of fluid velocity to particle position we implemented a conservative second-order interpolation scheme following the recommendations of Meyer & Jenny [18]. Evidently DNS and LES results differ somewhat but nevertheless LES gives qualitatively the same result, namely that preferential concentration is strongest around $St = 1$. This is in accordance with the findings of Wang & Squires [24] who also found that LES predicts well preferential concentration.

7 Computational Requirements

The computational requirements for the three testcases can be split into requirements for the flow and requirements for the particles. For the flow, the requirements are given by the Reynolds number. At the highest Reynolds number, the flow solver took 3750 CPU seconds per time step, at the lowest Reynolds number only 48CPU seconds per time step.

The requirements for the particles depend on the number of particles. For code validation it was sufficient to trace a comparatively small number of particles. Thus in this case the requirements for the particles are negligible in comparison to the requirements for the flow. At $Re_\lambda = 99$, detailed statistics were computed for several Stokes numbers. The number of particles must be sufficiently high to cover the complete domain. Here, the CPU time for the particles is comparable to the CPU time for the flow. For computing preferential concentration, the number density of the particles must be sufficiently high to resolve the particle clusters. The Reynolds number is low but the number density of particles is high. Therefore the CPU time was mainly determined by the particles in this case.

Additionally, the CPU time for the particles depends on Stokes number because for small Stokes numbers the adaptive solver chooses a smaller time step than for

Table 3 Table of computational requirements. Reynolds number Re_λ, Number of grid points N, Number of particles in millions N_p, Number of CPU cores N_c, CPU seconds per time step per million grid points α, CPU seconds per time step per million particles β. For the testcase $Re_\lambda = 52$ with 1.92 Mio. particles memory requirements were not an issue

Computational requirements				
Purpose	Code validation	SGS effects on kinetic energy		SGS effects on pref. concentration
Re_λ	265	99	52	52
N	1030^3	512^3	256^3	256^3
N_p in Mio.	1.92	1.92	1.92	40
St range	0-0.1	0.1-100	0.1-100	0.1-10
CPU seconds per time step for flow	3750	416	48	48
CPU seconds per time step for particles	125	64	68	2240
Memory (GB)	461	30	n.a.	21
N_c	250	64	8	64
α	3.3	3.1	2.9	2.9
β	65	33	35	56

large Stokes numbers. Table 3 lists all the requirements. α stands for CPU seconds for computation of flow without particles per time step per million grid points, β stands for CPU seconds per time step per million particles, excluding the computational time for the flow.

8 Conclusions

In this work we present results from simulations of particle-laden homogeneous isotropic turbulence at Reynolds numbers $Re_\lambda = 52$, 99 and 265. Stokes numbers based on the Kolmogorov time scale range from 0 to 100.

The analysis focusses on the effect of small-scale turbulence on particles, a modelling issue for LES. Published models are based on the reconstruction of the fluid velocity seen by the particles $\mathbf{u}_{f@p}$. The unresolved fluid velocity is usually modelled independently of Stokes number. In the present study we found that this assumption is questionable, leading to complications for modelling.

Furthermore, we found that particles with intermediate relaxation times tend to accumulate in regions with low turbulent kinetic energy. In other words, for intermediate relaxation times the locations for particle clusters correlate with locations where kinetic energy is small.

Besides locations for clustering we also analyzed the intensity for clustering. In accordance with previously published results, particle clustering is found to be maximal for particles with relaxation times around the Kolmogorov time scale. We found that the small-scale turbulence has little effect on the intensity of clustering.

References

1. Armenio, V., Fiorotto, V.: The importance of the forces acting on particles in turbulent flows. Phys. Fluids **13**(8), 2437–2440 (2001). DOI 10.1063/1.1385390. URL http://dx.doi.org/10.1063/1.1385390
2. Armenio, V., Piomelli, U., Fiorotto, V.: Effect of the subgrid scales on particle motion. Phys. Fluids **11**(10), 3030–3042 (1999)
3. Ayyalasomayajula, S., Gylfason, A., Collins, L.R., Bodenschatz, E., Warhaft, Z.: Lagrangian measurements of inertial particle accelerations in grid generated wind tunnel turbulence. Phys. Rev. Lett. **97**(14), 144507 (2006). DOI 10.1103/PhysRevLett.97.144507. URL http://dx.doi.org/10.1103/PhysRevLett.97.144507
4. Balachandar, S., Eaton, J.K.: Turbulence dispersed multiphase flow. Ann. Rev. Fluid Mech. **42**(1) (2010). DOI 10.1146/annurev.fluid.010908.165243. URL http://dx.doi.org/10.1146/annurev.fluid.010908.165243
5. Balachandar, S., Maxey, M.R.: Methods for evaluating fluid velocities in spectral simulations of turbulence. J. Comput. Phys. **83**(1), 96–125 (1989)
6. Biferale, L., Boffetta, G., Celani, A., Devenish, B.J., Lanotte, A., Toschi, F.: Multifractal statistics of Lagrangian velocity and acceleration in turbulence. Phys. Rev. Lett. **93**(6), 064502 (2004). DOI 10.1103/PhysRevLett.93.064502. URL http://dx.doi.org/10.1103/PhysRevLett.93.064502
7. Clift, R., Grace, J.R., Weber, M.E.: Bubbles, Drops and Particles. Academic Press, New York (1978)
8. Domaradzki, A.J., Saiki, E.M.: A subgrid-scale model based on the estimation of unresolved scales of turbulence. Physics of Fluids **9**(7), 2148–2164 (1997). URL http://scitation.aip.org/getabs/servlet/GetabsServlet?prog=normal&id=PHFLE6000009000007002148000001&idtype=cvips&gifs=yes
9. Fede, P., Simonin, O.: Numerical study of the subgrid fluid turbulence effects on the statistics of heavy colliding particles. Phys. Fluids **18**(4), 045103 (2006). DOI 10.1063/1.2189288. URL http://dx.doi.org/10.1063/1.2189288
10. Gobert, C., Schwertfirm, F., Manhart, M.: Lagrangian scalar tracking for laminar micromixing at high schmidt numbers. In: Proceedings of the 2006 ASME Joint U.S.-European Fluids Engineering Summer Meeting. Miami, Fl. (2006)
11. Guha, A.: Transport and deposition of particles in turbulent and laminar flow. Ann. Rev. Fluid Mech. **40**(1), 311–341 (2008). DOI 10.1146/annurev.fluid.40.111406.102220. URL http://dx.doi.org/10.1146/annurev.fluid.40.111406.102220
12. Hairer, E., Wanner, G.: Solving Ordinary Differential Equations II. Stiff and Differential-Algebraic Problems. Springer Series in Computational Mathematics. Springer, New York (1990)
13. Hogan, R.C., Cuzzi, J.N.: Stokes and reynolds number dependence of preferential particle concentration in simulated three-dimensional turbulence. Physics of Fluids **13**, 2938–2945 (2001). DOI 10.1063/1.1399292. URL http://dx.doi.org/10.1063/1.1399292
14. Kubik, A., Kleiser, L.: Forces acting on particles in seperated wall-bounded shear flow. Proc. Appl. Math. Mech. **4**, 512–513 (2004)
15. Kuerten, J.G.M., Vreman, A.W.: Can turbophoresis be predicted by large-eddy simulation? Phys. Fluids **17**, 011701 (2005)
16. Manhart, M.: A zonal grid algorithm for DNS of turbulent boundary layers. Comput. Fluids **33**(3), 435–461 (2004)

17. Meneveau, C., Lund, T.S., Cabot, W.H.: A Lagrangian dynamic subgrid-scale model of turbulence. J. Fluid Mech. **319**, 353–385 (1996)
18. Meyer, D.W., Jenny, P.: Conservative velocity interpolation for PDF methods. Proc. Appl. Math. Mech. **4**(1), 466–467 (2004)
19. Pope, S.B.: Turbulent Flows. Cambridge University Press, Cambridge, UK (2000)
20. Sagaut, P.: Large Eddy Simulation for Incompressible Flows. Springer (2006). URL http://books.google.de/books?hl=de&lr=&id=ODYiH6RNyoQC&oi=fnd&pg=PA1&dq=sagaut+Large+Eddy+Simulation+for+Incompressible+Flows&ots=XdayeamC04&sig=n8-WY6PRTKbN9_DvBAE473d0tgs
21. Stone, H.L.: Iterative solution of implicit approximations of multidimensional partial differential equations. SIAM J. Num. Anal. **5**(3), 530–558 (1968). URL http://scitation.aip.org/getabs/servlet/GetabsServlet?prog=normal&id=SJNAAM000005000003000530000001&idtype=cvips&gifs=yes
22. Sullivan, N.P., Mahalingam, S., Kerr, R.M.: Deterministic forcing of homogeneous, isotropic turbulence. Phys. Fluids **6**(4), 1612–1614 (1994)
23. Wang, L.P., Maxey, M.R.: Settling velocity and concentration distribution of heavy particles in homogeneous isotropic turbulence. J. Fluid Mech. **256**, 27–68 (1993). URL http://adsabs.harvard.edu/cgi-bin/nph-bib_query?bibcode=1993JFM...256...27W
24. Wang, Q., Squires, K.D.: Large eddy simulation of particle-laden turbulent channel flow. Phys. Fluids **8**(5), 1207–1223 (1996)
25. Williamson, J.H.: Low-storage Runge-Kutta schemes. J. Comput. Phys. **35**, 48–56 (1980)
26. Yang, Y., He, G.W., Wang, L.P.: Effects of subgrid-scale modeling on Lagrangian statistics in large-eddy simulation. J. Turbul. **9**, N8 (2008). DOI 10.1080/14685240801905360. URL http://dx.doi.org/10.1080/14685240801905360
27. Yeung, P.K., Pope, S.B.: An algorithm for tracking fluid particles in numerical simulations of homogeneous turbulence. J. Comput. Phys. **79**(2), 373–416 (1988)

On the Turbulence Structure in a Supersonic Diffuser with Circular Cross-Section

Somnath Ghosh and Rainer Friedrich

Abstract Effects of deceleration and mean compression on the turbulence structure of supersonic flow in a diffuser with an incoming supersonic fully-developed turbulent pipe flow are studied by means of DNS. Strong enhancement of the turbulence activity is observed when the flow undergoes deceleration. Turbulence production and pressure-strain terms in the Reynolds stress budgets are found to increase dramatically leading to increased Reynolds stresses. The central role of pressure-strain correlations in modifying the turbulence structure under these flow conditions is demonstrated.

1 Introduction

Decelerated compressible wall-bounded turbulent shear flows are still of great practical and theoretical interest and provide a challenge for improved turbulence modeling. Incompressible adverse pressure gradient (APG) shear flows have been studied experimentally among others by [8] and numerically by [3] and [6]. When the flow is compressible and undergoes an APG, the turbulence structure is affected not only by mean strain and shear, but also by mean compression. In his pioneering work, [2] studied effects of mean dilatation on the turbulence structure in wall-bounded flows in the context of engineering calculation methods. He found that mean dilatation effects have a greater impact on the turbulence structure than would be expected from the extra production terms in the Reynolds stress transport equations. He also mentioned indirect effects caused by the pressure-strain correlation tensor in such flows which can add 'overwhelmingly' to those of the extra production terms. In

Somnath Ghosh · Rainer Friedrich
Lehrstuhl für Aerodynamik, Technische Universität München, Boltzmannstr. 15, D-85748 Garching, Germany
e-mail: Somnath.Ghosh@aer.mw.tum.de
e-mail: r.friedrich@lrz.tum.de

S. Wagner et al. (eds.), *High Performance Computing in Science and Engineering, Garching/Munich 2009*, DOI 10.1007/978-3-642-13872-0_8,

their experimental investigation of APG supersonic turbulent boundary layers, [4] studied the behaviour of the Reynolds stresses and of the large-scale structures. Recently, [5] performed DNS/LES of supersonic axisymmetric nozzle flow and also LES of supersonic diffuser flow using a fully-developed pipe flow at the inlet in order to study, among other features, Bradshaw's indirect effect of the pressure-strain correlations on the turbulence structure. They found that the Reynolds stresses and pressure-strain correlations are dramatically increased in the diffuser even though the production by mean strain and mean compression is relatively small compared to that by mean shear. Now, supersonic internal flows subjected to APGs are more complicated than corresponding flows subjected to favourable pressure gradients (nozzle flow) so that their analysis needs greater care. When the Mach number of the incoming flow is at a rather low overall supersonic level and the flow contraction is moderate, substantial transonic regions can occur. It is the aim of this paper to gain insight into the complex dynamics of a supersonic diffuser flow which develops from an incoming fully-developed pipe flow. The results will provide a data-base for validation and improvement of Reynolds stress models.

2 Computational Details

The governing Navier-Stokes equations are solved in a characteristic-type pressure-velocity-entropy form on non-orthogonal curvilinear coordinates. Fifth order compact upwind schemes with low dissipation [1] have been used for the convection terms and sixth order compact central schemes [7] for the molecular transport terms. The flow field is advanced in time using a 3rd order low-storage Runge-Kutta scheme [9]. Fully-developed supersonic turbulent flow in a pipe serves as inflow condition for the diffuser flow. The walls are kept at the same constant temperature of $220K$ in both the flows. The centerline Mach number M_c and friction Reynolds number Re_τ of the incoming flow are 1.75 and 300. Re_τ is defined using the friction velocity $u_\tau = \sqrt{\tau_w/\rho_w}$, the pipe radius R and the kinematic viscosity at the wall, $\nu_w(T_w)$. The Mach number M_c is the ratio of the centerline velocity and local speed of sound. The domain length of each configuration (pipe or diffuser) is $L = 10R$. The adverse axial pressure gradient averaged over the first half of the diffuser and normalized with the local displacement thickness and the local wall shear stress (Clauser parameter) is 5. The ratio of diffuser radius to pipe radius at the end of the computational domain is 0.93. The number of grid points used to discretize the pipe domain is $256 \times 256 \times 140$ in streamwise, circumferential and radial directions while that for the diffuser is $384 \times 256 \times 140$. The higher resolution in the diffuser is required to capture the increased turbulence activity occurring due to deceleration of the flow. The pipe and diffuser flow simulations are coupled using MPI routines. The concept of characteristics is applied to set inviscid inflow conditions for the diffuser flow. The incoming characteristics are computed from the periodic pipe flow simulation and are received at every time-step in the diffuser computation through MPI. Partially non-reflecting outflow conditions are used in the subsonic region of the outflow plane. No sponge layer has been used [5].

The coupled DNS flow solver has a performance of about 400 Mflops/core on a high bandwidth partition of the Altix 4700. The DNS was run using 128 cores and required about 800,000 cpu-hours till the statistics reached a converged state.

3 Results

A snapshot of instantaneous axial velocity fluctuations, normalized with the local friction velocity, and presented in an (x,r)-plane that contains the axis, is shown in figure 1. An increase in near-wall 'sweep-ejection' activity as the flow is decelerated, can be observed in the first half of this carpet plot.

The carpet plot also shows the weak overall decrease in cross-sectional area. The diffuser has an inflow section of length $1.2R$ in which the radius is constant. The transition from constant to varying radius occurs along a short arch which is represented by 12 discrete points ($\Delta x/R = 0.25$) and has a radius of curvature of 0.25R. Since the Mach number of the incoming turbulent pipe flow is low supersonic ($M_c = 1.75$), substantial transonic regions appear after about 50% of the diffuser length (see figure 2, left), even though the contraction is weak. So, it is important to look at bulk and centerline quantities in order to better understand the complex flow behaviour and demarcate regions of deceleration and acceleration. Figure 2 (left) shows the axial development of the bulk values of Mach number, density, velocity, mass flux and Reynolds number. All bulk variables are obtained by averaging local mean variables over the circular cross-section and, except for the Mach number M_m, they are normalized by their inlet values. M_m drops from a value of 1.37 over 60% of the diffuser length and then shows a weak increase in the rest of the diffuser. The acceleration in the later part of the diffuser is caused by the fact that the flow, after reaching a subsonic state globally, sees the downstream part of the converging pipe as a subsonic nozzle in which the flow is, of course, accelerated. The behaviour of the bulk velocity clearly indicates deceleration in the first half (and acceleration in the second), while the bulk density first increases and then decreases. The bulk Reynolds number which is based on the local diffuser radius shows a decrease in

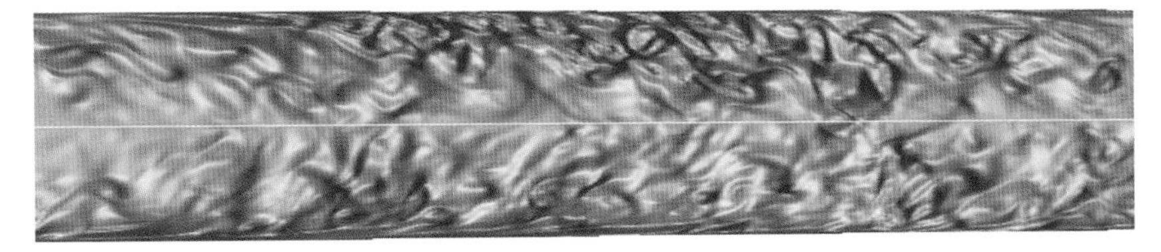

Fig. 1 Instantaneous axial velocity fluctuation, normalized with local friction velocity, u_τ, in a (x,r) plane of the diffuser

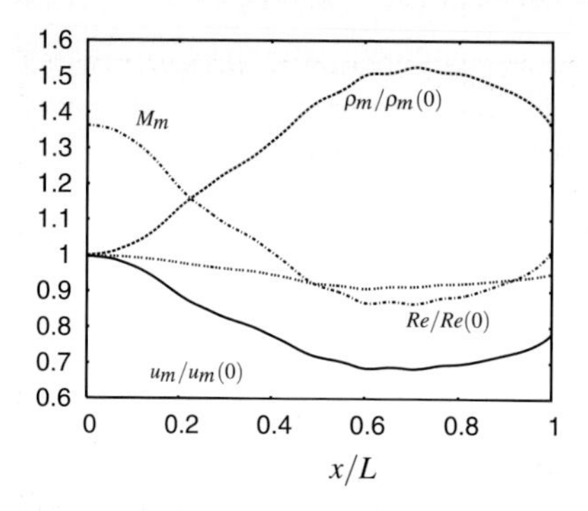

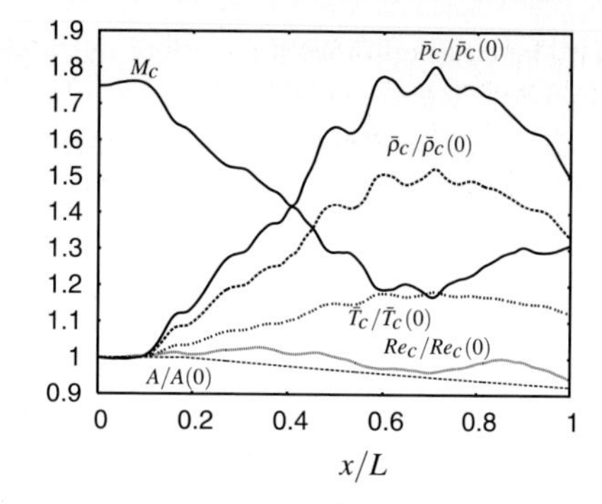

Fig. 2 Bulk(left) and centerline quantities in the diffuser

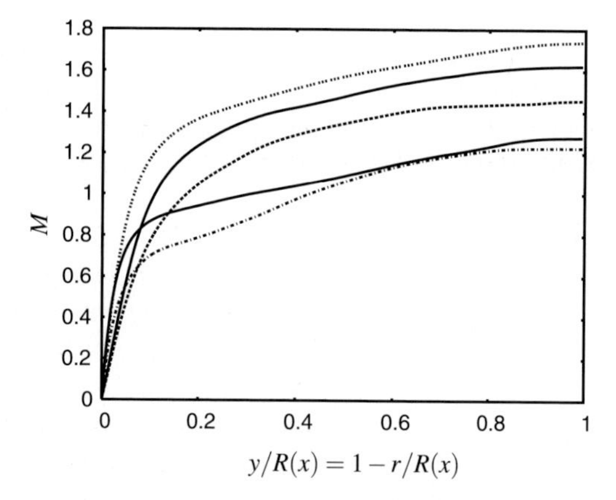

Fig. 3 Mean Mach number profiles in the diffuser at stations $x/L = 0$(... ...),0.2(—), 0.35(− − −), 0.6(−. − .−), 0.95(.... ....)

the first half and an increase in the second. Figure 2 (right) presents mean values at the centerline. While the Mach number M_c drops, the pressure, density and temperature increase up to $x/L = 0.6$, underlining the effect of deceleration in a supersonic diffuser whose cross-section A decreases by roughly 4% in the first half. For completeness we note the reference flow parameters: $Re_m(0) = 2766$, $Re_c(0) = 3857$.

In this paper, we are interested in effects of deceleration on the turbulence structure. So, we restrict our analysis to the first half of the diffuser where the flow is clearly decelerated. The mean extra strain rate is limited to less than 15% of the mean shear in the peak production zone. A look at the local Mach number profiles (fig. 3) reveals the deceleration at $x/L = 0.2, 0.35$ from the wall to the centerline. These profiles also give an impression of the growth of the near-wall subsonic layer. Profiles at positions $x/L = 0.6, 0.95$ show the acceleration of the flow first in the subsonic layer ($x/L = 0.6$) and then across the entire cross-section ($x/L = 0.95$). In the remaining part of the paper, we will discuss profiles up to $x/L = 0.35$ only and analyse the effects of mean compression on the turbulence structure. The mean density and mean temperature continue to show a roughly inverse proportionality in the radial direction (fig. 4) as in the incoming pipe flow since the radial variation of mean pressure remains negligible. The mean temperature is increased both

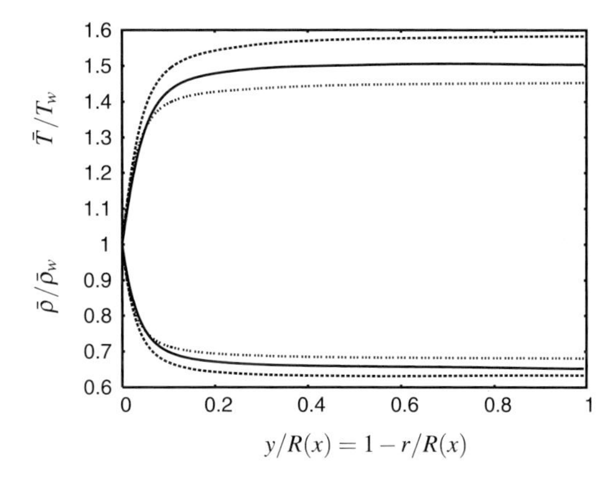

Fig. 4 Mean density, temperature profiles in the diffuser at stations $x/L = 0(\dots\ \dots)$, $0.2(—)$, $0.35(- - -)$

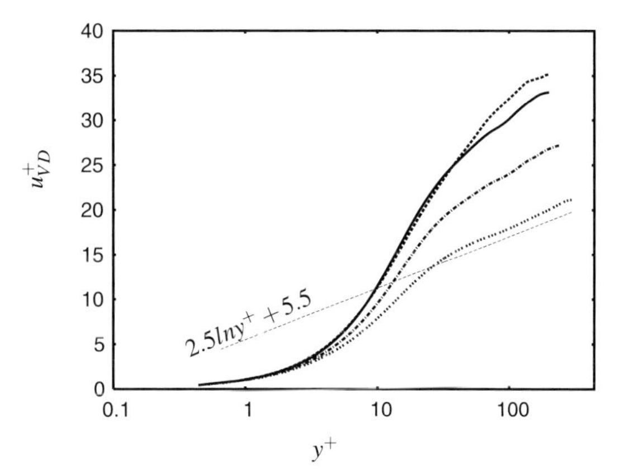

Fig. 5 Van Driest transformed velocity in the diffuser at stations $x/L = 0(\dots\ \dots)$, $0.15(-.-.-)$, $0.2(—)$, $0.35(- - -)$

by increase in mean and turbulent dissipation. The Van Driest transformed mean velocity profiles (fig. 5) show an overshoot over the log law for pipe flows, in a manner similar to incompressible flows under the influence of APG. However, here the variation of density in the axial direction is also reflected in the Van Driest transformed profiles.

Deceleration of the supersonic flow in the diffuser leads to an increase in turbulence intensities. As a result, both solenoidal and dilatational dissipation rates, $\bar{\rho}\varepsilon_s = \bar{\mu}\overline{\omega_i'\omega_i'}$, $\bar{\rho}\varepsilon_d = 4/3\bar{\mu}\overline{s_{kk}'s_{ll}'}$ are enhanced (not shown here). An increase in pressure-dilatation correlation $\overline{p'u_{i,i}'}$ is also observed. However, compressible dissipation rate as well as pressure-dilatation correlation continue to have negligible contributions to the TKE budget as in the incoming supersonic pipe flow. The Reynolds stresses increase monotonously through the compression region, both in the near-wall region as well as in the core. Here we show the axial Reynolds stress and the Reynolds shear stress scaled with the local wall shear stress (figs. 6, 7). It should be noted that while in incompressible boundary layers under the influence of APG, the turbulence intensities are decreased in the near-wall region and increased in the

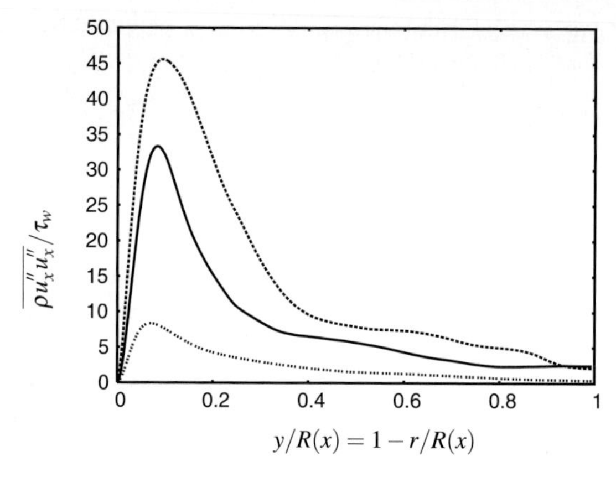

Fig. 6 Axial Reynolds stress in the diffuser normalized with local wall shear stress. x/L stations as in fig. 4

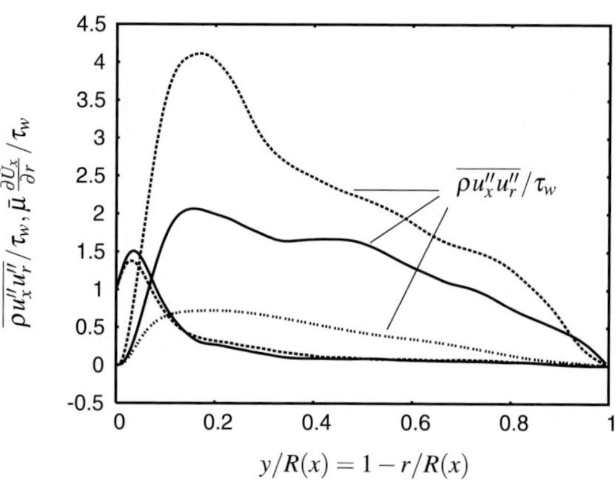

Fig. 7 Reynolds shear stress and mean shear in the diffuser. x/L stations as in fig. 4

outer layer [6], in the compressible diffuser flow, we see an increase in turbulence intensities both near the wall as well as in the core region.

The axial Reynolds stress production is increased and the production term is now decomposed into contributions from mean shear, extra strain rate and mean compression as in eq. (1). Extra strain rate and mean compression lead to small increases in the production of the axial stress as seen in fig. 8. But, the major increase in production is due to production by mean shear. The mean shear itself changes only marginally in the peak production region (as shown in fig. 7 at $x/L = 0.2$ and 0.35) which means that the increase in the Reynolds shear stress is the main reason for the increase in production by mean shear. This is contrary to findings in incompressible decelerated channel flows [3] where decreased mean shear leads to decreased production (and hence decreased turbulence intensities) in the near-wall region. The production term in the shear stress equation is similarly decomposed (eq. (2)) and again production due to mean shear (*shear1*) is the main reason for an increase in Reynolds shear stress (fig. 9). The term *shear2* has a small negative contribution, while production by mean dilatation and extra rate of strain remain equally small. The remarkable increase in production by mean shear is caused by

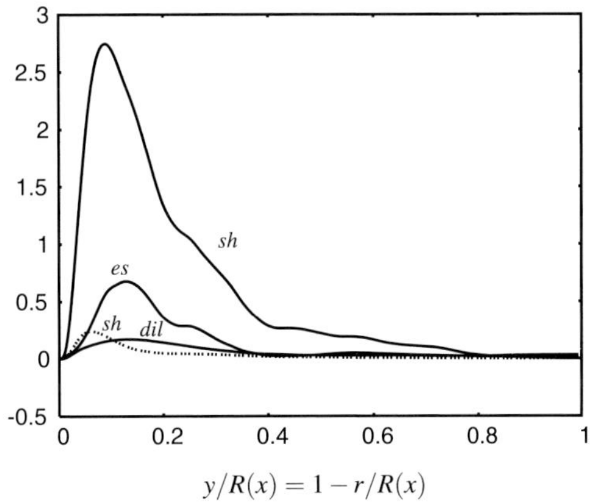

Fig. 8 Contributions to the production of the axial Reynolds stress in the diffuser at stations at $x/L = 0$(... ...) and 0.35 (—). sh : mean shear, dil: mean dilatation, es: extra rate of strain

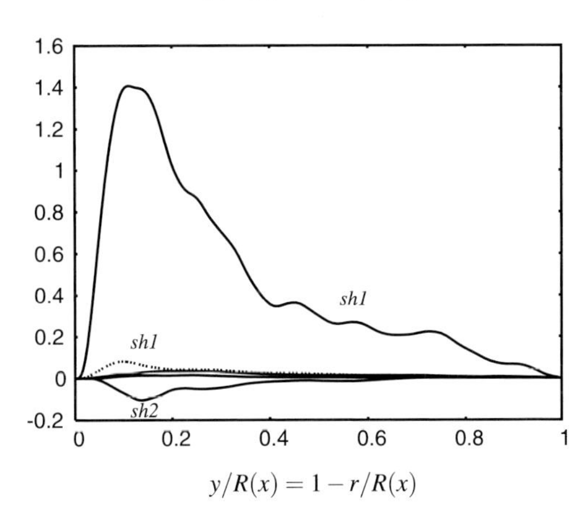

Fig. 9 Contributions to the production of the Reynolds shear stress in the diffuser at stations $x/L = 0$ (... ...) and 0.35 (—). sh1: mean shear (*shear1*), sh2: mean shear (*shear2*)

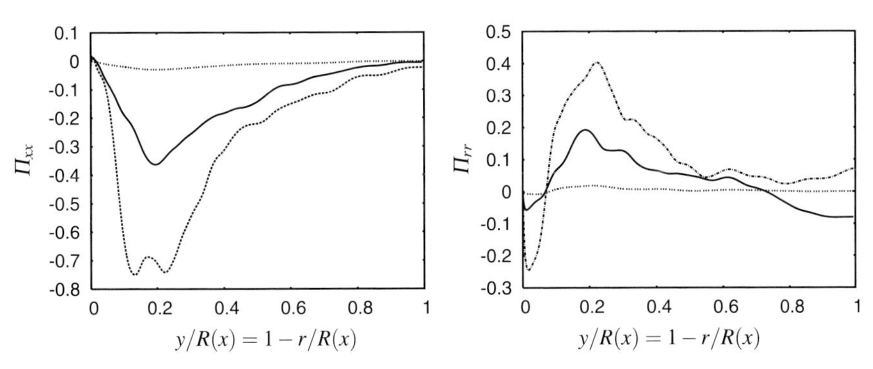

Fig. 10 Pressure-strain correlations: Π_{xx} (*left*), Π_{rr} (*right*) in the nozzle. x/L stations as in fig. 4

increased radial stress, which in turn increases due to the dramatic increase in the redistributive pressure-strain correlations (fig. 10). Thus the pressure-strain correlations play a significant role in controlling turbulence production in this flow. Both pressure fluctuations and strain rate fluctuations are found to increase.

$$P_{xx} = \underbrace{-\overline{\rho u_x'' u_r''}\frac{\partial \widetilde{u}_x}{\partial r}}_{shear} \underbrace{-\frac{1}{3}\overline{\rho u_x'' u_x''}\frac{\partial \widetilde{u}_l}{\partial x_l}}_{mean\ dilatation} \underbrace{-\overline{\rho u_x'' u_x''}(\frac{\partial \widetilde{u}_x}{\partial x} - \frac{1}{3}\frac{\partial \widetilde{u}_l}{\partial x_l})}_{extra\ rate\ of\ strain} \quad (1)$$

$$P_{xr} = \underbrace{-\overline{\rho u_r'' u_r''}\frac{\partial \widetilde{u}_x}{\partial r}}_{shear1} \underbrace{-\overline{\rho u_x'' u_x''}\frac{\partial \widetilde{u}_r}{\partial x}}_{shear2} \underbrace{-\frac{2}{3}\overline{\rho u_x'' u_r''}\frac{\partial \widetilde{u}_l}{\partial x_l}}_{mean\ dilatation} \underbrace{-\overline{\rho u_x'' u_r''}(\frac{\partial \widetilde{u}_x}{\partial x} + \frac{\partial \widetilde{u}_r}{\partial r} - \frac{2}{3}\frac{\partial \widetilde{u}_l}{\partial x_l})}_{extra\ rate\ of\ strain} \quad (2)$$

4 Conclusions

Effects of adverse pressure gradient on the turbulence structure of supersonic flow in a diffuser with incoming supersonic fully-developed pipe flow is investigated by means of DNS. Large increases in turbulence intensities occur when the flow undergoes deceleration. Turbulence production and pressure-strain correlations are increased. Analysis of the production terms shows that although the extra strain rate and mean compression increase turbulence production, their role is small compared to that of the redistributive pressure-strain correlations whose increase leads to a major increase in turbulence production by mean shear via an increase of the Reynolds stresses. Both pressure and strain rate fluctuations increase and a further analysis is currently being performed to explain this.

Acknowledgements The computations were performed on the SGI Altix 4700 at Leibniz Rechenzentrum. The excellent support of the LRZ staff is gratefully acknowledged.

References

1. Adams, N.A., Shariff, K.: A high-resolution hybrid compact-ENO scheme for shock-turbulence interaction problems. Journal of Computational Physics **127**, 27–51 (1996)
2. Bradshaw, P.: The effect of mean compression or dilatation on the turbulence structure of supersonic boundary layers. Journal of Fluid Mechanics **63**, 449–464 (1974)
3. Coleman, G.N., Kim, J., Spalart, P.R.: Direct numerical simulation of a decelerated wall-bounded turbulent shear flow. Journal of Fluid Mechanics **495**, 1–18 (2003)
4. Fernando, E.M., Smits, A.J.: A supersonic turbulent boundary layer in an adverse pressure gradient. Journal of Fluid Mechanics **211**, 285–307 (1990)
5. Ghosh, S., Sesterhenn, J., Friedrich, R.: Large-eddy simulation of supersonic turbulent flow in axisymmetric nozzles and diffusers. International J. of Heat and Fluid flow pp. 579–590 (2008)
6. Lee, J., Sung, H.J.: Effects of an adverse pressure gradient on a turbulent boundary layer. International J. of Heat and Fluid flow **29**, 568–578 (2008)
7. Lele, S.: Compact finite difference schemes with spectral-like resolution. Journal of Computational Physics **103**, 16–42 (1992)

8. Nagano, Y., Tsuji, T., Houra, T.: Structure of turbulent boundary layer subjected to adverse pressure gradient. Eleventh Symp. on Turbulent Sear Flows, Grenoble, France, September 8–10 (1997)
9. Williamson, J.K.: Low-storage Runge-Kutta schemes. Journal of Computational Physics **35**, 48–56 (1980)

Numerical Simulation of Supersonic Jet Noise with Overset Grid Techniques for Highly Parallelized Computing

J. Schulze and J. Sesterhenn

Abstract Supersonic jets with a complex shock pattern appear in numerous technical applications. Most supersonic jets, especially in modern military or civil aircraft, are not perfectly expanded. Thereby, shocks are appearing in the jet core and interacting with the turbulent mixing-layers and emanating shock induced noise. Under certain conditions this upstream traveling noise can be amplified due to a closed feedback loop. These so called *screech tones* can reach sound pressure levels of up to 160 dB (Tam et al., 1994) and hence lead to immense noise pollution and even structural fatigue. The focus of this research project lies in the numerical simulation of supersonic jet noise and finally the minimization of screech tones with an adjoint shape optimization approach of the nozzle geometry. To this end the nozzle geometry, based on a complex shape, has to be included in the computational domain. In the present paper the method of overset grid techniques in a highly parallelized environment is presented for the simulation of supersonic jet noise. Direct numerical simulations with a modeled nozzle inlet showed a good agreement of the screech frequency to a semi-empirical low found by Powell in 1953.

1 Introduction

Considering a supersonic jet, e.g. at the exit of a jet engine, in the over- or under-expanded case, a regular pattern of compression and expansion waves will be found within the supersonic part of the jet flow. A compression wave incident on the sonic line will be reflected as an expansion, and vice versa, cf. Fig. 1. At the location of interaction between the compression wave and the turbulent mixing layer, acoustic waves are generated. This shock-induced noise also plays an important role in what

J. Schulze · J. Sesterhenn
Department of Fluid Mechanics and Engineering Acoustics, TU – Berlin, 10623 Berlin, Germany
e-mail: jan.schulze@tnt.tu-berlin.de
e-mail: joern.sesterhenn@tnt.tu-berlin.de

S. Wagner et al. (eds.), *High Performance Computing in Science and Engineering, Garching/Munich 2009*, DOI 10.1007/978-3-642-13872-0_9,

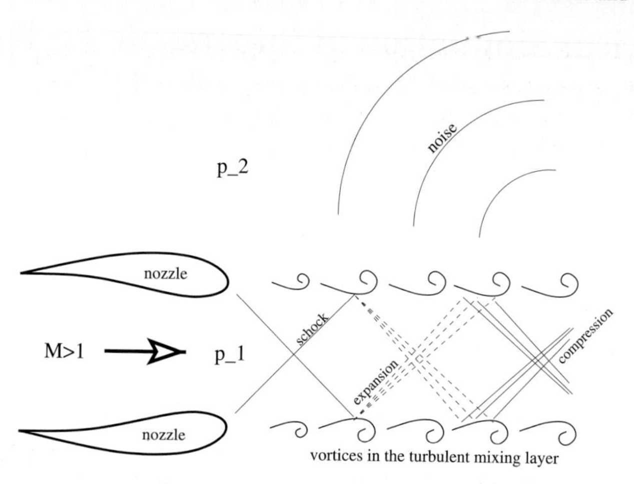

Fig. 1 Schematic view of the interaction of shock and turbulent mixing layer in a jet with emanated noise. Over-expanded $\Rightarrow p_2 > p_1$

is called *jet screech*. This phenomenon is caused by shock-induced acoustic waves traveling upstream and forcing the transitional shear-layer at the nozzle exit. At this point Kelvin-Helmholtz instabilities are growing to vortices, transported downstream and interacting with the shock tips which are emanating noise again and closing a feedback loop. The presence of a nozzle-geometry seems to be important since it has a leading effect for the sensitivity of the upstream traveling acoustics on the instabilities in the mixing layers.

Experimental results indicate particularly high sound pressure levels of up to 160 dB and even beyond [11]. Prediction and reduction of shock-induced noise, produced by modern civil aircraft with jet propulsion, traveling at high subsonic or supersonic Mach numbers is a matter of particular interest. Besides pollution of the environment by the radiated sound, the latter can also lead to high dynamic loads on parts of the aircraft causing structural fatigue and even destroy them.

In this project, the method of direct numerical simulation will be used to compute a supersonic, three-dimensional and rectangular jet that is not perfectly expanded, as it is found at the nozzle exit of jet engines for aircraft. Numerical methods of high order of accuracy are chosen for the direct solution of the compressible Navier-Stokes equations, which gives us the possibility to compute the sound field, generated by the supersonic jets, directly. The aim of this study lies in the minimization of supersonic jet-noise and in particular in the minimization of jet-screech. Since screech – a phenomenon which is not yet understood in all details – seems to be affected by the presence of the jet-nozzle, we will perform a shape optimization of the nozzle geometry to minimize the emanated noise. The optimization technique is based on the adjoints of the compressible Navier-Stokes equations. These equations have to be solved backwards in time to get the gradient information of the objective function. To this end the nozzle geometry has to be included in the computational domain. In this paper the method of overset grid techniques is presented to include the nozzle geometry for highly parallelized computations. Two-dimensional results for a high Reynolds-number supersonic jet are shown.

The organization of the paper is as follows: In section two the numerical method of overset grid techniques to include the nozzle geometry in the computational domain is presented. Results of direct numerical simulations with the presence of a nozzle will be found in section three, followed by some concluding remarks in section four.

2 Numerical Methods

To simulate the investigated flow case, the compressible Navier-Stokes equations are solved, based on a characteristic-type formulation [9] on an orthogonal grid stretched in both the stream-wise and the transverse directions. Along the span-wise direction, periodicity and statistical homogeneity are assumed. Spatial discretization is implemented using a finite difference compact scheme of sixth order [4] and a spectral like method in the periodic direction. The compact schemes preserve the dispersion relation even for higher wavenumbers which is an important issue for aeroacoustic applications. With an additional compact high order filter, spurious high frequent grid to grid oscillations are suppressed.

The time advancement is performed using an exponential integration based on Krylov subspaces. This integration is a low-dimensional approximation of the matrix exponential which represents the exact evolution operator for an autonomous linear system [7]. The main advantage of this method lies in its high numerical stability compared to other standard explicit techniques like the Runge-Kutta method. Hence, large time-steps can be used with CFL numbers of up to 50 and even beyond. Although, such a high CFL-number causes additional computational time, since one needs to compute more subspaces, it is still up to twice as fast as a comparable Runge-Kutta method.

A sponge region with a low-pass filter (applied spatially) and a grid stretching in the stream-wise direction is applied to avoid large structures interacting with the outflow boundary and scattering spurious energy back in the computational domain. Additionally, a boundary wave-acceleration is implemented to prevent acoustic reflections from the transverse boundaries. At the inflow two different boundary conditions are implemented. As a first approach the rectangular nozzle at the inlet is modeled with a laminar *tanh*-profile. The resulting shear-layers are forced slightly downstream of the inflow boundary with a model-spectrum including the most unstable frequency of the shear-layer and its first sub-harmonic with a stochastical phase-shifting (*inlet 1*). In a second approach the geometry of the nozzle is modeled by two backwards facing Joukowsky profiles (*inlet 2*, cf. Fig. 2). This geometry is included in the computational domain with overset grid techniques as explained in section 2.1.

The code is parallelized using the Massage Passing Interface (MPI). For the current setup up to 1020 CPU's are used on a SGI-Altix 4700. Approximately 12 TB of data are written to disk and 0.5 TB of main memory are used (cf. section 2.1).

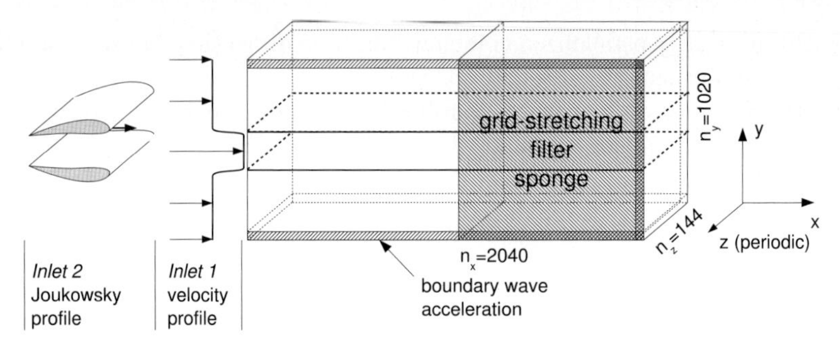

Fig. 2 Computational domain with some implemented numerical boundary conditions and the two different inlet conditions (velocity profile and Joukowsky-like nozzle geometry)

2.1 Nozzle Geometry with Overset Grid Techniques

A planar and rectangular jet is used for the investigation of screech tones. As it was mentioned before the presence of a nozzle is important. For this reason a nozzle geometry is being implemented in the present setup based on overset grid techniques [5, 8] and [10].

The nozzle geometry is modeled by two backwards facing Joukowsky profiles as shown in Fig. 2 (*inlet 2*). Each profile is surrounded by a curvilinear grid which is orthogonal due to the conformal mapping of the Joukowsky transformation. Both curvilinear grids are embedded in a cartesian grid where most of the jet is simulated. The data of the three different domains is divided through the Message Passing Interface (MPI). Each of the three domains is embedded in a MPI–Group, where each group has its own MPI–Intra-communicator. This intra-communicator is now used to decompose each group in several sub-domains to guarantee a further parallelization of the code (cf. Fig. 4).

The MPI–Routines are also used enable the communication needed for the interpolation of the curvilinear grid and the cartesian grid. All interpolated values need to be send from the curvilinear grids to the cartesian one and the other way round. For this communication MPI-Inter-communicators have to be initialized in the beginning of the code execution. Due to the massive parallelization of the code the interpolation zones overlap several sub-domains and the identification of the processes involved in the interpolation and intercommunication has to be set up carefully to avoid communication deadlock Additional communication at the boundaries of the sub-domains has to be implemented to receive ghost cells of the neighboring processes. The latter one are used to interpolate points close to the sub-domain boundaries where an interpolation stencil would extend the domain of the process. All the MPI-communication is embedded in a Multiple Program Multiple Data (MPMD) approach where each of the three grids has its own executable (cf. Fig. 4).

For the current setup, presented in section 3, a total number of 320 processes is used to decompose the two-dimensional domain for the parallelization of the code. The subdomains are partitioned as follows: cartesian grid in the background (green)

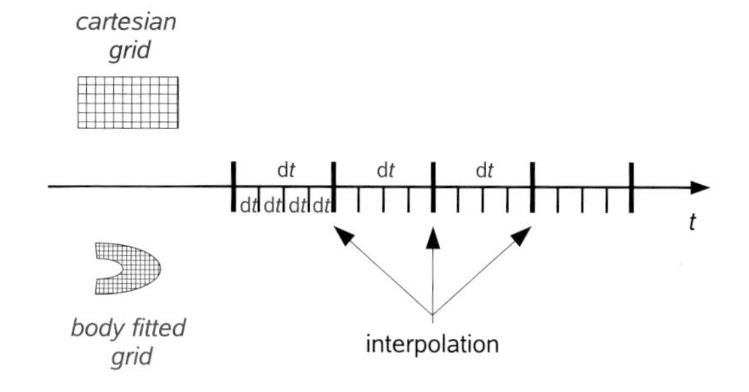

Fig. 3 Local time stepping to increase the performance of the code. An overall speedup of five is reachable

$16 \times 16 = 256$ processes; curvilinear domains (red) $4 \times 8 = 32$, (blue) $4 \times 8 = 32$ processes. Each process contains the same number of grid points for an optimally distributed load balancing.

2.1.1 Local Time-Stepping

More sophisticated approaches can distribute the number of points in a different way to enhance the overall performance of the code. Considering the fact, that each grid has its own local time step, determined by the CFL-number, and that the time-step of the body fitted domains can be much smaller than the one of the cartesian domain (since one needs to resolve the boundary layer). As an example, lets say that the local time-step of the curvilinear grid is one-fourth of the time-step of the comparable corse cartesian grid. Then one can set up a local time-stepping algorithm where the curvilinear grid is performing four time-steps while the cartesian grid is only doing one time-step in the same computational time. This imposes that the computational power of the curvilinear grid has to be four times as fast as the one of the cartesian grid. With other words, the curvilinear grid needs to have four times as many CPU's per grid point as the cartesian grid, so a nonuniform distribution of the grid points. Interpolation is then performed after all different domains reached the same physical time, which is after four iterations for the curvilinear grid or for one iteration for the cartesian grid, respectively (cf. Fig 3). Following this approach one can obtain an overall speedup of about a factor of five.

2.1.2 Interpolation

The interpolation is performed in an interpolation zone close to the boundary of the curvilinear grids. Depending on the order of interpolation, the width of this interpolation zone has to be at lest $n+1$ points, where n is the order of the interpolation. It is proposed by [3] that a fourth order Lagrange interpolation is a good compromise between accuracy and computational cost. Based on a Lagrange interpolation in three dimensions, the values in the interpolation zone are updated within each iteration of the time integration method. Due to the closed feedback loop in our application

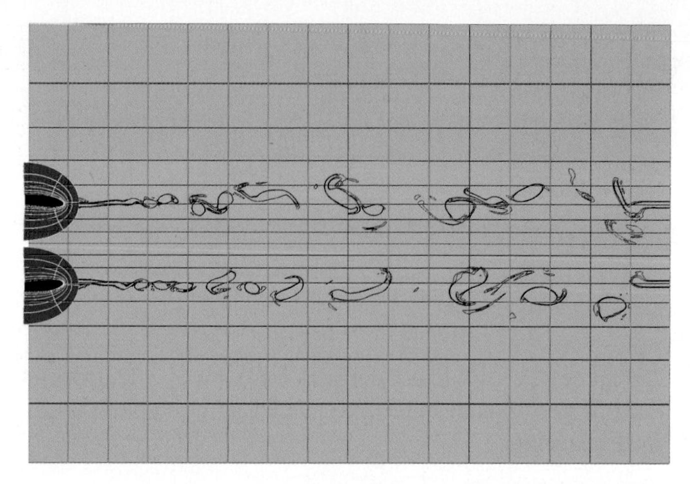

Fig. 4 Domain–decomposition and Multiple Program Multiple Data (MPMD) approach. Each of the three colored domains contains one MPI-Group an is running with its own executable sharing one communicator. Communication between the Groups is established with MPI-inter-communicators, while the communication inside one group is performed with MPI-Intra-communicators. The domain is decomposed as follows: Cartesian domain (green) $16 \times 16 = 256$ processes; Curvilinear domains (red): $4 \times 8 = 32$ processes; (blue): $4 \times 8 = 32$ processes

(screech) the information of the curvilinear grid and the cartesian grid needs to be interpolated in both directions. Thus, a two-way coupling is implemented.

To increase the numerical stability of the interpolation method the Lagrange interpolation is written in a barycentric formulation. A two-dimensional barycentric Lagrange interpolation can be written as:

$$p_{nm}(x,y) = \frac{\sum_{j=0}^{m} \sum_{i=0}^{n} f_{i,j} \mu_i(x) \mu_j(y)}{\sum_{j=0}^{m} \sum_{i=0}^{n} \mu_i(x) \mu_j(y)} \tag{1}$$

with $\mu_i(x) = \frac{\omega_i}{x - x_i}$, the interpolated value $p_{nm}(x,y)$ at the point (x,y) and the barycentric weights $\omega_i = \frac{1}{\prod_{k \neq i}(x_i - x_k)}$; (analogous for $\mu_j(y)$), and n, m the order of the interpolation in the x and y direction, respectively.

Since the geometry in our case is time independent, the values $\mu_i(x)$ and $\mu_j(y)$ need to be computed only once and can be saved throughout the simulation to reduce computational time.

Anyway, a direct interpolation of the values of the curvilinear grid on the cartesian grid is not possible since one needs a cartesian domain for the interpolation stencil. To interpolate a value from the curvilinear grid on the cartesian grid the stencil on the curvilinear grid needs to be in a cartesian domain (computational domain). These cartesian values are computed using an inverse Joukowsky transformation of the curvilinear grid. This is a different approach to [2] and [5] but fast and accurate

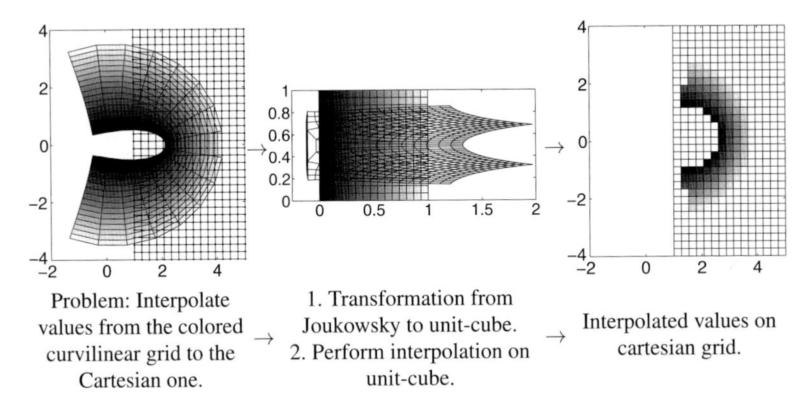

Fig. 5 Inverse Joukowsky transformation of the cartesian grid to apply a Lagrangian interpolation

since no system of equations has to be solved. Anyway, this approach is restricted to applications where an analytic solution to the inverse transformation of the domain exists. This is the case for a standard Joukowsky transformation. An example how the inverse transformed domain looks like is presented in Fig. 5. See also [8] for a more detailed description and some validation of the implemented interpolation method.

3 Results

In the present research project two different numerical inlet conditions are implemented (cf. Fig. 2). Within this paper we focus on inlet 2.

3.1 *Nozzle Geometry* (Inlet 2)

A two-dimensional supersonic simulation with included nozzle (inlet 2) is presented in Fig. 6 with a Reynolds number based on the jet diameter of $Re_D = 30000$, and a jet Mach number $M_J = 1.35$. To obtain oblique shocks in the jet core, a pressure ratio of $p_J/p_\infty = 1.2$ (under-expanded) is adjusted.

The cartesian grid in the background is decomposed in 256 sub-domains, bordered by weak white lines and contains in total 1024×1024 grid points. Each of the two computational domains around the nozzle-profiles is as well decomposed in each with 32 sub-domains containing 512×256 grid points. Hence, a total of 320 CPU-cores are used for this two-dimensional computation. The nozzle geometry is partly shown as two backwards facing Joukowsky profiles in black color.

In Fig. 7 the interpolation zones for the current two-dimensional simulation are visualized. Since there is a two-way coupling, two interpolation zones are needed.

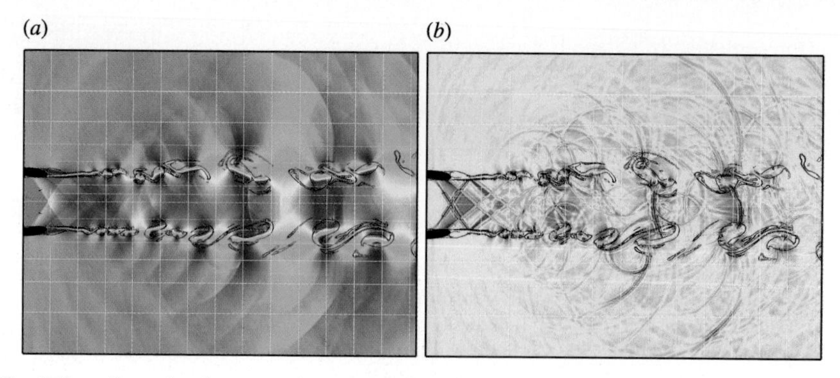

Fig. 6 Two-dimensional supersonic case for the jet computation with overset grid techniques. Mach number $M_J = 1.35$; Reynolds number based on the jet diameter $Re_D = 30000$. White lines border the sub-domains for the MPI domain decomposition. (*a*) Absolute velocity, Contour-lines: stream-wise vorticity. (*b*) Divergence of the velocity, Contour-lines: stream-wise vorticity

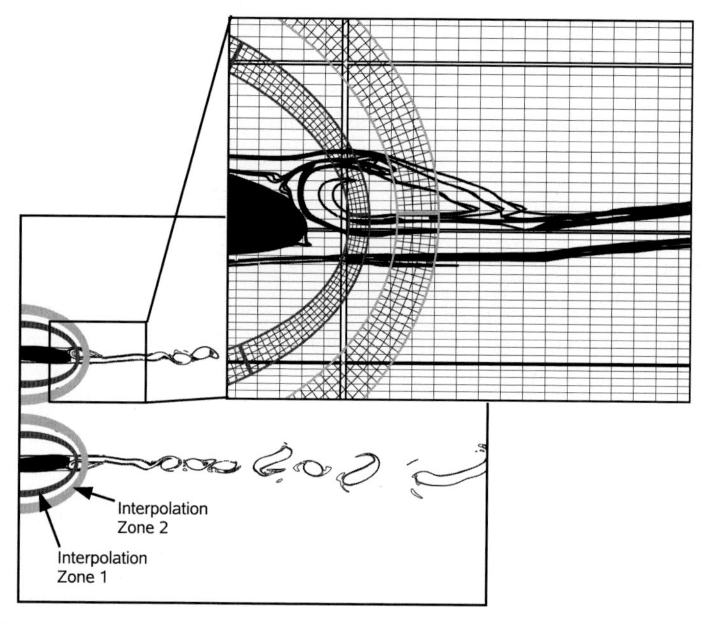

Fig. 7 Visualization of the interpolation zones and the cartesian grid in the background. Only every fourth point is shown

The interpolation zone close to the nozzle geometry interpolates the flow variables from the body fitted grid to the cartesian one. The second interpolation zone interpolates the values the other way round. Both on the body fitted grid and on the cartesian only every fourth point is shown. Due to a grid stretching on all grids one has to adjust the size and position of the interpolation zones carefully to guarantee the minimum number of required points in the interpolation zones.

The cartesian points inside the inner interpolation zone have no physical meaning since they belong to the simulation of the body fitted grid and can be neglected. Anyway, to obtain a numerically stable solution one needs to treat them in a special way. In the present study a simple but efficient method is used where the right hand side ($\partial \mathbf{u}/\partial t$) of all interior points is set to zero.

The contour-lines represent the stream-wise vorticity. One can see, that even large structures, like the recirculation bubble close to the nozzle, which extends both interpolation zones, are well resolved. In Fig. 6(b) we can see, that this also holds for steady shocks extending the interpolation zones.

In Fig. 6(a) the absolute velocity is visualized including the contour lines of the stream-wise vorticity. The emanated noise is clearly visible in Fig. 6(b) where the divergence of the velocity is shown again including the contour lines of the stream-wise vorticity. Acoustics directed in the downstream direction caused by the turbulence in the mixing layers and also shock induced noise directed in the upstream direction can be identified. In addition to this a complex shock pattern is visible in the jet core.

4 Conclusion

In the present paper the method of direct numerical simulation with overset grid techniques is used to simulate the noise of a not perfectly expanded supersonic jet. The nozzle is represented by two backwards facing Joukowsky profiles and included in the domain with overset grid techniques. All grids are decomposed with the message passing interface (MPI) to guarantee a maximum level of code parallelization. It could be shown, that the used interpolation method performs well together with the MPI-communication. The performance of the code can even be improved by the use of local time stepping.

Acknowledgements The authors like to thank the DEISA Consortium (co-funded by the EU, FP6 project 508830), for support within the DEISA Extreme Computing Initiative (www.deisa.org) and the DFG/CNRS research group FOR 508 (*Noise Generation in Turbulent Flows*) for support within this project.

References

1. Berland, J., Bogey, Ch. and Bailly, Ch.: Numerical study of screech generation in a planar supersonic jet. Physics of fluids **19**, DOI: 10.1063/1.2747225 (2007)
2. Desquesnes, G. et al.: On the use of a high order overlapping grid method for coupling in CFD/CAA. J. Comp. Phys. **220**, 355–382 (2006)
3. Guenanff, R., Sagaut, P., Manoha, E., Terracol, M., Lewandowski, R.: Theoretical aspects of a multi-domain high-order method for CAA. AIAA Paper 2003-3117 (2003)
4. Lele, S.: Compact finite difference schemes with spectral-like resolution. J. Comp. Phys. **103**, 16–42 (1992)

5. Marsden, O., Bogey, Ch. and Bailly, Ch.: High-Order Curvilinear Simulations of Flows Around Non-Cartesian Bodies. J. Comp. Aeroacoustics **13**(4), 731–748 (2004)
6. Powell, A.: On the mechanism of choked jet noise. Proc. Phys. Soc. London. **B66**, 1039–1056 (1953)
7. Schulze, J., Schmid, P. and Sesterhenn, J.: Exponential time integration using Krylov subspaces. Int. J. Numer. Meth. Fluids (2008) DOI: 10.1002/fld.1902
8. Schulze, J. et al.: Numerical Simulation of Supersonic Jet Noise. Num. Sim. Turbulent Flows & Noise Generation, NNFM **104**, 29–46 (2009)
9. Sesterhenn, J.: A characteristic-type formulation of the Navier-Stokes equations for high order upwind schemes. Computers & Fluids **30**(1), 37–67 (2001)
10. Sherer, S. and Scott, J.: High order compact finite-difference methods on general overset grids. J. Comp. Phys. **210**, 459–496 (2005)
11. Tam C., Ahuja, K. and Jones III, R.: Screech Tones from Free and Ducted Supersonic Jets. AIAA J. **32**(5), 917–922 (1994)

Vorticity Statistics in Fully Developed Turbulence

Michael Wilczek, Anton Daitche, and Rudolf Friedrich

Abstract We study the statistical properties of fully developed hydrodynamical turbulence. To this end, a theoretical framework is established relating basic dynamical features of fluid flows to the shape and evolution of probability density functions of, for example, the vorticity. Starting from the basic equations of motion, the theory involves terms, which presently cannot be calculated from first principles. This missing information is taken from direct numerical simulations. A parallel pseudospectral code is used to to obtain well-resolved numerical simulations of homogeneous isotropic turbulence. The results yield a consistent description of the vorticity statistics and provide insights into the structure and dynamics of fully developed turbulence.

1 Introduction

Turbulence is a ubiquitous phenomenon playing a key role in many natural and engineering environments. Although the basic equations of motion are known for almost two hundred years, the problem withstands a concise theoretical description from first principles. From a theoretical physicist's point of view it is desirable to develop a statistical theory of turbulent flows. The situation may be compared to classical thermodynamics, where macroscopic observables like pressure or temperature for an ideal gas are derived within a closed theoretical framework. In this sense, one of the main goals of theoretical turbulence research is to derive the statistics of dynamical variables like velocity, vorticity and acceleration right from the basic equations of motion. This, however, turns out to be a challenging problem, as one has to deal with a complex system far from equilibrium. From a mathematical point of view this

Michael Wilczek · Anton Daitche · Rudolf Friedrich
Institute for Theoretical Physics, University of Münster, Wilhelm-Klemm-Str. 9, 48149 Münster, Germany
e-mail: mwilczek@uni-muenster.de

S. Wagner et al. (eds.), *High Performance Computing in Science and Engineering, Garching/Munich 2009*, DOI 10.1007/978-3-642-13872-0_10,

Fig. 1 Volume rendering of vorticity (left) and velocity (right). The vorticity tends to organize into thin filaments, which are entangled forming a complicated global structure. The velocity appears more unstructured but displays long-range correlations. Volume rendering produced by VAPOR (`www.vapor.ucar.edu`)

problem is mirrored by the so-called closure problem of turbulence, which appears naturally in all theories of turbulence based on first principles.

Numerical simulations reveal that the turbulent velocity and vorticity field are everything else but purely random. While the velocity field displays complicated multi-scale correlations, the vorticity tends to be organized in thin filamentary structures, as can be seen in figure 1. Statistically these tubes lead to strong spatial correlations and non-Gaussian PDFs (probability density functions). Hence the understanding of the emergence and dynamics of these structures is at the heart of the turbulence problem. Within the current project we study the dynamical and statistical properties of these structures, as well as their impact on passive fluid particles. To allow for a sufficiently detailed presentation, we will focus on the statistical properties of the vorticity field in the current article.

The remainder of the article is structured as follows. After presenting some basic numerical results on the nature of the vorticity statistics in fully developed turbulence, we outline the theoretical framework, which allows to explain the observed statistics. We then turn to numerical and technical details on the code used to generate the numerical results. Finally, we present numerical estimates of terms identified within our statistical framework, which allow to describe the vorticity statistics in a consistent way.

2 Vorticity Statistics in Homogeneous Isotropic Turbulence

The dynamics of incompressible flows is described by the Navier-Stokes equations

$$\frac{\partial}{\partial t}\boldsymbol{u}(\boldsymbol{x},t)+\boldsymbol{u}(\boldsymbol{x},t)\cdot\nabla\boldsymbol{u}(\boldsymbol{x},t)=-\nabla p(\boldsymbol{x},t)+\nu\Delta\boldsymbol{u}(\boldsymbol{x},t)+\hat{\boldsymbol{F}}(\boldsymbol{x},t), \qquad (1)$$

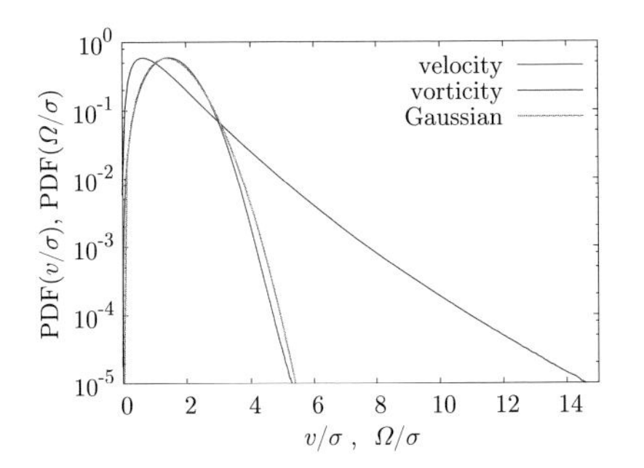

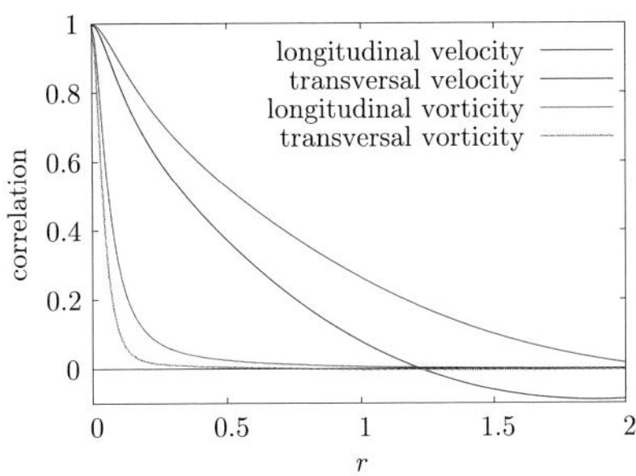

Fig. 2 Left: Vorticity PDF in fully developed turbulence compared with the velocity PDF and an angle-integrated Gaussian. The vorticity exhibits strongly non-Gaussian tails. Right: Correlation functions for the vorticity and velocity. The vorticity correlation decays rapidly, indicating the localized nature of the vorticity

where $\boldsymbol{u}$ is the velocity field, p denotes the pressure and $\hat{\boldsymbol{F}}$ is an external forcing driving the flow. The equations are completed by the continuity equation $\nabla \cdot \boldsymbol{u}(\boldsymbol{x},t) = 0$ and suitable boundary and initial conditions.

In case of fully developed turbulence, a statistical description of the field is necessary, where a complete characterization involves the statistical properties of the velocity field at N positions, $\boldsymbol{x}_1,\ldots,\boldsymbol{x}_N$ for every N. Due to incompressibility, the turbulent field is equivalently characterized by the vorticity field $\boldsymbol{\omega}(\boldsymbol{x},t) = \nabla \times \boldsymbol{u}(\boldsymbol{x},t)$, defined as the curl of the velocity field. By taking the curl of the Navier-Stokes equations (1), we obtain the vorticity equation

$$\frac{\partial}{\partial t}\boldsymbol{\omega}(\boldsymbol{x},t) + \boldsymbol{u}(\boldsymbol{x},t)\cdot\nabla\boldsymbol{\omega}(\boldsymbol{x},t) = \boldsymbol{S}(\boldsymbol{x},t)\cdot\boldsymbol{\omega}(\boldsymbol{x},t) + \nu\Delta\boldsymbol{\omega}(\boldsymbol{x},t) + \boldsymbol{F}(\boldsymbol{x},t), \quad (2)$$

where $S_{ij} = \frac{1}{2}\left(\frac{\partial u_i}{\partial x_j}(\boldsymbol{x},t) + \frac{\partial u_j}{\partial x_i}(\boldsymbol{x},t)\right)$ denotes the rate-of-strain tensor. Compared to the nearly normal statistics of the velocity, the single-point statistics of the vorticity exhibits strong deviations from Gaussian behavior, as can be seen from figure 2. The stretched exponential tails of the PDF indicate that the appearance of large values of vorticity is orders of magnitude more likely than in case of a Gaussian random field. These large values of the vorticity can clearly be related to the coherent structures present in the flow (see figure 1). Since these structures appear to be very localized, the two-point correlation of the vorticity decays rapidly as compared to the long-range correlated velocity field (see figure 2). The nontrivial single-point statistics, as well as the strong localization of the coherent structures, make the vorticity the central quantity to describe the fine structure of fully developed turbulence.

3 Theoretical Framework

3.1 The Lundgren-Monin-Novikov Hierarchy

There have been numerous phenomenological approaches in the past to characterize the statistical properties of turbulent fields (see e.g. [4] and references therein). However, a fully satisfactory theory of turbulence has to be derived from first principles, for example from the basic equations of motion. This can be achieved within the framework of the Lundgren-Monin-Novikov hierarchy, which was introduced by the authors independently into hydrodynamics in the late sixties [6–8]. Though this approach delivers a concise theoretical framework for the evolution of the PDFs, the mathematical complexity of the equations complicates a purely analytical solution of the problem. A number of efforts to approximate the resulting equations in terms of so-called closures yield mathematically tractable results, but result in serious physical defects. As a consequence, we follow a different strategy within the current project. Taking the LMN hierarchy as a starting point, we do not seek for an analytical closure in the first place, but use direct numerical simulations to give us information about the statistically unclosed terms. This enables us to characterize the statistical properties of the flow and reconstruct the statistics as a test for consistency of our approach.

To derive the theoretical framework, we start with the fine-grained PDF

$$\hat{f}^1(\boldsymbol{\Omega}_1;\boldsymbol{x}_1,t) = \delta(\boldsymbol{\omega}(\boldsymbol{x}_1,t) - \boldsymbol{\Omega}_1). \tag{3}$$

Here $\boldsymbol{\Omega}_1$ is the sample space variable, whereas $\boldsymbol{\omega}(\boldsymbol{x}_1,t)$ refers to an actual realization of the field and the superscript 1 indicates that the vorticity $\boldsymbol{\Omega}_1$ at a single point $\boldsymbol{x}_1$ in space is considered. By taking derivatives of this definition and using eq. (2), we obtain a kinetic equation for the fine-grained PDF $\hat{f}^1$. The evolution equation for the vorticity PDF $f^1(\boldsymbol{\Omega}_1;\boldsymbol{x}_1,t) = \langle \hat{f}^1(\boldsymbol{\Omega}_1;\boldsymbol{x}_1,t)\rangle$ is then obtained by ensemble averaging leading to

$$\frac{\partial}{\partial t} f^1 + \nabla \cdot \{\langle \boldsymbol{u}|\boldsymbol{\Omega}_1\rangle f^1\} = -\nabla_{\Omega_1} \cdot \{\langle \boldsymbol{S}\cdot\boldsymbol{\omega} + \nu\Delta\boldsymbol{\omega} + \boldsymbol{F}|\boldsymbol{\Omega}_1\rangle f^1\}. \tag{4}$$

The reader is referred to [6, 8, 9, 11, 13] for more details on the derivation. The interpretation of this kinetic equation is straight-forward; the probability density to find a vorticity $\boldsymbol{\Omega}_1$ is advected by the velocity conditioned on a given vorticity. The term on the right-hand side defines the divergence of a probability current involving the conditionally averaged right-hand side of equation (2). This basically identifies the two dynamical influences which determine the probability density function: vortex stretching and diffusion of vorticity. However, the appearing conditional averages cannot be calculated from first principles. As noted above, one of the goals of the present project is to determine these conditional averages from numerical experiments. Together with the evolution equation (4) and exploiting statistical symmetries, this determines the shape and evolution of the probability density function.

Once these conditional averages are known, a complete characterization of the one-point statistics of the vorticity field is given.

Instead of expressing the unknown (more mathematically speaking "unclosed") terms with the help of conditional averages, these terms may be coupled to the two-point PDF, $f^2(\boldsymbol{\Omega}_1, \boldsymbol{\Omega}_2; \boldsymbol{x}_1, \boldsymbol{x}_2, t)$. This may be exemplified with the advective term,

$$\begin{aligned} \langle \boldsymbol{u}(\boldsymbol{x}_1,t)\hat{f}^1\rangle &= \langle \frac{1}{4\pi}\int \mathrm{d}\boldsymbol{x}_2 \frac{\boldsymbol{\omega}(\boldsymbol{x}_2,t)\times(\boldsymbol{x}_1-\boldsymbol{x}_2)}{|\boldsymbol{x}_1-\boldsymbol{x}_2|^3}\hat{f}^1\rangle \\ &= \langle \frac{1}{4\pi}\int \mathrm{d}\boldsymbol{x}_2\,\mathrm{d}\boldsymbol{\Omega}_2 \frac{\boldsymbol{\omega}(\boldsymbol{x}_2,t)\times(\boldsymbol{x}_1-\boldsymbol{x}_2)}{|\boldsymbol{x}_1-\boldsymbol{x}_2|^3}\hat{f}^2\rangle \\ &= \frac{1}{4\pi}\int \mathrm{d}\boldsymbol{x}_2\,\mathrm{d}\boldsymbol{\Omega}_2 \frac{\boldsymbol{\Omega}_2\times(\boldsymbol{x}_1-\boldsymbol{x}_2)}{|\boldsymbol{x}_1-\boldsymbol{x}_2|^3} f^2. \end{aligned} \tag{5}$$

Here, the velocity is expressed in terms of the vorticity by Biot-Savart's law, additionally only fundamental properties of the delta distribution and ensemble averaging have been used. The important point here is to note the dependence on the two-point PDF $f^2(\boldsymbol{\Omega}_1, \boldsymbol{\Omega}_2; \boldsymbol{x}_1, \boldsymbol{x}_2, t)$, which also appears in the remaining unclosed terms. Proceeding in the same manner as above, an evolution equation may be obtained also for this PDF, however now coupling to the three-point PDF $f^3(\boldsymbol{\Omega}_1, \boldsymbol{\Omega}_2, \boldsymbol{\Omega}_3; \boldsymbol{x}_1, \boldsymbol{x}_2, \boldsymbol{x}_3, t)$. The resulting set of equations will in general take the form

$$\begin{aligned} \frac{\partial}{\partial t} f^1 &= \mathscr{F}[f^2] \\ \frac{\partial}{\partial t} f^2 &= \mathscr{F}[f^3] \\ &\vdots \\ \frac{\partial}{\partial t} f^{N-1} &= \mathscr{F}[f^N] \\ \frac{\partial}{\partial t} f^N &= -\sum_{i=1}^{N} \nabla_{\boldsymbol{x}_i}\cdot\left\{\langle \boldsymbol{u}_i|\boldsymbol{\Omega}_1 \ldots \boldsymbol{\Omega}_N\rangle f^N\right\} \\ &\quad -\sum_{i=1}^{N} \nabla_{\boldsymbol{\Omega}_i}\cdot\left\{\langle \mathrm{rhs}_i|\boldsymbol{\Omega}_1 \ldots \boldsymbol{\Omega}_N\rangle f^N\right\}. \end{aligned} \tag{6}$$

As indicated here, the hierarchy may formally be truncated on the N-th level by introducing conditional averages. As a result, the velocity and the right-hand side of the vorticity equation conditioned on the vorticity at N points enter the system of equations as unclosed terms. In this sense, eq. (4) corresponds to a truncation of the hierarchy on the level $N = 1$, i.e. the most simple case. This is the above-mentioned closure problem, which complicates a statistical treatment of turbulence. Hence a second major goal of the current project is to investigate the coupling of the different "levels" of this PDF hierarchy. The numerical computations give hints on the nature of the coupling, motivate reasonable closure approximations (in terms of realizability) and help to test analytical results.

3.2 Statistical Symmetries

The kinetic equations of the last paragraph display a quite complex mathematical structure. Furthermore, conditional averages with respect to a vectorial quantity are technically hard to estimate due to their dimensionality. As we want to study the vorticity statistics in fully developed stationary, homogeneous and isotropic turbulence, these statistical symmetries allow to simplify the problem. For example, the PDF $f^1(\boldsymbol{\Omega}_1;\boldsymbol{x}_1,t)$ is due to homogeneity independent from the position $\boldsymbol{x}_1$ and does not depend on time due to stationarity. On account of isotropy, it can only depend on the absolute value of $\boldsymbol{\Omega}_1$, such that there is a simple relation to the PDF of the absolute value, $\tilde{f}^1(\Omega_1) = 4\pi\Omega_1^2 f^1(\boldsymbol{\Omega}_1)$. The conditionally averaged terms of eq. (4) can also be simplified due to these statistical symmetries, so that for example the rate-of-strain tensor of the vortex stretching term and the diffusive term take the form

$$\langle \nu\Delta\omega_i|\boldsymbol{\Omega}\rangle = d(\Omega)\frac{\Omega_i}{\Omega} \qquad d(\Omega) = \left\langle \frac{\omega_i \nu\Delta\omega_i}{\omega} \middle| \Omega \right\rangle \tag{7}$$

$$\langle S_{ij}|\boldsymbol{\Omega}\rangle = \frac{1}{2}s(\Omega)\left(3\frac{\Omega_i\Omega_j}{\Omega^2} - \delta_{ij}\right) \qquad s(\Omega) = \left\langle \frac{\omega_i S_{ij}\omega_j}{\omega^2} \middle| \Omega \right\rangle \tag{8}$$

Homogeneity allows to re-express the diffusive contributions in terms of the conditionally averaged enstrophy dissipation tensor [11, 13]

$$\left\langle \nu\frac{\partial\omega_i}{\partial x_k}\frac{\partial\omega_j}{\partial x_k} \middle| \boldsymbol{\Omega} \right\rangle = \frac{1}{3}g(\Omega)\delta_{ij} + h(\Omega)\frac{\Omega_i\Omega_j}{\Omega^2}, \tag{9}$$

which is characterized by two scalar functions g and h related to the gradient and dissipation of enstrophy. These conditional averages will be estimated from our numerical simulations below.

4 Numerical Methods and Computations

4.1 Algorithm

The current project involves a direct numerical simulation of the three-dimensional vorticity equation (2) on a periodic domain. The method used is a standard pseudospectral method [1, 2], in which most of the computations (evaluation of derivatives etc.) are performed in Fourier space. Hence the equation takes the form

$$\frac{\partial}{\partial t}\tilde{\boldsymbol{\omega}}(\boldsymbol{k},t) + \nu k^2\tilde{\boldsymbol{\omega}}(\boldsymbol{k},t) = i\boldsymbol{k}\times\mathscr{F}\{\boldsymbol{u}(\boldsymbol{x},t)\times\boldsymbol{\omega}(\boldsymbol{x},t)\} + \tilde{\boldsymbol{F}}(\boldsymbol{k},t). \tag{10}$$

The nonlinearity on the right-hand side is evaluated in real space, reducing the computational efforts from $\mathscr{O}(N^2)$ (which an evaluation of the corresponding convolution would cost in Fourier space) to $\mathcal{O}(N \log N)$ (needed for the FFTs). The emerging aliasing errors are reduced with the help of a smooth Fourier filter [5]. Accurate time-stepping is achieved by a third-order, memory saving Runge-Kutta scheme [12]. The Laplace-term is treated exactly by an integrating factor technique [2]. Care has to be taken regarding the external forcing term, which is necessary to achieve a statistically stationary state. Different types of large-scale forcings have been implemented and tested for stationarity, homogeneity and isotropy.

In order to reach high Reynolds numbers as well as to gather a large amount of data we perform highly resolved simulations with up to 1024^3 grid points integrated for several thousands (up to tens of thousands) of time steps (corresponding to some large eddy turnover times), which requires an efficient parallel implementation of the numerical scheme. Hence our Fortran code is MPI parallelized and employs a slab domain decomposition, enabling us to effectively use up to 1024 cores at a resolution of 1024^3 grid points. Parallel IO is incorporated by the usage of MPI-IO. The Fourier transforms are performed by the free library FFTW [3]. Only involving these libraries the code is highly portable.

Furthermore, we implemented the possibility to follow trajectories of fluid particles (so-called Lagrangian tracer particles) through the flow fields. In addition to the Eulerian fields, the Lagrangian particle paths are evaluated according to

$$\frac{\mathrm{d}}{\mathrm{d}t}\boldsymbol{X}(t,\boldsymbol{y}) = [\boldsymbol{u}(\boldsymbol{x},t)]_{\boldsymbol{x}=\boldsymbol{X}(t,\boldsymbol{y})}, \tag{11}$$

i.e. the tracer particle starting at position $\boldsymbol{y}$ follows the velocity field interpolated at the current position of the particle. We use a tricubic interpolation scheme together with a third-order Adams-Bashforth-Moulton predictor-corrector method for the time-stepping. Additionally, more dynamical quantities of interest like for example the acceleration and the velocity gradients may be evaluated along the particle path.

4.2 Typical Runs

A typical simulation basically consists of two stages. First, a proper initial condition has to be obtained. To this end an artificial large-scale initial condition decays for some large eddy turnover times. During this time a physically reasonable solution develops. Then the forcing is applied and the system approaches statistical stationarity after some large eddy turnover times. This simulation stage is performed with adaptive time-stepping according to a Courant-Friedrichs-Levy condition. To save computational costs, the initial conditions for highly resolved simulations can be obtained from lower resolved simulations by up-sampling in Fourier space. The duration of the thermalization stage of the up-sampled field is shorter than for a random large-scale initial condition. It turns out that this thermalization stage of the simulation is rather costly, making up a considerable amount of the computational resources needed.

After the preparation of proper initial conditions the actual simulation is performed. Here, the flow field is advanced in the statistically stationary state. During this period fields (velocity, vorticity, velocity gradients etc.) are stored with a sampling rate sufficient to form a statistical ensemble. The statistical analysis is performed during the postprocessing stage. Optionally tracer particles are advected with the flow and stored frequently, giving the possibility to investigate spatio-temporal correlations. A typical 1024^3 run requires several tens of thousands of cpu hours for the preparation of initial conditions and the actual simulation. An Eulerian field here requires 12 GB of disk space, and the Lagrangian data produced per time step is of the order of 100 MB. In total, such a run easily produces a terabyte of data. Within the ongoing project runs with resolutions between 256^3 and 1024^3 grid points with Taylor-based Reynolds numbers ranging from about 75 to 250 have been performed, giving insight into the Reynolds number dependence of the statistical quantities under consideration as well as resolution issues. We would like to emphasize that the computational resources needed are not spent on a single run, our strategy rather is to perform a series of runs (differing in resolution, Reynolds number, type of forcing, initial condition etc.), accompanied by a continuous advancement and refinement of the presented theoretical framework.

4.3 Scaling and Performance

We close the technical section with a short overview of the scaling performance of our code. To this end we performed scaling tests with a grid resolution of 1024^3 and optionally ten millions of Lagrangian tracer particles, which represents a typical scenario for a highly resolved simulation. The tests were performed on the low density partitions of the HLRB II, the results are shown in figure 3. Both the simu-

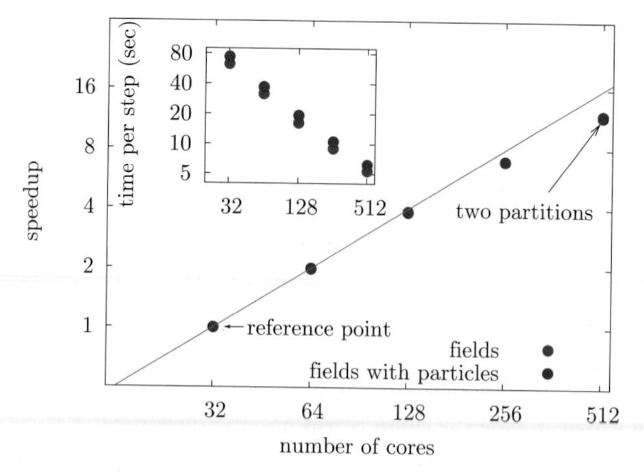

Fig. 3 Scaling results of a 1024^3 run with and without ten million tracer particles. The orange line indicates ideal scaling. The inset displays the time needed for a single simulation step

lations with and without particles perform very good. Slight performance penalties arise when using two partitions. As the calculation of the time evolution of the flow fields relies heavily on the use of Fourier transforms, the performance and scalability of this part of the code mirrors the effective implementation of the FFTW. The parallelization of the advection of tracer particles has been written from scratch, where care had to be taken in order to allow for a high sustained performance of the code. For example, as our computational domain is decomposed into slices along the z-direction, a particle may change the slice when passing through the simulation domain leading to communication overhead. As the scaling results show, the advection of the tracer particles performs rather well. The code scales just as well as without particles and the overall computational costs increase by 15 to 20 percent.

5 Merging Theory and Direct Numerical Simulations

The theoretical framework outlined above identifies a number of conditional averages necessary to characterize the single-point statistics of the vorticity. These are the conditional averages related to vortex stretching, vorticity diffusion, or equivalently, the terms from the conditional enstrophy dissipation tensor. Equation (4) takes the form of a conservation law, where the right-hand side may be interpreted as a divergence of a probability current. Under the imposed statistical symmetries, this current has to identically vanish, which we can check with our numerical data. Figure 4 shows the conditionally averaged vortex stretching term and the diffusive term. The vortex stretching term is positively correlated with the vorticity, such that strong vortices are subject to stronger vortex stretching than weak ones. The vorticity diffusion counterbalances the vortex stretching term, which can be seen from the

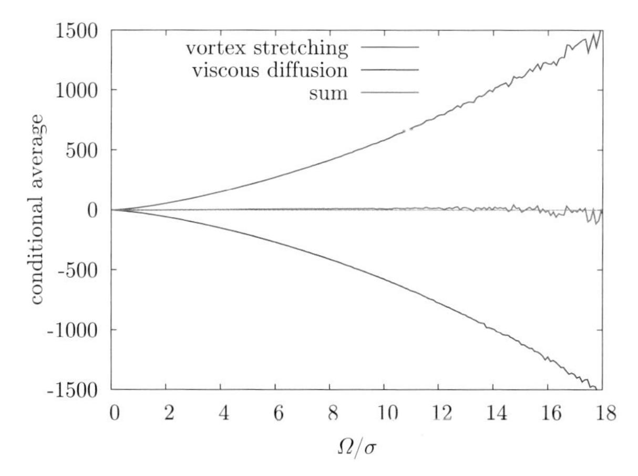

Fig. 4 Conditionally averaged vortex stretching $s(\Omega)\Omega$ and vorticity diffusion $d(\Omega)$. The two terms tend to cancel nearly identically, indicating that the external forcing has a negligible influence

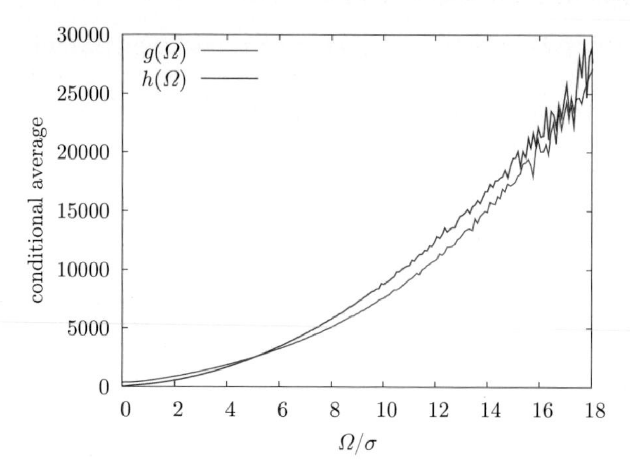

Fig. 5 Functions g and h characterizing the conditional enstrophy dissipation tensor. The functions depend strongly on the vorticity

fact that the sum almost identically vanishes. The consequence is that the external forcing has a negligible effect on the single-point vorticity statistics, it rather maintains a statistically stationary flow. This shows that the structure of the vorticity field is accountable to an internal mechanism inherent to the dynamics of the fluid flow. This conditional balance has been predicted and studied extensively by Novikov [9, 10] to characterize the local conditional structure of the vorticity field.

It turns out that a closed expression for the single-point PDF may be derived from the theoretical framework for homogeneous and stationary turbulence, when the diffusive term is re-expressed in terms of the conditional enstrophy dissipation tensor, characterized by the functions g and h (see eq. (9)). For example, the PDF of the absolute value in case of homogeneity and stationarity can be written as

$$\tilde{f}(\Omega) = \frac{\mathscr{N}}{\frac{1}{3}g(\Omega) + h(\Omega)} \exp - \int^{\Omega} \mathrm{d}\Omega' \, \frac{s(\Omega')\Omega' - \frac{2}{3\Omega'}g(\Omega')}{\frac{1}{3}g(\Omega') + h(\Omega')}. \tag{12}$$

The functions g and h have been evaluated numerically, the result is shown in figure 5. g and h both depend strongly on Ω, indicating, e.g., that the dissipation of enstrophy is strongly correlated with the local amplitude of vorticity. This is generically sufficient to yield strongly non-Gaussian vorticity PDFs. Furthermore, as the function h is non-vanishing, the conditional enstrophy dissipation tensor has non-vanishing off-diagonal elements, which seems to be an important statistical feature of the vorticity field. To check for the consistency of our theoretical framework, we use the numerically obtained conditional averages to evaluate equation (12) as well as the corresponding formula for homogeneous turbulence. The result is compared to the directly estimated vorticity PDF in figure 6. A good agreement is found within the whole range of vorticity values. While the homogeneous PDF collapses perfectly, slight deviations are found for the homogeneous and stationary case due to the minor influence of the neglected forcing term. One has to stress here that, once

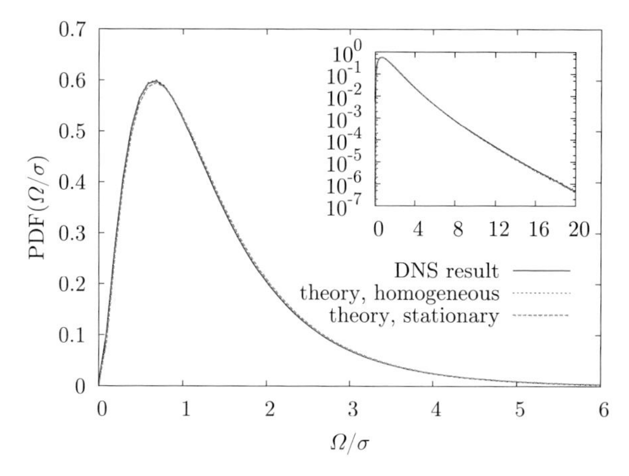

Fig. 6 Comparison of the theoretically and directly obtained PDF of vorticity. The good agreement demonstrates the consistent description of the vorticity statistics within the presented theoretical framework

the functional shape of the conditional avergages has been determined, no further parameters enter the evaluation of eq. (12).

6 Conclusions and Outlook

Based on works by Lundgren, Monin and Novikov [6–8], a theoretical framework has been developed within the current project to characterize the statistical properties of the vorticity field. This approach has to face the classical closure problem of turbulence. The idea here is to express the theory in such a way that the unclosed terms can be estimated from direct numerical simulations. To this end we apply a parallel pseudospectral method to simulate fully developed turbulence and obtain the unclosed terms from a statistical evaluation of the simulation output. As a result, the statistics of the vorticity can be related to dynamical effects like vortex stretching and dissipation of enstrophy. This not only allows to reconstruct the single-point vorticity statistics, it also sheds light on how, for example, the local structure of the vorticity field is correlated with the local self-amplification mechanisms. The present article focuses on the non-Gaussian nature of the vorticity PDF. There is, however, ongoing work on the two-point statistics of the vorticity, which will yield more information on the spatial organization of the vorticity field. Furthermore, the same approach has been applied to the velocity statistics, giving insights how the velocity statistics is related to pressure fluctuations and dissipation of kinetic energy. Beyond that, the statistics of Lagrangian tracer particles is studied, giving especially information on the temporal correlations of the flow.

The possibility to perform direct numerical simulations is of major importance for the development and refinement of these theoretical issues. The huge computational ressources necessary to obtain well-resolved direct numerical simulations of turbulence long for the power of modern supercomputers. As a closing remark, we would like to mention that the recently established remote visualization services of the LRZ have helped us to explore and analyze even large data sets, giving fascinating insight into the spatial structure and temporal evolution of turbulence.

References

1. Boyd, J.: Chebyshev and Fourier Spectral Methods. `www-personal.engin.umich.edu/~jpboyd/` (2000)
2. Canuto, C., Hussaini, M., Quarteroni, A., Zang, T.: Spectral Methods in Fluid Dynamics. Springer-Verlag, Berlin (1987)
3. `www.fftw.org`
4. Frisch, U.: Turbulence - The Legacy of A.N. Kolmogorov. Cambridge University Press, Cambridge, England (1995)
5. Hou, T.Y., Li, R.: Computing nearly singular solutions using pseudo-spectral methods. J. Comp. Phys. **226**(1), 379–397 (2007)
6. Lundgren, T.S.: Distribution functions in the statistical theory of turbulence. Physics of Fluids **10**(5), 969–975 (1967)
7. Monin, A.: Equations of turbulent motion. Prikl. Mat. Mekh. **31**(6), 1057 (1967)
8. Novikov, E.A.: Kinetic equations for a vortex field. Soviet Physics-Doklady **12**(11), 1006–1008 (1967)
9. Novikov, E.A.: A new approach to the problem of turbulence, based on the conditionally averaged Navier-Stokes equations. Fluid Dynamics Research **12**(2), 107–126 (1993)
10. Novikov, E.A., Dommermuth, D.: Conditionally averaged dynamics of turbulence. Mod. Phys. Lett. B **8**(23), 1395 (1994)
11. Pope, S.: Turbulent Flows. Cambridge University Press, Cambridge, England (2000)
12. Shu, C., Osher, S.: Efficient implementation of essentially non-oscillatory shock-capturing schemes. J. Comp. Phys. **77**(12), 379–397 (1988)
13. Wilczek, M., Friedrich, R.: Dynamical origins for non-Gaussian vorticity distributions in turbulent flows. Physical Review E **80**(1), 016316 (2009)

Assessment of CFD Predictive Capability for Aeronautical Applications

N. Thouault, F. Vogel, C. Breitsamter, and N.A. Adams

Abstract The present work focuses on assessing the predictive capabilities of state-of-the-art Computational Fluid Dynamics methods for aerospace engineering applications by comparison with experimental data. Two relevant cases are tested: a generic fan-in-wing configuration and a helicopter fuselage. Unsteady Reynolds Averaged Navier Stokes and Scale-Adaptive Simulations are carried out to predict the flow physics of such complex geometry arrangements. The simulation results show a satisfactory agreement with a wide range of experimental data.

1 Project Description

Computational Fluid Dynamics (CFD) methods are currently a widely spread tool for the design of industrial applications. Particularly in the field of aerospace engineering, numerical simulations have proven their value for a large variety of configurations (e. g. airfoil, wing, turbomachinery). Direct numerical simulations and large eddy simulations remain unaffordable to predict flows around complex geometries despite the increase of computational power in the last years. Therefore, Unsteady Reynolds Averaged Navier Stokes (URANS) or hybrid methods (e. g. Detached Eddy Simulations (DES) or Scale Adaptive Simulations (SAS)) are mostly used in the industry to investigate the flow physics of such configurations at reasonable computational cost.

The predictive capability of state-of-the-art URANS methods for complex geometries however requires a carefull evaluation by comparison with experimental data, particularly for cases with large separated areas. In the present work, two dif-

N. Thouault · F. Vogel · C. Breitsamter · N.A. Adams
Technische Universität München, Lehrstuhl für Aerodynamik, Boltzmannstr. 15, 85748 Garching, Germany
e-mail: nicolas.thouault@aer.mw.tum.de
e-mail: florian.vogel@aer.mw.tum.de

S. Wagner et al. (eds.), *High Performance Computing in Science and Engineering, Garching/Munich 2009*, DOI 10.1007/978-3-642-13872-0_11,

ferent wind tunnel models are investigated: a generic fan-in-wing configuration and a helicopter fuselage. Both cases feature complex, highly unsteady flow topologies and represent a challenge for an accurate CFD prediction.

After an overview of the background theory, the results obtained by URANS and SAS simulations will be compared to a wide array of experimental data. For each test case, details on the computational aspect will be provided.

2 Theory

The introduction of averaged and fluctuacting quantities in the original Navier Stokes equations leads to the RANS equations. This averaging procedure adds unknown terms in the equations, the Reynolds stresses, that need to be modeled. For this reason, a turbulence model is required to achieve closure of the RANS mean flow equations. In this section, the k-ω Shear-Stress-Transport (SST) model and the Scale Adaptive Simulation SAS-SST model are shortly described.

2.1 The k-ω SST Turbulence Model

The SST model is a two-equation eddy viscosity model based on the $k-\omega$ model from Menter [2]. The Reynolds stresses are related to the velocity gradient in the flow by a Boussinesq relationship (eq. 1). This model is a blend of the $k-\varepsilon$ and $k-\omega$ models implemented in the outer domain and in the near wall region respectively. The eddy viscosity is obtained from eq. 2 where F_2 is a blending function and S is an invariant measure of the strain rate. Additional information on the blending function can be found in ref. [2].

$$-\rho\overline{u_i u_j} = \mu_t\left(\frac{\partial \overline{U}_i}{\partial x_j} + \frac{\partial \overline{U}_j}{\partial x_i}\right) - \frac{2}{3}\rho\delta_{ij} \tag{1}$$

$$\nu_t = \frac{\mu_t}{\rho} = \frac{a_1 k}{max(a_1\omega, SF_2)} \tag{2}$$

2.2 The SAS-SST Turbulence Model

Similarly to a DES, the SAS is a blended model offering URANS capability near the solid wall and improved URANS capability in the outer domain. The k-ω SST model overestimates the eddy viscosity levels (as well as the length scale L which is proportional to the shear layer thickness) resulting in an unphysical damping of local instabilities. For this reason, the SAS-SST model introduces an additional term in the ω-equation of the standard SST model (eq. 3). This term is directly related to the Von Kármàn length scale, L_{VK}, which allows a dynamic adjusment of the length scale thereby improving the eddy viscosity levels. L_{VK} is related to the second derivative of the velocity, $L_{VK} = \kappa|\frac{\partial U}{\partial y}/\frac{\partial^2 U}{\partial^2 y}|$.

$$\frac{\partial \rho\omega}{\partial t} + \frac{\partial U_j \rho\omega}{\partial x_j} = \underbrace{\alpha\rho S^2 - \beta\rho\omega^2 + \frac{\partial}{\partial x_j}\left(\frac{\mu_t}{\sigma_\omega}\frac{\partial \omega}{\partial x_j}\right)}_{standard\ (k-\omega)-part} + \underbrace{\frac{2\rho}{\sigma_\Phi}\frac{1}{\omega}\frac{\partial k}{\partial x_j}\frac{\partial \omega}{\partial x_j}}_{SST-part} \tag{3}$$

$$\underbrace{-\frac{2\rho}{\sigma_\Phi}\frac{k}{\omega^2}\frac{\partial \omega}{\partial x_j}\frac{\partial \omega}{\partial x_j} + \tilde{\zeta}_2\kappa\rho S^2\frac{L}{L_{VK}}}_{=F_{SST-SAS}(SAS-part)}. \tag{4}$$

$L = \sqrt{k}/(c_\mu^{0.25}\omega)$ is the calculated length scale of the standard $k-\omega$ model.

The SAS model adjusts the eddy viscosity level only in the unsteady regions of the flow. In the regions without flow instabilities, the two terms included in $F_{SST-SAS}$ are of the same order of magnitude (eq. 5). Therefore $F_{SST-SAS}$ disappears and the turbulent model operates as the standard k-ω SST model. L_{VK} decreases in instability regions resulting in a increase of $F_{SST-SAS}$ (eq. 6). As the eddy viscosity is inversely proportional to ω, an increase of $F_{SST-SAS}$ leads to the desired relative reduction of eddy viscosity.

$$\frac{2\rho}{\sigma_\Phi}\frac{k}{\omega^2}\frac{\partial \omega}{\partial x_j}\frac{\partial \omega}{\partial x_j} \approx \tilde{\zeta}_2\kappa\rho S^2\frac{L}{L_{VK}} \qquad (SST\ \text{Regime}) \tag{5}$$

$$\frac{2\rho}{\sigma_\Phi}\frac{k}{\omega^2}\frac{\partial \omega}{\partial x_j}\frac{\partial \omega}{\partial x_j} < \tilde{\zeta}_2\kappa\rho S^2\frac{L}{L_{VK}} \qquad (SAS\text{-}SST\ \text{Regime}) \tag{6}$$

The $F_{SST-SAS}$ term has to modeled in order to perform a proper switch between SST and SAS-SST regimes (eq. 7).

$$F_{SST-SAS} = \rho F_{SAS} max\left[\tilde{\zeta}_2\kappa S^2\frac{L}{L_{VK}} - \frac{2}{\sigma_\Phi}k.max\left(\frac{1}{\omega^2}\frac{\partial \omega}{\partial x_j}\frac{\partial \omega}{\partial x_j}, \frac{1}{k^2}\frac{\partial k}{\partial x_j}\frac{\partial k}{\partial x_j}\right), 0\right] \tag{7}$$

F_{SAS} is used to calibtrate the SAS term in the SST environment ($F_{SAS} = 1.25$). Further information on the SST-SAS model can be found in [1] and [3].

3 Verification of URANS Predictions

The simulations have been performed at the HLRB2 on the SGI Altix 4700 platform, Leibniz Supercomputing center. The version 11 of ANSYS CFX has been used to conduct the numerical simulations. The parallelization is accomplished by means of the Single-Program-Multiple-Data (SMPD) model. For both test cases, an automatic decomposition of the computational domain has been used [1]. During the parallel run, the communication was realized using the MPICH message-passing libraries. The computational efficiency is, for the moment, not available due to the lack of access to the source code.

A second order discretization of the convective terms is used throughout all calculations. This discretization scheme is bounded and follows the boundedness principle used by Barth and Jesperson [6]. Considering the large aspect ratio of the near-wall cells, double precision is used throughout all computations to minimize round-off errors. The temporal discretization by means of a second order backward Euler scheme.

The experimental work has been performed in the Göttingen-type wind tunnel A at the Institute of Aerodynamics, Technische Universität München. The test section dimensions are 1.8 m in height, 2.4 m in width and 4.8 m in length. Maximum usable velocity is 65 m/s with a turbulence level of the freestream less than 0.4%.

3.1 *Fan-in-Wing*

Vertical and Short Take-Off and Landing aircraft (V/STOL) can be an alternative for short to medium range aircraft, to relieve congestion on airports and face the increase in air transportation traffic. For a fan-in-wing aircraft, additional lift is provided by a lift-fan being contained in the wing and operating perpendicularly to the freestream. The fan ingests part of the incoming flow on the wing upper side and, after compression, ejects the air at its nozzle thereby creating a jet-in-cross-flow problem on the wing lower side (Fig. 1). An extensive investigation has been performed on a generic wind tunnel model to analyse the fan-in-wing aerodynamic characteristics. In this section, results obtained from URANS simulations are compared to a wide range of experimental data. Additional information can be found in refs. [4] and [5].

3.1.1 Model Description

The model airfoil is of NACA 16-020 type. It provides enough space to insert two model fans inside the wing. The wing half model has an aspect ratio of 2.3, a semi-span area of 0.683 m^2, a taper ratio of 0.71 and 0° sweep of the 50% root chord line.

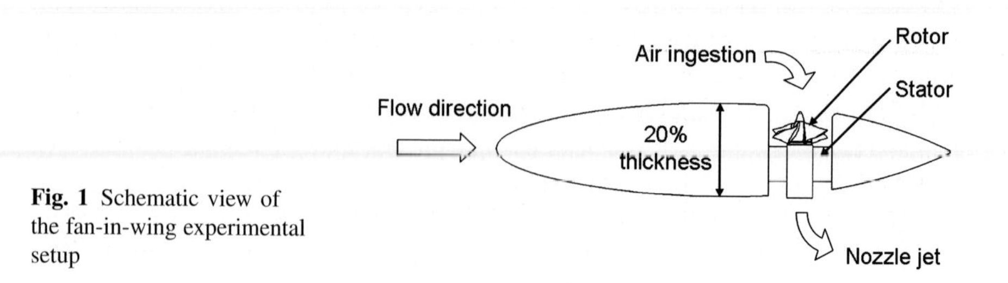

Fig. 1 Schematic view of the fan-in-wing experimental setup

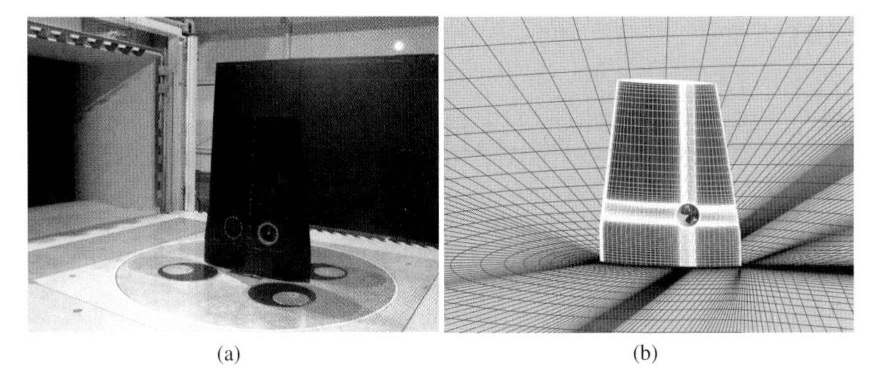

(a) (b)

Fig. 2 (a) Generic fan-in-wing model mounted on underfloor balance in wind tunnel test section and (b) grid topology

The model is designed to investigate several generic fan-in-wing configurations: (i) the closed wing without fan as reference, (ii) one single fan installed either at the rear (Fig. 2(a)) or in the front part of the wing, and (iii) two fans placed symmetrically with respect to the half chord. The geometry modeled in the numerical simulation is the one presented in Fig. 2(a) for a single fan at the wing rear part. The rear fan axis is located at $\frac{2}{3}$ c, and at 0.12 b from the root chord. The fan has a diameter of 0.13 c (c refers to the chord and b to the wing span). The model is installed on a peniche to raise the wing out of the wind tunnel wall boundary layer. The model fan is composed of two stages, a four-bladed rotor and four-bladed stator.

3.1.2 Numerical Setup

The entire wind tunnel model is reproduced numerically with minimum geometrical approximations (e.g. blade fillets, holes and gaps are removed). ANSYS ICEM CFD is used to generate the grid. A block structured mesh has been selected to obtain more accurate results especially in the near wall region. Special attention has been given to the nearest mesh node in the viscous sublayer for a good prediction of the separated regions on the wing and on the fan blades. The mesh is constructed to benefit from a low-Reynolds number near-wall formulation. A value of $y_+ < 1$ is achieved for the attached flow on the wing and fan blade surfaces. On the inlet lip and the rotor blade tips, the value of y_+ remains below 4. Overall, 939 blocks were created to obtain the mesh. The geometry requires an unconventional blocking strategy. An O-grid topology is set around each fan blade and around the wing. The overall computational topology is a C-H topology, commonly used for external aerodynamic numerical simulation. The mesh contains about 4×10^6 nodes. Additional details on the setup and a grid-dependence study can be found in ref. [5].

The initial solution for the unsteady run is obtained by performing a steady RANS calculation. The fan rotation during steady calculations is modeled with a reference frame transformation. The connection between the area surrounding the blades and the rest of the computational domain is done by frozen-rotor interfaces [1]. In this case, the results depend on the initial position of the blade and do not describe accurately the rotor/stator interaction. For the unsteady calculation, the area surrounding the blades is designated as sliding mesh. Sliding interfaces separate the rotating domain from the stationary domain. The interaction between the rotor and stator is thus taken into account and therefore provides a more realistic model. The root-mean-square residuals are less than 10^{-5} for continuity and less than 10^{-4} for all other residuals (heat transfer, momentum and turbulence quantities). The total energy model is used to address compressibility effects, especially significant at the rotor blade tip and on the inlet radius.

3.1.3 Computational Effort

A computational time of about 1000 CPU hours is necessary to obtain the initial solution. In addition, 3500 CPU hours have been consumed for the unsteady case for the baseline mesh (cf ref. [5]). 5 complete rotor revolutions have been simulated in approximately 8 days with 24 cores. After 5 revolutions, the time history of the loads acting on the wing show indicates that a periodic solution has been reached. A full revolution requires 286 time steps for the baseline mesh using a physical time step of 8×10^{-6} s. GGI Interfaces [1] between the rotating and stationary domain are necessary to ensure a proper connection. Also called sliding interfaces in the unsteady run, they increase significantly the computational time and the required memory. According to [1], each GGI connection means approximately 5% more CPU time and memory. In the fan-in-wing case, 2 GGI connections have been used. The increment of CPU time is hard to estimate because it can vary greatly depending on the nodes number in the GGI connections and the connection geometrical complexity.

3.1.4 Comparison to Experimental Data

Four different types of tests have been conducted in wind tunnel: force measurements, surface static pressure measurements, flowfield mapping using Stereo-PIV near the trailing edge and wool-tufts flow visualization. All the results presented in this section refer to a freestream velocity of $U_\infty = 40\ m/s$ (corresponding to a Reynolds number of $R_{e_c} = 2 \times 10^6$) and a fan rotation speed of N = 26200 rpm. In this section, the angle of attack is $\alpha = 0°$.

The drag coefficients predicted by the simulation show a very good agreement with the experimental data while the lift coefficients exhibit differences (Table 1). The differences observed on the lift coefficient are conveyed to the pitching moment coefficient. A small airfoil imperfection of the wind tunnel has been found to affect slightly the results. A carefull evaluation of the imperfection effect has been carried out in ref. [5].

Table 1 Comparison of aerodynamic coefficients based on URANS simulation and experiments ($\alpha = 0°$, $U_\infty = 40m/s$ and $N = 26,200rpm$)

	Experiment	Simulation
C_L	0.298	0.272
C_D	0.101	0.097
C_m	-0.055	-0.038

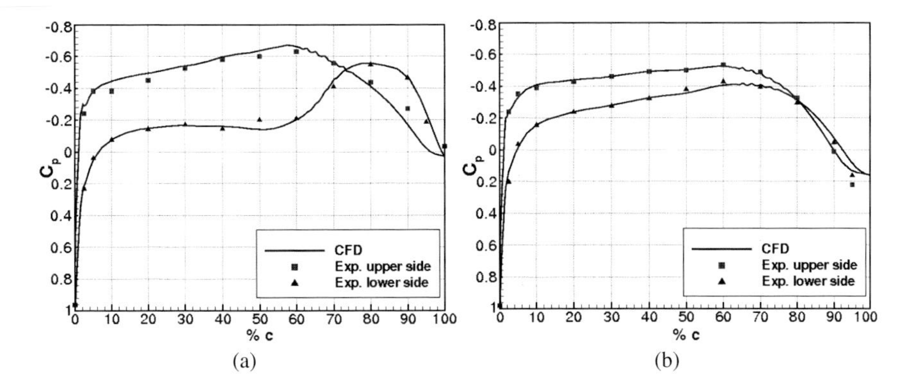

Fig. 3 Comparison of experimental and numerical chordwise pressure distribution at (a) $\frac{1}{3}$ b, (b) $\frac{2}{3}$ b ($\alpha = 0°$, $U_\infty = 40m/s$ and $N = 26,200rpm$)

Two chordwise pressure distributions located in the spanwise direction at $\frac{1}{3}$ b and $\frac{2}{3}$ b from the root chord are investigated. The suction peak on the rear part downstream of the jet, due to the jet mixing with the cross flow, shows a good agreement with the experimental measurements (Fig. 3(a)). The overall pressure distribution matches well with the experimental data for both chordwise distributions (Fig. 3(b)).

Wool-tuft flow visualization provides an overview of the near wall flow direction and indicates regions of attached and separated flow. In the simulation, surface streamlines are based on the time-averaged velocity. The reattachment of the boundary layer on the wing upper side and downstream of the fan can be seen for both numerical and experimental results (Fig. 4(a) and 4(c)). The separation at the trailing edge over this area is also relatively well described by the simulation at $\alpha = 0°$. However, a small separation area observed in the experimental picture, indicated by a circle in Fig. 4(c), is not reproduced by the simulation. On the wing lower side, the overall shape of the suction area appears to be well captured by the numerical simulation. The horseshoe vortex, clearly visible numerically, matches with the experimental picture (Fig. 4(b) and 4(d)). The separation occurring at the trailing edge and over the fan is also well described by the simulation at zero angle of incidence. The flow on the wing lower side and downstream of the fan nozzle is highly unsteady. In this area, separation and reattachement region vary rapidly in time. Therefore, the wool-tufts vizualisation cannot clearly indicate the attached and separated region.

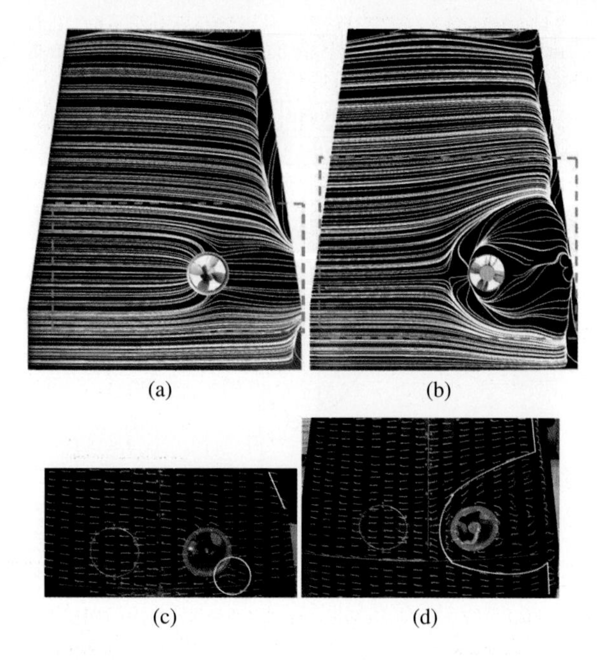

Fig. 4 Comparison of numerical surface streamlines (a) on the wing upper side and (b) on the wing lower side; and wool tuft flow visualization (c) on the wing upper side and (d) on the wing lower side at $\alpha = 0°$, $U_\infty = 40m/s$ and $N = 26,200rpm$

Flow field mapping based on Stereo-Particle Image Velocimetry (PIV) measurements provide quantitative velocity data for comparison downstream of the fan exit. The PIV measurements have been performed at 1.22 c near the trailing edge to capture the pair of counter-rotating vortices. The 'kidney shape' as well as the location of the velocity contours is well predicted by the numerical simulation (Fig. 5(a)). The position of the trailing edge wake is well described by the simulation. The velocity magnitude and the position of the vortex core are also in good agreement with the experimental results.

3.2 Flow Phenomena Around a Helicopter Fuselage

The aerodynamic design of the helicopter fuselage accounts for a significant part in the development process of a rotorcraft. In addition to a drag reduction in cruise, the fuselage design aims to decrease the undesireable structural dynamic loads caused by flow unsteadiness. In particular, flow separation and rollup of the separated boundary layer on the fuselage rear part cause a vortex shedding that can impinge downstream on the empennage.

Experimental investigations face limitation to bring precise information on the flow physics. Therefore, highly sophisticated numerical methods are required. In

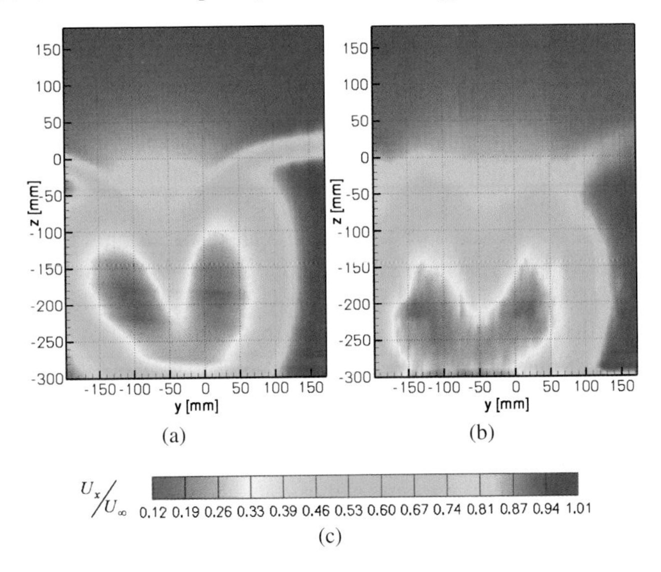

Fig. 5 Contours of streamwise velocity in a cross-flow plane at $x/c = 1.22$. (a) CFD results and (b): experimental results. The PIV sheet center ($y = 0\ mm$ and $z = 0\ mm$) is located at $x = 1.22\ c$, $y = 0.18\ b$, and $z = -0.14\ t$ (t, maximum wing thickness) from the wing root chord leading edge. $\alpha = 0°$, $U_\infty = 40m/s$ and $N = 26,200rpm$

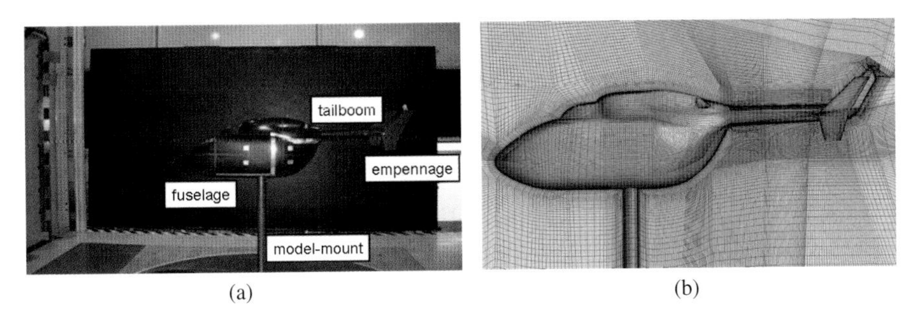

Fig. 6 Helicopter wind-tunnel model (a), grid topology (b)

this section, results obtained numerically with two different turbulence models are compared to experimental data. Unsteady simulations are carried out to model this unsteady and complex flow topology. Detailed information can be found in refs. [7] and [8].

3.2.1 Model Description

Here, the fuselage of a conventional light-weight transport helicopter (one main rotor and one tail rotor) is investigated. The model characteristic parts are indicated in Fig. 6(a) (e. g. tailboom, empennage). The empennage consists of two horizontal wings with endplates on each side. It ensures the helicopter stability in pitch and yaw and is used to control the position of the fuselage in trimmed cruise.

3.2.2 Numerical Setup

A structured mesh is generated with IcemCFD Hexa based on a block structured approach. The helicopter fuselage as well as the model mount are included in the computational domain. Overall, 643 blocks have been created. The near wall area is resolved by 20 mesh layers leading to a dimensionless wall distance less than $y_+ = 0.5$ on the fuselage body. Note that the mesh is written out as unstructured to be used by the code.

Two different turbulence models are used: the k-ω SST and the k-ω SAS-SST. The turbulence model comparison is performed with the same grid of 5×10^6 nodes. For the unsteady simulations, the time step is defined as 0.1 ms and the calculation run over 0.25 s. The unsteady calculations were initialized with the results of a steady state simulation to accelerate the convergence.

3.2.3 Computational Effort

For all simulations (including steady state runs and the unsteady calculation) a overall time of 3000 CPU hours was necessary. The calculations were performed in parallel mode with 20 to 40 cores to reduce the wall time.

3.2.4 Results and Validation

A wind tunnel campaign was conducted to provide data for comparison with the numerical simulations. Various experiments were carried out namely force and surface pressure measurements, flow field mapping using stereo-PIV and oil streamlines visualizations. In this section, the results are presented for the following flow conditions: an angle of attack $\alpha = 0°$ and a freestream velocity of U_∞ = 40m/s corresponding to a Reynolds number of Re = 6.4×10^5.

The experimental drag and lift coefficients are compared to numerical results obtained with different turbulence models in table 2. Both simulations show a good agreement with the experiment for the drag coefficient. The negative lift is strongly underpredicted by the k-ω SST model contrary to the SAS-SST model which matches the experiment very well.

The surface pressure distribution in the fuselage symmetry plane (Y=0) is presented in Fig. 7. Along the model upper side, the predicted surface pressure distribution show a good agreement with the experiment. The numerically obtained pressure distribution on the model lower side exhibits differences with the experiment downstream of the model mount. The suction peak close behind the model mount is not predicted by the k-ω SST model leading to a higher pressure level in

Table 2 Comparison of predicted aerodynamic coefficients to experimental data ($\alpha = 0°$, U_∞=40m/s)

	experiment	SST k-ω	SAS-SST k-ω
C_D	0.182	0.176	0.194
C_L	-0.185	-0.084	-0.183

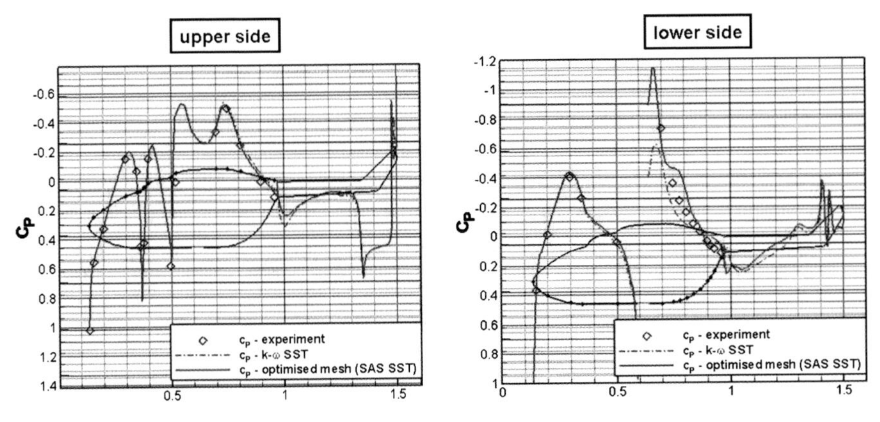

Fig. 7 Surface pressure distribution along the fuselage symmetry plane (Y=0)

this area and consequently to a lower down force. This effect explains the underprediction of negative lift by the k-ω SST model. The pressure distribution obtained with the SAS-SST model shows a better agreement with the experimental data and therefore results in an improved prediction of the lift coefficient.

4 Conclusions

The predictive capabilities of state-of-the-art Computational Fluid Dynamics have been assessed by comparison with experimental data for two different test cases using a commercially available flow simulation software (ANSYS CFX). An URANS simulation has been carried out with the k-ω SST model for a generic fan-in-wing configuration. In addition, a Scale Adaptive Simulation has been performed for a helicopter fuselage. The results show a satisfactory agreement with a wide range of experimental data including force measurements, surface pressure measurements, particle image velocimetry and flow visualization results. The improved prediction of unsteady complex flows, particularly wake regions, by the SAS-SST model has been highlighted.

Acknowledgements The authors would like to thank ANSYS CFX for providing the flow simulation software.

References

1. ANSYS CFX, *Ansys CFX-Solver Theory Guide*, Ansys Europe.
2. F. R. Menter, *Two-Equation Eddy-Viscosity Turbulence Models for Engineering Applications, AIAA Journal*, Vol. 32, 1994, pp. 1598–1605.
3. F. R. Menter and Y. Egorov, *A Scale-Adaptive Simulation Model using Two-Equation Models, Proceedings of the AIAA 43rd Aerospace Sciences Meeting*, AIAA 2005-1095, Reno, USA, 2005.
4. N. Thouault, C. Breitsamter, N. A. Adams, C. Gologan and J. Seifert, *Experimental Investigation of the Aerodynamic Characteristics of Generic Fan-in-Wing Configurations, The Aeronautical Journal*, Vol. 113, No. 1139, 2009, pa. 3338.
5. N. Thouault, C. Breitsamter and N. A. Adams, *Numerical and Experimental Analysis of a Generic Fan-in-Wing Configuration, Journal of Aircraft*, Vol. 46, No. 2, 2009, pp. 656–666.
6. T. J. Barth and D. C. Jesperson, *The Design and Application of Upwind Schemes on Unstructured Meshes*, AIAA Paper 89-0366, 1983.
7. F. Vogel, C. Breitsamter and N. A. Adams, *Unsteady Effects of the Separated Flow at the Tail Section of a Helicopter Fuselage, Proceedings of the 35th European Rotorcraft Forum*, Hamburg, Germany, 2009.
8. F. Vogel, C. Breitsamter and N. A. Adams, *Aerodynamic Analysis of a Helicopter Fuselage, Proceedings of the STAB symposium*, Aachen, Germany, 2008.

Computational Aspects of Implicit LES of Complex Flows

M. Meyer, S. Hickel, and N.A. Adams

Abstract Computational fluid dynamics for complex industrial applications up to now usually refers to RANS (Reynolds-Averaged Navier Stokes) simulations with appropriate statistical turbulence models. Existing RANS turbulence models, however, often fail to accurately predict separated and reattached flows. Better results but higher computational costs are expected from Large-Eddy Simulations (LES). It is therefore a common interest to develop efficient and robust LES approaches for the prediction of complex flows with industrial relevance. An efficient representation of physically complex flows can be achieved with an Implicit LES approach, where the truncation error of the numerical discretization itself functions as turbulence model. We employ an Adaptive Local Deconvolution Method (ALDM) implemented in a solver for the incompressible Navier-Stokes equations where the domain is discretized on a Cartesian grid. Boundaries of arbitrary shape are represented by a second-order accurate Conservative Immersed Interface Method (CIIM). In this report we point out the computational aspects of CIIM in the framework of ILES on the example of the flow over a circular cylinder at Re = 3,900. As benchmark an ILES with ALDM on body-fitted grids and with a usual formulation of the wall-boundary condition is taken. Two simulations with similar numerical resolution are compared with experimental reference data. Also compared are their computational cost.

1 Project Description

In aerospace engineering Computational Fluid Dynamics (CFD) is an indispensable tool commonly used in the development and optimization process of airfoils, aircraft components or complete airplanes. In CFD a continuous fluid is treated in a

M. Meyer · S. Hickel · N.A. Adams
Lehrstuhl für Aerodynamik, Technische Universität München, Boltzmannstr. 15, 85748 Garching, Germany
e-mail: michael.meyer@aer.mw.tum.de

S. Wagner et al. (eds.), *High Performance Computing in Science and Engineering, Garching/Munich 2009*, DOI 10.1007/978-3-642-13872-0_12,

discretized fashion on a computer. The most-common way is to discretize the spatial domain into computational cells to form a computational grid. Provided that on the grid all relevant length and time scales can be resolved, it is possible to directly solve the full unsteady Navier Stokes equations in so-called Direct Numerical Simulations (DNS).

As most flows are turbulent and exhibit a large range of length and time scales, DNS for industrial applications would require computational power that is not yet available. These applications are only feasible with a simplification of the Navier-Stokes equations, i.e. with approaches involving turbulence models. The most common approach is Reynolds averaging, where only the mean flow has to be resolved. The existing Reynolds Averaged Navier Stokes (RANS) turbulence models, however, often fail to accurately predict separated and reattached flow structures. Better results are obtained by Large-Eddy Simulations (LES), where large energetic flow structures are explicitly resolved the unresolved subgrid-scale structures are modeled. LES is computationally more expensive than RANS but less expensive than DNS. It is therefore of a great industrial interest and the objective of this project to develop efficient and robust LES approaches for the prediction of complex flows.

An efficient representation of physically complex flows can be achieved with Implicit Large Eddy Simulation (ILES) [9, 12]. In ILES the truncation error of the discretization of the convective terms acts as a subgrid-scale model which is thus an inherent feature of the discretization. In this context, the Adaptive Local Deconvolution Method (ALDM) [8–10] uses a discretization based on solution-adaptive deconvolution where parameters are determined specifically in such a way that the truncation error properly models the subgrid-scale energy transfer. Applications to canonical configurations [8, 11, 12] show excellent agreement with the reference results from theory, experiment, and DNS. Given this performance, our motivation is to use ALDM for investigations of flows around complex geometries with industrial relevance.

The generation of good-quality body-fitted grids for complex geometries is time-consuming and difficult. Contradictory requirements, namely adequate local resolution and minimum number of grid points can deteriorate the grid quality and therefore adversely affect accuracy and numerical convergence properties. For this reason we prefer Cartesian grids for the discretization of the domain. Cartesian grids also allow for an easy control of the truncation error and have superior computational efficiency. Flow boundaries and obstacles present in the flow are modeled by an Immersed Interface (II) approach. Note that II methods are a generalization of Immersed Boundary (IB) methods [22] as the formulation can be applied also to fluidic interfaces.

In this report we study the computational aspects of a novel second-order, Conservative Immersed Interface Method (CIIM) [20] in the framework of ALDM. As test case, we consider the flow over a circular cylinder at Reynolds number $Re = 3,900$ based on the cylinder diameter and the free-stream velocity. As a computational benchmark simulation we use the same ALDM turbulence model formulated on a curvilinear body-fitted grid with a usual formulation of the wall-boundary condition.

2 Numerical Method

2.1 Underlying Finite-Volume Discretization and ALDM

We consider a generic conservation equation in differential form on the domain Ω

$$\partial_t \mathbf{Q} + \nabla \cdot \mathbf{F} = 0, \tag{1}$$

with appropriate boundary conditions on $\partial\Omega$. The vector $\mathbf{Q}$ represents the conserved quantities and $\mathbf{F}$ is a flux function. The Navier-Stokes equations for incompressible fluid flow are recovered when $\mathbf{Q}$ equals the velocity $\mathbf{u} = [u, v, w]$ and $\mathbf{F}$ is the non-linear flux function

$$\mathbf{F} = \mathbf{u}\mathbf{u} + \mathbf{I}p - \nu \nabla \mathbf{u}, \tag{2}$$

where ν represents the kinematic viscosity, p is the pressure and $\mathbf{I}$ is the unity tensor. Mass conservation on Ω is ensured by the incompressible continuity equation

$$\nabla \cdot \mathbf{u} = 0. \tag{3}$$

Integrating Eq. (1) over the volume $V_{i,j,k} \cap \Omega$ of a computational cell (i,j,k) and a time step t_n to t_{n+1} with time step size $\Delta t = t_{n+1} - t_n$ yields

$$\frac{1}{V_{i,j,k}\Delta t} \int_{t_n}^{t_{n+1}} dt \int_{V_{i,j,k}\cap\Omega} dx\, dy\, dz \left(\frac{\partial \mathbf{u}}{\partial t} + \nabla \cdot \mathbf{F} \right) = 0. \tag{4}$$

A finite-volume discretization of Eq. (4) is obtained by

$$\frac{1}{V_{i,j,k}\Delta t} \int_{t_n}^{t_{n+1}} dt \left(\int_{V_{i,j,k}\cap\Omega} dx\, dy\, dz \frac{\partial \mathbf{u}}{\partial t} + \int_{\partial(V_{i,j,k}\cap\Omega)} dS\, \mathbf{F} \cdot \mathbf{n}_\perp \right) = 0, \tag{5}$$

where $\mathbf{n}_\perp$ is the normal vector and dS are the cell faces. In this paper, a fractional-step method is used and fluxes in Eq. (5) are computed with the Adaptive Local Deconvolution Method (ALDM).

The theoretical background of implicit LES with ALDM [1, 8] follows from the formal equivalence between cell-averaging and reconstruction in finite-volume discretizations, and top-hat filtering and deconvolution in explicit SGS-modeling. With ALDM, a local reconstruction of the unfiltered solution is obtained from a solution-adaptive combination of approximation polynomials. Flux calculations differ on curvilinear and Cartesian grids, as the geometry consideration of the body-fitted computational cells in general is significantly more complex. A suitable consistent numerical flux function operates on the approximately deconvolved solution. The solution-adaptive stencil-selection scheme and the numerical flux function contain free parameters which can be used to adjust the spatial truncation error of the discretization.

Unlike in classical numerical analysis, where the truncation error is analyzed in the limit of small grid size compared to the smallest flow scale and discretization coefficients are chosen in such a way that the formal order of accuracy of the discretization is maximum, free parameters of ALDM are selected such that the truncation error of the discretization functions as physically motivated SGS model. For the purpose of finding suitable discretization parameters Hickel et al. [8] performed an a posteriori analysis of the spectral numerical dissipation in LES of freely decaying homogeneous isotropic turbulence. Given that the primary purpose of an SGS model is to provide the correct spectral distribution of the dissipation of resolved scales by interactions with modeled SGS stresses, free discretization parameters were calibrated by constraining the numerical dissipation to the physical SGS dissipation obtained from the analysis of nonlinear interactions in turbulence [8]. With the calibrated discretization parameters, predictions of ALDM are in excellent agreement with theoretical results for isotropic turbulence at high Reynolds numbers. The general validity of the determined model is demonstrated for various canonical flow configurations [8, 11, 12].

2.2 Body-Fitted Grid Simulation Code LESOCC2

The simulation code Large-Eddy Simulation on Curvilinear Coordinates (LESOCC2) [14] employs non-orthogonal, boundary-fitted, block-structured grids. Boundary-fitted block-structured grids can offer considerable computational advantages for predicting wall-bounded turbulence: Imposing no-slip wall boundary conditions is straightforward for grid-aligned domain boundaries. The algorithm uses a cell-centered variable arrangement and Cartesian velocity components. The convective fluxes are discretized by ALDM in an algorithmically simplified version (SALD) [7]. SALD preserves the modeling capability while computational costs are reduced significantly. Another advantage of SALD is that it allows for a straight-forward application to curvilinear grids as the reconstruction of the unfiltered solution at cell faces by Harten-type deconvolution polynomials can be done dimension by dimension [13]. Viscous fluxes are discretized by centered differences. Numerical stability for the collocated grid arrangement is maintained by employing a momentum–interpolation technique.

2.3 Cartesian Grid Simulation Code INCA

The same implicit LES method has been implemented for the Incompressible Navier-Stokes equations on Cartesian Adaptive grids (INCA). Wall boundary conditions are imposed on the Cartesian grid by a second-order Conservative Immersed

Interface Method (CIIM) [20]. CIIM is based on the finite-volume discretization Eq. (5) applied to all cells of the entire computational domain, and a modification for cells that are cut by the immersed interface. CIIM operates on the fluxes of these cut cells only, and hence ensures mass and momentum conservation. Viscous forces on the immersed interface are accounted for by a friction term. The boundary condition of the normal velocity is satisfied by a momentum-exchange term and by imposing a homogeneous Neumann condition in the pressure projection. The main building blocks are outlined in the following subsections. For a more details regarding CIIM refer to Ref. [20].

2.3.1 Cut-Cell Volume Balance

The immersed interface is characterized by a level-set field, i.e. a signed distance function for each point of the domain with respect to the immersed surface. The zero-level-set contour describes the interface between the fluid and the obstacle. The intersection between the immersed boundary and the finite volume cell is approximated as planar interface separating fluid and solid part (see Fig. 1). The volume fraction of the fluid part of a cut cell is denoted $\alpha_{i,j,k}$, with $0 \leq \alpha_{i,j,k} \leq 1$. The volume $V_{i,j,k}$ of a cut cell (i,j,k) can be expressed as $V_{i,j,k} = \alpha_{i,j,k} \triangle x \triangle y \triangle z$, where $\triangle x$, $\triangle y$ and $\triangle z$ are the cell sizes in the respective coordinate directions. The cell faces S are represented by two parts: one is the combination of the six segments of cell faces cut by the interface, which can be written in the form of $A_{i\pm1/2,j,k} \triangle y \triangle z$, $A_{i,j\pm1/2,k} \triangle x \triangle z$, and $A_{i,j,k\pm1/2} \triangle x \triangle y$, where $1 \leq A_{k,l,m} \leq 0$ is the face aperture, the other is the part of the interface $\Gamma_{i,j,k}$ inside of the cell (i,j,k). For a cut cell the

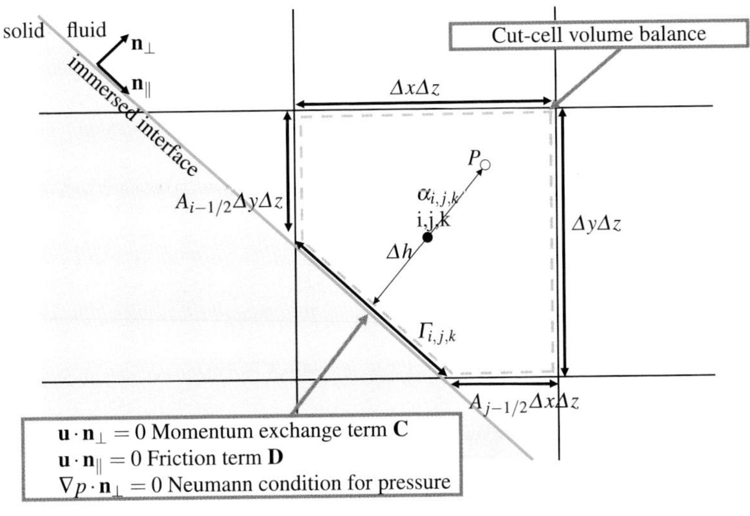

Fig. 1 Overview on the different steps of the conservative immersed interface method with regard to a cut cell (i,j,k)

general formulation of Eq. (5) yields

$$\begin{aligned}\mathbf{u}^{\mathbf{n+1}}_{\mathbf{i,j,k}} &= \mathbf{u}^{\mathbf{n}}_{\mathbf{i,j,k}}\\ &+ \frac{\triangle t\left[A_{i+1/2,j,k}\mathbf{F}_{i+1/2,j,k} - A_{i-1/2,j,k}\mathbf{F}_{i-1/2,j,k}\right]}{\alpha_{i,j,k}\triangle x}\\ &+ \frac{\triangle t\left[A_{i,j+1/2,k}\mathbf{F}_{i,j+1/2,k} - A_{i,j-1/2,k}\mathbf{F}_{i,j-1/2,k}\right]}{\alpha_{i,j,k}\triangle y}\\ &+ \frac{\triangle t\left[A_{i,j,k+1/2}\mathbf{F}_{i,j,k+1/2} - A_{i,j,k-1/2}\mathbf{F}_{i,j,k-1/2}\right]}{\alpha_{i,j,k}\triangle z}\\ &+ \frac{\triangle t\,(\mathbf{C}+\mathbf{D})}{\alpha_{i,j,k}\triangle x\triangle y\triangle z},\end{aligned} \tag{6}$$

where **C** and **D** are additional terms to account for the influence of the immersed interface on the fluid. A sketch of a cut cell in two dimensions for clarity is given in Fig. 1, explanations will follow below. For full fluid cells it is $\mathbf{C} = \mathbf{D} = 0$, and Eq. (6) degenerates to a standard finite-volume discretization on a Cartesian grid.

2.3.2 Friction Term D

In order to account for the shear stress on the interface Γ, we add the friction force **D** to the flux balance of the cut cell (see Fig. 1). We use a linear approximation of the wall-normal velocity gradient and only consider the tangential velocity $\mathbf{u}_{\parallel}$, which is determined by trilinear Lagrangian interpolation at point P, see Fig. 1, from the velocities at the the eight neighboring velocity-cell centers. The velocity at the boundary is zero. The distance between the foot point of the cell center on the immersed interface and the and point P is Δh, which is related to the cell volume.

2.3.3 Momentum Exchange Term C

For impermeable non-moving boundaries the interface normal velocity $\mathbf{u}_{\perp} = (\mathbf{u}\cdot\mathbf{n}_{\perp})\mathbf{n}_{\perp}$ of the fluid vanishes. This is achieved by a momentum exchange term which is added to the flux balance of Eq. (6).

2.3.4 Homogeneous Neumann Condition for the Pressure

The solution algorithm is based on a fractional-step method. The pressure of the previous time step is used to predict an intermediate velocity field, which is not divergence free. The incompressibility constraint is satisfied by a pressure correction which is obtained from solving a Poisson equation with a homogenous Neumann condition $\nabla p \cdot \mathbf{n}_{\perp} = 0$ on the interface Γ.

2.3.5 Mixing Procedure

Cartesian cut cell methods, such as the one employed in this paper, inevitably generate small fluid cells. A special treatment of these cells is mandatory since a stable time integration based on the time step calculated according to the full cell size CFL condition may not be possible. In the present method, the fluid of the small cells is mixed with that of the neighboring cells [15, 20]. We consider cells with a volume fraction $\alpha_{i,j,k} \leq 0.5$ as small cells. Note that discrete conservation is ensured by a flux formulation, where the conservative quantity obtained by a small cell corresponds to a loss of a target cell. This mixing is carried out before the final pressure projection, so that the solution after each time step is divergence free.

3 Test Case Round Cylinder

The flow over a circular cylinder at Re = 3,900 is a challenging test case for LES in general and for CIIM in particular. The laminar shear layers separate from the cylinder and undergo transition in the near wake of the cylinder. A good representation of flow physics in the vicinity of the cylinder surface is necessary to correctly predict the separation of the shear layers. Both, the transition region and the recirculation-bubble region require sufficient grid resolution for capturing turbulence phenomena.

Many Direct Numerical Simulations [3, 18] and Large-Eddy Simulations [2, 4, 5, 16, 19, 25, 26] have been conducted for this flow, so that this case is widespread in the literature and can be considered as established reference for IB methods. In order to study the computational aspects of CIIM, we compare the results of two implicit LES with ALDM, one on a curvilinear body-fitted grid with a usual formulation of the wall-boundary condition, and one on a Cartesian grid with CIIM. Both computations are compared with the particle-image velocimetry data of Lourenco and Shih [17] and the particle-image velocimetry data of Parnaudeau et al. [26].

3.1 Computational Setup

The computational domain is shown in Fig. 2. At the upper and lower boundaries free-slip boundary conditions are applied. In spanwise direction periodic boundary conditions are imposed. At the inlet a uniform velocity profile and at the outlet a pressure condition is defined. At the boundary of the obstacle a viscous-wall boundary condition is prescribed for the calculation on the curvilinear grid, and for the calculation on the Cartesian grid the immersed interface condition is imposed by CIIM.

An O-H topology is used for the curvilinear grid with a total number of 6 million computational cells. The Cartesian grid has a total number of 7 million computational cells and is locally refined at the cylinder, see Fig. 3. For both grids the first

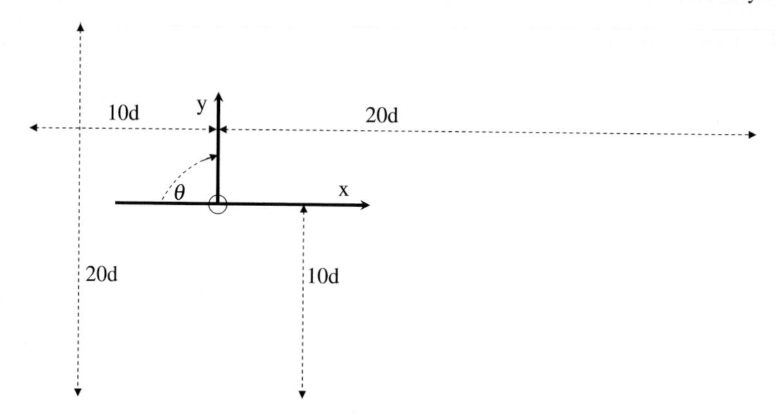

Fig. 2 Computational setup for the flow over a circular cylinder at Re = 3,900

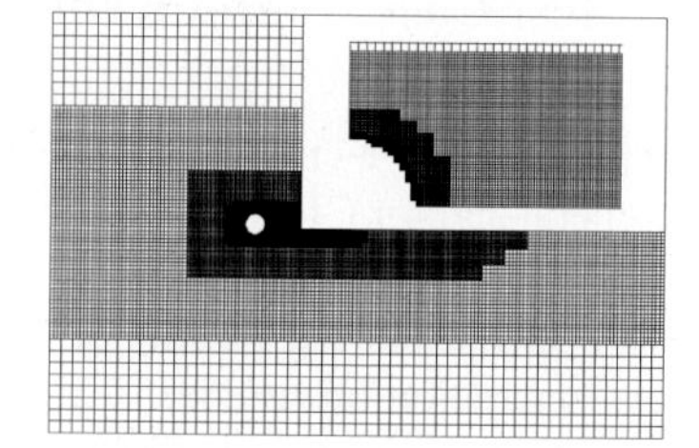

Fig. 3 Locally-refined Cartesian grid for the flow over a circular cylinder at Re = 3,900; right corner: zoom into the cylinder region

computational cell in the wall normal direction is located in the viscous sublayer of the cylinder. Both grids have a homogeneous spanwise resolution of $0.05d$.

For the discretization in space, ALDM is used for the convective terms, and a second-order accurate centered scheme for the viscous terms and the pressure gradient. Time advancement is performed using a third-order explicit three-step Runge-Kutta scheme [27]. The pressure Poisson equation is solved every Runge-Kutta substep. Computations are carried out with a CFL number of 1.0 based on the full cells on the Cartesian grid and with $CFL = 0.7$ on the curvilinear grid.

3.2 Simulation Results

Fig. 4 gives an impression of the instantaneous turbulent wake behind the cylinder. Note the thin boundary layer along the cylinder contour, which separates in form of a laminar shear layer on the upper and lower side of the cylinder before becoming unstable in the wake of the cylinder. These instabilities show an infolding of the shear layers (see Fig. 4) and lead to the production of streamwise-vorticity and long-elongated structures in the wake, shown in Fig. 5. Vortex shedding of larger structures is also visible. A good qualitative agreement is observed with Ref. [5].

Turbulence statistics are sampled over 30 shedding cycles after steady vortex shedding was established. Fig. 6 shows the velocity profiles at four different loca-

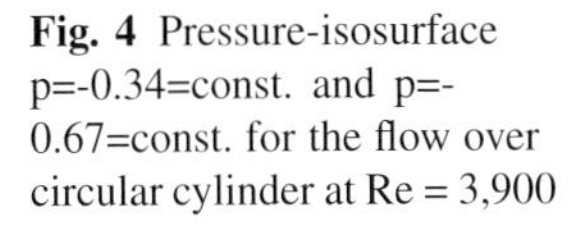

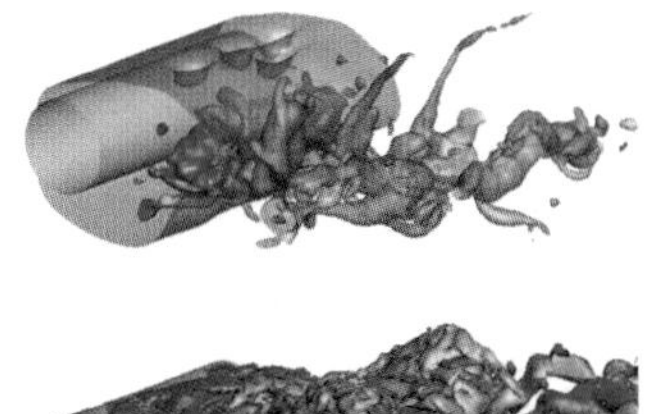

Fig. 4 Pressure-isosurface p=-0.34=const. and p=-0.67=const. for the flow over circular cylinder at Re = 3,900

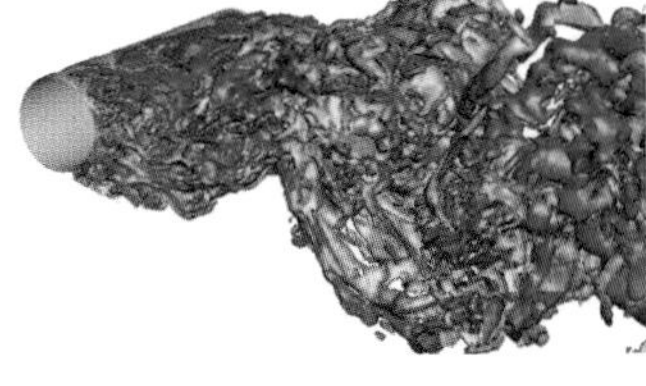

Fig. 5 Streamwise vorticity-isosurface $\pm 1.5 s^{-1}$ for the flow over circular cylinder at Re = 3,900

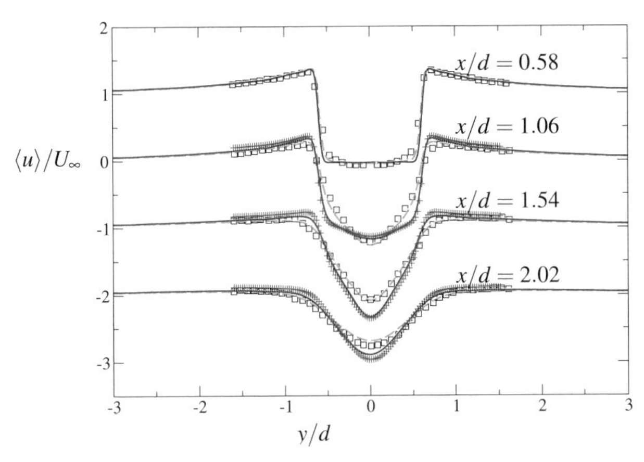

Fig. 6 Mean streamwise velocity at different locations in the wake of a circular cylinder at Re = 3,900 (□: experimental data Lourenco and Shih (extracted from Dröge 2007), +: experimental data Parnaudeau et al. [26], dashed line: LESOCC2 and solid line: INCA)

tions (x/d=0.58, 1.06, 1.54, 2.02) for the mean streamwise velocity $\langle u \rangle / U_\infty$. A strong velocity deficit occurs in the region of the recirculation bubble. The mean velocity profile shows a U-shape close to the cylinder which evolves toward a V-shape further downstream. Please note that at $x/d = 1.06$ the experiment of Lourenco and Shih [17] and the calculation on the curvilinear grid show a V-shape, whereas the recent experiment of Parnaudeau et al. [26] and the calculation on the Cartesian grid show U-shape. However, both calculations are in the range of experimental data.

In Fig. 7 the Reynolds normal stresses $\langle u'u' \rangle / U_\infty^2$ are shown. The two sharp peaks at $x/d = 1.06$ can be attributed to the transitional state of the shear layers, which show a flapping behavior due to primary vortex formation. Table 1 shows the mean drag coefficient, the pressure coefficient at the back of the cylinder (base suction coefficient), the separation angle of the shear layers and the Strouhal number of the vortex shedding. All quantities are well in the range of experimental data and results of previous DNS and LES computations. Further results and an assessment of the capability of CIIM to represent flow physics in the vicinity of the immersed

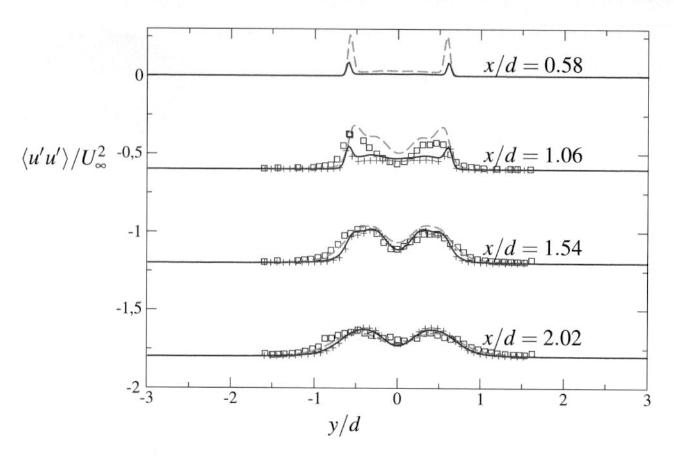

Fig. 7 Streamwise velocity fluctuations at different locations in the wake of a circular cylinder at Re = 3,900 (□: experimental data Lourenco and Shih (extracted from Dröge 2007), +: experimental data Parnaudeau et al. [26], dashed line: LESOCC2 and solid line: INCA)

Table 1 Drag coefficient, base suction coefficient, separation angle and Strouhal number for the flow over a circular cylinder at $Re = 3,900$

Case	C_D	C_p, b	θ_{sep}	St
Lourenco and Shih [17], experiment	0.99	-	86°	0.215
Ong and Wallace [24], experiment	-	-	-	0.21
Parnaudeau et al. [26], experiment	-	-	-	0.208
Dröge [3], DNS	1.01	-0.88	87.7°	0.210
Breuer [2], LES	1.016	-0.941	87.4°	-
Franke and Frank[4], LES	0.99	-0.85	88.2°	0.209
Fröhlich et al. [5], LES	1.08	-1.03	88.1°	0.216
Kravchenko and Moin [16], LES	1.04	-0.94	88°	0.210
Present curvilinear grid, LES	1.09	-1.05	88.9°	0.215
Present Cartesian grid, LES	1.05	-0.92	88°	0.210

interface is given in Ref. [21]. From the good agreement of the two calculations with the experimental results and with the results of other calculations we conclude that CIIM performs well.

4 Computational Aspects

4.1 Parallelization

In both numerical solvers parallelization is realized by means of the Single-Program-Multiple-Data (SMPD) model. Herein, identical copies of the code are running on a previously defined number of processors. For the communication during the parallel runs the native version of MPI (Message-Passing Interface) available for the

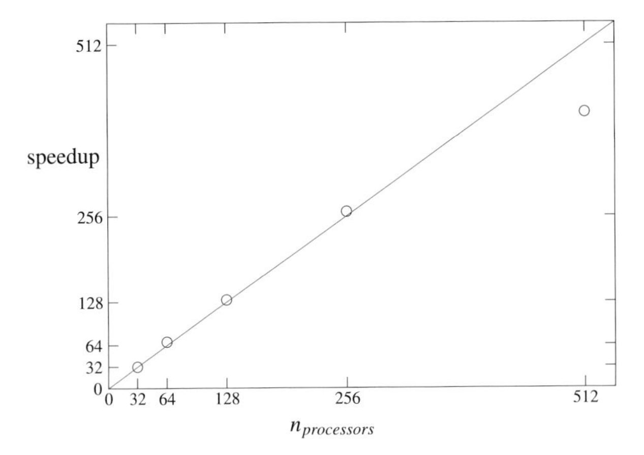

Fig. 8 Parallelisation of numerical solver INCA (speedup is defined with respect to simulated time)

SGI platform on the HLRB2 was used. A shared memory functionality based on Open Multi-Processing (OpenMP) is also implemented in INCA but not used for the simulations presented here.

In the case of the LESOCC2 simulation, block decomposition of the computational domain into 64 partitions has been realized in the ICEM grid generator tool of ANSYS. In the case of INCA block decomposition into 2052 partitions is realized in the automatic grid generation routine, which is directly incorporated in the solver.

Parallelization in the case of the LESOCC2 simulation shows a perfect load balance because the blocks are all of the same size. In the INCA simulations, block sizes differ and with the present grid the code nearly attains linear scaling up to 256 used cores, see Fig 8.

4.2 Computational Effort

The simulations have been carried out at the HLRB2 on the SGI Altix 4700 platform at the Leibniz Supercomputing Centre.

LESOCC2 uses a physical time step of $1.2 \times 10^{-3}s$, so that one vortex shedding period takes about 4000 timesteps. To simulate 30 periods for the statistical analysis, an overall amount of 120000 timesteps is required. This consumes approximately 50000 CPU hours, or, parallelized on 64 cores, nearly 33 days of computation. The efficiency can be expressed through the wall-clock time per timestep and per mesh-node or through the CPU time per timestep and per mesh-node. The current simulation consumed $3.73 \times 10^{-6}s$ wall-clock time per timestep and per mesh-node and $2.4 \times 10^{-4}s$ CPU time per timestep and per mesh-node.

INCA uses a physical time step 9.5×10^{-4}, so that one vortex shedding period takes about 5000 individual timesteps and for thc 30 periods an overall amount of

150000 timesteps is required. On the HLRB2 this consumes 120000 CPU hours running on 512 cores and 74000 CPU hours running on 256 cores, so that despite of using twice as much processors the computational time benefit is just 2 days (512 cores need 10days, 256 cores need 12 days). This phenomenon could be explained by the cores being located on different partitions in the case of the 512-core computation leading to an increased effort for inter-partition communication. The current INCA-simulation on 512 cores consumed $8.0 \times 10^{-7}s$ wall-clock time per timestep and per mesh-node and $4.1 \times 10^{-4}s$ CPU time per timestep and per mesh-node. For the same simulation on 256 cores we obtained $1.0 \times 10^{-6}s$ wall-clock time per timestep and per mesh-node and $2.5 \times 10^{-4}s$ for the CPU time per timestep and per mesh-node.

4.3 Computational Performance of CIIM

In order evaluate the computational performance of CIIM, which is used to impose the wall boundary condition on Cartesian grids in INCA, the CPU time used for CIIM has been measured and compared to the time used for other processes. Fig. 9 shows with respect to INCA and LESSOC2 the CPU time for solving the Poisson equation, the calculation of the fluxes and for CIIM (used in INCA) and, respectively, for the wall boundary condition (used in LESOCC2). The insert of the figure shows the relative CPU time needed for the cut-cell volume balance modification (see 2.3.1), the friction term (see 2.3.2), the imposition of the homogeneous Neumann condition for the pressure (see 2.3.4) and the mixing procedure (see 2.3.5). The CPU time needed for the momentum exchange term is included in the CPU time for homogeneous Neumann condition for the pressure. The CPU time of the insert is expressed with respect to the total CPU time needed for CIIM. A reference calculation was performed without any performance output for CIIM to ensure that the performance output does not lead to any significant increase in CPU time. The CPU time used for CIIM is about 3%, and thus significantly lower than the time needed for the Poisson-solver (80% of total CPU time) and comparable to the time needed for the ALDM flux calculation (2% of total CPU time).

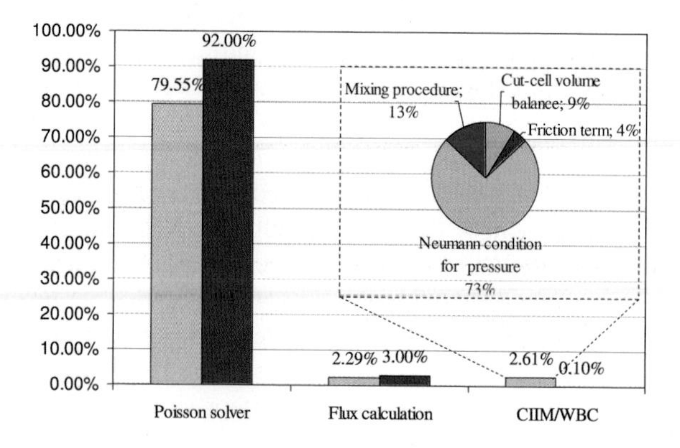

Fig. 9 CPU time needed for the the Poisson-solver, the flux calculation and CIIM/Wall Boundary Condition(WBC) (red: LESSOC2 and light blue: INCA). The insert shows the split-up for the CIIM subroutines used in INCA

5 Conclusions

Within this project, numerical simulations with two different LES solvers based on the same ALDM turbulence model have been performed: LESOCC2 uses curvilinear grids and an ordinary wall-boundary condition and INCA uses a Conservative Immersed Interface Method (CIIM) to impose the wall boundary condition on Cartesian grids. Results concerning flow physics of both simulations compare well with experiments and previous simulations. Results concerning the computational performance show that CIIM needs an insignificant amount of the total CPU time.

References

1. N. A. Adams, S. Hickel, S. Franz, Implicit subgrid-scale modeling by adaptive deconvolution. *J. Comput. Phys.* 200 412–431, 2004.
2. M. Breuer, Large eddy simulation of the sub-critical flow past a circular cylinder: numerical and modeling aspects, *Int. J. Numer. Meth. Fluids* 28 1281–1302, 1998.
3. M. Dröge, Cartesian Grid Methods for Turbulent Flow Simulation in Complex Geometries, *Ph.D thesis*, University of Groningen, 2007.
4. J. Franke, W. Frank, Large edy simulation of the flow past a circular cylinder at $Re_D = 3900$, *J. Wind Eng. Ind. Aerod.* 90 1191–1206, 2002.
5. J. Fröhlich, W. Rodi, P. H. Kessler, S. Parpais, J. P. Bertoglio, D. Laurence, Large eddy simulation of flow around circular cylinders on structured and unstructured grids, in CNRS DFG Collaborative Research Programme, vol. 66, Vieweg, Braunschweig, pp. 319–338, 1998.
6. A. Harten, B. Engquist, S. Osher, S. Chakravarthy, Uniformly high order accurate essentially non-oscillatory schemes, III. *J. Comput. Phys.* 71 231–303, 1987.
7. S. Hickel, N. A. Adams, Efficient implementation of nonlinear deconvolution methods for implicit large-eddy simulation, in High Performance Computing in Science and Engineering, *Springer*, Berlin, pp. 293–306, 2006.
8. S. Hickel, N. A. Adams, J.A. Domaradzki, An adaptive local deconvolution method for implicit LES, *J. Comput. Phys.* 213 413–436, 2006.
9. S. Hickel, T. Kempe, N. A. Adams, Implicit large-eddy simulation applied to turbulent channel flow with periodic constrictions. *Theor. and Comput. Fluid Dyn* 22 227–242, 2007.
10. S. Hickel, N. A. Adams, N. N.Mansour, Implicit subgrid-scale modeling for large-eddy simulation of passive-scalar mixing, *Phys. Fluids.* 19 095102, 2007.
11. S. Hickel, N. A. Adams, On implicit subgrid-scale modeling in wall-bounded flows, *Phys. Fluids.* 19 105106, 2007.
12. S. Hickel, N. A. Adams, Implicit LES applied to zero-pressure-gradient and adverse-pressure-gradient boundary-layer turbulence. *Int. J. Heat Fluid Flow* 29 626–639, 2008.
13. S. Hickel, D. V. Terzi, J. Fröhlich, An adaptive local deconvolution method for general curvilinear coordinate systems. In Advances in Turbulence XII, *Springer*, Berlin, 2009.
14. C. Hinterberger, Dreidimensionale und tiefengemittelte Large-Eddy-Simulation von Flachwasserströmungen. *PhD thesis*, University of Karlsruhe, 2004.
15. X. Y. Hu, B. C. Khoo, N. A. Adams, F. L. Huang, A conservative interface method for compressible flows, *J. Comput. Phys.* 219 553–578, 2006.
16. A. Kravchenko, P. Moin, Numerical studies of flow over a circular cylinder at $Re_D = 3900$, *Phys. Fluids.* 12 403–417, 2000.
17. L.M. Lourenco, C. Shih, Characteristics of the plane turbulent near wake of a circular cylinder, A particle image velocimetry study, extracted from [3].

18. X. Ma, G.-S. Karamanos, G. E. Karniadakis, Dynamics and low-dimensionality of a turbulent near wake, *J. Fluid Mech.* 410 29–65, 2000.
19. K. Mahesh, G. Constantinescu, P. Moin, A numerical method for large-eddy simulation in complex geometries, *J. Comp. Phys.*, 197 215–240, 2004.
20. M. Meyer, A. Devesa, S. Hickel, X.Y. Hu, N. A.Adams, A Conservative Immersed Interface Method for Large-Eddy Simulation of Incompressible Flows, submitted 2009.
21. M. Meyer, S. Hickel, N. A. Adams, Assessment of Implicit Large-Eddy Simulation with a Conservative Immersed Interface Method for Turbulent Cylinder Flow, (in press) *Int. J. Heat Fluid Flow* 31, 2010.
22. R. Mittal, G. Iaccarino, Immersed boundary methods, *Annu. Rev. Fluid Mech.* 37 239–261, 2005.
23. C. Norberg, Experimental investigation of the flow around a circular cylinder: influence of aspect ratio. *J. Fluid Mech.* 258, 287–316, 1994.
24. L. Ong, J. Wallace, The velocity field of the turbulent very near wake of a circular cylinder, *Exp. Fluids*, 20 441–453, 1996.
25. N. Park, J. Y. Yoo, H. Choi, Discretization errors in large eddy simulation: on the suitability of centered and upwind-biased compact difference schemes, *J. Comp. Phys.*, 198 580–616, 2004.
26. P. Parnaudeau, J. Carlier, D. Heitz, E. Lamballais, Experimental and numerical studies of the flow over a circular cylinder at Reynolds number 3900, *Phys. Fluids.* 20 085101-085101-14, 2008.
27. C.-W. Shu, Total-variation-diminishing time discretizations, *SIAM J. Sci. Stat. Comput.* 9 1073–1084, 1988.

Numerical Investigation of the Micromechanical Behavior of DNA Immersed in a Hydrodynamic Flow

Sergey Litvinov, Marco Ellero, Xiangyu Hu, and Nikolaus Adams

Abstract The topic of the project is relevant to the development of novel single-molecule manipulation techniques in biophysics and bionanotechnology where complex DNA-liquid interactions occur. The goals of the project are to verify the proposed numerical method by comparing the results for simple flow conditions with available numerical and analytical results, to analyze the dynamics of the DNA macromolecular exposed to an uniform and shear flow, perform simulations corresponding to the more complex experimental conditions.

1 Description of the Research Project

1.1 Stretching of DNA Molecules in a Hydrodynamic Flow

Recently the task attaching one DNA molecule specifically with both ends to different microscopic contact pads on surfaces has been carried out for the first time with respect to the construction of DNA-templated nanowires and networks [4, 25, 41]. These investigations show unambiguously that DNA molecules can be specifically integrated into microfluidic channel systems and be manipulated by hydrodynamic flows.

However, contrary to the process of DNA stretching, which requires solely the attachment of one end to substrate surfaces, the micromechanic response of stretched DNA molecules, tethered with both ends to the surface, is not yet understood. Up to now, the use of doubly-tethered DNA molecules for biophysical studies has never been reported in literature.

Sergey Litvinov · Marco Ellero · Xiangyu Hu · Nikolaus Adams
Lehrstuhl für Aerodynamik, Technische Universität München, 85748 Garching, Germany
e-mail: sergej.litvinov@aer.mw.tum.de
e-mail: marco.ellero@aer.mw.tum.de
e-mail: xiangyu.hu@tum.de
e-mail: nikolaus.adams@tum.de

S. Wagner et al. (eds.), *High Performance Computing in Science and Engineering, Garching/Munich 2009*, DOI 10.1007/978-3-642-13872-0_13,

1.2 Numerical Methods for DNA in a Hydrodynamic Flow

Liquid micro-flows with DNA [23, 39] or other macromolecules exhibit a considerable physical complexity. Due to the fact that mesoscopic length scales (from nm to μm) dominate the overall behavior of such systems [10, 20, 36], macroscopic approach such as finite-volume or finite-element based discretization of the continuum-flow equations, and microscopic approaches, such as molecular dynamics (MD), are not applicable since they can not resolve the mesoscopic structures, neglect thermal fluctuation effects, or are limited to characteristic length scales near the lower bound of microfluidics. Therefore a number of research groups pursues mesoscopic approaches which can resolve DNA molecular structure, and recover hydrodynamic forces and thermal fluctuation effects as well. Currently, three main approaches for DNA in microflow simulations can be distinguished. One is the Brownian dynamics method (BD) [28], one is the lattice Boltzmann method (LBM) [24], and the dissipative particle dynamics (DPD) method [16].

Extended BD simulations have been used to characterize the dynamics of long DNA molecules flowing in micro-channels [18, 37]. However, there are considerable limitations of such method. One limitation is the decoupling of the DNA-liquid interaction due to the implementation of a background flow field and a drag force on the DNA molecule with a simple viscous force model. The other limitation is that, the formulation of dissipative and random force terms in BD lacks the Galilean invariant and such violates a basic requirement.

DPD is a computational method that allows the simulation of simple and complex fluids on mesoscopic length and time scales [15, 30]. The DPD algorithm consists of two steps. First, the position of particles is updated in a streaming step, according to the particle momentum. Subsequently, the momentum of the particles is updated according to symmetric two-body forces acting between different particles. Dissipative and random forces are modeled by the Fluctuation-Dissipation theorem [7]. The method is strictly Galileian invariant and conserves momentum. The fully-Lagrangian nature of DPD facilitates the generalization to systems with DNA molecules in a direct way that LBM does not allow for [8]. On the other hand, in comparison with LBM, DPD is only suitable for smaller scale flows since the dissipative particles are oversimplified where no macroscopic thermodynamic state and parameters of the solute can be represented directly.

2 Model

In this section we describe a new model based on smoothed dissipative particle dynamics (SDPD) for the numerical simulation of polymers in dilute solutions, which is suitable for the numerical simulation of complex flows. The method inherits the favorable properties of Smoothed Particles Hydrodynamics (SPH) [27] for complex flows and the efficient representation of mesoscopic effects of SDPD [6]. Immersed polymers are taken into account in simulations by a straight-forward modification of the SDPD-particle interactions of particles containing parts of the polymer. The method is validated by the comparison with theoretical results of generic cases.

2.1 Mesoscopic Modeling of the Solvent

The key idea behind the SPH method is the approximation of the field gradient in the following form

$$\nabla F(\mathbf{r}) \approx -\sum_j F(\mathbf{r}_j)\nabla W(\mathbf{r}-\mathbf{r}_j,h)\frac{m_j}{\rho_j} \tag{1}$$

where $F(\mathbf{r})$ is field, m_j is particle mass, ρ is density, $W(\mathbf{r},h)$ is the kernel function with compact support and h is the smoothing length. Thus, the summation is a sum over particles j in the *neighborhood* of the point of interest $\mathbf{r}$. No assumption about particle positions is made and method is hence *mesh-free*.

By applying the approximation (1) to Navier-Stokes equations, one can obtain the second-order discredited form [17, 27]:

$$\rho_i = m_i \sum_j W_{ij}, \tag{2}$$

$$\frac{d\mathbf{v}_i}{dt} = -\frac{1}{m_j}\sum_j \left(\frac{p_i}{\sigma_i^2}+\frac{p_j}{\sigma_j^2}\right)\frac{\partial W_{ij}}{\partial r_{ij}}\mathbf{e}_{ij} + \frac{\mu}{m_i}\sum_j \left(\frac{1}{\sigma_i^2}+\frac{1}{\sigma_j^2}\right)\frac{\mathbf{v}_{ij}}{r_{ij}}\frac{\partial W_{ij}}{\partial r_{ij}}, \tag{3}$$

W_{ij} is a kernel function evaluated with $\mathbf{r}=\mathbf{r}_i-\mathbf{r}_j$, μ is the dynamic viscosity, σ_i is the inverse of a particle volume, $\mathbf{e}_{ij}$ and r_{ij} are the normalized vector and distance from particle i to particle j, respectively. Density and pressure are related by the equation of state $p = p_0\left(\frac{\rho}{\rho_0}\right)^\gamma + b$, where p_0, ρ_0, b and γ are parameters which may be chosen based on a scale analysis so that the density variation is less than a given magnitude. For $\gamma = 7$ the penetration of particles is precluded.

The Eq. (2) represents the deterministic part of the particle dynamics [6, 17]. Using the GENERIC formalism (General Equation for Non-Equilibrium Reversible-Irreversible Coupling) [14, 29], thermal fluctuations can be taken into account directly in (2) by introducing the following terms:

$$d\tilde{m}_i = 0, \quad d\tilde{\mathbf{P}}_i = \sum_j B_{ij}d\overline{\overline{W}}_{ij}\mathbf{e}_{ij}, \tag{4}$$

where $d\overline{\overline{W}}_{ij}$ is the traceless symmetric part of an independent increment of a Wiener process and B_{ij} is defined as

$$B_{ij} = \left[-4k_BT\mu\left(\frac{1}{\sigma_i^2}+\frac{1}{\sigma_j^2}\right)\frac{1}{r_{ij}}\frac{\partial W}{\partial r_{ij}}\right]^{1/2}. \tag{5}$$

Some important properties of this particular set of discrete equations are:

- total mass and momentum are exactly conserved;
- linear momentum is locally conserved due to the anti-symmetric form of the particle pair force;
- the total energy is conserved and the total entropy is a monotonically increasing function of time [32].

2.2 Mechanical Modeling of the Polymer Chain

The polymer is embedded into a number of special SDPD particles (denoted as polymer beads) which represent the segments of the polymer molecule. For typical particle sizes a bead would contain mainly solvent and the volume fraction of the polymer segment is small. Therefore, polymer beads interact hydrodynamically, with additional forces due to the chemical bond between the polymer segments contained in neighboring polymer beads. These additional forces are taken into account by a finite extensible non-linear elastic (FENE) potential is

$$U_{FENE} = -\frac{1}{2}kR_0^2 \ln\left(1 - \left(\frac{r}{R_0}\right)^2\right), \tag{6}$$

where r is the spring length between neighboring beads, R_0 is the maximum spring extension and k is the spring constant. The parameter R_0 is chosen as twice the average particle distance in such a way that the crossing of chain segments is avoided. The FENE force is superimposed on the hydrodynamic interaction of each polymer bead, represents the effect of the solvent. Note that by using the quasi-incompressible equation of state, the excluded volume effects between polymer beads are automatically taken into account. We emphasize that in the proposed model the hydrodynamic interaction are fully represented due to the fact that fluid particles and polymer beads (containing solvent) interact by hydrodynamic forces which are derived from a consistent Lagrangian discretization of the Navier-Stokes equations. The physical basis for this model is discussed in [21].

In this project the described model was implemented in two software packages. First software package was based on the GPL licensed code **sph2000** [11] and another one was build as a client for **PPM** library [31].

2.3 Object-Oriented Package sph2000

The software is an in-house development based on the following publicly available software packages and libraries:

- **sph2000**
  An object-oriented library for physical simulations with the SPH-method, written in C++ using design pattern [11]
  `http://www-ti.informatik.uni-tuebingen.de/~ganzenmu/sph2000/sph2000.html`
- **GSL — GNU Scientific Library**
  The GNU Scientific Library (GSL) is a numerical library for C and C++ [13]
  `http://www.gnu.org/software/gsl`
- **Blitz++**
  A C++ class library for matrix and vector operations [40]
  `http://www.oonumerics.org/blitz`
- **hdf5**
  A library for handling extremely large and complex data collections [9]
  `http://hdf.ncsa.uiuc.edu/products/hdf5`

The **sph2000** code allows for flexible design and the following features as required for the project are implemented

- Thermal fluctuation
- Periodic, moving wall, Lees-Edwards boundary conditions
- Polymer beads
- Fully implicit integration scheme
- S1 [34] integration scheme

Disadvantages of the **sph2000** implementation are the relatively poor performance for parallel runs. We attribute this issue to the memory management model based on pointers.

2.4 A Client for Parallel Particle Mesh (PPM) Library

Another implementation of the model with focus on performance was done in *FORTRAN 90* using **PPM** library which hides the complexity of the parallel implementation from the client author [31]:

- **PPM**
  Highly Efficient Parallel Particle-Mesh Library
  `http://www.mosaic.ethz.ch/Downloads/PPM`
- **H5Part**
  API to store particle data in hdf5 format [1]. Parallel I/O can be performed with implementation details hidden from the user of API.
  `http://vis.lbl.gov/Research/AcceleratorSAPP`
- **fgsl**
  Object-based FORTRAN interface to the GNU scientific library [2]
  `http://www.lrz-muenchen.de/services/software/mathematik/gsl/fortran`

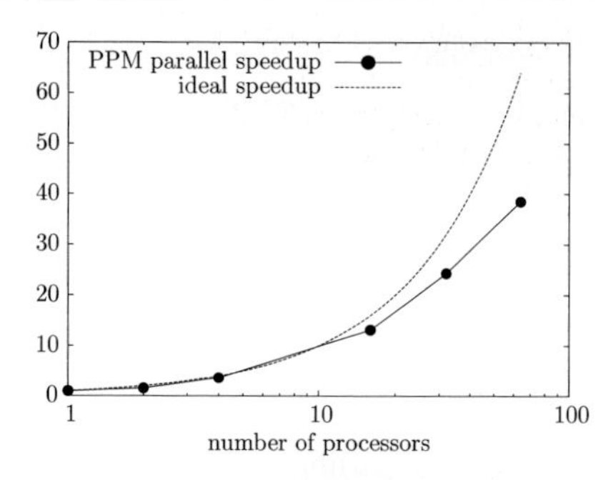

Fig. 1 The parallel speedup of the **ppm** client

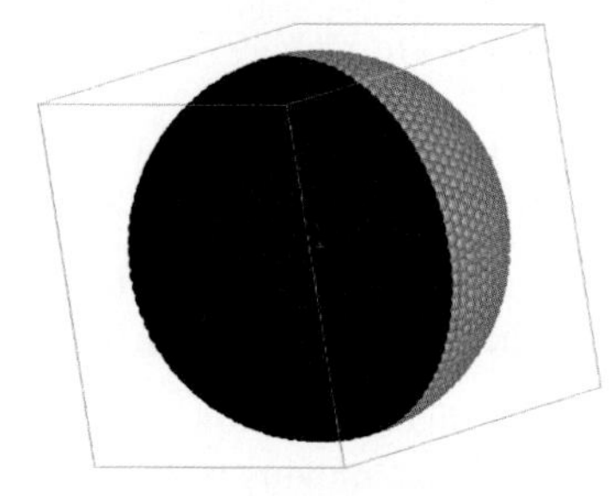

Fig. 2 Example of visualization with **punto**

The parallelization strategy for both implementations is based on domain decomposition. Communication is performed using *ghost* particle layer between neighboring sub-domains. In case of PPM-client we use only a subset of mapping options provided by the library.

The performance of single processor code achieved on HLRB is in the range of $10^4 - 3 \times 10^4$ particles processed per second. The parallel speedup of the client is shown in the Figure 1. Note that the degradation of performance for higher number of processor is partly caused by the increase of overall number of ghost particles.

Currently not all features of the **sph2000**-based implementation are present in the **PPM** client.

3 Visualization

There is a variety of commercial and open source tools for visualizing the particle data. For the current project we use **pv-meshless**. It is a customized version of ParaView [35] with some specific enhancements for visualization of the particle based data [3] (see Figure 3). The aim of the developers is to have it as a plug-in to mainstream version of ParaView.

Another tool used is **punto** (`http://www.fisfun.uned.es/~mrevenga/punto.html`): a command line driven program to do simple "online" visualization of particles (see Figure 2).

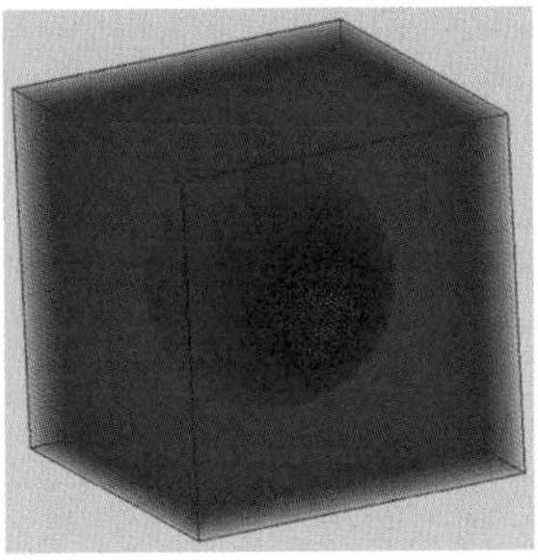

Fig. 3 Example of visualization with **pv-meshless**, number of particles is 10^5, solvent particles are half transparent

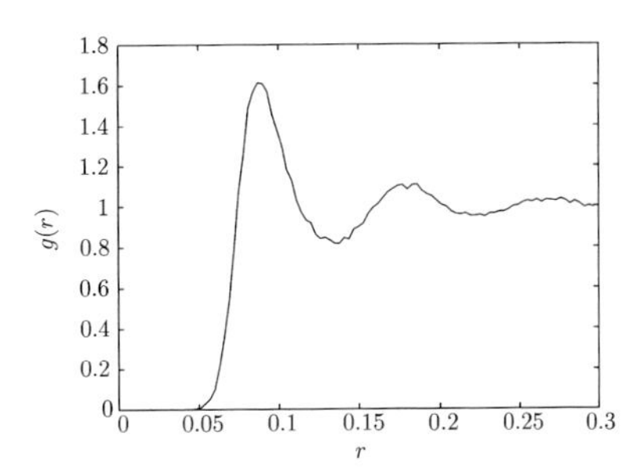

Fig. 4 Radial distribution function for SDPD particles

4 Summary of Results

In this section a brief summary of scientific findings is presented. For a more detailed description see [21].

4.1 Transport Properties of the Solvent

As it was shown in section 2, for the case of negligible thermal fluctuations the SDPD system of equations is equivalent to SPH. This is an important property of the method because it ensures the correct hydrodynamic behavior of the fluid for large scales. Hence we can focus our attention on the property of the fluid related to thermal fluctuations:

- Radial distribution function for SDPD particles found in simulations is shown in Figure 4. And it agrees with results from statistical mechanics.
- Important characteristic of thermal motion of SDPD particles is diffusion coefficient (D) which can be extracted from the average mean square displacement. The ratio of kinematic viscosity to diffusion coefficient is called *Schmidt number* defined by the

$$Sc = \frac{\mu}{\rho D} \tag{7}$$

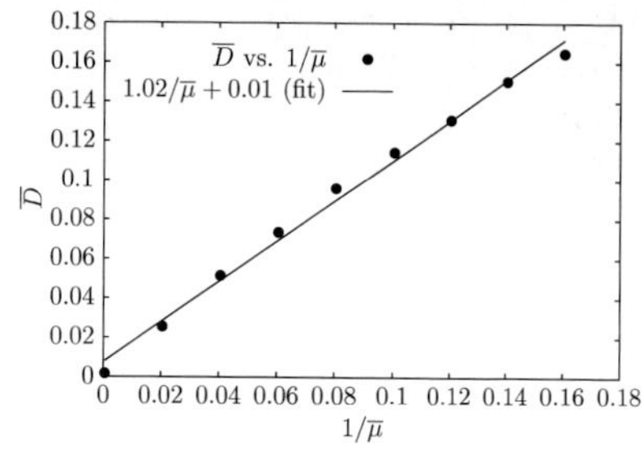

Fig. 5 Diffusion coefficient D vs. $1/\mu$, in SDPD viscosity is an input parameter

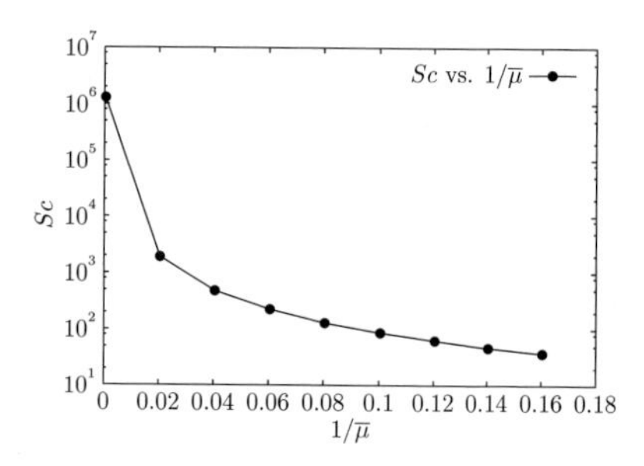

Fig. 6 Schmidt number Sc vs. $1/\mu$

It is important for various mesoscopic methods to realize Schmidt number typical for liquids ($\propto$ 100) [38]. We found that SDPD allows for simulating case of a wide range of Schmidt number (see Figure 5 and Figure 6) [22]).

4.2 Static Properties of Polymer

First, we validate the polymer model by recovering the static properties of the free polymer.

A typical simulation procedure is illustrated in Figures 7 and 8. The polymer chain is immersed in a domain full of SDPD particles. Boundary conditions are periodic. The "quantities of interest" are fluctuating in time and averaged values are extracted and compared with the predictions based on analytical models of polymer physics.

The following quantities are found to be in good agreement with the theoretical model where hydrodynamic interactions are taken into account (Zimm model). It is believed that Zimm model is a best available analytical model to describe experimental data for DNA in dilute solution [19].

- *Scaling* of average polymer size with number of beads constituting polymer (N). Widely used characterization of polymer size is the *radius of gyration*:

$$R_g^2 = \frac{1}{N} \sum_{k=1}^{N} (\mathbf{r}_k - \mathbf{r}_{mean})^2, \tag{8}$$

where $\mathbf{r}_k$ is the position of a bead, $\mathbf{r}_{mean}$ is the mean position of the beads.

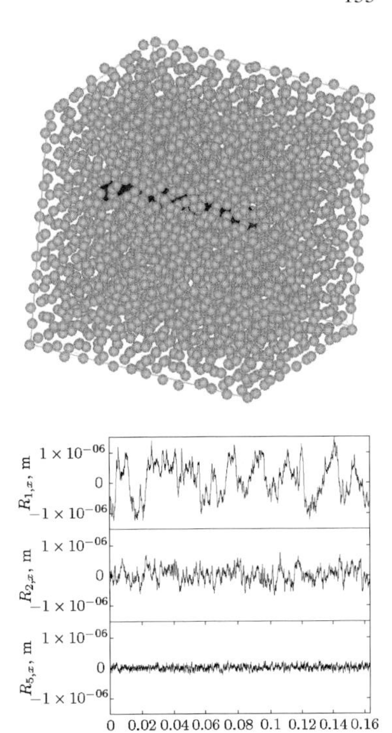

Fig. 7 Typical simulation configuration: polymer in the 'ocean' of SDPD particles

Fig. 8 Example of "quantities of interest" variation in time: normal modes of polymer chain, $p = 1,2,5$ (from top to bottom) corresponding to a polymer chain with $N = 90$

Gyration R_g for several chain lengths corresponding to $N = 5,10,20,30,40,60,70,$ $80,90,100$ beads is shown in the Figure 9, dotted line represents the best fit consistent with the theory ($R_g \propto N^{\nu}$) and redials a *static exponent* $\nu = 0.60 \pm 0.01$

- Another way to extract ν is via the *static structure factor* defined as:

$$S(\mathbf{k}) \equiv \frac{1}{N} \sum_{i,j} \langle \exp\left(-i\mathbf{k} \cdot (\mathbf{r}_i - \mathbf{r}_j)\right) \rangle \tag{9}$$

In the range of small wave vector $|\mathbf{k}|R_g \ll 1$, the structure factor is approximated by $S(\mathbf{k}) \approx N(1 - \mathbf{k}^2 R_g/3)$. For $|\mathbf{k}|R_g \gg 1$ the *structure factor* is approximated by $S(\mathbf{k}) \approx 2N/\mathbf{k}^2 R_g$. The intermediate regime $|\mathbf{k}|R_g \sim 1$ contains the information about the intramolecular spatial correlations. In the absence of external perturbations and close to equilibrium, $S(\mathbf{k})$ is isotropic and thus depends only on the magnitude of the wave vector $k = |\mathbf{k}|$. Therefore, $S(k)$ provides therefore different length scales even for a single polymer and in the intermediate regime is shown to behave as

$$S(k) \propto k^{-1/\nu} \tag{10}$$

Fig. 10 shows a log-log plot of linear part of $S(k)$ vs. R_g. From this figure it is shown that the curves evaluated from simulations with different chain lengths (N) collapse on a single master curve. The slope of the linear region of the master curve is $-1/\nu$.

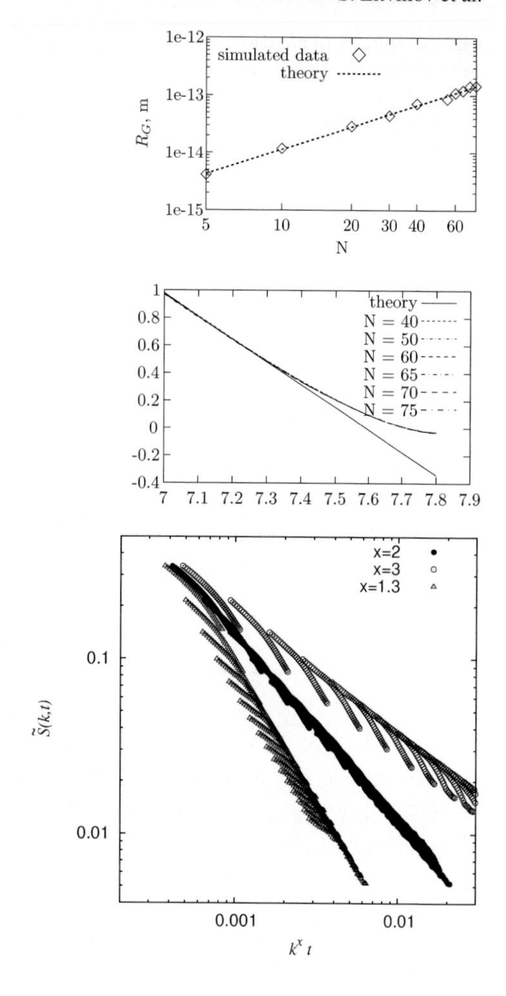

Fig. 9 Gyration radius characterizes the size of the polymer and depends on the number of beads, $\nu = 0.60 \pm 0.01$ (3D simulation)

Fig. 10 Static structure factor (3D simulation) shows prefect collapse for different polymer length, $\nu = 0.60$

Fig. 11 Scaling plot of the normalized dynamic structure factor, collapse at $x = 2$ (2D simulations)

4.3 Dynamic Properties of the Polymer

Dynamic polymer properties provide a stronger test of the model quality.

- *Dynamics structure factor* defined as

$$S(\mathbf{k},t) \equiv \frac{1}{N} \sum_{i,j} \langle \exp\left(-i\mathbf{k} \cdot [\mathbf{r}_i(t) - \mathbf{r}_j(0)]\right) \rangle \tag{11}$$

Fig. 11 shows the dynamical structure factor data plotted versus $k^x t$ for three different values of the dynamic scaling exponent, that is: $x = 1.3, 2, 3$. The figure suggests that in the appropriate scaling regime (that is $2 < R_G k < 8$) and best collapse is obtained for $x = 2$. This confirms the previous simulation results for 2D polymer [33].

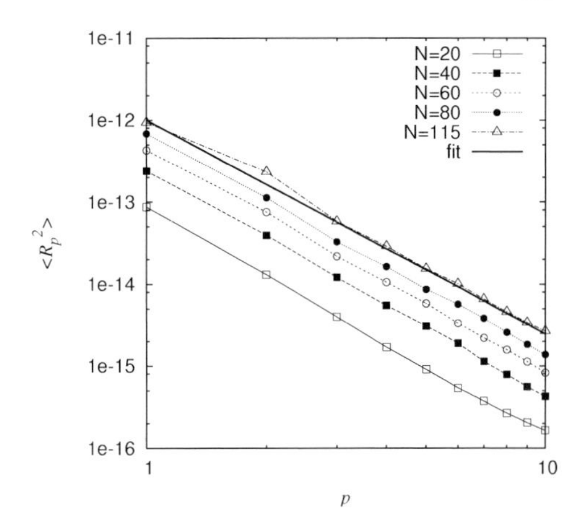

Fig. 12 Rouse mode mean-square amplitudes $< R_p^2 >$ vs mode p, for $N = 115$ the fit gives a slope equal to -2.61 ± 0.05

- In order to characterize dynamics of polymer on different length scales the set of Rouse coordinates are often employed.

$$\mathbf{R}_p = \frac{1}{N} \sum_{i=1}^{N} \cos\left(\frac{p\pi\left(i - \frac{1}{2}\right)}{N}\right) \mathbf{r}_i \tag{12}$$

For example, $\mathbf{R}_0$ is the chain's center of mass, while $\mathbf{R}_p$ for $p > 1$ represent high order modes describing the internal motion of the polymer molecule. In Figure 12 Rouse mode mean-square amplitudes $\langle R_p^2 \rangle$ vs mode p for different polymer chains are shown. The solid line represents the best fit of the data corresponding to $N = 115$ and gives a slope equal to -2.61 ± 0.05 is in good agreement with the theory [19].

4.4 Confined Polymer and Polymer in Complex Flows

- In order to investigate the interaction of SDPD polymer with the wall simulation of microchannel with varying gap H were performed. Normalized polymer stretch $X^* = X/X^\infty$ as a function of the inverse of the channel gap $H^* = H/R_G^\infty$ for several chain lengths ($N = 20, 60, 100$) is presented in Figure 13 (X^∞ and R_G^∞ are correspondingly extension and gyration radius in an unconfined environment). Deviations from the bulk behavior appear at $1/H^* \approx 0.2$. For $1/H^* > 0.3$ a universal scaling is realized [12].
- A study of polymer confined in the channel and exposed to a Poiseuille flow was also performed. In Figure 14 the span-wise distribution of polymer mass is shown. We found that the profile is affected by Schmidt number: for lower Schmidt number the depletion region is pronounced at the center of the channel; for high Schmidt number the polymer concentration tends to be higher in the center with smaller off-center peaks. Our simulations results are generally confirmed by the recent study of Millan et al. [26] and more detailed mesoscale simulation of Cannavacciuolo et al. [5].

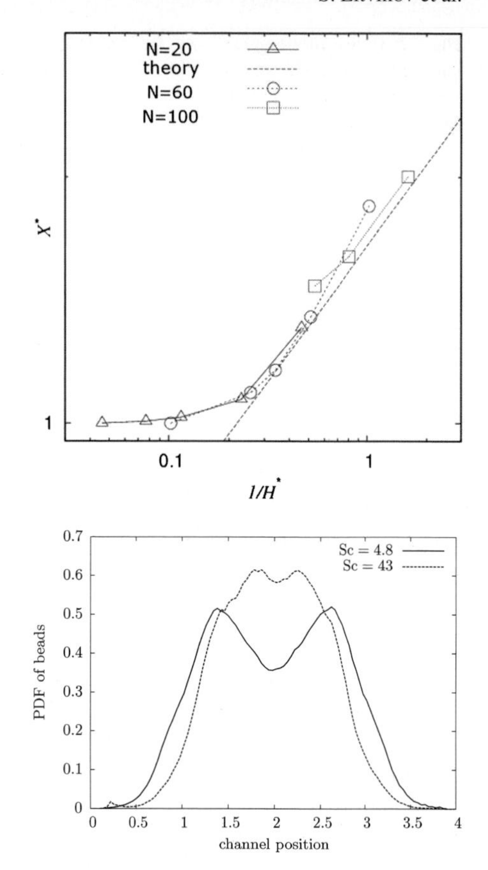

Fig. 13 Polymer in confinement 2D: the data correspond to different chain lengths ($N = 20, 60$) and values of the normalized channel gap H^* ranging from 1 to 22. All the quantities have been made dimensionless by the gyration radius R_G^∞ of the corresponding polymer molecule without confinement

Fig. 14 Polymer bead distribution in Poiseuille flow

5 Conclusions

We have developed and implemented a model for meshfree simulations of polymers. The polymer models are validated and found to be consistent with DNA behavior observed in experiments. Some cases of the interaction of the polymer with complex flows are considered.

References

1. Adelmann, A., Gsell, A., Oswald, B., Schietinger, T., Bethel, W., Shalf, J., Siegerist, C., Stockinger, K.: Progress on H5Part: a portable high performance parallel data interface for electromagnetics simulations. In: Particle Accelerator Conference PAC07, pp. 25–29 (2007)
2. Bader, R.: A Fortran binding for the GNU scientific library. In: ACM SIGPLAN Fortran Forum, vol. 26, p. 11. ACM (2007)
3. Biddiscombe, J., Graham, D., Maruzewski, P., Issa, R.: Visualization and analysis of SPH data. ERCOFTAC Bulletin, SPH special edition **76**, 9–12 (2008)

4. Braun, E., Eichen, Y., U.Sivan, Ben-Yoseph, G.: DNA-templated assembly and electrode attachment of a conducting silver wire. Nature **391**, 775–778 (1998)
5. Cannavacciuolo, L., Winkler, R.G., Gompper, G.: Mesoscale simulations of polymer dynamics in microchannel flows. Europhys. Lett. **83**(3), 34007 (2008)
6. Español, P.: Dissipative particle dynamics. In: V.M. Harik, M.D. Salas (eds.) Trends in Nanoscale Mechanics: Analysis of Nanostructured Materials and Multi-Scale Modeling. Kluwer (2003)
7. Español, P., Warren, P.: Statistical mechanics of dissipative particle dynamics. Europhys. Lett. **30**, 191–196 (1995)
8. Fan, X., Phan-Thien, N., Yong, N., Wu, X., Xu, D.: Microchannel flow of a macromolecular suspension. Phys. Fluids **15**, 11–21 (2003)
9. Folk, M., Cheng, A., Yates, K.: HDF5: A file format and I/O library for high performance computing applications. In: Proceedings of Supercomputing'99 (CD-ROM) (1999)
10. GadelHak, M.: The fluid mechanics of microdevices - the Freeman scholar lecture. ASME J. Fluids Eng. **121**, 5–33 (1999)
11. Ganzenmuller, S., Pinkenburg, S., Rosenstiel, W.: SPH2000: A parallel object-oriented framework for particle simulations with SPH. Lecture notes in computer science **3648**, 1275 (2005)
12. de Gennes, P.: Scaling concepts in polymer physics. Cornell Univ Pr (1979)
13. Gough, B.: GNU Scientific Library Reference Manual. Network Theory Ltd. (2009)
14. Grmela, M., Öttinger, H.: Dynamics and thermodynamics of complex fluids. I. Development of a general formalism. Phys. Rev. E **56**(6), 6620–6632 (1997)
15. Groot, R., Warren, P.: Dissipative particle dynamics: bridging the gap between atomistic and mesoscopic simulation. J. Chem. Phys. **107**, 4423–4435 (1997)
16. Hoogerbrugge, P., Koelman, J.: Simulating microscopic hydrodynamic phenomena with dissipative particle dynamics. Europhys. Lett. **19**, 155–160 (1992)
17. Hu, X., Adams, N.: A multi-phase SPH method for macroscopic and mesoscopic flows. J. Comput. Phys. **213**(2), 844–861 (2006). DOI 10.1016/j.jcp.2005.09.001
18. Jendrejack, R., Dimalanta, E., Schwartz, D., Graham, M., de Pablo, J.: Dna dynamics in a microchannel. Phys. Rev. Lett. **91**, 038102 (2003)
19. Jiang, W.H., Huang, J.H., Wang, Y.M., Laradji, M.: Hydrodynamic interaction in polymer solutions simulated with dissipative particle dynamics. J. Chem. Phys. **126**(4), 44901 (2007)
20. Karniadakis, G.E., Beskok, A.: Micro flows, fundamentals and simulations. Springer-Verlag, New York (2002)
21. Litvinov, S., Ellero, M., Hu, X., Adams, N.: Smoothed dissipative particle dynamics model for polymer molecules in suspension. Physical Review E **77**(6), 66703 (2008)
22. Litvinov, S., Ellero, M., Hu, X., Adams, N.: Self-diffusion coefficient in smoothed dissipative particle dynamics. J. Chem. Phys. **130**, 021101 (2009)
23. Maubach, G., Csaki, A., Seidel, R., Mertig, M., Pompe, W., Born, D., Fritzsche, W.: Controlled positioning of a DNA molecule in an electrode setup based on self-assembly and microstructuring. Nanotechnology **14**, 1055–1055 (2003)
24. McNamara, G., Zanetti, G.: Boltzmann equation to simulate lattice-gas automata. Phys. Rev. Lett. **61**, 2332–2335 (1988)
25. Mertig, M., Pompe, W.: Biomimetic fabrication of dna-based metallic nanowires and networks. In: C. Niemeyer, C. Mirkin (eds.) Nanobiotechnology-Concepts, Applications and Perspectives. WILEY-VCH, Weinheim (2004)
26. Millan, J.A., Laradji, M.: Cross-stream migration of driven polymer solutions in nanoscale channels: A numerical study with generalized dissipative particle dynamics. Macromolecules **42**(3), 803–810 (2009). URL `http://dx.doi.org/10.1021/ma8014382`
27. Monaghan, J.J.: Smoothed particle hydrodynamics. Rep. Prog. Phys. **68**(8), 1703–1759 (2005)
28. Öttinger, H.: Stochastic Processes in Polymeric Fluids. Springer (1996)
29. Öttinger, H., Grmela, M.: Dynamics and thermodynamics of complex fluids. II. Illustrations of a general formalism. Phys. Rev. E **56**(6), 6633–6655 (1997)

30. Ripoll, M., Ernst, M., Español, P.: Large scale and mesoscopic hydrodynamics for dissipative particle dynamics. J. Chem. Phys. **115**, 7271–7284 (2001)
31. Sbalzarini, I., Walther, J., Bergdorf, M., Hieber, S., Kotsalis, E., Koumoutsakos, P.: PPM–A highly efficient parallel particle–mesh library for the simulation of continuum systems. J. Comput. Phys. **215**(2), 566–588 (2006)
32. Serrano, M., Español, P.: Thermodynamically consistent mesoscopic fluid particle model. Phys. Rev. E **64**(4), 46115 (2001)
33. Shannon, S.R., Choy, T.C.: Dynamical scaling anomaly for a two dimensional polymer chain in solution. Phys. Rev. Lett. **79**(8), 1455–1458 (1997). DOI 10.1103/PhysRevLett.79.1455
34. Shardlow, T.: Splitting for dissipative particle dynamics. SIAM J. Sci. Comput. **24**(4), 1267–1282 (2003)
35. Squillacote, A., Lendel, M., Davis, J., Rashid, M., Rashid, H., Madenci, E., Guven, I., Scarpino, M., Holder, S., Ng, S., et al.: The ParaView guide: A Parallel visualization application. O'Reilly (2005)
36. Stone, H., Kim, S.: Microfluidics: basic issues, applications, and challenges. AIChE **47**, 1250–1254 (2001)
37. Streek, M.: Brownian Dynamics Simulation of Migration of DNA in structured Microchannels. PhD Thesis, Bielefeld University, Germany (2005)
38. Symeonidis, V., Karniadakis, G.E., Caswell, B.: Schmidt number effects in dissipative particle dynamics simulation of polymers. J. Chem. Phys. **125**(18), 184902 (2006)
39. Turner, S., M.Cabodi, Craighead, H.: Confinement-induced entropic recoil of single DNA molecules in nanofluidic structure. Phys. Rev. Lett. **88**, 128103:1–4 (2002)
40. Veldhuizen, T.: Blitz++: The library that thinks it is a compiler. Advances in Software tools for scientific computing **10**, 57–87 (2000)
41. Zhang, Y., Austin, R.H., Kraeft, J., Cox, E., Ong, N.: Insulating behavior of lamda-dna on the micron scale. Phys. Rev. Lett. **89**, 198102 (2002)

Comparing Frequency-Based Flow Solutions to Traditional Unsteady Fluid Dynamics Analysis in Turbomachinery

Michael Hembera, Florian Danner, Marc Kainz, and Hans-Peter Kau

Abstract Unsteady simulations are applied to resolve the time-dependent flow in turbomachinery. Consequently, the interaction between stationary and rotating parts requires an appropriate interface to transfer the flow quantities of the domains. Traditionally a so called domain scaling is used in combination with a sliding mesh approach. This method requires the pitch of the simulated blades to be equal in order to allow the use of periodic boundary conditions while cutting down the number of represented blade passages. Thus computational time can be saved. On the other hand, the flow solution is altered due to the scaling of the geometry. In order to overcome this problem, the non-linear harmonic (NLH) approach has been introduced. It allows for non identical blade pitches of the rows, since the flow solution is based on a Fourier-decomposition. Hence, the oscillating influence due to the perturbations of adjacent blade rows is approximated. Within the present study an unsteady Navier-Stokes-simulation is compared to a NLH computation in order to explain differences of the approaches. Therefore, an unsteady full-annulus simulation with about 40 million gridpoints as well a NLHs simulation with only one simulated blade passage per row with 1.38 million gridpoints was performed. Additionally, a NLHs simulation with a finer mesh resolution of 7 million gridpoints is included.

1 Introduction

The non-linear harmonic (NLH) approach (He et al. [1] and Chen et al. [2]) is a calculation method that can be used for unsteady turbomachinery simulations at low computation times. It is a hybrid method which is meant to combine the advantages

Michael Hembera · Florian Danner · Marc Kainz · Hans-Peter Kau
Technische Universität München, Institute for Flight Propulsion, 85748 Garching, Germany
e-mail: hembera@lfa.mw.tum.de
e-mail: danner@lfa.mw.tum.de
e-mail: kainz@lfa.mw.tum.de

S. Wagner et al. (eds.), *High Performance Computing in Science and Engineering, Garching/Munich 2009*, DOI 10.1007/978-3-642-13872-0_14,

of typical unsteady and steady calculations. For the investigations presented within this report, an implementation of the NLH-Method is applied, which was realised in the commercial solver EURANUS from Numeca. Details of the implementation can be found in Vilmin et al. [3].

The method is based on the assumption that the dominant disturbances are caused by the frequency that is determined by the rotational speed as well as the blade counts of subsequent rows (the blade-passing frequency – BPF). The periodically appearing perturbations are transformed into the frequency domain by a Fourier-transformation. Thus the flow quantities are divided into a time-mean value $(\bar{U}(\mathbf{r}))$ and time-dependent sinusoidal oscillating components $(U'(\mathbf{r},t))$:

$$U(\mathbf{r},t) = \bar{U}(\mathbf{r}) + \sum U'(\mathbf{r},t) \tag{1}$$

The time-dependent component can be decomposed into N harmonics with the position vector $\mathbf{r} = (x,y,z)$, flow variable $U = (\rho, \rho\mathbf{v}, \rho E)$ and relative velocity $\mathbf{v} = (v_x, v_y, v_z)$:

$$U'(\mathbf{r},t) = \sum_{k=1}^{N} \left(\tilde{U}_k(\mathbf{r}e^{I\omega_k t}) + \tilde{U}_{-k}(\mathbf{r}e^{I\omega_{-k} t}) \right) \tag{2}$$

The harmonic amplitudes $\tilde{U}_k$ and $\tilde{U}_{-k}$ are complex conjugates.

Every perturbation is represented by at least one sinusoidal basic oscillation and its associated harmonic frequencies, which are integer multiples of the basic blade passing frequency. In order to solve the flow field, one steady solution has to be computed and additionally, for every considered member of the Fourier series (representing the harmonics), two parts of conservation equations for the real and the imaginary part have to be solved. This leads to 10 additional equations for one harmonic of a specific perturbation which have to be solved supplementary to the steady solution. The accuracy of the solution is determined by the number of harmonics and, especially for multi-stage compressors, the number of perturbations or characteristic blade passing frequencies. For a single stage compressor, hence only one BPF exists. In this specific case the accuracy is only determined by the number of considered harmonics. In multi-stage compressors, one embedded row is influenced by the wake of the upstream row and the potential field of the downstream row. Assuming that the blade counts of the rows are non-identical, two BPFs exist. The method becomes non-linear through the implementation of harmonics into the time-averaged solution by deterministic stress terms.

In order to gain the final quasi-unsteady solution, the frequency based data has to be reconstructed in the time domain. Subsequently, they are superimposed with the steady solution. Accordingly, different time steps (meaning different relative positions of the rotating and steady parts) can be derived from one set of the solution which contains the steady solution, the amplitudes and the frequencies of the harmonics. Within the postprocessing procedure the solution of the different time steps is reconstructed in time with the according time-dependent values (equivalent to a phase-shift) of the oscillating terms.

The rotor-stator interface is implemented such that the unsteady solutions on both sides of the interface continuously match. As the method considers a limited number of frequencies only, the continuity of the unsteady flow cannot be stringently maintained. With higher numbers of harmonics, however, the continuity of the unsteady flow is preserved better.

2 Investigated Compressor Stage

For the present investigation the first 1.5 stages of a 4.5-stage transonic axial compressor for industrial gas turbines was chosen. These 1.5 stages consist of 40 inlet guide vanes (IGV), 23 rotor and 36 stator blades. The rotor is characterised by a maximum pre-shock Mach number of 1.45 at a design speed of 12960 rpm, corresponding to a blade tip velocity of approximately 440 m/s. Hence, at design speed the blade tip operates at supersonic conditions, while the hub runs in a subsonic environment. The blades are highly cambered at hub and show thin airfoils at blade tip. The mean aspect ratio of the rotor blades measures 1.4 with a hub-to-tip ratio of 0.57. The stage design total pressure ratio is $\Pi_t = 1.63$ at a corrected massflow rate of $\dot{m} = 46.0\,\text{kg/s}$. The considered compressor was already investigated within Hembera et al. [4, 5].

3 Numerical Model

To numerically investigate the first 1.5 stages, one full-annulus and one one-passage model was discretised with a block-structured grid using the commercial mesher AutoGrid from Numeca. The full-annulus mesh consisted of 40 million gridpoints and was solved with the fully turbulent 3D Favre-averaged-Navier-Stokes (FANS) equations with the turbulence model of Spalart-Allmaras (see Spalart and Allmaras [6]) applied. Thus one passage was characterised by relatively coarse mesh. This mesh is used for the 360° simulations as well as for a NLH simulation with only one simulated passage per row, which ends up in 1.38 million gridpoints. As only one passage had to be simulated with the NLH method, a refined mesh with 7 million gridpoints was considered additionally. This discrepancy in the relative mesh size of course influence the results, but it also points out the advantages and disadvantages of the two methods with the resulting limits in manageable mesh size. Full annulus simulations obviously require large meshes which come along with huge computational effort. Nevertheless, the relative resolution of a single blade passage is limited. As a consequence, on the one hand the mesh resolution is too low to adequately resolve some small-scale effects. On the other hand, due to discretization of the full annulus, large-scale effects – like rotating stall – can be simulated.

Fig. 1 Mesh topology of full annulus and NLH coarse mesh

3.1 Full-Annulus and NLH Coarse Grid

The first 1.5 stages of the compressor were entirely meshed, including all passages of the 40 IGV, 23 rotor and 36 stator blades. In total, 541 blocks were used resulting in about 40.28 million grid points. Each IGV passage consisted of 0.24 million points, each rotor passage of 0.79 million points and each stator passage of 0.35 million points. For the NLH coarse grid only one passage per row was used within the same relative resolution as the full annulus grid. The mesh ended up in about 1.38 Mio grid points. In radial direction 65 nodes were used, in axial direction 73 in the IGV, 97 in the rotor and 89 in the stator. The circumferential direction was discretised with 43 points in the IGV and in the stator and 107 points in the rotor. The mesh was refined toward the solid boundaries in order to meet the requirements for the low-Reynolds simulations of $y^+ < 3$. In figure 1 an extract of the mesh from the IGV, rotor and stator is shown.

3.2 NLH Fine Grid

The geometrical model of the 1.5 stages of the compressor for the NLH simulations on the fine grid consisted of one representative passage per blade row, which was discretised with a mesh of about 7.03 Mio grid points in 45 blocks. Therefore, 1.39 Mio points were used in the IGV passage, 4.24 Mio in the rotor passage and 1.4 Mio in the stator passage. The mesh resolution would correspond to 203 Mio mesh points for an equivalent full annulus mesh. In radial direction, 129 points were used, in axial direction 145 in the IGV, 181 in the rotor and 141 in the stator. The circumferential direction was discretised with 51 nodes in the IGV, 155 in the rotor and 59 in the stator.

3.3 Boundary Conditions and Numerical Settings

Identical boundary conditions were used for the two different methods in order to maintain comparability. The solid boundaries at hub, shroud and at the blades were treated as adiabatic Stokes-walls. Hence, the heat flux through the wall was zero and no slip as well as impermeability conditions were applied. The fluid was modelled as a perfect gas with constant heat capacity and constant isentropic coefficient. At the inlet, a boundary condition for constant total pressure $p_t = 101.325\,\text{kPa}$, constant total temperature $T_t = 293\,\text{K}$ and axial flow was applied. The boundary condition at the outlet was set to different values of constant static back pressure (using the radial equilibrium method) to obtain flow solutions for different operating points. For the NLH simulations, 2 perturbations or BPFs were used with 3 harmonics each. A steady simulation, using the frozen rotor approach, was carried out to serve as an initial solution for the unsteady simulations. The one-equation model from Spalart-Allmaras (see Spalart and Allmaras [6]) was used to model turbulence.

4 Comparison of Numerical Effort

The full annulus simulations required about 1500 physical time steps with 10 inner pseudo-time steps at a CFL-number of 3 to reach convergence. For the simulation of one inner pseudo-time step, about 3.3 CPUh on the used Itanium2 Montecito Dual Core 1.6 GHz processors were required. With 63 processors and 58 GB main memory, the full annulus simulation took about 30 days to reach convergence. In contrast to that, the quasi-unsteady NLH simulation with the same mesh and only one simulated passage per row required only 12 hours for one simulated operating point on 8 processors with 10 GB of memory. This resulted in less than 100 CPUh until the solution was convergenced. For the NLH simulation on the fine mesh, 24 hours were required to reach convergence on 29 processors with 34 GB main memory, resulting in a total of 750 CPUh. The results are summarized in table 1.

The comparison of the computational effort highlights distinct differences of the two methods. The FANS full annulus simulation required almost 500-times the CPUh of the NLH simulation. Regarding the memory requirements, only 5.8-times the amount of main memory was needed. If the required RAM per million gridpoints is compared this becomes clearer: The full annulus simulation required 1.45 GB per

Table 1 Comparison of the numerical effort

Effort	Full Annulus	NLH fine	NLH coarse
Mesh Size	40 Mio	7 Mio	1.38 Mio
Mesh Size per rotor passage	0.8 Mio	4.24 Mio	0.8 Mio
Processor Work	49500 CPUh	750 CPUh	96 CPUh
RAM	58 GB	34 GB	10 GB
Work per iteration	10*3.3 CPUh	1.5 CPUh	0.25 CPUh
Total time on processors	30d on 63 CPUs	24 h on 29 CPUs	12 h on 8 CPUs

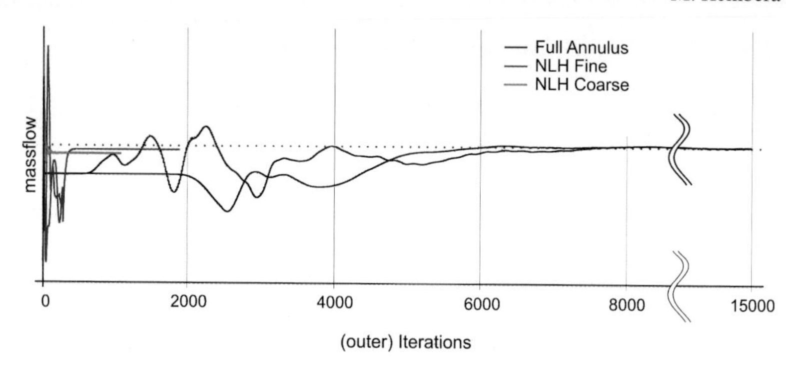

Fig. 2 Convergence behaviour

one million gridpoints whereas the NLH required 7.2 GB per one million gridpoints. This was characteristic for the NLH method: low requirements of CPUh at high requirements of RAM. This in turn resulted from the demand of solving 10 additional conservation equations per harmonic and perturbation. In the considered case, 60 additional equations for the 2 perturbations with 3 harmonics each had to be solved.

4.1 Convergence Behaviour

The convergence behaviour of the three different computations can be found in figure 2 on the basis of the massflow. In order to evaluate whether convergence was reached, other indicators like residuals, total pressure ratio or efficiency were used as well. The full annulus computation required almost 15000 iterations (including the inner iterations) to reach convergence. The coarse NLH simulation only required about 500 iterations, and the fine NLH simulation around 1000 iterations.

5 Numerical Results

The comparisons presented within this section are based on the afore-mentioned meshes. For the coarse mesh, the differences in the solutions are caused solely by the two different methods namely the NLH method and the approach based on the FANS equations. For the fine mesh however, the discrepancies mostly arise from the much finer resolution.

5.1 Compressor Maps

The compressor maps derived from the simulations are presented in figure 3(a) and 3(b). Due to the enormous amount of computational power for the full annulus simulations, only 2 operating points, controlled by the static back pressure of

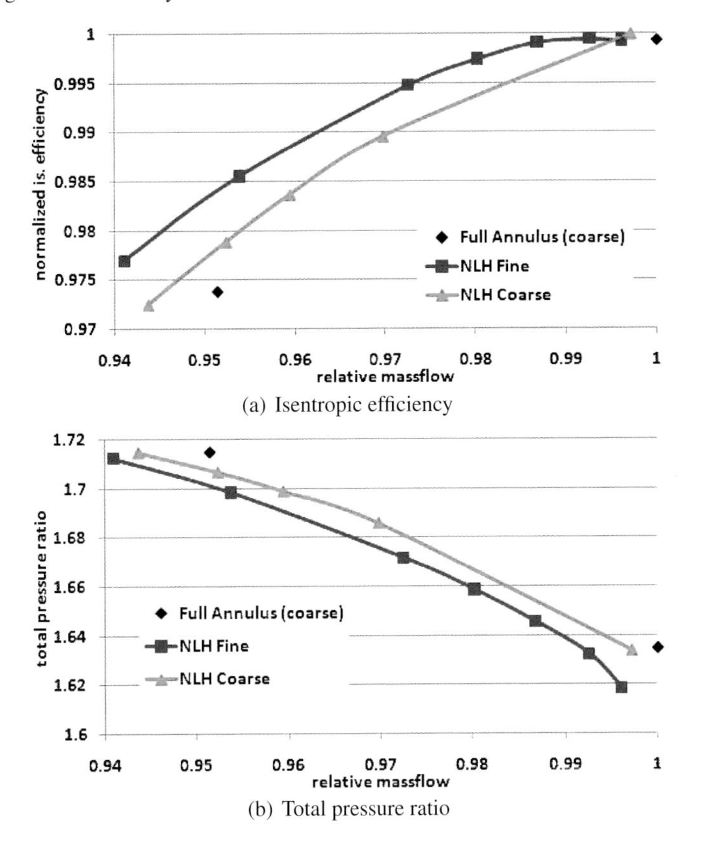

(a) Isentropic efficiency

(b) Total pressure ratio

Fig. 3 Compressor maps

122.5 kPa and 138 kPa at the outlet, were computed. These are represented by black diamonds.

In total, the characteristics of the full annulus simulation as well as of the NLH simulation with the coarse mesh were slightly shifted toward higher mass flows. The reason for that was the already mentioned lower resolution of the mesh and, as a consequence, the reduced capability of resolving secondary flow phenomena. Thus, blockage areas, especially at the throttled state, were reduced leading to a larger effective cross section for the flow. Nevertheless, the absolute values for the isentropic efficiency and the total pressure ratio are almost identical at design point for all the three different computations. Closer to surge however, unsteady flow phenomena increase so that this ends up in relatively large fluctuations in the full annulus simulation. The discrepancies in contrast to the NLH simulations might stem from the onset of rotating phenomena, which – per definition – cannot be resolved with the NLH simulations. Also possible are strengthened secondary flow phenomena, which, on the other hand, cannot be sufficiently resolved in the simulations with the coarse mesh for the NLH as well as for the full annulus simulation.

In total, the differences in the compressor maps at design point for the coarse mesh but also when compared to the fine mesh are very small, so that it can be said that very satisfying results can be achieved with the NLH simulation in a fraction of time in contrast to a full annulus simulation.

5.2 Blade-to-Blade-Flow

For both methods, a complete 360° blade-to-blade-view can be created. However, it is difficult to point out any similarities or differences in such a view. Because of this reason, only a small fraction of the complete 360° is shown.

For the comparison, two representative quantities at one single time step have been selected, the static pressure in figure 4(a) and the entropy in figure 4(b). All

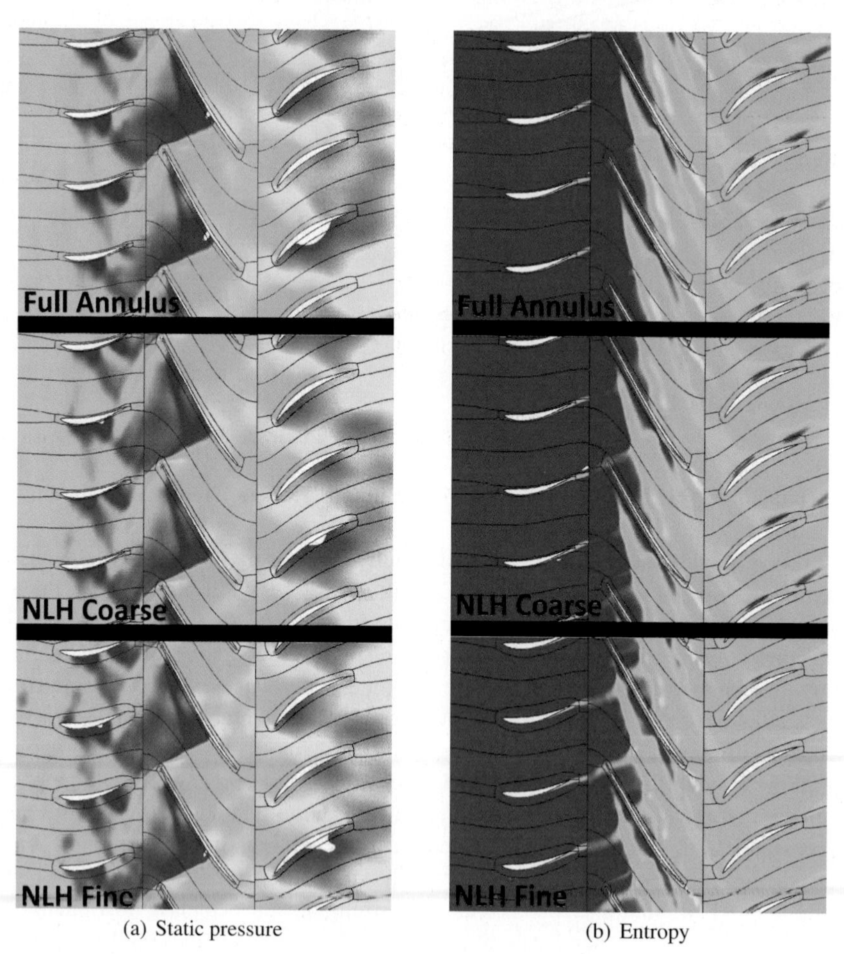

(a) Static pressure (b) Entropy

Fig. 4 Flow quantities of the different computations at 98.5% blade height

blade-to-blade views are at 98.5% radial blade height, where most of the interesting phenomena in transonic compressors like the vortex-shock interaction take place and where the gradients of certain flow variables become very high. The views are furthermore at design point and at an identical circumferential position, which has been arbitrarily chosen. The massflow is almost identical (difference below 0.7%), so that this should make no difference in these comparisons.

The basic phenomena in the static pressure plots of all three computations look fairly identical. The occurrence of the static pressure in the IGV as well as in the rotor and in the stator follows the same patterns. The only little differences are the absolute values of the static pressure, as can be seen for example in the stator. A merely more wave-like structure is visible in the NLH calculation with the fine grid, which has its origin in the method itself. This behaviour will be explained later.

In figure 4(b), the entropy levels of the two methods are compared. Here, the wave-like structure in the NLH method is even more visible, especially in the wake of the IGV. The scale of the quantity "entropy" has been chosen especially to make this phenomenon visible. It is a result from the instabilities based on sinusoidal oscillations. When considering only one harmonic, the disturbance resulting from the IGV and entering the rotor domain for example is approximated in circumferential direction by a sinusoidal wave. The spatial wave length of the disturbance is directly related to the blade passing frequency. When using more harmonics with smaller wave lengths (or two-times, three-times the frequency etc.), the real behaviour gets more and more approximated. The higher frequencies in nature bring about more local maxima and minima, which appear in the flow solution by these visible waves with, in contrast to the main disturbancc, much lower amplitudes.

This can be illustrated by a simple example. The step-function

$$f(x) = \begin{cases} -\frac{\pi}{2} & if \quad -\pi \leq x < 0 \\ \frac{\pi}{2} & if \quad 0 \leq x \leq \pi \end{cases} \tag{3}$$

can be approximated with a Fourier-series – here with 3 harmonics:

$$f(x) = 2\left(sin(x) + \frac{sin(3x)}{3} + \frac{sin(5x)}{5}\right) . \tag{4}$$

The result is shown in figure 5 with a different amount of harmonics.

Clearly visible are the local minima and maxima when approximating the step function with a Fourier series. As expected, these local extrema decrease with a higher number of harmonics. Only at the sharp edges a constant value of 9% remains (so called Gibbs effect), however in real flows no such "steps" in any quantities exist, so that this overshoot does not lead to any problems. This simple example describes the behaviour and the reason for the "wave-like" structures, especially in the entropy. This is visible very clearly in the entropy due to the fact that the wake of the IGV produces a characteristic trough.

This trough with a high gradient however is difficult to approximate with a limited number of harmonics, so that the entropy pattern, as described in figure 5 is represented by these waves. The pattern changes with the number of harmonics.

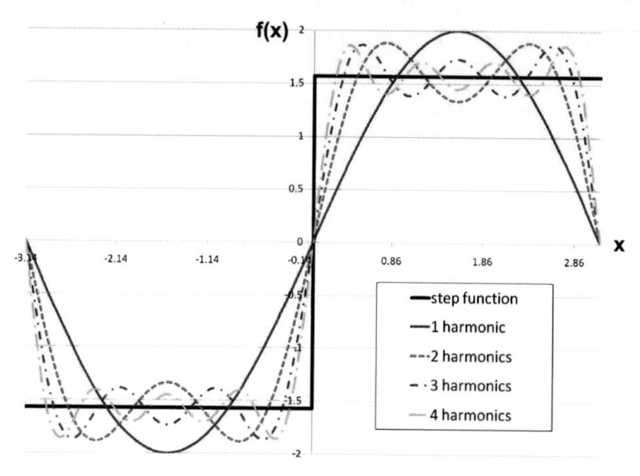

Fig. 5 Approximation of a step function with a fourier series

When using only one harmonic, no local maxima except the wake of the IGV is visible. However, as a consequence, the trough is much broader in the rotor domain and also more washed-out. The quality of the other quantities also decreases when reducing the number of harmonics. When using 4 harmonics, the local minima and maxima decrease and they almost disappear with 5 harmonics. When regarding the figure for entropy and the full annulus simulation, the wake seems to be dampened very strong when entering the rotor passage. This might be an effect from the coarse mesh resolution. In total, the main phenomena in the flow like the direction of the vortex trajectory, the position and strength of the passage shock of the coarse NLH mesh is represented in conformity with the full annulus simulations. The fine NLH mesh shows more details simply due to the higher mesh resolution but also because the used mesh resolution is required in order to be able to resolve the higher harmonics. This becomes clear when comparing the results of the coarse and the fine NLH mesh. In the fine mesh the gradients of the static pressure but also of the entropy patterns are much clearer. As a consequence, some structures like the wave-like structures between two IGV wakes become only visible when using the fine mesh.

5.3 Comparison with Fourier Analysis

In order to validate, whether the non-linear harmonics method represents the occurring frequency spectra correctly, a comparison of some representative signals by means of a Fast Fourier Transformation (FFT) is presented. For the comparison of the two methods with the Fast Fourier Analysis, the static pressure signal in the IGV-passage (point 1) and in the Stator 1-passage (point 2) close to the casing is analysed, as can be seen in figure 6.

In these points, it is possible to detect the signal from the rotor which travels through the rotor stator interface and is represented by – in this case – 3 harmon-

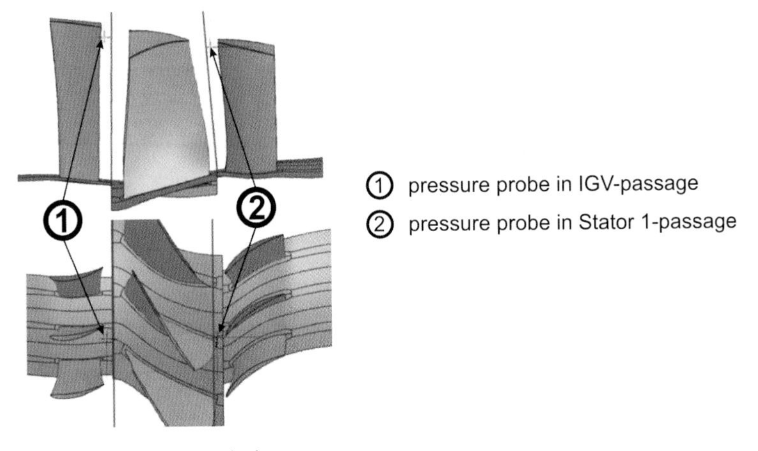

Fig. 6 Positions of the FFT analysis

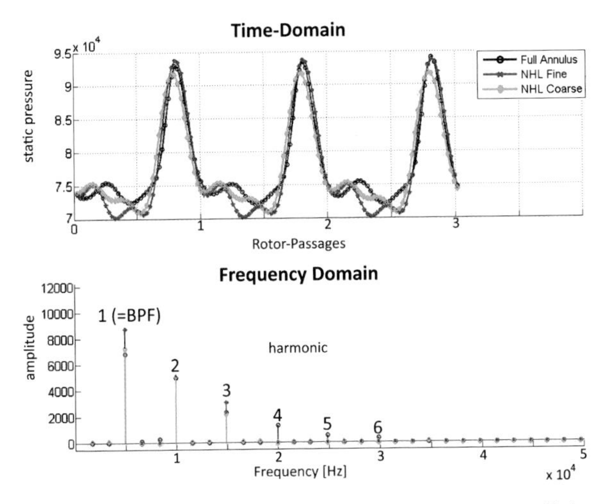

Fig. 7 Static pressure signal and related frequency spectrum at maximum efficiency in point 1

ics. Figure 7 shows the original static pressure signal for the point at maximum efficiency (ME) for all three configurations as well as the corresponding frequency spectrum after the FFT Analysis. The signals in the time domain are almost identical with very little discrepancies located between the definite pressure signal resulting from the shock wave of the rotor which travels upstream. It also seems as if the coarse grid with the NLH method does not reach the same absolute pressure level at the shock as do the other two simulations. In the frequency domain, the spectra of the full annulus signal (black) is almost completely represented by 3 harmonics. The amplitude of the fourth harmonic is very small and the one from the fifth almost disappeared. So in this case, the assumption of the NLH-method that the dominating disturbances are multiples of the BPF seems to be a very good approximation. The first harmonic, which is also the blade passing frequency BPF of 4968 Hz, has

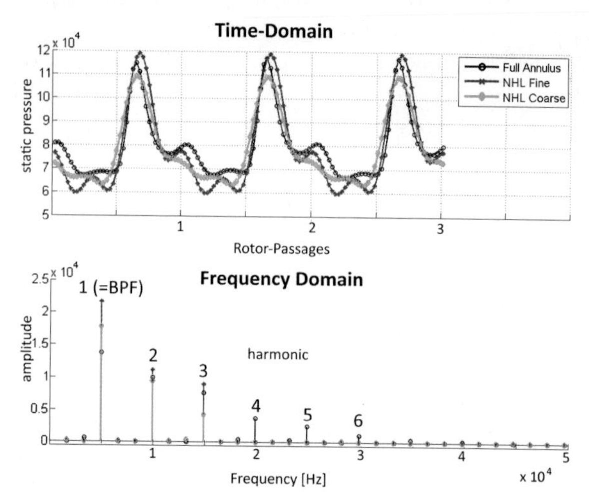

Fig. 8 Static pressure signal and related frequency spectrum close to the surge line in point 1

a higher amplitude for the NLH Fine case (red) and even a slightly higher amplitude for the NLH Coarse case (green). In the second harmonic, all three amplitudes lie on top of each other. The higher amplitudes for the harmonic computations might be a result of the only 3 computed harmonics, which have to end up in the same absolute value of static pressure in the time domain than the signal from the full annulus calculation with a (theoretical) infinite number of harmonics.

As already explained above, this behaviour of the method becomes clearer when the signal close to the surge line (SL), where the gradients of most of the flow values increase, is regarded. The static pressure fluctuations in figure 8 increased due to the throttled configuration. The passage shock which travels upstream is still the main effect which is visible. However, the signal has become stronger and, as a consequence, also the gradients became higher. From the theorie of Fourier transformation, it is known that a compressed signal in the time domain corresponds to an expanded signal in the frequency domain. As only 3 harmonics are computed with the NLH methods in this case, the signal of the NLH calculations cannot follow the steep gradients, so that the signal of the pressure wave appears wider. This can also be read off the frequency domain graph. For a better representation, 5 or 6 harmonics should be included in the NLH calculations close to the surge line. Again obvious is the overprediction of the amplitude of the first harmonic for both NLH grids, which is required to reach the same absolute static pressure value like in the full annulus calculations. However, the amplitude of the NLH Coarse grid computation seems to be too small, so that it cannot reach the same maxima as the other two configurations. Also the amplitudes of the higher harmonics are underpredicted which results in a reduced wave like structure in between two passage shocks.

The static pressure signal in front of the rotor is however dominated in both points (at ME and close to the SL) by the shock wave travelling upstream from the transonic rotor and, as a consequence, matches very well with the blade passing

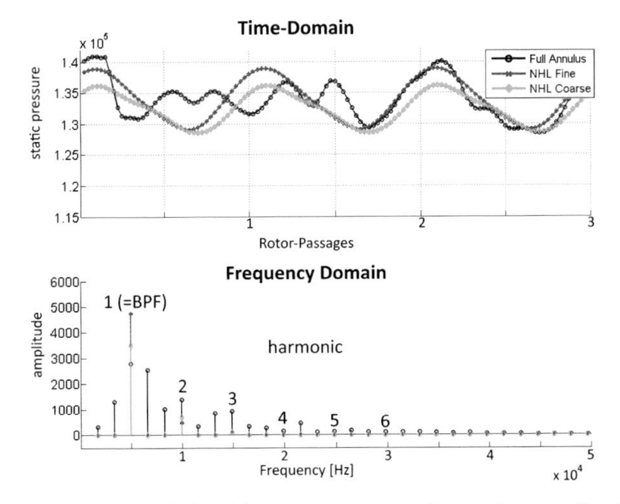

Fig. 9 Static pressure signal and related frequency spectrum close to the surge line in point 2

frequency. For this reason, a signal in point 2 of figure 6 in the stator 1 domain behind the rotor is regarded. Here, no such dominant effects as the shock exist. The signal matches pretty well for the case at maximum efficiency, so that again the assumption that the main disturbances are multiples of the blade passing frequency seems to be very accurate.

For the case close to the surge line, the result can be seen in figure 9. At this point, the signals do not match really well. Close to the surge line, the onset of rotating effects (such as rotating stall) can be discovered in the full annulus simulation. Of course – and per definition – the NLH method cannot represent these non-periodic effects. This is also visible when looking at the frequency domain. In the full annulus simulation, frequencies in between the harmonics show high amplitudes, so that this behaviour cannot be represented with a NLH simulation.

6 Conclusion

Within this report a full annulus simulation of a transonic compressor stage including IGV, Rotor 1 and Stator 1 with around 40 million gridpoints was compared to a NLH simulation with same resolution, but only one discretised passage per row. Additionally, a 7 million gridpoints NLH simulation was conducted. Due to the high requirements in mesh size for the full annulus computations, one blade passage had a relatively low resolution compared to a passage from the NLH with fine meshing. The mesh size of the NLH Fine calculations, however, were still almost 6 times smaller although the relative resolution was significantly higher. In order to reach convergence, the full annulus computation required 66 times the CPU time of the NLH Fine computation, and about 1.7 times more main memory.

At points around maximum efficiency both NLH grid calculations delivered accurate quasi-unsteady results compared to full annulus results. Due to the characteristics of the method, the number of harmonics and also the mesh size should be increased when throttling the compressor or if high flow gradients can be expected. The limits of the NLH method are reached as soon as rotating effects occur as they cannot be resolved. Nevertheless, the approach seems to be very promising for fast quasi-unsteady CFD-simulations in multi-stage compressors.

Acknowledgements The investigations presented in this report were conducted within the German collaboration programme "AG Turbo-COOREFF" and was funded by the Bundesministerium für Wirtschaft und Technologie (BMWi) and Rolls-Royce Deutschland Ltd. & Co. KG. Special thanks also goes to Numeca for providing the software licenses.

References

1. He, L.; Ning, W.: Efficient Approach for Analysis of Unsteady Viscous Flows in Turbomachines. AIAA J., **36**(11), 2005–2012 (1998)
2. Chen, T.; Vasanthakumar, P.; He, L.: Analysis of Unsteady Blade Row Interaction Using Nonlinear Harmonic Approach. AIAA JPP, **17**(3), 651–658 (2001)
3. Vilmin, S., Lorrain, E., Hirsch, C., Swoboda, M.: Unsteady Flow Modelling across the Rotor/Stator Interface using the Nonlinear Harmonic Method. ASME Paper GT2006-90210 (2006)
4. Hembera, M.; Kau, H.-P.; Johann, E.: Simulation of Casing Treatments of a Transonic Compressor Stage. ISROMAC Paper ISROMAC12-2008-20042 (2008)
5. Hembera, M.; Butzeck, C.; Danner, F.; Kau, H.-P.; Johann, E.: Development Of Circumferential Grooves For Axial Compressors Based On Flow Mechanisms. AIAA JPC Paper AIAA-2008-4988 (2008)
6. Spalart, P. R.; Allmaras, S. R.: A One-Equation Turbulence Model for Aerodynamic Flows. AIAA Paper AIAA-92-0439 (1992)

Capability of FDEM for Journal Bearings with Microstructured Surface

Torsten Adolph, Willi Schönauer, Roman Koch, and Gunter Knoll

Abstract For the numerical simulation of journal bearings, current software solutions use the Reynolds differential equation where inertia terms are not included. The Finite Difference Element Method (FDEM) is a black-box solver for nonlinear systems of elliptic and parabolic partial differential equations (PDEs). Based on the general black-box we implement the Reynolds equation with the nonlinear inertia terms for the simulation of a journal bearing. We can easily implement different models for the turbulence factors and the dynamic viscosity, and we also consider cavitation. We give results for grids with different microstructure for Reynolds number $Re = 24{,}500$, and we also give a global error estimate for each of the cases. This shows the quality of the numerical solution and is a unique feature of FDEM. Furthermore, we demonstrate the scalability of the code.

1 Introduction

This paper is the continuation of [1] where we first demonstrated the usefulness of FDEM for the numerical simulation of journal bearings. We want to compute the pressure in journal bearings considering turbulence and inertia effects. The theory and the derivation of the PDEs used in this simulation is explained in [4] and [10]. The simulation is based on the Reynolds differential equation which can be extended

Torsten Adolph · Willi Schönauer
Karlsruhe Institute of Technology, Steinbuch Centre for Computing, Hermann-von-Helmholtz-Platz 1, 76344 Eggenstein-Leopoldshafen, Germany
e-mail: torsten.adolph@kit.edu
e-mail: willi.schoenauer@kit.edu

Roman Koch · Gunter Knoll
Institut für Maschinenelemente und Konstruktionstechnik, University of Kassel, Mönchebergstr. 3, 34125 Kassel, Germany
e-mail: roman.koch@imk.uni-kassel.de
e-mail: gunter.knoll@imk.uni-kassel.de

S. Wagner et al. (eds.), *High Performance Computing in Science and Engineering, Garching/Munich 2009*, DOI 10.1007/978-3-642-13872-0_15,

Table 1 List of symbols

Symbol	Unit	Description
p	Pa	Pressure
h	m	Crack height
$\rho(p)$	kg/m^3	Density
$\eta(p)$	mPa s	Dynamic viscosity
u_m	m/s	Mean velocity in x-direction
v_m	m/s	Mean velocity in y-direction
w	m/s	Velocity in z-direction: $w = \partial h/\partial t$
V_μ	m/s	Bulk velocities
$I_{\mu\nu}$	m^3/s^2	Momentum fluxes
G_μ	–	Turbulence factors

by turbulence and inertia terms under specific restrictions, see [7, 9], resulting in

$$\begin{aligned}
&\frac{\partial}{\partial x}\left(G_x \frac{\rho h^3}{\eta}\frac{\partial p}{\partial x}\right) + \frac{\partial}{\partial y}\left(G_y \frac{\rho h^3}{\eta}\frac{\partial p}{\partial y}\right) \\
&\quad - \frac{\partial}{\partial x}(G_{Jx}\rho h u_m) - \frac{\partial}{\partial y}(G_{Jy}\rho h v_m) - \rho w \qquad (1) \\
&= -\frac{\partial}{\partial x}\left(G_x \frac{\rho h^2}{\eta}\left(V_x w + \frac{\partial I_{xx}}{\partial x} + \frac{\partial I_{xy}}{\partial y}\right)\right) \\
&\quad - \frac{\partial}{\partial y}\left(G_y \frac{\rho h^2}{\eta}\left(V_y w + \frac{\partial I_{yx}}{\partial x} + \frac{\partial I_{yy}}{\partial y}\right)\right)
\end{aligned}$$

with the notation given in Table 1. The turbulence factors G_μ give the turbulence effects, and the terms on the right hand side of the equal sign are the inertia terms.

In (1) we dropped the time derivative, so that we obtain a 2D elliptic PDE for the pressure. In comparison to the full model, this yields a considerable decrease of numerical complexity, however, stability of the numerical process is complicated to control.

For a detailed description of PDE (1) and its derivation, see [1]. There we also illustrate the computation of the bulk velocities, the momentum fluxes, the turbulence factors and the density as well as the implemented models of Barus [2] and Roelands [8] for the dynamic viscosity.

For the numerical simulation we use the Finite Difference Element Method (FDEM) program package. It is a black-box solver for the solution of arbitrary nonlinear systems of 2D and 3D elliptic and parabolic PDEs. We extended the Finite Difference Method to unstructured Finite Element grids, and the user may put into FDEM any system of PDEs on any unstructured domain that may even consist of several subdomains with different PDEs, and FDEM computes together with the solution an error estimate, which is a unique feature for such a general-purpose black-box solver. FDEM is described in detail in [11, Chap. 2]. The code is parallelized with MPI [6], and the large and sparse linear system of equations resulting from the discretization is solved by the SuperLU program package [5] that can also effectively use hundreds of parallel processors on sufficiently large matrices.

2 Topic

For the computation of the turbulence factors G_μ we need the Reynolds number that is computed from

$$Re = \frac{\rho V_x h}{\eta} . \qquad (2)$$

As V_x depends on the velocities u_1 and u_2 on the upper and on the lower side of the crack, respectively, we control Re by the variation of these two velocities.

In [1] we set $u_2 = 0$ and carried out eight computations with Reynolds numbers Re between 0.2 and 24,500 by choosing the value for u_1 between 1 m/s and 10^5 m/s. For all eight computations we used the same grid with a total number of nodes of 231. The purpose was to show that the CPU time on one processor is below 1 sec which was the prescribed CPU time limit. We wanted to have a smooth crack and modelled its height by a cosine function so that the crack becomes thinner in the left half of the domain and wider in the right half.

In this very paper all computations are carried out with Reynolds number $Re =$ 24,500, i.e. we choose $u_1 = 10^5$ m/s, $u_2 = 0$. To model the roughness of the crack walls, we choose a pseudorandom crack height, see Subsect. 3.1. We use six different grids with a total number of nodes between 231 and 205,761. Furthermore, we want to demonstrate the scalability of the FDEM code for up to 1,024 processors, see Subsect. 3.2.

3 Numerical Results

The implementation of the code and the computations have been carried out on the distributed memory parallel computer SGI Altix 4700 at the Leibniz Supercomputing Centre in Garching, Germany. We compute on Intel Itanium2 Montecito DualCore processors with 1.6 GHz clock rate (6.4 GFLOPS theoretical peak performance) and NUMAlink 4 interconnect.

The rectangular solution domain is $\Omega = [0, 2\pi] \times [0, 0.001]$, see Fig. 1, and we set $p = 0$ on the four boundaries, i.e. it holds

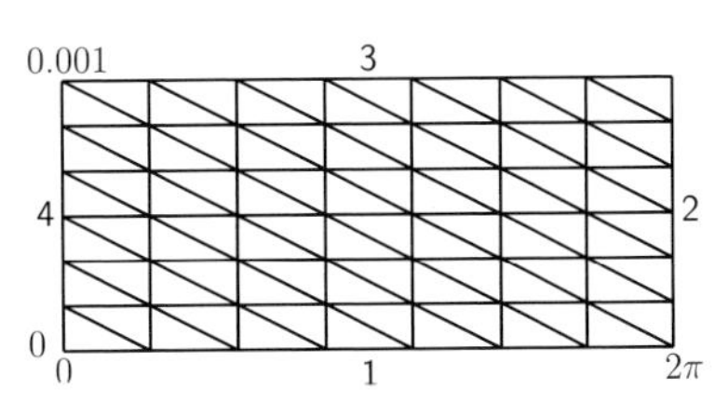

Fig. 1 Solution domain Ω and boundary numbering

$$p(x,0) = p(x,0.001) = p(0,y) = p(2\pi,y) = 0\,. \tag{3}$$

The given functions u_1, u_2, v_1, v_2 and w are constant, more precisely it holds

$$u_2 = v_1 = v_2 = w = 0\,. \tag{4}$$

By the choice of u_1 we control the Reynolds number as Re depends on the bulk velocity V_x (2) that is in turn dependent on $u_m = (u_1 + u_2)/2$. All computations are carried out with $u_1 = 10^5$ m/s which results in a maximum Reynolds number $Re = 24{,}500$.

We use the Barus viscosity equation [2] with $\eta_0 = 20\,\mathrm{mPa\,s}$ and the pressure coefficient $\alpha = 1.4 \times 10^{-8}$. Furthermore, it holds $\rho_0 = 890\,\mathrm{kg/m^3}$ for the density at ambient pressure [3].

In the derivatives of the momentum fluxes $I_{\mu\nu}$ occur third derivatives of the pressure p. As FDEM can cope with second derivatives at most, we have to introduce two auxiliary variables p_x and p_y (see [1]) so that we actually have to solve a system of three PDEs and therefore have three unknowns per node.

3.1 Results for Journal Bearings with Microstructured Surface

In spite of close machining tolerances, the surfaces of the journal bearing are rough in microscale. We want to consider this roughness in this paper. The flow direction is from left to right. The crack height h is chosen so that we get cavitation, i.e. there must be a downward trend of the crack height in the left half of the domain, and in the right half there must be an upward trend. h is a function of x, the influence of y is negligible. We depict the crack height for grid 1 and for grid 6 in Fig. 2.

The coarsest grid we use is a grid with 11 nodes in x-direction and 21 nodes in y-direction resulting in a total number of nodes of 231 and a number of elements of 400. Then we halve the mesh sizes in x- and y-direction from one grid to the other, see Table 2. Thus, the finest grid is a grid with $321 \times 641 = 205{,}761$ nodes and 409,600 elements. The elements of the grid have very small angles which causes

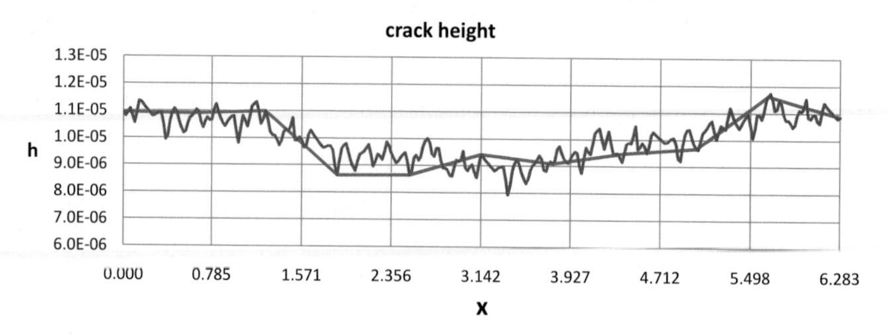

Fig. 2 Plot of the crack height $h(x)$ for grids 1 (*blue*) and 6 (*red*)

Table 2 Grids for the computations on one processor

Grid no.	no. of nodes $n_x \times n_y$	total n	no. of unknowns	no. of elements	bandwidths b_u	b_l
1	11× 21	231	693	400	173	186
2	21× 41	861	2,583	1,600	296	250
3	41× 81	3,321	9,963	6,400	698	664
4	81×161	13,041	39,123	25,600	1,409	1,344
5	161×321	51,681	155,043	102,400	2,870	2,784
6	321×641	205,761	617,283	409,600	5,750	5,664

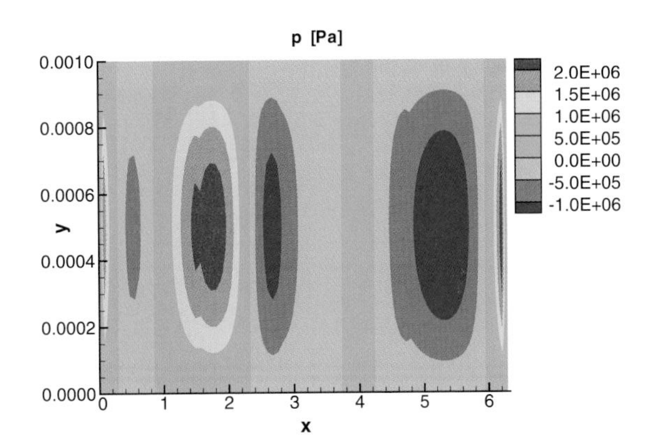

Fig. 3 Contour plot of the pressure p for the test step for grid 4

linear dependencies of the nodes that we collect for the difference stars. Here it is very important to have a robust algorithm for the selection of the nodes for the difference and error formulas.

In Fig. 3 we depict the solution resulting from the test step for grid 4. We see that the pressure is positive where the crack height h is decreasing and negative where h is increasing. For the computation step, the pressure is set to zero in each node where $p < 0$ holds. This is the usual way to cope with cavitation.

In Table 3 we present the results for the computation step of the six computations. In the second column you see the maximum pressure in the solution domain. Columns 3 and 4 give the maximum and the mean global relative estimated error, respectively. "Mean" means the arithmetic mean of all nodes. To get aware of the location of the maximum pressure and the maximum error you may have a look on Figs. 4 (for the pressure) and 5 (for the maximum error). The last 3 columns give the computation time on one processor for the generation of the difference and error formulas, for the computation of the large sparse matrix of the linear system of equations and of the r.h.s., and the CPU time for the SuperLU solver. You have to keep in mind that we have to generate the formulas only once whereas we have to solve the linear system of equations twice, once in the test and once in the computation step. For the computation of the error, we have to solve another linear system of equations in each step, but here we can use the previously factorized matrix.

Table 3 Maximum pressure, max. and mean relative estimated error (computation step) and CPU times for the six grids (comp. on one processor)

Grid no.	p_{max} (Pa)	rel. estimated error max.	rel. estimated error mean	CPU time (s) formula generation	CPU time (s) setup of linear sytem of eq.	CPU time (s) SuperLU
1	0.221×10^7	0.232×10^{-7}	0.250×10^{-8}	0.024	0.016	0.056
2	0.241×10^7	0.142×10^{-5}	0.861×10^{-7}	0.088	0.056	0.220
3	0.237×10^7	0.206×10^{-5}	0.977×10^{-7}	0.608	0.236	0.772
4	0.239×10^7	0.633×10^{-5}	0.224×10^{-6}	5.888	1.284	4.188
5	0.326×10^7	0.202×10^{-4}	0.325×10^{-6}	81.632	7.168	23.380
6	0.382×10^7	0.167×10^{-2}	0.573×10^{-5}	1301.848	32.748	219.232

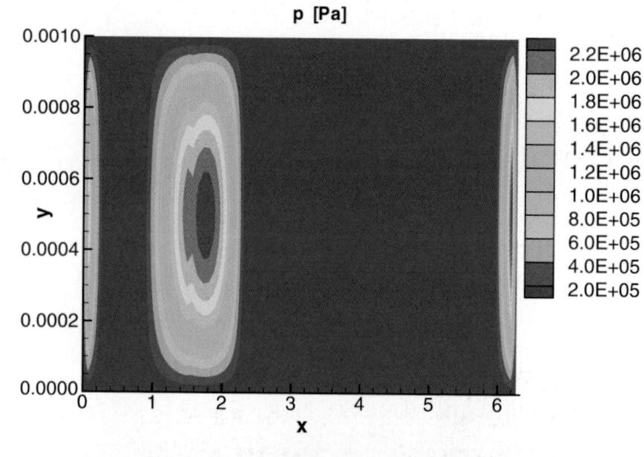

Fig. 4 Contour plot of the pressure p for the computation step for grid 4

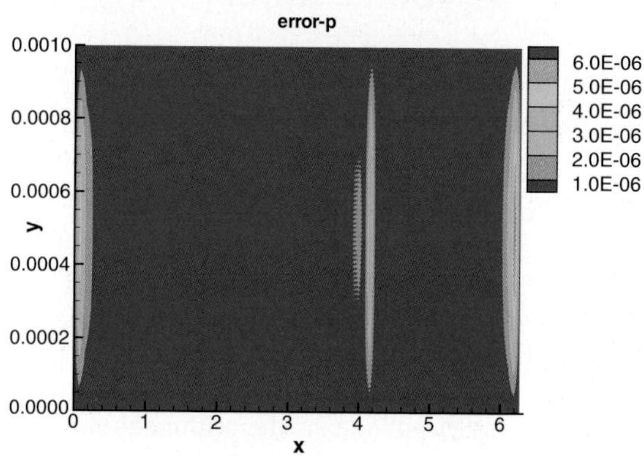

Fig. 5 Contour plot of the error of the pressure for the computation step for grid 4

The maximum errors occur for high pressures and at the boundaries of the cavitation domain, see Figs. 4 and 5. This is because we use nodes from the left and the right of the cavitation boundary for the error formulas.

From the figures we see that where the pressure equals zero because of cavitation after the computation step, the error equals zero in this part of the solution domain. We expect the error to become smaller as we use finer grids but in fact, we're finding quite the opposite. If we look at Fig. 2, we see that this must be a direct consequence of the crack height h. For finer grids, the crack gets much more rough than for the coarser ones. There is a constant swing between regions with and regions without cavitation. Therefore, it is much more difficult to get good error formulas. The grids and the values of h in each node have been provided by the IMK Kassel.

From Table 2 we also see that the bandwidth roughly doubles from one grid to the other. As we have banded matrices here, the work to compute the LU factorization is merely $O(n_m b_u b_l)$ with the number of unknowns n_m, the upper bandwidth of the matrix b_u and the lower bandwidth b_l. As it holds $n_{m,i} \approx 4n_{m,i-1}$ for two grids $i-1$ and i, the factor for the work for the LU factorization between two grids is 16. So we expect the 16-fold CPU time for the SuperLU solver. Fortunately, we notice that this factor is much smaller.

Instead, we see that the CPU time for the generation of the formulas increases by much more than the expected factor of 4. But this can be explained easily: The nodes around each central node are collected in rings, and for the determination of the next ring we use a logical vector of length n (no. of nodes). This vector must be completely searched for "true" entries. This becomes very expensive for larger n.

3.2 Scalability Tests

We conducted scalability runs and investigated weak scaling (problem size proportional to number of processors) up to 1,024 processors. The metric used for the scaling runs was the CPU time for the test and the computation step.

The basic grid for the one processor computation has 111×21 nodes. Then we halve the mesh size in x- and y-direction from one grid to the other which gives the fourfold number of nodes, and we increase the number of processors to the fourfold, see Table 4 which coincides with Table 2 (we only added a column for the number of processors).

In Table 5 we give the CPU time for the generation of the formulas, for the setup of the linear system of equations and for the SuperLU solver for the six computations. Again we have the 16-fold cost for the LU factorization from one grid to the other, but now we have the fourfold number of processors so that we expect the fourfold CPU time for the SuperLU solver. This is easily achieved for processor numbers up to 16, but then the factor increases to values between 5 and 5.5.

Table 4 Grids for the scalability runs

Grid no.	no. of proc.	no. of nodes $n_x \times n_y$	total n	no. of unknowns	no. of elements	bandwidths b_u	b_l
1	1	111× 21	2,331	6,993	4,400	173	186
2	4	221× 41	9,061	27,183	17,600	359	310
3	16	441× 81	35,721	107,163	70,400	695	688
4	64	881×161	141,841	425,523	281,600	1,409	1,390
5	256	1,761×321	565,281	1,695,843	1,126,400	2,789	2,860
6	1,024	3,521×641	2,256,961	6,770,883	4,505,600	5,609	5,761

Table 5 CPU times for the scalability runs for the six grids

Grid no.	CPU time (s) formula generation	setup of linear sytem of eq.	SuperLU	SuperLU ratio
1	0.376	0.156	0.488	—
2	0.392	0.180	1.504	3.1
3	0.440	0.196	5.368	3.6
4	0.696	0.300	26.072	4.9
5	0.896	0.556	138.720	5.3
6	1.708	0.916	752.557	5.4

The cost for the formula generation and for the linear system setup is supposed to remain the same for each computation as there are the same number of nodes on each processor. But evidently, this is not the case. We want to explain this here: We want to execute the generation and evaluation of the difference and error formulas, the generation of the r.h.s. and of the large sparse matrix purely local on the processors, without communication. So we must store on each processor also the data of the adjacent processor(s) that are needed for the rings of neighboured nodes. This is indicated as overlap. As we have a 1D domain decomposition (see [11, Sect. 2.8]), the width of the overlap depends on the x-coordinates of the nodes but not on the y-coordinate. As the number of nodes doubles from one grid to the other, the number of nodes in the overlap also increases. The logical vector for the ring search has the length that equals the number of nodes on a processor including the overlap. This is the reason for the observed increasing CPU times.

4 Concluding Remarks

We implemented an innovative code for the numerical simulation of a journal bearing which includes the Reynolds differential equation under consideration of the nonlinear inertia terms which are neglected in conventional software solutions. FDEM is a black-box solver for arbitrary nonlinear systems of PDEs. By entering the PDEs as Fortran Code we create a problem solver exactly for the requested problem, in this case for the journal bearing. As we use a Finite Difference Method, we get a direct access to an error estimate which is a unique feature under such general conditions. We demonstrated the usability of the FDEM code with the modifications for the numerical simulation of a journal bearing for Reynolds number $Re = 24{,}500$ with random crack height. Furthermore, we demonstrated the scalability of the code up to 1,024 processors.

Acknowledgements We are grateful for a generous alotment of computer time on the HLRB platform in the framework of project h1071.

References

1. Adolph, T., Schönauer, W., Koch, R., Knoll, G.: Application of FDEM on the numerical simulation of journal bearings with turbulence and inertia effects. In: W.E. Nagel, D.B. Kröner, M.M. Resch (eds.) High Performance Computing in Science and Engineering '09, Transactions of the High Performance Computing Centre, Stuttgart (HLRS) 2009, pp. 383–394. Springer Verlag (2010)
2. Barus, C.: Isothermals, isopiestics and isometrics relative to viscosity Am. J. of Science **45**, 87–96 (1893)
3. Dowson, D., Higginson, G.R.: Elastohydrodynamic Lubrication, The Fundamentals of Roller and Gear Lubrication. Pergamon Press, Oxford, Great Britain (1997)
4. Hirs, G.G.: A bulk-flow theory for turbulence in lubricant films. ASME Journal of Lubrication Technology **95**, 137–146 (1973)
5. Li, X.S., Demmel, J.W.: SuperLU_DIST: A scalable distributed-memory sparse direct solver for unsymmetric linear systems. ACM Trans. Mathematical Software **29**(2), 110–140 (2003)
6. Message Passing Interface (MPI) forum. URL `http://www.mpi-forum.org/`
7. Reynolds, O.: On the theory of lubrication and its application to Mr. Beauchamps tower's experiments, including an experimental determination of the viscosity of olive oils. Phil. Trans. R. Soc. Lond. **177**, 157–234 (1886)
8. Roelands, C.J.A.: Correlational aspects of the viscosity-temperature-pressure relationship of lubricating oils. Ph.D. thesis, Technical University Delft, The Netherlands (1966)
9. San Andrés, L.: Turbulent hybrid bearings with fluid inertia effects. ASME Journal of Tribology **112**, 699–707 (1990)
10. Schlichting, H., Gersten, K.: Grenzschicht-Theorie, 9th edn. Springer-Verlag, Berlin (1996)
11. Schönauer, W., Adolph, T.: FDEM: The evolution and application of the Finite Difference Element Method (FDEM) program package for the solution of partial differential equations (2005). URL `http://www.rz.uni-karlsruhe.de/rz/docs/FDEM/Literatur/fdem.pdf`. 239 pages

Numerical Investigation of a Transonic Axial Compressor Stage with Inlet Distortions

Jens Iseler, Andreas Lesser, and Reinhard Niehuis

Abstract This paper refers to a numerical study of a transonic compressor stage investigated for homogenous and disturbed inflow conditions. The simulations were conducted with a Navier-Stokes solver developed by the DLR Cologne. In the first part of the investigations, the flow behavior of the transonic compressor stage was predicted for undisturbed inflow conditions. For those simulations, periodicity was assumed and therefore only one blade passage was considered. The simulations were performed for several operating points at 100% and 85% rotational speed. The numerical results were compared to the experimental data from Dunker and revealed a satisfying agreement concerning the flow behavior. In the second part, steady and unsteady flow simulations for distorted inflow conditions were performed. Here, simulations of the entire compressor stage had to be done and continued.

1 Introduction

In order to decrease costs during the design and development process of modern aircrafts, high efficient and accurate design tools are necessary. These tools will allow the prediction of the flow behavior at the aircraft fuselage and its wings as well as inside of the jet engines for an entire flight mission. For simulating the flow physics correctly, the interaction of the inner jet engine flow and the outer flow around the aircraft has also to be taken into account. Accurate predictions are especially demanded during conditions where high aerodynamic loads are present. One of these critical phases is the take-off procedure, where highly turbulent air with strongly varying total pressure may enter into the jet engine. These inlet distortions increase

Jens Iseler · Andreas Lesser · Reinhard Niehuis
Universität der Bundeswehr München, Werner-Heisenberg-Weg 39, 85577 Neubiberg
e-mail: jens.iseler@unibw.de
e-mail: andreas.lesser@unibw.de
e-mail: reinhard.niehuis@unibw.dc

S. Wagner et al. (eds.), *High Performance Computing in Science and Engineering, Garching/Munich 2009*, DOI 10.1007/978-3-642-13872-0_16,

the risk of compression system instabilities. Inlet distortions maybe composed of total pressure-, angle- and total temperature distortions, depending on the particular configuration. All three disturbances have a reduction of the operation range in common. Despite many experimental investigations, there is still a high demand of research in the field of inlet distortions. A good overview of the effects of inlet distortions is given by Longley and Greitzer [1], as well as by Cousins [2]. Experimental investigations were carried out by Schmidt et al. [3], Peters et al. [4, 5], Reuss [6] and Wadia et al. [7, 8]. In order to predict accurately the creation and the migration of the inlet disturbances and its impact on the compressor, the outer aerodynamics in front of the engine and the flow into the jet engine have to be simulated simultaneously. At present, most of the numerical methods are specialized either for inner or for outer aerodynamics. One possibility to solve this problem is to couple numerically two codes- one for inner, the other for outer flow dynamics. This strategy is pursued by members of a DFG (German Research Foundation) project, where a coupling of two DLR codes TAU [9] and TRACE [10] is planned. Currently, validation tests of the TRACE-code are undertaken with regard to the impact of inlet distortions on the compressor performance. These validation tests refer to a data set of a transonic compressor stage, which was experimentally investigated for homogenous [11, 12] and disturbed [13] inflow conditions at the DLR's Institute of Propulsion Technology in Cologne.

2 Description of the Test Case

The compressor stage, as shown in Fig. 1, was designed for a spool speed of 20260 RPM with a total pressure ratio of 1.51 at an equivalent mass flow of 17.3 kg/s under standard reference conditions with 288 K and 101325 hPa. The rotor diameter is 398

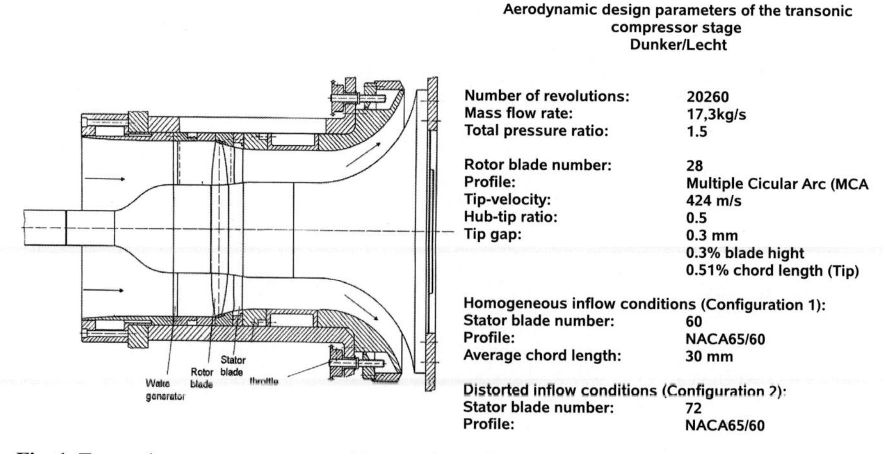

Fig. 1 Transonic compressor stage and its aerodynamic parameters

mm with a hub-to-tip ratio of 0.5 and the maximum blade tip speed is up to 424 m/s. Overall 28 blades with MCA profile and an average chord length of 50 mm were used.

Two different stator rows were used, depending on whether disturbed or undisturbed inflow conditions were given: For homogeneous inflow conditions, the blade row consists of 60 blades with NACA-65-profiles, while for inflow distortions overall 72 NACA-65-profiled blades were used. For the configuration of 28 rotor- and 60 stator blades, detailed temperature and pressure measurements as well as Laser2Focus measurements are available. For the second configuration, detailed measurements of temperature, pressure and flow direction exists only for disturbed inflow conditions. In Figure 1, the main parameters of the compressor stage are given. The inlet distortions were generated upstream of the compressor stage by a wake generator with non-rotating steel bars. The circumferential extent of these distortions was chosen as 60 and 120 degree. However, most of the published experimental data refer to a 120 degree-distortion.

3 Flow Solver Information

The CFD simulations are conducted with the RANS code TRACE developed by the DLR's Institute of Propulsion Technology in Cologne. This code allows the simulation of a multistage three-dimensional, periodically unsteady, and transitional turbomachinery flow on structured grids. TRACE is written in C and represents a time-accurate computational fluid dynamics solver specialized in turbomachinery flows. Integrated MPI-interfaces enable the code to run in parallel mode. In the used version, it solves the three-dimensional Reynolds-averaged viscous Navier-Stokes equations in the rotating frame of reference for compressible fluids on multi-block structured grids. The discretisation of the convective fluxes is based on the TVD-Upwind scheme by Roe, which is combined with a MUSCL extrapolation scheme in order to obtain second order accuracy in space. The viscous derivatives are discretised using a second order central-differences scheme. Furthermore, non-reflecting boundary conditions are implemented at the inlet and outlet boundaries. Steady-state simulations are realized by using the mixing plane concept proposed by Denton [14], where relative and absolute system are coupled by the transfer of circumferential averaged flow variables. Unsteady calculations are performed with an implicit dual time stepping method. The time-accurate coupling of the rotor and stator system is accomplished by the patched-cells-algorithm. In order to simulate the turbulent effects, the Wilcox k-ω-turbulence model [15] as well as an extended version of the Spalart and Allmaras model [16] is available. In the current investigation, a Kato-Launder-extension of k-ω-turbulence model is used. The CGNS (CFD General Notation System) file format is employed for the storage and retrieval of the flow field solution.

4 Type of Simulation

Four different types of numerical simulation were performed: Steady-state simulations for a one rotor and one stator blade passage, unsteady simulations for a configuration of one rotor blade passage and two stator blade passages [17] as well as steady and unsteady simulation for the entire compressor stage. The first two types were employed for homogeneous inflow conditions. For steady runs, the mixing plane concept was applied. Here, identical flow conditions in all rotor and stator blade passages are assumed. Therefore, only one rotor and stator passage has to be considered. The unsteady simulations were undertaken using the scaling technique. Here, blade numbers of rotor and stator row have to be unified. Thus, the stator geometry of configuration 1 had to be scaled down from 60 to 56 blades. Applying now a grid with one rotor and two stator passages, a one-to-one match at the interface is given. The last two simulation types refer to distorted inflow conditions (configuration 2). Here no simplification techniques are valid. Therefore, all 28 rotor and 72 stator passages have to be taken into account.

The simulations for homogeneous inflow conditions were conducted with a very high resolved computational grid in order to simulate accurately flow phenomena like the tip leakage flow, blade row interactions and shock-boundary layer interaction. The boundary layers of the blade surfaces are resolved with more than 20 elements in normal direction to the solid wall. Seven elements in radial and 155 elements in axial direction are used for the tip gap. The grid of the axial gap between rotor trailing edge and stator leading edge consists of 51 elements in axial direction, which permits an accurate simulation of the wake flow. In order to resolve the viscous sublayer, the mesh refinement near the walls was adapted to keep y+ in the order of 1.5 on the pressure and suction surface for all operating points. On the hub and in the tip clearance where wall functions were used the y+ values are in the order of 15. Steady (unsteady) simulations were performed with 8 (10) CPU's. With that amount, a minimum load of 89 % (92%) for each CPU was reached. The requested memory for those simulations was 31 (39) GB. 240 CPU hours were needed, to obtain a converged steady solution. The amount of stored data for a steady simulation was about 200 MB. For unsteady runs, the governing time dimension has to be highly discretised in order to predict accurately the transient nature of the tip leakage flow and wake flow. The ruling time dimension is represented by the blade passing frequency. Overall 512 physical time steps were used to resolve this passing frequency. A converged unsteady flow simulation of the blade passage was obtained after 8 blade passing periods and 14000 CPU-hours. In case of transient simulations, a significant higher amount of solution data was produced (approx. 60 GB per run).

For flow-simulations with distorted inflow conditions, the high grid resolution was maintained. The computational grid consists of 766 structured blocks and more than 50 million grid points. Steady and unsteady simulations are performed with 130 CPU's (94% minimum load). The requested memory is 495 GB. With that CPU-count, a performance of 3000 iterations per day is achieved. With 128 physical time steps, this leads to a runtime of approximately six months for an unsteady run. All important parameters are summarized in the following table.

Type	Gridsize	CPU	Memory (GB)	Stored Data (GB)	CPU-hours
Steady (passage)	1.5 *10^6	8	31	0.2	40
Unsteady (passage)	2.0*10^6	10	39	60	140000
Steady (entire stage)	51*10^6	130	495	70	13000
Unsteady (entire stage)	51*10^6	130	495	750	500000

5 Results of Steady and Unsteady Simulations

In the first part of the investigation, steady and unsteady simulations were performed for the homogeneous inflow conditions, using the periodic boundary technique. The numerical investigations were accomplished for the configuration of 28 rotor and 60 stator blades. This configuration was preferred, since the amount of experimental data for the second configuration with undisturbed inflow was very small. At the same time, a similar flow behaviour for both configurations can be assumed, as the same rotor geometry and the same stator profile was used.

Figure 2 and 3 display the radial profile of the total temperature at 100% rotational speed near peak efficiency (left hand side) and near stall downstream of the stator row. Steady as well as time-averaged unsteady numerical data are plotted. The temperature distributions show generally a good agreement, particularly between

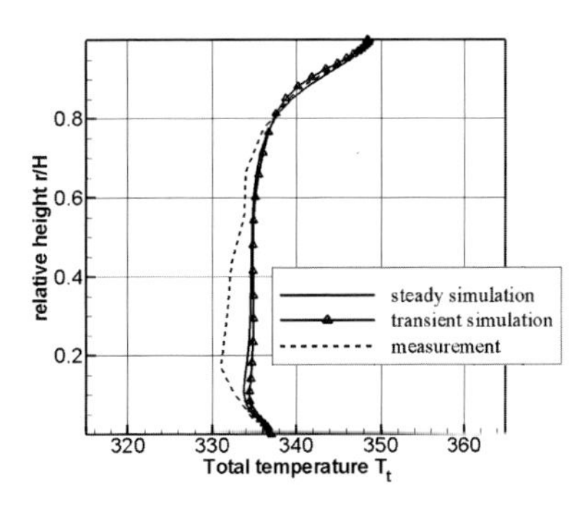

Fig. 2 Measured and predicted temperature profiles at design conditions at stator exit near peak efficiency

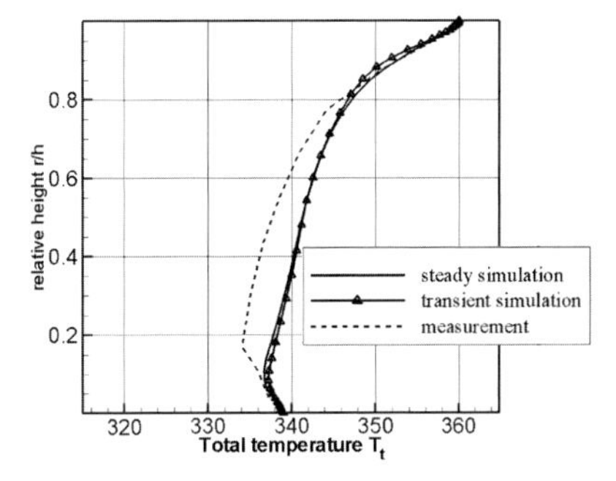

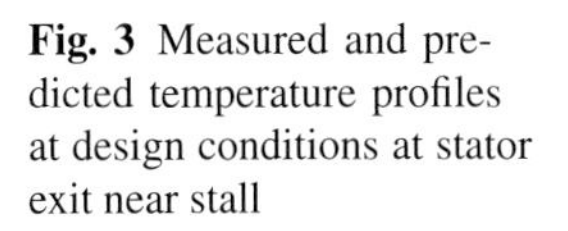

Fig. 3 Measured and predicted temperature profiles at design conditions at stator exit near stall

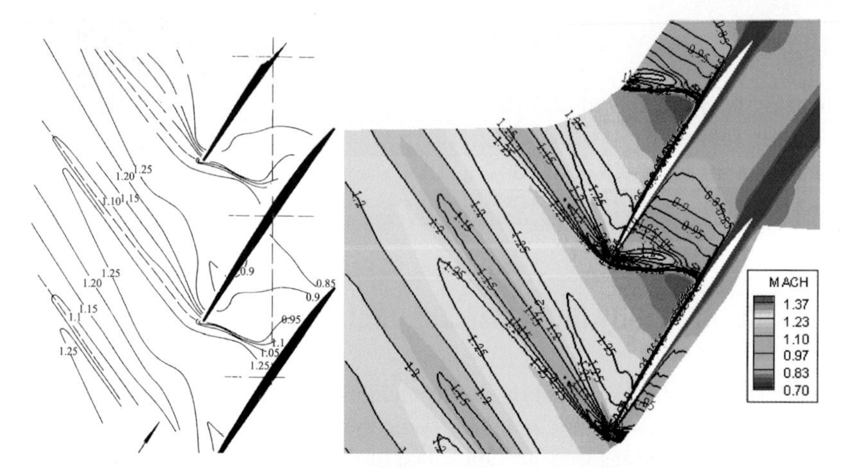

Fig. 4 Measured (left) and simulated Mach number distribution at 69% span near stall and 100% rotational speed; numerical results are derived from time-averaged transient data

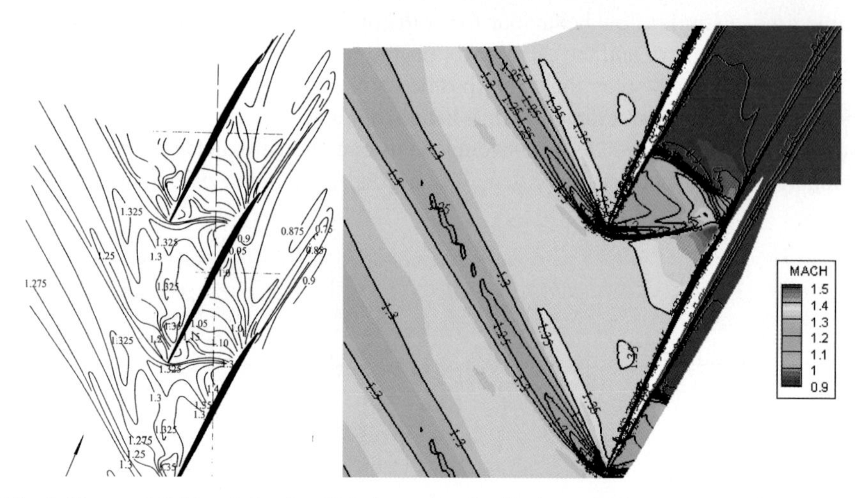

Fig. 5 Measured (left) and simulated Mach number distribution at 89% span near stall and 100% rotational speed; numerical results are derived from time-averaged transient data

80% and 100% span. The deviations generally do not exceed 1.5%. The differences between 20% and 80% are possibly due to a not activated transition model. According to the temperature plots, steady and time-averaged unsteady numerical data are nearly the same. Obviously, the unsteadiness of the rotor stator interaction plays only a small role at design speed.

The following four plots display measured and simulated Mach number distributions at 100% rotational speed. Figure 4 and 5 show the Mach number distribution for design condition, figure 6 and 7 refer to flow conditions at high aerodynamic load. For both operating points, the time-averaged numerical data and the experimental data (deduced from L2F-measurements) show a good agreement: Extent of

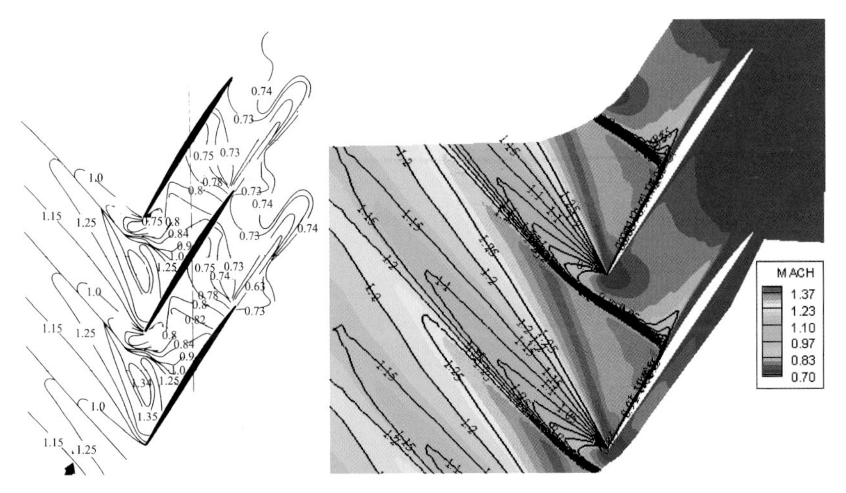

Fig. 6 Measured (left) and simulated Mach number distribution at 69% span near stall and 100% rotational speed; numerical results are derived from time-averaged transient data

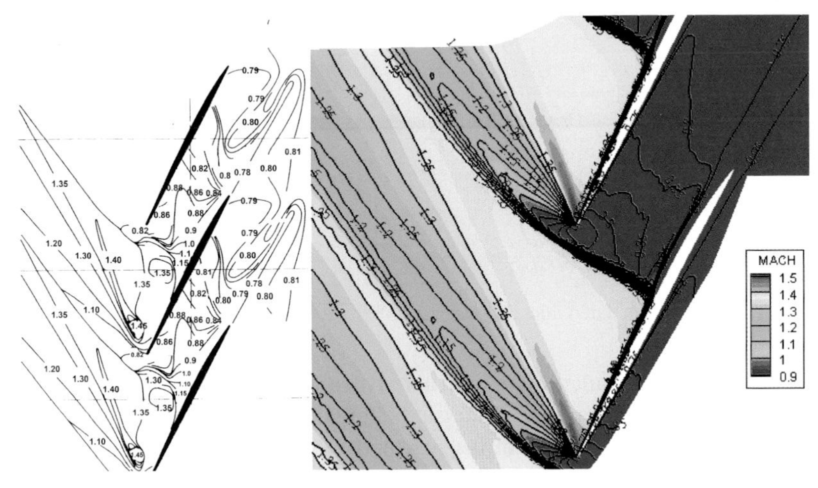

Fig. 7 Measured (left) and simulated Mach number distribution at 89% span near stall and 100% rotational speed; numerical results are derived from time-averaged transient data

the Prandtl-Meyer expansion, position and shape of the passage shock waves and Mach number levels for the whole rotor passage are well predicted. At design condition, both data exhibit an oblique passage shock wave followed by a normal shock wave. Mach numbers up to 1.45 are reached right in front of the passage shock at 89% span. Near instability onset, numerical and experimental data reveal a significant expansion on the suction side with Mach number values higher than 1.4. Furthermore, a detached normal shock has developed. Due to an intensive shock-boundary layer interaction, a remarkable growth of the friction layer can be stated downstream of the shock position. Near rotor tip, the boundary layer growth is un-

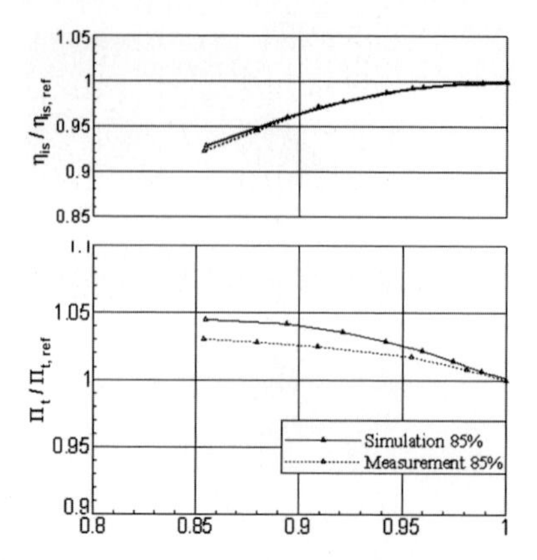

Fig. 8 Performance map for the 85% characteristic, configuration of 28 rotor and 72 stator blades(left) and front view of the rotor blade row (right)

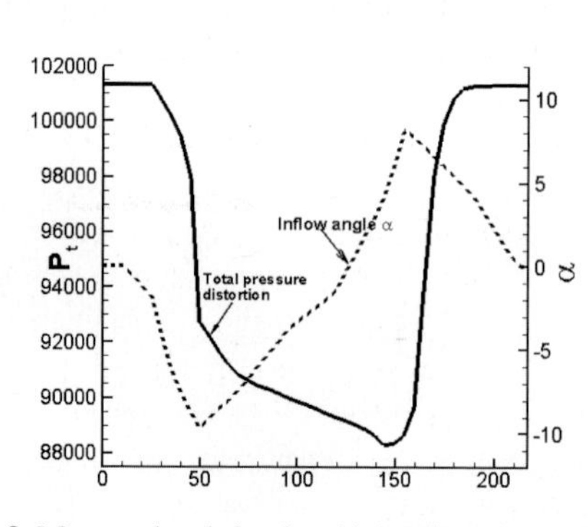

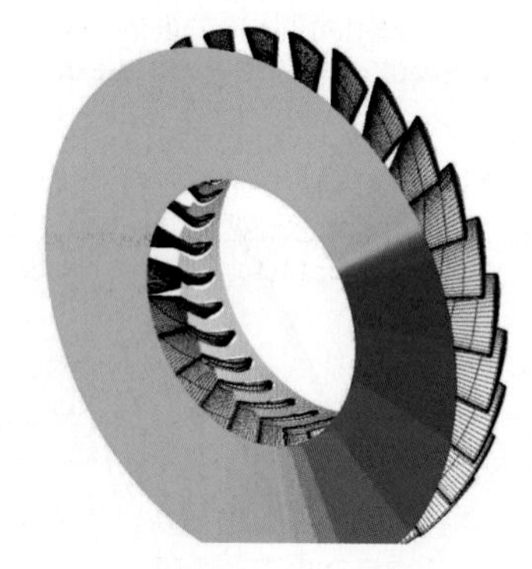

Fig. 9 Measured and simulated inlet distortion at 85% rotational speed

derestimated by the simulation for high aerodynamic load. None the less, an overall good prediction of the flow behaviour can be found in the rotor section.

With the experience gained in the first part of the project, further simulations were performed for the compressor configuration 2 with 72 instead of 60 stator blades. Since identical rotor geometry and the same stator-profile was used (NACA-65), the stage geometry could be adjusted without greater effort. At first, steady simulations for one blade passage and homogeneous inflow conditions were carried out. This was done in order to compare later the predicted and measured performance offset caused by the disturbed inflow.

Figure 8 shows the compressor performance map for 85% rotational speed deduced from numerical and experimental data. The plot contains performance values for several operating points between peak efficiency and instability onset. Efficiency and total pressure ratio are normalised with their respective values at peak efficiency. Overall, a similar behaviour with a decreasing mass flow rate can be found. As the performance data represent the only available experimental data without inlet distortions, a more detailed analysis cannot be performed. Figure 9 displays the computational grid for the flow simulation with inlet distortions. In order to pre-

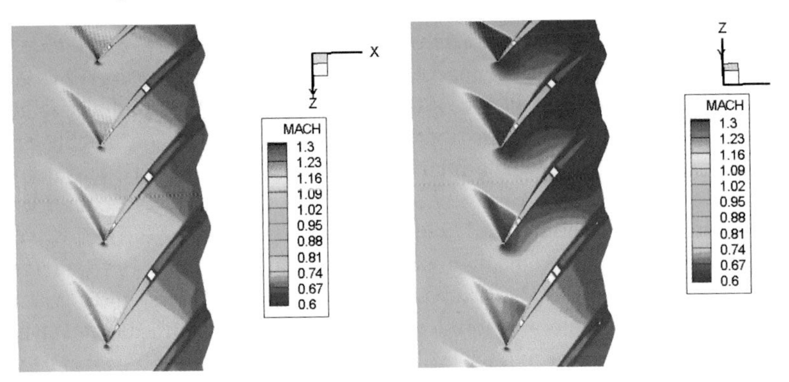

Fig. 10 Mach number distribution at rotor midspan for 85% characteristic for distorted inflow conditions

dict accurately the boundary layer development and the behaviour of the tip leakage vortex, the high resolution of the grid was maintained, which leads to more than 50 million grid points. The simulations are carried out at 85% rotational speed near peak efficiency and near instability onset. The 85% characteristic is preferred, since more experimental data are available for this speed, than for design speed. Measured and simulated inflow conditions are plotted in figure 10. At midspan, the total pressure distortion has an extent of almost 120° degree, the pressure loss is up to 13%. Moreover, a small variation of the inflow angle can be seen. The measuring plane of the inlet distortion is located 50 mm upstream of the compressor stage. As a consequence, this location also represents the inlet plane of the computational grid.

The plot on the right hand side shows the realisation of the total pressure distortion at the inflow boundary. Currently, converged steady-state results for peak efficiency are available. Figure 10 shows the Mach number distribution at mid span gained from the steady simulations. In the upper picture, the distribution outside of the distortion is displayed. Here, no significant changes relating to the blade passage simulations with homogenous inflow conditions are visible. In the picture below, the influence of the total pressure distortion is apparent: Inside of the distortion, the smaller axial velocities lead to a higher incidence angle at the rotor leading edge. Hence, an increase of the aerodynamic load combined with the amplified normal shock occurs. Unsteady simulations for disturbed inflow conditions are undertaken. Converged results are expected for November 2009.

6 Outlook

Detailed numerical simulations were performed on a transonic compressor stage for homogeneous and disturbed inflow conditions. For homogeneous inflow conditions, a comparison between predicted and measured flow variables was done at different operating conditions. The comparison of circumferentially averaged total temperature profiles indicates that the differences are of an acceptable magnitude. The differences observed are possibly according to a not activated transition model and therefore due to an underestimation of the secondary flow effects and profile losses. The simulated Mach number distributions at two different blade heights exhibit a good agreement with the experimental data deduced from L2F-measurements. Steady and transient simulations were accomplished with the RANS code TRACE, using a multi-block structured grid. 8 or 10 CPU's were used for the steady simulations and unsteady simulations, respectively. The requested memory for these simulations was below 40 GB. The steady simulations converged after 240 CPU-hours. According to a high temporal resolution, the unsteady runs took 14000 CPU-hours. After assuring a satisfying agreement between simulation and experiment for homogeneous inflow conditions, numerical simulations with total pressure distortions were carried out. Here, the assumption of periodicity between the blade passages is no more valid. Thus, the entire compressor stage had to be simulated. Overall 130 CPU's are used for these simulations. The required memory is up to 495 GB. Results of the steady-state simulations are already available. Transient simulations with distorted inflow conditions were undertaken for two operational points and are continued. First results are expected in the next period of the HLRB project, presumably in November 2009. In 2009 and 2010, a state-of-the-art compressor rig will be considered for further flow simulations with disturbed inflow conditions. The computational requirements will be very similar to the current test case.

References

1. Longley, J.P. and Greitzer, E.M. 1992: Inlet distortion effects in aircraft propulsion system integration. AGARD-LS-183
2. Cousins, W.T. 2004: History, philosophy, physics, and future directions of aircraft propulsion system inlet integration. ASME Paper GT2004-54210
3. Schmid, N.R., Leinhos, D.C., and Fottner, L. 2000: Steady performance masurements of a turbofan engine with inlet distortions containing co- and counter-rotating swirl from an intake diffuser for hypersonic flight. ASME Paper 2000-GT-11
4. Peters, T., Bürgener, T., and Fottner, L. 2001: Effects of rotating inlet distortions on a 5-stage HP compressor. ASME Paper 2001-GT-0300
5. Peters, T. and Fottner, L. 2002: Effects of co- and counter-rotating inlet distortions on a 5-stage HP-compressor. ASME Paper 2002-GT-30395
6. Reuss, N. 2005: Untersuchung von kombinierten Eintrittstotaldruckstörungen auf das instationäre Betriebsverhalten eines fünfstufigen Hochdruckverdichters. Dissertation, Universität der Bundeswehr
7. Wadia et al. 2002: Forward swept rotor studies in multistage fans with inlet distortion. ASME Paper GT-2002-30326
8. Wadia et al. 2008: High-fidelity numerical analysis of per-rev-type inlet distortion transfer in multistage fans - Part I: simulations with selected blade rows. ASME Paper GT-2008-50812
9. Klenner, J., Becker, K.; Cross, M., and Kroll, N. 2007: Future simulation concept. 1st European CEAS Conference, Berlin
10. Yang, H., Nürnberger, D., and Kersken, H.-P., 2006: Toward excellence in turbomachinery computational fluid dynamics: A hybrid structured-unstructured reynolds-averaged navier-stokes solver. ASME Transactions, Journal of Turbomachinery, pp. 390-402, Vol. 128

11. Dunker, R.J. and Hungenberg, H.G. 1980: Transonic axial compressor using laser anemometry and unsteady pressure measurements. AIAA Journal, Vol 18, No. 8
12. Dunker, R.J., Strinning, P., and Weyer, H.B. 1978. Experimental study of the flow field within a transonic axial compressor rotor by laser velocimetry and comparison with Through-Flow calculations, J. of Eng. For Power, Trans.ASME, Vol. 100, No. 2
13. Lecht, M., 1983: Beitrag zum Verhalten von Axialverdichterstufen bei stationärer Störung der Zuströmung. Forschungsbericht DFVLR Cologne
14. Denton, J.D., 1990: The calculation of three-dimensional viscous flows through multistage turbomachines, ASME Paper No. 90-GT-19, 1990
15. Wilcox, D.C., 1998: Turbulence modeling for CFD, 1998, Second Edition, DCW Industries, Anaheim
16. Spalart, P. and Allmaras, S., 1992: A one-equation turbulence model for aerodynamic flows, Technical Report AIAA-92-0439, 1992
17. Iseler, J. and Niehuis, R. 2008: Steady and unsteady flow-simulation of an axial transonic compressor stage, AIAA-2008-0083, 2008

A Parallel CFD Solver Using the Discontinuous Galerkin Approach

Christian Lübon, Manuel Keßler, and Siegfried Wagner

Abstract In the present paper a high-order Discontinuous Galerkin method is presented for the numerical simulation of the turbulent flow around complex geometries using unstructured grids. In order to close the Reynolds-averaged Navier-Stokes (RANS) system we use the Spalart-Allmaras turbulence model, the Wilcox K-ω turbulence model or a Detached Eddy Simulation technique. The paper includes some details of the code implementation. The excellent parallelisation characteristics of the scheme are demonstrated, achieved by hiding communication latency behind computation. Some results, like flows over a flat plate and around a sphere, which could not be predicted with an Unsteady Reynolds averaged Navier-Stokes calculation, are calculated with high accuracy and compared with theory and experiments.

1 Introduction

The Discontinuous Galerkin (DG) method combines ideas from the finite element and finite volume methods, the physics of wave propagation expressed by Riemann problems and the accuracy obtained by high order polynomial approxima tions within elements. It was originally developed for hyperbolic conservation laws in 2D flow [3] and 3D flow [16] using the Euler equations.

However, in real life applications the flow is in most cases turbulent and three dimensional. The original development of DG methods was devoted to the Euler equations that contain only derivatives of first order. The break through for solving the Navier-Stokes equations with derivatives of second order was presented by Bassi and Rebay [2]. Still another big step was to be done, namely to extend the algorithms

Christian Lübon · Manuel Keßler · Siegfried Wagner
Universität Stuttgart, Institut für Aerodynamik und Gasdynamik, Pfaffenwaldring 21, 70569 Stuttgart, Germany
e-mail: luebon@iag.uni-stuttgart.de

S. Wagner et al. (eds.), *High Performance Computing in Science and Engineering, Garching/Munich 2009*, DOI 10.1007/978-3-642-13872-0_17,

for the handling of turbulent flows. The RANS equations had to be solved, and thus, the algorithms had to be extended to include turbulence models [4]. Based on our previous experience [7–11] we have extended the algorithms to three-dimensional turbulent flow. This was done by solving the URANS equations and by implementing a Detached Eddy model.

2 Discontinuous Galerkin Schemes

2.1 Basic Equations

The Navier-Stokes equations can be written in the following compact differential form

$$\frac{\partial U}{\partial t} + \nabla \cdot \mathscr{F}_i(U) - \nabla \cdot \mathscr{F}_v(U, \nabla U) = 0 \tag{1}$$

Here U is the vector of conservative variables, $\mathscr{F}_i$ and $\mathscr{F}_v$ are the convective and diffusive flux functions, respectively. The next step is to handle high-order derivatives. According to Bassi and Rebay [2] we first reformulate this equation as a first-order system introducing the gradient of the solution ∇U as a new additional independent unknown Θ and get a second equation

$$\nabla U - \Theta = 0 \tag{2}$$

We now apply the DG approach resulting in the equations for an element E

$$\int_E v_k \Theta \, dE - \oint_{\partial E} v_k \cdot U_h \cdot \mathbf{n} \, d\sigma + \int_E \nabla v_k \cdot U_h \, dE = 0 \tag{3}$$

$$\int_E \left(v_k \frac{\partial U_h}{\partial t} \right) dE + \oint_{\partial E} v_k \cdot \mathscr{F}_i \cdot \mathbf{n} \, d\sigma - \int_E \nabla v_k \cdot \mathscr{F}_i \, dE$$
$$- \oint_{\partial E} v_k \cdot \mathscr{F}_v \cdot \mathbf{n} \, d\sigma + \int_E \nabla v_k \cdot \mathscr{F}_v \, dE = 0 \tag{4}$$

where $U_h = U(x,t) = \sum_{k=1}^{n} U_k(t) v_k(x)$ is the approximation for the numerical solution U_k, and the n shape functions v_k are a basis for the polynomial space P^k.

Our scheme can be advanced explicitly as well as implicitly in time. The explicit time integration, used in this paper, is performed with one-step Runge-Kutta type schemes of first to fourth order accuracy.

2.2 Turbulence Modeling

In the present URANS cases the one equation Spalart-Allmaras (SA) model is used [13]. The following equation shows the distribution of the transport variable $\widetilde{\nu}$

$$\frac{\partial \rho \tilde{\nu}}{\partial t} + \frac{\partial \rho \tilde{\nu} u_i}{\partial x_i} = \frac{1}{\sigma} \left[\frac{\partial}{\partial x_i} \left((\mu + \rho \tilde{\nu}) \frac{\partial \tilde{\nu}}{\partial x_i} \right) + \rho c_{b2} \frac{\partial \tilde{\nu}}{\partial x_i} \frac{\partial \tilde{\nu}}{\partial x_i} \right]$$

$$+ c_{b1}(1 - f_{t2}) \rho \tilde{S} \tilde{\nu} - \left(c_{w1} f_w - \frac{c_{b1}}{\kappa^2} f_{t2} \right) \frac{1}{\rho} \left(\frac{\rho \tilde{\nu}}{d} \right)^2 + \rho f_{t1} \Delta U^2 \tag{5}$$

with d as the distance to the closest wall, which represents the turbulent length scale, the vorticity $\tilde{S}$ and some calibration constants c_i and calibration functions f_i. The main flow equations and the turbulence model equations are solved in a fully coupled manner.

2.3 Detached Eddy Simulation

The Detached Eddy Simulation (DES) in this study is based on a modification of the S-A model, such that it reduces to RANS close to solid walls and to LES away from the wall [14]. In the S-A turbulence model, the turbulent length scale, the distance to the nearest wall d, is replaced by l_{DES}, defined as $l_{DES} = min\left(d, C_{DES}^{SA} \Delta\right)$ where Δ in this case is the largest cell spacing and $C_{DES}^{SA} = 0.65$ is an additional model constant, calibrated for homogeneous turbulence. Near solid boundaries Δ is larger than d and we get a model that acts as a S-A model. Far from walls $d >> \Delta$ a balance between the production and destruction term in the model equation shows that $\tilde{\nu} \sim S\Delta^2$ like a one-equation subgrid-scale model. Due to the LES part, a high-order spatial discretization scheme like the DG scheme with low dissipation is necessary to resolve the large eddies correctly, while the smaller ones are modeled.

3 Computational Aspects

3.1 Parallelisation and Object-Oriented Design

The method is completely coded in C++ with an object-oriented design, which among other things has been constructed for massively parallel simulations by using the MPI distributed memory paradigm from begin on. We carry out domain decomposition, this means that the grid is decomposed into subdomains (zones) and every processor is responsible for at least one subdomain. Clever splitting of the domain into a user specified number of zones is a complex issue and thus we are using the public domain software METIS [6] for the domain decomposition. METIS is a set of programs for partitioning graphs, partitioning finite element meshes, and for producing fill reducing ordering for sparse matrices. The algorithms implemented in METIS are based on multilevel graph partitioning. An example of a tetrahedral grid, the flow around a sphere, is shown in Figure 1.

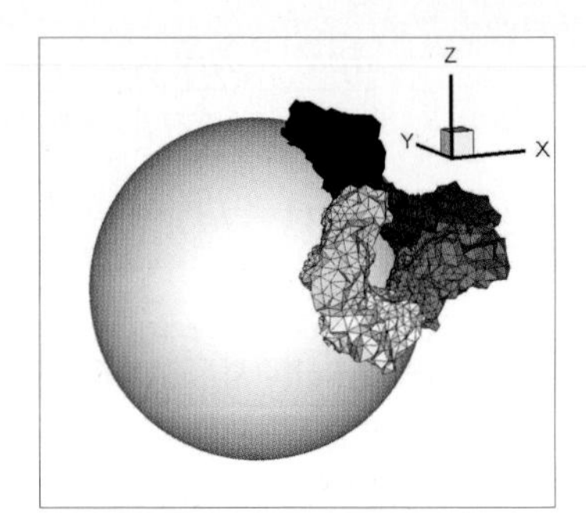

Fig. 1 Sphere grid with five computational domains

As most of the computation happens inside of the elements, the parallelization properties are very favorable. The only information transfer is on face boundaries and thus comparatively small, and no ghost cells, like in the Finite Volume method (FV), are required. The communication latency can be minimized, by executing zone local work during the communication. In contrast to the FV method this local work is not only the evaluation of the inner face integrals but also the evaluation of the element integrals.

3.2 Speedup and Performance

The objective of our work is to use unstructured Detached Eddy Simulations (DES) based on the DG discretization on complex geometries, like airplanes or helicopters. We use this discretization technique due to the nearly linear speedup for parallel computations, which cannot be reached with a high-order unstructured finite volume scheme because of the costly reconstruction schemes. The speedup is dependent on the spatial discretization order of the Discontinuous Galerkin scheme. If we use a high-order scheme we do a lot of local calculations with only small information transfer on the zone boundaries. In contrast, if we use a first order scheme, which is similar to a first order FV method, we get a lot of zone interactions and in the worst case the speedup efficiency drops to zero, see [8].

In order to assess the parallelization quality we used a tetrahedral mesh with $\approx 1 \cdot 10^6$ cells and split this mesh into up to 2048 regions. We use a high-order scheme, a fourth order method with 20 degrees of freedom per cell, and calculate the later shown sphere configuration. This means that we get $2 \cdot 10^7$ degrees of freedom for the entire mesh and $1 \cdot 10^4 - 6 \cdot 10^5$ degrees of freedom for each computational domain. We perform our computation on massively parallel systems, our 150 core instituts cluster Prandtl with Intel Woodcrest processors and 12 GFlops peak performance per core, the Nehalem cluster on the HLRS with 11.2 GFlops peak performance per core in Stuttgart, and finally the SGI Altix with Itanium II processors and 6.4 GFlops peak performance per core of HLRB II on the Leibniz Supercomputing Centre in Munich.

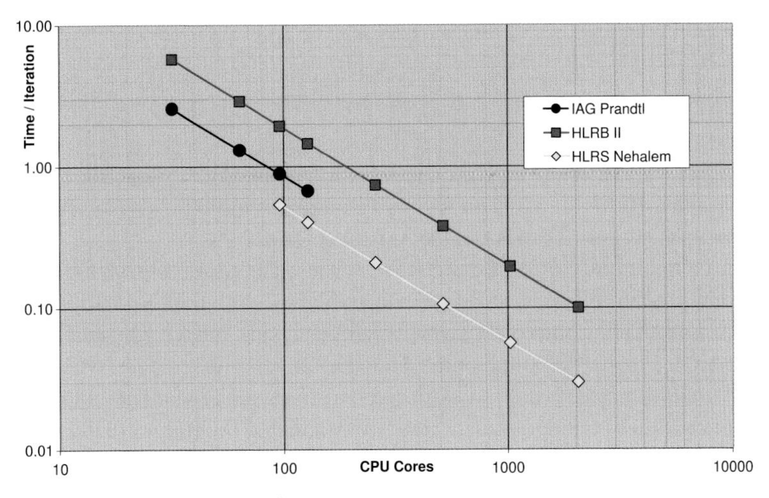

Fig. 2 Computational time per iteration

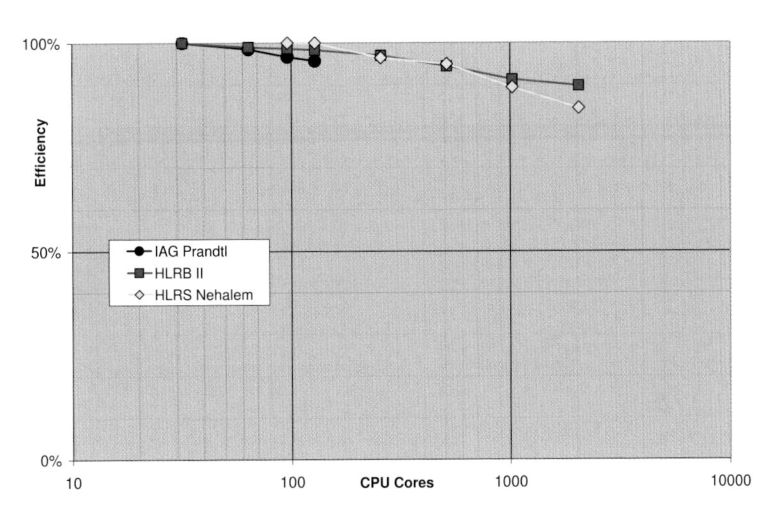

Fig. 3 Computational efficiency

For the medium calculation with 96 cores we get typically 550 MFlops on the HLRB II, this means $\approx 9\%$ of the peak performance. On the other platforms we get $\approx 10\%$ of the peak performance using the Prandtl cluster and $\approx 17\%$ of the peak performance on the new Intel Nehalem cluster. The computational time per iteration decreases nearly linear with the number of used cores (Figure 2), so we are able to reach a parallel efficiency up to 85% (Figure 3) and 3.8 TFlops on the Nehalem system.

4 Results

4.1 Flat Plate Flow

For a detailed validation of the laminar 3D implementation we calculated the flow over a flat plate ($\mathrm{Re} = 1 \cdot 10^5$ and $\mathrm{Ma} = 0.3$) and compared the velocity and skin friction profiles (Figure 4) with the theoretical solution.

For the calculations we use really coarse meshes. The results shown are generated on a mesh with 256 cells in total and 8 cells in normal direction. Looking at the first order P^0 solution, it is completely away from the physics, but we get good results for elements higher than P^2. Note that the boundary layer is discretized with only two cells in normal direction but up to 40 degrees of freedom dependent on the order.

For the validation of the turbulent flow we calculated the fully turbulent flat plate flow with a Reynolds number of $\mathrm{Re} = 3 \cdot 10^6$ and a Mach number of $\mathrm{Ma} = 0.3$ using the RANS approach based on the SA model on several grids with different sizes. The results shown in this paper (Figure 5) are calculated on an extremely coarse

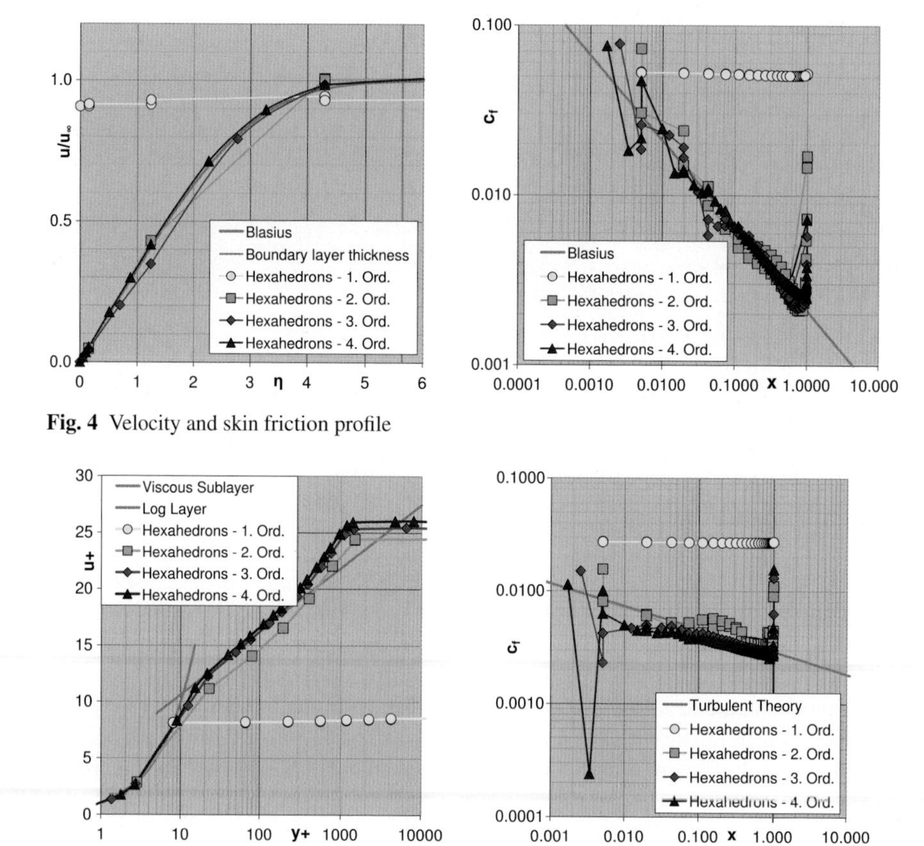

Fig. 4 Velocity and skin friction profile

Fig. 5 The law of the wall and skin friction profile

mesh like in the laminar case with a size of 32 cells in streamwise direction, 1 cell in spanwise direction and only 8 cells in normal direction. The near-wall resolution for elements higher than P^2 is $y^+ \approx 5$.

Due to the finite element approach in the DG cells we found a good agreement between the computed and theoretical normalized velocity profiles and are able to reach the law of the wall with this coarse grid as well as the skin friction profile compared with the theory [12]. The third-order hexahedral solution for example works with 10 degrees of freedom per cell which is comparable to 4 degrees of freedom in normal direction. For our mesh with only 8 cells in normal direction this means that we use 32 degrees of freedom in this direction.

4.2 Detached Eddy Investigation of the Flow Past a Sphere

The flow past a sphere belongs to the class of separated flows for which the location of flow detachment is not fixed by the geometry and it is not possible to use a classical URANS approach, because of the missing spectral gap between modeled frequencies from the turbulence model and transient frequencies calculated by the simulation. Because of these features it is a good application for DES.

The sphere is known for its drag crisis, which reflects the differences in separation between laminar and turbulent boundary layers. For our investigations we choose a sub-critical case with laminar separation of the boundary layer at $Re = 2 \cdot 10^4$ and a super-critical case with a turbulent boundary layer at $Re = 1.14 \cdot 10^6$. The difficult requirement, that the turbulence model would be suppressed in the laminar regions of the flow, is achieved by using the "tripless" approach of Travin et al. [15], so the transition takes place in the wake.

The DES prediction of the separation point at an azimuthal angle measured from the forward stagnation point of $\Theta \approx 81°$ for $Re = 2 \cdot 10^4$ agrees well with the experiment of Achenbach [1], as well as the separation point for the turbulent calculations of $\Theta \approx 115°$ (Figure 6).

A comparison of the pressure distribution with the simulations of Constantinescu et al. [5] and the experiments of Achenbach [1] is given in Figure 7 for both Reynolds numbers.

The mean values are calculated from averaging over $12 \cdot D/U_\infty$ units after an initialization time of $\approx 40 \cdot D/U_\infty$. The prediction of the sub-critical flow at $Re = 2 \cdot 10^4$ is in agreement with the simulation of Constantinescu at $Re = 1 \cdot 10^4$ and the measurements at $\mathrm{Re} = 1.62 \cdot 10^4$. The value of c_p is a little bit over predicted.

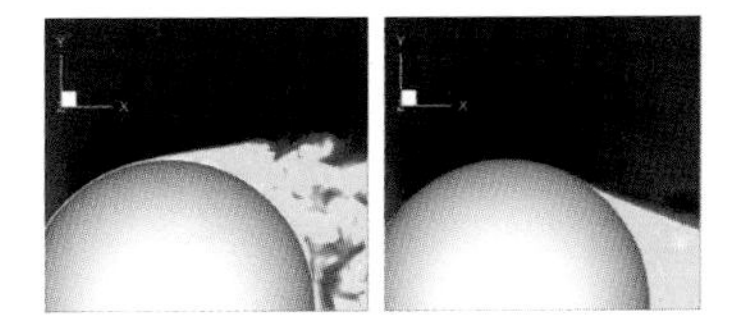

Fig. 6 Instantaneous velocity contours for laminar and turbulent separation

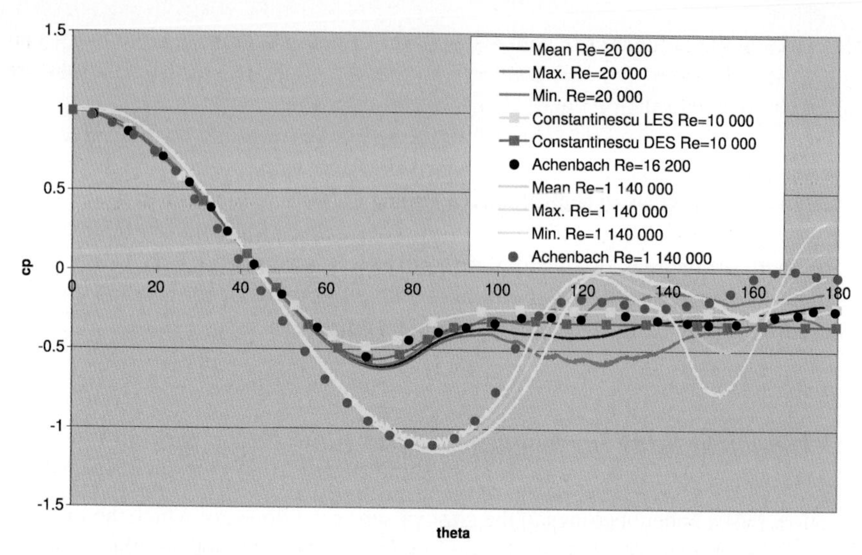

Fig. 7 Pressure distribution

A reason for this could be the slightly different Reynolds number. We found the same behavior for the super-critical case.

For our computations we used several grids and several discretization orders. The results shown are calculated on an unstructured grid with $\approx 1 \cdot 10^6$ cells and an order of four. Because of the polynomial ansatz functions, which are well-suited to discretize the viscous sublayer, we use a near-wall resolution for elements higher than P^2 of $y^+ \approx 3$ in the fully turbulent case. For time integration we use a fourth-order Runge-Kutta scheme with a time step from $10^{-4} - 10^{-6} \cdot D/U_\infty$.

5 Conclusion

We combined advanced numerical methods in an unstructured code for complex geometries with new promising methods in turbulence modeling. It has been shown that it is possible to do a Detached Eddy Simulation with an unstructured code and reach a high parallel efficiency.

It is necessary to increase our convergence rates by including multiple speedup schemes, like a matrix free implicit algorithm and an h/p multigrid scheme. This will be the focus of future research.

References

1. E. Achenbach. Experiments on the flow past spheres at very high Reynolds numbers. *Journal of Fluid Mechanics*, 54:565–575, 1972.

2. F. Bassi and S. Rebay. A High-Order Accurate Discontinuous Finite Element Method for the Numerical Solution of the Compressible Navier-Stokes Equations. *Journal of Computational Physics*, 131:267–279, 1997.
3. F. Bassi and S. Rebay. High-Order Accurate Discontinuous Finite Element Solution of the 2D Euler Equations. *Journal of Computational Physics*, 138:251–285, 1997.
4. F. Bassi and S. Rebay. A high order discontinuous Galerkin method for compressible turbulent flows. In B. Cockburn, G. E. Karniadakis, and C.-W. Shu, editors, *Discontinuous Galerkin Methods*, pages 77–88. Springer, 2000.
5. G. S. Constantinescu and K. D. Squires. LES and DES Investigations of Turbulent Flow over a Sphere. *38th AIAA Aerospace Sciences Meeting and Exhibit, Reno*, AIAA–Paper 2000-0540, 2000.
6. G. Karypis and V. Kumar. A fast and high quality multilevel scheme for partitioning irregular graphs. *SIAM J. Sci. Comput.*, 20:359–392, 1998.
7. C. Lübon, M. Keßler, and S. Wagner. *Turbulence Modeling and Detached Eddy Simulation with a High-Order Unstructured Discontinuous Galerkin Code*. Notes on Numerical Fluid Mechanics and Multidisciplinary Design. Springer. to appear.
8. C. Lübon, M. Keßler, and S. Wagner. *A Parallel CFD Solver Using the Discontinuous Galerkin Approach*, pages 291–302. High Perfomace Computing in Science and Engineering. Springer, 2008.
9. C. Lübon, M. Keßler, S. Wagner, and E. Krämer. High-order boundary discretization for discontinuous Galerkin codes. In *AIAA–Paper 2006-2822*. 25nd AIAA Applied Aerodynamics Conference, San Francisco, 2006.
10. C. Lübon, M. Keßler, S. Wagner, and E. Krämer. *Detached Eddy Simulation of Separated Flow on a High-Lift Device and Noise Propagation*, volume 97 of *Notes on Numerical Fluid Mechanics and Multidisciplinary Design*, pages 192–201. Springer, 2007.
11. C. Lübon and S. Wagner. *Three-Dimensional Discontinuous Galerkin Codes to Simulate Viscous Flow by Spatial Discretization of High Order and Curved Elements on Unstructured Grids*, volume 96 of *Notes on Numerical Fluid Mechanics and Multidisciplinary Design*, pages 145–153. Springer, 2007.
12. H. Schlichting. *Grenzschicht-Theorie*. Braun, Karlsruhe, 1982.
13. P. R. Spalart and S. R. Allmaras. A one-equation turbulence model for aerodynamic flows. *La Recherche Aérospatiale*, 1:5–21, 1994.
14. P. R. Spalart, W-H. Jou, M. Strelets, and S.R. Allmaras. Comments on the feasibility of LES for wings, and on a hybrid RANS/LES approach. In *Advances in DNS/LES*, 1997.
15. A. Travin, M. Shur, M. Strelets, and P. Spalart. Detached-eddy simulation past a circular cylinder. *Flow, Turbulence and Combustion*, 63:293–313, 2000.
16. J.J.W. van der Vegt and H. van der Ven. Space-time discontinuous Galerkin finite element method with dynamic grid motion for inviscid compressible flows I. General formulation. *Journal of Computational Physics*, 217:589–611, 2002.

Characterization of the Aeroacoustic Properties of the SOFIA Cavity and its Passive Control

Sven Schmid, Marius Kütemeyer, Thorsten Lutz, and Ewald Krämer

Abstract The American and the German Aerospace centers NASA and DLR are working together to build and operate the Stratospheric Observatory For Infrared Astronomy - SOFIA, a 2.5 m reflecting telescope housed in an open cavity on board of a Boeing 747 aircraft. The observatory operates in the stratosphere at an altitude above 13 km to observe astronomical objects in the infrared region of the electromagnetic spectrum. The flow over the open telescope port during the observation presents some challenging aerodynamic and aeroacoustic problems. Pressure fluctuations inside the cavity excite structural vibrations and deteriorate the image stability. NASA successfully optimized the shape of the cavity aperture by means of extensive wind-tunnel studies to mitigate the flow unsteadiness in the cavity. The simulations performed within the present project focus on different aeroacoustic aspects of the SOFIA cavity. Several URANS (Unsteady Reynolds Averaged Navier Stokes) simulations were carried out to study the impact of the cavity door position during opening and closing on the unsteady flow. The impact of the misalignment between the SOFIA telescope assembly and the aperture assembly was investigated with separately conducted Detached Eddy Simulations (DES). Moreover the remaining potential to improve the aeroacoustic characteristics by optimizing the aperture aft ramp geometry and installing a porous fence upstream of the cavity was analyzed with URANS and DES simulations as well.

Sven Schmid · Marius Kütemeyer
SOFIA Science Center, NASA Ames Research Center, Mail Stop N211-3, Moffett Field, CA 94035, USA
e-mail: Schmid@dsi.uni-stuttgart.de
e-mail: mkuetemyer@sofia.usra.edu

Thorsten Lutz · Ewald Krämer
Institut für Aerodynamik und Gasdynamik, Pfaffenwaldring 21, 70569 Stuttgart, Germany
e-mail: Lutz@iag.uni-stuttgart.de
e-mail: Kraemer@iag.uni-stuttgart.de

S. Wagner et al. (eds.), *High Performance Computing in Science and Engineering, Garching/Munich 2009*, DOI 10.1007/978-3-642-13872-0_18,

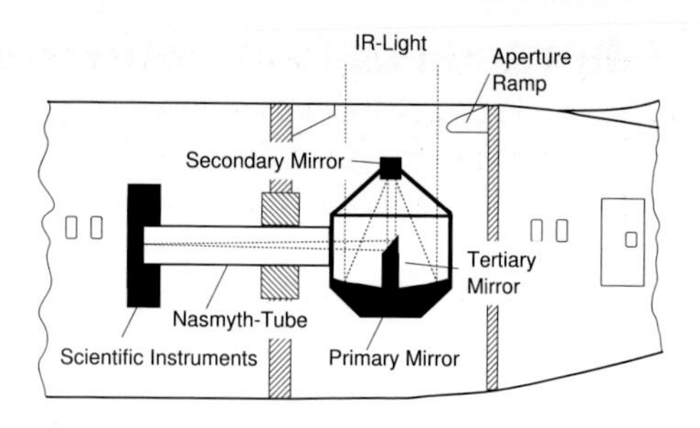

Fig. 1 Sectional view of the SOFIA telescope cavity

1 Introduction

SOFIA is a 2.5 m reflecting Cassegrain telescope with Nasmyth focus, housed in a Boeing 747-SP aircraft, see Fig. 1 [17]. Most of the infrared energy emitted from celestial objects is absorbed by water in the atmosphere, a phenomenon that limits ground-based astronomy. SOFIA will solve that problem by flying in the Stratosphere between 12 km and 14 km altitude, above 95% of atmospheric water. For the observation during flight a door in the rear part of the aircraft's fuselage will be opened to expose the telescope to the starry sky. The telescope assembly can be tilted around the elevation axis within the range from 20° to 60°. A movable aperture assembly that permanently follows the movements of the telescope keeps the size of the actual cavity opening at a minimum and stabilizes the free shear layer spanning the aperture [6]. Flow over open ports is in general characterized by instabilities and unsteady flow phenomena [8]. At high Mach numbers, sound pressure levels can become exceedingly high [3], leading to unwanted structural vibrations or even fatigue problems. In case of SOFIA, any perturbation of the structure deflects the telescope assembly from its target position and hence deteriorates the pointing stability.

The aeroacoustic properties of the SOFIA telescope cavity has been characterized with DES and URANS in several previous studies by Schmid et al. [9, 10, 12]. The comparison of URANS and DES results to experimental data of a 7% wind-tunnel model that was tested in the 14ft by 14ft transonic wind tunnel at the NASA Ames Research Center showed an excellent agreement. Both the amplitudes as well as the frequency content of the unsteady pressure fluctuations on the surface of the telescope assembly were well predicted by the CFD simulation.

The same authors demonstrated the effectiveness of the present flow control approach comparing the results to a generic cavity configuration without aperture ramp on the downstream wall [13]. Different means for passive flow- and acoustic control, like vortex generators, a porous fence upstream of the cavity, a modified cavity geometry and baffle plates have been investigated with CFD- and acoustic simulations by Schmid et al. as well [11, 14, 15]. The objective of the CFD-investigations performed within the framework of the present project is the support of a profound comprehension of the prevailing aeroacoustic phenomena and the analysis of possible approaches for flow control in order to improve the telescope's performance in long term.

The shear layer spanning the opening of the cavity amplifies convected flow disturbances that are scattered into acoustic waves at the cavity rear wall. Acoustic waves propagate upstream inside and outside the cavity and excite further shear-layer disturbances closing a feedback loop. Frequencies with a phase lag of a multiple of 2π are being amplified in particular, yielding a selection of discrete modes. Rossiter [7] found that the frequencies can be estimated by the semi-empirical equation

$$F = \left(\frac{U}{L}\right) \cdot \frac{m-\gamma}{\frac{1}{K}+Ma}, \qquad m = 1,2,3... \tag{1}$$

where f is the frequency of the mode m, L is the reference length and Ma the Mach number. γ and K are empirical constants, γ represents the phase delay of disturbances that are scattered at the rear wall, K is the average convection speed (relative to free-stream velocity) of disturbances in the shear layer. The existence and the magnitude of these so called Rossiter modes depend basically on the stability characteristics of the shear layer that evolves from the boundary layer upstream of the cavity [2, 8]. The relevant boundary layer parameter is the momentum thickness δ_2. Small values in general lead to higher shear layer disturbance amplification and hence to higher pressure fluctuations inside the cavity.

2 Numerical Methods

In the present study DES computations are performed with the Finite-Volume CFD-code TAU that was developed by the Institute of Aerodynamics and Flow Technology of DLR [5]. The code bases on a dual-grid cell-vertex formulation and solves the unsteady, compressible, three-dimensional Reynolds-averaged Navier-Stokes equations on unstructured or hybrid grids.

The simulations presented here are performed applying the central-difference algorithm with second- and fourth-order numerical dissipation according to Jameson [1]. Time accurate simulations are carried out by a dual time-stepping scheme (DTS) that allows for convergence acceleration techniques like multigrid and residual smoothing. In the present RANS simulations, the Menter SST turbulence model was used. The SA-DES model accounts for subgrid turbulence in the DES mode. In order to guarantee that the LES-region is far enough from viscous walls to avoid problems resulting from grid induced separation, the DDES-approach was applied [16].

Hybrid grids around the complete SOFIA aircraft were created with the commercial Software GRIDGEN by Pointwise. The number of cells varies between $23.0E+06$ and $50.0E+06$ cells for each grid in total. Prisms were extruded on viscous walls to resolve boundary layers, tetrahedra fill up the rest of the computational domain. In order to resolve boundary layers appropriately, a y^+ value of 1 was aimed at the first cell vertex above the surfaces. The prism stacks contain between 28 and 38 prism layers in wall-normal direction. The boundary layer is resolved by 25-35 cells in average. For the DES simulations, a structured block containing $22.5E+06$ nearly isotropic hexahedra was placed in the region of the cavity shear-layer as this zone's correct representation impacts the whole simulation quality crucially. The structured approach guarantees a minimum of numerical dissipation that is necessary to resolve the small scale turbulent structures in the LES mode.

The present computations were carried out on the Linux-Cluster SGI Altix 4700 (HLRB II). For parallelization, the domain was equally decomposed into 1020 subdomains to utilize 1020 processor cores in parallel. A typical DES-computation consisted of 16000 physical timesteps with 40 inner iterations per step. The average CPU time consumption for one single inner iteration (3v-multigrid cycle) is about $0.9\,\mu$s per cell. The performance of the TAU code on the HLRB II as well as the scaling behavior has already been studied prior to this investigation and documented in the previous paper by Schmid et al. [11].

3 Impact of Cavity Door Position on the Aeroacoustic Characteristics

The SOFIA telescope port will be closed during the aircraft's takeoff and landing. The partial covering of the aperture by the door changes the flow and the acoustics of the cavity significantly. The a-priori prediction of the impact of the door opening degree on the fluctuation levels inside the cavity is not possible because several different physical effects counteract and partly compensate each other. Closing the door decreases the surface between the interior and the exterior where kinetic energy of the external flow is fed into the cavity to drive and sustain the unsteady pressure fluctuations. Regarding this effect, decreasing the opening leads to a reduction of the overall sound pressure level. But as decreasing the door opening alters the cavity acoustical properties, especially the amount of dissipation associated with acoustic radiation through the opening, this effect has the opposite consequence on the sound pressure level. This phenomenon was proven with a linear acoustic cavity model, based on the solution of the homogeneous Helmholtz equation [4]. Furthermore when the door is partially closed, a part of the aperture ramp is covered and the device will be in off-design conditions what may reduce the stabilizing effect on the shear layer and increase the flow induced pressure fluctuations in the cavity as a consequence.

The impact of the door position on the cavity aeroacoustics was investigated by URANS simulations in the present study. Three different configurations with 10%, 50% and 70% partial open aperture were considered herein and compared to the configuration with a fully opened aperture [10, 12]. According to the URANS data, the main flow inside the cavity is altered significantly due to the partial aperture opening. Figure 2 shows instantaneous snapshots of the Mach number distribution on the plane perpendicular to the direction of flight through the center of the pri-

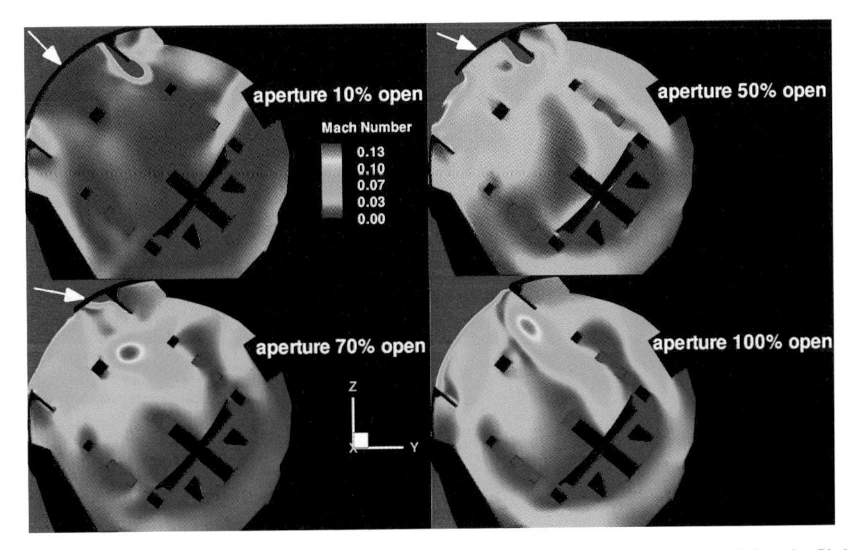

Fig. 2 Sectional view of Mach number along the primary mirror symmetry plane (view in flight direction); cavity door (marked by white arrows) opens from left to right

mary mirror for the different configurations. The 10%-configuration shows the lowest Mach number average, revealing that the amount of kinetic energy entrained from the external flow is lower compared to the other configurations, what was already stated as an alleviating factor earlier in this document. The simulation data also reveals that the partial open door configurations with aperture openings of 50% and the 10% show a prominent swirl flow inside the cavity as can be seen by the region with high Mach number at the upper aperture sidewall. The present simulations of the partially opened SOFIA telescope port revealed a distinct impact of the door position on the amplitudes of the unsteady pressure fluctuations. The average SPL value over the telescope assembly surface is 126 dB for the full open cavity configuration. The 10%, 50% and the 70% open cavities yield 123 dB, 132 dB and 128 dB respectively.

4 Shear-Layer Control by Means of a Porous Fence

In a separate analysis the capability of a porous fence as a means of passive flow control has been studied. Detached Eddy Simulations show that the shear-layer fence upstream of the cavity has a significant impact on the steady and unsteady flow features in and around the telescope port. Pressure fluctuations are mitigated by three main effects. First, the fence directs the flow away from the aperture. The interaction of the shear-layer disturbances with the aperture ramp is weakened, less acoustic energy is scattered back into the cavity that perpetuates the disturbance feedback. Second, the fence thickens the shear layer and increases its defect. Thicker shear

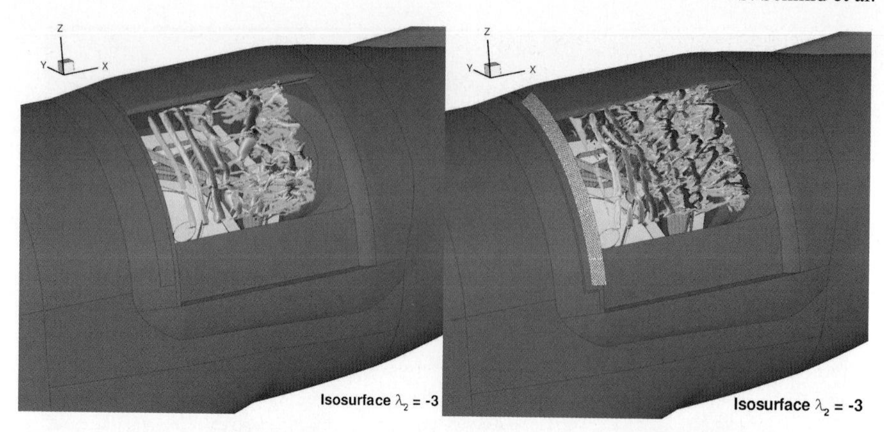

Fig. 3 Turbulent shear-layer structures (baseline configuration left, fence configuration right)

layers are generally characterized by more benign stability characteristics. Third, the fence decreases the shear layer's spatial coherence by breaking the large scale Kelvin Helmholtz vortices into smaller and more random organized turbulent structures as can be seen in Fig. 3. As the pictures show the transversal structures are ruptured by the fence, promoting bursting and creating chaotic three-dimensional vortices already at an earlier stage further upstream. The shear-layer fence yields a significant reduction of the unsteady loads on the telescope assembly [14].

5 Impact of TA/AA Misalignment on Aeroacoustics

The elevation angle (EL) of the telescope can be changed during flight within a range between 20° and 60°. Furthermore, in order to compensate the movement of the aircraft caused by air turbulence, the telescope can be tilted around the cross elevation axis (CE) and the line of sight axis (LOS) by $\pm 3°$. During normal operation the aperture assembly (AA) of the cavity will be aligned with the telescope assembly (TA). However, in some of the test flight points there will be a misalignment between the aperture and the telescope. The impact of this misalignment has been investigated with DES in the present study. Two different configurations have been simulated. In configuration A the telescope has an elevation angle $EL = 50°$, a cross elevation angle of $CE = +3°$ and a line of sight angle of $LOS = +3°$. Configuration B is characterized by $EL = 30°$, $CE = -3°$ and $LOS = -3°$. The results of these configurations are compared to the reference configuration with $EL = 40°$, $CE = 0°$ and $LOS = 0°$. The aperture assembly angle is $AA = 40°$ in all configurations. Figure 4 shows the different telescope positions relative to the aperture.

The CFD grids of the configurations A and B were built identically to the reference configuration except that the telescope assembly was tilted around the three telescope axes. The DES simulations reveal that the misalignment between the aper-

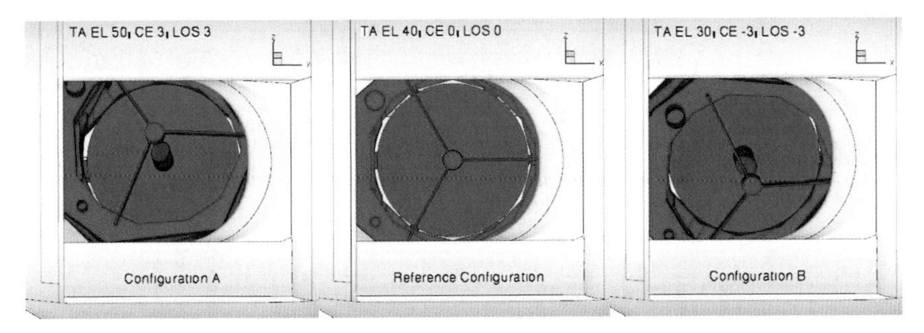

Fig. 4 TA/AA misalignment configurations

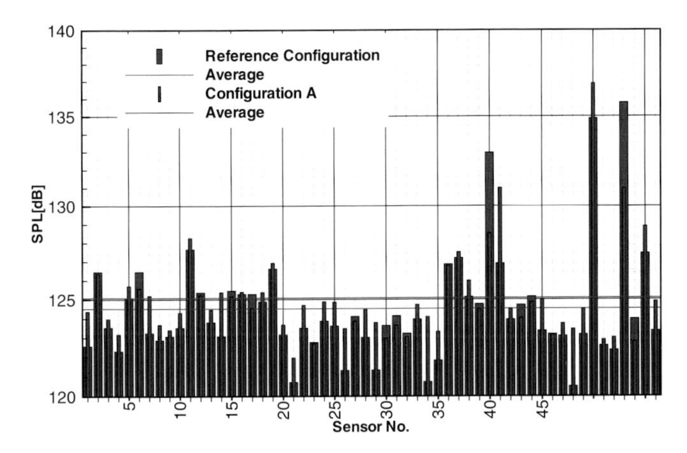

Fig. 5 Overall SPL values of configuration A compared to reference configuration

ture and the telescope does not significantly affect the flow and the acoustics on the SOFIA telescope port. Figures 5 and 6 show the impact of the misalignment on the sound pressure levels on the TA surface.

Fig. 7 shows the arrangement of the 56 locations where pressure transducers were mounted on the telescope assembly of the 7%-model in wind tunnel.

For both configurations A and B the change of the SPL-average (indicated as red, blue and magenta horizontal lines in Figs. 5 and 6) over all 56 locations is less than 0.5 dB. The majority of the pressure sensors of the misaligned configurations differed by less than 1 dB. Only few sensors show prominent differences, in particular the sensors No. 40, 50 and 53. Sensor No. 40 is located on the TA headring, sensor No. 50 and 53 are located on the aft SMA spiderarm. The distinct differences in those three pressure sensors are caused by the interaction of the internal cavity mean flow with the telescope structure. It was observed both in wind tunnel as well as in CFD simulations that a flow separation on the aperture ramp causes a prominent downstream into the cavity. In configuration B the sensor No. 40 is more exposed to this high velocity region, yielding an SPL value of 140dB. In contrast, configuration A shows the opposite effect, lowering the pressure fluctuations at sensor No. 40 as

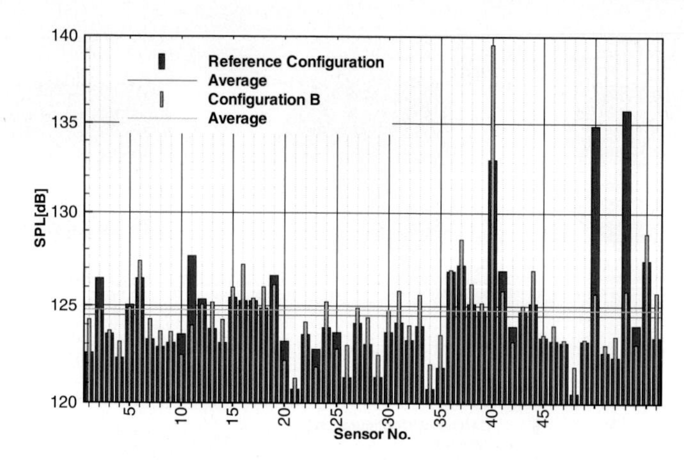

Fig. 6 Overall SPL values of configuration B compared to reference configuration

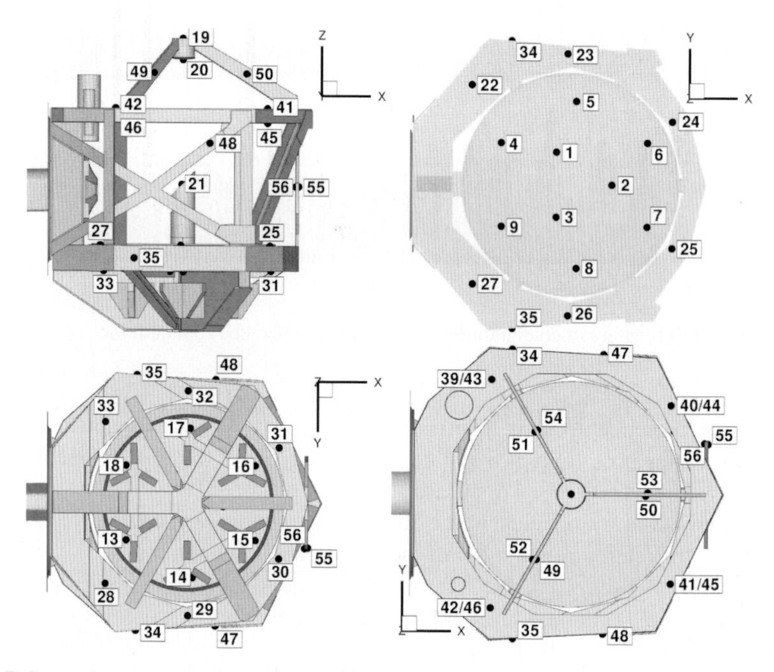

Fig. 7 Sensor locations on telescope assembly

the flow velocity in the vicinity of this location is significantly lower. In configuration B the aft SMA spiderarm faces significantly lower flow velocities compared to configuration A, what can be seen in the lowered SPL values of sensor No. 50 and 53.

The spatial distribution of the overall sound pressure levels on the TA surface is visualized in Fig. 8. As already mentioned, the only difference between the three

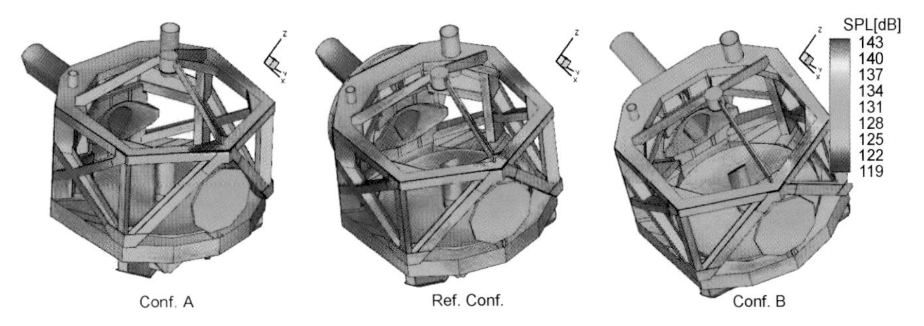

Fig. 8 SPL contour plots on the TA surface

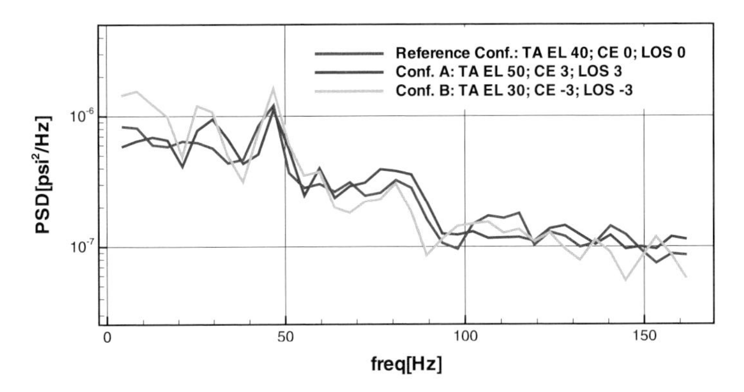

Fig. 9 Impact of TA/AA misalignment on PSD of unsteady pressure fluctuations

configurations is basically the different location of the impingement point of the separated flow on the TA structure. The energy spectra of the acoustic cavity resonances are not significantly changed by the TA/AA-misalignment. Figure 9 shows the mean average of all 56 PSD lines. The prominent peak at a frequency of about 47 Hz, a footprint of an acoustic resonance mode dominates the PSD plots of all three configurations. A slight amplitude increase can be observed at the 30 Hz acoustic resonance peak, both for configuration A and B. Acoustic simulations revealed that this mode is a standing wave in x-direction of the cavity. Furthermore the peak at around 10 Hz is slightly lower for configuration A and slightly increased for configuration B. This mode is the first organ pipe resonance of the cavity, characterized by fluctuations parallel to the line of sight.

6 Parametric Optimization of the Aperture-Ramp Geometry

As a further approach to improve the aeroacoustic environment under which the telescope has to work, the geometry of the aperture ramp has been optimized in a separate study. Both wind tunnel data as well as CFD simulations show a flow sep-

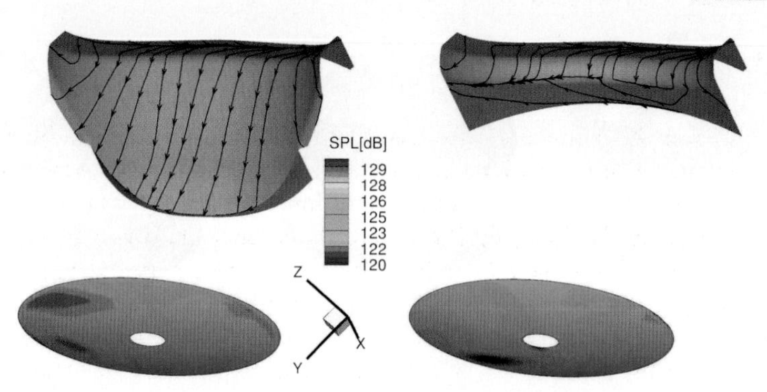

Fig. 10 Bottomview of time averaged streamtraces on aperture ramp and SPL contour plots on primary mirror surface; right: baseline configuration; left: modified configuration

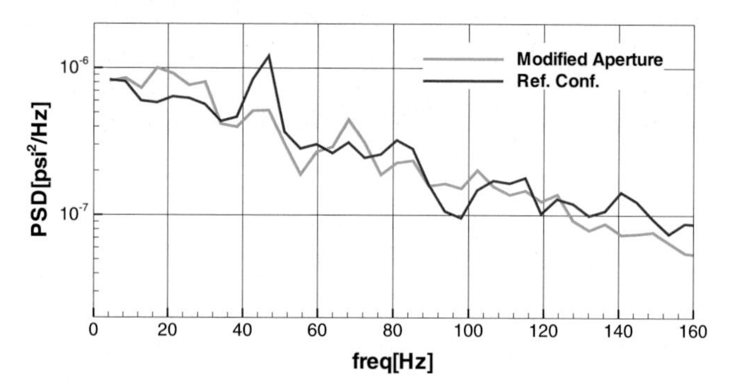

Fig. 11 Impact of aperture ramp modification on pressure PSD plots

aration on the aperture ramp which leads to an impingement of fluid on the primary mirror causing fluctuating loads [15]. The geometry of the cavity internal part of the aperture ramp has been modified and optimized parametrically by means of 2D RANS simulations with the objective to suppress separation. The optimized shape that showed an entirely attached flow in 2D has then been applied to the $3D$ configuration and a DES simulation was performed.

Figure 10 shows the original aperture ramp geometry (right) in comparison to the optimized configuration (left). The time averaged streamlines on the aperture ramp surface show that the flow remains fully attached on the modified aperture ramp. The comparison between the SPL contour plots on the primary mirror surface reveals that the pressure fluctuations are lowered due to the modification, which is prone to improve the telescope's pointing-stability. Figure 11 reveals the impact of the modification on the average PSD of all telescope pressure sensors. The modified aperture ramp reduces the spike between $40Hz$ and $50Hz$ which is the footprint of an acoustic resonance mode [4].

7 Conclusions

The flow over the partial and full open SOFIA telescope cavity has been characterized by means of URANS simulations. It turned out that the door position has a significant impact on the aeroacoustics. According to the URANS model the sound pressure levels of the 50% partial open cavity will in average be 6dB higher compared to the baseline standard configuration. Two different approaches for passive flow control turned out to be possible means to improve the cavity environment. DES simulations show that the aeroacoustic loads on the SOFIA telescope assembly are reduced significantly by a porous fence, installed upstream of the telescope port. DES simulations predict that modifying the aperture ramp geometry can decrease the average sound pressure level by $1dB$ on the telescope what can be attributed to the reduction of the $47Hz$ acoustic mode. DES simulations show furthermore that the misalignment between the telescope assembly and the aperture assembly has only a minor impact on the cavity aeroacoustics. The SPL-values predicted by the CFD simulations on the TA surface change locally due to the displacement of sensor positions, but the average SPL over the telescope surface remains nearly unchanged.

Acknowledgements SOFIA, the "Stratospheric Observatory for Infrared Astronomy" is a joint project of the Deutsches Zentrum für Luft- und Raumfahrt e.V. (DLR; German Aerospace Centre, grant: 50OK0901) and the National Aeronautics and Space Administration (NASA). It is funded on behalf of DLR by the Federal Ministry of Economics and Technology based on legislation by the German Parliament, the state of Baden-Württemberg and the Universität Stuttgart. Scientific operation for Germany is coordinated by the German SOFIA-Institute (DSI) of the Universität Stuttgart, in the USA by the Universities Space Research Association (USRA).

References

1. Technical documentation of the DLR TAU-code. Tech. rep., Institute of Aerodynamics and Flow Technology, Deutsches Zentrum für Luft- und Raumfahrt e.V. Braunschweig (2007)
2. Ahuja, K.K., Mendoza, J.: Effects of cavity dimensions, boundary layer and temperature on cavity noise with emphasis on benchmark data to validate computational aeroacoustic codes. NASA CR-4653 (1995)
3. Atwood, C.A.: Selected computations of transonic cavity flows. ASME **147**, 7–18 (1993)
4. Düring, M.: Investigation of Methods to Influence Resonance Characteristics of the SOFIA Telescope Cavity. Studienarbeit, IAG, Universität Stuttgart (2009)
5. Gerhold, T.: Overview of the Hybrid RANS Code TAU. In: N. Kroll, J.K. Fassbender (eds.) Notes on Numerical Fluid Mechanics and Multidisciplinary Design, vol. 89, pp. 81–92. Springer (2005)
6. Rose, W.: SOFIA V design validation test final report. Tech. rep., ROSE Engineering & Research, INC., P.O. Box 5146, Incline Village, NV 89450 (1998)
7. Rossiter, J.E.: Wind-tunnel experiments on the flow over rectangular cavities at subsonic and transonic speed. R&M No. 3438, Aeronautical Research Council, London, UK (1964)
8. Rowley, C.W.: Modeling, Simulation, and Control of Cavity Flow Oscillations. PhD-Thesis, California Institute of Technology Pasadena, California (2001)
9. Schmid, S.: Simulation der instationären Strömung um das Stratosphärenobservatorium SOFIA. Ph.D. thesis, Institut für Aerodynamik und Gasdynamik, Universität Stuttgart (2009)

10. Schmid, S., Lutz, T., Krämer, E.: Numerical Simulation of the Flow Field Around the Stratospheric Observatory For Infrared Astronomy. In: C. Tropea, S. Jakirlic, H.J. Heinemann, R. Henke, H. Hönlinger (eds.) Notes on Numerical Fluid Mechanics and Multidisciplinary Design, vol. 96, pp. 364–371. Springer Berlin Heidelberg New York (2007)
11. Schmid, S., Lutz, T., Krämer, E.: Simulation of the Flow Around the Stratospheric Observatory For Infrared Astronomy SOFIA Using URANS and DES. In: Third Joint HLRB and KONWIHR Result and Reviewing Workshop, Garching, Germany (2007)
12. Schmid, S., Lutz, T., Krämer, E.: DES Simulations of the Unsteady Flow Field Around the Stratospheric Observatory For Infrared Astronomy SOFIA. In: XXII International Congress of Theoretical and Applied Mechanics, Adelaide, Australia (2008)
13. Schmid, S., Lutz, T., Krämer, E.: IUTAM Symposium on Unsteady Flows and their Control, vol. 14, chap. Simulation of the Unsteady Cavity Flow of the Stratospheric Observatory For Infrared Astronomy. Springer Berlin Heidelberg New York (2009)
14. Schmid, S., Lutz, T., Krämer, E.: Passive control of the unsteady flow inside the SOFIA telescope port by means of a porous fence. In: Third Symposium on Hybrid RANS-LES Methods, Gdansk, Poland (2009)
15. Schmid, S., Lutz, T., Krämer, E., Kühn, T.: Passive Control of the Flow Around the Stratospheric Observatory For Infrared Astronomy. Journal of Aircraft **46**(4), 1319–1325 (2009)
16. Spalart, P., Deck, S., Shur, M., Squires, K., Strelets, M., Travin, A.: A New Version of Detached-Eddy Simulation, Resistant to Ambiguous Grid Densities. Theoretical and Computational Fluid Dynamics (2005)
17. Titz, R., Röser, H.P.: Astronomy and Technology in the 21st Century. Wissenschaft & Technik Verlag (1998). ISBN 3-89685-558-1

Towards the Numerical Simulation of a Scram Jet Intake at High Mach Number

Claus-Dieter Munz, Gregor Gassner, Christoph Altmann, Arne Taube, and Marc Staudenmaier

Abstract We describe the current progress of our project towards a numerical simulation of a scram-jet intake at high Mach number. We will outline why we have not yet reached our goals and need more computational time and resources. Nevertheless we present results of complex large-scale computations already performed on the HLRB II supercomputer that will hold as pioneering computations to enable the code for massively parallel computations on a very large number of processors and also describe discovered obstacles as well as the implemented solutions towards successful calculations.

1 Introduction

For propulsion systems of future hypersonic aircrafts but also reusable space transportation systems, air-breathing scramjets can be an alternative to the classical rocket technology at Mach numbers greater than 5. Nowadays, the main research interests are to provide both experimental as well as numerical fundamentals in order to design and build a scramjet demonstrator. Several problems in the area of aero- and gasdynamics, such as flow at the compression ramp, shock-boundary interactions, thermodynamics of the supersonic combustion, but also material sciences for high-strength ceramic composites will be addressed. Within the presented project, this scramjet will be investigated numerically.

Claus-Dieter Munz · Gregor Gassner · Christoph Altmann · Arne Taube · Marc Staudenmaier
Institut für Aerodynamik und Gasdynamik, Universität Stuttgart, Pfaffenwaldring 21, 70569 Stuttgart, Germany
e-mail: munz@iag.uni-stuttgart.de

S. Wagner et al. (eds.), *High Performance Computing in Science and Engineering, Garching/Munich 2009*, DOI 10.1007/978-3-642-13872-0_19,

2 Goals of the Project

The main goal of this project is the numerical simulation of a scram-jet intake at a high Mach number of M=8. This project is part of the DFG Graduiertenkolleg GRK 1095 - conception of a scram-jet demonstrator - and its partners at TU München, RWTH Aachen, Universität Stuttgart and the DLR. The numerical as well as experimental setup contains interactions of strong shocks with the developed boundary layer. This shock-boundary layer interaction can cause transition to turbulence and/or flow detachment yielding a multi scale problem with strong anisotropic gradients and structures. For the numerical discretization, the so-called 'Space-Time Expansion discontinuous Galerkin scheme' (STE-DG) is used, which is implemented into the IAG code HALO. This fully explicit numerical scheme features isotropic/anisotropic hp-adaptation, time consistent local time stepping and arbitrary high order in space and time on general unstructured grids. One main advantage of explicit DG schemes is that they are able to scale optimally for a very high number of processors, which makes them suitable for very large computations. An essential element achieving this scalability is a dynamic load balancing, as the computational load strongly varies during the computation due to the local time stepping feature and the automatic hp adaptation. The scheme has already proven its large-scale capabilities with computations on more than 4000 processors on the HLRB II supercomputer. In addition, calculations on up to now 1000 processors were performed to test all code components and to determine parts that do not scale. We are currently in the progress of completing these calculations and implement the 'Variational Multi Scale' (VMS) framework based 'Large Eddy Simulation' (LES) for turbulent flows. Once the implementation is done and validated, the VMS-LES STE-DG framework is combined with shock capturing techniques of the HALO code to finally simulate the scram jet intake. To get to know the intake geometry, Figure 1 shows a two dimensional calculation, performed with an earlier version of the HALO code on desktop PC. The mesh as well as the Mach number distribution are shown. Performing a LES calculation of a complex framework like a scramjet is not an easy task. It is highly complex not only in the field of numerical schemes/algorithms and code developement but also when it comes to computing hardware and requirements. We therefore decided to follow a step-by-step approach to structure and sharpen the design steps and lead the project to success. These steps consist of different calculations targetting at different features that are necessary to reach our final goal. After some basic information about the numerical scheme, we will describe already performed preparatory calculations.

3 Numerical Code

Since all presented numerical problems contain strongly time dependent instationary phenomena, we are using a special unstructured explicit discontinuous Galerkin code, called HALO (Highly Adaptive Local Operator). This code is of high order

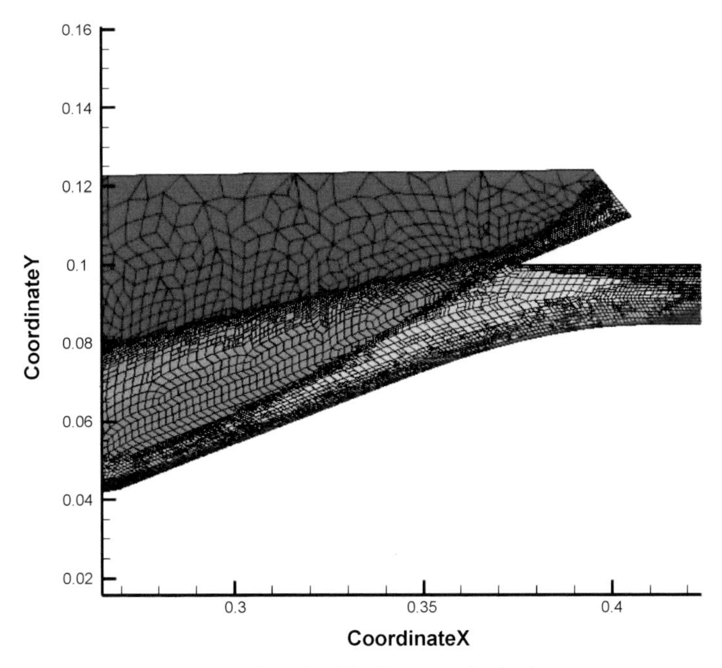

Fig. 1 Mach number distribution and mesh of the 2D scram-jet-intake

of accuracy in both space and time and can handle unstructured hybrid grids with hanging nodes, consisting of tetrahedra, hexahedra, prisms and pyramids in 3D and even polygons in 2D. It is now almost fully parallelized and requires only a minimum of inter-processor communication due to its explicit character. A main feature of this code is its time discretization method [1, 2] and [3] that allows a high order time consistent local time stepping mechanism, where each gridcell can run with its own maximum possible time step. Various equation systems are implemented, such as Euler or Navier-Stokes equations as well as viscous Magnetohydrodynamic (MHD) equations.

We will first summarize the basics of the discontinuous Galerkin scheme inside HALO and its local time stepping functionality. For a detailed view on these topics the reader is referred to the corresponding articles [3] or [2]. We are starting the DG discretization as usual with subdividing our domain Ω into into non-overlapping spatial grid cells Q_i and introduce our numerical DG solution. To obtain the weak formulation of the STE-DG scheme, we multiply by a spatial test function $\phi = \phi(\mathbf{x})$ and integrate over an arbitrary space-time cell $Q_i^n := Q_i \times [t^n, t^{n+1}]$. Please note that performing a space-time integration is substantially different to the classical purely spatial dependent DG formulation. In addition to the now following classical integration by parts to derive the weak formulation, Gassner et al. [3] introduced a new variational formulation for diffusion problems, the keystone of which is the application of a second integration by parts. This generates a new diffusion surface integral which depends on the solution itself. With the definition of suitable numerical fluxes and a suitable numerical state for the second diffusion surface integral we get an adjoint consistent formulation and thus a discretization with optimal order of convergence. This is accomplished by solving the so-called diffusive generalized Riemann problem (dGRP). See [3] for more detail on this method of diffusion flux treatment.

Finally, the discrete variational formulation of the advection diffusion equation results in

$$\begin{aligned}\int_{Q_i^n} (u_i)_t \phi \, d\mathbf{x}dt - \int_{Q_i^n} \mathbf{f}^a \cdot \nabla \phi \, d\mathbf{x}dt + \int_{Q_i^n} \mu \nabla u_i \cdot \nabla \phi \, d\mathbf{x}dt + \\ \int_{\partial Q_i^n} \mathbf{g^a} \cdot \mathbf{n} \phi \, dsdt - \int_{\partial Q_i^n} \mathbf{g^d} \cdot \mathbf{n} \phi \, dsdt + \int_{\partial Q_i^n} g^s \left[\nabla \phi \cdot \mathbf{n}\right]^- dsdt = 0,\end{aligned} \tag{1}$$

where the test function ϕ runs through all the basis functions. Here, the term $\mathbf{g^a}$ denotes the numerical advection flux, the term $\mathbf{g^d}$ denotes the numerical diffusion flux and $g^s := \mu u_i - [\mu u_i]^-$ denotes the additional scalar diffusion flux. u_i denotes the numerical solution in grid cell Q_i without any restrictions on the equation system type.

For nonlinear flux functions the space-time integrals in (1) have to be computed in an approximate way. While this could be done using Gaussian quadrature formulae in space and time, we are using a call-local predictive solution to get approximate values at this space-time Gauss points and will correct later with neigboring information by the inter-cell fluxes. This is done in an explicit way, since we are interested in an explicit scheme.

3.1 Space-Time Expansion

The concept of the STE-DG scheme is a Taylor expansion at the barycenter to get a predictive approximate solution in the space-time cell [2]:

$$u(\mathbf{x},t) = u(\mathbf{x}_i,t_n) + \sum_{j=1}^{N} \frac{1}{j!} \left((t-t_n)\frac{\partial}{\partial t} + (\mathbf{x}-\mathbf{x}_i)\cdot\nabla \right)^j u \Big|_{(\mathbf{x},t)=(\mathbf{x}_i,t_n)}, \tag{2}$$

about the barycenter $\mathbf{x}_i$ of the grid cell Q_i at time t_n. This Taylor expansion provides approximate values for u and ∇u at all space-time points $(\mathbf{x},t) \in \Omega_i^n$. Since no neighbor information has gone into this formulation, we can consider this an predictor of our solution that will be corrected later with neigboring information by the inter-cell fluxes. While the pure space derivatives at $(\mathbf{x}_i,t_n)$ are readily available within the DG framework, the time and mixed space-time derivatives have to be computed using the Cauchy-Kowalevsky (CK) procedure. This procedure will replace these with pure spatial derivatives by the differential equation directly. More information about the CK procedure can be found in [2]. This derived framework now also allows for a natural consistent explicit local time stepping.

3.2 Local Time Stepping

A major disadvantage of a conventional explicit DG-scheme is its severe time step restriction to ensure stability. While this restriction is often a natural condition of consistency in approximating unsteady solutions, it becomes obstructive for unstructured grids with small grid cells. The grid cell with the most restrictive local time step defines the time step for all grid cells. But, this drop in efficiency can be avoided: Due to the locality of the explicit STE-DG scheme, each grid cell may run with its own time step in a time-consistent manner, while the high order accuracy of the numerical scheme is preserved. The local time step is determined exclusively by the in-cell time step restriction and is completely independent of time steps of neigboring cells. Please note that this local time stepping algorithm minimizes the total number of time steps for a computation with fixed end time. However, when the difference of time levels of adjacent grid cells are very small compared to the local time steps, we locally synchronize the time levels of those cells to gain efficiency. Common global time levels as needed e. g. at the end of the computation can easily be introduced. This procedure has absolutely no influence on the accuracy of the underlying numerical scheme, as convergence tests e.g. in [2] verify.

All presented calculations deal with Navier-Stokes equations. To prevent discontinuities - e.g. shocks - to destroy the calculation, artificial viscosity is used to smear the shock profile.

4 Scale-up Efficiency

For testing the scaling capabilities of the HALO code, we chose an example where a perfect load balance was to achieve. We were using the so-called manufactured solution technique for the 3D compressible instationary Navier-Stokes equations: When forcing an exact solution, this results in an inhomogeneous source term on the right hand side of the Navier-Stokes equations that is put into the code. The problem was set up with periodic boundaries so that the boundary communication will not differ from the inter-processor communications. The size of the computational problem was increased exactly the same way as the number of processors for calculation was increased. This way, we kept a constant load in computation as well as in communication. Table 1 shows the good scale-up efficiency of the HALO code for up to 4080 processors with a constant load per processor. The efficiency when calculating on N processors is calculated as the calculation time on one processor divided by the time needed for a calculation on N processors.

Table 1 Scale-up efficiency of the HALO code

Number of processors	1	1000	2197	4080
Efficiency [%]	-	99.1	97.8	98.8

5 Preparatory Calculations

In order to be able to perform a large calculation of the scram-jet with features like hp-adaption, shock capturing and VMS, preparatory calculations were set up to test certain features and prepare the code for efficient computation on a large number of processors. These calculations do also target physical problems and will be described next.

5.1 3D Flow Around a Sphere

The laminar time periodic flow around a sphere was set up to test the codes ability to handle unstructured hybrid grids and a p-adaption mechanism in three space dimensions. We were solving the unsteady compressible Navier-Stokes equations with a free-stream Mach number of $Ma_{\text{inf}} = 0.3$ and a Reynolds number of $Re = 300$. The problem was discretized with a block-unstructured grid consisting of prisms for the boundary layer, tetrahedra and hexahedra. Figure 2 shows the different grid blocks and dimensions of the computational domain. P-adaption was arranged to vary the polynomial degree p between 1 and 5 where each grid cell was allowed to adapt every 500 time steps. To demonstrate the calculation results, Figure 3 shows a 3D view of the instantaneous vortex measure λ_2 for this calculation. The color indicates the velocity magnitude. Here, one can easily see that the very large cells at the end of the wake cannot provide the necessary resolution and are therefore producing large inter cell jumps of the solution. Finally, Figure 4 on the next page shows the distribution of the local polynomial degree p at end time $t_{end} = 1000$.

5.2 3D Freestream Injector

This calculation especially targets shock capturing and p-adaption capabilities of the scheme, as we simulate a $Ma = 1.4$, $Re = 30000$ injection. The injection nozzle is designed according to devices used for gas injection used in automotive industry. This problem also contains complex curved geometries that are challenging for high order schemes. The calculation aims at an aeroacoustic simulation of the instationary injection process, including the start up of the process. Results can be validated with calculations performed with other codes, both at our institute as well as in industry. Preliminary 2D calculations already provided insight regarding the necessary grid resolutions and shock capturing strategies. In our case, the problem was calculated on a grid with over 16 million DG degrees of freedom (4 million of hexahedral elements) on 500 to 1000 processors. Figure 5 shows a 2D slice plane of the calculation (density distribution and velocity streamlines) together with an iso-surface plot of the density that brings out the development of the flow. To illustrate the flow development, several streamlines have been added. The picture presents

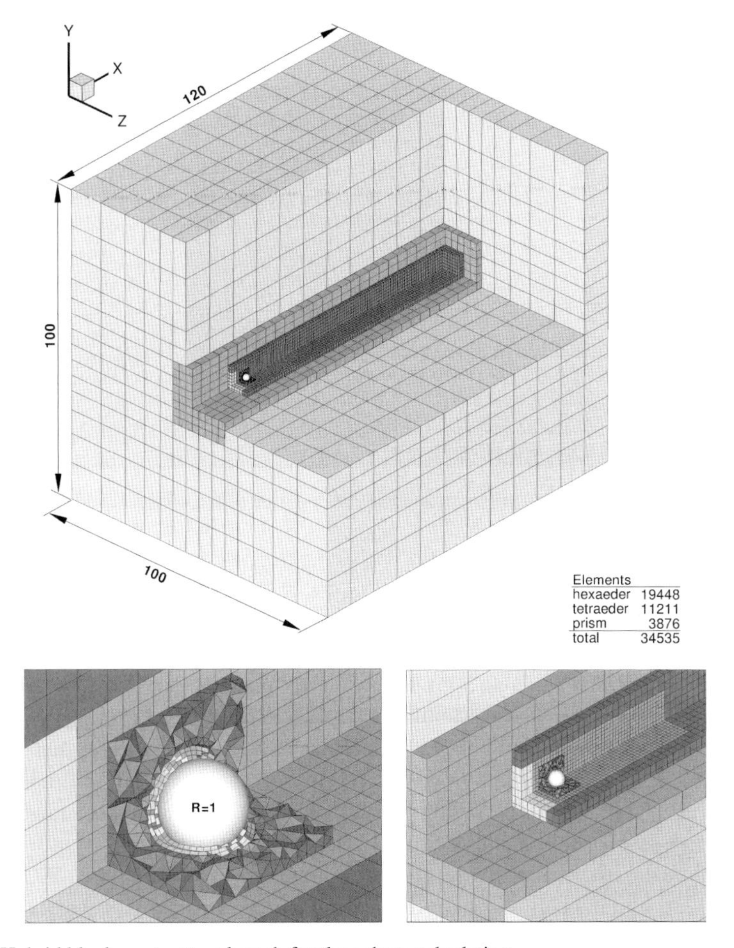

Fig. 2 Hybrid block-unstructured mesh for the sphere calculation

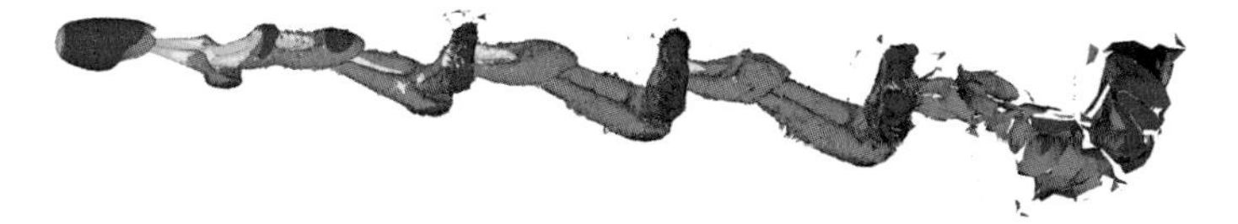

Fig. 3 Instantaneous λ_2 isosurface of the sphere

an early phase of simulation where the injection process was just started. The geometry itself is rather complex, consisting of four kidney-shaped injection 'nozzles' within the cylindric injector and is assembled with unstructured hexahedra, allowing hanging nodes and polygons at connection surfaces.

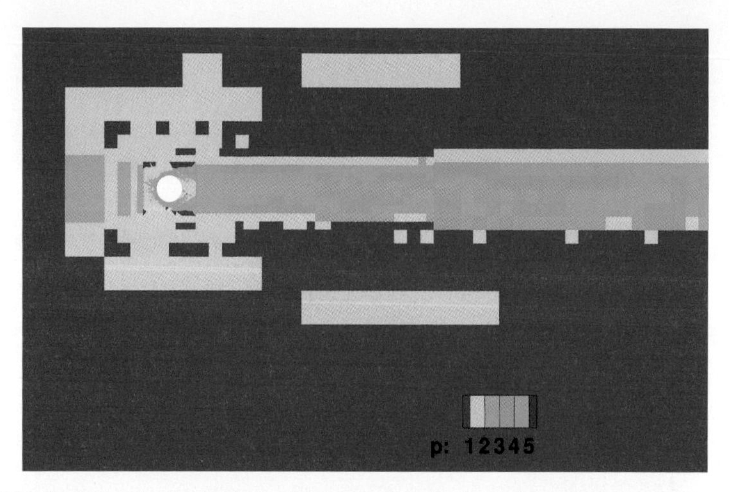

Fig. 4 Distribution of the local polynomial degree at end time of the sphere calculation

5.3 3D Turbulent Flow Around a Cylinder

This example directly aims as a preparatory step towards LES calculations. To do so, first a " 'Direct Numerical Simulation" ' (DNS) is to be performed in order to obtain reference data for the following VMS-LES calculations. Together with the local time stepping, p-adaptivity and zonal grids will help to reduce computational costs for this large-scale calculation. To simplify the difficult task of mesh generation, an originally 2D mesh can be expanded into the third dimension within the code. This feature may also be used to speed up the 3D flow initialization. It is here considered to initialize an originally 2D flow as initial condition for the 3D calculation. This technique will then also be used for the scram-jet intake, as its geometry is also suitable for such an initialization. In this example, the turbulent flow around a cylinder at Reynolds number $Re = 3900$ is calculated on a hybrid zonal mesh with over 700.000 cells, resulting in 10.5 million DG degrees of freedom. For this computation, 1000 processors on the HLRB II were used. Figure 6 shows the p-adapted domain, where one can easily see that the wake of the cylinder that contains the small-scale turbulent structures needs a high order polynomial representation. Figure 7(a) shows a 2D slice plane of the velocity magnitude while Figure 7(b) displays a 3D view of the instantaneous vortex measure λ_2. This example had also shown up to be well suited for a precise optimization of the dynamic load balancing. Due to the high order explicit local time stepping of the HALO code, this task is of top importance for us, even more when calculating on thousands of processors. In addition, adaption strategies will also challenge this algorithm.

Currently, the DNS calculations are being performed on the HLRB II to improve the robustness of the code and its features. Once that step is completed, we will focus on the VMS-LES calculations of the cylinder to compare them with the DNS results.

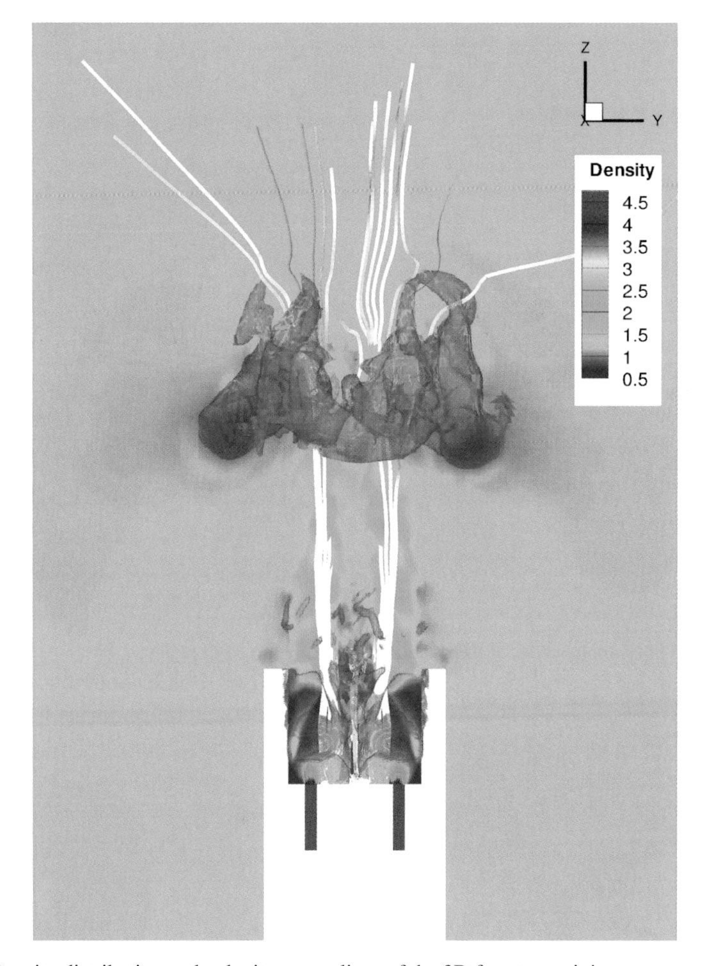

Fig. 5 Density distribution and velocity streamlines of the 3D freestream injector

To do so, several LES related postprocessing tools are also being implemented, like a spatial and temporal averaging of the solution.

6 Code Performance

The HLRB II provides a quick job performance summary directly on command line. This tool can provide a first insight into the performance of the calculations and help to determine the computational efficiency of the numerical code. We hereby discovered strong performance differences, depending on the example to calculate. Especially the scale-up tests as well as the sphere calculation where able to perform with up to 725 MFLOPS per processor. The latter performance was discovered for

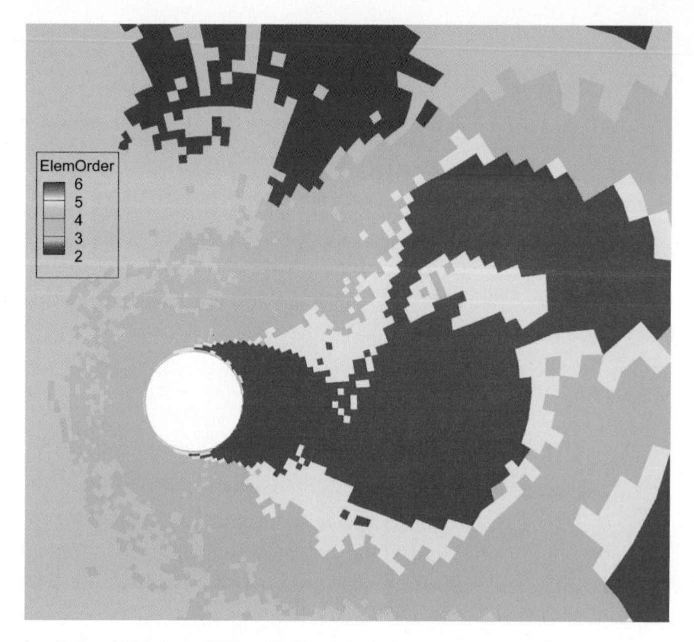

Fig. 6 Local polynomial order of the cylinder calculation on a cut plane

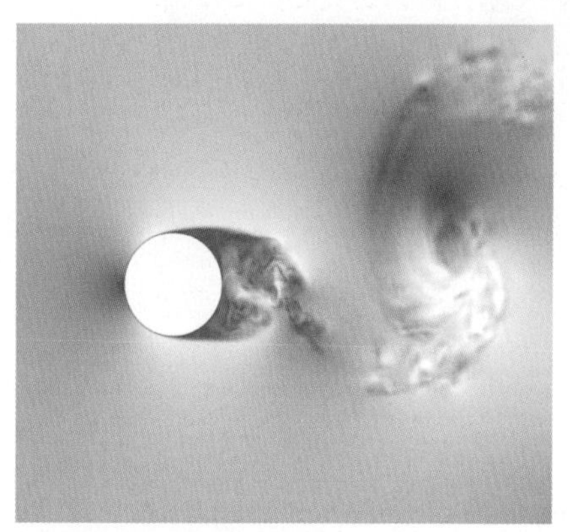

(a) Velocity magnitude of the flow around a cylinder on a cut plane.

(b) Instantaneous λ_2 isosurface of the cylinder calculation.

Fig. 7 Plots of the cylinder calculation

the 4080 processor scale-up calculation resulting in a total of more than 3 TFLOPS. Unfortunately, the efficiency dropped down especially for the cylinder calculation. Certain settings of p-adaptivity and mesh structures as well as output-stages do have a significant influence there. Since the reason for that behavior has not been determined yet, we are still in the process of investigation but it seems that these areas could further be improved.

7 Outlook

Our main task, a three-dimensional LES calculation of a scram-jet-intake is not accomplished yet. Since it has very high requirements not only on the computational environment (large numbers of processors, as provided by the HLRB) but also on the numerical method and practical implementation, this is not a simple venture. We therefore chose the way of incrementally improving code functionalities via selected calculations only targetting parts of the necessary features. Thus, overhasty steps are mostly avoided. These already performed preparatory calculations did not only show the potential of the HALO code but also gave an insight into the physics we are addressing and helped improving the code as desired.

Acknowledgements This project is kindly supported by the Deutsche Forschungsgemeinschaft (DGF) and the Graduiertenkolleg (GRK).

References

1. F. Lörcher, G. Gassner, and C.-D. Munz, "A discontinuous Galerkin scheme based on a space-time expansion. I. Inviscid compressible flow in one space dimension," *Journal of Scientific Computing*, vol. 32, no. 2, pp. 175–199, 2007.
2. G. Gassner, F. Lörcher, and C.-D. Munz, "A discontinuous Galerkin scheme based on a space-time expansion. II. Viscous flow equations in multi dimensions," *Journal of Scientific Computing*, vol. 34, no. 3, pp. 260–286, 2008.
3. G. Gassner, F. Lörcher, and C.-D. Munz, "A contribution to the construction of diffusion fluxes for finite volume and discontinuous Galerkin schemes," *J. Comput. Phys.*, vol. 224, no. 2, pp. 1049–1063, 2007.

Direct Numerical Simulations of Turbulent Mixed Convection in Enclosures with Heated Obstacles

Olga Shishkina, Matthias Kaczorowski, and Claus Wagner

Abstract To simulate turbulent forced and mixed convection flows in complicated three-dimensional domains, a fast finite-volume high-order method based on the Chorin ansatz is developed. The Poisson solver, which is applied to compute the pressure, uses the separation of variables together with capacitance matrix technique as suggested in [Shishkina, Shishkin & Wagner, J. Comput. & Appl. Maths (2009) 226, 336–344]. The developed numerical method generally allows to use hexahedral computational meshes, which are non-equidistant in all three directions and non-regular in any two directions. By means of Direct Numerical Simulations (DNS) we investigate instantaneous and statistical characteristics of turbulent forced and mixed convection flows which develop in parallelepiped convective cells with heated parallelepiped obstacles inside. Cold fluid comes into the cell through thin slits close to the top. The outlet slits are located close to the bottom. The working fluid is air with Prandtl number $\mathscr{P}r = 0.714$. The considered Grashof number $\mathscr{G}r = 4.22 \times 10^8$ and Reynolds number based on the velocity of the inlet flow $\mathscr{R}e = 2.37 \times 10^4$ and 1.18×10^4. It is shown that in the cases of forced and mixed convection principally different large-scale circulations of air are developed inside the domain, although the same geometry and the same Reynolds numbers are considered. In particular, for $\mathscr{R}e = 2.37 \times 10^4$ a downward flow is developed in the case of forced convection, while mixed convection leads to an upward flow in the central part of the domain. Distribution of the mean heat fluxes at the surfaces of the obstacles is shown to be very irregular and strongly dependent on the positions of the surfaces (vertical or horizontal) as well as on their locations inside the domain.

Olga Shishkina · Matthias Kaczorowski · Claus Wagner
DLR - Institute for Aerodynamics and Flow Technology, Bunsenstr. 10, 37073 Göttingen, Germany
e-mail: Olga.Shishkina@dlr.de
e-mail: Matthias.Kaczorowski@dlr.de
e-mail: Claus.Wagner@dlr.de

S. Wagner et al. (eds.), *High Performance Computing in Science and Engineering, Garching/Munich 2009*, DOI 10.1007/978-3-642-13872-0_20,

1 Introduction

Investigation of turbulent mixed convection flows past heated obstacles has physical as well as engineering objectives, since problems such as climate control in buildings, cars or aircraft, where the temperature must be regulated to maintain comfortable and healthy conditions, can be formulated as mixed convection problems [6]. In this type of convection the flows are determined both by the buoyancy force like in natural convection and by inertia forces like in forced convection, while neither of these forces dominates.

Mixed convection is characterised by Grashof number $\mathscr{G}r = \alpha g \hat{H}^3 \Delta\hat{T}/\nu^2$ and Reynolds number $\mathscr{R}e = \hat{H}\hat{u}_{inlet}/\nu$, where $\mathscr{G}r$ is of the same order as $\mathscr{R}e^2$ and α denotes the thermal expansion coefficient, ν the kinematic viscosity, g the gravitational acceleration, $\Delta\hat{T}$ the difference of the temperatures at the heated obstacles and in cold inlet flows, $\hat{H}$ the height of the container and $\hat{u}_{inlet}$ is the velocity of the inlet flow. So far large-scale problems of mixed convection in air have been investigated mainly in theoretical and experimental studies and in underresolved numerical simulations, utilising turbulence modelling (see, for example, [3]). The objective of the present study is to investigate instantaneous and statistical characteristics of turbulent mixed convection flows past heated obstacles, compared to forced convection, for different $\mathscr{G}r$ and $\mathscr{R}e$, by means of Direct Numerical Simulations (DNS). The chosen computational domain, which is a parallelepiped with four parallelepiped obstacles, can be assumed as a generic room in indoor ventilation problems. The considered Prandtl number is $\mathscr{P}r = 0.714$, which corresponds to air at the temperature about 22°C.

The paper is organised as follows. §2 is devoted to the description of the governing equations and computational domain. In §3 some details on the numerical method and mesh resolution requirements for DNS of mixed convection are discussed. In §4 we present our numerical results on forced and mixed convection. §5 demonstrates the parallel performance of the code and §6 summarises our conclusions.

2 Governing Equations and Computational Domain

The system of the governing momentum, energy and continuity equations for the velocity field $\hat{\mathbf{u}}$, temperature $\hat{T}$, and pressure $\hat{p}$ for mixed convection flows in the framework of the Boussinesq approximation reads as follows

$$\begin{aligned} \hat{\mathbf{u}}_t + \hat{\mathbf{u}} \cdot \nabla \hat{\mathbf{u}} + \rho^{-1} \nabla \hat{p} &= \nu \Delta \hat{\mathbf{u}} + \alpha g (\hat{T} - \hat{T}_0) \hat{\mathbf{e}}_x, \\ \hat{T}_t + \hat{\mathbf{u}} \cdot \nabla \hat{T} &= \kappa \Delta \hat{T}, \\ \nabla \cdot \hat{\mathbf{u}} &= 0, \end{aligned} \tag{1}$$

where $\hat{\mathbf{u}}_t$ and $\hat{T}_t$ denote the time derivatives of the velocity and temperature fields, respectively, ρ is the density, κ the thermal diffusivity, $\hat{T}_0 = 0.5(\hat{T}_H + \hat{T}_C)$, $\hat{T}_H$ the

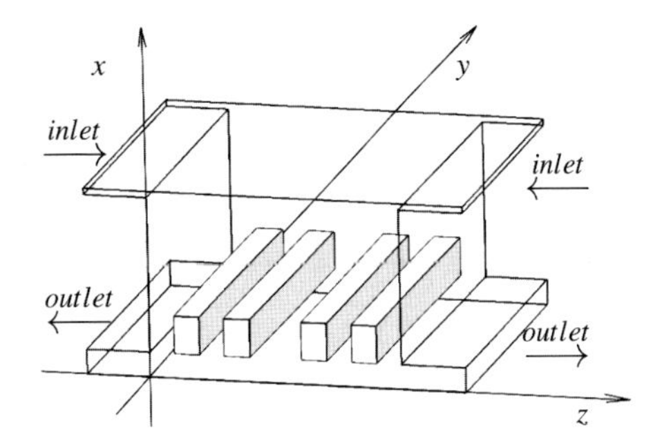

Fig. 1 Sketch of the domain. The obstacles are heated, while the inlet flows are cooled

temperature of the heating obstacles, $\hat{T}_C$ the temperature of the cooling inlet flows and $\hat{\mathbf{e}}_x$ the unit vertical vector.

The temperatures of the cold inlet flows and heated obstacles are fixed. At the outlets and outer rigid adiabatic walls $\partial \hat{T}/\partial \mathbf{n} = 0$, where $\mathbf{n}$ is the normal vector. The velocities at the outer boundaries of the inlet ducts are fixed; one of the horizontal components of the velocity is put to $\hat{u}_{inlet}$, while the others are equal to zero. At the outer boundaries of the outlet ducts $\partial \hat{\mathbf{u}}/\partial \mathbf{n} = 0$. At all solid walls $\hat{\mathbf{u}} = 0$ due to impermeability and no-slip conditions. The considered fluid is air and all material properties are taken at 22°C.

The considered computational domain is a box with 4 parallelepiped ducts for the inlet and outlet flows, which are connected to the box (fig. 1). The length, height and width (without ducts) of the domain are $\hat{L}$, $\hat{H}$ and $\hat{D}$, respectively. Through 2 thin ducts, located close to the top, cold inlet flows come inside the box. The length, height and width of the ducts for the inlet flows are $\hat{L}$, $\hat{H}/150$ and $\hat{W}$, respectively. 2 ducts with the sizes $\hat{L} \times \hat{H}/20 \times \hat{W}$ for outlet flows are situated close to the bottom. There are 4 heated parallelepiped obstacles inside the box, which are raised at the distance $\hat{H}/20$ from the bottom and are located in a parallel manner to each other. The length, height and width of the obstacles are $\hat{L}$, $\hat{H}/5$ and $\hat{D}/10$, respectively. The distance between 2 central obstacles equals $\hat{D}/5$. The other distances between the obstacles or between the walls with the connected inlet/outlet ducts and the nearest obstacle are equal to $\hat{D}/10$.

To non-dimensionalize the governing equations, we use the following reference constants: $\hat{x}_{ref} = \hat{D}$ for distance, $\hat{u}_{ref} = \left(\alpha g \hat{D} \Delta \hat{T}\right)^{1/2}$ for velocity, $\hat{t}_{ref} = \hat{x}_{ref}/\hat{u}_{ref}$ for time, $\hat{T}_{ref} = \Delta \hat{T} \equiv \hat{T}_H - \hat{T}_C$ for temperature and $\hat{p}_{ref} = \hat{u}_{ref}^2 \rho$ for pressure. Thus we obtain the following system of the governing dimensionless equations in the Boussinesq approximation

$$\begin{aligned}\mathbf{u}_t + \mathbf{u} \cdot \nabla \mathbf{u} + \nabla p &= \Gamma^{-3/2} \mathscr{G}r^{-1/2} \Delta \mathbf{u} + T \mathbf{e}_x, \\ T_t + \mathbf{u} \cdot \nabla T &= \Gamma^{-3/2} \mathscr{G}r^{-1/2} \mathscr{P}r^{-1} \Delta T, \\ \nabla \cdot \mathbf{u} &= 0.\end{aligned}$$

Here $\mathbf{u}$ is the dimensionless velocity vector-function, T the temperature, $\mathbf{u}_t$ and T_t their time derivatives, p the pressure, $\Gamma = \hat{D}/\hat{H}$ the aspect ratio of the box, $\mathscr{P}r = \nu/\kappa$ Prandtl number. All boundary conditions are taken in accordance with the above discussed. The dimensionless temperature T equals 0.5 at the obstacles and -0.5 at the inlets.

Within the presented work we conducted DNS of forced and mixed convection in the above discussed geometry for $\mathscr{G}r = 4.22 \times 10^8$, $\mathscr{R}e = 2.37 \times 10^4$ and 1.18×10^4.

3 Numerical Method and Mesh Resolution Requirements

To simulate turbulent mixed convection flows in enclosed domains, we use the fourth-order finite-volume discretization schemes in space and the explicit Euler-leapfrog discretization scheme in time, which are similar to those used in our DNS of turbulent Rayleigh–Bénard convection [5, 12–15]. The principle changes in the numerical method and the corresponding computational code were done within the Poisson solver, which is needed to couple the pressure and the velocity fields. The new Poisson solver uses the capacitance matrix technique together with the separation of variables method, which was developed and described in details in [11]. The developed numerical method generally allows to conduct DNS of turbulent natural, forced or mixed convection flows in enclosures using computational meshes, which are non-equidistant in all 3 directions and irregular in any 2 directions.

To conduct DNS of turbulent convective flows, one should use a fine computational mesh [4] with the mean width $\hat{h}_m$, which is at least as small as

$$\hat{h}_m \leq \pi \hat{\eta}, \quad \hat{\eta} = \left(\nu^3 / < \bar{\varepsilon}_u >_{\Upsilon}\right)^{1/4}, \tag{2}$$

where $\hat{\eta}$ is the smallest Kolmogorov scale, ε_u the rate of turbulent kinetic energy dissipation and the bar-sign denotes averaging in time and $< \cdot >_{\Upsilon}$ averaging over the whole domain Υ.

In our DNS of mixed and forced convection flows for $\mathscr{P}r = 0.714$, $\mathscr{G}r = 4.22 \times 10^8$, $\mathscr{R}e = 2.37 \times 10^4$ and 1.18×10^4 and $\hat{D}/\hat{H} = 4/3$ and $\hat{L}/\hat{H} = 4/15$ we use a mesh, which is non-equidistant in all 3 directions and consists of 320 nodes in the vertical direction, and 64×600 nodes in two horizontal directions. Below we shortly discuss, how one can estimate the resolution requirement for the mixed convection flows and show that the constructed mesh is applicable for the DNS of the above described problems.

To obtain a lower-bound estimate of the number of the required nodes in a computational mesh for DNS of mixed convection flows, we first consider the case of natural convection (without any inlet/outlet flows). In the case of classical Rayleigh-Bénard convection, where the fluid is heated from below and cooled from above, for the rate of turbulent kinetic energy dissipation $\varepsilon_{u,nat}$ one has

$$\varepsilon_{u,nat} = \nu^3 \hat{H}^{-4} (\mathscr{N}u - 1) \mathscr{G}r \mathscr{P}r^{-1}, \tag{3}$$

where $\mathscr{N}u$ is the mean vertical heat flux, i.e. the Nusselt number. From relations (2), (3), we obtain the following requirement for the mean mesh width $\hat{h}_m$:

$$\hat{h}_m / \hat{H} \leq \pi \mathscr{P}r^{1/4} (\mathscr{N}u - 1)^{-1/4} \mathscr{G}r^{-1/4}. \tag{4}$$

For $\mathscr{P}r \sim 0.7$ one can estimate $\mathscr{N}u$ using the relation $\mathscr{N}u \approx 0.124\mathscr{R}a^{0.309}$ [8], where $\mathscr{R}a$ is the Rayleigh number, $\mathscr{R}a = \mathscr{G}r\mathscr{P}r$. Thus, from this relation and (4) for $\mathscr{P}r = 0.714$ and $\mathscr{G}r = 4.22 \times 10^8$ we obtain $\hat{h}_m/\hat{H} \leq 7.55 \times 10^{-3}$. This means that in order to conduct DNS for these Prandtl and Grashof numbers we need a mesh with at least $\hat{H}/\hat{h}_m \sim 135$ nodes in the vertical direction. For the case $\hat{D}/\hat{H} = 4/3$ and $\hat{L}/\hat{H} = 4/15$ the number of the required nodes in the horizontal direction is at least 177 and 36, respectively.

Further, to conduct DNS, one should also provide a good resolution within the thermal boundary layers around heated obstacles. The thickness of the thermal boundary layers is formally defined as $\lambda_\vartheta = \hat{H}/(2\mathscr{N}u)$. For the particular case $\mathscr{G}r = 4.22 \times 10^8$ and above estimated $\mathscr{N}u$ one obtains $\lambda_\vartheta/\hat{H} \approx 9.67 \times 10^{-3}$. Thus if one used an equidistant mesh with the mesh width $\hat{h}_m$, one would have only one or two nodes within the thermal boundary layer, which is definitely not enough to resolve the flow within the layer. In our DNS we take up to 14 nodes within each thermal boundary layer and the mesh width $\hat{h}_{obs}$ close to the heated obstacles is determined as follows: $\hat{h}_{obs}/\hat{H} = 6.86 \times 10^{-4}$.

Finally, close to the inlet and outlet slits the flow should be also well resolved. For this, the requirement $x^+ < 1$ must be fulfilled. Here $x^+ = \hat{u}_\tau \hat{h}_{in}/\nu$ and $\hat{h}_{in}$ denotes the mesh width close to the inlet slits and $\hat{u}_\tau \equiv \sqrt{\nu \partial \hat{u}_z/\partial \hat{x}|_{\hat{x}=\hat{H}}}$ is the friction velocity. The latter can be estimated from $\hat{u}_{inlet} \sim C\hat{u}_\tau$ with some constant C. Thus we obtain the following requirement for mesh width close to the inlet slits: $\hat{h}_{in}/\hat{H} \leq C\nu/(\hat{H}\hat{u}_{inlet}) = C/\mathscr{R}e$, which for $\mathscr{R}e = 2.37 \times 10^4$ and $C \approx 20$ is reduced to $\hat{h}_{in}/\hat{H} \leq 8.43 \times 10^{-4}$. In our DNS $\hat{h}_{in}/\hat{H} = 2.67 \times 10^{-4}$.

4 Results

DNS of mixed and forced convection flows were conducted for $\mathscr{P}r = 0.714$, $\mathscr{G}r = 4.22 \times 10^8$, $\mathscr{R}e = 2.37 \times 10^4$ and 1.18×10^4, and $\Gamma = \hat{D}/\hat{H} = 4/3$ and $\hat{L}/\hat{H} = 4/15$. These Prandtl and Reynolds numbers correspond, in particular, to convection in air by the temperature 22°C, with $\nu = 1.55 \times 10^{-5}\mathrm{m}^2/\mathrm{s}$, $\kappa = 2.17 \times 10^{-5}\mathrm{m}^2/\mathrm{s}$, $\alpha = 3.40 \times 10^{-3}\mathrm{K}^{-1}$ and the following geometrical parameters of the domain: $\hat{H} = 3\mathrm{m}$, $\hat{D} = 4\mathrm{m}$, $\hat{L} = 0.8\mathrm{m}$. Taking $\hat{u}_{ref} = 0.123\mathrm{m/s}$ together with $\hat{u}_{inlet} = \hat{u}_{ref}$ or $\hat{u}_{inlet} = 0.5\hat{u}_{ref}$, one obtains $\mathscr{R}e = 2.37 \times 10^4$ or 1.18×10^4, respectively. Further, if the obstacles are heated and we deal with mixed convection, we further assume that $\hat{u}_{ref} = \sqrt{\alpha g \Delta \hat{T} \hat{D}}$. From this and the definitions of the Grashof number, Reynolds number and the aspect ratio we obtain $\hat{u}_{inlet} = \hat{u}_{ref}\mathscr{R}e\mathscr{G}r^{-1/2}\Gamma^{-1/2}$. Taking $\mathscr{G}r = 4.22 \times 10^8$, which for $\mathscr{R}e = 2.37 \times 10^4$ corresponds to $\mathscr{G}r = \mathscr{R}e^2\Gamma^{-1}$, one obtains $\hat{u}_{inlet} = \hat{u}_{ref}$. For $\mathscr{R}e = 1.18 \times 10^4$ this value of the Grashof number corresponds to $\mathscr{G}r = 4\mathscr{R}e^2\Gamma^{-1}$ and $\hat{u}_{inlet} = 0.5\hat{u}_{ref}$. Finally, we can determine the difference between the temperatures of the obstacles and the inlet flow for $\mathscr{G}r = 4.22 \times 10^8$ as follows: $\Delta\hat{T} = \hat{u}_{ref}^2/(\alpha g \hat{D}) = 0.113\mathrm{K}$.

In figure 2 the instantaneous distributions of the temperature, velocity magnitude and heat flux magnitude are presented for $\mathscr{P}r = 0.714$, $\mathscr{G}r = 4.22 \times 10^8$ and

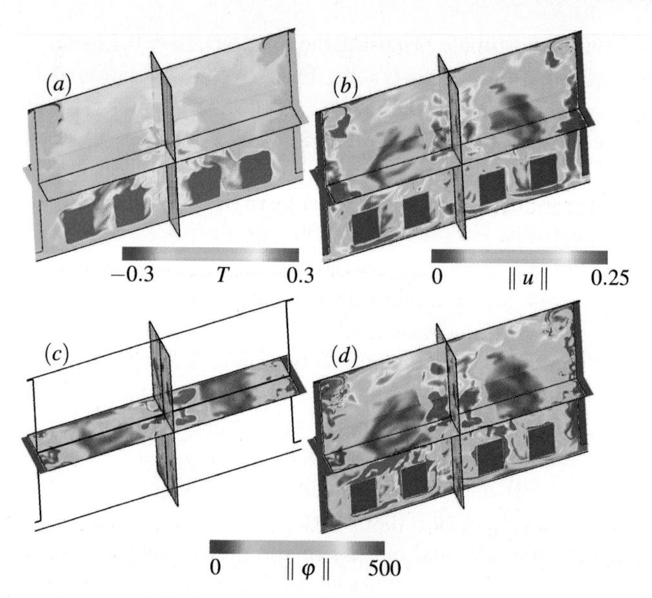

Fig. 2 Snapshots of the temperature T (a), velocity magnitude $\| u \|$ (b) and heat flux magnitude $\| \varphi \|$ for $\mathscr{P}r = 0.714$, $\mathscr{G}r = 4.22 \times 10^8$, $\mathscr{R}e = 2.37 \times 10^4$

$\mathscr{R}e = 2.37 \times 10^4$. The largest absolute velocities are observed close to the inlet slits, along the vertical walls, the bottom and the central vertical cross-section between two central obstacles. Although the computational domain is relatively thin, $\hat{L}/\hat{H} = 4/15$, the flow has a well-pronounced three-dimensional structure (see, for example figure 2 c). The heat flux magnitude reaches the highest values close to the inlet slits. Relatively large values of the heat flux magnitude are observed around the obstacles and along the central vertical cross-section.

Let us consider the heat flux vector-function, which is defined as follows:

$$\hat{\varphi} = (\hat{\varphi}_x, \hat{\varphi}_y, \hat{\varphi}_z), \quad \hat{\varphi}_\beta = \frac{\hat{u}_\beta \hat{T} - \kappa \partial \hat{T}/\partial \hat{x}_\beta}{\kappa \Delta \hat{T}/\hat{H}}, \quad \beta = x, y, z.$$

Averaging in time the energy equation (1), one obtains the following relation

$$\nabla \cdot \bar{\hat{\varphi}} = \mathbf{0}. \tag{5}$$

This means, in particular, that averaged in time heat, which is gained from the heated obstacles, equals to that, which is lost owing to the inlet/outlet flows.

In figure 3 distributions of the mean heat flux magnitude $\overline{\varphi}_{abs}$ in all 3 central cross-sections are presented for the case $\mathscr{P}r = 0.714$, $\mathscr{G}r = 4.22 \times 10^8$ and $\mathscr{R}e = 2.37 \times 10^4$. The heat flux vectors $(\varphi_x, \varphi_y, \varphi_z)$, which are also shown in figure 3, illustrate the heat flow from the surfaces of the obstacles towards the central vertical cross-section between two central obstacles. In the bulk the strongest upward heat flow is observed along this vertical cross-section, while close to the top the heat

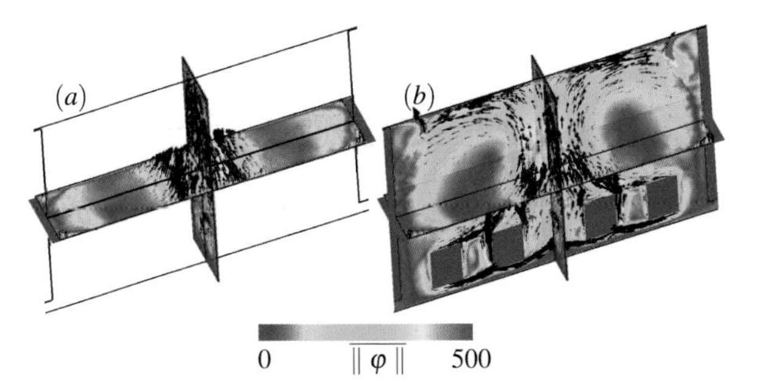

Fig. 3 Distributions of the mean heat flux magnitude $\overline{\| \varphi \|}$ with superimposed heat flux vectors (for $z \in [0.095, 0.905]$), $\mathscr{P}r = 0.714$, $\mathscr{G}r = 4.22 \times 10^8$ and $\mathscr{R}e = 2.37 \times 10^4$. The largest values of the heat flux magnitude are observed close to the inlet ducts (the corresponding heat flux vectors are not shown in Fig. 3 *b*)

flow is oriented towards the inlet slits. The extremely high absolute values of the heat flux are obtained in the vicinity of the inlet slits, therefore the heat flux vectors $(\varphi_x, \varphi_y, \varphi_z)$ in figure 3 (*b*) are shown only for the interval $z \in [0.095, 0.905]$. The fact that the highest values of $\bar{\varphi}_{abs}$ are observed close to the inlets can be explained as follows. According to (5), the mean heat flux through all surfaces of the heated obstacles, multiplied with the area of these surfaces, should be of the same order as the mean heat flux through the inlet slits, multiplied with the area of these slits. The considered geometry of the domain is such that the total area of the obstacle surfaces is 200 times larger than the area of the inlet slits, therefore the mean heat flux near the slits is much higher than that around the obstacles or in the bulk of the convective cell.

Distributions of the temperature, averaged in time, are presented in figure 4 together with superimposed mean velocity vectors. One can see that two large-scale rolls develop in the bulk. The fluid rises up mainly in the centre of the domain, moves along the top boundary towards the inlet slits, falls down along the vertical walls, at which the inlet and outlet ducts are attached, and further moves along the bottom boundary, partly below and partly above the obstacles, towards the centre of the convective cell.

Comparing figures 3 and 4, one can conclude that the structures of the mean fluid flow and mean heat flow look very similar in the bulk and differ significantly in the regions near the inlet slits. In contrast to the bulk flow, where the vectors of mean velocity field and mean heat flux are aligned, the vectors of both fields point in opposite directions near the inlets. In the inlet ducts the vectors of the mean heat flux point outside, while the vectors of the mean fluid velocity point inside. The latter causes the development of 2 additional smaller rolls near the inlet slits.

Although the considered Grashof number $\mathscr{G}r = 4.22 \times 10^8$ in the cases presented in figures 4 (*a*) and (*b*) is the same, the temperatures of the mean flows are different and equal $T_{mean} = 0.0714$ and $T_{mean} = 0.195$ in the cases $\mathscr{R}e = 2.37 \times 10^4$ and $\mathscr{R}e = 1.18 \times 10^4$, respectively. Thus, for stronger cold inlet flows one obtains lower mean temperature inside the container in comparison to weaker inlet flows. The

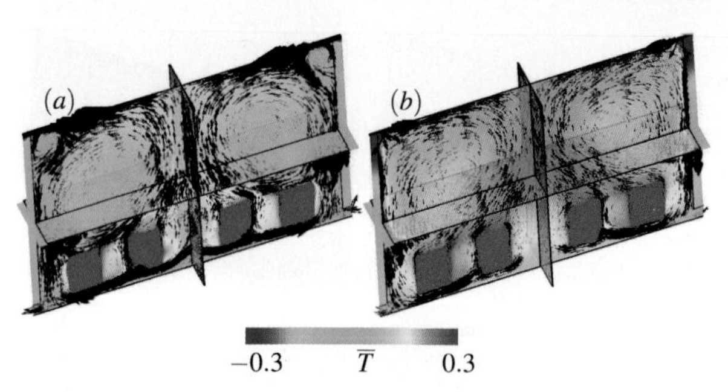

Fig. 4 Mean temperature $\overline{T}$ with superimposed velocity vectors for $\mathscr{P}r = 0.714$, $\mathscr{G}r = 4.22 \times 10^8$, $\mathscr{R}e = 2.37 \times 10^4$ (a) and $\mathscr{R}e = 1.18 \times 10^4$ (b)

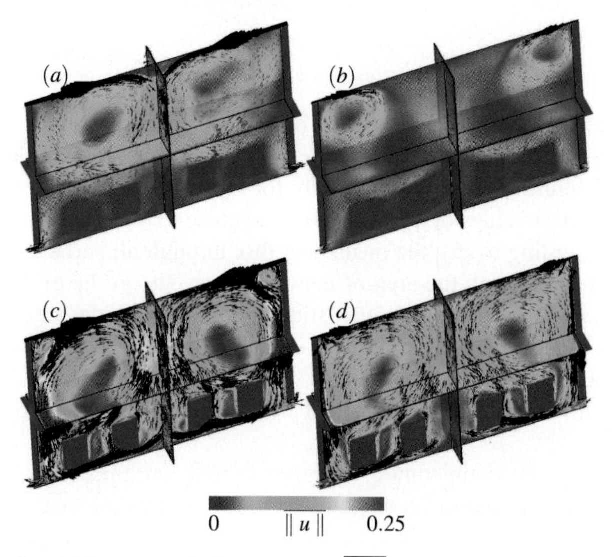

Fig. 5 Distributions of the mean velocity magnitude $\overline{\| u \|}$ as obtained in the DNS of forced (a,b) and mixed (c,d) convection for $\mathscr{P}r = 0.714$ and $\mathscr{R}e = 2.37 \times 10^4$ (a,c), $\mathscr{R}e = 1.18 \times 10^4$ (b,d) and $\mathscr{G}r = 4.22 \times 10^8$ (c,d)

mean flow structures in the cases $\mathscr{R}e = 2.37 \times 10^4$ and $\mathscr{R}e = 1.18 \times 10^4$ are similar in principle. As expected, the flow becomes generally weaker (i.e. the absolute velocity of the mean flow decreases) and the sizes of the 2 additional rolls near the inlet slits become smaller with decreasing $\mathscr{R}e$.

In figure 5 distributions of the mean absolute velocity are presented for all considered cases: forced (without heating of the obstacles) and mixed convection for $\mathscr{G}r = 4.22 \times 10^8$ and 2 different Reynolds numbers $\mathscr{R}e = 2.37 \times 10^4$ and 1.18×10^4. One can see that the flow structures in the forced convection and mixed convection cases are totally different. For example, for $\mathscr{R}e = 2.37 \times 10^4$ the structure of the mean flow has 2 large-scale rolls in both forced and mixed convection cases, but in the pure forced convection the fluid moves predominantly downwards in the centre of the cell, while in mixed convection it moves generally upwards. No ad-

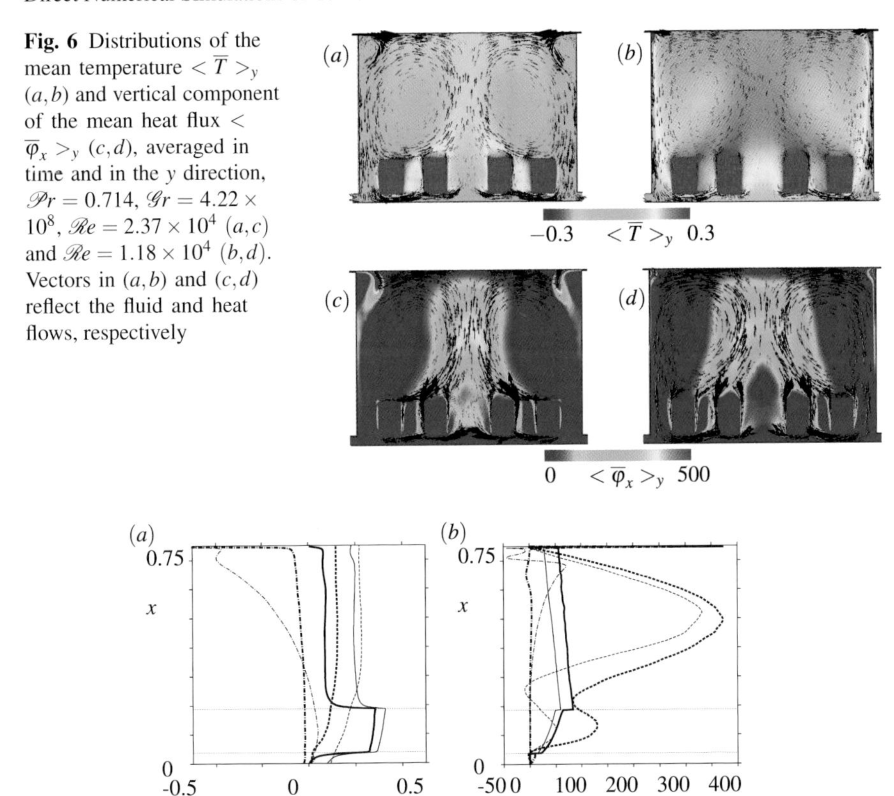

Fig. 6 Distributions of the mean temperature $<\overline{T}>_y$ (a,b) and vertical component of the mean heat flux $<\overline{\varphi}_x>_y$ (c,d), averaged in time and in the y direction, $\mathscr{P}r = 0.714$, $\mathscr{G}r = 4.22 \times 10^8$, $\mathscr{R}e = 2.37 \times 10^4$ (a,c) and $\mathscr{R}e = 1.18 \times 10^4$ (b,d). Vectors in (a,b) and (c,d) reflect the fluid and heat flows, respectively

Fig. 7 Profiles of the mean temperature $<\overline{T}>_y$ (a) and vertical component of the mean heat flux $<\overline{\varphi}_x>_y$ (b), averaged in time and in the y-direction, as functions of the vertical coordinate x, for $\mathscr{P}r = 0.714$, $\mathscr{G}r = 4.22 \times 10^8$ and $z = 0$ $(-\cdot-)$, $z = 0.5$ $(--)$ and averaged over horizontal cross-sections at a height x from the bottom (—). Thick and thin curves correspond to the cases $\mathscr{R}e = 2.37 \times 10^4$ and $\mathscr{R}e = 1.18 \times 10^4$, respectively. Horizontal lines reflect the lower and upper surfaces of the heated obstacles

ditional rolls near the inlet slits are observed in the forced convection case. For a fixed Reynolds number the absolute velocities of the mean flow are higher in mixed convection compared to forced convection.

Additionally, in figure 6 we present the distributions of the mean temperature $<\overline{T}>_y$ and vertical component of the mean heat flux $<\overline{\varphi}_x>_y$, averaged in time and in the horizontal direction along the obstacles, for the mixed convection case with $\mathscr{G}r = 4.22 \times 10^8$ and $\mathscr{R}e = 2.37 \times 10^4$ and 1.18×10^4. One can see that the vertical heat flux is significantly higher close to the central vertical cross-section between two central obstacles and this structure of the mean heat flux is stable.

More details one can see in figure 7, where the profiles of the mean temperature $<\overline{T}>_y$ and vertical component of the mean heat flux $<\overline{\varphi}_x>_y$, averaged in time and in the y-direction, are presented as functions of the vertical coordinate x, for mixed convection cases with $\mathscr{G}r = 4.22 \times 10^8$ and 2 different Reynolds numbers. The con-

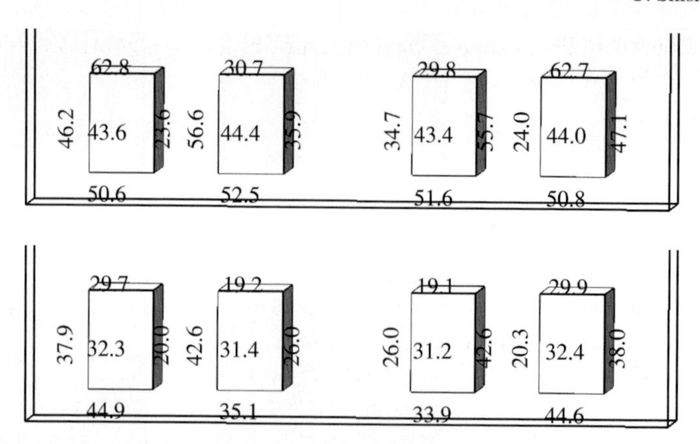

Fig. 8 Time- and surface-averaged heat fluxes at different boundaries of the heated obstacles for $\mathscr{P}r = 0.714$, $\mathscr{G}r = 4.22 \times 10^8$ and $\mathscr{R}e = 2.37 \times 10^4$ (*a*) and $\mathscr{R}e = 1.18 \times 10^4$ (*b*). The heat fluxes, averaged over the whole surfaces of the obstacles are given at their centres

sidered locations are the following: close to the inlet slits ($z = 0$) and in central vertical cross-section between two central obstacles ($z = 0.5$). The profiles, averaged over horizontal cross-sections at a height x from the bottom, are also presented in figure 7. One can see that the mean temperature in the central vertical cross-section ($z = 0.5$) is slightly higher than the mean temperature, averaged over horizontal cross-sections. Thus, these dimensionless temperatures are $< \overline{T} >_{y,mean} = 0.0714$ and $< \overline{T} >_{y,c} = 0.122$ in the case $\mathscr{R}e = 2.37 \times 10^4$ and $< \overline{T} >_{y,mean} = 0.195$ and $< \overline{T} >_{y,c} = 0.226$ in the case $\mathscr{R}e = 1.18 \times 10^4$, respectively.

Comparing the area-averaged and central profiles of the vertical component of the mean heat flux $< \overline{\varphi}_x >_y$, averaged in time and in the y-direction (figure 7 *b*), one concludes that the mean heat flux in the centre of the cell can be 5-7 times higher than the area-averaged mean heat flux. For $\mathscr{R}e = 2.37 \times 10^4$ the mean heat flux in the cell centerline $< \overline{\varphi}_x >_{y,c}$ can be as high as 372 at distance $x = 0.49$, while the area-averaged mean heat flux in the bulk $< \overline{\varphi}_x >_{y,mean}$ is about 73. For $\mathscr{R}e = 1.18 \times 10^4$ the corresponding quantities achieve the following values: $< \overline{\varphi}_x >_{y,c} = 332$ at distance $x = 0.52$ and $< \overline{\varphi}_x >_{y,mean} = 49$.

Finally, in figure 8 we present the time- and surface-averaged heat fluxes separately for different boundaries of the heated obstacles and for both considered mixed convection cases. The heat fluxes, averaged over the whole surfaces of each obstacle are also given in figure 8. One can see that different surfaces of a considered obstacle provide significantly different mean heat flux. Although all heated obstacles provide only slightly different mean (averaged over the whole surfaces of each obstacle) heat fluxes, the values of the mean heat flux and its distribution depend strongly on position of the considered surface (level or upright) and on its location in the convective cell. Thus, the most irregular mean heat flux distribution is observed at the surfaces of the obstacles, which are situated closer to the vertical walls with attached inlet and outlet ducts. At the vertical surfaces of the obstacles, which are located closer to the centerline of the container, the mean heat fluxes are significantly (about 1.6 times) smaller than the mean heat fluxes at the opposite vertical surfaces of the obstacles.

5 Details on Performance on the HLRB System

All the reported simulations were conducted on the HLRB system, using 32 processors. During a singe run for each particular case of $\mathscr{G}r$ and $\mathscr{R}e$ we simulate 150.000 time steps, which cover approximately 1.95 dimensionless time units. So far for each case up to 1.3×10^7 time steps have been conducted.

The numerical code is parallelized using the MPI technique. Despite the fact that the pressure solver using the Capacitance matrix technique requires significantly more inter processor communications, our computations showed a very high parallel efficiency of up to 99.875%.

Below a typical example of the resources used during a single run is shown:

```
resources-used.cpupercent=3196
resources-used.cput=1267:55:46
resources-used.mem=20342464kb
resources-used.ncpus=32
resources-used.vmem=5523926816kb
resources-used.walltime=39:47:17
```

6 Conclusions

To simulate turbulent forced and mixed convection flows in three-dimensional domains with rigid boundaries in all 3 directions, a fast finite-volume high-order method was developed, which allows the usage of hexahedral computational meshes, which are non-equidistant in all directions and could be non-regular in any 2 directions.

DNS of forced and mixed convection were conducted and the instantaneous and mean flow fields were analysed for $\mathscr{P}r = 0.714$ and $\mathscr{G}r = 4.22 \times 10^8$ and $\mathscr{R}e = 2.37 \times 10^4$ and 1.18×10^4. It was shown that in the cases of forced and mixed convection principally different large-scale circulations of air are developed inside the container, although the same geometry and the same Reynolds numbers are considered. In particular, for $\mathscr{R}e = 2.37 \times 10^4$ a downward flow is developed in the case of forced convection, while mixed convection leads to an upward flow in the central part of the domain. Distribution of the mean heat fluxes at the surfaces of the obstacles is shown to be very irregular and strongly dependent on the positions of the surfaces (vertical or horizontal) as well as on their locations inside the domain. Thus, at those vertical surfaces of the obstacles, which are located closer to the centerline of the container, the mean heat fluxes are about 1.6 times smaller than those at the opposite sides of the obstacles.

Acknowledgements The authors are grateful to the Deutsche Forschungsgemeinschaft (DFG) and Leibniz Supercomputing Centre (LRZ) for supporting this work.

References

1. G. Ahlers, S. Grossmann & D. Lohse. 2009. Heat transfer and large-scale dynamics in turbulent Rayleigh–Bénard convection. *Rev. Mod. Phys.* 81: 503–537.
2. M.S. Emran, J. Schumacher. 2008. Fine-scale statistics of temperature and its derivatives in convective turbulence. *J. Fluid Mech.* 11: 13–34.
3. S.D. Fitzgerald, A.W. Wood. 2007. Transient natural ventilation of a room with a distributed heat source. *J. Fluid Mech.* 591: 21–42.
4. G. Grötzbach. 1983. Spatial resolution requirements for direct numerical simulation of Rayleigh-Bénard convection. *J. Comput. Phys.* 49: 241–264.
5. M. Kaczorowski, C. Wagner. 2009. Analysis of the thermal plumes in turbulent Rayleigh–Bénard convection based on well-resolved numerical simulations. *J. Fluid Mech.* 618: 89–112.
6. P.F. Linden. 1999. The fluid mechanics of natural ventilation. *Annu. Rev. Fluid Mech.* 31: 201–238.
7. D. Lohse, K.-Q. Xia. 2010. Small-scale properties of turbulent Rayleigh–Bénard convection. *Ann. Rev. Fluid Mechanics*, submitted.
8. J.J. Niemela, L. Skrbek, K.R. Sreenivasan, R.J. Donnely. 2000. Turbulent convection at very high Rayleigh numbers. *Nature* 404: 837–841.
9. U. Schumann, J. Benner. 1982. Direct solution of the discretized Poisson-Neumann problem on a domain composed of rectangles. *J. Comput. Phys.* 46: 1–14.
10. U. Schumann, R.A. Sweet. 1976. A direct method for the solution of Poisson's equation with Neumann boundary conditions on a staggered grid of arbitrary size. *J. Comput. Phys.* 20: 171–182.
11. O. Shishkina, A. Shishkin, C. Wagner. 2009. Simulation of turbulent thermal convection in complicated domains. *J. Comput. & Appl. Maths* 226: 336–344.
12. O. Shishkina, C. Wagner. 2006. Analysis of thermal dissipation rates in turbulent Rayleigh–Bénard convection. *J. Fluid Mech.* 546: 51–60.
13. O. Shishkina, C. Wagner. 2007. Local heat fluxes in turbulent Rayleigh–Bénard convection. *Phys. Fluids* 19: 085107.
14. O. Shishkina, C. Wagner. 2008. Analysis of sheet-like thermal plumes in turbulent Rayleigh–Bénard convection. *J. Fluid Mech.* 599: 383–404.
15. O. Shishkina, A. Thess. 2009. Mean temperature profiles in turbulent Rayleigh–Bénard convection of water. *J. Fluid Mech.* 633: 449–460.
16. E.D. Siggia. 1994. High Rayleigh number convection. *Ann. Rev. Fluid Mech.* 26: 137–168.
17. R.J.A.M. Stevens, R. Verzicco, D. Lohse, 2009. Radial boundary layer structure and Nusselt number in Rayleigh–Bénard convection. *J. Fluid Mech.*, submitted.
18. R. Verzicco, R. Camussi. 2003. Numerical experiments on strongly turbulent thermal convection in a slender cylindrical cell. *J. Fluid Mech.* 477: 19–49.

Determination of Acoustic Scattering Coefficients via Large Eddy Simulation and System Identification

S. Föller, R. Kaess, and W. Polifke

Abstract The scattering of acoustic waves at a sudden change in cross section in a duct system is investigated. In particular, a method is presented which allows to determine the coefficients of transmission and reflection of plane acoustic waves by combining large eddy simulation (LES) of turbulent compressible flow with system identification. The complex aero-acoustic interactions between acoustic waves and free shear layers are captured in detail, such that the scattering coefficients can be determined accurately from first principles. The method works as follows: First, a LES with external, broadband excitation of acoustic waves is carried out. Time series of acoustic data are extracted from the computed flow field and analyzed with system identification techniques to determine the acoustic scattering coefficients for a range of frequencies. The combination of broadband excitation with highly parallelized LES make the overall approach quite efficient, despite the difficulties associated with simulation of low-Mach number compressible flows.

1 Introduction

In contemporary urban environments, the emissions of noise from, e.g., transportation systems or air conditioning equipment have become detrimental to the well-being and health of millions of people. Consequently, the field of aero-acoustic analysis and design, which is decisive for noise prediction and reduction, has become of increasing importance. The development of novel methods, which aim for

Dipl.-Ing. Stephan Föller · Prof. Wolfgang Polifke, Ph.D.
Lehrstuhl für Thermodynamik, TU München, Boltzmannstr. 15, 85747 Garching, Germany
e-mail: foeller@td.mw.tum.de
e-mail: polifke@td.mw.tum.de

Dipl.-Ing. Roland Kaess
Astrium GmbH, Space Transportation, 81663 München, Germany
e-mail: roland.kaess@astrium.eads.de

S. Wagner et al. (eds.), *High Performance Computing in Science and Engineering, Garching/Munich 2009*, DOI 10.1007/978-3-642-13872-0_21,

an efficient design process that meets strict acoustical design objectives, has become a challenging focus of modern engineering sciences.

For the design of mufflers, after treatment devices in exhaust systems, or ventilation systems, a common approach to study the propagation and dissipation of sound in such duct systems utilizes low-order acoustic network models (e.g. [1, 15]). This approach allows for a quick assessment of the acoustic behavior under different operating conditions and for different geometries. The models consist of an assembly of discrete acoustic elements, i.e. "acoustic two-ports", which are represented mathematically by an acoustic scattering or transfer matrix. Of course, in order to predict the overall system behavior correctly, the matrix coefficients of the individual acoustic elements have to be known. Very simple elements like ducts can be described analytically. However, the accurate representation of geometries involving complex turbulent acoustic interactions requires advanced numerical methods. For this reason, a methodology has been developed to determine the acoustic scattering behavior of elements involving compressible turbulent wall-bounded flows. It combines acoustically excited, compressible large eddy simulations and acoustic signal analysis methods to obtain accurate low-order element descriptions.

2 Background

In the following, a brief introduction to the acoustic element notation is given and the possible approaches to determine their acoustic transfer behavior are discussed.

In linear aero-acoustics the transmission, reflection and absorption of plane, harmonic waves in ducts at frequencies below the first cut-on frequency are described completely by the so-called "scattering matrices" [1, 19]

$$\begin{pmatrix} f_d \\ g_u \end{pmatrix} = \begin{pmatrix} t_u & r_d \\ r_u & t_d \end{pmatrix} \begin{pmatrix} f_u \\ g_d \end{pmatrix}. \tag{1}$$

Here, the characteristic wave amplitudes f and g represent plane acoustic waves traveling along the duct in the downstream and upstream direction, respectively. They can be related to the acoustic velocity fluctuations u' and the acoustic pressure fluctuations p' by the following relations:

$$f = \frac{1}{2}\left(\frac{p'}{\rho c} + u'\right), \tag{2}$$

$$g = \frac{1}{2}\left(\frac{p'}{\rho c} - u'\right). \tag{3}$$

The coefficients of the scattering matrix (1) can be interpreted as transmission and reflection coefficients t and r for the characteristic wave amplitudes f and g in the down- and upstream direction (subscripts d and u), respectively. The coefficients depend strongly on the involved geometry, flow velocities, speed of sound and on the frequency of the impinging waves (e.g. for a sudden area expansion [2, 23]).

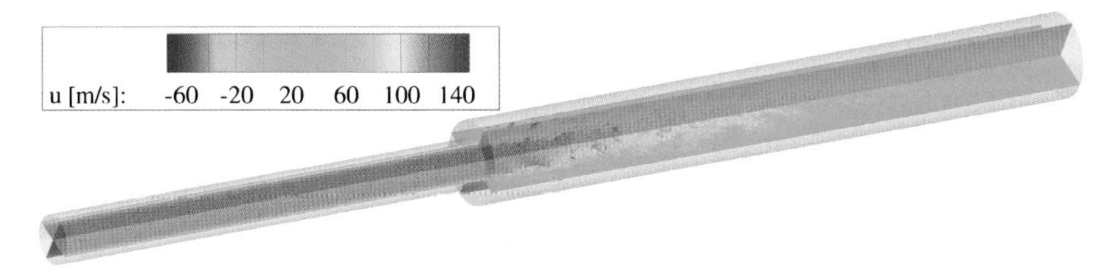

Fig. 1 Geometry of the cylindrical backward-facing step with contour plots of instantaneous axial velocity

The coefficients of the scattering/transfer matrix can be determined by analytical, experimental or numerical methods. Since analytical methods [2, 3, 11] are restricted to very simple configurations, like ducts or discrete area changes, only experimental approaches (e.g. the two-source method [1, 4]) or numerical, CFD-based techniques [7, 13] are applicable to more complex geometries and flow fields.

Polifke and co-workers [7, 20, 21] have proposed a hybrid CFD/SI approach, where time series data from a single CFD simulation are post-processed with system identification tools to determine transfer matrix coefficients over a broad frequency range. Until recently, applications of CFD/SI were based on unsteady Reynolds-averaged Navier-Stokes (URANS) simulations [7], where it is known that some turbulent flows cannot be treated accurately (e.g. flow separations, free shear layers, swirled vortex break-down or turbulent combustion). Hence, it must be expected that the CFD/SI approach based on URANS can not provide accurate results for the acoustic transmission behavior in the complex turbulent flows, because of the interaction between acoustical waves and an erroneous mean flow field. Furthermore, direct interactions of higher frequency waves and turbulent vortices are omitted by relying on a time and space averaged turbulence model. For this reason, a method that combines large eddy simulations (LES) and system identification methods (SI) has been developed. The LES approach resolves the large-scale, energy containing turbulent flow structures, whereas the small-scale turbulent fluctuations and their effects on the resolved scales are represented by a Sub-Grid-Scale (SGS) model.

In this project, the cylindrical backward-facing step (Fig. 1) has been selected to validate the LES/SI method [5, 6], because there are analytical [2, 3, 11] as well as experimental [23] results for the acoustic transfer behavior available. The use of LES/SI for the identification of flame transfer functions (of crucial importance for the analysis of thermo-acoustic flame transfer functions) is discussed in [8, 10, 24].

3 Method: LES/SI

The numerical determination of the acoustic transfer behavior consists of a two step process, a compressible LES of the acoustic element under consideration, followed by a system identification process based on recorded CFD data. In this case, the Wiener-Hopf-Inversion method is applied [14]. Figure 2 illustrates the procedure.

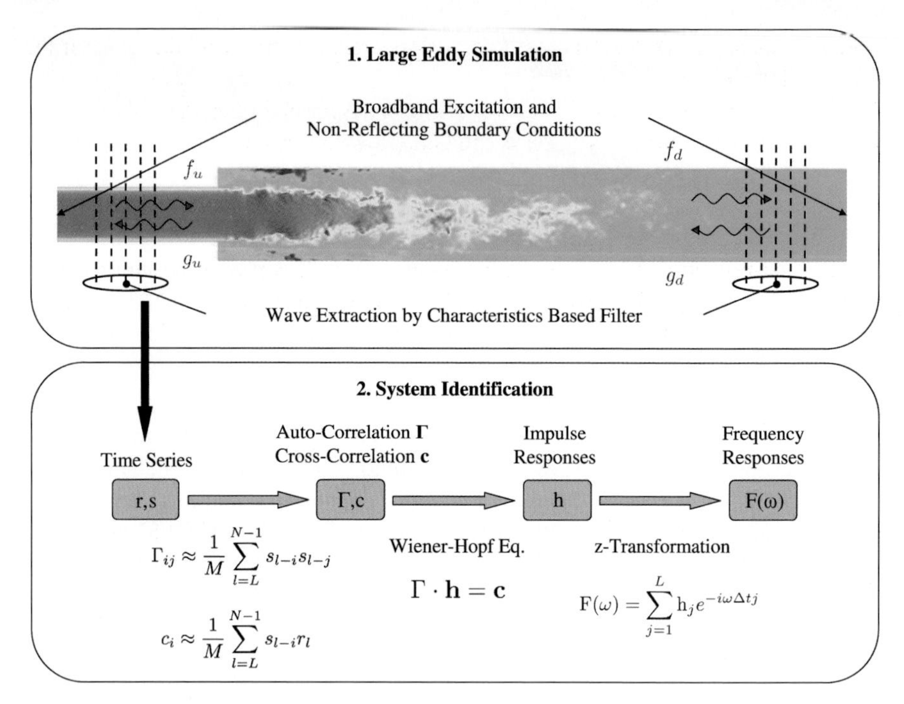

Fig. 2 Fundamental approach of the identification of acoustic scattering matrices by applying the Wiener-Hopf-Inversion method based on LES raw data

At first, a LES is set up. During the transient simulation, broadband, low-amplitude noise is superimposed on the mean flow at the in- and outflow boundaries of the computational domain. More precisely, the temporal variations of the wave amplitudes $\partial f/\partial t$ at the inlet condition and $\partial g/\partial t$ at the outlet condition are chosen as the excitation signals, according to Poinsot and Lele [17]. Hence, during the simulation, acoustic waves are travelling through the CFD domain simultaneously in the up- and downstream direction, while being reflected and transmitted at the sudden change in cross-section. The generation of vortices at the trailing edge of the area change in response to impinging acoustic waves – a significant mechanism for dissipation of acoustic energy – is also considered.

For system identification, ingoing characteristic wave amplitudes f_u and g_d are considered as signals s, while outgoing wave amplitudes f_d and g_u are considered as responses r. In URANS simulations, these signals and responses can easily be retrieved by subtracting the mean values of flow variables from the instantaneous values. On the other hand, the resolved turbulence fluctuations of a LES make it difficult to differentiate between acoustic and turbulent fluctuations of pressure and velocity. In order to extract acoustic signals from LES data, a characteristics based filter (CBF) has been developed by Kopitz et. al [12]. To identify the acoustic plane wave component, the fundamental effect that the acoustic waves propagate with the speed of sound, while the turbulent fluctuations are convected with the order of the average flow velocity, is utilized.

The accuracy of the identification procedure described above depends strongly on the reflection coefficients at the in- and outlet boundary conditions, as shown by Yuen et al. [25]. If the reflection coefficients are too large, the signals f_u and g_d will be correlated to some extent, which results in a significant error in the inversion of the Wiener-Hopf equation (the auto-correlation matrix Γ becomes ill-conditioned). Hence, non-reflecting boundary conditions are applied, ensuring that the acoustic reflections are kept at the lowest possible level. In the present simulations, a modified version of the characteristics-based boundary conditions (NSCBCs) of Poinsot and Lele [17] was used. As stated by Polifke et al. [22], the appearance of partial reflection at low frequencies makes it necessary to extend the original NSCBCs formulation with a correction term. This corresponds to the acoustic plane wave amplitudes leaving the domain. In order to further improve the boundary treatment proposed in [22], Kaess et al. [9] coupled the NSCBC with the CBF by utilizing the latter not only for extracting the acoustic signals, but also for allowing the determination of the required correction term very accurately. Beside the compliance with the strict, acoustic requirements, the boundary conditions are utilized to impose the operating conditions. Therefore, a velocity inlet and a pressure outlet are applied. In case of the inlet condition, the mean velocity profile of a fully developed turbulent pipe flow at the selected Reynolds number is imposed. Artificial turbulent fluctuations are not injected in the present simulations.

The post-processing is done by analyzing the recorded CFD time series of signals and responses (see Figure 2 right). Therefore, the correlation based Wiener-Hopf-Inversion (WHI) has been chosen to identify the acoustic transfer behavior of the simulated system [14, 18]. At first, the cross-correlation vector **c** and the auto-correlation matrix Γ are calculated. Afterwards, the Wiener-Hopf equation is set up and solved with respect to the unknown unit-impulse-response vector **h**. Finally, a z-transformation of the latter is performed to obtain the scattering coefficients of the system in the frequency domain.

4 Results

In the present project, the LES/SI method has been used to determine the scattering coefficients of a cylindrical backward-facing step (Fig. 1). At the area expansion, flow separation occurs and a shear layer is formed in the downstream pipe. In contrast to the simplicity of the geometry, a complex aero-acoustic interaction can be observed where the impinging acoustic waves directly interact with the unstable shear layer [3]. The introduced LES/SI approach should be able to capture these phenomena, since it allows for the direct interaction of the acoustic waves and the resolved vortical structures. Furthermore, experimental [23] and analytical data [2] are available to validate the LES/SI results. Therefore, the backward-facing step has been selected as a validation test case for the novel methodology.

Table 1 Geometrical information and operating conditions

Variable name	Symbol	Expression	Value
pipe diameter, upstream	D_u	-	0.05m
pipe diameter, downstream	D_d	-	0.085m
area ratio	A_R	$(D_u/D_d)^2$	0.35
pressure, ambient	p_0	-	101300Pa
temperature, ambient	T_0	-	298.15K
speed of sound, ambient	c_0	$\sqrt{\kappa R T_0}$	346.177m/s
Mach number, inlet	M	-	0.05...0.2

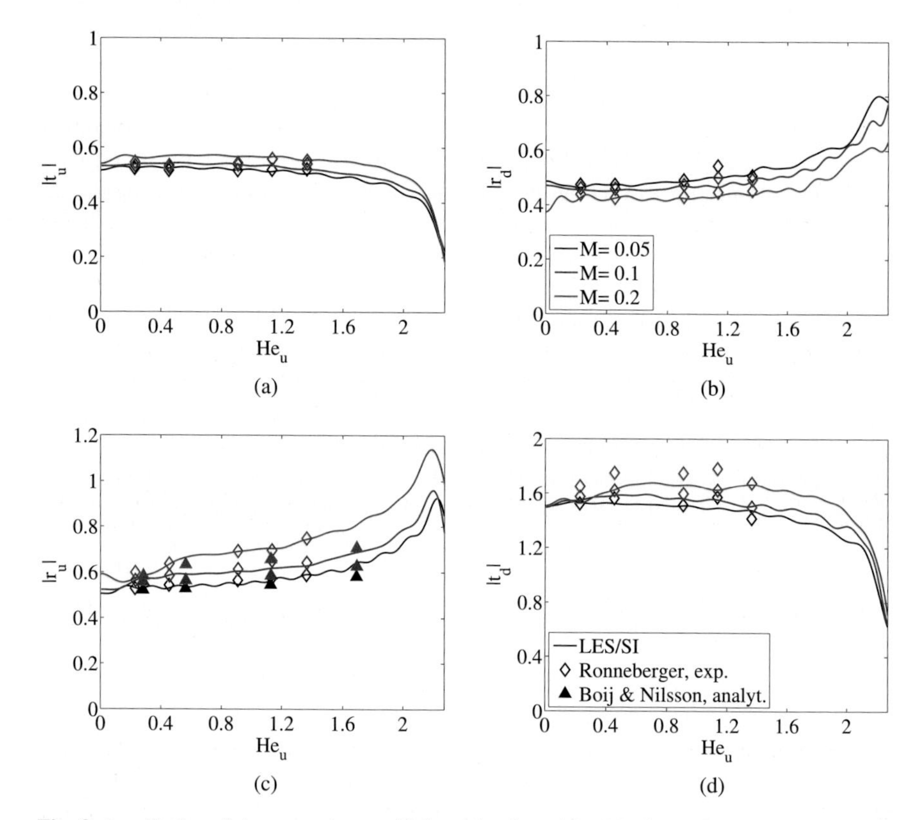

Fig. 3 Amplitudes of the scattering coefficients for three inlet Mach numbers, plotted over the upstream Helmholtz number and compared with results of Ronneberger [23] and Boij et al. [2]

The acoustical transfer behavior has been analyzed at different Mach numbers and ambient conditions which are summarized in Table 1. The area ratio has been set to 0.35 because for this ratio the majority of the analytical solutions and experimental measurements are available.

For a broad range of frequencies, the obtained results of the scattering coefficients are shown in Figures 3 and 4.

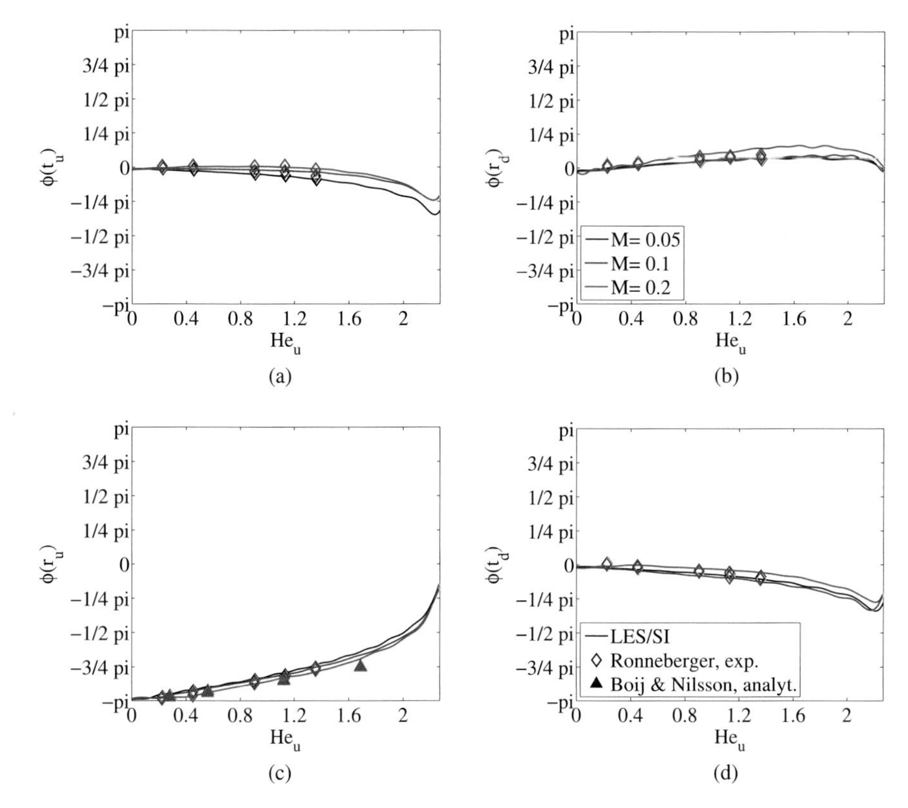

Fig. 4 Phase angle of the scattering coefficients for three inlet Mach numbers, plotted over the upstream Helmholtz number and compared with results of Ronneberger [23] and Boij et al. [2]

Here, the coefficients are plotted versus the upstream Helmholtz number

$$\mathrm{He}_u = \frac{\pi f D_u}{c_0}. \tag{4}$$

The good agreement of the LES/SI results with analytical solutions and experimental data demonstrates clearly that the LES/SI approach can be applied to characterize accurately the scattering behavior of compressible, turbulent wall-bounded flows in the linear acoustic regime.

At the moment, an improved version of the LES/SI method is being developed, since the currently used WHI post-processor is not able to provide accurate results for systems having strong eigenfrequencies in the analyzed frequency range (e.g. combustion chambers) or generating flow induced sound of high intensity (e.g. shear flows at high Reynolds numbers). The influence of this high intensity sound is to some extent visible in the amplitudes of the scattering coefficients, see Fig. 3. The small periodic fluctuations in the amplitudes of the LES/SI results are caused by the turbulence induced sound waves. Thus, the improved post-processing method will provide a much higher level of robustness against self-induced turbulent sound and strong eigenfrequencies. Towards the end of the present project, the LES/SI will be applied to more challenging aero-acoustic configurations like the cylindrical orifice in a pipe and the T-joint of two cylindrical pipes.

5 Numerical Methods and High Performance Computing

For the aero-acoustical LES, the flow solver AVBP developed by CERFACS[1] is used, an explicit solver for the compressible, three dimensional Navier-Stokes equations on unstructured grids. The spatial and temporal derivatives are expressed by discretization schemes of second or third order, i.e. the Finite Volume based Lax-Wendroff scheme (2nd order accuracy) or the Finite Element based Two-Step Taylor-Galerkin scheme (3rd order accuracy). The unresolved small scale turbulent fluctuations can be modeled by different SGS models. Here, the so-called Wall Attached Layer Eddy (WALE) model developed by Nicoud and Ducros [16] and the classical Smagorinsky model are applied. For the flow configuration analyzed in this paper, the two SGS models cause only negligible differences in the results of the acoustical scattering coefficients. However, the classical Smagorinsky model should be used with care for the analyzed aero-acoustical configurations, since the SGS modelling approaches influence the initial develop of coherent turbulent structures in the shear layer and, therefore, have an impact on possible interactions of acoustic waves and coherent turbulent structures. From this point of view, the WALE model is more preferable due to its superior ability in modelling the subgrid-scale fluctuations of shear flows.

In terms of parallelization, the AVBP solver can be optimally adapted to a wide variety of processor architectures. Tests have shown that it can be highly parallelized with an optimal load average per CPU, especially if high performance network interconnections are available. A version of AVBP optimized for the HLRB II has been provided by CERFACS and is in use by other projects of the Lehrstuhl für Thermodynamik, TU München. In order to test the scaling behavior of the solver on the HLRB II, a LES with 4.3 million grid nodes has been set up and jobs with different size classes have been conducted. In Figure 5, the results for the scaling behavior are shown, indicating an almost ideal speedup.

The required CPU time on the cluster systems of the Leibniz-Rechenzentrum, Garching (LRZ) is mainly governed by the grid size and the time step of the aero-acoustical LES. These parameters depend on two requirements. On the one hand, the adequate resolution of the large-scale turbulent vortices has to be ensured to yield accurate results of the flow field. On the other hand, the sound waves have to be accurately captured. Thus, the mesh resolution depends on the Reynolds number and the upper frequency limit for which the acoustical transfer behavior is to be determined. The typical mesh consists only of hexahedral cells and its size varies

[1] `http://www.cerfacs.fr/4-26334-The-AVBP-code.php`

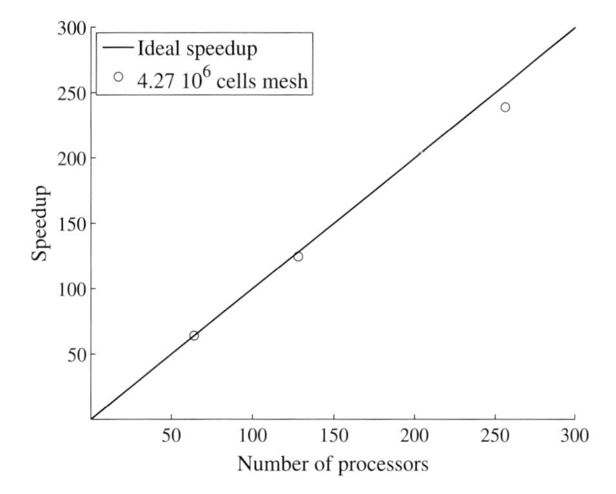

Fig. 5 Scaling behavior of the flow solver AVBP on high bandwidth partitions of HLRB II

Table 2 Averaged mesh resolution, normalized by wall units (r: radial direction, φ: circumferential direction, z: axial direction)

resolution in direction	M=0.05	M=0.1	M=0.2
r^+, z^+	55	70	120
φ^+	60	90	150
$r^+_{\text{near-wall region}}$	35	35	35

between 2 and 8 million cells. For the analyzed configurations, the number of cells result in an averaged mesh resolution which is summarized in Table 2.

In combination with the mesh refinement in the radial direction close to the walls, a two-layer wall function is applied which is based on the wall-shear velocity and dynamically switches between the linear or the logarithmic law in dependence of r^+.

The corresponding time step Δt_{CFD} is calculated utilizing the Courant-Friedrichs-Lewy (CFL) stability criterion. In combination with the smallest cell size, this yields values of the order of $5 \cdot 10^{-7}$ s. Beside the upper frequency limit, there is a lower frequency limit which determines the minimal duration of the LES. Experience shows that the time series length should correspond to at least 10 cycles of the lowest frequency from the simulation. Moreover, a statistically stationary initial solution for each analyzed flow condition has to be obtained prior to the LES run with external excitation. In combination with the explicit time stepping scheme, the sufficient mesh resolution, the associated time step and the simulated flow time, the number of the iterations per simulation is determined (see Table 3). The given iteration numbers take into account both the iterations required to obtain the initial solution and the iterations for the simulation with acoustic excitation.

The CPU time per iteration and cell has been obtained from simulations which are representative for the analyzed configurations and have been run on the HLRB II. With this information, the overall required CPU time is estimated and one obtains

Table 3 Averaged numerical parameters and CPU time

Simulation parameters	Value
Δt_{CFD}	$5 \cdot 10^{-7}$s
t_{CFD}	0.35s
No. of cells	$5 \cdot 10^{6}$
No. of iterations	$7 \cdot 10^{5}$
CPU time/(iteration cell)	$1.79 \cdot 10^{-5}$s
CPU time/iteration	89.3s
CPU time/simulation	$1.74 \cdot 10^{4}$h

in total an amount of $1.74 \cdot 10^4$ CPU hours per LES. In order to obtain results in a reasonable time frame, the selected degree of parallelization varies between 32 and 256 processors depending on the size of the computational mesh and time step. Up to now, mainly the Linux-Cluster of the LRZ has been used to perform the first LES runs. The results in Figures 3 and 4 have been obtained utilizing the queue segments x64_mpi_8 and x64_mpi_32 of this cluster. But due to the practical parallelization limit of 32 to 64 nodes on these segments, it is intended to shift the majority of future computations to the HLRB II in order to decrease the duration of the large-scale simulation to less than one week.

6 Summary

An efficient hybrid approach of broadband excited LES combined with system identification methods makes it possible to study aero-acoustic interactions in turbulent wall-bounded flows. Results obtained to date for the reflection and transmission coefficients of plane acoustic waves at a sudden change of cross section in a duct exhibit excellent agreement with analytical and experimental data.

The approach benefits clearly from the high parallelization capabilities of the flow solver AVBP, which exhibits an almost ideal speedup on the high performance cluster systems of the LRZ. In combination with broadband identification, it is possible to determine acoustic scattering coefficients for a relevant range of frequencies within a few days.

In future work, it is planned to increase the robustness of the approach against turbulent noise ("pseudo-sound") by introducing novel identification techniques. Once this is achieved, it will be possible to determine the aero-acoustic properties of exhaust system or ventilation duct components from first principles.

Acknowledgements Financial support was provided by the Deutsche Forschung Gemeinschaft (DFG), project numbers PO 710/5. The authors would like to thank the CFD team of CERFACS, which provided the AVBP flow solver, and the LRZ, for the access to the high performance cluster systems.

References

1. Åbom, M., Bodén, H.: Modelling of fluid machines as sources of sound in duct and pipe systems. Acta Acustica **3**(6), 549–560 (1995)
2. Boij, S., Nilsson, B.: Reflection of sound at area expansions in a flow duct. Journal of Sound and Vibration **260**(3), 477–498 (2003). URL `http://www.sciencedirect.com/science/article/B6WM3-47RRW1S-6/2/58ba97cbe5a42cbf75f1f6e9c5bb867c`
3. Boij, S., Nilsson, B.: Scattering and absorption of sound at flow duct expansions. Journal of Sound and Virbration **289**, 577–594 (2006)
4. Durrieu, P.: Quasisteady aero-acoustic response of orifices. Journal of the Acoustical Society of America **110**, 1859–1872 (2001)
5. Föller, S., Kaess, R., Polifke, W.: Reconstruction of acoustic transfer matrices from large-eddy-simulations of complex turbulent flows. In: 14th AIAA/CEAS Aeroacoustics Conference (29th AIAA Aeroacoustics Conference), AIAA-2008-3046. AIAA/CEAS, Vancouver, Canada (2008)
6. Föller, S., Kaess, R., Polifke, W.: Determination of acoustic scattering coefficients via LES and system identification. In: Euromech Colloquium 540 "Large-Eddy Simulation for Aerodynamics and Aeroacoustics". Euromech, Munich, Germany (2009)
7. Gentemann, A.M.G., Fischer, A., Evesque, S., Polifke, W.: Acoustic transfer matrix reconstruction and analysis for ducts with sudden change of area. In: 9th AIAA/CEAS Aeroacoustics Conference, AIAA 2003-3142. AIAA, Hilton Head, S.C., U.S.A. (2003)
8. Giauque, A., Poinsot, T., Nicoud, F.: Validation of a Flame Tranfer Function Reconstruction Method for Complex Turbulent Configurations. In: 14th AIAA/CEAS Aeroacoustics Conference (29th AIAA Aeroacoustics Conference), AIAA-2008-3046. AIAA/CEAS, Vancouver, Canada (2008)
9. Kaess, R., Huber, A., Polifke, W.: A time-domain impedance boundary condition for compressible turbulent flow. In: 14th AIAA/CEAS Aeroacoustics Conference (29th AIAA Aeroacoustics Conference), AIAA-2008-2921. AIAA/CEAS, Vancouver, Canada (2008)
10. Kaess, R., Poinsot, T., Polifke, W.: Determination of the stability map of a premix burner based on flame transfer functions computed with transient CFD. In: 4th European Combustion Meeting. The Combustion Institute, Vienna, Austria (2009)
11. Kooijman, G.: Acoustical response of shear layers. Ph.D. thesis, Technische Universiteit Eindhoven (2007)
12. Kopitz, J., Bröcker, E., Polifke, W.: Characteristics-based filter for identification of acoustic waves in numerical simulation of turbulent compressible flow. In: 12th Int. Congress on Sound and Vibration (ICSV12), 389. IIAV, Lisbon, Portugal (2005)
13. Kopitz, J., Polifke, W.: Stability analysis of thermoacoustic systems by determination of the open-loop-gain. In: 12th Int. Congress on Sound and Vibration (ICSV12). IIAV, Lisbon, Portugal (2005)
14. Ljung, L.: System Identification - Theory For the User. Prentice Hall (1999). 2nd Edition
15. Munjal, M.L.: Acoustics of Ducts and Mufflers. John Wiley & Sons (1986)
16. Nicoud, F., Ducros, F.: Subgrid-scale stress modelling based on the square of the velocity gradient tensor. J. of Flow, Turbulence and Combustion **62**, 183–200 (1999)
17. Poinsot, T., Lele, S.K.: Boundary conditions for direct simulation of compressible viscous flows. J. of Comput. Phys. **101**(1), 104–129 (1992)
18. Polifke, W.: Numerical techniques for identification of acoustic multi-poles. In: Advances in Aeroacoustics and Applications, VKI LS 2004-05. Von Karman Institute, Brussels, BE (2004)
19. Polifke, W.: System modelling and stability analysis. In: Basics of Aero-Acoustics and Thermo-Acoustics, VKI LS 2007-02. Von Karman Institute, Brussels, BE (2007)
20. Polifke, W., Gentemann, A.M.G.: Order and realizability of impulse response filters for accurate identification of acoustic multi-ports from transient CFD. Int. J. of Acoustics and Vibration **9**(3), 139–148 (2004)

21. Polifke, W., Poncet, A., Paschereit, C.O., Döbbeling, K.: Reconstruction of acoustic transfer matrices by instationary computational fluid dynamics. J. of Sound and Vibration **245**(3), 483–510 (2001)
22. Polifke, W., Wall, C., Moin, P.: Partially reflecting and non-reflecting boundary conditions for simulation of compressible viscous flow. J. of Comp. Phys. **213**(1), 437–449 (2006)
23. Ronneberger, D.: Theoretische und experimentelle Untersuchung der Schallausbreitung durch Querschnittssprünge und Lochplatten in Strömungskanälen. Tech. rep., DFG-Abschlussbericht, Drittes Physikalisches Institut der Universität Göttingen (1987)
24. Tay Wo Chong, L., Kaess, R., Komarek, T., Föller, S., Polifke, W.: Identification of flame transfer functions using les of turbulent reacting flows. In: High Performance Computing in Science and Engineering, Garching 2009. Springer, LRZ, Garching, Germany (2009)
25. Yuen, S.W., Gentemann, A.M.G., Polifke, W.: Investigation of the influence of boundary conditions on system identifiability using real time system modeling. In: 11th Int. Congress on Sound and Vibration (ICSV11), pp. 3501–3508. IIAV, Saint-Petersburg, Russia (2004)

Identification of Flame Transfer Functions Using LES of Turbulent Reacting Flows

L. Tay Wo Chong, R. Kaess, T. Komarek, S. Föller, and W. Polifke

Abstract Combustion systems can exhibit detrimental thermoacoustic instabilities, driven by unsteady heat release of the flame. For analysis and control of such instabilities, a characterization of the flame dynamics in terms of the so -called Flame Transfer Function (FTF), i.e. the normalized ratio of heat release to velocity fluctuations, is required. In the project reported here, LES simulations of a turbulent premixed swirl burner were carried out using the LES Solver AVBP developed at CERFACS. The system was excited using broadband perturbations at the inlet and then system identification methods were used to identify the FTF. Combustion was modeled using the Thickened Flame model with one-step kinetics. Results were compared to experiments and favorable agreement was obtained. Calculations required significant numerical resources and were carried out on the Linux Cluster and on the HRBLII Supercomputers of the Leibniz-Rechenzentrum in Garching using up to 256 cores.

1 Introduction

Stringent emission regulations (for NOx, CO, etc.) have been established for gas turbines. In order to comply with these regulations, the lean premixed combustion technology, which offers a number of advantages in controlling emission levels,

L. Tay Wo Chong · T. Komarek · S. Föller · W. Polifke
Lehrstuhl für Themodynamik, TU München, Boltzmannstr. 15, 85748, Garching, Germany
e-mail: tay@td.mw.tum.de
e-mail: komarek@td.mw.tum.de
e-mail: foeller@td.mw.tum.de
e-mail: polifke@td.mw.tum.de

R. Kaess
Astrium GmbH, Space Transportation, 81663, München, Germany
e-mail: kaess@astrium.eads.de

S. Wagner et al. (eds.), *High Performance Computing in Science and Engineering, Garching/Munich 2009*, DOI 10.1007/978-3-642-13872-0_22,

has been introduced for stationary engines. Using this technology, the combustors operate with excess air to reduce the combustion temperature. However, this mode of operation makes the combustor prone to blow-out or flashback and in particular thermoacoustic instabilities.

Thermoacoustic instabilities, i.e. a self-excited coupling between fluctuations of pressure or velocity and heat release, can lead to very high levels of pressure pulsations in the combustor, possibly resulting in structural damage. For this reason, the correct description of the flame dynamics has significant technological importance. To analyze the stability of a combustion system, it is necessary to know how the heat release of a flame responds to perturbations of velocity by the determination of the Flame Transfer Function. The Flame Transfer Function (FTF) describes the relevant dynamic properties of a flame in terms of amplitude and phase. Unfortunately, the experimental determination of FTFs for configurations of technical interest is very difficult and prone to measurement errors.

Alternatively, it is possible to determine the FTF from computational fluid dynamics simulation: First an unsteady CFD simulation is performed to generate time series of fluctuating velocity and heat release rate. Then the FTF is reconstructed from the data, using methods from system identification (SI) [5, 6, 8]. For the numerical simulation of turbulent flows, Large Eddy Simulation (LES) is now established as a powerful, albeit computationally expensive, tool. It is a technique intermediate between Direct Numerical Simulations and the solution of the Reynolds-averaged equations (RANS). Recent numerical predictions of turbulent reacting flows using LES have shown the potential of the approach for laboratory and industry scale configurations [4]. Nevertheless, LES demands high computational resources with massively parallel computations in order to achieve results in short time (*LES is not slow, it is just expensive!*[1]). In the present work, LES simulations of a perfectly premixed axial swirl burner were carried out using the compressible solver AVBP on the Linux Cluster and on the HRBLII Supercomputers of the Leibniz-Rechenzentrum in Garching, Germany using up to 256 cores with high parallelization efficiency. The system was excited using broadband perturbations at the inlet and then system identification methods were used to identify the FTF.

2 Background

2.1 Large Eddy Simulations

In LES, the turbulent large scales are calculated explicitly, whereas the effects of smaller ones are modeled using subgrid closures. To separate the large from the small scales, LES is based on the definition of a filtering operation considering a filter width, i.e., the wavelength of the smallest scale retained by the filtering operation. The filter function determines the size and structure of the small scales. A complete review on the filtering approach for LES can be found in [17, 21].

[1] to quote T. Poinsot

2.2 Combustion Model

In LES the thickness of a premixed flame is typically smaller than the mesh size. Due to this, the reaction source term of the species transport equations needs to be modeled. The thickened Flame Model [3] is based on artificially thickening the flame by a factor F, while maintaining the laminar flame speed of an unthickened flame. The transport equations of the species from the chemical scheme are solved and the reaction rate is expressed using an Arrhenius law, as in Direct Numerical Simulations. The thickening can be achieved based on laminar flame premixed theory, multiplying the species and thermal diffusivities by the factor F, while dividing the Arrhenius pre-exponential factor of the reaction rate by the same factor. In this way, the flame can be resolved on a LES mesh. Due to the thickening of the flame, the interaction between turbulence and chemistry is changed, having a reduction on the flame surface and reaction rate. To take into account this effect, an efficiency function E, corresponding to a subgrid scale wrinkling factor, is included in the model [3, 17]. Then the species and energy filtered transport equations have to be modified for reacting flow simulations, while the mass and momentum filtered equations are unmodified.

2.3 The Flame Transfer Function

The dynamic response of a flame to a perturbation is represented by its flame transfer function $F(\omega)$. It relates the acoustic velocity fluctuations u' upstream of the flame with the flame heat release fluctuations $\dot{q}'$ in the frequency domain by the following expression:

$$F(\omega) = \frac{\dot{q}'(\omega)/\bar{\dot{q}}}{u'(\omega)/\bar{u}} \tag{1}$$

The fluctuations are normalized with their mean values of heat release $\bar{\dot{q}}$ and velocity $\bar{u}$.

2.4 LES/SI

The hybrid approach presented in this paper is a combination of large eddy simulation with System Identification (SI) methods, as they have been developed in digital signal processing. The flame is considered as a "black-box" input-output system, with the velocity just upstream of the flame as input, and the total heat release rate as output. For sufficiently small levels of perturbation, the response may be assumed to be linear (and time-invariant). Then the "unit impulse response" $\mathbf{h}$ describes the relation between the input and output signals in the time domain. From the unit impulse response, the flame transfer function $F(\omega)$ is obtained by

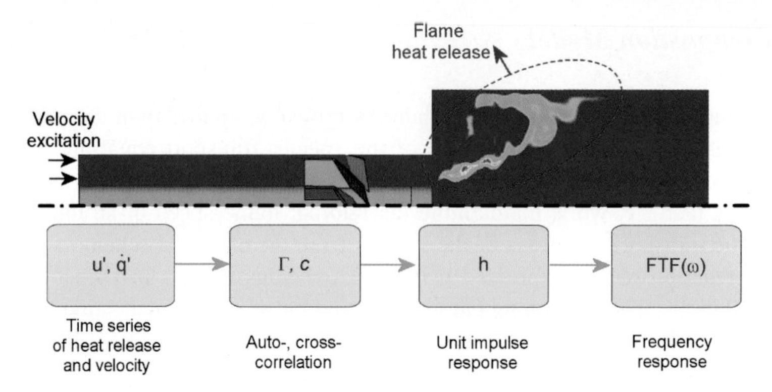

Fig. 1 LES/System Identification process

z-transform (the discrete time equivalent of the Fourier transform). A flow chart of the LES/SI method is given in Fig. 1. First, a LES simulation of the system under consideration is set up. After obtaining a statistically stabilized solution, the system is excited with a broadband perturbation superimposed on the mean flow. Variations of the characteristic ingoing wave at the inlet are chosen as excitation perturbations. These perturbations will propagate to the flame front and create a response in the heat release of the flame. A plane is defined upstream of the flame, and area averaged instantaneous axial velocity values are exported every time step from this plane. The global heat release is obtained by a volume integration of the heat release over the computational domain and exported also every time step. These time series of the simulations will be imported into a post-processor. The mean values of axial velocity and heat release are subtracted from the exported time series to obtain their fluctuations. These values are normalized with the mean values of velocity and heat release. After this procedure, the auto-correlation Γ and cross-correlation $\mathbf{c}$ of the signals are calculated. Then, the Wiener-Hopf equation, an optimal linear least square estimator, defined by [18]:

$$\Gamma \mathbf{h} = \mathbf{c}, \tag{2}$$

is inverted to obtain the unit impulse response $\mathbf{h}$ of the signals. Through a z-transformation of the unit impulse response, the flame transfer function can be obtained.

3 Numerical Set Up

The BRS burner is a swirl stabilized premix burner with an axial swirl generator mounted on a central bluff body. It was developed and investigated within the framework of the KW 21 project GV 6 Premixed Flame Dynamics by Komarek and Polifke [12]. The methane air mixture enters the plenum which is followed by the

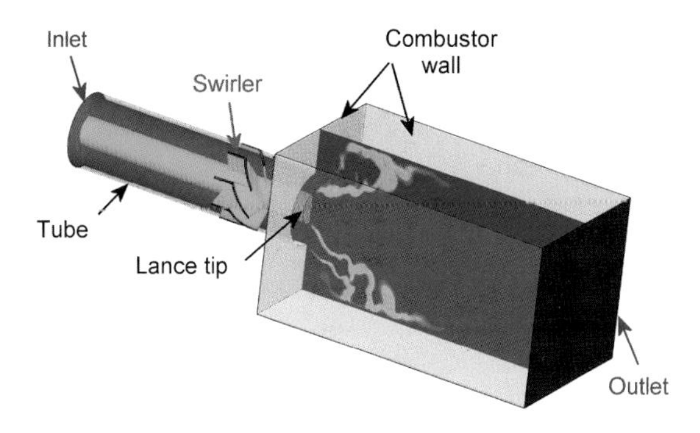

Fig. 2 Scheme of the numerical set up of the BRS burner

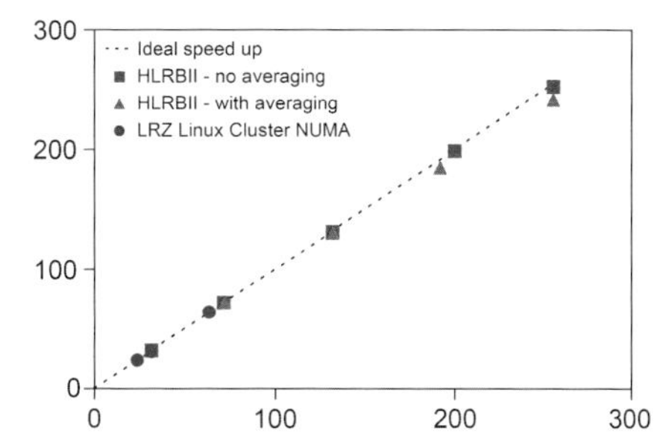

Fig. 3 Parallelization behavior for different number of CPUs. 7.5 million cells mesh. Reference for ideal speed up: 32 CPUs for HLRBII and 24 CPUs for the Linux Cluster

burner with an axial swirl generator. The combustion chamber has a length of 300 mm and a quadratic cross section of 90×90 mm. The burner is operated in "perfectly premixed" mode with a completely homogeneous mixture of natural gas and air to eliminate any equivalence ratio fluctuations. Details about the experimental settings and the FTF measurements can be found in [12]. For the simulations, the plenum was removed to introduce an excitation signal with more uniform power spectral distribution for the identification of the FTF. A full 3D unstructured mesh consisting of around 7.5 million cells was created using the program Gambit. In Fig. 2, the scheme of the numerical set up of the BRS burner is presented.

For the present work, the Finite Volume based LES solver AVBP [2, 4, 22] from CERFACS was used. AVBP exhibits a very good parallelization scaling behavior, this it is well suited for massively parallel computations. Calculations were carried out on the Linux cluster and the HRLBII supercomputers of LRZ München. For a run with broadband excitation, 256 CPUs were used. The speed-up for this amount of CPUs is almost linear as shown in Fig. 3, with a loss of performance of only about 5% when the averaging option is activated.

The solver produces the grid partitions using standard algorithms like the recursive inertial method [1]. The aim is to minimize the interfaces generated by the partitioning process. It uses the Message Passing Interface (MPI) for the communication between the processes; and for the application, a Master-Slave parallelization scheme is chosen. Apart from the generation of complete output files, an interval of iterations can be selected after which the master process collects certain data from the slave processes and writes it to an output file, in order to provide e.g. convergence data. This process was modified in order to collect acoustic information at certain predefined monitor planes [9, 11]. Additionally, as already used in [11], a domain integrated reaction rate is written out "on the fly", allowing to collect the necessary time series data during the calculations and saving in this way a huge amount of disk space.

In the calculations, the full compressible multi-species Navier-Stokes equations are solved using a methane-air mixture with equivalence ratio of 0.77 at atmospheric conditions for a power rating of 50 kW. The subgrid stresses are modeled using the WALE model [15]. Reacting flow calculations were carried out using the Thickened Flame combustion model with one step kinetics and a thickening factor of 5. The Lax-Wendroff second order numerical scheme [14] is chosen as discretization scheme. The temporal integration is carried out using a single step explicit scheme.

3.1 Boundary Conditions

For the inlet and outlet, Kaess et al. [10] developed a modified version of the non-reflecting boundary conditions of Polifke et al. [20], which in turn are based on the characteristics-based boundary conditions of Poinsot and Lele [16]. The technique of "plane wave masking" was combined with a characteristics based filter as proposed by Kopitz et al. [13] to differentiate between acoustic and turbulent fluctuations. LES simulations were carried out considering adiabatic and non-adiabatic conditions on the combustor walls. The non-adiabatic combustor walls and the lance tip are no-slip isothermal walls with a temperature of 600K, which is an estimated value. The simulations with excitation were done only for non-adiabatic conditions. The boundary conditions are indicated in Table 1 and shown in Fig. 2.

In compressible LES simulations, the acoustic waves need to be properly resolved while they are traveling through the CFD domain. Due to the necessary fine resolution of the flow field, flame and acoustics, the resulting time step size of the CFD is 1.25e-7 s, which is based on an acoustic CFL number [14] lower than 0.7. Thus the use of high performance computational resources is necessary. For example, for the calculation of a FTF, the overall excited simulation length should be higher than 0.2 seconds to identify low frequency waves. Then a simulation using LES/SI will need at least 1.6e6 iterations. For this study, a reacting flow calculation

Table 1 Boundary Conditions

Boundary Condition	Type	Details
Inlet	Non-reflective velocity inlet	V_{inlet}=19 m/s
Outlet	Non-reflective pressure outlet	P_{outlet}=101325 Pa
Combustor wall (Adiabatic)	Adiabatic no slip wall	-
Combustor wall (Non-adiabatic)	Isothermal no slip wall	T_{Wall_isot}=600 K
Tube/swirler	Adiabatic no slip wall	-
Lance tip	Isothermal no slip wall	T_{Wall_isot}=600 K

takes approximately 55 minutes to perform 10000 iterations with 256 CPU's in the HLRBII supercomputer from LRZ. Then, a complete simulation will take approximately 6 days.

4 Results and discussion

As a first step, the averaged flow field from non-reacting and reacting flow simulations without excitation are compared with experiments using particle image velocimetry. The averaging time was 37.5 ms for the non-reacting and reacting case. This represents 300000 iterations with a time step of 1.25e-7 s. In Fig. 4, axial velocity profiles at the middle cross plane for different positions of the combustor are shown. The agreement with experiment for non-reacting flow is very good. The size of the recirculation zone and velocity magnitudes are well reproduced. The presence of a precessing vortex core was observed. For the reacting case, the agreement is also good, although the velocities inside of the recirculation zone are not well reproduced. In both cases, inner and outer recirculation zones are created due to the strong swirl and the confinement created by the combustion chamber.

For the reacting flow case, in order to compare the heat release from simulations with the one from experiments, the heat release was integrated over the depth of the combustion chamber, in order to determine the distribution in correspondence to the line-of-sight integrated measurements of the OH Chemiluminescence. The agreement is very good as shown in Fig. 5. The axial heat release distribution from experiments and from simulations at non-adiabatic conditions exhibits also good agreement. Considering adiabatic walls, the flame predicts an incorrect stabilization in both shear layers, producing a flame much shorter than in the experiments. The

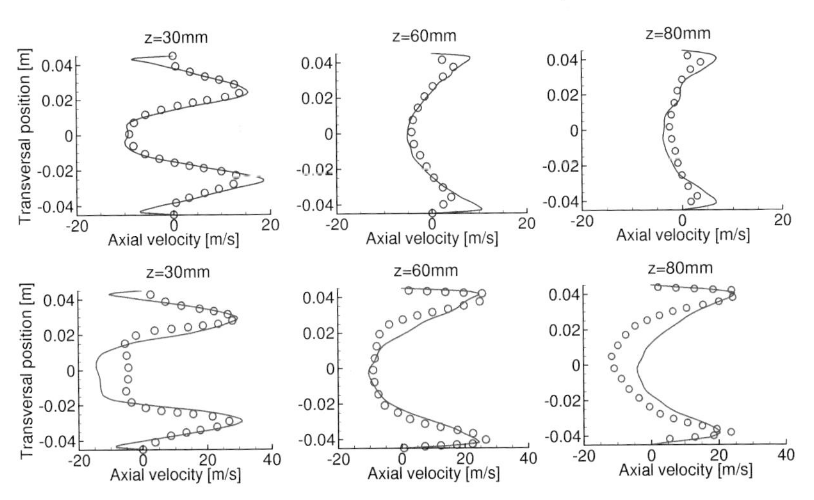

Fig. 4 Axial velocity profiles with non-reacting flow (up) and with reacting flow (down) at the middle cross plane shown in Fig. 2. o Experiments, - LES

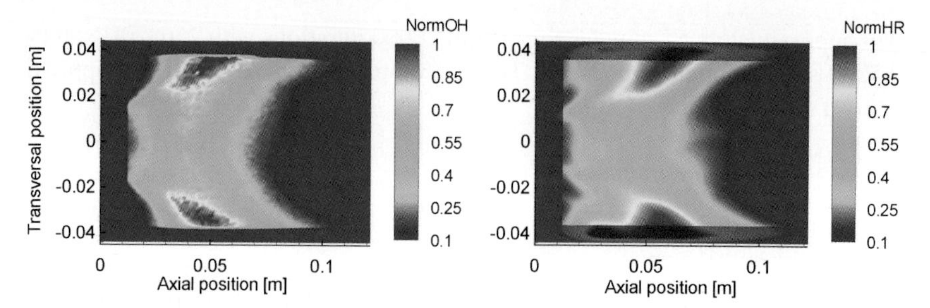

Fig. 5 Spatial distribution of heat release: OH chemiluminescence from experiments (left), line-of-sight integration of heat release from LES (right). Both are normalized with their maximum value

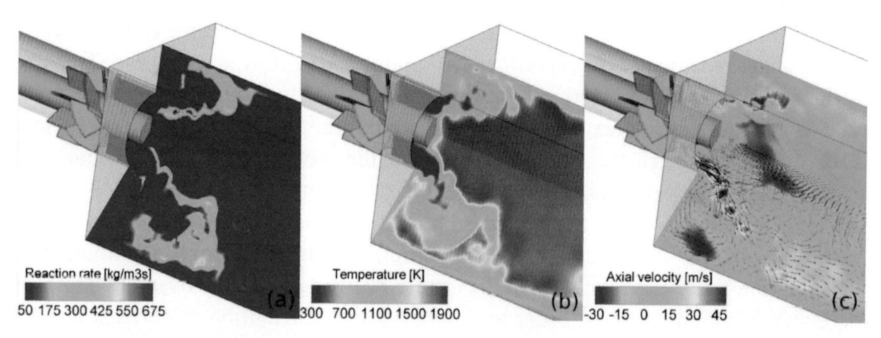

Fig. 6 Instantaneous reaction rate (a), temperature (b) and axial velocity (c) of the BRS burner with non-adiabatic conditions

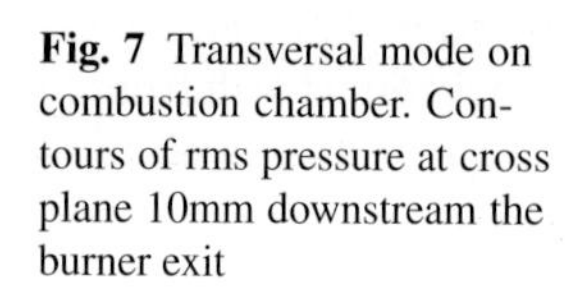

Fig. 7 Transversal mode on combustion chamber. Contours of rms pressure at cross plane 10mm downstream the burner exit

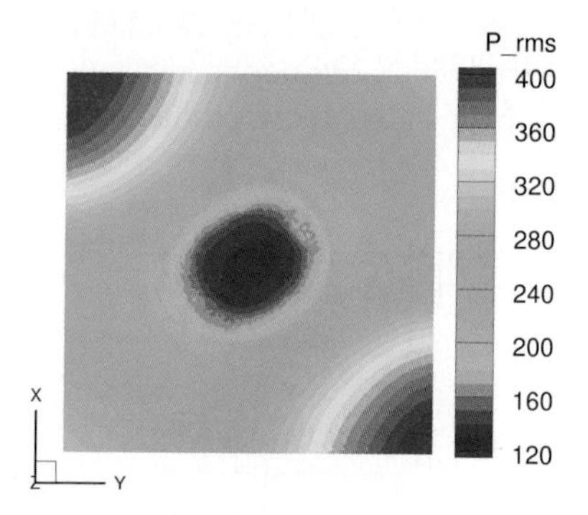

non-adiabatic walls cool down the combustion products, which are transported to the outer shear layer by the outer recirculation zone, quenching the flame in the outer shear layer. In Fig. 6, the instantaneous reaction rate, temperature and axial velocity are shown. The influence of non-adiabaticity on the flame stabilization is clearly seen.

During the simulations, a transversal eigen-mode at 3920Hz developed in the combustion chamber, which matches with the transversal eigen-frequency of the geometry. The diagonal mode shown in Fig. 7 is due to a combination of the two degenerate transversal modes.

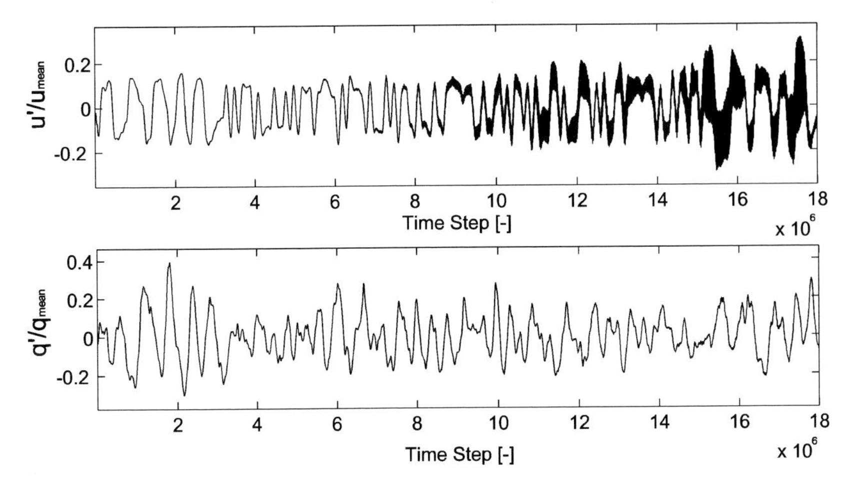

Fig. 8 Heat release (q') and axial velocity (u') fluctuations normalized by their mean values

4.1 Identification of Flame Transfer Function

Perturbations on the characteristic ingoing wave were imposed at the inlet using a broadband excitation with a frequency-limited discrete random binary signal [7, 8] and an amplitude of 6.5% of the inlet mean velocity. The waves travel through the connecting tube and swirler and reach the flame, creating, e.g. a variation of the flame surface. This variation produces a change in the turbulent flame speed and consumption rate, which depends on the excitation frequency. Other effects like swirl number variation [12] also influence the flame response. The simulation was run for 225ms in real time. This simulation time gives a frequency resolution of 4.4Hz, which nominally is also the lowest resolved frequency. The signals were filtered to remove the frequency content higher than 600Hz and the Wiener-Hopf inversion was applied to obtain the FTF. The FTF identified from a single time series of length 1800000 (STS) is presented in Fig. 9. The level of agreement with the experimental data is not satisfactory. It is expected that the significant disagreements are due to the influence of "noise", i.e. turbulent "pseudo sound", fluctuations of the heat release produced by resolved turbulent fluctuations, and perturbations introduced by the transversal modes, which grow to significant amplitudes after about 750000 iterations, as shown in Fig. 8. These perturbations introduce spurious contributions to the auto- and cross-correlations Γ and $\mathbf{c}$ of signal and response, respectively, and thus deteriorate the quality of the identification scheme. In general, the identification process can be improved using a longer time series [7]. This was not possible in the present case due to the unstable transversal mode, which kept growing in amplitude as the run continued.

Attempts to suppress the transversal mode were not successful, therefore a "work-around" was devised: In order to obtain a more robust estimate of auto- and cross-correlations Γ and $\mathbf{c}$, time series generated from several LES runs were com-

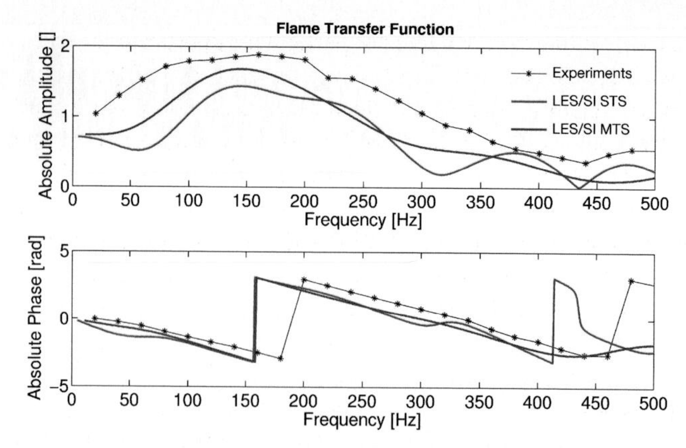

Fig. 9 Flame transfer function from experiments and from LES/SI with a single signal (STS) and with multiple signals (MTS)

bined. Each run started from the same initial condition, but used different excitation time series, generated from different seeds for the random number generator. The auto- and cross-correlation coefficients from different runs were then combined into multiple time series (MTS) auto- and cross-correlation, thus suppressing the relative magnitude of uncorrelated noise. Results for the flame transfer function from multiple time series (MTS) are also shown in Fig. 9. In that case, four different excitation signals of length 750000 with the same statistical characteristics as the single signal excitation were used. It can be seen that the agreement with experiment is noticeably improved.

The amplitude of the Flame Transfer Function should reach 1 in the limit of low frequency [19]. Nevertheless, the identified FTF exhibits an amplitude of 0.75 at the lowest frequency identified (10Hz). Huber and Polifke [7] have shown the influence of noise on the identification of the FTF, having also lower amplitudes than 1 at low frequencies. The Wiener-Hopf inversion presents some limitations in the presence of high levels of noise. The transversal mode develops perturbations inside the combustion chamber, which create a certain level of response and are not captured as input signal, not allowing for a "clean" identification process to identify the FTF.

5 Summary

LES of turbulent reacting flow in a premixed swirl burner were carried out using HPC resources of the Linux Cluster and the HRBLII Supercomputers of the Leibniz-Rechenzentrum in Garching. Calculations required significant numerical resources and were carried out using up to 256 cores. Results for averaged distributions of velocity and heat release rate were compared to experiments, and good quantitative agreement was obtained. For the flame dynamics, results were less satisfactory: The phase of the transfer function, which is of particular importance for thermo-acoustic instability, was reproduced quite well. However, the gain of the transfer function was significantly under-predicted and shows incorrect asymptotic behaviour in the limit of low frequencies. A high-frequency transversal mode, which appeared in the reacting flow simulations, has been found to have a noticeable detrimental impact on the identification of the FTF. Simulations performed after this study, using the same burner at a lower power rating of 30 kW, did not develop the transversal mode. This allowed robust identification and yielded very satisfactory agreement between experimental and identified flame dynamics [23].

Acknowledgements The authors gratefully acknowledge financial support for L. Tay, R. Kaess, S. Föller and T. Komarek by DAAD, DFG (Projects Po 710-3 and Po 710-5) and KW21 (project GV-6), respectively. We are indebted to CERFACS for making available the solver AVBP; to Leibniz Rechenzentrum for providing access to HPC resources, to T. Sattelmayer for making available lab infrastructure and diagnostic equipment and to A. Winkler, J. Wäsle and M. Konle for their support in the PIV measurements.

References

1. CERFACS (2007) AVBP 6.0 Handbook
2. Boudier G, Gicquel LYM, Poinsot T (2008) Effects of mesh resolution on large eddy simulation of reacting flows in complex geometry combustors. Combust. Flame 155:196–214
3. Colin O, Ducros F, Veynante D and Poinsot T (2000) A thickened flame model for large eddy simulations of turbulent premixed combustion. Phys. Fluids 12:1843–1863
4. Cuenot B, Sommerer Y, Gicquel LYM, Poinsot T (2006) LES of reacting flows. In: 3rd International workshop on Rocket Combustion Modeling, Paris
5. Gentemann A, Hirsch C, Kunze K, Kiesewetter F, Sattelmayer T and Polifke W (2004) Validation of flame transfer function reconstruction for perfectly premixed swirl flames. In: Proceedings of ASME Turbo Expo, Vienna
6. Giauque A (2007) Fonctions de transfert de flamme et energies des perturbations dans les ecoulements reactifs. PhD. Thesis. Institut National Polytechnique, Toulouse
7. Huber A and Polifke W (2009) Dynamics of practical premixed flames, part I: Model structure and identification. International journal of spray and combustion dynamics 1(2):199–228
8. Huber A and Polifke W (2009) Dynamics of practical premixed flames, part II: Identification and interpretation of CFD data. International journal of spray and combustion dynamics 1(2):229–250
9. Kaess R (to be published) Computation of Open Loop Gain Curves with Large Eddy Simulation. PhD. Thesis, Technische Universität München
10. Kaess R, Huber A and Polifke W (2008) Time-domain impedance boundary condition for compressible turbulent flows. In: 14th AIAA/CEAS Aeroacoustics Conference, Vancouver
11. Kaess R, Polifke W, Poinsot T, Noiray N, Durox D, Schuller T and Candel S (2008) CFD-based mapping of the thermo-acoustic stability of a laminar premix burner. In: Proceedings of the Summer Program CTR, Stanford
12. Komarek T and Polifke W (2009) Impact of swirl fluctuations on the flame response of a perfectly premixed swirl burner. In: Proceedings of ASME Turbo Expo, Orlando
13. Kopitz J, Broecker E and Polifke W (2005) Characteristics-based filter for identification of planar acoustc waves in numerical simulations of turbulent compressible flow. In: 12th Int. Congress on Sound and Vibrations, Lisbon

14. Moureau V, Lartigue G, Sommerer Y, Angelberger C, Colin O and Poinsot T (2005) Numerical methods for unsteady compressible multi-component reacting flows on fixed and moving grids. J. Comput. Phys. 202:710–736
15. Nicoud F and Ducros F (1999) Subgrid-scale stress modeling based on the square of the velocity gradient tensor. Flow, Turbulence and Combustion 62:183–200
16. Poinsot T and Lele SK (1992) Boundary conditions for direct simulation of compressible viscous flows. J. Comput. Phys. 101:104–129
17. Poinsot T and Veynante D (2001) Theoretical and numerical combustion. R T Edwards, Philadelphia
18. Polifke W (2004) Numerical Techniques for Identification of Acoustic Multi-Ports. In: Advances in Aeroacoustics and applications, VKI lecture series, Brussels
19. Polifke W and Lawn C (2007) On the low-frequency limit of flame transfer functions. Combust. Flame 151(3):437–451
20. Polifke W, Wall C and Moin P (2006) Partially reflecting and non-reflecting boundary conditions for simulation of compressible viscous flow. Journal of Computational Physics 213:437–449
21. Sagaut P (2004) Large eddy simulation for incompressible flows. Springer-Verlag, Berlin
22. Selle L, Lartigue G, Poinsot T, Koch R, Schilmacher, Krebs W, Prade B, Kaufmann P and Veynante D (2004) Compressible large eddy simulation of turbulent combustion in complex geometry on unstructured meshes. Combust. Flame 137:489–505
23. Tay Wo Chong L, Komarek T, Kaess R, Föller S and Polifke W (2010) Identification of flame transfer functions from LES of a premixed swirl burner. In: Proceedings of ASME Turbo Expo, Glasgow (To be published)

Computational Modelling of the Respiratory System for Improvement of Mechanical Ventilation Strategies

Andrew Comerford, Sophie Rausch, Lena Wiechert, Michael W. Gee, and Wolfgang A. Wall

Abstract This paper is concerned with a brief outline of our computational models of the respiratory system against the background of acute lung diseases and mechanical ventilation. We divide the lung into two major subsystems, namely the conducting airways and the respiratory zone represented by lung parenchyma. Due to their respective complexity, both parts are themselves out of range for a direct numerical simulation resolving all relevant length scales. Therefore, we develop detailed individual models for parts of the subsystems as a basis for novel multi-scale approaches taking into account the unresolved parts appropriately. In the tracheo-bronchial region, CT-based geometries up to a maximum of approximately seven generations are employed in fluid-structure interaction simulations, considering not only airway wall deformability but also the influence of surrounding lung tissue. Physiological outflow boundary conditions are derived by considering the impedance of the unresolved parts of the lung in a fully coupled 3D-1D approach. In the respiratory zone, an ensemble of alveoli representing a single ventilatory unit is modeled considering not only soft tissue behavior but also the influence of the covering surfactant film. Novel nested multi-scale procedures are then employed to simulate the dynamic behavior of lung parenchyma as a whole and local alveolar ensembles simultaneously without resolving the alveolar micro-structure completely.

Andrew Comerford · Sophie Rausch · Lena Wiechert · Michael W. Gee · Wolfgang A. Wall
Institute for Computational Mechanics, Technische Universität München, Boltzmannstr. 15, D-85747 Garching, Germany
e-mail: comerford@lnm.mw.tum.de
e-mail: rausch@lnm.mw.tum.de
e-mail: wiechert@lnm.mw.tum.de
e-mail: gee@lnm.mw.tum.de
e-mail: wall@lnm.mw.tum.de

S. Wagner et al. (eds.), *High Performance Computing in Science and Engineering, Garching/Munich 2009*, DOI 10.1007/978-3-642-13872-0_23,

1 Introduction

Mechanical ventilation is a vital supportive therapy for critical care patients suffering from Acute Respiratory Distress syndrome (ARDS). However, although many therapeutic approaches have been developed, the mortality associated with ARDS remains around 40 % [17]. Heterogeneity of the ARDS lung predisposes patients towards a number of further associated complications which are collectively termed ventilator induced lung injuries (VILI) and deemed one of the most important factors in the pathogenesis of ARDS. VILI mainly occurs at the alveolar level of the lungs in terms of primary mechanical and secondary inflammatory injuries [3]. Primary injuries are consequences of alveolar overexpansion or frequent recruitment and derecruitment inducing high shear stresses. Since mechanical stimulation of cells can result in the release of proinflammatory mediators – a phenomenon commonly called mechanotransduction – secondary inflammatory injuries often directly follow, possibly starting a cascade of events leading to sepsis or multi-organ failure. Understanding the reason why the lungs still become damaged or inflamed despite recent developments towards "protective" ventilation protocols [1] is a key question sought by the medical community.

Computational models of the respiratory system can provide essential insights into the involved phenomena and open up new vistas towards improved patient-specific ventilation protocols. However, due to the complexity of the lung comprising 23 generations of dichotomously bifurcating airways ending in approximately 300 million alveoli, a direct numerical simulation resolving all levels of the respiratory system is impossible from the onset. Therefore, we first established detailed models for distinct parts of the lung being of particular interest with respect to VILI, i.e. the tracheo-bronchial and the alveolar region. In order to take into account the unresolved parts of the lung appropriately, novel multi-scale approaches are developed yielding overall models for the respiratory zone as well as for the conducting airways as a whole.

2 Research Software Platform BACI

All our computational models are implemented at the Lehrstuhl für Numerische Mechanik, TU München in our in -house research software platform BACI. Among others, this research platform can be used for simulating convection-diffusion problems, laminar and turbulent incompressible and variable-density flow as well as nonlinear solid mechanics. Particular capabilities are, for instance, turbulent non-reactive and reactive flow, combustion as well as contact and thermo-mechanics. Furthermore, it is one of the most advanced codes for the simulation of coupled problems such as FSI and transport of various properties within flow domains. It also has a variety of advanced multi-scale features. The Finite-Element-Method (FEM) is the premier approach to spatial discretization. State-of-the-art element technology such as enhanced assumed strain (EAS), assumed natural strain (ANS) and discrete strain gap (DSG) methods are used to develop efficient solid and solid-shell elements. Furthermore, a uniform nodal-strain tetrahedron with isochoric stabilization has been devised, see [5]. For flow simulation, residual-based variational multiscale techniques as well as state-of-the-art stabilization approaches are used within BACI. Additional techniques are available for turbulent flow simulation. BACI has been and is developed within an object-oriented C++ environment. Parallelization is based on domain decomposition methods using MPI (Message Passing Interface). State-of-the-art solution techniques for nonlinear and linear system of equations (e.g. algebraic multigrid) as well as for coupling of several fields (e.g. monolithic fluid-structure interaction nonlinear solvers) are incorporated in BACI and are continuously further developed in our group. BACI makes use of and contributes to Sandia National Laboratories parallel software framework 'Trilinos' for the implementation of efficient parallel sparse linear algebra operations. Also, the institute is an accredited contributor to Sandia's algebraic multigrid solver framework 'ML' which is used to solve linear systems of equations of more than 10^9 unknowns.

3 Computational Model of the Central Airways

Modeling the tracheo-bronchial region can provide in-depth knowledge into how the human lung functions mechanically under artificial ventilation. We are particularly interested in investigating the influence of different protocols on pressure and flow distributions, thereby revealing possible lung regions at risk of overdistension. Furthermore, we want to quantify stresses and strains in airway walls being also of direct relevance for the development of VILI.

We have shown that the consideration of specific geometric features is essential, i.e. severe differences in flow characteristics occur between artificial and CT-based models of the airways [20]. Therefore, we utilize realistic patient-specific geometries usually up to a maximum of seven generations [2]. These geometries are extracted from standard CT-images (with a resolution of 0.6mm) that are routinely performed in the hospital. High quality meshes are generated, typically containing approximately 2×10^6 tetrahedral elements equating to about 1.6×10^6 degrees of freedom.

When modeling the airway tree, investigations are in general limited to the very first generations due to a restricted resolution of CT imaging. To obtain correct pressure levels in the resolved tracheo-bronchial region, peripheral impedance of unresolved parts of the lung must be taken into account. For this purpose, we developed a fully coupled 3D-1D approach [2] based on a previously derived method for arterial impedance modeling [13].

Briefly, we attach artificial asymmetric airway trees made up of 1D flexible tubes to each outlet of the resolved 3D tracheo-bronchial model. The impedance of each individual branch is then summed in series and parallel in order to obtain the total impedance of a specific tree. The coupling of the 1D models to the 3D model is achieved utilizing a Dirichlet to Neumann approach introduced in [19] for arterial

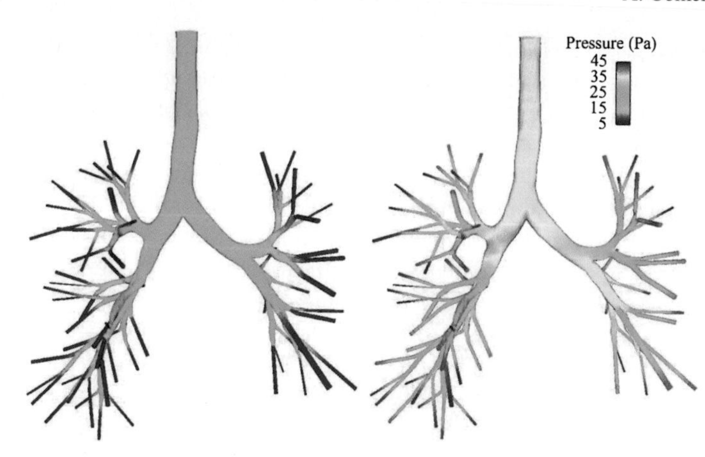

Fig. 1 Comparison of pressure contours at maximum inspiration with traction free (left) and impedance boundary conditions (right). Using impedance conditions, the pressure is approximately 27% higher

blood flow. Apart from being mathematically well posed and numerically stable, this method has the advantage that the solution of the 1D and 3D domain is achieved simultaneously.

Simulation results show that flow remains relatively disturbed throughout the model, reinforcing the need for 3D simulations of the bronchial tree. In higher generations, the flow is relatively uniform meaning the application of the 1D model in these locations seems acceptable. The flow distribution within the 3D domain is effected only marginally by the impedance of the downstream domain suggesting that the resistance to flow is mainly due to the larger airways rather than the impedance of the lower tree. However, the change in pressure distribution between traction free and impedance conditions is evident (confer Fig. 1). Between the two different simulations there is a significant change in computational effort (1750 and 3500 CPUh for traction free and impedance respectively). We have also performed simulations under diseased state conditions which resulted in considerable changes in flow distributions.

In [20], we considered the influence of deformable airway walls on flow phenomena for the first time. The coupling of incompressible flows and soft tissue is a very challenging task. Many existing approaches are either unstable or very inefficient in such situations. Therefore, we have developed new robust and efficient coupling schemes for fluid-structure interaction (FSI) simulations (confer e.g. [4, 8–10]). We figured out that monolithic schemes with block preconditioning are most suitable for our complex simulations in the tracheo-bronchial region [7], see Fig. 2. The shown simulation performed with traction free boundary conditions has a calculation time of approximately 13000 CPUh.

Compared with previous pure computational fluid dynamics (CFD) simulations, we found that the influence of FSI on normalized flow distributions and secondary flow intensities is moderate [20]. However, airflow patterns, both axial and in-plane,

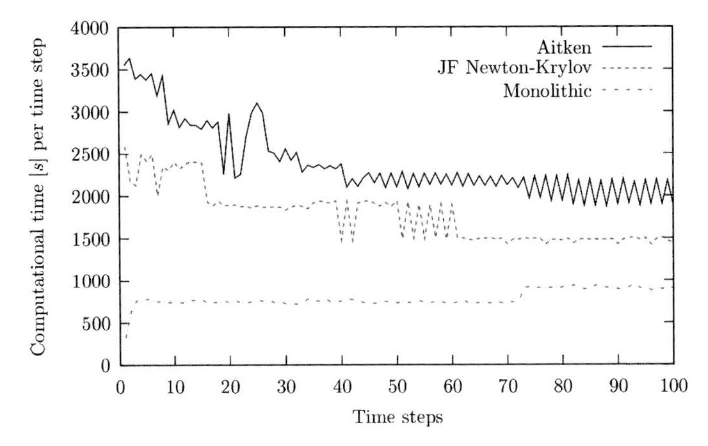

Fig. 2 Computational time per time step for a seven generation airway model. The computation was performed on 24 processors

Fig. 3 Embedding of tracheobronchial model in simplified model of surrounding lung tissue taking into account fluid-structure interaction. Colors from blue to red indicate increasing overall displacements

were quite different although the changes in cross sectional areas were only around 2%. Even more importantly, CFD simulations are not capable of capturing stresses in lung tissue. Hence, consideration of airway wall deformability is crucial for investigating mechanisms of VILI.

As a first step towards realization of an overall lung model, we embed our FSI-models in a homogenized tissue block in order to consider the influence of surrounding parenchyma and also the interdependence of neighboring airways as illustrated in Fig. 3. More realistic shapes of the lung lobes will be extracted from standard CT data sets in the future.

We also incorporate nanoparticle mass transfer into our elaborate model. The transport of such particles in the lung is a very important topic as they have both beneficial (nanodrug delivery) and adverse effects (irritation of the airway wall due to pollutant inhalation). From our study it was found that surface deposition is effected by near wall airflow separation and attachment induced by the realistic lung geometry (see Fig. 4). This pure fluid simulation with traction free boundary conditions had a calculation time of 13400 CPUh.

Fig. 4 Flux to the surface at maximum inspiration, Sc=10 and Sc=100, left and right, respectively. This non-dimensional number characterizes the relative thickness of the fluid to mass transport boundary layer. Non-uniform mass transfer results directly from the realistic surface topology, where the flux is effected by the growing and shrinking of the boundary layers

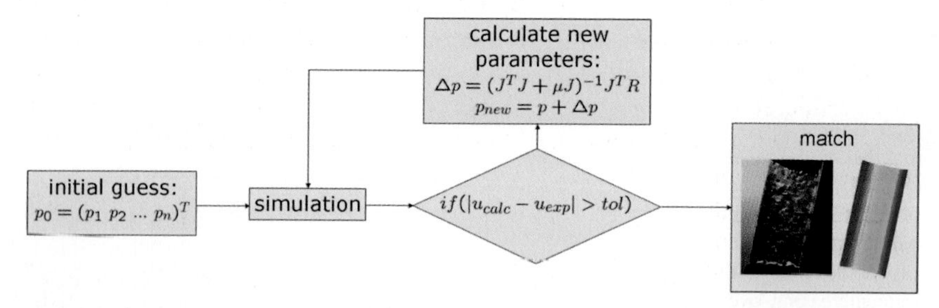

Fig. 5 Simulation of the experiment during an inverse analysis used for parameter identification of lung parenchyma

4 Computational Model of Lung Parenchyma

Since pulmonary alveoli are the main site of VILI, a detailed modeling of all involved phenomena is necessary in order to obtain insights in the underlying mechanisms.

For modeling alveolar tissue behavior, several complex non-linear material models have to be combined. For more details see [16, 22]. The corresponding parameters are determined utilizing an inverse analysis. This involves performing uniaxial tension tests on living rat lung slices, to determine stress-strain curves. The resulting stress curves are utilized as input parameters for the simulation, while the strain curves are utilized to determine the termination criteria. Within the simulations we vary the material parameters, according to the Levenberg-Marquardt Algorithm [11, 12], until we reach the required tolerance ie. the termination criteria, see Fig. 5 and [15].

Pulmonary alveoli are covered by a continuous film of surface active agents. To model the influence of this liquid lining, we established a surface coupling of structural and interfacial mechanics [22]. Surface active agents – also known as surfactant – exhibit dynamically varying surface stresses. In order to capture this behavior, a non-linear and time-dependent constitutive model is employed [14].

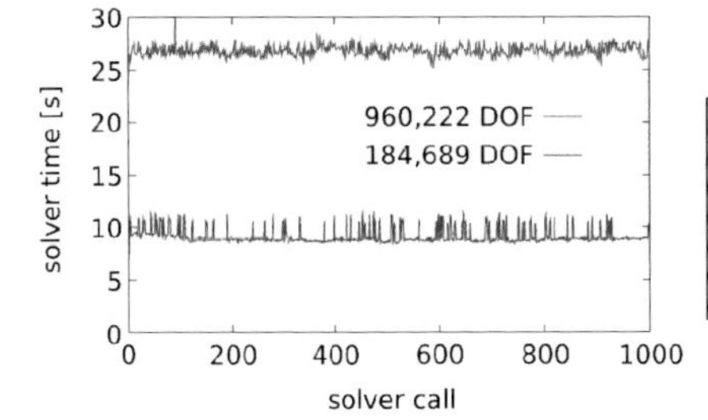

DOF	184,689	960,222
processors	16	32
CG iterations per solve	170	164
solver time per call	8.9 s	25.8 s
AMG setup time	1.5 s	2.2 s

Fig. 6 Comparison of two discretizations of artificial alveolar geometries. Left: Solver time per call. Right: Solver details

The computational model generally comprises an assemblage of alveoli. In a first step we tested our solvers on artificially generated ventilatory units. We used an algebraic multi-grid preconditioned conjugate gradient method [6] with four grids on two artificial alveolar geometries. Chebychev smoothers are employed on the finer grids, whereas an LU decomposition is applied on the coarsest grid. For more details see Fig. 6 and [23]. It is noteworthy that $O(n)$ overall scalability is achieved, as theoretically predicted by multigrid theory, with the presented approach for these examples.

Recently, micro-CT data for isolated fixed rat lung tissue became available [18]. This allows us to investigate realistic alveolar assemblages in detail for the first time. The overstraining of the alveolar tissue, which causes the inflammation of the tissue during VILI, cannot be quantified globally. That is why it is especially important to look at the local deformation in the single alveolar walls. The geometry of the alveolar tissue has a very irregular open-foam like structure. Due to the complex shape very fine tetrahedral meshes are needed. To investigate the influence of the mesh density and the cube size several different discretizations are compared within the simulations (not all meshes are shown here), see Fig. 7. As the first step we simulated uniaxial tension and simple shear deformation of the cubes, see Fig. 8. The results for the larger cube under 10% uniaxial tension are shown in Fig. 9. This deformation results in up to 40% strain in the alveolar walls. Consequently, resolving the realistic alveolar morphology is crucial when investigating local overstretching of lung tissue in case of VILI. These relatively small alveolar ensembles already require the solution of very large systems of equations. The corresponding simulation to the shown results consumes about 33600 CPUh even though highly efficient multigrid solvers are applied. For the determination of material parameters of alveolar walls, via an inverse analysis, it is necessary to perform the simulation several times.

These calculation times clearly indicate that it is impossible to resolve the whole lung parenchyma for a computational simulation. One promising possibility to overcome this problem is a newly developed dynamic multi-scale approach [24]. This procedure enables us to investigate the time-dependent behavior of lung parenchyma as a whole and local alveolar ensembles simultaneously without necessitating to resolve the alveolar micro-structure completely. A schematic representation of the

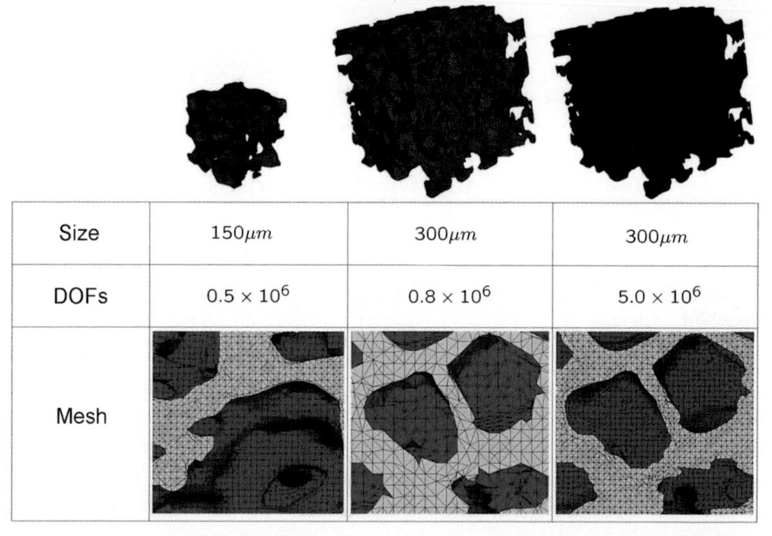

Size	$150\mu m$	$300\mu m$	$300\mu m$
DOFs	0.5×10^6	0.8×10^6	5.0×10^6
Mesh			

Fig. 7 Different discretizations for CT-based alveolar geometries to study the influence of cube size and mesh density

Fig. 8 Different deformations simulated for the CT-based geometry: Left hand side uniaxial tension, right hand side simple shear

Fig. 9 Deformed CT-based geometry under uniaxial tension, the colors show the largest eigenvalue of the principal strain from low (blue) to high (red)

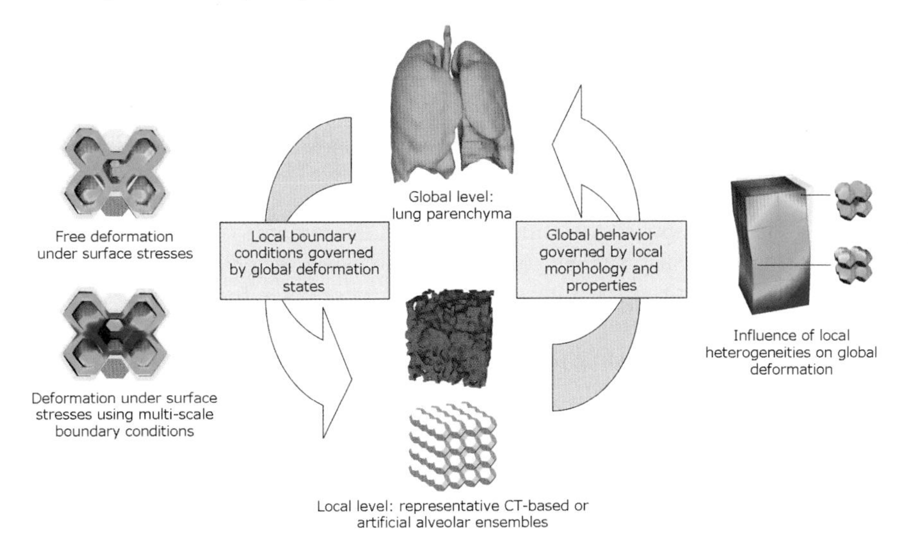

Fig. 10 Schematic overview on nested dynamic multi-scale method with illustrative first simulation results

dynamic multi-scale method along with some first simulation results can be seen in Fig. 10. On the right hand side, the overall heterogeneous deformation of an idealized lung tissue strip due to locally different alveolar behavior is shown. The influence of the global behavior on the alveolar deformation (micro scale) is illustrated on the left hand side. Due to the fully coupled nature of the developed multi-scale approach, also the alveolar micro-structure is affected by the global parenchyma model by means of the prescribed boundary conditions.

5 Summary and Outlook

Substantial progress towards an overall computational lung model has been presented. We have developed robust FSI algorithms that allow the efficient solution of large biomedical problems. New boundary conditions bridging the scale between large airways and the acinar region of the lung have been introduced. Moreover, we are now able to simulate transport of nanoparticles in the central airways, hence understanding how drugs and pollutants are distributed through the lungs.

For the lung parenchyma we developed a sophisticated model, including both nonlinear surfactant and tissue mechanics. Corresponding material formulations were fitted to experiments on living lung tissue. Furthermore simulations on CT-based alveolar geometries were performed, demonstrating the importance of resolving to this level. In particular the local strains were four times as high as the global ones. Novel multi-scale approaches were used to establish an overall lung parenchyma model while resolving alveolar structures locally. For a detailed presentation of our computational models, we refer to [21].

Currently we are working on combining the presented models for the conducting airways and the respiratory zone to form a single virtual lung model. This model will represent one of the most sophisticated simulations of the human lung utilizing our advanced computational approaches and our regional lung models widely published in the international community. Although the research is developed against the background of VILI, our approaches are by no means restricted to this particular application. Hence, we believe that our models can promote further understanding of the lung under healthy and diseased conditions and will be valuable for investigating a number of interesting problems.

Acknowledgements Support by the German Science Foundation/Deutsche Forschunsgemeinschaft (DFG) through projects WA1521/6-2, WA1521/8-1 and WA1521/9-1 within the priority program "Protective Artificial Respiration" (PAR) is gratefully acknowledged. We also would like to thank our medical partners, i.e. the Guttmann workgroup at University Hospital Freiburg (Division of Clinical Respiratory Physiology) and the Uhlig workgroup at University Hospital Aachen (Institute for Pharmacology and Toxicology).

References

1. Amato, M.B., Barbas, C.S., Medeiros, D.M., Magaldi, R.B., Schettino, G.P., Lorenzi-Filho, G., Kairalla, R.A., Deheinzelin, D., Munoz, C., Oliveira, R., Takagaki, T.Y., Carvalho, C.R.: Effect of a protective-ventilation strategy on mortality in the acute respiratory distress syndrome. New England Journal of Medicine **338**, 347–354 (1998)
2. Comerford, A., Foerster, C., Wall, W.A.: Structured tree impedance outflow boundary conditions for 3d lung simulations. Journal of Biomechanical Engineering (submitted, 2009)
3. DiRocco, J., Carney, D., Nieman, G.: The mechanism of ventilator-induced lung injury: role of dynamic alveolar mechanics. Yearbook of Intensive Care and Emergency Medicine 2005 pp. 80–92 (2005). URL `http://dx.doi.org/10.1007/0-387-26272-5_9`
4. Gee, M., Küttler, U., Wall, W.: Truly monolithic algebraic multigrid for fluid-structure interation. International Journal for Numerical Methods in Engineering **submitted** (2009)
5. Gee, M.W., Dohrmann, C.R., Key, S.W., Wall, W.A.: A uniform nodal strain tetrahedron with isochoric stabilization. International Journal for Numerical Methods in Engineering **78**, 429–443 (2009)
6. Gee, M.W., Hu, J.J., Tuminaro, R.S.: A new smoothed aggregation multigrid method for anisotropic problems. Numerical Linear Algebra with Applications **16**(1), 19–37 (2009). URL `http://dx.doi.org/10.1002/nla.593`
7. Kuettler, U., Gee, M., Foerster, C., Comerford, A., Wall, W.A.: Coupling strategies for biomedical fluid-structure interaction problems. International Joufnal for Numerical Methods in Biomedical Engineering (accepted, 2009)
8. Kuettler, U., Wall, W.A.: Fixed-point fluid-structure interaction solvers with dynamic relaxation. Computational Mechanics **43**(1), 61–72 (2008)
9. Kuettler, U., Wall, W.A.: Vector extrapolation for strong coupling fluid-structure interaction solvers. Journal of Applied Mechanics **76**(2), 012205–1–7 (2009)
10. Kuettler, U., Wall, W.A.: Overview of strong coupling fluid-structure interaction schemes (in preparation, 2009)

11. Levenberg, K.: A method for the solution of certain non-linear problems in least squares. Quarterly of Applied Mathematics **2**, 164–168 (1944)
12. Marquardt, D.W.: An algorithm for least-squares estimation of nonlinear parameters. Journal of the Society for Industrial and Applied Mathematics **11**(2), 431–441 (1963)
13. Olufsen, M.S., Peskin, C.S., Kim, W.Y., Pedersen, E.M., Nadim, A., Larsen, J.: Numerical simulation and experimental validation of blood flow in arteries with structured-tree outflow conditions. Annals of Biomedical Engineering **28**, 1281–1299 (2000)
14. Otis, D.R., Ingenito, E.P., Kamm, R.D., Johnson, M.: Dynamic surface tension of surfactant TA: experiments and theory. Journal of Applied Physiology **77**(6), 2681–2688 (1994)
15. Rausch, S., Dassow, C., Uhlig, S., Wall, W.A.: Determination of material parameters of lung parenchyma based on living precision cut lung slice testing (in preparation, 2009)
16. Rausch, S., Wall, W.A.: Parameter identification for modelling alveolar tissue on different scales. In: 7th EUROMECH Solid Mechanics Conference (2009)
17. Rubenfeld, G.D., Caldwell, E., Peabody, E., Weaver, J., Martin, D.P., Neff, M., Stern, E.J., Hudson, L.D.: Incidence and outcomes of acute lung injury. New England Journal of Medicine **353**, 1685–1693 (2005)
18. Schittny, J.C., Mund, S.I., Stampanoni, M.: Evidence and structural mechanism for late lung alveolarization. American Journal of Physiology – Lung Cellular and Molecular Physiology **294**(2), L246–L254 (2008)
19. Vignon-Clementel, I., Figueroa, C., Jansen, K., Taylor, C.: Outflow boundary conditions for three-dimensional finite element modeling of blood flow and pressure in arteries. Computer Methods in Applied Mechanics and Engineering **195**, 3776–3796 (2006)
20. Wall, W.A., Rabczuk, T.: Fluid structure interaction in lower airways of ct-based lung geometries. International Journal for Numerical Methods in Fluids **57**, 653–675 (2008)
21. Wall, W.A., Wiechert, L., Comerford, A., Rausch, S.: Towards a comprehensive computational model for the respiratory system. Communications in Numerical Methods in Engineering – With Biomedical Applications **submitted** (2009)
22. Wiechert, L., Metzke, R., Wall, W.A.: Modeling the mechanical behavior of lung tissue at the micro-level. Journal of Engineering Mechanics **135**, 434–438 (2009)
23. Wiechert, L., Rabczuk, T., Gee, M.W., Metzke, R., Wall, W.A.: Coupled problems in computational modeling of the respiratory system. In: High Performance Computing on Vector Systems, pp. 145–166 (2007). URL `http://dx.doi.org/10.1007/978-3-540-74384-2_12`
24. Wiechert, L., Wall, W.A.: A nested dynamic multi-scale approach for 3d problems accounting for micro-scale multi-physics. Computer Methods in Applied Mechanics and Engineering (accepted, 2009)

Part III
Geo Sciences

SeisSol – A Software for Seismic Wave Propagation Simulations

Martin Käser, Cristobal Castro, Verena Hermann, and Christian Pelties

Abstract We present important extensions of the software SeisSol based on the arbitrary high-order Discontinuous Galerkin (DG) Finite-Element method to model seismic wave propagation using non-conforming hybrid meshes for model discretization and to account for highly heterogeneous material. In these new approaches we include a point-wise integration of numerical fluxes across element interfaces for the non-conforming boundaries preserving numerical accuracy while avoiding numerical artifacts due to the mesh coupling. We apply the proposed scheme to a scenario test case of the Grenoble valley to demonstrate the methods capability. Furthermore, we present the extension to space-variable coefficients to describe material variations inside each element using the same numerical approximation strategy as for the velocity-stress variables in the formulation of the wave equation. The combination of the DG method with a time integration scheme based on the solution of Arbitrary accuracy DErivatives Riemann problems (ADER) still provides an explicit, one-step scheme which achieves arbitrary high-order accuracy in space and time. We confirm the accuracy of the proposed scheme a numerical experiments considering randomly heterogeneous material and compare our results to independent reference solutions. Finally, we apply the scheme to an earthquake scenario considering the effect of a sedimentary basin and compare the efficiency of the classical global time stepping technique and the local time stepping for different partitioning strategies and show results for a strong scaling test.

Martin Käser · Cristobal Castro · Verena Hermann · Christian Pelties
Department für Geo- und Umweltwissenschaften, Geophysik, Ludwig-Maximilians-Universität München, Theresienstrasse 41, 80333 München, Germany
e-mail: kaeser@geophysik.lmu.de

S. Wagner et al. (eds.), *High Performance Computing in Science and Engineering, Garching/Munich 2009*, DOI 10.1007/978-3-642-13872-0_24,

1 Introduction

The numerical computation of complete and sufficiently accurate wave fields for complex subsurface models is getting increasingly important in seismology, as full wave form inversion techniques become feasible with modern supercomputers. However, there is a large variety of numerical schemes to choose from when computing synthetic seismograms and this choice might be problem dependend. A method called the Discontinuous Galerkin (DG) Finite-Element method has first been introduced for the solution of hyperbolic partial differential equations (PDE) by Reed and Hill [15] in the context of the neutron transport equation. Since that a variety of DG methods have been developed in other research areas and its convergence properties on different mesh types have been analyzed by many authors. For a comprehensive overview of the history of DG-related developments we refer to Chapter 1.1 in [7] and the book of Cockburn *et al.* [3]. Dumbser and Munz [6] introduced an arbitrary high-order DG scheme in space and time by applying the Cauchy-Kowalewski procedure that makes extensive and recursive use of the governing equations. The original idea of constructing such explicit one-step time integration schemes that automatically provide the same time accuracy as space accuracy was presented in [17, 18] based on Arbitrary accuracy DErivatives Riemann problems (ADER).

Recently, this approach, now termed ADER-DG, has been introduced in the field of computational seismology in a series of papers [4, 8, 9, 13, 14] using a triangular and tetrahedral discretization of the computational domain for the seismic wave equation and different rheologies. Furthermore, the approach allows for p-adaptation as well as local time-steeping [5] enhancing its flexibility and applicability for realistic, large-scale problems.

In this paper we show recent extensions of the ADER-DG method and its large-scale applications to seismic wave propagation problems typically encountered in computational geophysics. These cover seismic wave propagation using non-conforming hybrid meshes, highly heterogeneous material, and sedimentary basins.

2 The Numerical Approximation

We solve the elastic wave equation in two space dimensions using its velocity-stress formulation leading to the first order hyperbolic system of partial differential equations (PDE)

$$\frac{\partial Q_p}{\partial t} + A_{pq}\frac{\partial Q_q}{\partial x} + B_{pq}\frac{\partial Q_q}{\partial y} + C_{pq}\frac{\partial Q_q}{\partial z} = S_p\,, \tag{1}$$

where $Q = (\sigma_{xx}, \sigma_{yy}, \sigma_{zz}, \sigma_{xy}, \sigma_{yz}, \sigma_{xz}, u, v, w)^T$ is the vector of unknown stresses and velocities, A, B and C are the square Jacobian matrices including the material properties, and S is a source term. Note that the space-time dependency of $Q(x,y,t)$ and $S(x,y,t)$ as well as the space dependency of $A(x,y,z)$, $B(x,y,z)$ and $C(x,y,z)$ are omitted to simplify the notation. The solution Q_p of the p-th component is numerically approximated by $(Q_h^{(m)})_p$ inside each element via a linear combination of time-independent orthogonal polynomial basis functions $\theta_l(\xi,\eta,\zeta)$ of degree N and only time-dependent degrees of freedom $\hat{Q}_{pl}^{(m)}(t)$, i.e.,

$$\left(Q_h^{(m)}\right)_p(\xi,\eta,\zeta,t) = \hat{Q}_{pl}^{(m)}(t)\,\theta_l(\xi,\eta,\zeta), \tag{2}$$

where index (m) accounts for the element. For a detailed derivation of the fully discrete ADER-DG scheme the reader is referred to the references given in the introduction.

3 Non-conforming Hybrid Meshes

The ADER-DG method has the advantage of working with arbitrary high-orders of accuracy and unstructured meshes that can be used to model complex geometries. However, regarding the CPU time, it is faster to compute on regular meshes instead of unstructured ones to reach a desired error level. Therefore, the performance of the ADER-DG method can be increased by combining different mesh types, i.e., creating hybrid meshes, as proposed similarly Ichimura *et al.* [16, 19] for the FE method, or for a combination of the FE and FD method by Moczo *et al.* [12].

Furthermore, the mesh spacing - and therefore the limiting time-step - is usually determinated by the lowest wave length. As the wave length changes for the propagation through different materials it is suitable to adapt the mesh spacing to the local velocity in order to save runtime. Therefore, we suggest a new possibility to refine or coarsen a purely quadrilateral mesh. However, the element interfaces then become non-conforming. In the following, we will treat both boundary types, hybrid meshes and non-conforming interfaces, in the same way and refer to them as non-conforming boundaries.

As we are using hybrid meshes in 2D, the general basis functions $\theta_l(\xi,\eta)$ defined in reference space can be different depending on the element type of the particular element $E^{(m)}$ and we introduce the notation to distinguish them as

$$\theta_l(\xi,\eta) = \begin{cases} \Phi_l(\xi,\eta) \text{ if } & E = E_T \quad \text{(element type is triangular)}, \\ \Psi_l(\xi,\eta) \text{ if } & E = E_Q \quad \text{(element type is quadrilateral)}. \end{cases} \tag{3}$$

In the derivation of the ADER-DG method the so-called flux term occurs due to the integration by parts of the governing equation (1). This flux term represents a line integral along the boundary ∂E, i.e., along the edges, of an element E over the product of the test function θ_k and the numerical flux F_p. The line integral can be decomposed into a sum of integrals over the element edges such that

$$\int_{\partial E} \theta_k F_p \, dS = \sum_{j=1}^{n} |S^{(j)}| \int_0^1 \theta_k F_p^{(j)} \, d\chi \,. \tag{4}$$

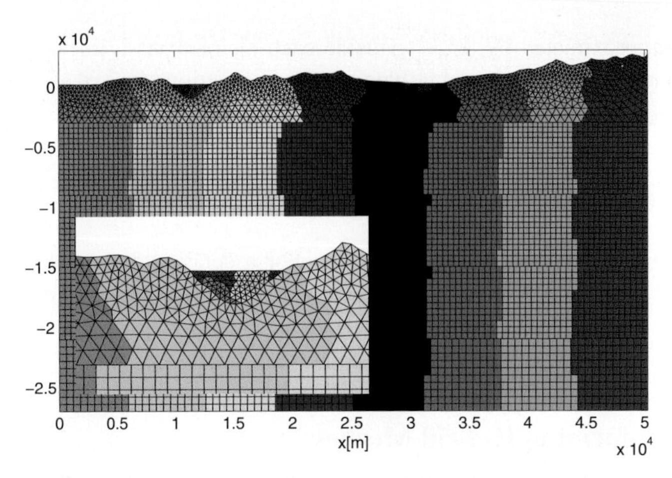

Fig. 1 Computational mesh for the Grenoble testcase showing the colored partitions of a hybrid non-conforming mesh

Here we integrate over the unit interval $\chi \in [0,1]$ and therefore introduce the length of the j-th edge $|S^{(j)}|$ as a scaling fractor due to the interval transformation from the physical into the reference space. The number n of edges again depends on the element type, i.e., $n = 3$ for triangles and $n = 4$ for quadrilaterals. The flux term $F_p^{(j)}$ also depends on the particular edge j and can be decomposed into an outgoing and an incoming part as shown in Chapter 4.14 of the book of LeVeque [11] or in [8]. On hybrid meshes, we have to take care of the different basis functions for different element types when computing the incoming flux. Treating non-conforming boundaries we solve the integrals numerically by using the Gauss-Legendre procedure, which is given by

$$\int_0^1 \theta_k^{(m)} \theta_l^{(m_j)} dS = \sum_{i=1}^{N+1} \theta_k^{(m)} \theta_l^{(m_j)} w_i \,. \tag{5}$$

with w_i being the Gaussian weights and N being the polynomial degree.

In the following example we use an east-west cross-section north of the city of Grenoble in the French Alps. The valley in which Grenoble is located contains a sedimentary basin in form of a belt. Wave speeds are slow there compared to its surrounding by solid bedrock. The computational domain of 50 km length, 27 km depth and up to 3 km height is presented in Fig. 1 with a zoomed section of the western basin. To account for the very slow wave speeds in the basin, we apply a very fine mesh on it, which is partitioned separately. Building the surface, the two parts of the basin as well es the uppermost layer obtain a triangular mesh. With increasing velocities in the lower zones we proportionally adapt the mesh spacing of the quadrilaterals. Fig. 2 illustrates snapshots for a simulated time of 1.5 s, 3.0 s, 4.5 s and 6.0 s. The wave field get strongly scattered by the surface topography and the ground motion shows high values and a long duration inside the basins.

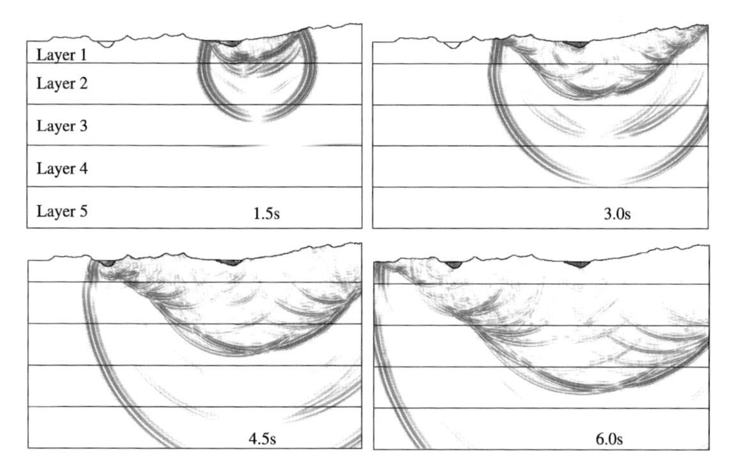

Fig. 2 Snapshots of the seismic wave field for the Grenoble testcase for different times. The strong scattering due to the surface topography is clearly visible, while the mesh transition between different mesh types does not produce numerical artifacts

4 Highly Heterogeneous Material

In the following we extended the ADER-DG method for seismic wave propagation problems incorporating the high-order polynomial approximation of variable material inside each element to handle highly heterogeneous material. The variable material introduces additional terms in the form of volume integrals and adds further computational complexity to the calculation of flux and stiffness matrices. For mathematical details the reader is referred to [1]. In contrast to previous formulations of the ADER-DG scheme the incorporation of the variable coefficient enables us now to treat also the material variations inside an element and its effect on the seismic wave field more accurately. Therefore, simulations can be carried out with coarser meshes, as sub-cell information about the material is properly considered inside each element. Furthermore, the approximation order for the material variation and the wave propagation can be chosen independently to enhance computational efficiency.

We apply the scheme to a challenging problem, where seismic waves propagate through a highly heterogeneous medium of a random perturbation of mean material properties representing a layered material variation typically encountered in sedimentary environments. The resulting heterogeneity of the P-wave speed is shown in Fig. 3 where a number of thin, lense-shaped high- and low-velocity zones are visible. Snapshots of the obtained horizontal velocity field u are shown in Fig. 4, which displays a strong scattering of the seismic waves and deformation of the wave fronts due to the highly heterogeneous material. To validate our results we use a finite difference (FD) and spectral element method method (SEM) to obtain independent reference solutions. The comparison of recorded seismograms at the free surface is plotted in Fig. 5. The visual fit between all three solutions is perfect even for the late arrivals in the coda of the seismograms.

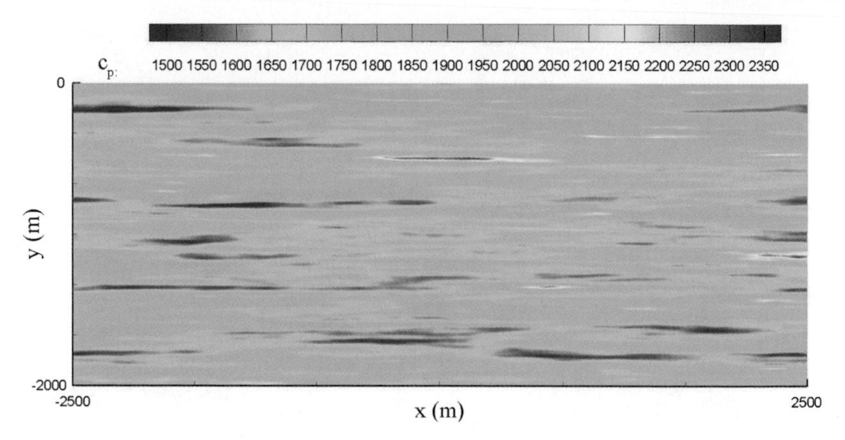

Fig. 3 Spatial variation of the P-wave velocity model with extreme values ranging from 1450 m/s to 2400 m/s. The variation of the S-wave speed has an identical spatial distribution, however, with extreme values ranging from 830 m/s to 1370 m/s

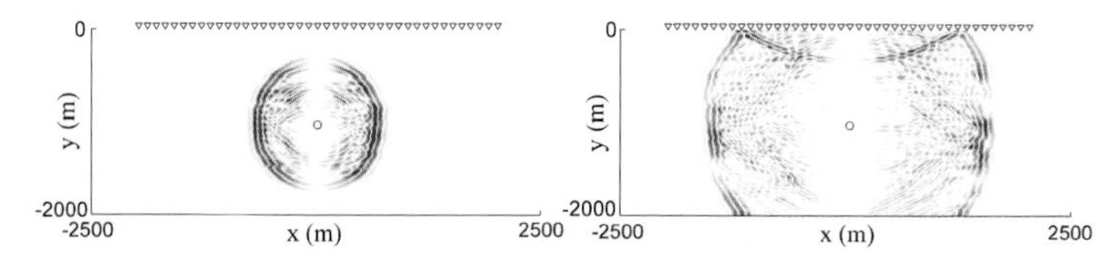

Fig. 4 Snapshots of the horizontal velocity wave field u in the randomly heterogeneous model at times 0.5 s and 1.0 s. The circle represents the location of the explosive source while the triangles are receiver in the free surface

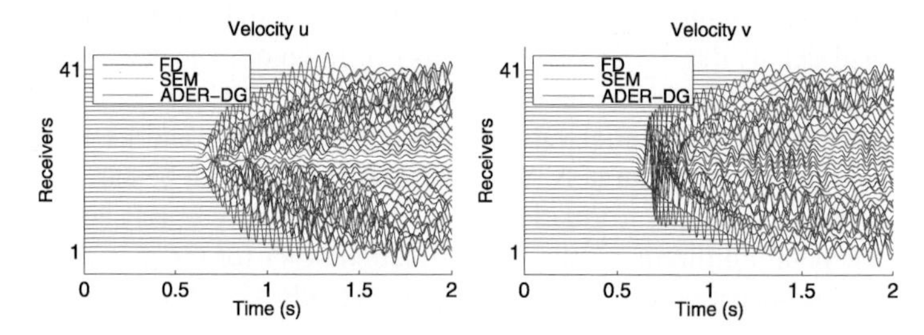

Fig. 5 Comparison of the numerical solution of the particle velocity u (left) and v (right) on each of the 41 receivers placed on the free surface for the randomly heterogeneous material test case

5 Large-Scale Basin Application

Here we present a large-scale application of the proposed ADER-DG scheme to a ground motion simulation problem in the continental area of San Antonio on the west coast of Chile. This is done by solving the set of linear hyperbolic equations describing seismic wave propagation in three space dimensions as presented by [4]. We want to remark, that this application represents a simplified case of the real geophysical situation, but demonstrates the flexibility and feasibility of the proposed

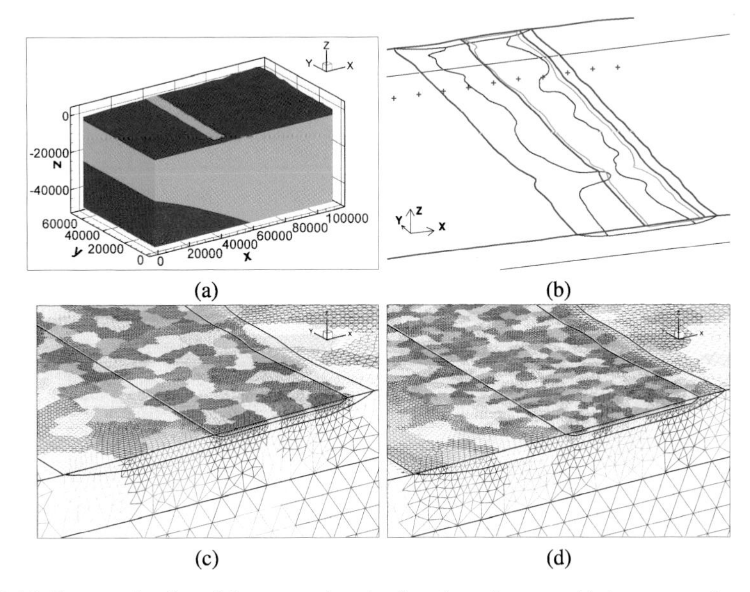

Fig. 6 (a) Computational model representing the San Antonio area with its topography and the basin and subduction geometry. (b) Zoomed view of the geometry of the two thin basins embedded in a near-surface layer. Crosses show the surface locations of the ten seismic receivers. (c) Color-coded MPI domains obtained by a standard partition of Metis. (d) Improved partitioning strategy by subdividing each geological unit (zone) separately in the desired number of MPI domains

methodology for realistic scenarios where detailed knowledge of the earthquake source and the geological properties of the subsurface are included. As the accuracy and correctness of the adaptive DG scheme for seismic wave propagation problems was already validated in previous work of [5, 10], we concentrate on the comparison of the efficiency of the method. To this end, we first show a simulation using the classical ADER-DG scheme using a global time step (GTS) defined by the smallest time step in the entire computational domain following a CFL stability criterion. Then we show the results of the same simulation using local time stepping (LTS) with the same standard mesh partition. Finally, we improve the efficiency of that scheme by modifying the partitioning strategy to get a better load balance for multi-processor calculations.

The computational domain with the coordinate origin $(x,y,z) = (0,0,0)$ centered at longitude 72°W, latitude 34°S, and elevation 0 m above see level is shown in Fig. 6(a), where five different geological units with distinct material properties (for details see [2]) are shown: (1) a 200 m thick sedimentary basin at the surface embedded into (2) a wider and 500 m deep basin structure, (3) a near-surface layer of 2000 m thickness over (4) a large block of continental crustal material, and (5) a oceanic wedge indicating the subducting slab along a dipping curved interface. The discrete model consists of about 1.5 Mio. tetrahedral elements. Synthetic seismograms are registered at ten receivers with a regular distance of 3 km in increasing x-direction across the basins at the model surface as shown in Fig. 6(b). The seismo-

Table 1 CPU-time reduction due to more efficient time stepping or mesh partitioning

scheme	partition	CPU-time [s]	number of cores
ADER-DG GTS	Metis (standard)	43256	1024
ADER-DG LTS	Metis (standard)	29444	1024
ADER-DG LTS	Metis (grouped)	26976	1024

grams are used to check the consistency of the results obtained by the three different simulations.

The simulations are summarized in Table 1 with CPU-times obtained on the LRZ Intel Itanium2 Montecito Dual Core 1.6 GHz cores. We see, that the standard mesh partition as shown in Fig. 6(c) is not adequate anymore. Whereas for GTS schemes it is only important that each subdomain consists of the same number of elements in oder to assure well-adjusted load balance, for LTS this strategy leads to problems. In fact, for LTS schemes it is important that the number of element updates in each subdomain is equal. This is a highly non-trivial task, as the number of element updates per cycle in each subdomain also changes in time. Therefore, a dynamic load balancing technique should provide the best load balance. Nevertheless, our preliminary and effective solution is the partition of the mesh respecting the different geological units (zones). This way, each zone is partitioned into the desired number of subdomains as shown in Fig. 6(d). This is similar to the case of the tsunami wave simulation above where each cluster is partitioned separately. Finally, each processor obtains one subdomain from each geologic zone which provides an improved load balance due to the fact that computationally expensive (small) and cheap (large) elements are distributed more evenly. Using the LTS scheme with the modified partition of Fig. 6(d) in a third simulation reduces the computational time further as shown in Table 1.

The resulting ground motion components u, v, and w in $x-$, $y-$, and $z-$direction, respectively, are shown as seismograms in Fig. 7 at four selected receivers obtained by the three different simulations. Receiver 6 inside the inner basin shows a long duration due to trapped waves. Trapped waves are a well-known basin effect, if the impedance contrast between a hard basement rock and a soft sedimentary basin rock allows only a small fraction of the wave energy inside the basin to escape from it. Receiver 9 on the east side of the basin again shows the direct wave arrivals more clearly, however, followed by a surface wave coda mainly caused by the basin.

Plotting the seismograms of all stations as a cross-section from W to E in Fig. 8(a), the ground motion amplification and longer duration due to the basins is clearly visible. To emphasize the importance of considering the low velocity basins even if extremely thin in their spatial extent, we also show a comparison to results obtained by neglecting the basins. Fig. 8(b) displays the comparison with and without the basin at receiver 6. Without the basin there is only weak ground motion after about 16 sec, i.e. after the direct S-wave has passed the receiver. Therefore, considering such fine, low-velocity structures is currently a key issue in realistic ground motion modeling and seismic hazard analysis, but remains a challenge for many non-adaptive numerical schemes.

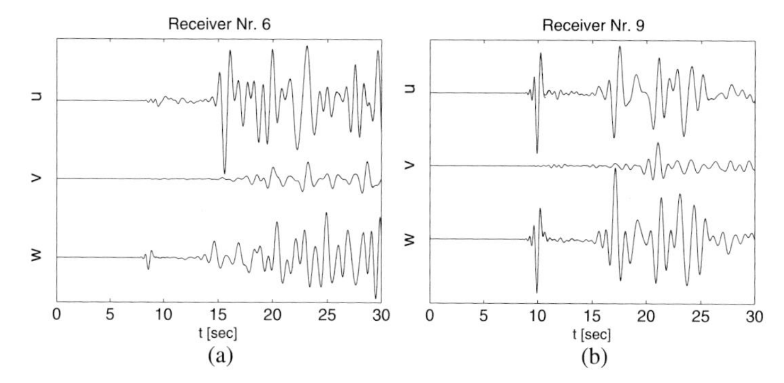

Fig. 7 Seismograms of the ground motion velocities u, v, and w at two selected receivers. The seismograms of the three different simulations are superimposed and coincide visually in one single line, which confirms that there is no difference in the obtained results. The seismogram scaling is the same within each plot, but might differ from one plot to the other to improve visualization. (a) Receiver 6 is inside the inner basin, and (b) receiver 9 is positioned east of the basins

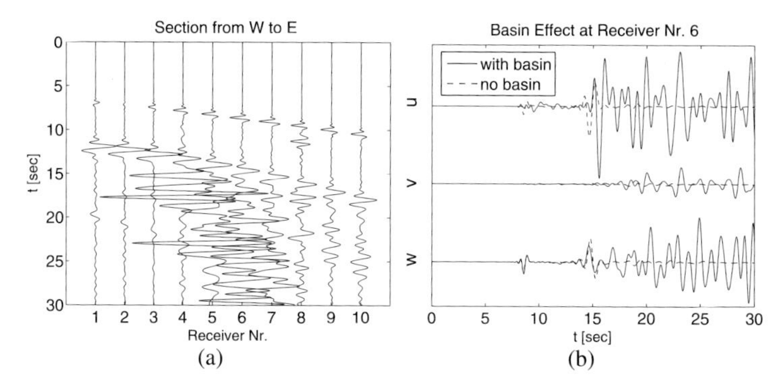

Fig. 8 (a) Equally scaled seismograms of the vertical ground motion velocity plotted as a cross-section from W to E (left to right) show a clear amplification and prolongation of the signals within the basin structure. In particular, it is obvious that the strong ground motion is caused by the arrival of S-waves. (b) Comparison of the velocity seismograms at receiver 6 inside the inner basin considering or neglecting the basin. It is obvious how the basin causes strong wave amplification and prolongation of the shaking due to wave trapping

6 Scalability

A major advantage of the ADER-DG scheme is its locality as an element updated in time only requires information from the direct neighbors in form of polynomial coefficients. This property does not change with increasing approximation order as only the amount of coefficients grows. Therefore, the scheme is well-suited for parallelization. In fact, communication between elements takes place only once per time step and represents much less than 1% of the CPU-time. The results of a strong-scaling test are shown in Fig. 9. The CPU-time reduction remains still close to the ideal case and the speed-up is satisfactory up to 1024 cores with an efficiency of

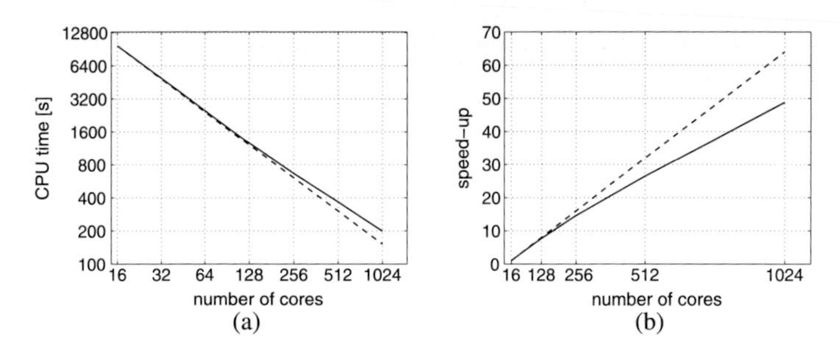

Fig. 9 CPU time decrease (top) and speed-up (bottom) for an increasing number of cores for the strong-scaling test (solid) in comparison to the theoretically ideal case (dashed)

76%. The crucial issue for the scaling properties of the algorithm is the load balance which depends on the mesh partitioning strategy. The LTS approach yields more difficulties as each element runs its own time step. Therefore, element updates happen asynchronously as smaller elements update more often than larger ones. The grouping of elements into zones and partitioning of each zone separately as shown in Fig. 6(c) seems to be an preliminary solution.

7 Concluding Remarks

We presented results of important extensions of the arbitrary high-order Discontinuous Galerkin (DG) Finite-Element method to model seismic wave propagation using non-conforming hybrid meshes for model discretization and to account for highly heterogeneous material. We use a point-wise Gauss-Legendre integration of numerical fluxes across element interfaces for the non-conforming boundaries preserving numerical accuracy while avoiding numerical artifacts due to the mesh coupling. We validate this approach by a scenario test case in the Grenoble valley, where the mesh spacing is adapted to the given surface and basin geometry and to the velocity structure. We briefly presented the extension to space-variable coefficients to describe material variations inside each element using the same numerical approximation as for the velocity-stress variables in the seismic wave equation. Finally, we apply the scheme to an earthquake scenario and compare the efficiency of the classical global time stepping technique and the local time stepping for different partitioning strategies and show results of a strong scaling tests.

The simulations using the SeisSol software package typically require 256 to 1024 cores with run-times between a few hours up to two days. Furthermore, we realized that the scalability of SeisSol depends on the size of the problem to solve. As a rule of thumb, the mesh partitioning process should produce subdomains, where the number of elements per core is not less than 1000 to keep the efficiency sufficiently high. Furthermore, writing output in the sense of full wave field data, i.e. all degrees of freedom for all elements have to be written into files, unfortunately has to be minimized to avoid considerably enhanced CPU-time usage.

Acknowledgements The authors thank the Deutsche Forschungsgemeinschaft (DFG), as the work was financed by the Emmy Noether Programm (KA 2281/1-1). Many thanks also to the Leibniz Rechenzentrum in München, Germany, for providing the GAMBIT and ICEM CFD mesh generators and the necessary support and computational resources to run all development tests, validations and demanding 3D applications.

References

1. Castro, C.E., Käser, M., Brietzke, G.B.: Seismic wave propagation modeling through highly heterogeneous material using the discontinuous galerkin finite element scheme with sub-cell resolution (submitted)
2. Castro, C.E., Käser, M., Toro, E.F.: Space-time adaptive numerical methods for geophysical applications. Roy. Soc. Phil. Trans. (in press)
3. Cockburn, B., Karniadakis, G.E., Shu, C.W.: Discontinuous Galerkin Methods. Lecture Notes in Computational Science and Engineering. Springer (2000)
4. Dumbser, M., Käser, M.: An arbitrary high order discontinuous Galerkin method for elastic waves on unstructured meshes II: The three-dimensional case. Geophys. J. Int. **167**(1), 319–336 (2006)
5. Dumbser, M., Käser, M., Toro, E.F.: An arbitrary high order discontinuous Galerkin method for elastic waves on unstructured meshes V: Local time stepping and p-adaptivity. Geophys. J. Int. **171**(2), 695–717 (2007)
6. Dumbser, M., Munz, C.: Building blocks for arbitrary high order discontinuous Galerkin schemes. J. Sci. Comput. **27**, 215–230 (2006)
7. Hesthaven, J., Warburton, T.: Nodal Discontinuous Galerkin Methods: Algorithms, Analysis, and Applications. Springer, New York (2008)
8. Käser, M., Dumbser, M.: An arbitrary high order discontinuous Galerkin method for elastic waves on unstructured meshes I: The two-dimensional isotropic case with external source terms. Geophys. J. Int. **166**(2), 855–877 (2006)
9. Käser, M., Dumbser, M., de la Puente, J., Igel, H.: An arbitrary high order discontinuous Galerkin method for elastic waves on unstructured meshes III: Viscoelastic attenuation. Geophys. J. Int. **168**(1), 224–242 (2007)
10. Käser, M., Hermann, V., de la Puente, J.: Quantitative Accuracy Analysis of the Discontinuous Galerkin Method for Seismic Wave Propagation. Geophys. J. Int. **173**(3), 990–999 (2008)
11. LeVeque, R.: Finite Volume Methods for Hyperbolic Problems. Cambridge University Press, Cambridge (2002)
12. Moczo, P., Bystricky, E., Kristek, J., Carcione, J., Bouchon, M.: Hybrid modeling of P-SV seismic motion at inhomogeneous viscoelastic topographic structures. Bull. Seism. Soc. Am. **87**, 1305–1323 (1997)
13. de la Puente, J., Dumbser, M., Käser, M., Igel, H.: Discontinuous Galerkin methods for wave propagation in poroelastic media. Geophysics **73**(5), T77–T97 (2008)
14. de la Puente, J., Käser, M., Dumbser, M., Igel, H.: An arbitrary high order discontinuous Galerkin method for elastic waves on unstructured meshes IV: Anisotropy. Geophys. J. Int. **169**(3), 1210–1228 (2007)
15. Reed, W.H., Hill, T.R.: Triangular mesh methods for the neutron transport equation. Tech. Rep. LA-UR-73-479, Los Alamos Scientic Laboratory (1973)
16. T. Ichimura, M. Hori, Bielak, J.: A hybrid multiresolution meshing technique for finite element three-dimensional earthquake ground motion modelling in basins inclding topography. Geophys. J. Int. **177**, 1221–1232 (2009)
17. Titarev, V.A., Toro, E.F.: ADER: Arbitrary high order Godunov approach. J. Sci. Comput. **17**, 609–618 (2002)

18. Toro, E.F., Millington, R.C., Nejad, L.A.M.: Towards very high order godunov schemes. In: E.F. Toro (ed.). Godunov methods; Theory and applications, pp. 907–940. Kluwer Academic Plenum Publishers, Oxford, (2001). International Conference
19. Tsuyoshi Ichimura, Hori, M., Kuwamoto, H.: Earthquake motion simulation with multiscale finite-element analysis on hybrid grid. Bull. Seism. Soc. Am. **97**, 1133–1143 (2007)

Advances in Modelling and Inversion of Seismic Wave Propagation

V. Hermann, N.D. Pham, A. Fichtner, S. Kremers, Lianjie Huang, Paul Johnson, Carène Larmat, H.-P. Bunge, and H. Igel

Abstract We report on progress in modelling and inversion of seismic waveforms. This involves in particular the simulation of wave propagation through Earth models with complex geometries (i.e., internal interfaces or topography) using numerical solutions based on tetrahedral meshes. In addition, efficient solvers in 3-D based on a regular-grid spectral element method allow for the simulation of many Earth models and for the inversion (i.e., for the fit) of observed seismograms using adjoint techniques. We present an application of this approach to the Australian continent. Furthermore results are presented on exploiting ideas from reverse acoustics to estimate finite source properties of large earthquakes and to constrain crustal scattering through modeling joint observations of rotational and translational ground motions.

1 Introduction

Computational seismology has become an increasingly important discipline of seismology and will become even more relevant as the observed data volumes increase. After several years of code developments in the field of computational wave propagation, now the focus is on solving scientific problems with the tested and benchmarked research codes. Here we report results based on two flavours of numerical methods: the spectral element method and the discontinuous Galerkin method.

Within the last few decades a number of different numerical methods for modelling and inversion of seismic waveforms has been developed, including basic fi-

V. Hermann · N. D. Pham · A. Fichtner · S. Kremers · H.-P. Bunge · H. Igel
Department of Earth and Environmental Sciences, Ludwig-Maximilians-University of Munich, München, Germany
e-mail: igel@geophysik.uni-muenchen.de

Lianjie Huang · Paul Johnson · Carène Larmat
Geophysics Group, Earth and Environmental Sciences Division, Los Alamos National Laboratory of the University of California, Los Alamos, USA

S. Wagner et al. (eds.), *High Performance Computing in Science and Engineering, Garching/Munich 2009*, DOI 10.1007/978-3-642-13872-0_25,

nite difference schemes ([28]), Fourier pseudospectral methods ([5]), finite element approaches ([20]) and spectral element methods (SEM) ([13]). The Discontinuous Galerkin (DG) method ([6]) has been developed within the last few years and has already become a well-established numerical method. It is suitable for the simulation of seismic wave propagation in general and able to handle complex geometries as well as heterogeneous media using an arbitrary high approximation order in space and time (ADER-DG). Many simulations have been accomplished on the SGI Altix 4700 machine (HLRB II) of Leibniz-Rechenzentrum. One of the models, e.g., serves as a demonstration for handling quite complex structures. It contains different materials of strongly varying velocities including a salt dome which is of great interest in exploration industry. In this example, the computational domain is composed of 3.1 million elements. For a simulated time of 10*s*, the program was running about 20*h* on 512 processors. In this study, the ADER-DG method is used to examine topographic effects on wave amplification. This subject was first introduced over 30 years ago and is still an ongoing topic as shown by recent publications (e.g. [18]).

In the field of seismic tomography we are now moving into a new era: while so far seismic data were reduced to a few travel time observations of some seismic phases (e.g., P and S waves) the aim is now to use the information contained in the complete waveforms of the observations. In the section on seismic tomography we show the first application of this concept to tomography on a continental scale. The tomographic inversion of the subsurface structure under Australia demonstrates for the first time that - given the current computational power - inversion based on complete 3-D wave propagation solvers is feasible and leads to much improved tomographic models ([8, 9]).

We use a simular concept to investigate whether it is possible to recover earthquake source characteristics by time reversing seismic wavefields. In theory, time reversing the complete seismograms as source injected at the receiver locations constitutes the adjoint field and leads to a first update for the seismic source process ([27]). The results shown indicate that the concept works in principle but that a very high station density is necessary to quantitatively recover source characteristics.

Finally, the DG method is used to directly model an entirely new type of observation: rotational motions (in seismology we usually observe three components of translational motions). It turns out that this new observable - in combination with collocated translational observations - allows the recovery of information on the scattering properties of the Earth's crust ([22]).

All the results presented indicate that the 3-D modelling tools for seismic wave propagation can now be used to model actual observations. The parallelized algorithms based on implementations with Fortran-MPI have been used on O(100) processors, and show good scaling behaviour. However, for the next generation hardware, more tests and optimization is required to render the algorithms performant on O(100k)-core hardware. We intend to accomplish this through collaborations within the MAC (Munich Centre for Advanced Computing) project.

2 Topographic Effects on Seismic Waves

Amplification effects on seismic waves due to strong topographic gradients have been the subject of thorough analysis for more than 30 years. A number of notable earthquakes have led to repeated and consistent observations of amplified ground motion on top of rising topographic features such as mountains, ridges and cliffs. According to Boore [4], surface topography can be neglected when seismic wavelengths are much larger than topographic irregularities and the slopes of the irregularities are small. If these conditions do not hold, considerable amplification can be observed.

At Mount Hochstaufen (see [15]), SE-Germany, we observe topographic amplification up to a factor of 4.4 compared to a station at the mountain base using the spectral ratio method. The obtained amplification increases with rising receiver altitude.

In order to examine this effect more intensely, we try to model this amplification pattern. Therewith we are prepared to extend the study theoretically by simulating constructed topographies. For the simulation we use high-performance computational methods: The ADER-DG method offers the possibility of handling very complex geometries and achieving arbitrary high orders of accuracy within the framework of numerical wave propagation ([6]). This ability is needed for such a rugged topography as the Staufen Massif displays. Our simulations were running as a 128-processor partition on the Intel Itanium2 Montecito Dual Core machine (called HLRB II) of the Leibniz-Rechenzentrum in Garching.

For the model setup we employ a digital elevation model of the area around the Staufen Massif ($\Omega = [0;18.5km] \times [0;17.5km] \times [-6km;1.5km]$) kindly provided by the Bavarian Geologic Survey. The three dimensional unstructured tetrahedral mesh contains about 1.6million elements with a spacing of $200m$ at the surface and growing with depth to $350m$ (see Fig. 1). Throughout the whole domain we keep the material properties constant ($v_p = 4.7\frac{km}{s}$). As we want to involve only teleseismic events we choose an inflowing plane p-wave in z-direction as a source term.

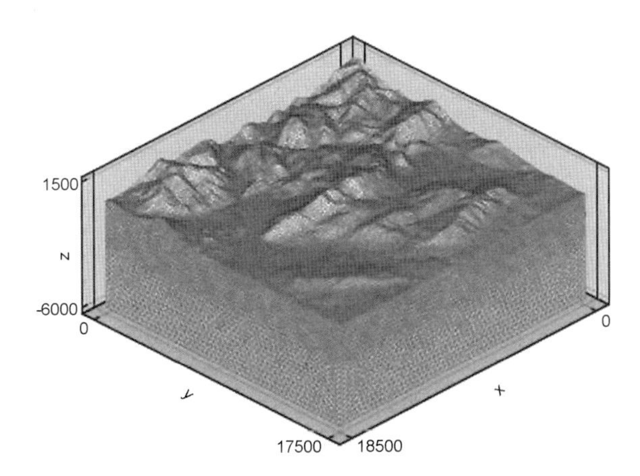

Fig. 1 Unstructured tetrahedral mesh for the area around Mt. Hochstaufen. The unit is meter

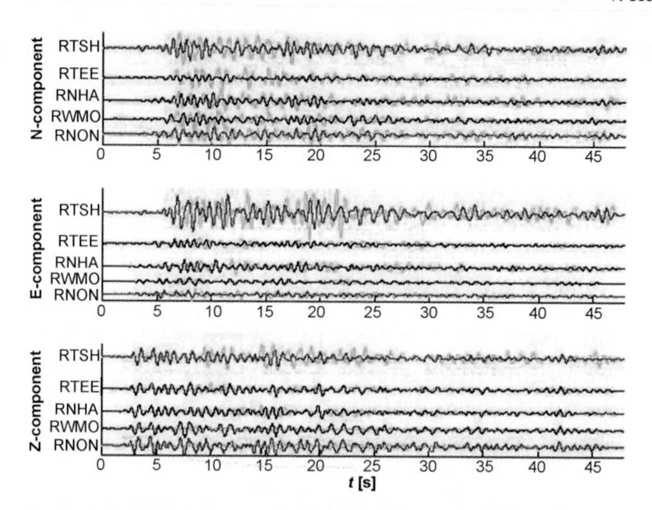

Fig. 2 Comparison of observed (blue) and synthetic (red) seismogams for Vanuatsu Islands event at different stations around Hochstaufen (RTBE, RNHA, RWMO, RNON) and on top of the massif (RTSH)

Having finished the simulation we can deconvolve the input from the computed seismograms and obtain the Green's function of our model at some defined receiver location which can be used for ongoing analysis.

As an example, Fig. 2 includes some of the results for a comparison of observed and synthetic data at Mt. Hochstaufen. The components are shown for the Vanuatsu Islands event ($M = 6.8$) which happend on 7th of August, 2006. Most of the stations are around Hochstaufen, whereas station RTSH is located almost on top of te mountain. Comparing observed to synthetic data in general, the time domain amplification factors are within the same range whereas due to the many simplifications, data were not utilizable for a profound spectral analysis. We examine the amplification factors as ratios of groundmotions at a receiver location in the valley around Hochstaufen and one on top of the mountain. Depending on the component (N, E and Z) we obtain values of 1.25 (N), 3.81 (E) and 3.28 (Z) for the observed seismograms. Averaging over only 7 events the standard deviation of 25% is acceptable. The spectral distribution of amplification shows a larger difference in the amplification of the E- and Z-components compared to the N-component with rising frequency. The E-component is amplified most, the Z-component has only slightly smaller values, and the N-component suffers least amplification, which might be related to the east-west elongated shape of Mt. Hochstaufen.

For the numerically computed data time domain amplification shows average values of 1.88 (N), 7.94 (E) and 1.21 (Z). The standard deviation on average accounts for 22%. The trend in amplification could be partially extracted from the simulation, as the E-component displays the largest amplification for both data sets. There appears rising amplification with altitude and stronger topographic gradients.

Concluding, one can say, that this study was a first step towards a very interesting subject of topographic effects at Mt. Hochstaufen. Further applications incorporate higher topographic resolution, different slopes and altitudes. We intend to analyze the frequency dependence of the amplification factor in detail.

3 Seismic Tomography Using Waveforms

Recent progress in numerical methods and computer science allows us today to simulate the propagation of seismic waves through realistically heterogeneous Earth models with unprecedented accuracy (e.g. [6, 14]). These new capabilities must now be used to further our understanding of the Earth's 3D structure. Detailed knowledge of subsurface heterogeneities is essential for studies of the Earth's dynamics and composition, for reliable tsunami warnings, the monitoring of the Comprehensive Nuclear Test Ban Treaty and the exploration for resources including water and hydro-carbons.

Full waveform tomography is a tomographic technique that takes advantage of numerical solutions of the elastic wave equation (e.g. [8, 9, 26]). Numerically computed seismograms automatically contain the full seismic wavefield, including all body and surface wave phases as well as scattered waves generated by lateral variations of the model Earth properties. The amount of exploitable information is thus significantly larger than in tomographic methods that are based, for example, on measurements of surface wave dispersion or the arrival times of specific seismic phases. The accuracy of the numerical solutions and the exploitation of complete waveform information result in tomographic images that are both more realistic and better resolved [9, 26].

We developed a novel technique for full 3D waveform tomography for radially anisotropic Earth structure. This is based on the combination of spectral-element simulations of seismic wave propagation and adjoint techniques. The misfit between observed seismograms and spectral-element seismograms is reduced iteratively using a pre-conditioned conjugate-gradient scheme.

The application of our method to the upper mantle in the Australasian region allows us to explain the details of seismic waveforms with periods between $30s$ and $200s$ (Fig. 3), and it provides tomographic images with unprecedented resolution (Fig. 4). In the course of 19 conjugate-gradient iterations the total number of exploited waveforms increased from 2200 to nearly 3000. The final model, *AMSAN.19*, thus explains data that were not initially included in the inversion. This is strong evidence for the effectiveness of the inversion scheme and the physical consistency of the tomographic model. Our model of the shear wave speed in the upper mantle (Fig. 4) reveals the deep structure of the Australasian region: The Coral and Tasman Seas, located east of mainland Australia, are characterized by a pronounced low-velocity zone, centred around $140km$ depth. It indicates the presence of a flow channel where temperatures are several hundred degrees higher than in the surrounding mantle. A similar low-velocity zone is not present under continental Australia. A low-velocity band extends along the eastern continental margin down to at least $200km$ depth. This is associated with high surface heat flow, re-

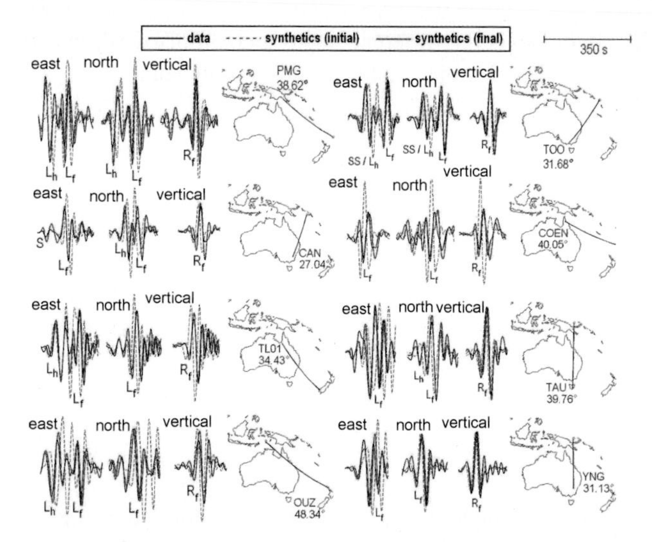

Fig. 3 Exemplary waveform comparisons for a variety of source-receiver geometries. Black solid lines are data, black dashed lines are synthetics for the initial model (a smoothed version of a previous tomography) and red solid lines are synthetics for the final model shown in Fig. 1. The dominant period is 30 s. A time scale is plotted in the upper-right corner. While significant discrepancies exist between data and the initial synthetics, the final synthetics accurately explain both the phases and the amplitudes of the observations

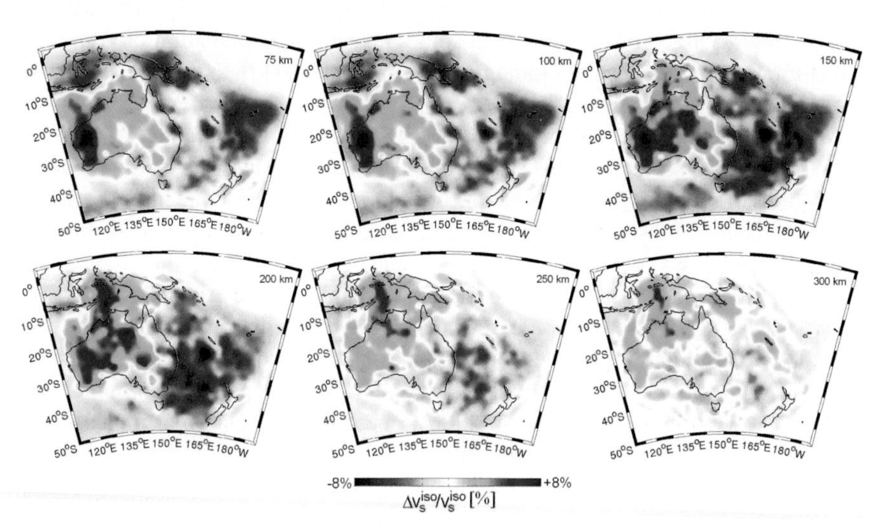

Fig. 4 Horizontal slices through the relative lateral variations of the shear wave speed at different depth in the Australasian region

cent volcanism and some mild seismic activity. The continental lithospheric mantle is confined to depths above $250km$, where the velocity perturbations exceed 5%. A region of high wave speeds under the Arafura and Timor Seas is evidence for a northward continuation of the North Australian craton.

Currently our research focusses on the application of our method to the European upper mantle and regions where more accurate Earth models are needed for the improved assessment of seismic risk.

4 Time Reversal of Seismic Waves

Time Reversal is a promising method to determine earthquake source characteristics without any a-priori information (except the earth model and the data). It consists of injecting flipped-in-time records from seismic stations within the model to create an approximate reverse movie of wave propagation from which the location of the source point and other information might be inferred. The backward propagation is performed numerically using a spectral element code. We investigate the potential of time reversal to recover finite source parameters, test it on a synthetic data set, the SPICE Tottori benchmark model, and on point source moment tensor sources.

Time Reversal has been largely used in acoustics in the last twenty years for locating sound source and scatterers in laboratory experiments (e.g. [10]); however the basic ideas date back to the early 1960's ([21]). In its simplest form, Time Reversal is used to reconstruct a sound source. This is accomplished in the laboratory by measuring acoustic waves emanating from an acoustic source. The waves are reversed-in-time, and re-emitted from the detector locations. Due to reciprocity and invariance of the wave equation to time-reversal, they back-propagate following their forward wave paths back onto the source point. When the backward step is numerically performed, the method is often referred as Time Reversal Imaging and relies on the accuracy of both the numerical scheme and the velocity model mimicking the actual media. In the past several years Time Reversal has begun to be applied to seismological questions first related to seismic migration ([2]) and then to source imaging but limited to simple velocity models ([23]). In the last decade, the drastic increase of compute power combined with the development of new methods (beside the classical finite difference method) such as spectral element has made the modeling of wave propagation in complex velocity models feasible. Application to actual seismic recordings was first conducted by [16] for the 24 December 2004, M9.0 Sumatra–Andaman earthquake. Most recently, several glacial earthquakes involving sudden motion of ice-mass downhill were located with the method [17].

Time Reversal versus adjoint methods Time Reversal is the first step in the iterative inversion problem of source location if the problem is formulated using the adjoint technique (e.g., [27]). Therefore it is important to note that when time reversing seismograms we do not actually image the source but estimate a source update (it would be the source only in the case of an exactly linear problem). An example is shown in [27] for a point source of which the location and source time function are known and the moment and mechanism are the unknowns. It is shown that to calculate the gradients the elements of the strain tensor are used. In this formulation the adjoint sources are simply the time-reversed data.

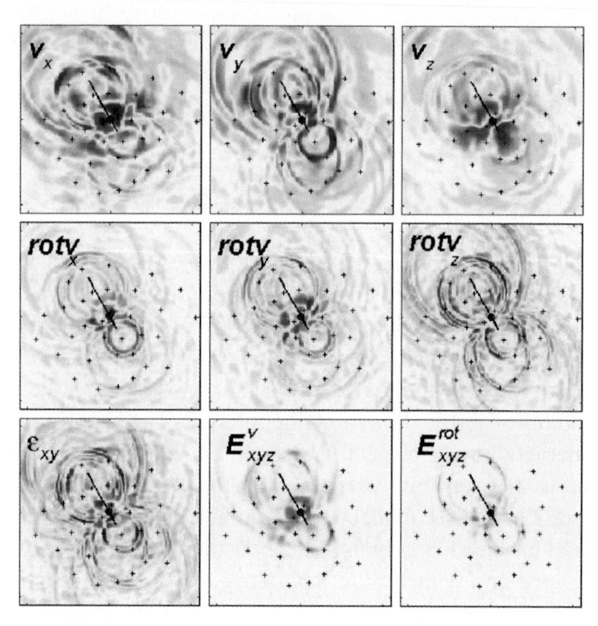

Fig. 5 Snapshots of a horizontal plane through the source depth (12.5 km) at source time for various properties of the reverse field of the SPICE finite source. Receiver locations are indicated by '+', the location of the hypocenter by the red circle. The fault is indicated by the black line

Synthetic Tests The concepts for time-reversing seismograms recorded close to large seismic sources are first illustrated using synthetic modelling. The forward modelling is carried out using a spectral-element approach (e.g., [7]) in Cartesian coordinates. To investigate the potential to recover quantitative information on finite-source properties (e.g., source extent, rupture velocities, location of asperities, etc.) we time-reverse synthetic seismograms calculated within the kinematic source inversion blind test organized within the European SPICE project ([19]). Results can be seen in Fig. 5. While focusing can be observed in all cases, there is also a substantial amount of additional energy propagating in the medium originating from phases generated by the adjoint sources (e.g., surface waves, also called ghost waves in the time reversal community).

Synthetic test with a point source moment tensor The initial tests with a point source moment tensor served as a control whether the code is setup correctly and the time-reversed input is working. In Fig. 6 snapshots of the rotational energy for a time series ending at the expected focusing point in time are shown in a plane through the source for the SPICE setup with 33 injected time-reversed signals. Injection at the stations can be observed, leading to a sharp focus at source time.

How to improve the results in terms of focussing and source location estimation is matter of ongoing discussion, but valuable approaches seem to be weighting of the adjoint sources and choosing the right field (and component) for visualization. Further studies will deal with the problem of time reversing real earthquake data.

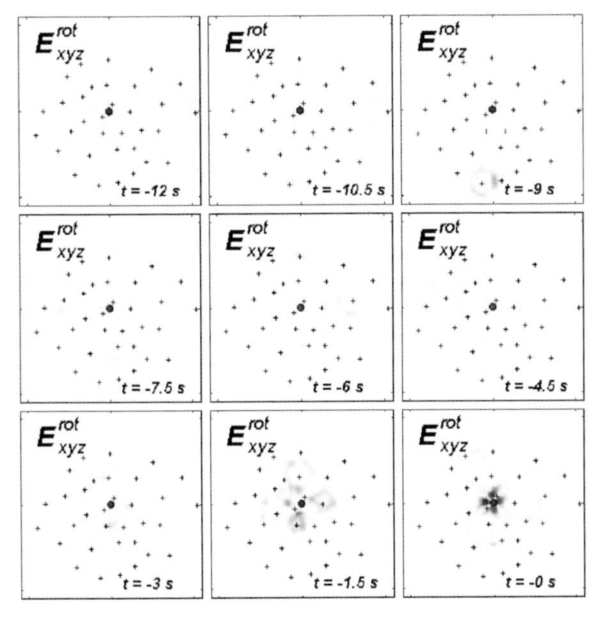

Fig. 6 Snapshots in source depth (12.5 km) at various times for the sum of squared amplitudes (rotations) of the reverse field. Receiver locations are indicated by '+', the source location by the red circle

5 Scattering in the Earth's Crust

It was a surprise to discover considerable rotational energy in a time window containing the P coda in the teleseismic seismometer records of [11]. To understand the original of the observations, we study the generation of P coda rotations under the assumption of P-SH scattering in the crust by modeling complete theoretical seismograms created by a plane P wave of dominant period 1s (the same as the predominant period of the observed P waves), propagating upward from the bottom of a random medium.

A random medium can be described through a spatial distribution $u(x)$ of material parameters u. This distribution can be expressed as [12]

$$u(x) = u_0(x) + U(x) \tag{1}$$

where $u_0(x)$ is the mean value of $u(x)$ and $U(x)$ is a realization of the random quantity. In our study, to simplify the modeling, we perform simulations with random media in which the wave velocities are randomly perturbed in space, but the mass density ρ, the ratio between P and S wave velocities $\frac{V_p}{V_s}$, and the mean values of the velocities are kept constant. The realization of the random velocities is calculated in terms of white noise filtered by a spectral filter (see [12]). Since we just consider high frequencies (i.e., small heterogeneities), we choose the spectral filter

Table 1 The modeling parameters used in this study

Mesh type	Hexahedral
Element edge length	$950m$
Total number of elements	176128
Polynomial order inside elements	4
Number of processors	64
Length of seismograms	$60s$
Boundary conditions	Free surface (top), inflow (bottom), periodic (sides)
Average time step	$8.13008 \times 10^{-3}s$
Run time per simulation	$\approx$ 2 hours

corresponding to the Zero Von Kármán correlation function mentioned in [12]:

$$\hat{f}(k) = \kappa \left(a^{-2} + k^2\right)^{-d/4}, \tag{2}$$

where a is the Von Kármán correlation length, d is the Euclidean dimension of the space, and κ is a constant corresponding to the given value of perturbation.

To be as realistic as possible, the used model setup is based on the crust model of [3] at the Wettzell area. The model is $60800m$ long, $60800m$ wide and $40850m$ deep. The mass density is taken $\rho = 2.9g/cm3$, and the mean values of P and S wave velocities (respectively) are $V_p = 6600m/s$, $V_s = 3700m/s$. In order to produce significant scattering energy, we use correlation lengths between $1000m$ and $15000m$ (see [1, 25]). Furthermore, the root mean square perturbation of wave velocities is taken in the range from 0% (homogeneity) up to about 11%.

Seismograms are calculated using the ADER-DG method (the combination of a Discontinuous Galerkin finite element method and an Arbitrary high-order DERivative time integration approach developed by [6]) that was extended to allow outputting the three components of rotation rate, in addition to the three components of translational velocity. The modeling parameters are detailed in Tab. 1. Fig. 7 illustrates schematically a three-dimensional Von Kármán random medium used in this study with correlation length $a = 2000m$, root-mean square perturbation of 6.51%. In addition, synthetic seismograms obtained at three receivers located at different sites on the surface of the model are shown. For each set of translational velocities (V_x, V_y, V_z) or rotation rates $(\Omega_x, \Omega_y, \Omega_z)$ the amplitudes are scaled. The figure shows that, as waves travel through the random medium, both rotational and horizontal translation components are generated by scattering. The delayed arrival of the vertical rotation compared to the onset of the vertical velocity that we noticed in the observations [22] can also be clearly seen in our synthetics. Moreover, the simulated seismograms differ at each station because the random field is completely three-dimensional. This allows us to stack aspects of the data from different receivers rather than multiple simulations.

The effects of the perturbation and correlation length on P coda rotations are systematically investigated. The partitioning of P and S energy and the stabilization of the ratio between the two are used to compare the simulated P coda rotations with the observed ones (see [22] for details). For each pair of scattering parameters

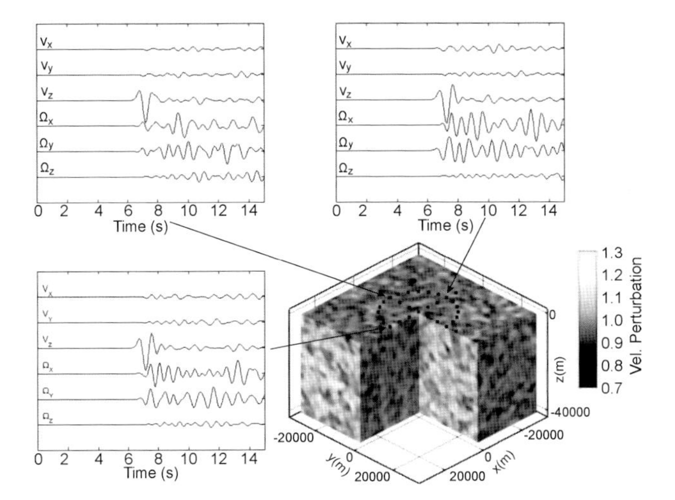

Fig. 7 Schematic illustration of a 3D Von Kármán random medium used in this study (correlation length $a = 2000m$, root-mean square perturbation of V_s is 6.51%) and 6-component seismograms obtained at three different receivers for a plane P wave propagating upward (in a vertical direction) from the bottom of the model. For each set of translational velocities (V_x, V_y, V_z) or rotation rates $(\Omega_x, \Omega_y, \Omega_z)$ amplitudes are scaled

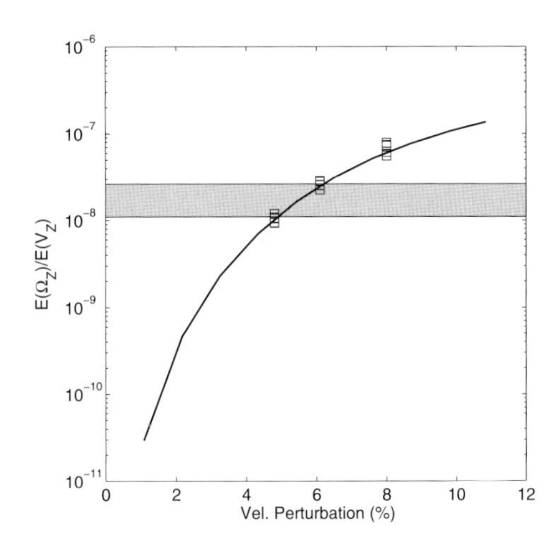

Fig. 8 Comparison of observed and simulated energy ratios. Black curve: the (average) energy ratio as a function of velocity perturbation calculated from simulated seismograms (correlation length fixed at $2000m$); Squares: the (average) energy ratio as a function of correlation length (from $1000m$ up to $15000m$) obtained from simulations; Horizontal gray band: range of the energy ratios obtained from observations

(a velocity perturbation and a correlation length), we obtain a value of the energy ratio, averaged through all 24 receivers in circular configuration (Fig. 7). Initially, we take a fixed correlation length of $2000m$ and calculate the average energy ratios as a function of velocity perturbation. The results are shown as a black line in Fig. 8. As expected, the energy ratio increases as we increase the perturbation. The curve shows a non-linear behavior, with the slope decreasing as the perturbation increases. Similarly, we can calculate the average energy ratio as a function of correlation length at some fixed values of velocity perturbation. We cover a range of correlation lengths from $1000m$ up to $15000m$ and fixed perturbation values of 4.8%, 6.1% and 8%. The results are plotted as squares in Fig. 8 and strongly suggest that for our model geometry the energy ratio is much more sensitive to the velocity perturbation than to the correlation length.

The results (Fig. 8) demonstrated that the observed signals can be explained with P-SH scattering in the crust with scatterers of roughly $5km$ correlation length (not well-constrained) and root mean square perturbation amplitude of 5% (well-constrained). This result further illustrates the efficacy of rotation measurements, for example, as a filter for SH-type motion. Similar processing steps will be possible for the horizontal components of rotation and the corresponding components of translation. It is conceivable that the combination of these various components might lead to tight constraints on near-receiver structure, results otherwise only available from array measurements.

6 Conclusions

We report several examples of 3-D wave propagation algorithms to recover either the Earth's structure or the earthquake source characteristics. 1) Using the capability of the Discontinuous Galerkin method to simulate waves through models with complex geometries we attempt to model sensational observations made at the Mt. Hochstaufen showing amplifications due to local topography by a factor of more than 10. Or results suggest that we cannot reproduce the amplification indicating that either more topographic details or additional internal 3-D strucrure is required. 2) We show a first tomographic inversion using complete 3-D modeling. This indicates that we enter a new era in seismic tomography in which we can expect to use much more information from the observed seismograms and recover sharper images of Earth's interior. 3) The time reversal study indicates that source information can be recoverd be re-injecting time-flipped seismograms at the receiver locations but the recovery of finite source characteristics is difficult. Finally, 4) we are able to model observations of rotational motions using ring laser technology with the 3-D discontinous Galerkin method, which allows putting constraints on crustal scattering properties. The key message is that 3-D wave simulation agorithms are about to become standard research tools for seismic data processing. The recent funding of a large EU project in this field (QUEST, www.quest-itn.org, coordinated by LMU seismology) supports the view that we are now looking into a much wider usage of 3-D simulation technology for real data analysis.

Acknowledgements We acknowledge funding from the Bavarian Elite Network THESIS, KONWIHR, and the LRZ for providing the computational resources. We also acknowledge the European Human Resources Mobility Programme (SPICE project), the DFG Emmy-Noether Programme, and the DAAD (N.D. Pham) and the Vietnamese Government. HI and SK also acknowledge support from the Los Alamos National Laboratory, New Mexico.

References

1. Aki, K., Richards, P. G.: Quantitative Seismology: theory and methods, 2nd. University Science Books, California (2002).
2. Baysal, E., Kosloff, D. D., Sherwood, J. W. C.: Reverse time migration. Geophysics **48**, 11 (1983).
3. Bassin, C., Laske, G., Masters, G.: The Current Limits of Resolution for Surfaace Wave Tomography in North America, EOS Trans. AGU **81**, F897 (2002).
4. Boore, D. M.: A Note on the Effect of Simple Topography on Seismic SHWaves. Bull. Seism. Soc. Am., **62**(1), 275–284 (1972).
5. Carcione, J. M.: The wave equation in generalised coordinates, Geophysics, **59**, 1911–1919 (1994).
6. Dumbser, M., Käser, M.: An arbitrary high order discontinuous Galerkin method for elastic waves on unstructured meshes II: The three-dimensional isotropic case. Geophysical Journal International **167**, 319–336 (2006).
7. Fichtner, A., Igel, H.: Efficient numerical surface wave propagation through the optimization of discrete crustal models - a technique based on non-linear dispersion curve matching (DCM). Geophys. J. Int. **173**(2), 519–533 (2008).
8. Fichtner, A., Kennett, B. L. N., Igel, H., Bunge, H.-P.: Spectral-element simulation and inversion of seismic waves in a spherical section of the Earth. Journal of Numerical Analysis Industrial and Applied Mathematics **4**, 11–22 (2009).
9. Fichtner, A., Kennett, B. L. N., Igel, H., Bunge, H.-P.: Full seismic waveform tomography for upper-mantle sturcture in the Australasian region using adjoint methods. Geophysical Journal Internatioal, in pess.
10. Fink, M.: Time reversed acoustics. Phys. Today **50**(3), 34–40 (1997).
11. Igel, H., Cochard, A., Wassermann, J., Flaws, A., Schreiber, U., Velikoseltsev, A., Pham, N. D.: Broadband Observations of Earthquake Induced Rotational Ground Motions, Geophys. J. Int. **168**, 182–196 (2007).
12. Klimeš, L.: Correlation Functions of Random Media, Pure Appl. Geophys., **159**, 1811–1831 (2002).
13. Komatitsch, D., and J. P. Vilotte: The spectral-element method: an efficient tool to simulate the seismic response of 2D and 3D geological structures, Bull. Seism. Soc. Am., **88**, 368–392 (1998).
14. Komatitsch, D., Tromp, J.: Spectral-element simulations of global seismic wave propagation-I. Validation. Geophysical Journal International **149**, 390–412 (2002).
15. Kraft, T., Wassermann, J., and Igel, H.: High-precision relocation and focal mechanism of the 2002 rain-triggered earthquake swarms at Mt. Hochstaufen, SEGermany. Geophys. J. Int., **167**(3), 1513–1528 (2006).
16. Larmat, C., Montagner, J.-P., Fink, M., Capdeville, Y., Tourin, A., Clévédé, E.: Time reversal imaging of seismic sources and application to the great Sumatra earthquake. Geophys. Res. Lett. **33**, (2006).
17. Larmat, C., Tromp, J., Liu, Q., Montagner, J.-P.: Time reversal location of glacial earthquakes. J. Geophys. Res. **B09314**, (2008).
18. Lee, Shiann-Jong, Chan, Yu-Chang, Komatitsch, Dimitri, Huang, Bor-Shouh, and Tromp, Jereon: Effects of Realistic Surface Topography on Seismic Ground Motion in the Yangminshan Region of Taiwan Based Upon the Spectral-Element Method and LIDAR DTM, Bull. Seism. Soc. Am., **99**, 681–693 (2009).
19. Mai, P., Burjanek, J., Delouis, B., Festa, G., Francois-Holden, C., Monelli, D., Uchide, T., Zahradnik, J.: Earthquake Source Inversion Blindtest: Initial Results and Further Developments. AGU Fall Meeting (2007).
20. Moczo, P., J. Kristek, M. Galis, P. Pazak, and M. Balazovjech: The finite-difference and finite-element modeling of seismic wave propagation and earthquake motion, Acta physica slovaca, **57**(2), 177–406 (2007).

21. Parvulescu, A. and Clay, C. S.: Stability of Propagated Sound I. Geophys. J. Acoust. Soc. Am. **37**, 6 (1965).
22. Pham, N. D., Igel, H., Wassermann, J., Käser, M., de la Puente, J., Schreiber, U.: Observations and Modeling of Rotational Signals in the P Coda: Constraints on Crustal Scattering, Bull. Seismol. Soc. Am. **99**(2B), 1315–1332 (2009).
23. Rietbrock, A., Scherbaum, F.: Acoustic imaging of earthquake sources from the chalfant valley, 1986, aftershock series. Geophys. J. Int. **119**, 260–268 (1994).
24. Somerville, P. G., Smith, N. F., Graves, R. W., Abrahamson, N.: Modification of empirical strong ground motion attenuation relations to include the amplitude and duration effects of rupture directivity. Seism. Res. Lett. **68**, 199–222 (1997).
25. Stein, S., Wysession, M.: An Introduction to Seismology, Earthquake, and Earth Structure, Blackwell Publishing, Oxford (2003).
26. Tape, C., Liu, Q., Maggi, A., Tromp, J.: Adjoint tomography of the southern California crust. Science **325**, 988–992 (2009).
27. Tromp, J., Tape, C., Liu, Q.: Seismic tomography, adjoint methods, time reversal and banana-doughnut kernels. Geophys. J. Int. **160**, 195–216 (2005).
28. Virieux, J.: P-SV wave propagation in heterogeneous media: Velocity-stress finite-difference method, Geophysics, **51**, 889–901 (1986).

Part IV
Astrophysics

Constrained Local UniversE Simulations (CLUES)

Stefan Gottlöber, Yehuda Hoffman, and Gustavo Yepes

Abstract The local universe is the best known part of our universe. Within the CLUES project (http://clues-project.org — Constrained Local UniversE Simulations) we perform numerical simulations of the evolution of the local universe. For these simulations we construct initial conditions based on observational data of the galaxy distribution in the local universe. Here we review the technique of these constrained simulations. In the second part we summarize our predictions of a possible Warm Dark Matter cosmology for the observed local distribution of galaxies and the local spectrum of mini-voids as well as a study of the satellite dynamics in a simulated Local Group.

1 Introduction

During the last decade cosmological parameters have been determined to a precision of just a few percent giving rise to the standard model of cosmology: a flat Friedmann universe whose mass-energy content is dominated by a cosmological constant (the Λ term), a cold dark matter (CDM) component and baryons. The convergence to a standard model of cosmology sets the framework for studying the formation of both the large scale structure and of galaxies within that model. The main goal

Stefan Gottlöber
Astrophysical Institute Potsdam, An der Sternwarte 16, 14482 Potsdam, Germany
e-mail: sgottloeber@aip.de

Yehuda Hoffman
Racah Inst. of Physics, Hebrew University, Jerusalem 91904, Israel
e-mail: hoffman@huji.ac.il

Gustavo Yepes
Departamento de Física Teorica C-15, Universidad Autonoma de Madrid, Madrid 28049, Spain
e-mail: gustavo.yepes@uam.es

S. Wagner et al. (eds.), *High Performance Computing in Science and Engineering, Garching/Munich 2009*, DOI 10.1007/978-3-642-13872-0_26,

of this study is to achieve a physical understanding of how the observed structure emerged within the context of the given cosmological model. This model specifies the cosmological expansion history, the initial conditions and the material content of the universe. These plus the knowledge of the physical laws governing the dynamics of the dark matter, baryons and radiation provide the framework within which a successful model of galaxy formation can be developed.

The basic paradigm of structure formation in the ΛCDM was formulated more than 30 years ago [19]. It suggests that dark matter (DM) clusters to form DM halos, within which galaxy formation takes place via complex baryonic physics. Intensive numerical efforts of the last decade have validated the basic White & Rees picture. Yet, the process of galaxy formation is much more complicated than that simple picture. DM halos grows via process of mergers of substructures and galaxy formation proceeds by the combined action of clumpy and anisotropic gas accretion and mergers dwarf galaxies. As dwarf galaxies cross the virial radius of the DM halo they become satellites, subject to tidal forces exerted by the central potential. This transforms the satellites into tidal streams and eventually phase mixing dissolves the streams into the stellar halo of galaxies. This process manifests itself in the combined color and dynamical phase space of galactic stella halos. This is the universal mode of galaxy formation but it can be observed and analyzed only in the very local universe, resulting in the so-called near-field cosmology. This motivates cosmologists to turn their attention and study the archeology of the Local Group (LG) in their quest for understanding galaxy formation. This also motivates us to simulate the formation of the LG in the most realistic possible way.

Cosmological simulations may cover a large dynamical and mass range. A representative volume of the universe should be large, but this comes at the expense of the mass resolution, which must be decreased in proportion to the simulated volume size. One may reduce the box size at the expense of possibly not being representative. To overcome this problem we use smaller simulation boxes specifically designed to represent the observed local universe. The algorithm of constrained realizations of Gaussian field provides a very attractive method of imposing observational data as constraints on the initial conditions and thereby yielding structures which can closely mimic those in the actual universe. The prime motivation of the CLUES project is to construct simulations that reproduce the local cosmic web and its key 'players', such as the Local Supercluster, the Virgo cluster, the Coma cluster, the Great Attractor and the Perseus-Pisces supercluster. The main drawback of current constrained simulations is that they do not directly constrain the sub-megaparsec scale structure, yet they enable the simulation of these scales within the correct environment. Such simulations provide a very attractive possibility of simulating objects with properties close to the observed LG and situated within the correct environment. Their random origin, however, limits the predictive power of the simulations and in particular the constraining power of our 'near field cosmology' experiments. It is one of our main goals to improve the constraining power and ability of the simulations to reproduce the actual observed LG.

2 Constrained Simulations

During the past few years we have performed a series of constrained simulations of the local universe. We continue this research to answer within the next generation of constrained simulations the question how unique the LG is. To this end we will use more and better observational data of the local Universe. Our main motivation is to use the constrained simulations as a numerical near-field cosmological laboratory for experimenting with the complex DM and gastrophysics processes.

2.1 Observational Data

Observational data of the nearby universe is used as constraints on the initial conditions and thereby the resulting simulations reproduce the observed large scale structure. The implementation of the algorithm of constraining Gaussian random fields [4] to observational data and a description of the construction of constrained simulations was described already elsewhere [8, 10]. Here we briefly describe the observational data used. Two different observational data is used to set up the initial conditions. The first is made of radial velocities of galaxies drawn from the MARK III [20], SBF [17] and the Karachentsev [6] catalogs. Peculiar velocities are less affected by non-linear effects and are used as constraints as if they were linear quantities [21]. The other constraints are obtained from the catalog of nearby X-ray selected clusters of galaxies [13]. Given the virial parameters of a cluster and assuming the spherical top-hat model one can derive the linear overdensity of the cluster. The estimated linear overdensity is imposed on the mass scale of the cluster as a constraint. For the CDM cosmogonies the data used here constrains the simulations on scales larger than $\approx 5h^{-1}$Mpc [8]. It follows that the main features that characterise the local universe, such as the Local Supercluster, Virgo cluster, Coma cluster and Great attractor, are all reproduced by the simulations. The small scale structure is hardly affected by the constraints and is essentially random.

2.2 Constrained Initial Conditions

The Hoffman-Ribak algorithm [4] is used to generate the initial conditions as constrained realizations of Gaussian random fields on a 256^3 uniform mesh, from the observational data mentioned above. Since these data only constrain scales larger than a very few Mpc, we need to perform a series of different realizations in order to obtain one which contain an LG candidate with the correct properties (e.g. two halos with proper position relative to each-other, mass, negative radial velocity, etc). Hence the low resolution, 256^3 particle simulations were searched and the one with the most suitable Local Group like objects were chosen for follow up, high resolution re-simulations. We generated more than 200 realizations of these simulations and all of them were evolved since the starting redshift ($z = 50$) till present time

($z = 0$) using the TREEPM N-body algorithm in GADGET2. We used 64 processors of HLRB2 Altix supercomputer and the typical wall clock time for each of them was of the order of 10 hours.

High resolution extension of the low resolution constrained realizations were then obtained by creating an unconstrained realization at the desired resolution, FFT-transforming it to k-space and substituting the unconstrained low k modes with the constrained ones. The resulting realization is made of unconstrained high k modes and constrained low k ones. In order to be able to zoom into the desired Local Group object, we set up the unconstrained realization to the maximum number of particles we can allocate in one node of the HLRB2 Altix cluster. Since our initial conditions generator code is OpenMP parallel, we can only run it in shared memory mode. Thus, thanks to the cc-NUMA architecture of the Altix, we can address as much as 2 Tbytes of RAM for an OpenMP program. This means we can accommodate as many as 4096^3 particles in this memory. We used up to 500 processors in one node and 1.5 Tbytes of memory for that. The typical cpu time per run is of the order of 5-6 hours wall clock. Most of this time was taken to generate the sequence of Gaussian random numbers, which was done in serial mode. We would not have been able to proceed in this project if we had not have access to this computer. In fact, there are very few architectures in the world with this huge amount of shared memory per node. We are now rewriting the initial condition generator code to MPI to overcome these limitations.

We did two different kind of initial conditions using the procedure described above. On one hand we were interested in simulating the large scale structures of the whole simulation box of $64h^{-1}$Mpc with enough resolution to resolve halos which can be related with the dwarf galaxies in our real Universe. To this end, once we generated the initial conditions for the largest number of particles, (i.e. 4096^3) then we degraded them down to a maximum of 1024^3 in the whole box.

On the other hand, we were also interested in re-simulating with very high resolution the formation of the LG candidates found in the low resolution simulations. To this end, we re-simulate the evolution of this region of interest, using the full resolution (equivalent to 4096^3 effective particles) only within a sphere of just $2h^{-1}$Mpc. Outside this region, the simulation box is populated with lower resolution (i.e higher mass) particles. We were thus able to achieve high resolution in the region of interest, while maintaining the correct external environment. These initial conditions were set up at very high redshift ($z = 100$) to avoid spurious effects due to cell crossing in the high resolution volume. We then populate this volume with two different species, dark matter and SPH gas particles, for the case of running hydrodynamical simulations of these objects.

2.3 *Description of Simulations*

Using the above initial conditions, we carried out the simulations using the MPI N-body + SPH code GADGET2, developed by V. Springel [15]. The total amount of computing time that has been invested in all the experiments done within the CLUES collaboration has been tremendous. We had to distribute the work among different supercomputers in Europe, mainly the HLRB2 at LRZ, the MareNostrum at BSC, and JUMP and more recently Juropa at Jülich. In all cases, the initial conditions were always generated in HLRB2 at LRZ for the reasons explained above.

In total we have run 3 big collisionless (N-body only) simulations with 1 billion particles each (i.e. 1024^3), and 2 more are currently running. Two of these simulations correspond to one realization with WMAP3 cosmological parameters, that was done both assuming CDM and WDM with $m_{WDM} = 1\,\mathrm{keV}$ and the other one is a CDM realization with WMAP5 cosmological parameters.

The simulations started at $z = 50$ and were evolved until $z = 0$. We stored of the order of 190 snapshots, 60 of them with very fine time interval of 15 million years between them, until $z = 6$. Then, we enlarged the time interval to 100 million years between consecutive snapshots. Since each snapshot is 32 Gbytes in size, we stored of the order of 6.1 Tbytes of data in each run. We used 500 MPI processors and the typical computing time needed for these runs was of the order of 130-150k CPU hours.

The other set of simulations we have also performed correspond to the re-simulations of the LG object found in two of the constrained realizations. As explained above, we have re-simulated a sphere with a radius of $2h^{-1}$Mpc around the object position at present time, with the maximum resolution allowed by our initial condition setting (i.e. 4096^3 effective particles in the re-simulate volume). We use the same set of initial conditions to run two simulations, one with dark matter only and another one with dark matter, gas dynamics, cooling, star formation and supernovae feedback. In our SPH simulation, each high resolution dark matter particle was replaced by an equal mass, gas - dark matter particle pair with a mass ratio of roughly 1:5.

Due to the strong clustering of the objects formed in these simulations, the GADGET code does not scale very well with number of processors. Thus, we decided to use the minimum necessary to allocate all the data into memory. Total number of particles in these simulations are of the order of 130-150 million. On HLRB2 Altix at LRZ we have used 64 to 128 processors for these runs. The total amount of cpu time exceeded the computational resources allocated in normal projects, so we used also most of the time that was granted to us from DEISA Extreme Computing Initiative (DECI) in two consecutive projects named, SIMU-LU and SIMUGAL-LU. Each of them consistent of 1 million cpu hours divided among MareNostrum and HLRB2. We also benefited from the DEISA high speed network in order to transfer the snapshots between the two centres.

Simulations of the Full Box

To find a LG candidate we first identify in the constrained simulation the Virgo cluster. Then we search for an object which closely resembles the Local Group and is in the right direction and distance to Virgo. Based on the locations of the LG and Virgo we now define a coordinate system similar to the super-galactic coordinate

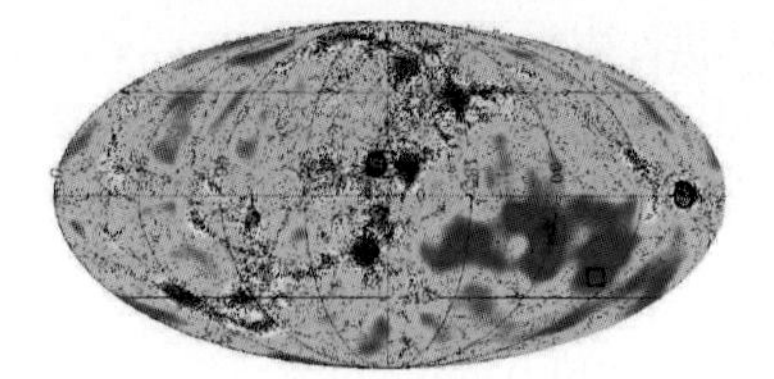
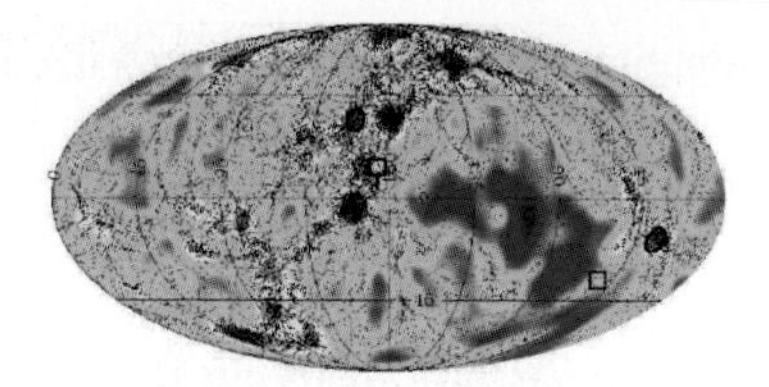

Fig. 1 The left and the right plot show two projected sky maps with halos up to a distance of $20h^{-1}$Mpc form the simulated Milky Way (see text for an explanation of the used coordinate system)

system. We assume that the equatorial plane of the super-galactic coordinate system lies in the super-galactic plane, which is spanned by the Local Group and the Local Supercluster. Thus we need another object besides Virgo to define this plane. As shown by Zavala et al [22] the Ursa Major cluster is a natural choice. Rotating the coordinate system until the simulated Virgo cluster is at the same angular position as the observed one we have fixed our coordinate system. This coordinate system is visualised in the left panel of Fig. 1 which shows a sky map with the angular distribution of halos within $20h^{-1}$Mpc from the simulated Milky Way in equatorial coordinates (RA, DEC) in a Mollweide projection. Here the value of each pixel is given by a mass-weighted count of all halos located in that pixel (high density: red, low density: blue) and all halos with masses larger than $5 \times 10^9 h^{-1}\mathrm{M}_\odot$ appear as black points (see [22] for details).

The simulated Local Supercluster can be seen as prominent vertical structure in the left panel of Fig. 1. By construction the simulated Virgo cluster (black circle) and the real one (black square) are at the same position in the centre of the plot. However, the simulated position (black circle on the right edge of the plot) of a second nearby cluster, Fornax, differs from its observed position (the black square). Such deviations are within the expected variations of the constrained simulations. In fact, small scale structure (as the position of the local group) are not constrained by the observational data. The sky maps shown in Fig. 1 depend obviously of the chosen origin (MW). The situation can be improved by an additional adjustment of the coordinate system. One can relax the requirement that the angular position of the simulated and real Virgo has to coincide and use a coordinate system that minimises the quadratic sum of the distances between the simulated and real clusters Virgo and Fornax (Fig. 1, right panel).

Zoomed Simulations of the Local Group

On the left panel of Fig. 2 the gas distribution in the Local Group is shown. The size of the plot is about $2h^{-1}$Mpc across, viewed from a distance of $3.3h^{-1}$Mpc. On the three right panels we show the gas disks of the three main galaxies as seen from a distance of $250h^{-1}$kpc, the size of the plot is about $50h^{-1}$kpc. Since the three disks are seen from the same distance and direction one can evaluate the relative orientation of the three gas disks. Note, that the M31 and MW disks are smaller than the disk of M33 due to major mergers which these objects had recently ($\sim z = 0.6$).

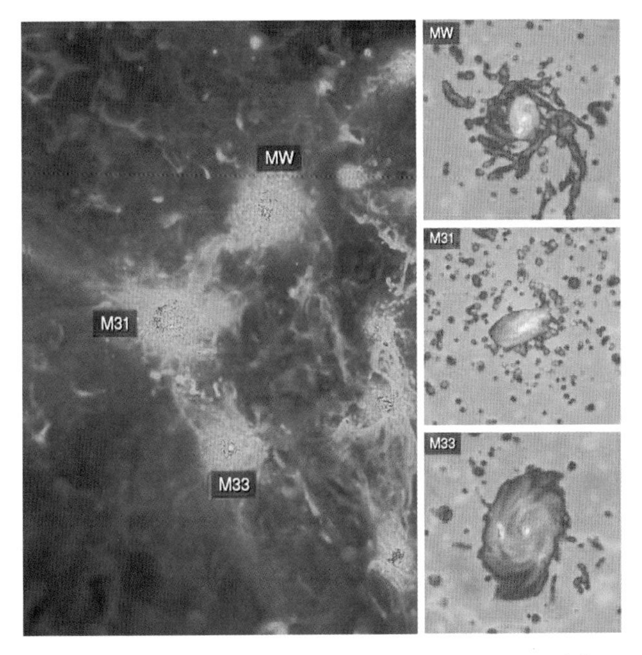

Fig. 2 The left hand side of the plot shows the gas distribution in the Local Group. On the right hand side we zoom on the three objects (MW, M31, M33)

In the section 2.4 we have described the new realizations of the local group which do not show such recent major mergers.

The images were rendered with A. Khalatyan's SPMViewer using the remote visualisation server RVS1 at LRZ. This allowed us to render the particles directly on the GPU, which took one to two minutes per frame when using all 52 million gas particles of the high resolution region. We chose a logarithmic colour table for visualising the density distribution of the gas: dark colours for low density and bright colours for high density regions. The colours and density range are adjusted such that the faint structures and filaments connecting the three main galaxies become visible. For the zoomed images we shifted the density range for colour mapping by a factor of $10^{0.5}$ in order to improve the contrast and to to enhance the spiral arm features of the gas disks.

2.4 An Ensemble of Constrained Simulations

As we explained above, we generated an ensable of 200 constrained simulations within the framework of the WMAP5 cosmological model. These low resolution simulations have been inspected to find suitable Local Group like objects. Typically 10 such objects fulfil our selection criteria [12] in a volume of $(64\ h^{-1}\mathrm{Mpc})^3$. However, these objects are not necessarily located close to the actual location of the

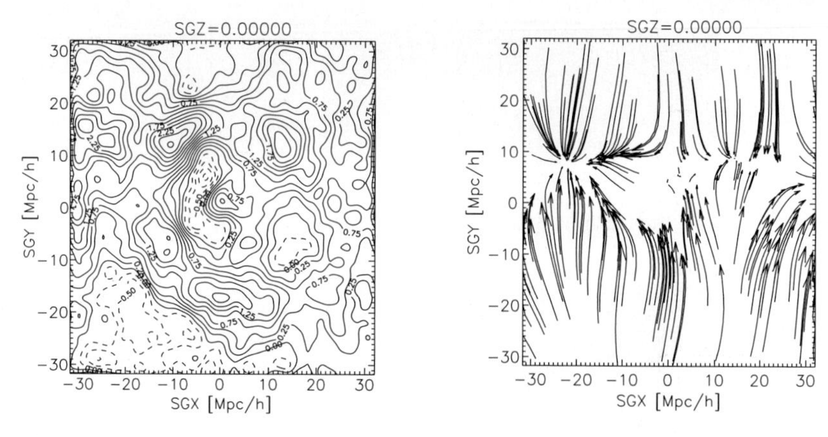

Fig. 3 The mean linear density and velocity fields. Left: linear density field smoothed by a Gaussian kernel with $R_g = 1h^{-1}$Mpc, solid, solid thick and the dashed lines correspond to positive, zero and negative values. Right: velocity field

Local Group, namely nearby and in the right direction of the the simulated Virgo cluster and the Local Supercluster. In addition to the LG identified in the WMAP3 simulation (see Fig. 2) we have identified five more objects which allow us to study the scatter in the properties of these "Local Groups".

Moreover, the ensemble of 200 constrained simulations can be used to study the statistical properties of the linear constrained realizations, the non-linear density and velocity fields extracted from the simulations, and the relation between the constrained initial conditions and the final simulations. A first step in this direction is presented here. A sub-ensemble of 60 linear constrained realizations is studied and the mean and the variance of the linear density and velocity fields are studied. Fig. 3 presents the mean, taken over the 60 constrained realizations, of the density (left panel) and the three-dimensional velocity fields. The presented velocity field corresponds to the divergent component of the velocity field which is constructed by removing the tidal component which is produced by the mass distribution outside the computational box [5]. The cosmography exhibited by the constrained realizations is dominated by the Local Supercluster, running at roughly $SGY \approx 15h^{-1}$Mpc across the box, and in particular by the Virgo cluster at $[SGX, SGY, SGZ] \approx [-8, 15, 0]h^{-1}$Mpc. Note that linearly recovered Virgo cluster is already shifted by a few Mpc to the left on the Supergalactic plan. The velocity field itself converges close to $\approx [-20, 10, 0]h^{-1}$Mpc, and therefore we expect to see a drift of the simulated Virgo in the negative SGX direction.

Fig. 3 shows the mean density and velocity fields, however individual constrained realizations deviate from the mean. To study the nature of the scatter of the constrained realizations from the mean field we calculate the scatter, i.e. the standard deviation, of the density field. It is interesting to see how robust are the various features uncovered by the constrained realizations against the scatter exhibited by the constrained realizations. Fig. 4 presents the mean density field divided by the local scatter. This normalization of the field by "sigma" is strongly scale dependent.

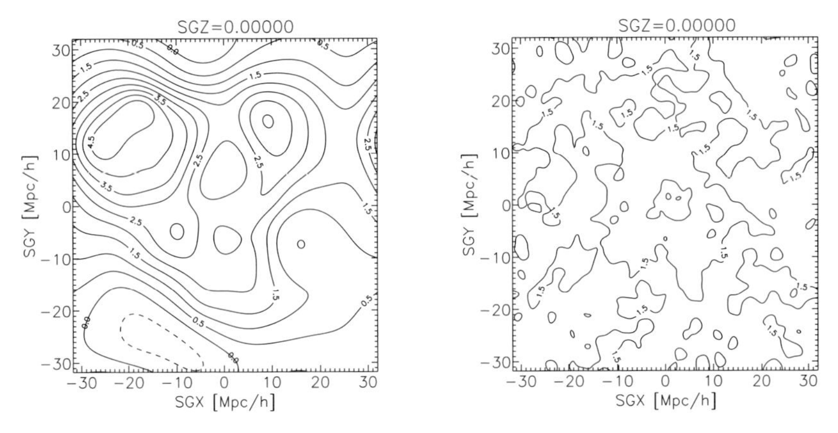

Fig. 4 Left:The scatter, i.e the standard deviation, around the mean density field smoothed by a Gaussian kernel with $R_g = 5h^{-1}$Mpc is evaluated. The figure shows the mean density field normalised by it scatter. Right: The same but for $R_g = 1h^{-1}$Mpc

A Gaussian kernel of $R_g = 5\ h^{-1}$Mpc is used in the left panel and $R_g = 1\ h^{-1}$Mpc in the right one. Normalizing the mean field by the local scatter provides a local measure of the statistical significance of any feature recovered by the constrained realizations. The figure shows that indeed the Local Supercluster is the most robust feature of the constrained realizations. Yet, the normalized $R_g = 1h^{-1}$Mpc map is almost featureless with typical values of around unity. It follows that the $\approx 1h^{-1}$Mpc structure is virtually unconstrained by the imposed data, within the WMAP5 cosmology adopted here, whereas scales larger than roughly $5h^{-1}$Mpc are strongly constrained by the data.

3 Latest Results from CLUES Simulations

Within our project at the Leibniz Rechenzentrum we have performed a series of constrained simulations with 1 billion particles in boxes of 160 h^{-1}Mpc as well as 64 h^{-1}Mpc side length using both GADGET and ART. Moreover, we have re-simulated an object which closely resembles our local group with different resolutions as well as different physics. In the following two subsections we summarise a few of the most interesting results.

3.1 Warm Dark Matter in the Local Universe

We have performed a series of constrained simulations with 1024^3 dark matter particles assuming the WMAP3 normalization of the power spectrum [14]: $\Omega_m = 0.24$, $\Omega_\Lambda = 0.76$, $H_0 = 100h\,\mathrm{km\,s^{-1}Mpc^{-1}}$ with $h = 0.73$, $n = 0.95$ and $\sigma_8 = 0.75$. In order to test the effects of different dark matter candidates in the structure of the local groups, we also generated the same initial conditions but assuming that the dark matter is made of a warm, low mass candidate with a mass per particle of $m_{\rm WDM} = 1\,{\rm keV}$. The effects of Warm Dark Matter (WDM) on the structure formation is to remove power from short scales, due to the large thermal velocities of the particles. For such WDM particles the free-streaming length is $350h^{-1}$kpc which corresponds to a filtering mass of $\sim 1.1 \times 10^{10} h^{-1} M_\odot$ [1]. Thus, we just had to change the initial power spectrum, from the standard CDM, to the WDM which contains a sharp cut-off at the free-streaming length. In order to test the effect of the WDM on the structures formed, we need to have enough mass resolution to resolve those structures well below the cut-off imposed by the free streaming of WMD particles. This translates into a minimum number of particles in a simulation box such that the Nyquist Frequency imposed by the discreteness of our sampling of particles, be always smaller than the frequency at the cut-off in the WDM spectrum. Using 1024^3 particles in the simulation box, we ensure that this conditions fulfils. Thus, we generated the same realization for both CDM and WDM constrained initial conditions.

In a universe filled with Warm Dark Matter the exponential cut-off in the power spectrum leads naturally to a reduction of small scale structure. Our choice of $m_{\rm WDM} = 1\,{\rm keV}$ is close to the lower limit for the mass of the WDM particle predicted by observations. We have chosen this extreme case to study the maximal possible effect of the Warm Dark Matter on the local structure of the universe as discussed in more detail in Zavala et al [22] and Tikhonov et al [16]. We summarize here the main results from the analysis of the big simulations with 1024^3 dark matter particles assuming the WMAP3 normalization of the power spectrum [14] and two model WDM and CDM that were described in the previous sections.

In the ΛCDM case the mass function of halos in the whole simulated box follows closely the estimates from the Sheth & Tormen formalism, except in the highest mass end due to the influence of the Local Supercluster, a massive structure in this small box which comes in due to the observational constraints. In the Warm Dark Matter universe the mass function lies close to the CDM one for halos with masses higher than the filtering mass, however for lower masses it flattens and then rise due to spurious numerical fragmentation for masses $\lesssim 3 \times 10^9 h^{-1} M_\odot$ [18]. The mass resolution of $m_{\rm DM} = 1.63 \times 10^7 h^{-1} M_\odot$, allows us to derive robust results for halos with masses larger than this limiting mass, corresponding to maximum rotation velocities of $24\,{\rm km\,s^{-1}}$.

From the Warm and Cold Dark Matter simulations we have calculated the velocity functions of haloes which has the advantage over the mass function that it can be compared more directly with observational data. From our simulations we have obtained predictions for the velocity function of halos in the field of view, within $20h^{-1}$Mpc, that is being surveyed by the ongoing Arecibo Legacy Fast ALFA (ALFALFA) HI blind survey [2, 3]. This survey is divided into two regions on the sky, one includes the Virgo cluster, the other points into the opposite direction. Additionally constrained to distances less than $20h^{-1}$Mpc we call these here the "Virgo-direction region" and "anti-Virgo-direction region", respectively. Zavala et al. [22] compared the constrained simulations with the ALFALFA observational data. The results are summarised in Fig. 5.

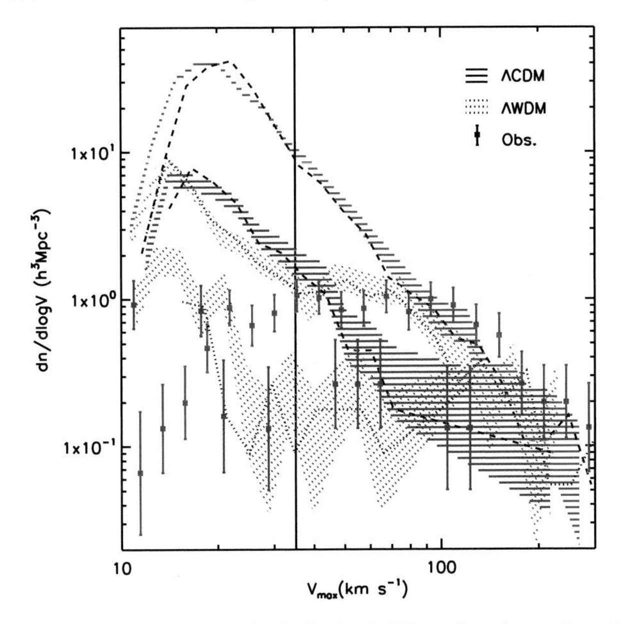

Fig. 5 Velocity function for the sample of galaxies in the Virgo-direction region taken taken from the ALFALFA catalogs (red square symbols with error bars). Predictions from our constrained simulations for the observed field of view appear as the dashed (ΛCDM) and dotted (ΛWDM) red areas, delimited by Poisson error bars. In blue the same is shown for the anti-Virgo-direction. The vertical solid line marks the value of V_{max} down to which the simulations and observations are both complete

For velocities in the range between $80\,\mathrm{km\,s^{-1}}$ and $300\,\mathrm{km\,s^{-1}}$, the velocity functions predicted for the ΛCDM and ΛWDM simulations agree quite well with the observed velocity functions. This result is encouraging because it confirms that our constrained simulations properly reflect our local environment. In fact, the simulations are able to predict the shape and normalization of the velocity function in the high velocity regime, in particular in the Virgo direction, where the normalization is an order of magnitude larger than for the universal velocity function.

The minimum mass for which we can trust the simulations and observations corresponds to a maximum circular velocity of $\sim 35\,\mathrm{km\,s^{-1}}$ (solid line in Fig. 5). Between this velocity and about $V_{max} \sim 80\,\mathrm{km\,s^{-1}}$ the predictions agree well for the ΛWDM cosmogony. On the contrary, the ΛCDM model predicts a steep rise in the velocity function towards low velocities. Thus, ar our limiting circular velocity, $V_{max} \sim 35\,\mathrm{km\,s^{-1}}$, it forecasts ~ 10 times more sources both in Virgo-direction (red) as well as in anti-Virgo-direction (blue) than the ones observed by the ALFALFA survey. These results indicate a potential problem for the cold dark matter paradigm [22]. The spectrum of mini-voids also points to a potential problem of the ΛCDM model [9]. The ΛWDM model provides a natural solution to this problem, however, the late formation of halos in the ΛWDM model might be a problem for galaxy formation [16].

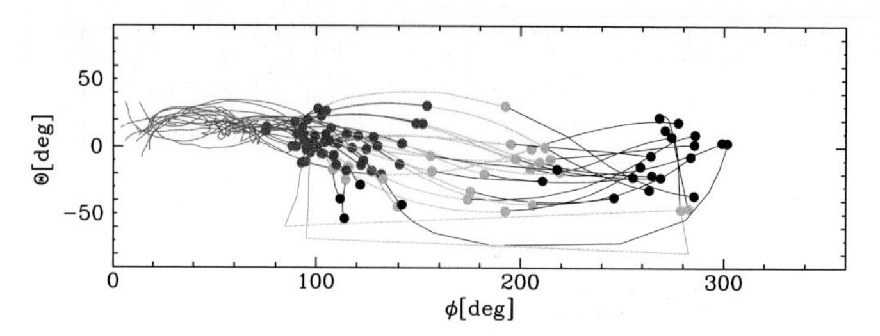

Fig. 6 Orbits of the large in-falling group of sub-haloes as seen by the observer placed at the centre of M31. The spherical coordinate system was aligned to match the angular momentum vector of the host which is pointing along the z axis. Different colours correspond to different times. Red circles: 3 Gyrs ago, blue: 2 Gyrs ago, green: 1 Gyr ago, black: present time

3.2 Satellites in the Local Group

Klimentowksi et al. used the dark matter only zoomed simulation of the Local Group with an effective resolution of 4096^3 particles as a numerical laboratory for studying the evolution of the population of its sub-haloes [7]. In M31 they have detected a large group of in-falling satellites which consists of 30 haloes. Fig. 6 shows the orbits of the satellites projected on to the sky. The observer was placed in the centre of M31 and the coordinate system was chosen so that the z-axis is aligned with the angular momentum of M31. About 4 Gyrs ago the satellites fell into M31 from one well-defined region in the sky but the spread increases with time. This example suggests that the alignment of angular moments of in-falling satellites is not well conserved, even though the group has not yet decayed. One should thus be very careful when trying to reproduce the histories of dwarf galaxies using their present proper motions.

Since we were running simulations of the LG with and without baryons modelled hydrodynamically we can quantify the effect of gas physics on the $z = 0$ population of sub-haloes. We found [11] that above a certain mass cut, $M_{\rm sub} > 2 \times 10^8 h^{-1} M_\odot$ sub-haloes in gas-dynamic simulations are more radially concentrated than those in in dark matter only simulations. The increased central density of such sub-halos results in less mass loss due to tidal stripping than the same sub-halo simulated with only dark matter. The increased mass in hydrodynamic sub-haloes with respect to dark matter ones causes dynamical friction to be more effective, dragging the sub-halo towards the centre of the host. In Fig. 7 we show as an example the mass and radial position of an in-falling satellite as a function of time. The same satellite has been identified in the dark matter only simulation (blue) and the gas-dynamical simulation (red). In the left panel one can see that the sub-halo's mass in the dark matter only simulation closely follows the SPH sub-halo's mass for a short period directly after accretion. However, by $z = 0.4$ the DM halo has already lost a good $\sim 10\%$ more then its SPH counterpart. By $z = 0$ the DM sister has lost $\sim 65\%$ of its infall mass, while the SPH sub-halo has only lost $\sim 45\%$ of its infall mass. In the right panel we show the orbit of this sub-halo as a function of redshift which diverge

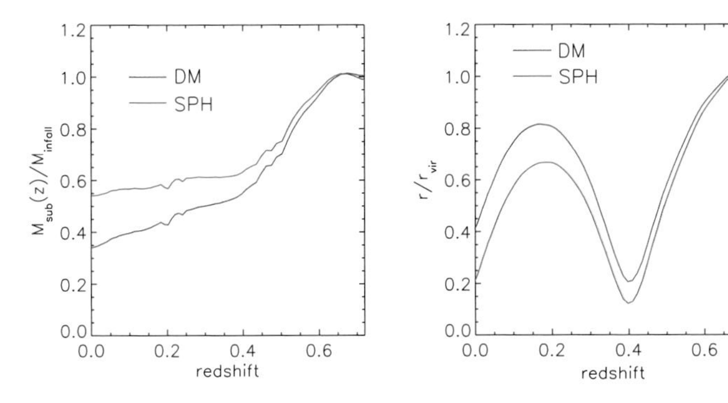

Fig. 7 Left: The mass of an SPH sub-halo (red line) and its DM sister (blue line) as a function of redshift, normalised to the mass each sub-halo had at infall, around $z = 0.72$. Right: The distance from the main halo of an SPH sub-halo and its DM sister as a function of redshift

shortly after infall. By the first pericentric passage, the SPH sub-halo penetrates further in by about a factor of 2 than the DM sister. The sub-halo from the dark matter only simulation is consistently further behind the SPH sub-halo at all stages of the orbit, resulting in the SPH sub-halo being nearly a factor of 2 closer to the host halo's centre then the DM sub-halo at $z = 0$.

4 Summary

Within the CLUES project (`http://clues-project.org`) we have performed a series of constrained simulations of the local universe. These simulations reproduce the observed large scale structures around the Local Group whereas small scale structures below scales of $\sim 1h^{-1}$Mpc are essentially random. In four simulations we have identified Local Groups situated at the right position but having different merging histories. Constrained simulations are a useful tool to study the formation and evolution of our Local Group in the right cosmological environment and the best possible way to make a direct comparison between numerical results and observations, minimizing the effects of the cosmic variance.

Acknowledgements We thank Jesus Zavala (Munich), Noam Libeskind (Potsdam), Jarek Klimentowski (Warsaw) and Kristin Riebe (Potsdam) for providing plots from our common research projects. The computer simulations described here have been performed at LRZ Munich and BSC Barcelona. We acknowledge support by ASTROSIM, DFG and DEISA for our collaboration. GY would like to thank MICINN for financial support under project grants FPA 2009-08958, AYA 2009- 13875-C03 and the SyeC Consolider CSD 2007-0050. YH would like to thank ISF (13/08) for financial support.

References

1. Bode, P., Ostriker, J. P., & Turok, N., 2001, ApJ, 556, 93
2. Giovanelli, R., Haynes, M. P., Kent, B. R., Perillat, P., Catinella, B., Hoffman, G. L., Momjian, E., Rosenberg, J. L., et al., 2005, AJ, 130, 2613
3. Giovanelli, R., Haynes, M. P., Kent, B. R., Saintonge, A., Stierwalt, S., Altaf, A., Balonek, T., Brosch, N., et al., 2007, AJ, 133, 2569
4. Hoffman, Y. & Ribak, E., 1991, ApJL, 380, L5
5. Hoffman, Y., Eldar, A., Zaroubi, S., & Dekel, A., 2001, astro-ph/0102190
6. Karachentsev, I. D., Karachentseva, V. E., Huchtmeier, W. K., & Makarov, D. I., 2004, AJ, 127, 2031
7. Klimentowski, J., Lokas, E. L., Knebe, A., Gottlöber, S., Martinez-Vaquero, L. A., Yepes, G., & Hoffman, Y., 2010, MNRAS, 402, 1899
8. Klypin, A., Hoffman, Y., Kravtsov, A. V., & Gottlöber, S., 2003, ApJ, 596, 19
9. Klypin, A. & Tikhonov, A., 2009, MNRAS, 395, 1915
10. Kravtsov, A. V., Klypin, A., & Hoffman, Y., 2002, ApJ, 571, 563
11. Libeskind, N. I, Yepes, G., Knebe, A., Gottlöber, S., Hoffman, Y., & Knollman, S. R., 2010, MNRAS, 401, 1889
12. Martinez-Vaquero, L.A., Yepes, G., Hoffman, Y., Gottlöber, S., & Sivan, M., 2009, MNRAS, 397, 2070
13. Reiprich, T. H. & Böhringer, H., 2002, ApJ, 567, 716
14. Spergel, D. N., Bean, R., Doré, O., Nolta, M. R., Bennett, C. L., Dunkley, J., Hinshaw, G., Jarosik, N., et al., 2007, ApJS, 170, 377
15. Springel, V. 2005, MNRAS, 364, 1105
16. Tikhonov, A. V., Gottlöber, S., Yepes, G., & Hoffman, Y., 2009, MNRAS, 399, 1611
17. Tonry, J. L., Dressler, A., Blakeslee, J. P., Ajhar, E. A., Fletcher, A. B., Luppino, G. A., Metzger, M. R., & Moore, C. B., 2001, ApJ, 546, 681
18. Wang, J. & White, S. D. M., 2007, MNRAS, 380, 93
19. White, S. D. M. & Rees, M. J., 1978, MNRAS, 183, 341
20. Willick, J. A., Courteau, S., Faber, S. M., Burstein, D., Dekel, A., & Strauss, M. A., 1997, ApJ, 109, 333
21. Zaroubi, S., Hoffman, Y., & Dekel, A., 1999, ApJ, 520, 413
22. Zavala, J., Jing, Y. P., Faltenbacher, A., Yepes, G., Hoffman, Y., Gottlöber, S., & Catinella, B., 2009, ApJ, 700, 1779

The Core Helium Flash: 3D Hydrodynamic Models

Miroslav Mocák and Ewald Müller

Abstract We continue our study of turbulent convection during the core helium flash close to its peak by comparing the results of 2D and 3D hydrodynamic simulations performed with our multidimensional Eulerian hydrodynamics code HERAKLES on the ALTIX 4700 computer of the LRZ. In our previous study based on well resolved 2D and coarsely resolved 3D models covering only the onset of convection we found that (i) the temporal evolution and properties of the convection are similar to those predicted by quasi-hydrostatic stellar evolutionary calculations, and that (ii) the core helium flash does not lead to any explosive behavior. These results are confirmed by our new well resolved 3D models covering longer evolutionary periods. They also show that the flow patterns developing in the 3D models are structurally different from those of the corresponding 2D ones, and that the typical convective velocities in these models are smaller than those in their 2D counterparts. The 3D models also tend to agree better with the predictions of mixing length theory. Our hydrodynamic simulations further show the presence of turbulent entrainment that results in a growth of the convection zone on a dynamic time scale.

1 Introduction

The core helium flash is the most violent event in the life of a star with an initial mass between $0.7M_{\odot}$ and $2.2M_{\odot}$ [33]. The pre-flash stellar core has a white dwarf-like structure with an off-center temperature maximum followed by a super-adiabatic temperature gradient (convection zone) resulting from plasma and photo-neutrino cooling. When helium burning temperatures are reached, the liberated energy can-

Miroslav Mocák · Ewald Müller
Max-Planck-Institut für Astrophysik, Karl-Schwarzschild-Str.1, 85748 Garching bei München, Germany
e-mail: mmocak@mpa-garching.mpg.de
e-mail: emueller@mpa-garching.mpg.de

S. Wagner et al. (eds.), *High Performance Computing in Science and Engineering, Garching/Munich 2009*, DOI 10.1007/978-3-642-13872-0_27,

not be used to expand and cool the layers at the temperature maximum due to the electron degeneracy but rather leads to a heating of the core and strongly increasing nuclear energy release. Therefore, the core helium ignition leads to a runaway situation due to the extreme dependence of the triple α helium burning rate on the temperature ($\dot{\varepsilon} \sim T^{30}$). At this stage, the gas gets rapidly heated on timescale of $\sim$ 10000 yr but does not expand. The temperature rises until the degeneracy of the electron gas is lifted.

For a solar mass star, the total energy production rate at the peak of the core helium flash is almost 10^{42} ergs/s. Nevertheless, the energy release only results in a slow expansion of the helium core ($\sim$m/s) as energy transport by a convection, heat conduction and radiation seems to be always able to deliver most of the flash energy quiescently from the stellar interior to the upper layers. While the pre-flash evolution proceeds on the nuclear time scale $\sim 10^8$ yr, typical e-folding times for the energy release from helium burning become as low as hours at the peak of the flash, and therefore are comparable to convective turnover times. Thus, the usual assumptions used in simple descriptions for convection in hydrostatic stellar evolution modeling (e.g. instantaneous mixing) are no longer valid. Previous attempts to overcome these assumptions by allowing for hydrodynamic flow remained inconclusive predicting either severe explosions [5–7, 11, 12, 14, 15, 35, 38, 39] or a quiescent behavior [10, 13, 22, 25, 26].

Here we discuss our temporally extended and better resolved 2D and 3D simulations of turbulent convection within the helium core during the peak of the core helium flash performed with the HERAKLES code [19, 20] which solves the hydrodynamic equations coupled to nuclear burning and thermal transport in up to three dimensions. We employ a detailed equation of state, and a time-dependent gravitational potential. Our hydrodynamic models are characterized by a convectively unstable layer (the convection zone) embedded in between two stable layers composed of several chemical nuclear species and of a partially degenerate electron gas.

2 Initial Model

The initial model was obtained with the stellar evolution code GARSTEC [36]. Some of its properties are listed in Tab. 1. The model possesses a white dwarf like degenerate structure with an off-center temperature maximum resulting from plasma and photo-neutrino cooling. Its central density is $\sim 7 \times 10^5$ g cm^{-3}. At the outer edge of the isothermal region in the center of the helium core the temperature jumps up to $T_{max} \sim 1.7 \times 10^8$ K, and the adjacent convection zone has a super-adiabatic temperature gradient. The core is predominantly composed of ^{4}He (mass fraction X(^{4}He)$>$ 0.98), and some small amounts of ^{1}H, ^{3}He, ^{12}C, ^{13}C, ^{14}N, ^{15}N and ^{16}O, respectively. For our hydrodynamic simulations we adopt the abundances of ^{4}He, ^{12}C and ^{16}O as the triple-α reaction dominates the nuclear energy production during the flash. The remaining composition is assumed to be adequately represented by a gas with a mean molecular weight equal to that of ^{20}Ne.

Table 1 Some properties of the initial model M: Total mass M, stellar population, metal content Z, mass M_{He} and radius R_{He} of the helium core ($X(^4He) > 0.98$), nuclear energy production in the helium core L_{He}, temperature maximum T_{max},radius r_{max}, density ρ_{max} at the T_{max}

Model	M [M$_\odot$]	Pop.	Z	M_{He} [M$_\odot$]	R_{He} [10^9cm]	L_{He} [10^9L$_\odot$]	T_{max} [10^8K]	r_{max} [10^8cm]	ρ_{max} [10^5g cm^{-3}]
M	1.25	I	0.02	0.47	1.91	1.03	1.70	4.71	3.44

The initial stellar model is spherically symmetric (1D), hydrostatic, and was evolved on a Lagrangian grid of 2294 zones using the OPAL equation of state [30]. For our 2D and 3D hydrodynamic simulations we use the equation of state of Timmes & Swesty [34], and an Eulerian spherical polar grid, i.e. we had to interpolate the stellar model onto our Eulerian grid. As the equations of state employed in the GARSTEC code and ours are different, the pressure differs by about 1% for the density and temperature of the stellar model. Stabilization of the initial hydrodynamics model resulted in a small decrease of the temperature compared to the stellar model.

3 Code

The HERAKLES code solves the Euler equations coupled to source terms describing thermal transport, self-gravity and nuclear burning. The hydrodynamic equations are integrated with the piecewise parabolic method [9] involving a Riemann solver for real gases [8]. The evolution of the chemical species is described by a set of additional continuity equations [28].

The source terms in the momentum and energy equation arising from self-gravity and nuclear burning are treated by means of operator splitting. Every source term is calculated separately at the end of the integration step. Using dimensional splitting [32], the code is of 2nd order integration accuracy in spacetime. In the simulations discussed below transport of energy by conduction and radiation was neglected, as it is roughly seven orders of magnitude smaller than that resulting from convection.

The gravitational potential is approximated by a 1D Newtonian potential obtained from the spherically averaged mass distribution. The nuclear network consists of first order nonlinear differential equations describing the amount of energy released by nuclear burning. We solve this stiff system by means of the semi-implicit Bader-Deufelhard method which utilizes the Richardson extrapolation approach and sub-stepping techniques [1].

The code performs well on many computer systems including massively parallel ones with vector processors. The included computational kernels are fully vectorized, while the vector length is adjusted to the memory architecture. This allows for optimal performance on both vector and superscalar, cache-based machines. The current version of the code is parallelized with OpenMP. On the SGI ALTIX 4700 of the LRZ and for a 3D simulation with $400 \times 180 \times 360$ zones the speedup on 64 and 128 processors amounts to 42 and 43, respectively. The unsatisfying scaling behavior beyond 64 processors is due to (i) an improper distribution of data across the local processor memories (some data reside far from the processors, where they are being processed), and (ii) due to the nuclear reaction network. The latter prevents a proper load balancing and thus a proper scaling, as only a small amount of computational zones are burning. Running the code with burning switched off we obtain a satisfactory speedup of 80 on 128 processors.

4 Results

We performed two 2D and one 3D simulation, whose properties are summarized in Table 2. Models helf.2d.a and helf.3d cover about 6000 s of evolutionary time during the peak of the core helium flash. The computational domain consists of a wedge of 120° in both angular directions centered at the equator ($\theta = \pi/2$) to properly capture the largest vortices ($\sim 40°$) found in previous 2D simulations [25]. In radial and angular direction we used a grid resolution of $\Delta r = 5.55 \times 10^6$ cm and $\Delta\theta = \Delta\phi = 1.5°$, respectively. The additional 2D model, hefl.2d.b, was evolved for almost 34 hrs (130 000 s) on a grid covering 180° in θ-direction and having more than twice the angular resolution of the other models.

All models posses a convection zone which spans 1.5 density scale heights and which is enclosed by convectively stable layers extending out to a radius of 1.2×10^9 cm. We used reflective boundary conditions in every coordinate direction except for model hefl.2d.a, where it turned out to be necessary to impose periodic boundary conditions in angular direction, because reflective boundaries in combination with the 120 °wedge size affected the large scale convective flow adversely leading to disproportionately high convection velocities.

We cut off the inner part of the grid at a radius of 2×10^8 cm (sufficiently distant from the temperature inversion at $r \sim 5 \times 10^8$ cm) to allow for a larger CFL time step. To trigger convection, we imposed a random flow field with a maximum (absolute) velocity of 10 cm s^{-1}, and a random density perturbation $\Delta\rho/\rho \leq 10^{-2}$. This procedure is necessary, as a spherically symmetric model evolved with Herakles on a grid in spherical polar coordinates will remain spherically symmetric forever.

The 3D model hefl.3d and the corresponding 2D model hefl.2d.a evolve initially ($t < 1200$ s) similarly. Convection sets in after roughly 1000 s (the thermal relaxation time), and hot rising bubbles appear in the region where helium burns in a thin shell ($r \sim 5 \times 10^8$ K). After another ~ 200 s, the bubbles cover the complete height of the convection zone ($R \sim 4.8 \times 10^8$ cm). The flow eventually approaches a quasi-steady state consisting of several upstreams (or plumes) of hot gas carrying the liberated nuclear energy off the burning region, thereby inhibiting a thermonuclear runaway.

The models exhibit a sandwich-like stratification: two stable layers enclose the convection zone at the top and bottom, respectively. The convection zone is characterized by a large kinetic energy density, and the adjacent convectively stable layers

Table 2 Some properties of the 3D and 2D simulations: number of zones in r (N_r), θ (N_θ), and ϕ (N_ϕ) direction, spatial resolution in r (Δr in 10^6 cm), θ ($\Delta\theta$), and ϕ ($\Delta\phi$) direction, characteristic velocity v_c (in 10^6 cm s^{-1}) of the flow during the first 6000 s, expansion velocity at temperature maximum v_{exp} (in m s^{-1}), typical convective turnover time $t_o = 2R/v_c$ (in *sec*) where R is the height of the convection zone, and maximum evolutionary time t_{max} (in s), respectively

run	grid	Δr	$\Delta\theta$	$\Delta\phi$	v_c	v_{exp}	t_o	t_{max}
hefl.2d.a	180×90	5.55	1.5°	-	1.44	24.	650	6000
hefl.2d.b	360×240	2.77	0.75°	-	1.84	92.	510	130000
hefl.3d	$180 \times 90 \times 90$	5.55	1.5°	1.5°	0.85	10.	1105	6000

show waves induced by convection. The models reach a steady state after roughly 2000 s, and the steep increase of the total kinetic energy from 10^{39} erg to 10^{45} erg marks the onset of convection. The kinetic energy density shows small fluctuations in the fully evolved convection zone, and is by an order of magnitude larger in the 2D model hefl.2d.a than in 3D model hefl.3d. This is in agreement with other studies, as it is well know that 2D turbulence is more intensive (see, e.g. [27]). The total energy production is about 10% higher in the 3D model due to its slightly higher temperature and the strong dependence of the triple-α reaction rate on temperature.

Fully evolved convection ($t > 2000$ s) in the 3D model hefl.3d differs significantly from that in the corresponding 2D model hefl.2d.a. The convective flow of the 2D model is dominated by vortices having (angular) diameters ranging from 30°to 50° (Fig. 1), and an aspect ratio of close to one. The vortices are qualitatively similar to those found in other 2D simulations [2, 17, 18, 29]. This vortex structure of 2D turbulence is quite typical, and arises from the self-organization of the flow [16, 23] The convective flow in the 3D model hefl.3d consists of column-shaped plumes (Fig. 1 and 2), and contrary to the 2D model does not show any dominant angular mode. Power spectra of angular velocity fluctuations show that turbulent elements have an almost time-independent characteristic angular size of 30° - 50°in the 2D case, while the spectra computed for the 3D model change with time and exhibit no dominant angular mode. We find turbulent flow features across the whole convection zone resulting from the interaction of convective up and downflows. Close to the edges of the convective zone we observe the smallest turbulent flow features that form when compact turbulent plumes are decelerated and break-up [3]. Shear instabilities likely play an important role in the development of the turbulent flow as well [4].

Temperature and density fluctuations are 30-40 % larger in the 2D model than in the 3D one. This is expected, because vortices are stable in 2D flows but decay in 3D ones (Fig. 1). The fluctuations in the composition (^{4}He and ^{12}C) are larger by 10-30% in the 3D model which is a result of the more non-uniform mixing of chemical elements. Auto-correlation functions of the radial velocity that provide a measure of the radial size and lifetime of flow patterns show that the convective flow extends across the whole convective region as determined by the Schwarzschild criterion in the initial stellar model.

Convection may induce mixing in convectively stable layers adjacent to convectively unstable regions. Following Meakin & Arnett [24], we prefer to call this

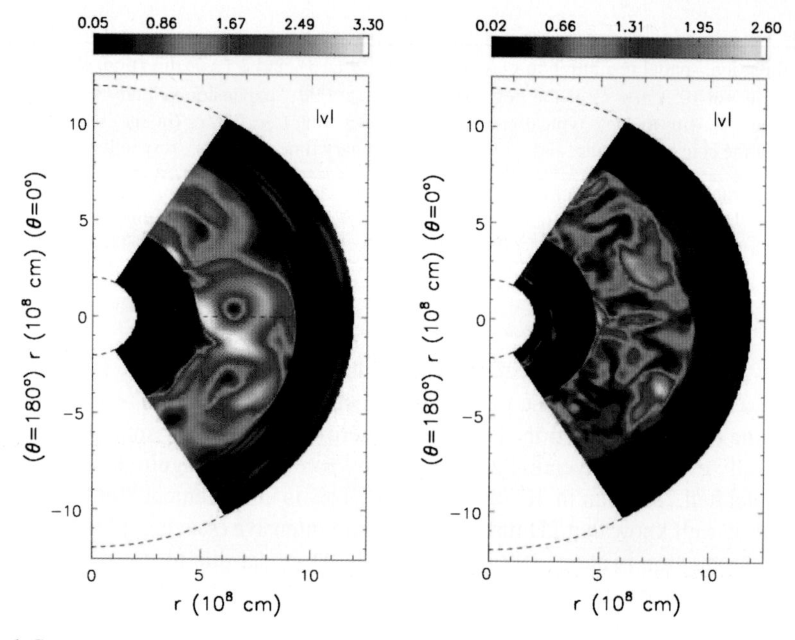

Fig. 1 Snapshots taken at ~ 4754 s showing contour plots of the absolute value of the velocity (in units of 10^6 cm s^{-1}) for the 2D model hefl.2d.a (left), and for the 3D model hefl.3d in a meridional plane of azimuth angle $\phi = 50°$ (right)

process turbulent entrainment (or mixing). The term (convective) overshooting commonly used in astrophysics accounts only for localized ascending or descending plumes crossing the edge of the convection zone. If the filling factor of these plumes or their crossing frequency is high, they can change the entropy in convectively stable regions surrounding convection zones, a process that is known as penetration [3]. Turbulent entrainment accounts for both overshooting and penetration.

Contrary to [18] we determine the depth of the entrainment neither by the radius where the kinetic energy carried into the stable layers is zero, nor by the radius where the kinetic energy has dropped to a certain fraction of its maximum value [3]. We find both conditions insufficient, because the kinetic energy flux approaches zero (at the convective boundaries) much faster than the convective flux. Instead, we rather use the ^{12}C mass fraction, as it is low outside the convection zone during the flash and can rise there only due to turbulent entrainment. Here we use the condition $X(^{12}C) = 2 \times 10^{-3}$ to define the edges of the convection zone. Due to the turbulent entrainment, these edges are pushed towards the stellar center and surface, i.e. the width of the convection zone increases on the dynamic timescale. This is in contradiction with the predictions of hydrostatic stellar modeling, where the width of the convection zone is determined by the local Schwarzschild or Ledoux criterion [37].

The speed, at which the radius of the outer edge of the convection zone increases with time due to entrainment, is estimated for models hefl.2d.a and hefl.3d to be at most 14 m s^{-1}. The radius of the inner edge of the convection zone changes at

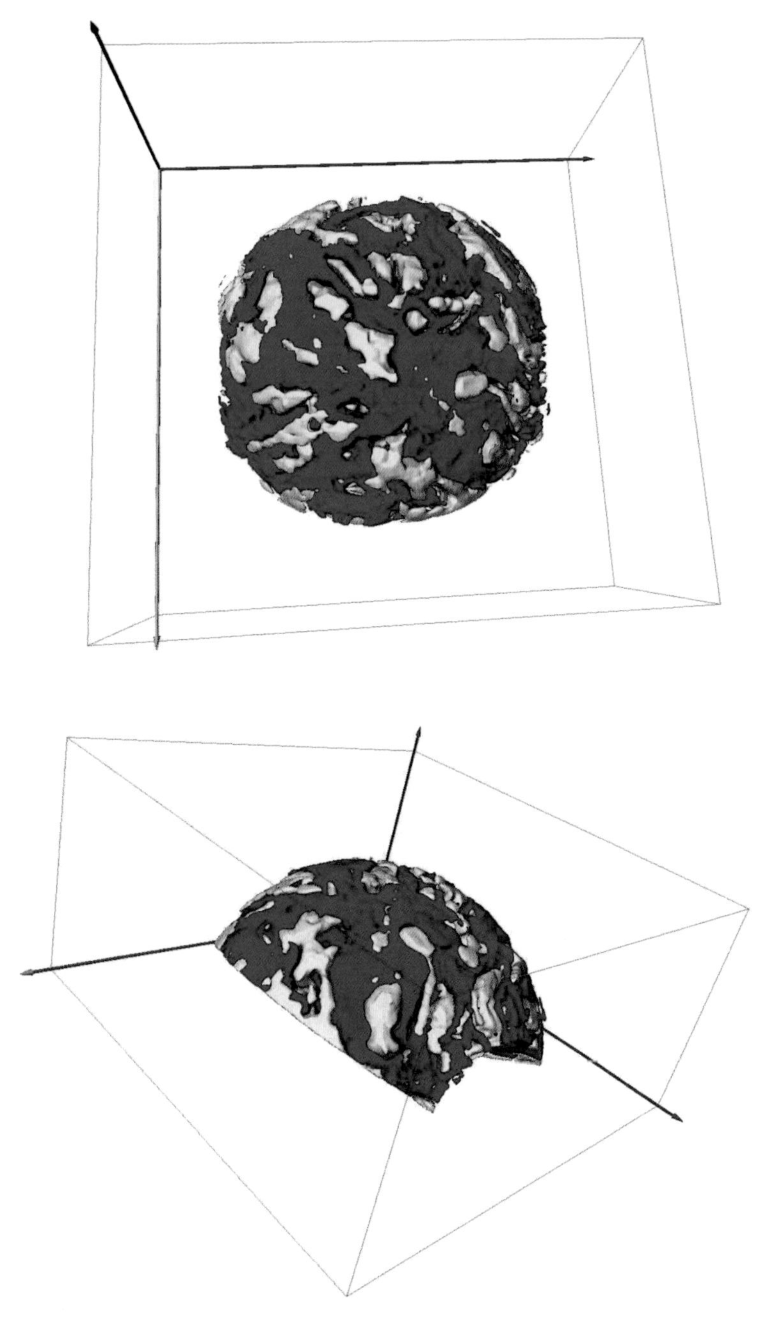

Fig. 2 Different views of the velocity field of the 3D model hefl.3d at $t = 3000$ s. The blue isosurfaces correspond to radial downflows ($v_r = -6 \times 10^5$ cm s^{-1}), and the yellow isosurfaces to radial upflows ($v_r = 6 \times 10^5$ cm s^{-1}), respectively. The small and long edges of the boxes have a size of $12,000$ km and $24,0000$ km, respectively

a much smaller rate [2, 24], as the region interior to the convection zone is more stable against convection and has a higher density than the region exterior to the convection zone [31].

Entrainment is more efficient in the 2D model hefl.2d.a than in the 3D model hefl.3d. In 2D, the observed convective flow structures are large and fast rotating vortices that due to the imposed axisymmetry are actually tori [2]. They have a high filling factor near the edge of the convection zone where they overshoot or penetrate [3]. 3D structures crossing the edge of the convection zones are smaller (localized) plumes with a lower filling factor and smaller velocities.

Mixing length theory (or MLT) commonly used for treating convection in stellar evolutionary calculations relies on assumptions and parameters that are often chosen based on convenient ad-hoc arguments about the convective flow, like e.g. the value of the mixing length, the amount of upflow-downflow symmetry or the position where, within the convection zone, convective elements start to rise [21]. MLT assumes that the temperature of a convective element (blob) is the same as that of the ambient medium surrounding it when it starts to rise. However, as a blob will not rise until it is hotter than the surroundings, this MLT assumption is contradictory. MLT further assumes that once the blobs begin to rise they carry their surplus of heat lossless over a distance given by the mixing length before they release it to the surrounding gas instantaneously at the end of their path. These assumptions are also not fulfilled in general.

Our simulations show that convective elements typically start their rise deep inside the star from the region of dominant nuclear burning where they are accelerated by buoyant forces. The assumptions of MLT that convective blobs form and begin their motion at different depths of the convection zone, and that the average convective blob propagates a distance equal to half of the assumed mixing length before dissolving with the surrounding gas [21], therefore do not hold. MLT finally also assumes a correlation between the thermodynamic variables and the velocity of the flow in a convection zone. However, the results of our simulations falsify this assumption [26].

According to results discussed above 2D simulations are biased due to the imposed symmetry restriction which tends to overestimate the activity in the convection zone. However, qualitative similarities with geometrically unconstrained 3D models exists. Hence, we find it justified to explore the long-term evolution (i.e. covering a few tens of hours instead of a few hours) of our core helium flash model by performing long-term cost-effective 2D instead of very costly 3D simulations.

The early evolution (first 8 hrs) of the 2D model [25] is characterized by a very dynamic flow involving typical convective velocities of 1.8×10^6 cm s^{-1}. Our long-term hydrodynamic simulation of this model covering 36 hrs revealed that the global and angle-averaged maximum temperatures continue to rise at the initial rate of 40 K s^{-1}. This is 60% lower than the rate predicted by stellar evolutionary calculations. As a consequence, the typical convective velocities increase by about 50% and reach a level of 2.8×10^6 cm s^{-1} at the end of the simulation.

Hydrodynamic simulations of convection driven by nuclear burning covering several convective turnover times show a rapid growth of the convection zone due to

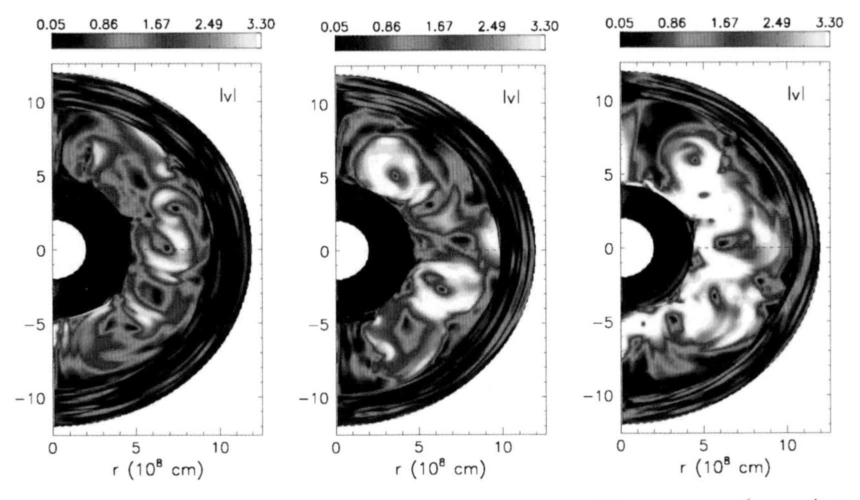

Fig. 3 Snapshots of the spatial distribution of the velocity modulus $|v|$ (in units of 10^6 cm s^{-1}) for the 2D model hefl.2d.b at 24 000 s (left), 60 000 s (middle), and 120 000 s (right), respectively

the turbulent entrainment [24]. An analysis of our simulations shows a similar behavior. Turbulent motion near the upper edge of the convection zone pumps material into the convectively stable layer at an entrainment rate of 14 m s^{-1} without any significant slowdown over the whole duration of the simulation covering more than 250 convective turnover times. The entrainment rate at the inner convective boundary is about a factor of six smaller (2.3 m s^{-1}) slightly increasing during the second half of the simulation ($t > 60000$ s). These entrainment rates have to be considered as upper limits because of the imposed axisymmetry which leads to exaggerated convective velocities and large filling factors for the penetrating plumes. The turbulent entrainment causes a growth of the convection zone on a dynamic timescale, in agreement with the oxygen shell burning hydrodynamic models [24].

5 Conclusions

We performed, analyzed and compared 2D axisymmetric and 3D hydrodynamic simulations of the core helium flash. In agreement with our previous study of 2D hydrodynamic models of the core helium flash we find that the core helium flash in three dimensions neither rips the star apart, nor that it significantly alters its structure.

The evolved convection of the 3D models differs qualitatively from that of the axisymmetric ones. The typical convective structure in the 2D simulations is a vortex with a diameter roughly equal to the width of the convection zone, whereas the 3D structures are smaller in extent and have a plume-like shape. The typical convective velocities are much higher in the 2D models than in the 3D ones. In the latter models the convective velocities tend to fit those predicted by the mixing length theory better. Both 2D and 3D models are characterized by an upflow-downflow asymmetry, where downflows dominate.

Our hydrodynamic simulations show the presence of turbulent entrainment at both the inner and outer edge of the convection zone, which results in a growth of the convection zone on a dynamic timescale. While entrainment occurs at an almost constant speed at the outer boundary of the convection zone, it tends to accelerate at the inner one. The entrainment rates are higher at the outer edge than at the inner edge of the convection zone, as the latter is more stable against entrainment.

The upper part of the evolved convection zone is characterized by a sub-adiabatic temperature gradient, where buoyancy breaking takes places, i.e. rising convective plumes start to slow down in this region and eventually descend back into deeper layers of the star. Convection should not exist in that mass layer according to mixing length theory.

The fast growth of the convection zone due to entrainment has some potentially interesting implications. As entrainment is not considered in canonical stellar evolutionary calculations, stars evolving towards the core helium flash may never reach a state as the one given by our initial stellar model. Hence, this may influence the growth of the convection zone observed in our hydrodynamic simulations, as the thermodynamic conditions at the edges of the convection zone may differ.

If a rapid growth of the convection zone indeed occurs, the main core helium flash studied here will never be followed by subsequent mini-flashes, as convection will lift the electron degeneracy in the helium core within 10 days. In addition, the helium core will likely experience an injection of hydrogen from the stellar envelope within a month and undergo a violent nuclear burning phase powered by the CNO cycle. However, the growth of the convection zone within the core that is simulated in our models does not have to continue until it will reach the outer convection zone extending up to the surface of the star. Hence, mixing of nuclear ashes to the stellar atmosphere does not necessarily take place. But a fast dynamic growth of the inner convection zone will lead to a change of the composition of the stellar core (less carbon and oxygen), and consequently of the luminosities of low-mass stars on the horizontal branch.

We found significant differences between the properties and the evolution of 2D and 3D models having a convection zone powered by (semi-)degenerate helium burning. However, as our 3D models are likely not yet fully converged and as they cover only a relatively short period of evolutionary time ($< 6000\,$s), long-term 3D simulations using a higher grid resolution are needed to obtain a better and more reliable understanding of the hydrodynamics of the core helium flash.

Acknowledgements The computational resources and support provided by the LRZ where the simulations were performed are gratefully acknowledged. We would like to thank Konstantinos Kifonidis for his help improving the performance of our code on the SGI ALTIX 4700, and Achim Weiss for providing the initial model for our simulations. One of the authors (M.Mocak) also acknowledges financial support from the CommunautT francaise de Belgique - Actions de Recherche ConcertTes.

References

1. G. Bader and P. Deuflhard. A semi-implicit mid-point rule for stiff systems of ordinary differential equations. *Numer. Math.*, 41:373–398, January 1983.
2. G. Bazan and D. Arnett. Two dimensional Hydrodynamics of Pre–Core Collapse: Oxygen Shell Burning. *Astrophys. J.*, 496:316, March 1998.
3. N. H. Brummell, T. L. Clune, and J. Toomre. Penetration and Overshooting in Turbulent Compressible Convection. *Astrophys. J.*, 570:825–854, May 2002.
4. F. Cattaneo, N. H. Brummell, J. Toomre, A. Malagoli, and N. E. Hurlburt. Turbulent compressible convection. *Astrophys. J.*, 370:282–294, March 1991.
5. P. W. Cole, P. Demarque, and R. G. Deupree. Convective heating of the inner core of red giants prior to the peak of the core helium flash. *Astrophys. J.*, 291:291–296, April 1985.
6. P. W. Cole and R. G. Deupree. The core helium flash with two-dimensional convection. *Astrophys. J.*, 239:284–291, July 1980.
7. P. W. Cole and R. G. Deupree. The violent phase of the core helium flash. *Astrophys. J.*, 247:607–613, July 1981.
8. P. Colella and H. H. Glaz. *J. Comput. Phys.*, 59:264, January 1984.
9. P. Colella and P. R. Woodward. *J. Comput. Phys.*, 54:174, January 1984.
10. D. S. P. Dearborn, J. C. Lattanzio, and P. P. Eggleton. Three-dimensional Numerical Experimentation on the Core Helium Flash of Low-Mass Red Giants. *Astrophys. J.*, 639:405–415, March 2006.
11. R. G. Deupree. Two- and three-dimensional numerical simulations of the core helium flash. *Astrophys. J.*, 282:274–286, July 1984.
12. R. G. Deupree. Two- and three-dimensional numerical simulations of the hydrodynamic phase of the core helium flash. *Astrophys. J.*, 287:268–281, December 1984.
13. R. G. Deupree. A Reexamination of the Core Helium Flash. *Astrophys. J.*, 471:377, November 1996.
14. R. G. Deupree and P. W. Cole. A survey of the core helium flash with dynamic convection. *Astrophys. J.*, 269:676–685, June 1983.
15. A. C. Edwards. The hydrodynamics of the helium flash. *Mon. Not. Roy. Soc.*, 146:445 1969.
16. B. Fornberg. *J. Comput. Phys.*, 25:1–31, 1977.
17. N. E. Hurlburt, J. Toomre, and J. M. Massaguer. Nonlinear compressible convection penetrating into stable layers and producing internal gravity waves. *Astrophys. J.*, 311:563–577, December 1986.
18. N. E. Hurlburt, J. Toomre, J. M. Massaguer, and J.-P. Zahn. Penetration below a convective zone. *Astrophys. J.*, 421:245–260, January 1994.
19. K. Kifonidis, T. Plewa, H.-T. Janka, and E. Müller. Non-spherical core collapse supernovae. I. Neutrino-driven convection, Rayleigh-Taylor instabilities, and the formation and propagation of metal clumps. *Astron. Astrophys.*, 408:621–649, September 2003.
20. K. Kifonidis, T. Plewa, L. Scheck, H.-T. Janka, and E. Müller. Non-spherical core collapse supernovae. II. The late-time evolution of globally anisotropic neutrino-driven explosions and their implications for SN 1987 A. *Astron. Astrophys.*, 453:661–678, July 2006.
21. R. Kippenhahn and A. Weigert. *Stellar Structure and Evolution*. Stellar Structure and Evolution, XVI, 468 pp. 192 figs. Springer-Verlag Berlin Heidelberg New York. Also Astronomy and Astrophysics Library, 1990.
22. J. Lattanzio, D. Dearborn, P. Eggleton, and D. Dossa. Three Dimensional Simulations of the Core Helium Flash - with Rotation. *ArXiv Astrophysics e-prints*, December 2006.
23. J. C. McWilliams. *J. Fluid. Mech.*, 146:21–43, 1984.
24. C. A. Meakin and D. Arnett. Turbulent Convection in Stellar Interiors. I. Hydrodynamic Simulation. *Astrophys. J.*, 667:448–475, September 2007.
25. M. Mocák, E. Müller, A. Weiss, and K. Kifonidis. The core helium flash revisited. I. One and two-dimensional hydrodynamic simulations. *Astron. Astrophys.*, 490:265–277, October 2008.

26. M. Mocák, E. Müller, A. Weiss, and K. Kifonidis. The core helium flash revisited. II. Two and three-dimensional hydrodynamic simulations. *Astron. Astrophys.*, 501:659–677, July 2009.
27. H. J. Muthsam, W. Goeb, F. Kupka, W. Liebich, and J. Zoechling. A numerical study of compressible convection. *Astron. Astrophys.*, 293:127–141, January 1995.
28. T. Plewa and E. Müller. The consistent multi-fluid advection method. *Astron. Astrophys.*, 342:179–191, February 1999.
29. D. H. Porter and P. R. Woodward. High-resolution simulations of compressible convection using the piecewise-parabolic method. *Astrophys. J. Suppl.*, 93:309–349, July 1994.
30. F. J. Rogers, F. J. Swenson, and C. A. Iglesias. OPAL Equation-of-State Tables for Astrophysical Applications. *Astrophys. J.*, 456:902, January 1996.
31. H. P. Singh, I. W. Roxburgh, and K. L. Chan. Three-dimensional simulation of penetrative convection: penetration below a convection zone. *Astron. Astrophys.*, 295:703, March 1995.
32. G. Strang. *SIAM J. Numer. Anal.*, 5:506, October 1968.
33. A. V. Sweigart and P. G. Gross. Evolutionary sequences for red giant stars. *Astrophys. J. Suppl.*, 36:405–437, March 1978.
34. F. X. Timmes and F. D. Swesty. The Accuracy, Consistency, and Speed of an Electron-Positron Equation of State Based on Table Interpolation of the Helmholtz Free Energy. *Astrophys. J. Suppl.*, 126:501–516, February 2000.
35. K. R. Villere. *A helium flash with superadiabatic convection*. PhD thesis, AA (California Univ., Santa Cruz.), 1976.
36. A. Weiss and H. Schlattl. Age-luminosity relations for low-mass metal-poor stars. *Astron. Astrophys. Suppl.*, 144:487–499, June 2000.
37. Achim Weiss, Wolfgang Hillebrandt, Hans-Christoph Thomas, and Ritter. *Cox and Giuli's Principles of Stellar Structure*. Gardners Books, February 2004.
38. A. J. Wickett. Convection in the helium flash. In E. A. Spiegel and J.-P. Zahn, editors, *Problems of Stellar Convection*, volume 71 of *Lecture Notes in Physics, Berlin Springer Verlag*, pages 284–289, 1977.
39. R.G. Zimmermann. PhD thesis, University of California at Los Angeles, (1970), 1970.

3D Simulations of Large-Scale Mixing in Core Collapse Supernova Explosions

N. Hammer, H.-Th. Janka, and E. Müller

Abstract We present the first 3D simulations of the large-scale mixing that takes place in the shock-heated stellar layers ejected in the explosion of a blue supergiant star. The blast is initiated and powered by neutrino-energy deposition behind the stalled shock by means of choosing sufficiently high neutrino luminosities from the contracting, nascent neutron star, whose high-density core is excised and replaced by a retreating inner grid boundary. The outgoing supernova shock is followed beyond its breakout from the stellar surface about two hours after the core collapse. Violent convective overturn in the post-shock layer causes the explosion to start with significant large-scale asphericity, which acts as a trigger of the growth of Rayleigh-Taylor instabilities at the composition interfaces of the exploding star. Deep inward mixing of hydrogen is found as well as fast-moving, metal-rich clumps penetrating with high velocities far into the hydrogen envelope of the star. Comparing with corresponding 2D (axially symmetric) calculations, we determine the growth of the Rayleigh-Taylor fingers to be faster, the deceleration of the dense metal-carrying clumps in the helium and hydrogen layers to be reduced, the asymptotic clump velocities in the hydrogen shell to be higher, and the outward radial mixing of heavy elements and inward mixing of hydrogen to be more efficient in 3D than in 2D.

Nicolay J. Hammer
Max-Planck Institut für Astrophysik, Karl-Schwarzschild-Str.1, 85748 Garching, Germany
present address: Max-Planck-Institut für Plasmaphysik, Boltzmannstr. 2, 85748 Garching, Germany
e-mail: `hammer@mpa-garching.mpg.de`

H.-Thomas Janka · Ewald Müller
Max-Planck Institut für Astrophysik, Karl-Schwarzschild-Str.1, 85748 Garching, Germany
e-mail: `thj@mpa-garching.mpg.de`
e-mail: `emueller@mpa-garching.mpg.de`

S. Wagner et al. (eds.), *High Performance Computing in Science and Engineering, Garching/Munich 2009*, DOI 10.1007/978-3-642-13872-0_28,

1 Introduction

Besides an experimental confirmation of a core collapse event through the detection of supernova neutrinos [14], the second most important insight provided by observations of Supernova 1987A was the occurrence of large-scale non-radial flow and extensive mixing of chemical species in the envelope of the progenitor star during the explosion (see e.g. [2]). While SN 1987A still remains the most prominent and thoroughly-observed example, observations of many other core-collapse supernovae have meanwhile provided ample evidence that large-scale extensive mixing seems to occur generically in such events (see, e.g. [32]).

Two dimensional hydrodynamic simulations of non-radial flow and mixing in the stellar envelopes of core collapse supernova progenitors have been performed by several groups. The first simulations were started by artificially seeding Rayleigh-Taylor instabilities (RTIs) in the mantle and envelope of the progenitor and following their evolution until shock breakout from the stellar surface (see e.g. [2, 10, 11, 13, 21, 22]. More recent simulations consistently connect the seed asymmetries arising from convective flow in the neutrino-heated bubble, and by the standing accretion shock instability (SASI) in the "supernova engine" during the first second of the explosion (see, e.g. [15]) to the later Rayleigh-Taylor and Richtmyer-Meshkov instabilities after shock passage through the outer stellar layers with application to SN 1987A [18, 19]. Very recently Rayleigh-Taylor induced mixing and the amount of fallback occurring during artificially triggered (piston model), and initially spherically symmetric supernova explosions of zero- and solar-metallicity 15 and $25\,M_{\odot}$ stars have been studied [16], and the observational characteristics of jet-driven supernovae in a red supergiant progenitor have been explored [9]. The latter simulations show the development of fluid instabilities that produce helium clumps in the hydrogen envelope, and nickel-rich jets that may account for the non-spherical excitation and substructure of spectral lines.

Only a few attempts to 3D simulations have been made up to now. [24] simulated the adiabatic point explosion of an $n = 3$ polytrope, and claimed to have found pronounced Rayleigh-Taylor instabilities in that event. However, this claim was not confirmed by the 3D simulations of [23] and [3]. Both studies found only a weak instability, i.e., no extensive clumping and mixing. Subsequently, 3D simulations by [21] and [34] using relatively coarse resolution and/or considering only a wedge-shaped fraction of the star showed that seed perturbations grow strongly at the He-H interface when a realistic presupernova star model is used instead of an $n = 3$ polytrope. Later [17] studied the difference in the growth of 2D versus 3D single-mode perturbations at the He-H and O-He interfaces of SN 1987A. They found that the 3D single-mode perturbation grows about 30% faster than a corresponding 2D one. This finding is also supported by laser experiments and comprehensive Rayleigh-Taylor growth simulations [1, 8, 25].

2 Simulation Setup

We use a computational grid in spherical polar coordinates consisting of $1200 \times 180 \times 360$ zones in r, θ, and ϕ-direction. The equidistant angular grid has a resolution of $0.935°$ in θ-direction and $1°$ in ϕ-direction covering the whole sphere except for a cone with a half opening angle of $5.8°$ around the symmetry axis of the coordinate system. Reflective boundary conditions are imposed in θ-direction, and periodic ones in ϕ-direction. The clipping of the computational grid around the symmetry axis avoids a too restrictive CFL time step condition. We have not observed any numerical artifacts as consequences of this grid constraint.

The radial mesh is logarithmically spaced between a time-dependent inner boundary, initially located at a radius of 200 km, and a fixed outer boundary at 3.9×10^{12} cm. It has a maximum resolution of 2 km at the inner boundary, and a resolution of 4×10^{10} cm at the outer one (corresponding to a roughly constant relative radial resolution $\Delta r/r \approx 1\%$). We allow for free outflow at the outer boundary, and impose a reflective boundary condition at the inner edge of the radial grid. During the simulation we move (approximately every 100th timestep) the inner radial boundary to larger and larger radii to relax the CFL condition. This cutting of the computational grid reduces the number of radial zones from 1200 at the beginning to about 400 towards the end of the simulation, but involves only the innermost few percent of the initial envelope mass. We convinced ourselves by means of 2D test calculations that this removal of mass has no effect on the dynamics and mixing occurring at larger radii (the cut radius exceeds at no time a few percent of the radius of the Rayleigh-Taylor finger tips).

The initial model used for our 3D simulations is based on the explosion models of Scheck et al. [26], who followed the onset and early development of neutrino-driven explosions in 3D with the same code, numerical setup, and input physics as described in [28]. The stellar progenitor considered here is a spherically symmetric model of a $15.5 M_\odot$ blue supergiant with a radius of 4×10^{12} cm ([18, 33]. Since the models of [26] included only the central part of the star out to the middle of the C/O shell at $r = 1.7 \times 10^9$ cm, the envelope of the progenitor had to be added for the long-time simulations presented here.

Scheck et al. [26] simulated one 3D explosion model with $2°$ angular resolution, which reached a final time of 0.58 s after bounce, and another one with $3°$ angular resolution, which covered an evolution time of 1 s after core bounce. Although both models were exploited for our study, we report here only on simulations using the former model that has an (unsaturated) explosion energy of 0.6×10^{51} erg [26]. We also performed a 3D simulation where we artificially increased the explosion energy of this initial model to 1.0×10^{51} erg by extending the thermal energy input as accomplishable by ongoing neutrino heating.

We used a tabulated EOS [30] that considers contributions of an arbitrarily degenerate and relativistic electron-positron gas, of a photon gas, and of a set of ideal Boltzmann gases, consisting of the eight nuclear species (n, p, α, ^{12}C, ^{16}O, ^{20}Ne, ^{24}Mg, and ^{56}Ni) included in our initial model. Nuclear burning is not taken into account in our simulations. Both in the 3D and 2D simulations we neglected the influence of gravity on the motion of the ejecta. While this has no important impact on the dynamics of the expanding ejecta, the amount of fallback is underestimated. However, that way we could avoid the accumulation of mass near the inner (reflecting) radial boundary which would have implied a considerably more restrictive CFL condition.

We ran ten 2D and three 3D simulations, but here we focus on one 3D model that has an explosion energy of 10^{51} ergs. The main findings for this particular 3D model are qualitatively similar to those we obtained for less energetic 3D explosions, but the resulting maximum values for the asymptotic clump velocities vary with the explosion energy roughly as $v_{\text{clump}} \propto \sqrt{E_{\text{exp}}}$. The 2D simulations were started from different meridional slices of the 3D initial explosion model. This ensures as much similarity of the initial conditions as possible. The differences between the chosen slices give rise to some variation of the initial asymmetry and explosion energy among the various 2D models, which manifests itself both in different clump structures and velocities, and the amount of mixing in the stellar envelope.

3 Code Performance

For our simulations we used a variant of the PROMETHEUS hydrodynamics code [10], which is optimised for LRZ's HLRB II SGI Altix 4700 system. Several changes in the OpenMP parallelisation of the code were made to make optimal use of a large (>32) number of CPUs and the ccNUMA bus system.

We ran several benchmarks using up to 510 CPUs, i.e. one complete node of the SGI Altix 4700 system, to study the parallel performance of our code and to determine the optimal number of CPUs for our 3D simulations. The benchmark runs were performed on two computational grids of $1200 \times 180 \times 360$ and $400 \times 90 \times 180$ zones (in r, Θ, and ϕ-direction), respectively. The unsatisfying scaling behavior of the code beyond 256 processors (see Fig. 1) is caused by memory bandwidth limitations. Accordingly, we used 256 CPUs for our 3D simulations, which required a total of about 650 000 CPU hours.

4 Results

4.1 Dynamic Evolution

The evolution of the explosion after its launch by neutrino energy deposition and the growth of instabilities in the considered $15.5\,M_{\odot}$ progenitor were described in much detail by [18]. It was shown there that the asymmetries associated with the convective activity developing in the postshock region during the neutrino-heating phase act as seeds of secondary Rayleigh-Taylor instabilities at the composition interfaces of the exploding star. At about 100 s dense Rayleigh-Taylor fingers containing the

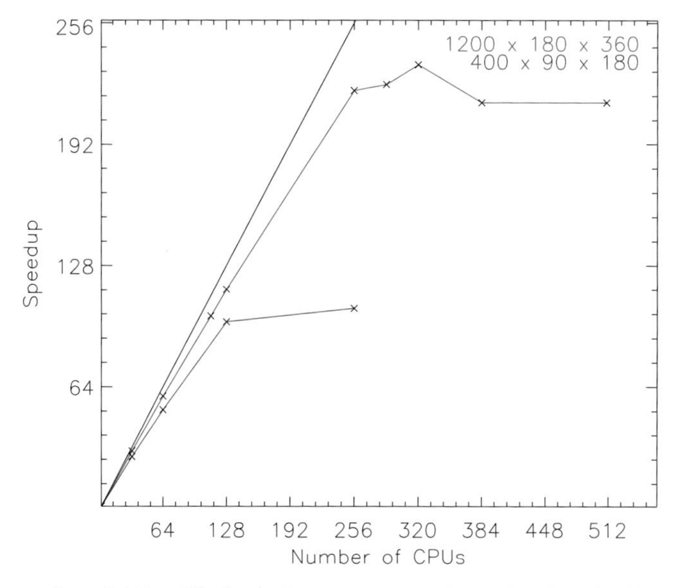

Fig. 1 Measured parallel SpeedUp for the PROMETHEUS code as a function of grid size and numbers of CPUs used on LRZ's HLRB II SGI Altix 4700 system. The grid used for our 3D simulation is the larger one (red curve).

metals (C, O, Si, iron-group elements) have grown out of a compressed shell of matter left behind by the shock passing through the Si/O and (C+O)/He interfaces. These fingers grow quickly in length and while extending into the helium shell, fragment into ballistically moving clumps and filaments that propagate faster than the expansion of their environment.

While for a 2D model with explosion energy around 1.8×10^{51} erg the Si and Ni containing structures still move with nearly 4000 km/s at 300 s, a strong deceleration occurs when the metal clumps enter the relatively dense He-shell that forms after the shock passage through the He/H interface. At $t \gtrsim 10,000$ s the metal carrying clumps have dissipated their excess kinetic energy and propagate with the same speed as the helium material in their surroundings. In the presence of a hydrogen envelope and the corresponding deceleration of the shock as it propagates through the inner regions of the hydrogen layer (in which case the helium "wall" below the He/H interface builds up), [18] could not observe any metal clumps that achieve to penetrate into the hydrogen envelope, in contrast to what was observed in the case of SN 1987A. The 2D calculations performed in the course of the present work confirm these findings.

Kifonidis et al. [19], investigating explosions also with $E_{\text{exp}} \sim 2\times10^{51}$ erg, proposed a cure of the metal-H mixing problem by invoking a sufficiently large dipolar or quadrupolar asymmetry of the beginning explosion with a globally aspherical shock wave. This was motivated by recent hydrodynamic studies that show that large-amplitude, low-ℓ mode oscillations (ℓ defining the order of an expansion in spherical harmonics) due to the standing accretion shock instability (SASI) can be

Fig. 2 Surfaces of the radially outermost locations with constant mass fractions of $\sim 3\%$ for carbon (green), and oxygen (red), and of $\sim 7\%$ for nickel (blue). The two panels display the hydrodynamic instabilities at about 9000 s shortly after the supernova shock has broken out of the stellar surface. The side length of the panels is about 7.5×10^{12} cm

an important ingredient in the explosion mechanism and in the sequence of processes that leads to observed asymmetries of supernova explosions and neutron star kicks [4–7, 20, 26–29]. This on the one hand led to higher initial maximum velocities of the metal-rich clumps and thus their faster propagation through the He core on a timescale shorter than the build-up of the thick He "wall" that prevented their penetration into the hydrogen shell in the older calculations. On the other hand it triggered the growth of a Richtmyer-Meshkov instability at the He/H interface due to the deposition of vorticity by the shock. This led to strong mixing of hydrogen inward in mass space and down to low velocities, an effect that has turned out to be necessary to explain the shape of the light curve maximum in the case of SN 1987A [31].

These conclusions, however, were based on 2D models. Therefore the question remained whether 3D might yield differences and whether also in 3D one would have to advocate a global low-ℓ mode asphericity of the beginning explosion as an explanation of the high velocities of nickel clumps in the hydrogen shell and of the strong inward mixing of hydrogen observed in SN 1987A.

4.2 Radial Element Mixing

When we speak about "nickel" here and in the following, we mean all elements heavier than ^{16}O. Thus "nickel" includes the contributions of ^{20}Ne and ^{24}Mg of the set of nuclear species considered in our simulations. This makes sense because the asymptotic radial distributions of iron-group elements turn out to be very similar to those of neon and magnesium.

At 350 s oxygen and nickel rich, dense fingers, some of which display the typical mushroom shape of Rayleigh-Taylor structures, penetrate through the outer boundary of the carbon layer of the exploding star (mass fraction of ^{12}C equal to 3%). The most prominent features at that time develop to the largest mushrooms visible later in the evolution. These carry predominantly nickel and may contain a narrow spine of oxygen. In contrast, many of the far more numerous, less extended and more narrow fingers are composed of a mixture of nickel and oxygen. Also fingers with a dominant mass fraction of oxygen can be identified (see Fig. 2).

The metal distributions in both 2D and 3D models peak at the location of the metal core with similar velocities (around 2000–3000 km/s at 350 s, around 1500–2500 km/s at 2600 s, and around 1000 km/s at 9000 s). In 3D, however, the nickel and oxygen distributions possess wider high-velocity tails up to velocities of more than 6000 km/s at 350 s. In the long run this causes the large discrepancies of the radial composition mixing. Two effects appear to contribute to the formation and continued presence of these high-velocity tails. Firstly, in 3D the Rayleigh-Taylor instabilities at the composition interfaces grow more rapidly, which leads to the formation of extremely fast (mostly nickel and less oxygen) clumps with velocities in excess of 5000 km/s at 350 s. These dense bullets penetrate deep into the hydrogen layer even before the helium wall has formed. The more rapid growth of the

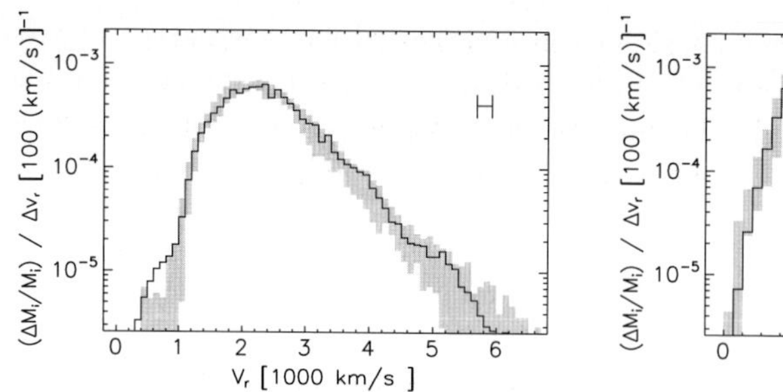

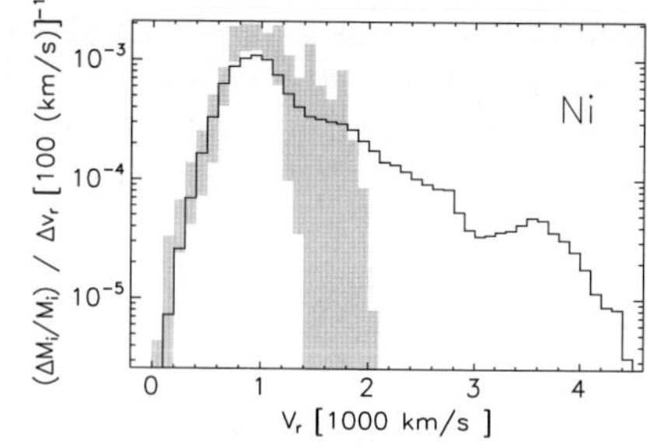

Fig. 3 Normalized mass distributions of hydrogen (upper panel) and of elements heavier than oxygen (collectively denoted as "nickel") versus radial velocity v_r for the 3D simulation (solid line) compared to the corresponding 2D results (grey region) at about 9000 s. The binning is done in intervals of $\Delta v_r = 100$ km/s and the distributions $\Delta M_i/M_i$ with i being the element index are given per unit length of velocity. The grey-shaded area indicates the significant variation between the ten 2D simulations, which were started from different meridional slices of the initial 3D explosion model

Rayleigh-Taylor instabilities also seems to be the reason for the efficient mixing of hydrogen into the metal core, which is clearly different from the 2D case already at $\sim$1000 s. Secondly, a larger fraction of the nickel and oxygen with a velocity of $\sim$ 4000 km/s at 350 s, which is in the ballpark of the maximum nickel and oxygen velocities of the 2D models, experiences less decelaration in the 3D simulation. This is obviously linked to differences in the propagation of the dense metal-containing clumps as they move quasi-ballistically through the helium shell and inner part of the hydrogen envelope under the influence of drag forces exerted by the surrounding medium. While the fastest bullets with a speed of 5000–6000 km/s at 350 s still move with 3000–4500 km/s at $t > 2600$ s, the clumps with initially $\sim$ 4000 km/s are decelerated to 2000–3000 km/s. In contrast, in 2D none of the material of the metal core has retained velocities of more than 2000 km/s at the time when the shock reaches the stellar surface ($t \sim 8000$ s).

While the maximum hydrogen velocities are very similar in 2D and 3D and reach up to 6000 km/s, a clearly bigger hydrogen mass attains the lowest velocities (below 1000 km/s) in the 3D case (see Fig. 3). Correspondingly, a significant fraction of the hydrogen is mixed deep into the metal core of the star to regions with enclosed masses of $M(r) \lesssim 2\,M_\odot$. These findings are very similar to the results obtained in 2D by Kifonidis et al. [19], who considered cases with already initially largely aspherical explosions. In 3D similarly strong inward mixing of hydrogen does not require a very pronounced global asymmetry of the early blast. Richtmyer-Meshkov instabilities at the shock-distorted composition interfaces play no important role in our 3D simulations, because the outgoing supernova shock is significantly less deformed than in the 2D models discussed by [19].

Even more impressive than the hydrogen differences is the large discrepancy between the 2D and 3D nickel distributions in velocity space and the maximum nickel velocities. While in 2D even the fastest nickel clumps move with less than 2000 km/s at $t \sim 9000$ s, a large fraction of the nickel retains velocities of more than 3000 km s^{-1} and some even more than 4000 km/s in 3D (see Fig. 3). Similarly large 2D vs. 3D differences are obtained for oxygen, whereas the hydrogen and helium distributions of the 2D case are close to that of the 3D run (except for the low-velocity tail of the hydrogen distribution mentioned above). Interestingly, a significantly larger fraction of the nickel than of the oxygen propagates with the highest speeds of clumps. This is compatible with the fact that the biggest and most extended Rayleigh-Taylor structures contain mostly nickel and only thin, shorter cores of oxygen.

In the 2D case nickel and oxygen are mixed outward to an enclosed mass of about $5 M_\odot$ (in rough agreement with the results of [18]), which is close to the base of the hydrogen envelope. In 2D the metal-containing clumps thus get stuck in the dense helium wall as pointed out by [18]. In contrast, in the 3D run nickel and oxygen are mixed deep into the hydrogen layer. The fastest nickel and oxygen clumps are able to penetrate nearly to the surface of the exploding star (in the mass space).

It is amazing that in our 3D simulations the fastest metal clumps have achieved to overtake most of the hydrogen at the time the shock breaks out (at $\sim$ 8000 s). The nickel and oxygen mixing into the hydrogen envelope is clearly more efficient than even in the globally asymmetric explosions studied by Kifonidis et al. [19], where at 10000 s the metals are spread out only up to an enclosed mass of about $10 M_\odot$, although the peak metal velocities are roughly the same at $\sim$300 s.

We refrain here from drawing far reaching conclusions on the observational consequences of this finding, e.g. concerning the early visibility of X-ray and γ-ray signatures of such strong mixing. While we consider the 2D/3D differences for the same explosion model and the same and fixed numerical resolution of our first explorative study as enlightening, we think that more simulations with different explosion energies, different progenitor stars, and in particular also with finer grid zoning are desirable before the results can be interpreted quantitatively with respect to observations of supernovae.

5 Conclusions

Our 2D/3D comparison revealed that the asymptotic velocities of metal-rich clumps are much higher in 3D than in the corresponding 2D cases. In the latter the iron-group and oxygen carrying dense fingers get stuck in the massive, dense helium shell ("wall") that forms below the base of the hydrogen envelope, and stay co-moving with the helium matter there at a velocity of $\lesssim$1500 km/s, a fact that was pointed out by [18], whose results are in agreement with our 2D simulations. In contrast, in the 3D runs we obtained "nickel" ($\equiv$elements heavier than ^{16}O) and oxygen bullets propagating at maximum velocities of 4500 km s^{-1} and a large fraction ($\sim$10–20%) of the metal core of the progenitor star was seen to reach velocities of more than 2000 km/s. The most extended and fastest Rayleigh-Taylor structures, some of which have a mass of several $10^{-3} M_\odot$, contain mostly nickel. These nickel "bullets" expand significantly more rapidly than the longest fingers with dominant or appreciable oxygen content, which are smaller on average but much more numerous than the large nickel features. Iron-group nuclei, neon, and magnesium are thus carried far into the hydrogen layer. They move well ahead of oxygen-rich knots, and both iron and oxygen overtake the material from the carbon layer in the ejecta. The onion-shell stratification of the progenitor is thus partially turned over during the explosion.

Besides strong mixing of heavy-elements into the hydrogen envelope, we also observed hydrogen being transported deep into the metal core of the star to regions with velocities of <1000 km/s and an enclosed mass of $<2M_{\odot}$.

A small fraction of the metal core receives velocities of more than 6000 km/s initially, which is significantly higher than in our 2D runs (but similar to the peak metal velocities seen by [19]). These clumps penetrate into the hydrogen envelope faster than the helium wall builds up. They keep their speed of more than 4000 km/s until the time of shock breakout and experience much less deceleration than in the 2D models of [19]. Moreover, a larger fraction of the core metals start out with velocities of about 4000 km/s, which is of the order of the speed of the fastest clumps in our 2D models, but again the deceleration of these dense bullets is less strong than in the 2D case. This suggests that the final differences can be attributed only partly to 2D/3D differences in the growth rate of perturbations at the composition interfaces, which were pointed out by [17]. Instead, the discrepancies between 3D and 2D results that develop when the majority of the metal-rich clumps begins to enter the dense helium wall below the He/H boundary after $\sim$1000 s, and the increase of these discrepancies over the first 1–2 hours after the onset of the explosion while the clunps make their way through the helium layer, require a different explanation.

Hammer et al. [12] therefore hypothesized that these differences are a consequence of differences in the drag forces of the surrounding medium, which affect the propagation of the metal-rich clumps and lead to their deceleration as they plow through the helium layer. By means of a simple analytic model they could demonstrate that the different geometry of the bullets in 2D and 3D — toroidal (axisymmetric) structures versus quasi-spherical bubbles — can explain the differences observed in the simulations.

The drag force acting on the dense metal clumps is probably not independent of the grid resolution and insufficient zoning could underestimate the drag effects on the clump propagation. Although our 2D results are in reasonably good agreement with 2D AMR calculations [18, 19], we still think that simulations with varied mesh size or an AMR code would be useful to confirm our findings in 3D.

While the 2D/3D differences found in our investigation are most likely generic, the extent of the radial mixing, the size and mass distribution of the metal fingers, their velocities and spatial distribution, and the composition of individual clumps probably depends strongly on the structure of the progenitor star, the explosion energy, and the initial asymmetry of the blast. Having advanced the modeling of mixing instabilities in supernova explosions to the third dimension, we therefore plan to direct our focus to the exploration of a wider variety of progenitor stars with the goal to establish a link between explosion models and observations.

Acknowledgements We thank L. Scheck for providing us with his 3D explosion models as initial data for our studies. This work was supported by the Deutsche Forschungsgemeinschaft through the Transregional Collaborative Research Centers SFB/TR 27 "Neutrinos and Beyond" and SFB/TR 7 "Gravitational Wave Astronomy", and the Cluster of Excellence EXC 153 "Origin and Structure of the Universe" (`http://www.universe-cluster.de`). The computations were performed on the SGI Altix 4700 of the Leibniz-Rechenzentrum (LRZ) in Munich.

References

1. Anuchina, N., Volkov, V., Gordeychuk, V., Eskov, N., Ilyutina, O., Kozyrev, O.: Numerical simulations of Rayleigh-Taylor and Richtmyer-Meshkov instability using MAH-3 code. J. Comp. Appl. Math. **168**, 11–20 (2004)
2. Arnett, W.D., Bahcall, J.N., Kirshner, R.P., Woosley, S.E.: Supernova 1987A **27**, 629–700 (1989). DOI 10.1146/annurev.aa.27.090189.003213
3. Benz, W., Thielemann, F.K.: Convective instabilities in SN 1987A **348**, L17–L20 (1990). DOI 10.1086/185620
4. Blondin, J.M., Mezzacappa, A.: Pulsar spins from an instability in the accretion shock of supernovae **445**, 58–60 (2007). DOI 10.1038/nature05428
5. Blondin, J.M., Mezzacappa, A., DeMarino, C.: Stability of Standing Accretion Shocks, with an Eye toward Core-Collapse Supernovae **584**, 971–980 (2003). DOI 10.1086/345812
6. Burrows, A., Livne, E., Dessart, L., Ott, C.D., Murphy, J.: A New Mechanism for Core-Collapse Supernova Explosions. Astrophys. J. **640**, 878–890 (2006). DOI 10.1086/500174
7. Burrows, A., Livne, E., Dessart, L., Ott, C.D., Murphy, J.: Features of the Acoustic Mechanism of Core-Collapse Supernova Explosions. Astrophys. J. **655**, 416–433 (2007). DOI 10.1086/509773
8. Cabot, W.: Comparison of two- and three-dimensional simulations of miscible Rayleigh-Taylor instability. Physics of Fluids **18**, 045101 (2006)
9. Couch, S.M., Wheeler, J.C., Milosavljević, M.: Aspherical Core-Collapse Supernovae in Red Supergiants Powered by Nonrelativistic Jets. Astrophys. J. **696**, 953–970 (2009). DOI 10.1088/0004-637X/696/1/953
10. Fryxell, B., Arnett, D., Müller, E.: Instabilities and clumping in SN 1987A. I - Early evolution in two dimensions **367**, 619–634 (1991). DOI 10.1086/169657
11. Hachisu, I., Matsuda, T., Nomoto, K., Shigeyama, T.: Nonlinear growth of Rayleigh-Taylor instabilities and mixing in SN 1987A. Astrophys. J. Lett. **358**, L57–L61 (1990). DOI 10.1086/185779
12. Hammer, N.J., Janka, H., Mueller, E.: Three-Dimensional Simulations of Mixing Instabilities in Supernova Explosions. ArXiv e-prints (2009)
13. Herant, M., Benz, W.: Hydrodynamical instabilities and mixing in SN 1987A - Two-dimensional simulations of the first 3 months **370**, L81–L84 (1991). DOI 10.1086/185982
14. Hirata, K., Kajita, T., Koshiba, M., Nakahata, M., Oyama, Y.: Observation of a neutrino burst from the supernova SN1987A. Physical Review Letters **58**, 1490–1493 (1987)
15. Janka, H.T., Langanke, K., Marek, A., Martínez-Pinedo, G., Müller, B.: Theory of core-collapse supernovae **442**, 38–74 (2007). DOI 10.1016/j.physrep.2007.02.002
16. Joggerst, C.C., Woosley, S.E., Heger, A.: Mixing in Zero- and Solar-Metallicity Supernovae. Astrophys. J. **693**, 1780–1802 (2009). DOI 10.1088/0004-637X/693/2/1780
17. Kane, J., Arnett, D., Remington, B.A., Glendinning, S.G., Bazán, G., Müller, E., Fryxell, B.A., Teyssier, R.: Two-dimensional versus Three-dimensional Supernova Hydrodynamic Instability Growth. Astrophys. J. **528**, 989–994 (2000). DOI 10.1086/308220
18. Kifonidis, K., Plewa, T., Janka, H.T., Müller, E.: Non-spherical core collapse supernovae. I. Neutrino-driven convection, Rayleigh-Taylor instabilities, and the formation and propagation of metal clumps **408**, 621–649 (2003). DOI 10.1051/0004-6361:20030863

19. Kifonidis, K., Plewa, T., Scheck, L., Janka, H.T., Müller, E.: Non-Spherical Core-Collapse Supernovae II. Late-Time Evolution of Globally Anisotropic Neutrino-Driven Explosions and Implications for SN 1987A **453**, 661–678 (2006)
20. Marek, A., Janka, H.T.: Delayed Neutrino-Driven Supernova Explosions Aided by the Standing Accretion-Shock Instability. Astrophys. J. **694**, 664–696 (2009). DOI 10.1088/0004-637X/694/1/664
21. Müller, E., Fryxell, B., Arnett, D.: High Resolution Numerical Simulations of Instabilities, Mixing, and Clumping in Supernova 1987A. In: I.J. Danziger, K. Kjaer (eds.) European Southern Observatory Astrophysics Symposia, *European Southern Observatory Astrophysics Symposia*, vol. 37, p. 99 (1991)
22. Müller, E., Fryxell, B., Arnett, D.: Instability and clumping in SN 1987A **251**, 505–514 (1991)
23. Müller, E., Hillebrandt, W., Orio, M., Hoflich, P., Monchmeyer, R., Fryxell, B.A.: Mixing and fragmentation in supernova envelopes. Astron. Astrophys. **220**, 167–176 (1989)
24. Nagasawa, M., Nakamura, T., Miyama, S.M.: Three-dimensional hydrodynamical simulations of type II supernova - Mixing and fragmentation of ejecta. Proc. Astron. Soc. Pacific **40**, 691–708 (1988)
25. Remington, B.A., Drake, R.P., Ryutov, D.D.: Experimental astrophysics with high power lasers and Z pinches. Reviews of Modern Physics **78**, 755–807 (2006). DOI 10.1103/RevModPhys.78.755
26. Scheck, L.: PhD thesis, Technical University Munich (2007)
27. Scheck, L., Janka, H.T., Foglizzo, T., Kifonidis, K.: Multidimensional supernova simulations with approximative neutrino transport. II. Convection and the advective-acoustic cycle in the supernova core **477**, 931–952 (2008). DOI 10.1051/0004-6361:20077701
28. Scheck, L., Kifonidis, K., Janka, H.T., Müller, E.: Multidimensional supernova simulations with approximative neutrino transport. I. Neutron star kicks and the anisotropy of neutrino-driven explosions in two spatial dimensions **457**, 963–986 (2006). URL 10.1051/0004-6361:20064855
29. Scheck, L., Plewa, T., Janka, H.T., Kifonidis, K., Müller, E.: Pulsar Recoil by Large-Scale Anisotropies in Supernova Explosions. Physical Review Letters **92**(1), 011103 (2004)
30. Timmes, F.X., Swesty, F.D.: The Accuracy, Consistency, and Speed of an Electron-Positron Equation of State Based on Table Interpolation of the Helmholtz Free Energy **126**, 501–516 (2000)
31. Utrobin, V.P.: The Light Curve of Supernova 1987A: The Structure of the Presupernova and Radioactive Nickel Mixing. Astronomy Letters **30**, 293–308 (2004). DOI 10.1134/1.1738152
32. Wang, L., Wheeler, J.C.: Spectropolarimetry of Supernovae. Ann. Rev. Astron. Astrophys. **46**, 433–474 (2008). DOI 10.1146/annurev.astro.46.060407.145139
33. Woosley, S.E., Pinto, P.A., Ensman, L.: Supernova 1987A - Six weeks later **324**, 466–489 (1988). DOI 10.1086/165908
34. Yamada, Y., Nakamura, T., Oohara, K.: Three Dimensional Simulations of Supernova Explosion. I. Progress of Theoretical Physics **84**, 436–443 (1990). DOI 10.1143/PTP.84.436

Relativistic Simulations of Neutron Star and Strange Star Mergers

A. Bauswein and H.-Th. Janka

Abstract Compact stars made of strange matter, though still a hypothetical energetic ground state of matter, may be an alternative to neutron stars in accordance with the observed properties of the known compact stars. Binary systems of these so-called strange stars or of ordinary neutron stars do not exist infinitely long, but their orbits shrink due to gravitational-wave emission so that after some 100 million years of evolution the two compact objects merge. We investigate in our project how observations of such merger events, primarily by upcoming gravitational-wave experiments that will have the capability to detect such sources out to the Virgo galaxy cluster in 20 Mpc distance, could help to decide on the existence of strange matter stars. By performing three-dimensional relativistic hydrodynamical simulations on the HLRB II SGI Altix 4700 machine at the Leibniz-Rechenzentrum we identify fundamental differences between mergers of neutron stars and strange stars. The analysis of the simulated models focuses on observable signatures of the collision events, in particular on gravitational-wave measurements and the detection of strange matter clumps ("strangelets") ejected during the merging and making a contribution to the cosmic ray flux. The results suggest that once the experiments reach the required sensitivity, a decision on the existence of strange stars may become possible.

1 Introduction

Stars more massive than 8 $M_\odot$ end their hydrostatic evolution with a supernova explosion. The remnants of such events are either a black hole for the most massive progenitor stars or a compact star, commonly referred to as neutron star. The true

A. Bauswein · H.-Th. Janka
Max-Planck-Institut für Astrophysik, Karl-Schwarzschild-Straße 1, Postfach 1317, 85741 Garching bei München, Germany
e-mail: bauswein@mpa-garching.mpg.de
e-mail: thj@mpa-garching.mpg.de

S. Wagner et al. (eds.), *High Performance Computing in Science and Engineering, Garching/Munich 2009*, DOI 10.1007/978-3-642-13872-0_29,

nature of the latter has been a mystery since the discovery of these objects. Besides a composition of mainly neutrons, more exotic phases of matter have been proposed to exist in compact stars. There is even the possibility that these objects are actually quark stars. According to the strange matter hypothesis [5, 21] a mixture of up, down and strange quarks may form the true and absolute ground state of matter and if this hypothesis was true, compact stars are so-called strange stars consisting of strange quark matter rather than ordinary nucleonic matter [7, 8].

There are no theoretical arguments against the viability of the strange matter hypothesis and the existence of strange stars and theoretically a wide range of possibilities exists for the properties of the equation of state of high-density matter. Also observations cannot exclude the possibility that the known compact stars are strange stars, because astronomical measurements do not reach the required precision.

It is known that compact stars exist also in binaries and that the orbits of these systems decay as predicted by general relativity due to the emission of gravitational waves, which carry away energy and angular momentum. The decrease of the orbital separation on a time scale of 100 to 1000 Myr finally leads to the merging of the two stars. As the stars approach each other, the gravitational-wave signal is expected to become stronger, which is why the inspiral and merging of compact stars is one of the most promising sources for ground-based gravitational-wave detectors like LIGO (`http://www.ligo.org`) and VIRGO (`http://www.virgo.infn.it`).

In this work we try to explore by simulations if the merging process of two strange stars yields observational features that allow one to distinguish them from colliding neutron stars. In particular the implications for cosmic ray measurements and for the detection of gravitational waves are considered. Neutron stars as well as strange stars are very compact objects, comparable in size and mass, and therefore need to be considered in a general relativistic framework. The details of the underlying model and the implementation are discussed in the next section. Section 3 gives information on performance and scaling aspects and we present the results of our simulations in Sect. 4 together with a discussion of our findings. Finally, we finish with a summary, our conclusions and an outlook in Sect. 5.

2 Mathematical Model and Numerical Implementation

The mathematical model and its numerical implementation consist mainly of two parts. One part solves the relativistic three-dimensional Euler equations describing the hydrodynamical evolution of the system. The other module provides a solution of the Einstein field equations to yield the gravitational field in which the fluid moves. Details of the model and the implementation can be found in [16, 17] and references therein.

The hydrodynamics are treated within the relativistic Smoothed Particle Hydrodynamics (SPH) formalism, which is a relativistic version of the classical SPH method (see e.g. [3]) to solve the Lagrangian Euler equations. This approach assumes the fluid to be represented by a number of particles with constant rest mass

m_i and coordinates r_i. The particles themselves are considered not to be pointlike but to be spread out over a small spatial domain. This is described by the so-called kernel function $W(r - r_i; h)$, which peaks at a particle's position r_i and decreases with the distance $r - r_i$ from the particle's center. The extension of the kernel function and thus the particle is characterized by the smoothing length h. In our specific implementation a spherically symmetric spline kernel with a compact support is used [14]. Given this, any physical quantity can be expressed as an average value $\langle A(r)\rangle = \int d^3r' A(r') W(r - r'; h)$, which can be discretized by $\sum_i m_i \frac{A(r_i)}{\rho(r_i)} W(r - r_i; h)$ with the sum running over neighboring particles. For instance the density reads $\rho(r) = \sum_i m_i W(r - r_i; h)$. This representation has the advantage that spatial derivatives via application of the product rule can be expressed as derivatives of the kernel function, which is analytically given. So it avoids a finite difference approximation of spatial derivatives. With this approach the relativistic hydrodynamical equations result in a set of ordinary differential equations instead of partial differential equations as in an Eulerian treatment. In this way an initial value problem for the conserved rest mass density, the conserved momentum and the relativistic energy is formulated (see for instance [17]), which can be solved easily with the classical fourth order Runge-Kutta method. In the current code an adaptive time step is used. So the essential idea of the SPH approach can be summarized as follows. All quantities are represented by their values at moving interpolation points, the particles, which corresponds to a Lagrangian view of hydrodynamics, and the time evolution of the hydrodynamical quantities is computed "on the particles".

Finally an equation of state describing the thermodynamics of the stellar fluid has to be implemented in order to close the system of hydrodynamical equations. Details on this will be given at the end of this section.

In SPH several methods can be imposed in order to treat hydrodynamic shock waves. In the present code a scheme developed by [13] is implemented, which solves a local Riemann problem given by the two states of neighboring particles.

SPH has been and is successfully used also for other applications in astrophysics like the simulation of planet formation, star formation, supernovae, accretion disks, cosmological structure and galaxy formation.

As mentioned in the introduction of this section, the code solves the Einstein field equations as a relativistic treatment of gravitation. This is done in the so-called conformal flatness approximation, where the spatial part of the line element is assumed to be $\gamma_{ij} = \psi^4 \delta_{ij}$ with the conformal factor ψ as a function of the coordinates and with the Kronecker symbol δ_{ij} ([9, 20]). Using this, the Einstein field equations simplify to five coupled, elliptical, nonlinear partial differential equations. The source terms of these equations contain noncompact terms besides the matter contributions (for details see [17]). This approach has the advantage to be on the one hand much simpler, numerically more stable and more efficient than solving the full general relativistic equations. On the other hand it contains nevertheless relativistic effects and from comparisons one concludes that the differences to a full treatment seem acceptable [16]. The elliptic equations for the gravitational field are solved by a multigrid method. These equations are discretized by finite differencing on a fixed grid assuming boundary conditions given by a multipole expansion of the source terms. This expansion is also used for estimating the values of the metric components beyond the grid if particles leave the region covered by the computational domain of the metric solver. Because these equations are coupled and nonlinear, they are solved iteratively until convergence.

The time evolution of the whole system is achieved by the coupling of the two modules described above, the hydrodynamics solver and the metric solver. This is done by the successive evolution of the hydrodynamical equations through the SPH scheme and the solution of the elliptical partial differential equations, which do not contain an explicit time dependence. Thus the matter distribution determines the gravitational field through a solution of the field equations. With this gravitational field the fluid can be evolved for the next time step by the Runge-Kutta method providing an updated new matter distribution for the subsequent time step.

In the conformal flatness approximation of general relativity, gravitational waves are neglected, since the spatial part of the metric is assumed to be conformally flat. For this reason the gravitational waves generated by the fluid motion have to be extracted in another way. This is done by the quadrupole formula. Furthermore the backreaction of the gravitational wave emission has to be considered to account for the loss of energy and angular momentum from the fluid. For a given time evolution of the matter and the mass quadrupole of the source, the backreaction can be computed by means of a set of elliptical partial differential equations [4]. This method results in some corrections to the elements of the metric tensor. The detailed implementation of this scheme is presented in [16].

The equation of state (EoS) of high-density neutron star matter as well as of potential strange quark matter is uncertain because of insufficient constraints by laboratory experiments and astronomical measurements. Also the theoretical approaches of describing high-density matter are ambiguous and do not yield a reliable prediction. Therefore, for astrophysical simulations one refers to various models for the EoS and tries to explore the different consequences. The EoSs for our project are given as three-dimensional tables. They contain the pressure and the energy density as functions of the rest mass density, the temperature and the electron fraction. Several EoSs are available. Beside simplified EoSs as polytropes and zero-temperature descriptions, four microphysical EoSs are implemented. There is the model of [19], which was derived within a relativistic mean field theory (below referred to as "Shen"). Another EoS employs the liquid drop model of nuclear physics [10] (below referred to as "LS"). Both EoSs describe neutron star matter. Furthermore two EoSs of strange quark matter are available for simulating strange stars. For these EoSs (below referred to as MIT60 and MIT80) the MIT bag model was used (for details see [1, 2]).

3 Scalability of the Code

The code has run successfully on different supercomputers like the IBM p575 P5 system at the Rechenzentrum Garching (RZG) of the Max Planck Society and the IBM p690 P4+ "JUMP" system of the Jülich Supercomputing Centre (JSC). The

Table 1 Speed up (reduction factor of runtime) of the code for a model with 7 million particles

Number CPUs	Speed up	Efficiency [%]
1	1	100
2	2	100
4	4	100
8	8	100
16	15	94
32	29	91
64	53	83
128	80	63

parallelization with OpenMP shows good scalability for up to 32 CPUs on these machines in the case of small particle numbers (about 500,000). The scaling behavior of the code on the HLRB II SGI Altix 4700 machine of the LRZ was tested for a model with 7 million particles. The results are listed in Table 1.

The actual scaling behavior is somewhat better than the listed numbers, because the influence of serial work, which could not be avoided for these performance measurements, will be smaller in real production runs. For instance the serial part of the multigrid solver contributed relatively too much.

The scaling behavior improves with particle number. For the current study of compact star mergers smaller particle numbers of about 700,000 were used. For the strange star models even lower particle numbers of about 150,000 were sufficient because strange stars have a more homogeneous structure in comparison to neutron stars. While the scaling behavior for these models is slightly worse, this can be balanced by running several setups in parallel with an appropriate number of threads per single model resulting in a very good efficiency. It is important to note here that for the described project many model runs for varied EoSs and system parameters (system masses and stellar mass ratios) are essential. For bunches of our models the workload of the machine and the efficiency of the code are comparable to high resolution runs.

4 Simulations and Results

The merging of two compact stars, neutron stars or potentially strange stars, proceeds via different stages, which are shown exemplarily in Fig. 1 for a neutron star coalescence described with the LS EoS. During the inspiral phase the stars slowly approach each other due to gravitational-wave emission (upper left panel) and the orbital period decreases. When the revolution time reaches about 2 ms, the actual merger takes place. In the cases with mass ratios of the binary components unequal to unity, the lighter star is tidally disrupted by the more massive one as can be seen in the plots for a binary with 1.2 $M_\odot$ and 1.35 $M_\odot$ (upper right panel). The subsequent evolution depends on the total mass M_{tot} of the system. For high M_{tot} the stars collapse to a black hole shortly after they came into contact (prompt collapse). If M_{tot} falls below some threshold value, which depends on the EoS, the colliding stars form a differentially rotating remnant as in Fig. 1 (lower

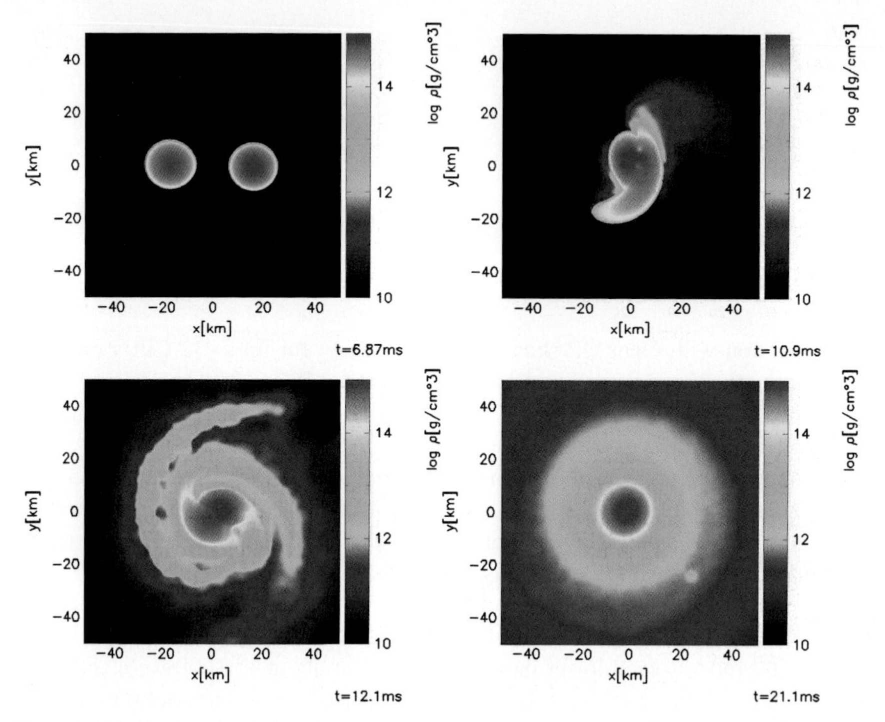

Fig. 1 Evolution of the rest mass density in the orbital plane of a merging neutron star binary with 1.2 $M_{\odot}$ and 1.35 $M_{\odot}$ components for the LS EoS. The system rotation is counterclockwise. The figures were created with the visualization tool SPLASH [18]

left panel). The fast differential rotation supports the object against gravitational collapse. Typical total masses of compact binaries exceed the maximum mass of single, non-rotating compact stars and even of uniformly rapidly rotating compact stars. The so-called hypermassive remnant finally settles to a axisymmetric configuration (lower right panel). However, also this object collapses to a black hole after angular momentum redistribution, which typically does not occur within the simulation time.

Figure 2 shows a coalescence of strange stars for the same binary setup (1.2 $M_{\odot}$ and 1.35 $M_{\odot}$ stars) as in Fig. 1. Although the same stages of the evolution can be identified, one easily recognizes some striking differences to the neutron star case. First, tidal disruption is hampered and no primary massive spiral arm is formed. This behavior can be explained by the higher compactness of strange stars in comparison to neutron stars. Additionally, no dilute halo structure develops as in the neutron star case, where matter is streaming off from the whole remnant. Instead, the remnant of the strange star merger forms two filigree spiral arms after about two revolutions, and only from these features matter spreads into the surroundings. Fragmentation of the spiral arms leads to a disk consisting of clumps of high-density strange quark matter orbiting the compact remnant. Only a small amount of matter from the tips of the tidal tails can escape from the merger site and become gravitationally unbound.

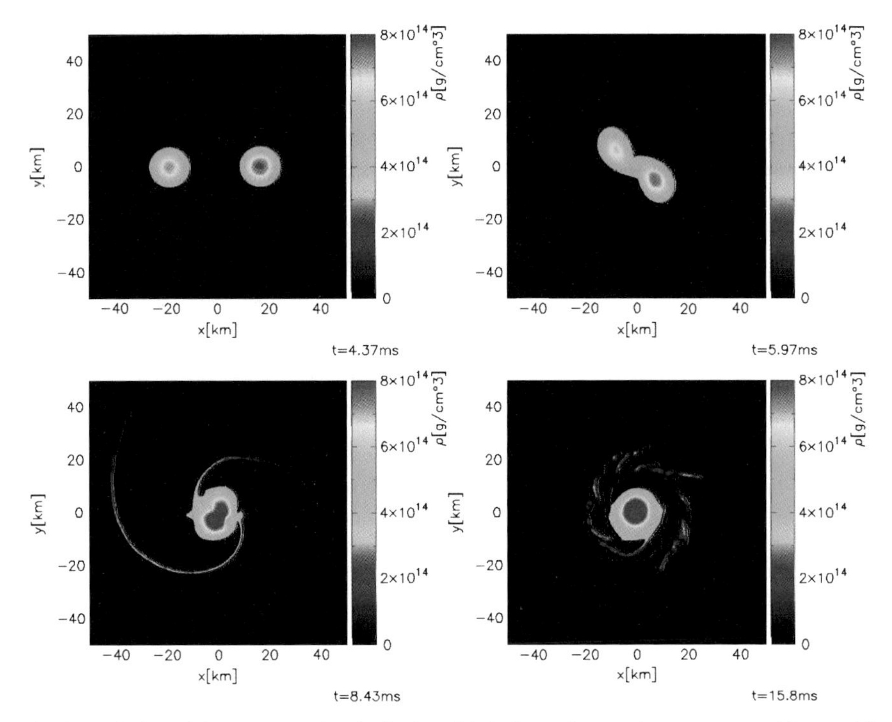

Fig. 2 Evolution of the rest mass density in the orbital plane of a merging strange star binary with 1.2 $M_{\odot}$ and 1.35 $M_{\odot}$ components for the MIT60. note the linear scale of the rest mass density in contrast to Fig. 1. The figures were created with the visualization tool SPLASH [18]

The fundamentally different dynamics of the strange star coalescence is a consequence of the higher compactness of strange stars and of the selfbinding of strange quark matter, meaning that as the pressure goes to zero the density remains finite (even above nuclear density). Pictorially spoken this results in a liquid-like behavior of strange quark matter.

Within the current project we performed more than 90 simulations of strange star and neutron star mergers, varying the individual masses of the binary components and the EoSs. This allows not only for a comparison of the dynamical behavior but also a for an assessment of the question whether the mergers of each class yield observational signatures that distinguish them as members of different object classes. In the following we will discuss in particular the consequences for cosmic ray experiments detecting strange matter clumps ("strangelets"), and we will address the question whether gravitational-wave measurements can discriminate between the two forms of compact stars.

If the strange matter hypothesis was correct, also small lumps of quark matter are absolutely stable, and basically any size of strange quark matter from atomic nucleus dimensions to compact star radii can exist in the universe. Microscopic nuggets of strange quark matter, so-called strangelets, will be charged and therefore detectable by cosmic ray experiments like the AMS-02 instrument, which is supposed to be installed on the International Space Station in 2010. So the detection of a strangelet would decide on the strange matter hypothesis and would prove the existence of strange stars.

It is assumed that mergers of strange stars would be crucial sources of the strangelet contribution to the cosmic ray flux because small amounts of strange quark matter become gravitationally unbound from the merger site. Therefore a computation of the amount of ejecta is essential for an estimate of the strangelet flux expected at cosmic ray detectors.

Our runs are the first simulations modeling the merging process of strange stars and therefore permit the first estimation of the ejecta mass on grounds of source models yielding input data for strangelet acceleration and propagation models [12]. The estimated amount of unbound matter per merging event is shown in Fig. 3 as a function of the mass ratio $q = M_1/M_2$ and total binary mass $M_{tot} = M_1 + M_2$ for compact stars described by the MIT60 and MIT80 EoSs. Above a certain M_{tot} value we cannot determine any amount of ejecta (see white lines in Fig. 3). For these binary configurations a prompt collapse to a black hole occurs and therefore the remnant has no time to develop spiral arms from which matter could become unbound. If M_{tot} is below a certain limit, we obtain a steep rise of the ejecta mass in a narrow region of the M_{tot}-q-plane in both EoS cases for $q < 0.85$. For MIT60 the region where more than $0.01 M_{\odot}$ of matter become unbound is located around a total mass of about 2.5 $M_{\odot}$. For MIT80 significantly lower total masses are required to obtain unbound matter and the ejected masses are lower as well. This dependence is explained by the fact that MIT80 leads to more compact stars with correspondingly smaller radii, which impedes the tidal disruption.

Folding the results of these simulations for the ejecta masses with binary population estimates for merger rates of systems with different q and M_{tot} based on theoretical models of compact star binary evolution, one can compute the integral strangelet production rate of a galaxy. It is found that the strangelet production rate depends sensitively on the mass-radius relation of strange stars. In general from EoSs resulting in more compact stars one obtains lower strangelet production. The two microphysical models used in our study represent two extreme cases in terms of the mass-radius relation and the compactness of the resulting strange stars. For the MIT60 EoS the abundance of strangelets is higher by a factor of 100 compared to previous estimates guided by neutron star merger models with cruder treatment of the dynamics (e.g. nonrelativistic simulations). On the other hand, for the MIT80 EoS the integrated strangelet flux vanishes, because for this EoS only very light binary configurations eject matter. These, however, are nonexistent in the model calculations for the compact star population. From these results we conclude that the strange matter hypothesis cannot be ruled out as previously assumed, if cosmic ray experiments will not find any evidence of strange quark matter in cosmic rays.

In this case there is another interesting consequence of our results. If a cosmic ray strangelet hit an ordinary neutron star, the whole star is expected to be converted to a strange star, because it is energetically favorable to adopt the ground state and not to remain in a metastable state of neutron star matter (although with an extremely long life time). Simple estimates reveal that already a tiny abundance of strangelets in the

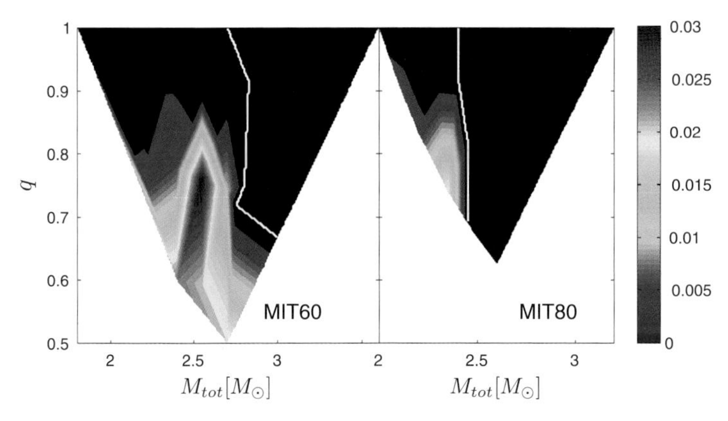

Fig. 3 Ejected mass per merger event, color-coded and measured in $M_\odot$, as function of the mass ratio $q = M_1/M_2$ and the total system mass $M_{tot} = M_1 + M_2$ of the binary configurations for the MIT60 and MIT80 EoSs. The white line separates binary mergers with and without ejecta. The shape of the colored areas is determined by the choice of the masses of the compact stars within reasonable ranges

cosmic ray flux is sufficient to convert all neutron stars in a galaxy to strange stars [6, 11]. Therefore it was argued that the unambiguous detection of a neutron star would rule out the strange matter hypothesis [6, 11]. However, our results for the MIT80 EoS have the implication that even if the strange matter hypothesis was correct, it would not necessarily lead to a measurable flux of strangelets. Consequently, neutron stars and strange stars could be in coexistence and the unambiguous identification of a neutron star cannot exclude the strange matter hypothesis.

The analysis of gravitational-wave signals from compact star mergers can be considered as a complementary approach to distinguish neutron star mergers from strange star collisions. Typical wave forms are plotted in Fig. 4, which displays the gravitational-wave amplitude measured at a distance of 20 Mpc for the mergers shown in Figs. 1 and 2. The characteristic inspiral phase with an increasing amplitude and frequency is followed by the relatively short and more complex pattern of the actual merger and then by the ringdown of the oscillating hypermassive remnant. The oscillations of the postmerger remnant are clearly dominated by a single frequency, which suggests to use it as a characteristic quantity together with the maximal frequency reached during the inspiral phase. Both frequencies are plotted versus each other in Fig. 5 for different binary configurations and for all EoSs used in our study.

In general the frequencies of the strange star mergers (MIT60 and MIT80) are located at higher frequencies. There is an overlap between the results of the MIT60 EoS and the LS EoS, which describes neutron star matter and implies relatively compact stars. However, this degeneracy can be broken by considering additional characteristics of the gravitational-wave signal as discussed in [2]. For the interpretation of this diagram one should bear in mind that both, the strange star EoSs and the nuclear EoSs have been chosen to bracket the range of possibilities in terms of

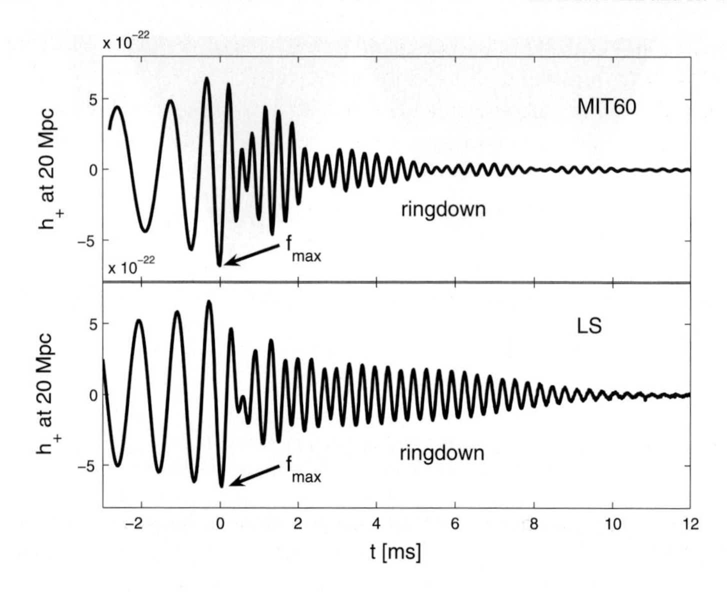

Fig. 4 Gravitational-wave amplitudes for the plus polarization measured perpendicular to the orbital plane at a distance of 20 Mpc for binary mergers with $M_1 = 1.2 M_\odot$ and $M_2 = 1.35 M_\odot$, using the MIT60 EoS (upper panel) and the LS EoS (lower panel)

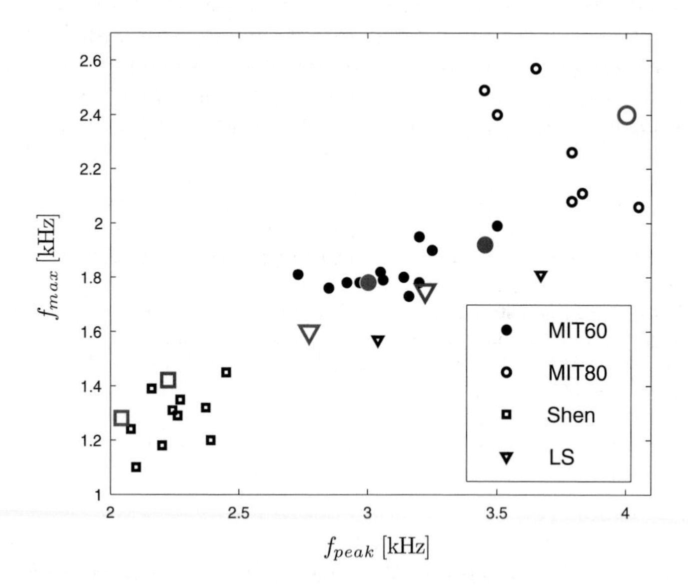

Fig. 5 The maximal frequencies during inspiral versus the dominant frequencies of the ringdown of the postmerger remnant for all configurations that form a hypermassive object. Results for all EoSs used in this study are plotted. The blue symbols denote the binary configurations with two 1.35 $M_\odot$ stars, the red ones mark the frequencies for the merger events of two stars with 1.2 $M_\odot$

the compactness of the resulting compact stars in both of the classes. As argued in [15] and [2], the compactness of the stars is crucial for the location of the frequency pair in this plot. Therefore we conclude that certain regions in this plane are excluded for nuclear EoSs and strange quark matter EoSs, respectively. Considering these arguments ultimately leads to the conclusion that a decision about the viability of the strange matter hypothesis becomes possible once gravitational waves from compact binary mergers can be measured and the characteristic features of the signal can be extracted. In those cases where gravitational-wave detections may not be finally conclusive, as in the overlapping region in Fig. 5, cosmic ray experiments could provide crucial information.

5 Summary, Conclusions and Outlook

We presented a comparison of neutron star mergers and strange star coalescence. An extraction of observable quantities revealed that cosmic ray experiments and gravitational-wave detectors may have the ability to decide on the strange matter hypothesis and on the existence of strange stars once instruments with sufficient measurement sensitivity have become available.

In the future we plan to consider a larger sample of microphysical EoSs, both for neutron star and for strange quark matter. We also plan to explore the consequences of neutron star mergers for nucleosynthesis aspects considering especially their potential as sites of r-process element production in the ejecta of the collisions. Moreover, we currently develop and test a code for simulating the coalescence of black holes with neutron stars or alternatively with strange stars.

Acknowledgements It is a pleasure to thank R. Oechslin, G. Pagliara, I. Sagert, J. Schaffner-Bielich, M. M. Hohle and R. Neuhäuser for contributing to this project in various ways. This work was supported by the Sonderforschungsbereich Transregio 7 "Gravitational Wave Astronomy", by the Sonderforschungsbereich Transregio 27 "Neutrinos and Beyond", and the Cluster of Excellence EXC 153 "Origin and Structure of the Universe" of the Deutsche Forschungsgemeinschaft, and by CompStar, a research networking programme of the European Science Foundation. The computations were performed at the Rechenzentrum Garching of the Max-Planck-Gesellschaft and at the Leibniz-Rechenzentrum Garching.

References

1. Bauswein, A., Janka, H.T., Oechslin, R., Pagliara, G., Sagert, I., Schaffner-Bielich, J., Hohle, M.M., Neuhaeuser, R.: Mass Ejection by Strange Star Mergers and Observational Implications. Phys. Rev. Lett. **103**(1), 011101 (2009)
2. Bauswein, A., Oechslin, R., Janka, H.T.: submitted to Phys. Rev. D (2009)
3. Benz, W.: Smooth Particle Hydrodynamics - a Review. In: J.R. Buchler (ed.) Numerical Modelling of Nonlinear Stellar Pulsations Problems and Prospects (1990)

4. Blanchet, L., Damour, T., Schaefer, G.: Post-Newtonian hydrodynamics and post-Newtonian gravitational wave generation for numerical relativity. Mon. Not. R. Astron. Soc. **242**, 289–305 (1990)
5. Bodmer, A.R.: Collapsed nuclei. Phys. Rev. D **4**(6), 1601–1606 (1971). DOI 10.1103/PhysRevD.4.1601
6. Caldwell, R.R., Friedman, J.L.: Evidence against a strange ground state for baryons. Phys. Lett. B **264**, 143–148 (1991)
7. Glendenning, N.: Compact Stars. Springer-Verlag, New York (1996)
8. Haensel, P., Potekhin, A.Y., Yakovlev, D.G.: Neutron Stars 1. Springer-Verlag, New York (2007)
9. Isenberg, J., Nester, J.: Canonical Gravity. In: General Relativity and Gravitation - Berne, Switz p. 23, 1980 (1980)
10. Lattimer, J.M., Swesty, F.D.: A generalized equation of state for hot, dense matter. Nuclear Physics A **535**, 331–376 (1991). DOI 10.1016/0375-9474(91)90452-C
11. Madsen, J.: Astrophysical limits on the flux of quark nuggets. Phys. Rev. Lett. **61**, 2909–2912 (1988)
12. Madsen, J.: Strangelet propagation and cosmic ray flux. Phys. Rev. D **71**(1), 014026 (2005). DOI 10.1103/PhysRevD.71.014026
13. Monaghan, J.J.: SPH and Riemann Solvers. Journal of Computational Physics **136**, 298–307 (1997)
14. Monaghan, J.J., Lattanzio, J.C.: A refined particle method for astrophysical problems. Astron. Astrophys. **149**, 135–143 (1985)
15. Oechslin, R., Janka, H.T.: Gravitational Waves from Relativistic Neutron-Star Mergers with Microphysical Equations of State. Phys. Rev. Lett. **99**(12), 121102 (2007). DOI 10.1103/PhysRevLett.99.121102
16. Oechslin, R., Janka, H.T., Marek, A.: Relativistic neutron star merger simulations with non-zero temperature equations of state. I. Variation of binary parameters and equation of state. Astron. Astrophys. **467**, 395–409 (2007). DOI 10.1051/0004-6361:20066682
17. Oechslin, R., Uryū, K., Poghosyan, G., Thielemann, F.K.: The influence of quark matter at high densities on binary neutron star mergers. Mon. Not. R. Astron. Soc. **349**, 1469–1480 (2004). DOI 10.1111/j.1365-2966.2004.07621.x
18. Price, D.J.: splash: An Interactive Visualisation Tool for Smoothed Particle Hydrodynamics Simulations. Publications of the Astronomical Society of Australia **24**, 159–173 (2007). DOI 10.1071/AS07022
19. Shen, H., Toki, H., Oyamatsu, K., Sumiyoshi, K.: Relativistic Equation of State of Nuclear Matter for Supernova Explosion. Progress of Theoretical Physics **100**, 1013–1031 (1998)
20. Wilson, J.R., Mathews, G.J., Marronetti, P.: Relativistic numerical model for close neutron-star binaries. Phys. Rev. D **54**, 1317–1331 (1996)
21. Witten, E.: Cosmic separation of phases. Phys. Rev. D **30**(2), 272–285 (1984). DOI 10.1103/PhysRevD.30.272

The Physics of Galactic Nuclei

Marc Schartmann, Christian Alig, Martin Krause, and Andreas Burkert

Abstract Our current numerical work realised with the help of the HLRB II supercomputer focuses on the so-called Seyfert phenomenon. Whenever enough gas is accreted onto the central supermassive black holes of otherwise normal galaxies, their centres light up to become as bright as the stars of the whole galaxy. The main aim of this joint project is to understand the origin and evolution of the complex stellar, gaseous and dusty structures around supermassive black holes and how such activity cycles might be triggered and what consequences this has on the surrounding distribution of gas and dust. We mainly follow four areas of research: (i) star formation in the direct vicinity of supermassive black holes, (ii) radiative interaction of dust clouds in tori, (iii) multi-phase turbulence in obscuring tori and (iv) studying the connection between nuclear star formation and black hole feeding. This is done with the help of up-to-date numerical methods: high resolution grid based hydrodynamics codes (PLUTO, NIRVANA), highly adaptive SPH-codes (GADGET2) are developed further to include radiation pressure effects and a new code is developed to treat the radiation pressure interactions in an ensemble of dust clouds.

1 Introduction

Active nuclei are powered by supermassive black holes. Whenever enough gas reaches the centre, a fast rotating, thin and hot gaseous accretion disc is able to form, which receives mass from a ring-like, dusty, geometrically thick gas reservoir. Anisotropically blocking the light, this gives rise to two characteristic observational signatures, depending whether the line of sight is obscured (edge-on view) or not

M. Schartmann · Ch. Alig · M. Krause · A. Burkert
Max-Planck-Institut für extraterrestrische Physik, Giessenbachstrasse, D-85748 Garching, Germany
Universitäts-Sternwarte München, Scheinerstrasse 1, D-81679 München, Germany
e-mail: schartmann@mpe.mpg.de

S. Wagner et al. (eds.), *High Performance Computing in Science and Engineering, Garching/Munich 2009*, DOI 10.1007/978-3-642-13872-0_30,

(face-on view). These nuclear regions of nearby Seyfert galaxies, as well as our own galactic centre have been observed in great detail with the most up-to-date instruments at the largest available telescopes and interferometers, yielding unprecedented resolution. But despite this great wealth of recent data, the physical processes occuring in galactic nuclei are still poorly understood. The main aim of our work is to obtain a deep understanding of the origin and evolution of the complex stellar and gaseous structures and their relation to the observed phases of nuclear activity and how such activity cycles relate to nuclear star formation as observed in nearby active galaxies. This goal is achieved with state-of-the-art numerical methods and in close collaboration with observers. In the following sections, the results we obtained so far in the single projects are discussed in more detail.

2 Nuclear Disc Formation in Galactic Nuclei

Recent observations reveal two sub-parsec-scale rings of about 100 young and massive stars near the radio source SgrA* in our Galactic Centre (GC) [2, 5, 6, 11, 18]: a warped clockwise rotating disc with a mean eccentricity of 0.36 ± 0.06 and a second inclined counter-clockwise rotating disc extending from 0.04 pc to 0.5 pc. Due to the hostility of the environment, they cannot be explained by normal means of star formation, as tidal forces would disrupt typical molecular clouds in the vicinity of the central black hole. Therefore, it was proposed that they formed by fragmentation of self-gravitating, eccentric accretion discs ([1, 20] and references therein). [4, 8, 12] investigate the formation of such accretion discs by the rapid deposit of gas around the central black hole through the tidal disruption of an infalling cloud with an impact parameter larger than the cloud radius. [21] proposed a model in which the infalling cloud covers the black hole in parts during the passage, which they show is a more likely scenario. This enables efficient redistribution of angular momentum and dissipation of kinetic energy due to material with opposite angular momentum streaming around the black hole from opposite sides. We present the first realisation of this scenario using high-resolution numerical simulations, which enables us to show that this is a possible formation mechanism for the observed stellar discs in the Galactic Centre.

2.1 *Numerical Method and Model Setup*

For our simulations we are using the N-body Smoothed Particle Hydrodynamics (SPH) Code GADGET2 with a total number of 10^6 SPH particles and a softening length of $\varepsilon = 10^{-3}$ pc. The number of neighbours is set to $n_{\rm neigh} = 50 \pm 5$. The SPH particle mass is $m_{\rm SPH} = 8.81 \times 10^{-2}\,{\rm M}_\odot$ and the corresponding minimum mass that can be resolved is $m_{\rm min} = n_{\rm neigh} \times m_{\rm SPH} = 4.4\,{\rm M}_\odot$.

GADGET2 uses a modified version of the standard Monaghan Balsara artificial viscosity:

$$\Pi_{ij} = -\frac{\alpha}{2}\frac{[c_i + c_j - 3w_{ij}]\,w_{ij}}{\rho_{ij}} \quad \text{with} \quad w_{ij} = \mathbf{v}_{ij}\cdot\mathbf{r}_{ij}/|\mathbf{r}_{ij}| \tag{1}$$

c_i the sound speed of particle i and v_{ij},r_{ij} the relative particle velocity, separation.

The GADGET2 Code is written in C and uses MPI for parallelisation. Fig. 1 shows a study we performed for our simulations of the time needed between two iteration steps. We gain roughly a factor of 1.8 in speed each time we double the number of CPUs. All simulations were done on 128 CPUs. The total runtime depends on the number of particles with small time step. Simulations with small initial cloud black hole separation (leading to a large number of particles with small time step) can take up to 4 times the full queue running time of 48 hours, leading to a total number of 24600 CPU hours. Simulations with large impact parameter usually take around 2 times the full queue running time, leading to a total number of 12300 CPU hours. Due to the huge possible parameter space for the infall, a number of simulations need to be done exploring a range of initial conditions. In addition, tests for the numerical stability of the results have been conducted including a very high resolution simulation with 5×10^6 SPH particles on 256 CPUs.

Our simulations start with a spherical and homogeneous molecular cloud of radius 3.5 pc ([14]), H_2 gas density of 10^4 cm^{-3} ([7]) and temperature of $T_{\text{cloud}} =$ 50 K ([7]), which is placed at a distance of 5 pc from the black hole on the x-axis (the direction of motion) and an offset b on the y-axis. The black hole is included as a constant static, Newtonian potential of a point mass of $M_{\text{BH}} = 3.5\times10^6$ M$_\odot$ ([5]). The (isothermal) hydrodynamical evolution is characterized by the fraction $\frac{b}{R}(<1)$. All SPH particles within an accretion radius $r_{\text{acc}} = 2\times10^{-2}$ pc are removed from the simulation and treated as being accreted by the black hole. The evolution is then characterized by the ratio $\frac{v_c}{v_b}$ where v_c is the initial cloud velocity and $v_b^2 = \frac{2GM_{bh}}{b}$ is the black hole's escape velocity at distance b.

Our parameter study comprises of three simulations with fixed initial cloud velocities of 50 km/s and varying offsets b, set to 1, 2 and 3 pc, respectively. In addition we did two simulations with b fixed to 3 pc and varying initial cloud velocities, set to 30 and 80 km/s (in concordance with observations of [14]), respectively. The

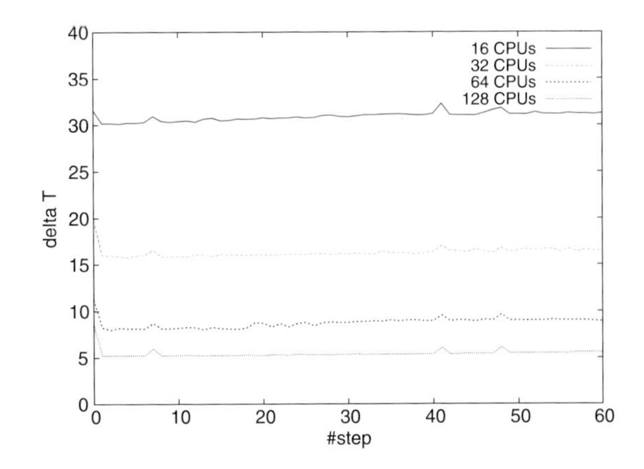

Fig. 1 Time (arbitrary units) needed between two iteration steps plotted against iteration step number. As can be seen, doubling the number of CPUs used gives roughly a factor of 1.8 in speed

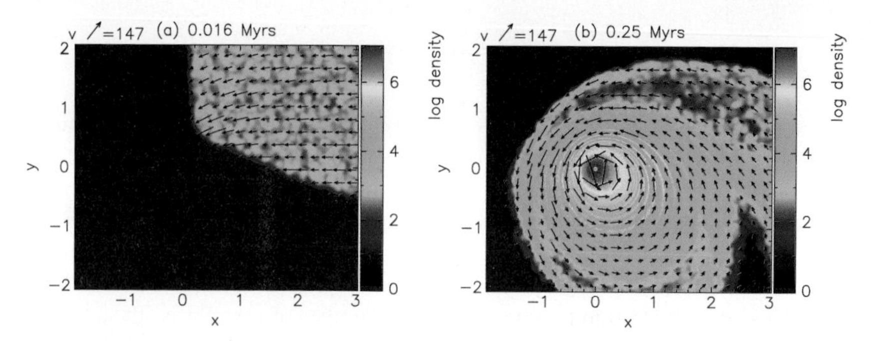

Fig. 2 Logarithmic density in $\frac{M_{\odot}}{pc^3}$ in the xy-plane with z=0 pc after **(a)** 0.016 and **(b)** 0.25 Myr. Velocity is given in km/s with the unit length of the arrow corresponding to 147 km/s. Material with opposite angular momentum collides, a density wave builds up and spirals into the forming disc

simulations were typically run for an evolutionary time of 0.25 Myrs which corresponds roughly to the situation when all parts of the cloud have crossed the black hole. In this report, we concentrate on our standard model, which results in the best comparison with observations, having the following parameters: $\frac{b}{R} = 0.85$, $\frac{v_c}{v_b} = 0.8$, $v_c = 80$ km/s and $b = 3$ pc.

2.2 Density Evolution

Fig. 2 shows an early and the final state of the evolution of the cloud for our standard model. The cloud approaches the black hole from the right side along the x-axis. Due to tidal forces the parts closest to the black hole quickly start to form a finger-like extension stretching towards the black hole. This material passes the black hole on orbits corresponding to its initial angular momentum and will subsequently collide with gas from the opposite side. A density wave forms (green area) which thickens and moves around the disc, which starts to build up. After 0.25 Myrs, an inner, relaxed equilibrium disc has formed while the outer region represents a 1-armed infalling spiral in the disc plane (Fig. 2b). In all simulations, the SPH particles in the inner 0.1 pc reach the minimum smoothing length so that we are limited by numerics in this region and the physics is not resolved anymore due to strong smoothing over the forces and thus we cannot make any reliable predictions for the size of the inner disc radius.

2.3 Accretion

Accretion rates are calculated from material which falls below our accretion radius of $r_{acc} = 2 \times 10^{-2}$ pc. This does not necessarily mean black hole accretion since the

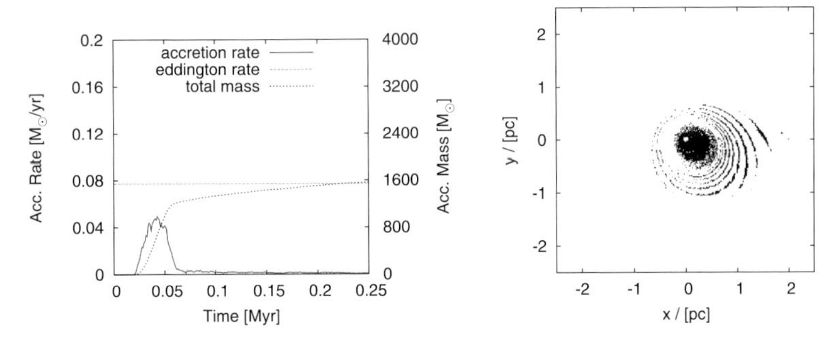

Fig. 3 **(a)** Accretion rate towards the black hole (red solid line, left y-axis). The dashed green line shows the Eddington limit at 10% accretion efficiency. Also plotted is the total accreted mass in blue (dotted line) in M$_\odot$ (right y-axis). **(b)** Plot of all grid cells with a Toomre parameter smaller than one, indicating instability and leading to star formation. Each grid cell has a size of 0.02×0.02 pc

formation of a small and hot black hole accretion disc evolving viscously is beyond the resolution limit and outside the scope of our current simulation. Nevertheless, it is interesting to compare our accretion rates for our simulations to the Eddington rate of the black hole which is shown in Fig. 3a for the case of our standard model. The Eddington mass accretion rate is defined as $\dot{M}_{edd} = \frac{4\pi G M_{BH} m_p}{\varepsilon \sigma_T c}$, with m_p being the proton mass, ε the accretion efficiency and σ_T the Thompson scattering cross-section. From this we get an Eddington rate of $\dot{M}_{edd} = 0.0775\ \frac{M_\odot}{\mathrm{yr}}$ using an accretion efficiency of 10% and a 3.5×10^6 M$_\odot$ black hole. At the end of all simulations we reach very low accretion rates indicating that we have reached a nearly stationary state. The accretion rates of our standard model stays below the Eddington rate at all times. The impact parameter plays the most important role in determining the final accreted mass while the cloud velocity has a much smaller effect.

2.4 Disc Properties

To analyse disc stability, we calculate the Toomre parameter $Q = \frac{c_s \Omega}{\pi \Sigma G}$ on a 250×250 grid applied to an area of 5×5 pc in the xy-plane. Fig. 3b shows the resulting disc defined by all grid points with a Toomre parameter smaller than one ($Q < 1$). We compare the properties of the unstable discs defined in this way to the observed stellar disc. Simple elliptical fits were made to the discs and the masses were calculated by counting gas which is unstable according to the Toomre criterion. When comparing the disc masses resulting in our parameter study for fixed velocity, variable impact parameter, with the simulations with fixed impact parameter, variable velocity, we find that in contrast to the accretion rates, for which the impact parameter was the dominating factor, the cloud velocity is the most important factor in determining the disc masses (the lower the velocity, the higher the unstable disc mass). The initial specific angular momentum is unimportant for determining the final disc properties. According to the data from [18] the stars in the observed discs have specific angular momentum reaching from roughly 30 to 70 $\frac{pc^2}{Myr}$. However, our simulations within or closest to this range do not reproduce the properties of the observed discs which again demonstrates the highly nonlinear evolution of a cloud engulfing the black hole. On average we find disc eccentricities of order $e = 0.4$ and an outer radius of $r \sim 1.35$ pc. The observed eccentricities of the stellar rings in the GC are $e = 0.34 \pm 0.06$ [2], the estimated mass of the gaseous disc that formed the stars is expected to be around $M \sim 10^{4-5}$ $M_\odot$ [15, 16]. The observed outer radius of the disc is roughly $r = 0.5$ pc [5], with the observed disc size still increasing due to the addition of newly observed stars belonging to the disc structure. These properties are very similar to our gaseous discs, which have a bit too large radii ($r = 1 - 1.7$ pc) but already show the same low eccentricity ($e = 0.24 - 0.51$) as the stellar discs and the required mass in gas ($M = 0.7 - 6.95 \times 10^4$ $M_\odot$) to form the observed stellar disc. For details we refer to Alig et al. 2010, submitted to MNRAS.

3 Radiation Pressure Driven Dust Cloud Interactions

It is currently a matter of active research, how AGN tori are able to sustain their vertical scale height against gravity – as needed by the Unified Scheme of Active Galactic Nuclei (AGN). Several mechanisms have been proposed recently. Two promising solutions are the effects of infrared radiation pressure mutually exerted by members of an ensemble of dusty clouds (e. g. [10, 19]) and gravitational heating due to accreted clouds (e. g. [3]). Toy models of clumpy dust tori are currently vitally discussed (mainly treated with the help of radiative transfer codes) and excellent agreement is found with spectroscopical as well as interferometric observations. In line with these findings, we build up on this success and develop a numerical N-body-like method to assess the dynamical, equilibrium structure of clumpy tori under gravitational and radiation pressure interaction of the dust clouds, taking direct collisions between them into account.

Our newly developed SIROCCO[1]-code is entirely written in the C-language and relies on a second order accurate leapfrog algorithm with a dynamic timestep in a direct N-body approach, where radiation pressure forces are taken into account. The code is prepared for simulations of local boxes as well as for global torus simulations. A shared memory parallelisation with the help of the OpenMP library enables us to do calculations of ensembles of the order of a few times 10^5 clouds, as necessitated by recent radiative transfer models [9, 17]. Furthermore, elastic, inelastic and partially elastic collisions between clouds can be taken into account. This is an important feature, as cloud collisions were estimated to be as frequent as approximately one per orbit and will lead to the redistribution of angular momentum and

---

[1] SIROCCO stands for **S**imulating the **I**nteraction of **R**adiating and **O**bscuring **C**loud **CO**nglomerations.

transport of material towards the centre. The parallel performance of the code is shown in Fig. 6b (red dots and line). It scales almost ideal up to a total number of 16 processors, being limited by the percentage of serial code.

So far, only the N-body part and the cloud collisions have been tested. We are currently working on an approximate treatment of the cloud-cloud shadowing processes, via the implementation of an efficient ray-tracing scheme. With the help of such N-body simulations, we will not only be able to calculate the evolution of radiatively driven clumpy tori, but also test the dynamical stability of existing clumpy torus models.

4 Multi-Phase Turbulence in the Tori of Active Galactic Nuclei

As a further step in our investigations of clumpy and dusty tori around Active Galactic Nuclei, we carry out hydrodynamics simulations including a frequency resolved treatment of radiative transfer. Infrared radiation is transported through the torus via scattering, absorption and re-emission by dust grains. The transmitted forces raise the temperature of the separately computed gas component. This gas component is permitted to cool via ultraviolet line emission, which is fed back into the radiation field at the appropriate frequency.

We solve the following system of equations:

$$\frac{\partial \rho_i}{\partial t} + \nabla \cdot (\rho_i \mathbf{v}) = 0 \tag{2}$$

$$\frac{\partial \rho \mathbf{v}}{\partial t} + \nabla \cdot (\rho \mathbf{v}\mathbf{v}) = -\nabla p + \frac{\rho}{c} \int \kappa_\nu \mathbf{I}_\nu \, \mathrm{d}\nu \tag{3}$$

$$\frac{\partial e}{\partial t} + \nabla \cdot (e \mathbf{v}) = -p \nabla \cdot \mathbf{v} - n^2 \Lambda(T_\mathrm{g}) \tag{4}$$

$$\frac{\partial I_{\nu,j}}{c \partial t} + \frac{\partial I_{\nu,j}}{\partial x_j} = \kappa_\nu \rho_\mathrm{d} \left(\frac{4\pi}{6} B_\nu(T_\mathrm{d}) - I_{\nu,j} \right) + \frac{1}{6} n^2 \Lambda(T_\mathrm{g}) \delta(\nu - \nu_0) \, . \tag{5}$$

The first equation is the separate conservation of gas mass (ρ_g), dust mass (ρ_d) and dust internal energy density (ρ_e). Sputtering and evaporation would result in source terms on the right-hand side of this equation. The second one is the Euler equation, where we use the total mass density $\rho = \rho_\mathrm{g} + \rho_\mathrm{d}$. The radiation pressure appears as the last term on the right-hand side. We evolve six separate radiation fields $I_{\nu,j}$ (erg/s/cm^2/Hz), forward and backwards propagating radiation for each direction. Next comes the internal energy equation for the gas, including the cooling function. For each of the six coordinate directions, j, we evolve the radiation field, using the radiative transfer equation directly. $B_\nu(T_\mathrm{d})$ is the Planck function at the dust temperature T_d. The source terms for the radiative transfer equations describe in this order: emission and absorption/scattering by the dust particles and radiation produced by the gas at the fiducial UV-wavelength ν_0. We relate the thermodynamic quantities by equations of state. For the gas, $p = (\gamma - 1)e$ and $n = \rho / \mu m_\mathrm{p}$, where we

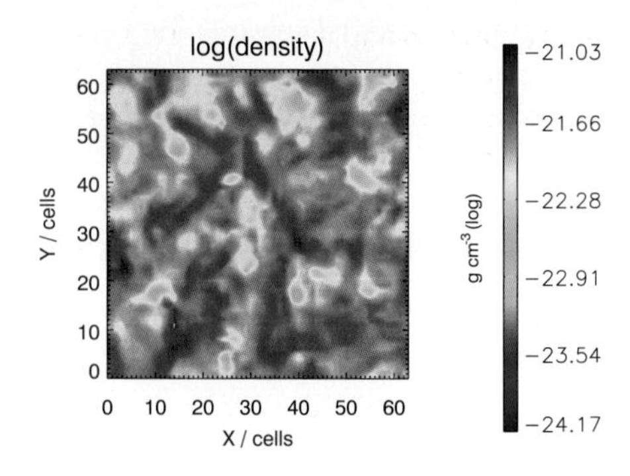

Fig. 4 Density distribution for our periodic box simulation. We start from decayed hydrodynamic turbulence. When the light is switched on, the turbulent decay carries on almost undisturbed. The radiation forces are strong in this example, but cancel largely. The code is required to pass this test to avoid spurious accelerations later in the production runs

calculate μ according to collisional ionisation equilibrium, and $T_{\mathrm{g}} = p/nk_{\mathrm{B}}$ with the Boltzmann constant k_{B}. The temperature and internal energy density for the dust are related by the heat capacity $C(T)$: $\rho_{\mathrm{e}} = \int_0^{T_{\mathrm{d}}} C(T)\mathrm{d}T$. The shortest timescale in the system is the one for dust heating/cooling. Therefore, we always use the equilibrium dust temperature, determined by the implicit equation:

$$0 = \int \mathrm{d}\nu \kappa \Big(\sum_{j=1}^{6} I_{\nu,j} - 4\pi B_\nu(T_{\mathrm{d}})\Big) \,. \tag{6}$$

All other equations are evolved explicitely. As the light propagation timestep is often much smaller than the gas cooling and hydrodynamics timesteps, the code can subcycle the radiative transfer within the same hydrodynamics timestep. We have implemented this radiative transfer module as a module for the MPI-parallelised 3D magneto-hydrodynamics code NIRVANA. The new module is also paralellised by MPI. Test runs at LRZ showed that the code scales well down to a domain size of 16 cells. The code conserves the total energy (radiation, kinetic and thermal) very well ($< 10^{-4}$ over the simulation time).

A first consistency check is a periodic equal temperature box (see Fig. 4). The radiation forces are high in this example, but they should cancel almost exactly since, for a given cell, the sky is equally bright in all directions. Using 16 processors at LRZ for two days, we are able to compute a periodic box of 60×64^2 cells for almost 15,000 timesteps. The only exception regarding the boundary conditions is for the radiation field: we include a displacement similar to the shearing box model. Our box represents a volume of $(0.1\ \mathrm{parsec})^3$. We prepare the initial conditions using large scale sinusoidal density fluctuations, and random velocities, which we let decay, without using the radiative transfer module, for a 1000 years. The result of this decayed turbulent cascade is a Kolmogorov power spectrum for the hydrodynamic variables with small velocity. This is the initial condition when we switch on the radiation field. We use a black body spectrum in equilibrium with dust at a temperature of 1500 K. In agreement with our expectations, we notice only small accelerations and the turbulent decay proceeds almost unchanged. A remaining problem is that clouds sometimes get preferentially compressed into diagonal shapes, due to the forces being applied in the directions of the Cartesian coordinates, only. We are working to improve this feature. The code will then be applied to more realistic setups with net radiation fluxes through the computational domain.

5 Feeding Supermassive Black Holes with Nuclear Star Clusters

Recently, high resolution observations with the help of the near-infrared adaptive optics integral field spectrograph SINFONI at the VLT proved the existence of massive and young nuclear star clusters in the centres of a sample of Seyfert galaxies. With the help of three-dimensional high resolution hydrodynamical simulations with the PLUTO code, we follow the evolution of such clusters. The gas ejection of their stars provide both, material for the obscuration within the so-called unified scheme of AGN and a reservoir to fuel the central, active region and it additionally drives turbulence in the interstellar medium on tens of parsec scales. This leads to a vertically wide distributed clumpy or filamentary inflow of gas on tens of parsec scale (see Fig. 5a,b,c), whereas a turbulent and very dense disc builds up on the parsec scale (see Fig. 5d). In order to capture the relevant physics in the inner region, we treat this disc separately in a one-dimensional simulation, which also enables a direct comparison with observations. Due to the unknown physical mechanism, we concentrate on the effects of a parametrised turbulent viscosity to generate angular momentum and mass transfer and additionally take star formation into account. Fig. 6a shows the comparison of the mass contributions of the various components from the 1D modelling for an assumed alpha viscosity parameter of $\alpha = 0.05$ and a gas temperature of 400 K. The input parameters of the simulations have been constrained by observations of the nearby Seyfert 2 galaxy NGC 1068. At the current

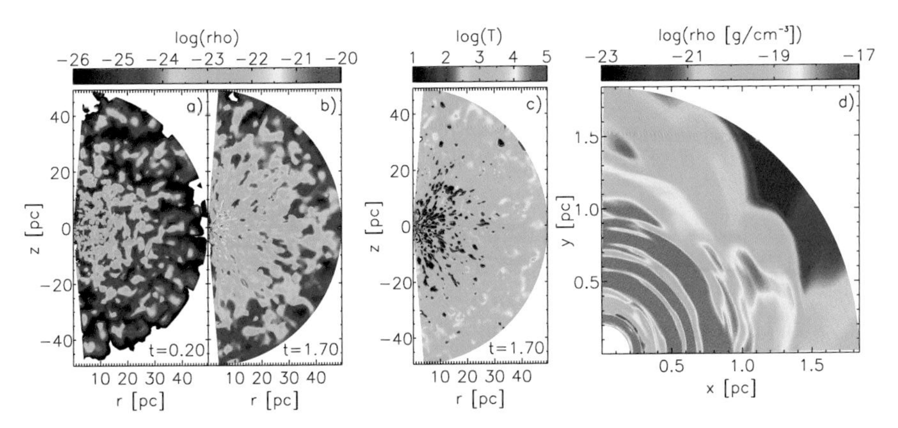

Fig. 5 Snapshots of the density distribution in a meridional plane of our 3D hydrodynamical standard model after (a) 0.2 orbits (corresponding to $7 \cdot 10^4$ yr) and (b) 1.7 orbits (approximately $6 \cdot 10^5$ yr). Panel (c) shows the temperature distribution in the same meridional plane after 1.7 orbits. Panel (d) shows a zoom into the innermost part of the equatorial plane, displaying the nuclear disc component

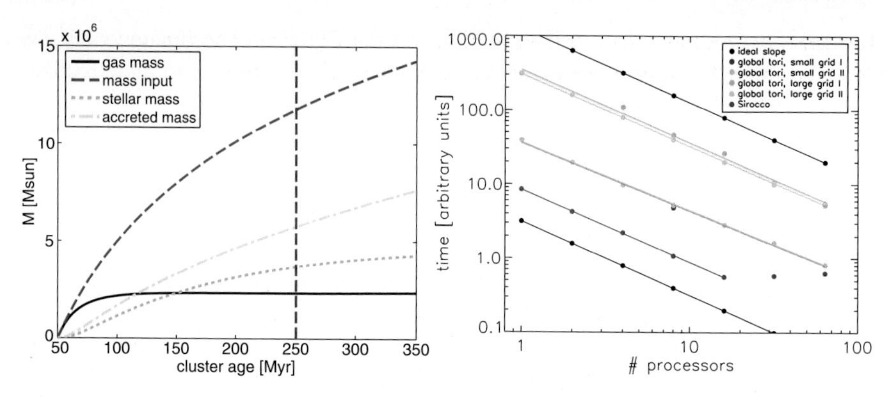

Fig. 6 **(a)** Comparison of mass contributions of the various components of our 1D effective disc model for an alpha viscosity parameter of 0.05 and a gas temperature of 400 K. The blue dashed line denotes the estimated current age of the nuclear star cluster in NGC 1068. **(b)** Parallel performance of our codes on the SGI Altix 4700 supercomputer at the LRZ. Different grid sizes and multiple measurements are shown for the PLUTO code and one example for the OpenMP version of the SIROCCO code (red). Plotted is the time for a certain number of steps in arbitrary units against the number of processors used. The black lines and dots indicate the ideal scaling $\propto 1/N_{\mathrm{Proc}}$

age of its nuclear starburst of 250 Myr, our simulations yield disc sizes of the order of 0.8 to 0.9 pc, gas masses of $10^6\,M_\odot$ and mass transfer rates of $0.025\,M_\odot/\mathrm{yr}$ through the inner rim of the disc. This shows that our large scale torus model in combination with the 1D effective disc model is able to approximately account for the disc size as inferred from interferometric observations in the mid-infrared and compares well to the extent and mass of a rotating disc structure as inferred from water maser observations. Several other observational constraints are in good agreement as well. On basis of these comparisons, we conclude that the proposed scenario seems to be a reasonable model and shows that nuclear star formation activity and subsequent AGN activity are intimately related.

5.1 Numerics and Code Performance

PLUTO[2] [13] is a multiphysics, multialgorithm modular code which was especially designed for the treatment of supersonic, compressible astrophysical flows in the presence of strong discontinuities in a conservative form. Thereby, it relies on a modern Godunov-type High-Resolution Shock-Capturing (HRSC) scheme. For the simulations presented in this report, we use the two-shock Riemann solver and a 3D spherical geometry. The code is completely written in the C language, parallel performance is granted via the Message Passing Interface (MPI). The parallel performance of the PLUTO code on the SGI Altix 4700 supercomputer at the LRZ shows almost ideal scaling properties, also after adding optically thin line cooling

[2] http://plutocode.to.astro.it/

for arbitrary cooling functions, as well as mass input due to stellar evolutionary processes (see Fig. 6b). For details we refer to Schartmann et al. (2010, in press, arXiv:0912.4677).

References

1. Alexander, R.D., Armitage, P.J., Cuadra, J., Begelman, M.C.: Self-Gravitating Fragmentation of Eccentric Accretion Disks **674**, 927–935 (2008). DOI 10.1086/525519
2. Bartko, H., Martins, F., Fritz, T.K., Genzel, R., Levin, Y., Perets, H.B., Paumard, T., Nayakshin, S., Gerhard, O., Alexander, T., Dodds-Eden, K., Eisenhauer, F., Gillessen, S., Mascetti, L., Ott, T., Perrin, G., Pfuhl, O., Reid, M.J., Rouan, D., Sternberg, A., Trippe, S.: Evidence for Warped Disks of Young Stars in the Galactic Center **697**, 1741–1763 (2009). DOI 10.1088/0004-637X/697/2/1741
3. Beckert, T., Duschl, W.J.: The dynamical state of a thick cloudy torus around an AGN **426**, 445–454 (2004). DOI 10.1051/0004-6361:20040336
4. Bonnell, I.A., Rice, W.K.M.: Star Formation Around Supermassive Black Holes. Science **321**, 1060 (2008). DOI 10.1126/science.1160653
5. Genzel, R., Schödel, R., Ott, T., Eisenhauer, F., Hofmann, R., Lehnert, M., Eckart, A., Alexander, T., Sternberg, A., Lenzen, R., Clénet, Y., Lacombe, F., Rouan, D., Renzini, A., Tacconi-Garman, L.E.: The Stellar Cusp around the Supermassive Black Hole in the Galactic Center **594**, 812–832 (2003). DOI 10.1086/377127
6. Ghez, A.M., Salim, S., Hornstein, S.D., Tanner, A., Lu, J.R., Morris, M., Becklin, E.E., Duchêne, G.: Stellar Orbits around the Galactic Center Black Hole **620**, 744–757 (2005). DOI 10.1086/427175
7. Güsten, R., Philipp, S.D.: Galactic Center Molecular Clouds. In: S. Pfalzner, C. Kramer, C. Staubmeier, A. Heithausen (eds.) The Dense Interstellar Medium in Galaxies, p. 253 (2004)
8. Hobbs, A., Nayakshin, S.: Simulations of the formation of stellar discs in the Galactic Centre via cloud-cloud collisions **394**, 191–206 (2009). DOI 10.1111/j.1365-2966.2008.14359.x
9. Hönig, S.F., Beckert, T., Ohnaka, K., Weigelt, G.: Radiative transfer modeling of three-dimensional clumpy AGN tori and its application to NGC 1068 **452**, 459–471 (2006). DOI 10.1051/0004-6361:20054622
10. Krolik, J.H.: AGN Obscuring Tori Supported by Infrared Radiation Pressure **661**, 52–59 (2007). DOI 10.1086/515432
11. Lu, J.R., Ghez, A.M., Hornstein, S.D., Morris, M.R., Becklin, E.E., Matthews, K.: A Disk of Young Stars at the Galactic Center as Determined by Individual Stellar Orbits **690**, 1463–1487 (2009). DOI 10.1088/0004-637X/690/2/1463
12. Mapelli, M., Hayfield, T., Mayer, L., Wadsley, J.: In situ formation of the massive stars around SgrA*. ArXiv e-prints p. arXiv:0805.0185 (2008)
13. Mignone, A., Bodo, G., Massaglia, S., Matsakos, T., Tesileanu, O., Zanni, C., Ferrari, A.: PLUTO: A Numerical Code for Computational Astrophysics **170**, 228–242 (2007). DOI 10.1086/513316
14. Miyazaki, A., Tsuboi, M.: Dense Molecular Clouds in the Galactic Center Region. II. Statistical Properties of the Galactic Center Molecular Clouds **536**, 357–367 (2000). DOI 10.1086/308899
15. Nayakshin, S., Cuadra, J.: A self-gravitating accretion disk in Sgr A* a few million years ago: Is Sgr A* a failed quasar? **437**, 437–445 (2005). DOI 10.1051/0004-6361:20042052
16. Nayakshin, S., Dehnen, W., Cuadra, J., Genzel, R.: Weighing the young stellar discs around Sgr A* **366**, 1410–1414 (2006). DOI 10.1111/j.1365-2966.2005.09906.x
17. Nenkova, M., Sirocky, M.M., Nikutta, R., Ivezić, Ž., Elitzur, M.: AGN Dusty Tori. II. Observational Implications of Clumpiness **685**, 160–180 (2008). DOI 10.1086/590483

18. Paumard, T., Genzel, R., Martins, F., Nayakshin, S., Beloborodov, A.M., Levin, Y., Trippe, S., Eisenhauer, F., Ott, T., Gillessen, S., Abuter, R., Cuadra, J., Alexander, T., Sternberg, A.: The Two Young Star Disks in the Central Parsec of the Galaxy: Properties, Dynamics, and Formation **643**, 1011–1035 (2006).
19. Pier, E.A., Krolik, J.H.: Radiation-pressure-supported obscuring tori around active galactic nuclei **399**, L23–L26 (1992). DOI 10.1086/186597
20. Rice, W.K.M., Lodato, G., Armitage, P.J.: Investigating fragmentation conditions in self-gravitating accretion discs **364**, L56–L60 (2005). DOI 10.1111/j.1745-3933.2005.00105.x
21. Wardle, M., Yusef-Zadeh, F.: On the Formation of Compact Stellar Disks around Sagittarius A* **683**, L37–L40 (2008). DOI 10.1086/591471

Numerical Models of Turbulence in Isothermal and Thermally Bistable Interstellar Gas

Wolfram Schmidt, Daniel Seifried, Markus Niklaus, and Jens C. Niemeyer

Abstract We present a novel subgrid scale model that is applicable to highly compressible turbulence. In large eddy simulation of forced supersonic turbulence, this subgrid scale model reduces the bottleneck effect in the turbulence energy spectra and it allows for a prediction of the local rate of turbulent energy dissipation as well as the turbulent pressure. The latter has an impact, for instance, on the gravitational support of turbulent gas and the efficiency of star formation in elaborate large-scale simulations of disk galaxies. We also report results from turbulence simulations with parameterized radiative cooling, in which the effects of thermal processes on the scale of star forming clouds was investigated.

1 Introduction

The interstellar gas in galaxies is in a highly turbulent state, which is maintained by several physical processes such as gravitational instabilities, thermal instabilities or blast waves originating from supernovae [3]. We aim at numerical simulations of the production of turbulence by these processes and its impact on star formation in disk galaxies. Our approach to the problem is similar as in [16]. In addition, we have developed the method of adaptively refined large eddy simulation [9], which

Wolfram Schmidt · Jens C. Niemeyer
Institut für Astrophysik, Universität Göttingen, Friedrich-Hund-Platz 1, 37077 Göttingen, Germany
e-mail: schmidt@astro.physik.uni-goettingen.de

Daniel Seifried
Institut für Theoretische Astrophysik, Universität Heidelberg, Albert-Ueberle-Str. 2, 69120 Heidelberg, Germany

Markus Niklaus
Deutsches Fernerkundungsdatenzentrum, Deutsches Zentrum für Luft- und Raumfahrt, Oberpfaffenhofen, Münchner Straße 20, 82234 Weßling, Germany

S. Wagner et al. (eds.), *High Performance Computing in Science and Engineering, Garching/Munich 2009*, DOI 10.1007/978-3-642-13872-0_31,

allows us to predict the energy and the corresponding pressure of turbulence on numerically unresolved length scales [2]. After having carried out prototype simulations of disk galaxies with this method [13], we identified two key challenges: Firstly, a subgrid scale (SGS) model applicable to the supersonic regime is required. Secondly, the role of thermal processes on length scales ~ 10pc, which are only marginally resolved or unresolved in simulations of a whole disk galaxy, has to be investigated. Our SGS model for supersonic turbulence has reached a mature stage now, and we report results from large eddy simulations (LES) of supersonic isothermal turbulence in Sect. 3. The basic concept of LES is briefly reviewed in the next Section. We have also completed an extensive parameter study of turbulence in thermally bistable gas [1]. Some of the results are presented in Sect. 4. In the last Section, we comment on the implications for advanced disk galaxy simulations.

2 Large Eddy Simulation

The basic idea of large eddy simulations is to decompose the equations of incompressible fluid mechanics into a resolved part and a subgrid-scale part using a suitable filter operator [5]. This is also possible for compressible fluid dynamics [15], although SGS models have been developed almost exclusively for incompressible or weakly compressible turbulence. Recently, however, attempts have been made to extend this approach to the highly compressible regime [11, 17].

To understand the method of scale separation that is applied in LES, let us consider the compressible Navier-Stokes equation for the momentum density of the fluid. The exact solutions for the density, $\overset{\infty}{\rho}$, the velocity, $\overset{\infty}{\mathbf{u}}$, and other fluid dynamical fields, are approximated by filtered variables $\rho := \langle \overset{\infty}{\rho} \rangle_\Delta$, $\mathbf{u} := \langle \overset{\infty}{\rho}\overset{\infty}{\mathbf{u}} \rangle_\Delta / \rho$, which are smooth on length scales ℓ smaller than the grid resolution Δ. For the filtered variables, dynamical equations similar to the original equations can be formulated. Because of the non-linear term in the momentum equation, however, filtering introduces an additional stress term that accounts for the interaction between the numerically resolved flow and subgrid-scale velocity fluctuations:

$$\frac{\partial}{\partial t}\rho\mathbf{u} + \nabla\cdot\rho\mathbf{u}\otimes\mathbf{u} = -\nabla P + \nabla\cdot\left(\langle\overset{\infty}{\sigma}\rangle_\Delta + \tau_{\mathrm{sgs}}\right), \tag{1}$$

where the SGS turbulence stress tensor is defined by

$$\tau_{\mathrm{sgs}} = -\langle \overset{\infty}{\rho}\overset{\infty}{\mathbf{u}}\otimes\overset{\infty}{\mathbf{u}} \rangle_\Delta + \rho\mathbf{u}\otimes\mathbf{u}. \tag{2}$$

For this tensor, which is *a priori* uncomputable, we use the following closure [11]:

$$\tau_{ij} = 2C_1\Delta\rho^{1/2}K_{\mathrm{sgs}}^{1/2}S_{ij}^{*} - 2C_2K_{\mathrm{sgs}}\frac{2u_{i,k}u_{j,k}}{|\nabla\otimes\mathbf{u}|^2} - \frac{2}{3}(1-C_2)K_{\mathrm{sgs}}\delta_{ij}, \tag{3}$$

where $|\nabla \otimes \mathbf{u}| := (2u_{i,k}u_{i,k})^{1/2}$, $S^*_{ij} = S_{ij} - \frac{1}{3}d\delta_{ij}$, S_{ij} is the symmetic part of $u_{i,j}$, and $d = u_{i,i}$. The trace of τ_{sgs} is determined by the SGS turbulence energy, which specifies the kinetic energy density of unresolved velocity fluctuations: $\tau_{ii} = -2K_{\mathrm{sgs}}$, For the computation of K_{sgs}, a dynamical equation has to be solved in addition to the filtered equations for the resolved gas dynamics [15].

Due to the microscopic viscosity ν of the gas, the viscous stress in Eq. (1) dissipates kinetic energy on the smallest dynamical length scales $\ell \sim \eta$ of the physical flow:

$$\overset{\infty}{\sigma}_{ij} = 2\nu\overset{\infty}{\rho}\,\overset{\infty}{S}{}^*_{ij} = 2\nu\overset{\infty}{\rho}\left(\overset{\infty}{S}_{ij} - \frac{1}{3}\overset{\infty}{d}\delta_{ij}\right). \tag{4}$$

The rate of energy dissipation, filtered on the grid scale, is given by

$$\rho\varepsilon_\Delta = \langle \overset{\infty}{\sigma}_{ij}\overset{\infty}{u}_{i,j}\rangle_\Delta = \langle 2\overset{\infty}{\rho}\,\overset{\infty}{S}{}^*_{ij}\overset{\infty}{S}{}^*_{ij}\rangle_\Delta = \langle \nu\overset{\infty}{\rho}|\overset{\infty}{S}{}^*_{ij}|^2\rangle_\Delta. \tag{5}$$

It is important to note that $\rho\varepsilon_\Delta \neq \sigma_{ij}u_{i,j}$, where σ_{ij} and $u_{i,j}$ are the filtered viscous stress tensor and velocity gradient, respectively. In the case of incompressible turbulence, the Kolmogorov scale η can be related to the Reynolds number, $\eta/L \sim \mathrm{Re}^{3/4}$, where $\mathrm{Re} := VL/\nu$ for an integral length L and characteristic velocity V of the flow. In computational astrophysics, the Reynolds number is usually very high and, consequently, η is much smaller than any feasible grid resolution Δ. In this case, the viscous stress term in the filtered momentum equation (1) is negligible, which means that the *physical* energy dissipation occurs entirely on subgrid scales $\ell \ll \Delta$. However, the rate of energy dissipation (5) is asymptotically constant in the limit $\eta/\Delta \to 0$, because the arbitrarily steep velocity gradients give rise to a finite product of the viscosity times the squared rate of strain.

3 Supersonic Isothermal Turbulence

In the following, we present various results from LES of forced supersonic isothermal turbulence, where the SGS model outlined in Sect. 2 was applied. For the implementation, we used the code Enzo 1.5 developed by the Laboratory for Computational Astrophysics at the University of California in San Diego (http://lca.ucsd.edu). Statistically stationary homogenous turbulence with given characteristic velocity V and integral time scale T is produced by spatially periodic stochastic forcing on a length scale $L = TV$, with adjustable mixtures of solenoidal (divergence-free) and compressive (rotation-free) modes [12, 14]. Setting the adiabatic exponent γ equal to 1.001 [12], the energy dissipated per integral time is small compared to the kinetic energy, i. e., the gas is pseudo-isothermal. This approximate treatment of isothermality enables us to observe energy conservation. With our implementation of the SGS model, the sum of resolved kinetic energy, SGS turbulence energy, and internal energy minus the power of the forcing integrated over time is conserved for the whole computational domain to a relative precision better than 10^{-8}. The fraction of computational time consumed by the SGS model is in the percent range.

3.1 Turbulence Energy

The SGS turbulence energy is an intermediate reservoir of energy that exchanges energy with the resolved flow and loses energy by dissipation into heat. A visualization of K_{sgs} prepared from an LES with 512^3 grid cells is shown in Fig. 1. In this simulation, the weighing of solenoidal to compressive forcing modes is $2:1$ and the root mean square Mach number of the flow in is about 5.5 in the statistically stationary regime. In the reddish regions, K_{sgs} is higher than the average value, while it is lower in bluish regions. For comparison, Fig 2 shows the local denstrophy $\Omega_{1/2} = \frac{1}{2}\left|\nabla \times (\rho^{1/2}\mathbf{u})\right|$, which is an indicator of compressible turbulent velocity fluctuations [7]. It appears that high SGS turbulence energy is concentrated in regions of intense denstrophy. However, $\Omega_{1/2}$ exhibits a more pronounced small-scale structure, and the instantaneous values of K_{sgs} and $\Omega_{1/2}$ can differ greatly at particular spatial positions. This is a consequence of the different effects contributing to SGS dynamics.

With regard to the numerically resolved flow dynamics, K_{sgs} introduces a turbulent pressure in addition to the thermal pressure. This can be seen by substituting Eq. (3) for the SGS turbulence stress tensor into the momentum equation (1). The trace of τ_{sgs} gives rise to the term $-\frac{2}{3}\nabla K_{\mathrm{sgs}}$, which can be absorbed into the pressure gradient if the thermal pressure P is replaced by the the effective pressure

$$P_{\mathrm{eff}} = P + \frac{2}{3}K_{\mathrm{sgs}}. \tag{6}$$

The relative contribution of the turbulent pressure $P_{\mathrm{sgs}} = \frac{2}{3}K_{\mathrm{sgs}}$ depends on the Mach number of the flow, the scaling of the turbulent velocity fluctuations, and the nu-

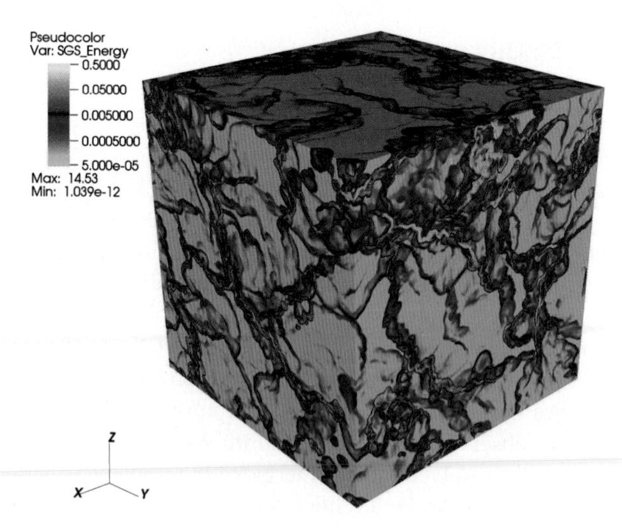

Fig. 1 Visualization of the SGS turbulence energy density K_{sgs} in a 512^3 LES with $2/3$ solenoidal and $1/3$ compressive forcing modes

merical resolution. For example, the spatial average of P_{sgs} is about $0.06P_0$ for the LES mentioned above. This suggests that the contribution from turbulent pressure is small in this case. However, a different picture emerges from the phase diagram shown in Fig. 3, where the two-dimensional probability density of the effective pressure vs. the mass density is plotted. One can see that the mean effective pressure for a given mass density (dotted line) closely follows the isothermal relation $P \propto \rho$, with a trend toward higher pressure in overdense gas, Nevertheless, intermittent turbulent velocity fluctuations can produce an effective pressure that exceeds the thermal pressure by one oder of magnitude. In Sect 4, it is shown that this effect is more pronounced in the cold phase of thermally bistable turbulence.

3.2 *Turbulent Viscosity*

The trace-free part of the SGS turbulence stress tensor (3) is composed of a viscous stress with the turbulent viscosity $\nu_{sgs} = C_1\Delta(K_{sgs}/\rho)^{1/2}$ and the non-diagonal elements of the non-linear term. While the first term has a quasi-dissipative effect on the resolved flow (kinetic energy is converted into SGS turbulence energy), the second, non-linear term modifies this effect and can also backscatter energy to the resolved flow. Both effects, in turn, compete with the intrinsic numerical dissipation of PPM (conversion of kinetic energy into internal energy). What is the outcome of this interplay?

In a variety of compressible turbulence simulations with dissipative numerical schemes such as PPM, the so-called bottleneck effect, i. e., an unphysical enhancement of spectral power in the high-wave number range, has been observed [4, 7, 12].

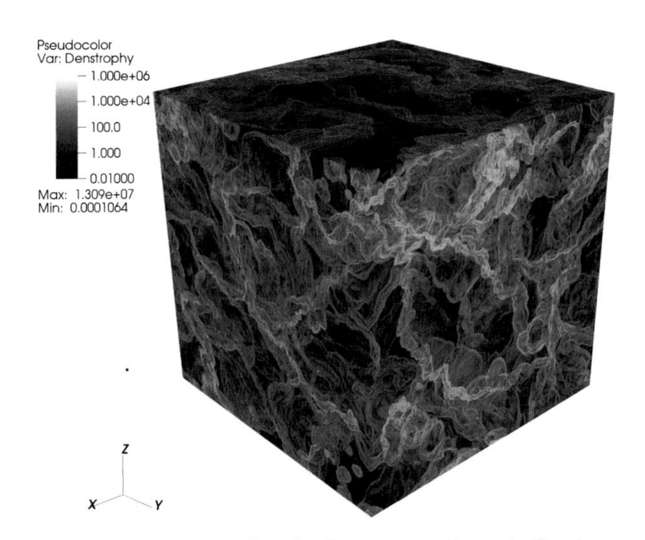

Fig. 2 Visualization of the denstrophy $\Omega_{1/2}$ for the same snapshot as in Fig. 1

On the other hand, it was reported that coupling an SGS model to PPM in simulations of decaying transonic turbulence largely suppressed the bottleneck effect [17]. To investigate whether this can be achieved with our SGS model in LES of supersonic turbulence, we computed turbulence energy spectrum functions with fractional density weighing,

$$E_{1/3}(k) = \oint \frac{1}{2}|(\widehat{\rho^{1/3}\mathbf{u}})^2(\mathbf{k})|k^2 \ \Omega_{\mathbf{k}}, \tag{7}$$

where $\widehat{\rho^{1/3}\mathbf{u}}$ is the Fourier transform of $\rho^{1/3}\mathbf{u}$. This variable stems from the dissipation rate per unit volume, which is $\sim \rho u^3$ [7].

Figure 4 compares time-averaged compensated spectrum functions, $k^{5/3}E_{1/3}(k)$, for plain PPM simulations and LES with 256^3 grid cells. As in [4], we consider the two limiting cases of purely solenoidal (divergence-free) and compressive (rotation-free) forcing. The bumps caused by the bottleneck effect at high wavenumbers noticeably flatten the spectra obtained from the PPM simulations without SGS model

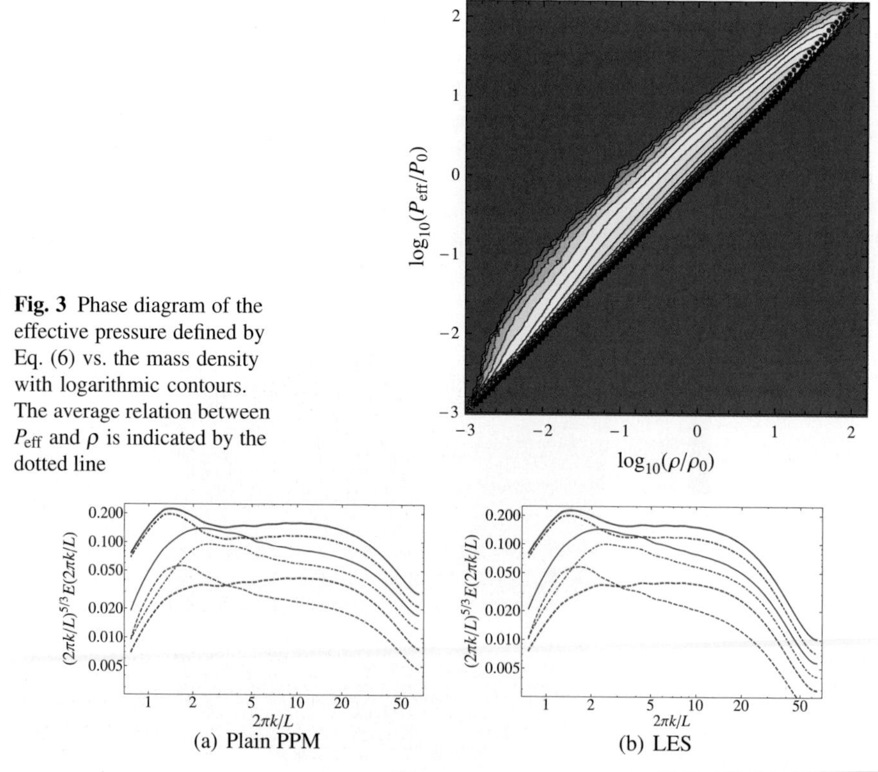

Fig. 3 Phase diagram of the effective pressure defined by Eq. (6) vs. the mass density with logarithmic contours. The average relation between P_{eff} and ρ is indicated by the dotted line

Fig. 4 Compensated spectrum functions, $k^{5/3}E_{1/3}(k)$, for the mass-weighted velocity field $\rho^{1/3}\mathbf{u}$ in plain PPM simulations (a) and LES (b) on 256^3 grid. The spectra for solenoidal and compressive forcing are shown in blue and red, respectively. The dot-dashed lines show transversal components, and the dashed lines show longitudinal components of $E_{1/3}(k)$

(a). The LES spectrum, on the other hand, has a well-defined plateau, which indicates approximate Kolmogorov scaling, in the case of solenoidal forcing (b). However, the spectra of compressively driven turbulence are significantly stiffer, as reported in [4, 12]. Results from 512^3 LES are presented in [11].

3.3 Turbulent Dissipation

In the limit of infinite Reynolds number, $\mathrm{Re} = VL/\nu \to \infty$, the viscous stress term in the filtered momentum Eq. (1) vanishes (see Sect. 2), and the rate of energy dissipation on the grid scale is determined by the SGS turbulence energy [15]:

$$\rho \varepsilon_\Delta \simeq \rho \varepsilon_{\mathrm{sgs}} = C_\varepsilon \frac{K_{\mathrm{sgs}}^{3/2}}{\rho^{1/2}\Delta}. \tag{8}$$

In contrast, an extrapolation of the expression for the microscopic dissipation rate on length scales $\ell \sim \eta$ to the grid scale Δ was proposed in [10]:

$$\rho \varepsilon_\Delta = \rho \nu_\Delta |S^*|^2. \tag{9}$$

The grid-scale viscosity ν_Δ in the above expression is assumed to be a constant controlled by the numerical dissipation of PPM. Defining the Reynolds number of the resolved flow by $\mathrm{Re}_\Delta = 2L^2\langle |S^*|^2\rangle / u_{\mathrm{rms}}^2$, where u_{rms} is the root mean square velocity [12][1], the viscosity can be evaluated from $\nu_\Delta = VL/\mathrm{Re}_\Delta$. The problem with this approach is that the viscosity on the grid scale, which is effectively an

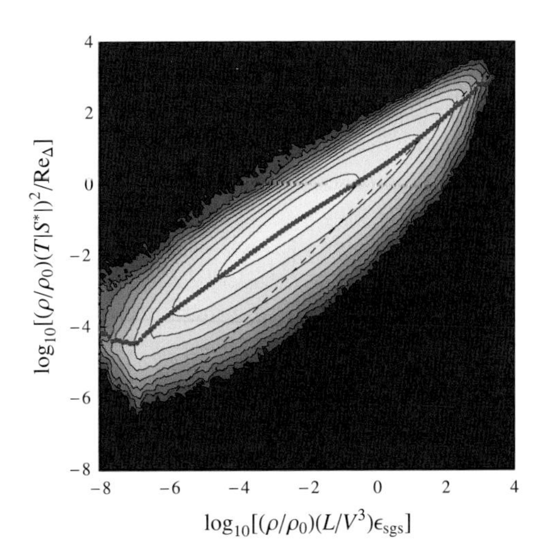

Fig. 5 Correlation diagram of the normalized rate of energy dissipation defined by Eq. (9) vs. the rate of SGS energy dissipation (8). The averages of expression (9) for given values of $\rho\varepsilon_{\mathrm{sgs}}$ are given by the thick dotted line

[1] For compressible turbulence, the mean rate of energy dissipation is proportional to $\langle |S^*|^2\rangle = \langle \omega^2 + \frac{4}{3}d^2\rangle$.

SGS turbulent viscosity, is *not* a constant. This follows from *any* non-trivial SGS model. Neglecting compressibility and the non-linear term in the SGS turbulence stress (3), equilibrium between production and dissipation of SGS turbulence energy implies $K_{\text{sgs}} \sim (C_1/C_\varepsilon)\rho\Delta^2|S^*|^2$ [15]. On the other hand, substituting the Eq. (9) into Eq. (8), we obtain the relation $K_{\text{sgs}} \sim \rho(\nu_\Delta\Delta)^{2/3}|S^*|^{4/3}/C_\varepsilon$, which is inconsistent with equilibrium value of SGS turbulence energy.

This discrepancy becomes apparent in Fig. 5, which shows a correlation diagram for the 512^3 LES discussed at the beginning of this Section. Calculating $\rho\varepsilon_\Delta$ from Eq. (9) and averaging for fixed values of $\rho\varepsilon_{\text{sgs}}$, yields the dotted line, which roughly corresponds to the relation $|S^*|^2 \propto \varepsilon_{\text{sgs}}^{2/3}$ following from the equilibrium between SGS turbulence production and dissipation. The values of $\rho\varepsilon_{\text{sgs}}$ scatter by several orders of magnitude around this line because of intermittency, SGS transport effects and the influence of the non-linear term. As a consequence, Eq. (9) is an erroneous approximation to the dissipation rate on the grid scale. Although Eqs. (8) and (9) yield about the same *mean* dissipation rate, the former determines the *local* rate of energy dissipation on the footing of a physically well motivated scale-separation of fluid dynamics, while the latter is based on a putative analogy between the numerical and the microscopic viscosity.

4 Thermally Bistable Turbulence

In this Section, we present results from numerical simulations of turbulence subject to parameterized radiative cooling [1]. Apart from the cooling term and a heating term corresponding to a constant UV background in the energy equation, we applied the same simulation setup as explained at the beginning of Sect 3. The consumption of computational resources was between 1 an 2 kCPU-hrs for 256^3, and about 30 kCPU-hrs for 512^3 grids.

Figure 6 shows the evolution of statistics for a 512^3 simulation with an initial temprature $T_0 = 4908\,\text{K}$, and a mean number density $n_0 = 1\,\text{cm}^{-3}$. The forcing is purely compressive, with a characteristic velocity V equal to sound speed c_0 for the initial state. Plotted are the mean values as well as the minima and maxima of the temperature T and the local Mach number u/c_s. The onset of the thermal instability can be seen at time $t \approx 0.7T$, where the minimal temperature rapidly drops to the $10\,\text{K}$ floor of the cooling function (a). The systems evolves into a two-phase state, where the bulk of the computational domain is filled by warm gas at densities lower than $1\,\text{cm}^{-3}$, while the cold phase with densities up to $\sim 1000\,\text{cm}^{-3}$ occupies only a small volume fraction. Thus, the volume-weighted mean Mach number, is maintained at a value of the order unity by the forcing (b). The turbulent flow in the cold phase, on the other hand, is highly supersonic. Since the mass-fraction of cold gas is about one half, the mass-weighted mean Mach number rises to values around 5.

The thermal instability is seeded on small scales [1]. For this reason, one would expect a substantial dependence on the numerical resolution. In a resolution study, we computed probability density functions (PDFs) of several quantities from sim-

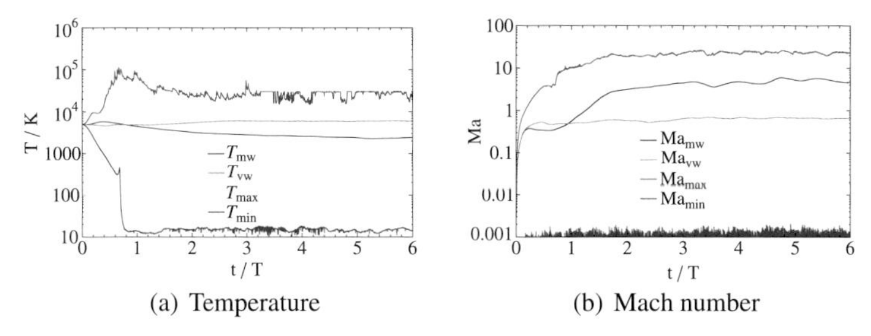

(a) Temperature (b) Mach number

Fig. 6 Temporal evolution of the global averages (mass-weighted and volume-weighted), minima and maxima of the temperature (a) and the Mach number (b) in a 512^3 simulation of thermally bistable turbulence

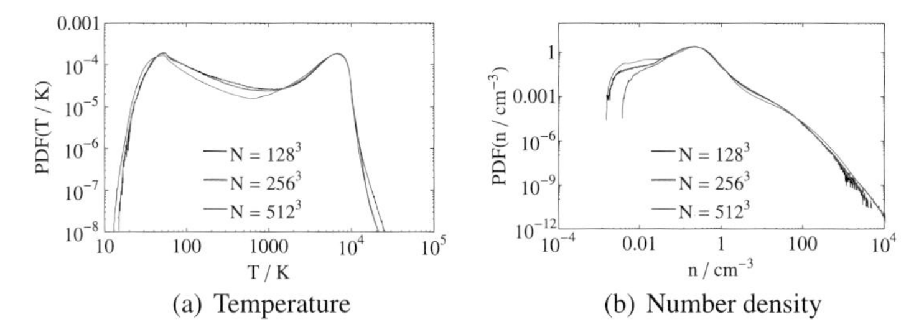

(a) Temperature (b) Number density

Fig. 7 PDFs of the temperature (a) and the number density (b) for different numerical resolutions. The highest resolution corresponds to Figs. 6

ulations performed on grids with 128^3, 256^3, and 512^3 cells. The results for the temperature and number density PDFs are shown in Fig. 7. Actually, the PDFs are remarkably robust with regard to the numerical simulation. The peaks in the temperature PDF indicate the cold and the warm phase (a). In the distribution of the mass density (b), the cold gas gives rise to a power-law tail toward high densities, while the maximum of the PDF is shaped by the warm phase. Nevertheless, resolution-dependent differences can be seen. In particular, high-density tail is affected by the numerical resolution and the phase separation is more pronounced in the 512^3 simulation in comparison to the low-resolution runs.

To study the basic trends, we chose different initial states and varied the forcing in 256^3 simulations. For this parameter study, we did not make use of the SGS model.[2] As an example, the dependence of the mass fractions of the gas phases on the mean kinetic energy of the turbulent flow is plotted in Fig. 8. Here, we use the following definitions: The cold gas phase encompasses temperatures lower than 200 K, warm gas has temperatures higher than 5000 K, and the gas in between is considered as unstable. The most important observation is that the mass fraction of cold gas *decreases* with increasing turbulence intensity, and more gas resides in the unstable phase, although higher turbulence intensity implies stronger compression. However, the variation is relatively small and it is even less for mean number den-

[2] Numerical tests and the implementation of the model were not completed when the study was carried out

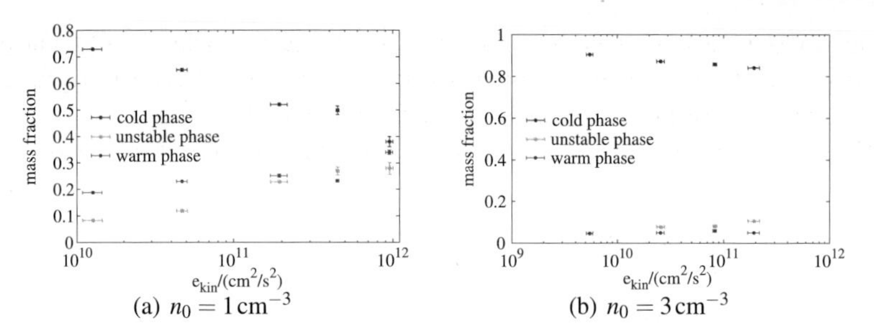

Fig. 8 Mass fractions of warm (red) and cold (blue) gas as a function of the mean kinetic energy per unit mass for two different mean number densities. The mass fractions of gas in an intermediate, unstable state are indicated by green dots

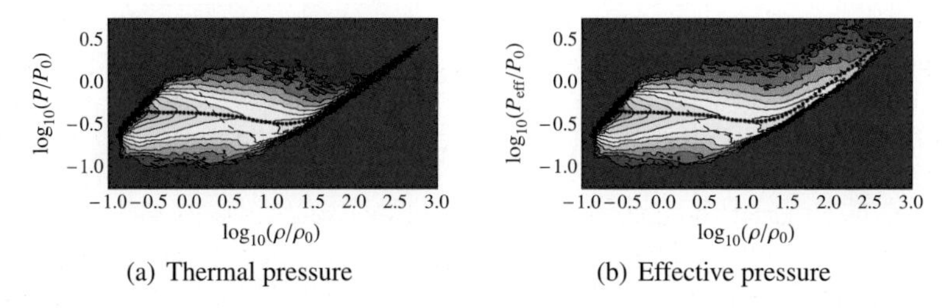

Fig. 9 Phase diagram for the thermal pressure (a) and the effective pressure (b) vs. the mass density in a 512^3 simulation of thermally bistable turbulence. The equilibrium curve, for which cooling is balanced by heating, is plotted as dashed black line. The dotted lines indicate the mean pressure for a given density

sities higher than $1\,\mathrm{cm}^{-3}$. As a consequence, the formation of a large mass fraction of cold gas is a very robust effect if the forcing is compressive. However, we found that solenoidal forcing may suppress the phase separation.

Figure 9 shows phase diagrams for one of our first LES of thermally bistable turbulence. In this case, compressive forcing with $V = 0.5c_0$ was applied. In diagram (a), one can see that the pressure of the unstable gas is roughly constant, while the cold gas is nearly isothermal. The scatter is caused by turbulence on resolved length scales. In diagram (b), the effective pressure is related to the density. The gas is unaffected by SGS turbulence pressure for $n/n_0 \lesssim 10$. The pressure of the cold gas at higher densities, however, is significantly enhanced by unresolved velocity fluctuations.

5 Conclusion

In this report, we outlined the status of our SGS model for supersonic turbulence, and we presented results from simulations of thermally bistable turbulence. Both the development of the SGS model and the parameter study of turbulence in interstellar gas with radiative cooling are precursors to our final project goal: Performing an adaptively refined large eddy simulation of a disk galaxy, with a fully consistent numerical cutoff in terms of the Truelove criterion, the floor of the cooling function, and the influence of SGS turbulence on the fluid dynamics.

The gain of the SGS model in simulations of compressible turbulence is:

1. The determination of scaling laws benefits from the reduction of the bottleneck effect in turbulence energy spectra.
2. The rate of turbulent energy dissipation is predicted on the basis of SGS dynamics rather than numerical viscosity.
3. The effective pressure, i. e., the sum of thermal and turbulent pressure on the grid scale, can be computed from the SGS model.

For galaxy simulations, item 3 is of particular importance, because a cutoff length that satisfies the Truelove criterion (i. e., the local Jeans length is resolved at least by 4 grid cells) will only be feasible on scales, for which unresolved turbulent velocity fluctuations are a significant fraction or even a multiple of the speed of sound. Therefore, SGS turbulence pressure will have an impact on to the support against gravitational collapse and, consequently, to the local efficiency of star formation [6, 8, 13]. As regards the unresolved thermal processes, the situation appears to be quite favorable, because the results from our parameter study suggest that, in particular grid cells, the bulk of the gas will cool down to molecular cloud temperatures, once the number density becomes large compared to $1\,\mathrm{cm}^{-3}$, and turbulence is not predominantly solenoidal. In the planned disk galaxy simulations, both criteria will be met for the density peaks at high refinement levels, where sink particles are inserted to capture collapsing overdense gas. Tracking down star-forming regions by appropriate refinement criteria [12] and modeling the efficiency of star formation as well as the feedback from stars to the interstellar medium will be challenges, where the SGS model is the key to link the production of turbulence on larger scales to turbulent dissipation and feedback processes on small scales.

References

1. E. Audit and P. Hennebelle. Thermal condensation in a turbulent atomic hydrogen flow. *Astron. & Astrophys.*, 433:1–13, 2005.
2. A. Burkert, R. Genzel, N. Bouche, G. Cresci, and S. Khochfar et al. High-Redshift Star-Forming Galaxies: Angular Momentum and Baryon Fraction, Turbulent Pressure Effects and the Origin of Turbulence. 2009. E-print arXiv:0907.4777.
3. B. G. Elmegreen. Turbulence and Galactic Structure. In D. L. Block, I. Puerari, K. C. Freeman, R. Groess, and E. K. Block, editor, *Penetrating Bars Through Masks of Cosmic Dust*, volume 319 of *Astrophysics and Space Science Library*, p. 561, 2004.
4. C. Federrath, J. Duval, R. Klessen, W. Schmidt, and M. M. Mac Low. Comparing the statistics of interstellar turbulence in simulations and observations. E-print arXiv:0905.1060, 2009.
5. M. Germano. Turbulence - The filtering approach. *Journal of Fluid Mechanics*, 238:325–336, 1992.
6. R. S. Klessen, F. Heitsch, and M.-M. Mac Low. Gravitational Collapse in Turbulent Molecular Clouds. I. Gasdynamical Turbulence. *ApJ*, 535:887–906, 2000.

7. A. G. Kritsuk, M. L. Norman, P. Padoan, and R. Wagner. The Statistics of Supersonic Isothermal Turbulence. *ApJ*, 665:416–431, 2007.
8. M. R. Krumholz and C. F. McKee. A General Theory of Turbulence-regulated Star Formation, from Spirals to Ultraluminous Infrared Galaxies. *Astrophys. J.*, 630:250–268, 2005.
9. A. Maier, L. Iapichino, W. Schmidt, and J. C. Niemeyer. Adaptively refined large eddy simulations of clusters. ApJ accepted. E-print arXiv:0909.1800, 2009.
10. L. Pan, P. Padoan, and A. G. Kritsuk. Dissipative Structures in Supersonic Turbulence. *Physical Review Letters*, 102(3):034501, 2009.
11. W. Schmidt. Large Eddy Simulations of Supersonic Turbulence. To appear in Proceedings of ASTRONUM 2009, 4th International Conference on Numerical Modeling of Space Plasma Flows in Chamonix, E-print arXiv:0910.0183, 2009.
12. W. Schmidt, C. Federrath, M. Hupp, S. Kern, and J. C. Niemeyer. Numerical simulations of compressively driven interstellar turbulence. I. Isothermal gas. *Astron. & Astrophys.*, 494:127–145, 2009.
13. W. Schmidt, C. Federrath, M. Hupp, A. Maier, and J. C. Niemeyer. Star Formation in the Turbulent Interstellar Medium and its Implications on Galaxy Evolution. In S. Wagner, M. Steinmetz, A. Bode, and M. Brehm, editors, *High Performance Computing in Science and Engineering, Garching/Munich 2007*. Springer, 2008.
14. W. Schmidt, W. Hillebrandt, and J. C. Niemeyer. Numerical dissipation and the bottleneck effect in simulations of compressible isotropic turbulence. *Comp. Fluids.*, 35:353–371, 2006.
15. W. Schmidt, J. C. Niemeyer, and W. Hillebrandt. A localised subgrid scale model for fluid dynamical simulations in astrophysics. I. Theory and numerical tests. *Astron. & Astrophys.*, 450:265–281, 2006.
16. E. J. Tasker and G. L. Bryan. Simulating Star Formation and Feedback in Galactic Disk Models. *Astrophys. J.*, 641:878–890, 2006.
17. P. R. Woodward, D. H. Porter, S. Anderson, T. Fuchs, and F. Herwig. Large-scale simulations of turbulent stellar convection flows and the outlook for petascale computation. *Journal of Physics Conference Series*, 46:370–384, 2006.

Turbulence Modeling and the Physics of the Intra-Cluster Medium

Luigi Iapichino, Jens C. Niemeyer, Surajit Paul, and Wolfram Schmidt

Abstract The effective modeling of the stirring and development of turbulent flows in grid-based hydrodynamical simulations is computationally challenging. Here we present two possible ways to tackle the problem: first, we consider the use of the adaptive mesh refinement (AMR), applying novel refinement criteria which are optimized to follow the evolution of a turbulent flow. In a second step, the AMR is combined with a subgrid scale (SGS) model for the unresolved turbulence, thus resulting in a new numerical technique called *FEARLESS* (Fluid mEchanics with Adaptively Refined Large Eddy SimulationS). *FEARLESS* performs both the adaptive refinement of the regions where turbulent flows develop and a consistent coupling of the SGS turbulence with the resolved scales, and is argued to be a suitable tool in simulations of turbulent clumped flows. The results of galaxy cluster simulations, performed with the new tool, give rise to several interesting implications with regard to the physics of these objects, and to the numerical methods employed for their exploration in computational cosmology.

Luigi Iapichino
Zentrum für Astronomie der Universität Heidelberg, Institut für Theoretische Astrophysik, Albert-Ueberle-Str. 2, D-69120 Heidelberg, Germany
e-mail: luigi@ita.uni-heidelberg.de

Jens C. Niemeyer · Wolfram Schmidt
Institut für Astrophysik, Universität Göttingen, Friedrich-Hund-Platz 1, D-37077 Göttingen, Germany
e-mail: niemeyer@astro.physik.uni-goettingen.de
e-mail: schmidt@astro.physik.uni-goettingen.de

Surajit Paul
The Inter-University Centre for Astronomy and Astrophysics, Pune University Campus, Pune 411 007, India
e-mail: surajit@iucaa.ernet.in

S. Wagner et al. (eds.), *High Performance Computing in Science and Engineering, Garching/Munich 2009*, DOI 10.1007/978-3-642-13872-0_32,

1 Introduction

The build-up of the cosmic structure, which proceeds through the process of hierarchical clustering [1], plays a key role for the conversion of potential gravitational energy into kinetic and internal energy in galaxy clusters and groups. In this process, cluster mergers induce bulk motions in the intra-cluster medium (henceforth ICM), with velocities of the order of $1000\,\mathrm{km\ s^{-1}}$ [2]. In this flow the gas bulk velocity is transonic and shocks are ubiquitous, thus the development and dissipation of turbulent gas motions can provide a non-negligible contribution to the energy budget.

The turbulent nature of the flow in the ICM will be directly probed with the help of high-resolution X-ray spectroscopy of emission line broadening by the next generation of X-ray telescopes [3–6]. At the present time, there is a number of observational clues that have been interpreted as evidence for the turbulent state of the ICM (see [7], for a review). In addition to the observations, there are many astrophysical problems that could benefit from a deeper theoretical knowledge of turbulence in the ICM such as the amplification of magnetic fields [8–10], the non-thermal emission in clusters [11], and the acceleration of cosmic rays [12–14], to name a few.

From the computational viewpoint, the evolution of the cosmic structure is the typical problem whose prominent features (complexity, nonlinearity and lack of symmetry) call for the use of numerical simulations as a central investigation tool. The wide range of length scales involved in the process makes the Smoothed Particle Hydrodynamics (SPH) technique very suitable and popular in the field, but grid codes, especially when coupled with the Adaptive Mesh Refinement (AMR), are also able to catch a significant part of the scales needed for properly modeling the clustering [15], and allow a complementary and less problematic approach to the baryonic processes [16, 17].

The formation and evolution of the cosmological large scale structure is therefore a case of turbulence generation in a strongly clumped medium. Some authors of this report (L.I., J.C.N., W.S.) are part of the research team which developed a new numerical method, called *FEARLESS* (Fluid mEchanics with Adaptively Refined Large Eddy SimulationS), that combines AMR with Large Eddy Simulations (LES) for the modelling of subgrid scale (SGS) turbulence. With the combined use of grid refinement and SGS model, *FEARLESS* is very suitable for simulations of intermittent turbulent flows in clumped media.

FEARLESS has been developed as part of this research project *h0973: Modelling of turbulent flows applied to numerical simulations of galaxy clusters*, focused on the physics of galaxy clusters, and of the companion project *h0972: Star formation in the turbulent interstellar medium and its implications on galaxy evolution*, devoted to the galactic scales. In this report we will briefly present the basic ideas of turbulence modelling through AMR and LES (Sect. 2), and then the results of our project *h0973* will be presented. In particular, we will review our works, and the related technical and computational issues, on mesh refinement techniques for resolving turbulent flows (Sect. 3), and the first application of the full FEARLESS approach to cluster simulations (Sect. 4). Finally in Sect. 5 the conclusion will be drawn, and some upcoming work will be sketched.

2 Numerical Tools for the Modeling of Turbulent Flows

FEARLESS arises from the combined use of adaptive mesh refinement (AMR) and a subgrid scale (SGS) model for the unresolved scales. This novel tool has been successfully implemented into the ENZO[1] code, v. 1.0 [18], an AMR, grid-based hybrid (hydrodynamics plus N-Body) code based on the PPM solver [19] modified for the study of cosmology [20], which provides the necessary infrastructure for performing numerical simulations of cosmological structure formation.

In the first phase of the project (cf. [21]) the development work was aimed on the AMR, in particular on refinement criteria which are best suited for refining turbulent flows. One such criterion was introduced in [22], and is based on the regional variability of structural invariants of the flow, i.e. variables related to the spatial derivatives of the flow velocity. An examples is the modulus of the vorticity $\omega = \nabla \times \mathbf{v}$ (the curl of the velocity field), expected to become high in regions where the flow is turbulent. The regional threshold for triggering the refinement is expressed in the comparison of the cell value of the variable $q(\mathbf{x},t)$ and the average and the standard deviation of q, calculated on a local grid patch:

$$q(\mathbf{x},t) \geq \langle q \rangle_i(t) + \alpha \lambda_i(t) \tag{1}$$

where λ_i is the maximum between the average $\langle q \rangle$ and the standard deviation of q in the grid patch i, and α is a tunable parameter.

This technique has been used in the simulations in [7, 23, 24], and we refer to Sect. 3 for their most relevant results.

In the case of astrophysically relevant Reynolds numbers, even using AMR one cannot resolve all the relevant length scales down to the dissipative one [25]. This is the case also in the case of galaxy clusters, which have to be followed in their evolution in large cosmological volumes, although the Reynolds number in the ICM is supposed to be only moderately high ([7, 26] for a review). In engineering applications as well as other fields of computational fluid dynamics, SGS models have been developed in order to account for the influence of unresolved turbulence on the resolved scales. This technique is often referred to as Large Eddy Simulations (LES) [27].

As shown by [25], the dynamics equations of a compressible, viscous, self-gravitating fluid can be decomposed into large-scale (resolved) and small-scale (unresolved) parts, using the filter formalism proposed by [28]. By means of filtering, any field quantity a can be split into a smoothed part $\langle a \rangle$ and a fluctuating part a', where $\langle a \rangle$ varies only at scales greater than the prescribed filter length. Following this procedure one can define the turbulent energy e_{t}, which can be visualized as an subgrid energy buffer between the resolved kinetic energy and the internal energy (graphically shown in Fig. 1). The turbulent energy is governed by an equation in the following form:

$$\frac{\partial}{\partial t}\langle \rho \rangle e_{\mathrm{t}} + \frac{\partial}{\partial r_j}\hat{v}_j \langle \rho \rangle e_{\mathrm{t}} = D + \Sigma + \Gamma - \langle \rho \rangle (\lambda + \varepsilon)\,, \tag{2}$$

[1] ENZO homepage: http://lca.ucsd.edu/software/enzo/

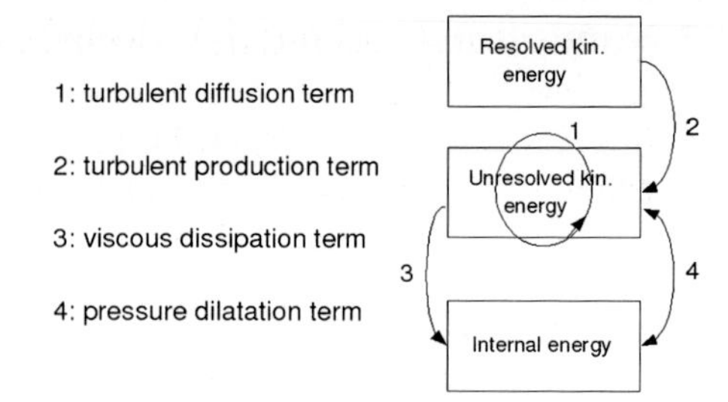

Fig. 1 Graphical description of the energy budget in *FEARLESS*, showing the three energy components and the exchange terms between them

where ρ is the gas density and $\hat{v}_j$ is the density-weighted filtered velocity component, according to Favre [29]. The quantities at the right-hand side of Eq. 2 determine the evolution of e_{t} and are the turbulent diffusion term D, the turbulent production term Σ, the pressure dilatation term λ and the viscous dissipation term ε. Their role in the energy budget is visually described in Fig. 1, and the way they are modelled (i.e. their closures) represent the SGS model. A more detailed and rigorous description can be found in [25, 30].

A severe limitation on the use of the SGS model is that the chosen closure relations and, in fact, the very concept of SGS turbulence energy only applies if the velocity fluctuations on subgrid scales are nearly isotropic. This limits the LES methodology to flows where all anisotropies stemming from large scale features, like boundary conditions or external forces, can be resolved. In the *FEARLESS* method, the grid resolution is locally adjusted by adaptive mesh refinement (AMR) in order to ensure that the anisotropic, energy-containing scales are resolved everywhere. In this way, it is assumed that turbulence is asymptotically isotropic on length scales comparable to or less than the grid resolution. It is very difficult to justify the latter assumption *a priori*, because there are no refinement criteria that would guarantee asymptotic isotropy on the smallest resolved length scales. By a case-by-case careful analysis of simulation results, however, one can gain confidence whether AMR resolves turbulent regions appropriately.

3 Resolving the Turbulent Flow with AMR

In a first group of works, only the novel AMR criteria described in Sect. 2 were tested and applied to hydrodynamical simulations relevant for the physics of galaxy cluster evolution.

Firstly, 3D hydrodynamical simulations were performed in an idealized setup, representing a moving subcluster during a merger event [23]. Whereas AMR simulations performed with the usual refinement criteria based on local gradients of temperature and density do not properly resolve the production of turbulence in the subcluster wake, the new criteria provide a better resolution of this flow, allowing to follow the onset of the shear instability, the evolution of the turbulent wake and the subsequent back-reaction on the subcluster core morphology.

The merging events, studied in the previous work in a simplified fashion, have been further investigated by means of full cosmological simulations, starting from realistic initial conditions [7]. In this case, the refinement criteria based on regional variability of control variables of the flow are used together with the customary overdensity criteria for baryons and dark matter. This approach is effective in resolving turbulent flows in the ICM, with a turbulent velocity in the cluster core larger than 200 km s^{-1} and a turbulent pressure contribution at the percent level. Further details, especially from the computational side, were already given in our project intermediate report [21].

The recent study of major merger events performed in the framework of this project [24] is somehow complementary to the minor merger scenario presented above. Although this problem has been already addressed several times in the last years [17, 31–37], the novel tools developed for the study of turbulent flows allowed to focus on the propagation of merger shocks and their role in injecting turbulence in the ICM, in particular in the post-shock region. Emphasis will be thus mainly put on quantitative and morphological features of merger shocks and the subsequently generated turbulence in ICM. Turbulence in the post-shock region is likely to be related with magnetic field amplification and CR acceleration, and consequently can be linked with cluster radio observations.

We performed a series of cosmological simulations of cluster mergers, based on a flat ΛCDM background cosmology with parameters $\Omega_\Lambda = 0.7$, $\Omega_{\rm m} = 0.3$, $\Omega_{\rm b} = 0.04$, $h = 0.7$, $\sigma_8 = 0.9$, and $n = 1$. The simulations have been initialized at redshift $z_{\rm in} = 60$ using the [38] transfer function, and evolved to $z = 0$. Cooling physics, feedback and transport processes are neglected. An ideal equation of state was used for the gas, with $\gamma = 5/3$.

The simulation box has a comoving size of 128 Mpc h^{-1}. It is resolved with a root grid (AMR level $l = 0$) of 64^3 cells and 64^3 N-body particles. The mass of each particle in this grid is $8.3 \times 10^{11}\ M_\odot$. A static grid ($l = 1$) was nested inside the root grid. It has a size of 64 Mpc h^{-1} and was resolved in 64^3 cells and 64^3 particles (particle mass $1.03 \times 10^{11}\ M_\odot$). Inside this grid, in a volume with side of 38.4 Mpc h^{-1}, a further static grid ($l = 2$) is added, and grid refinement from level $l = 3$ to $l = 6$ is enabled. The linear refinement factor N is set to 2, allowing an effective spatial resolution of 31.25 kpc h^{-1} at the maximum refinement level. The static grids and the region where AMR is allowed are nested on the location of a cluster merger, identified in a DM-only low-resolution run. Seven representative major mergers were chosen in our volume, focusing on events occurring at $0.25 < z < 0.7$ between halos with mass $M > 10^{13}\ M_\odot$, and with a mass ratio between the merging clumps larger than 0.5.

The morphological evolution of a representative merger is shown in Fig. 2 through the evolution of temperature. The two subclumps approach each other almost head-on, with a relative velocity of 980 km s^{-1}, and merge at $z \simeq 0.3$. A web of filaments is visible around the forming structure, as density contours. The gas in the ICM is severely attracted in the newly formed potential well, generating a shock wave which propagates through the cluster. Part of the kinetic and gravitational energy of the merger event is thus dissipated into the ICM. As the shock propagates

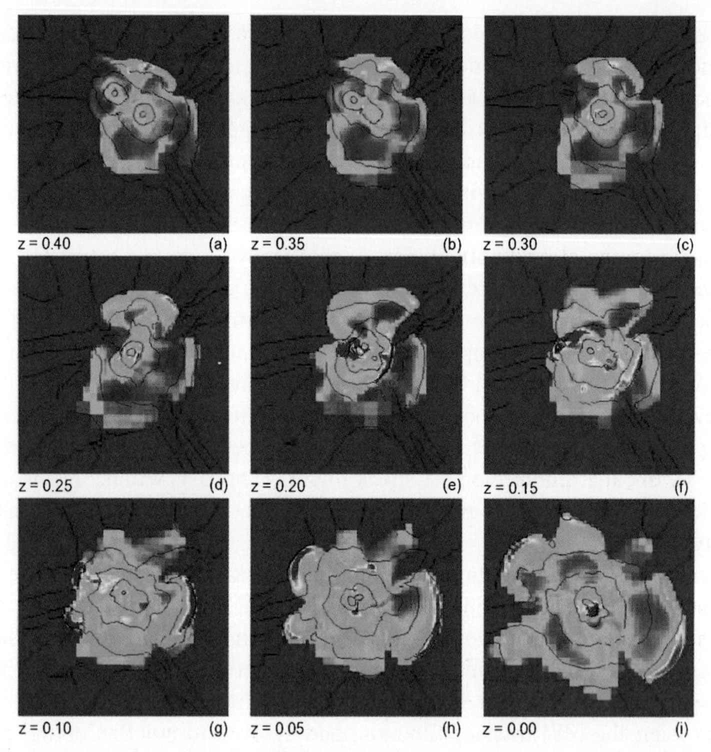

Fig. 2 The evolution of a merger event is shown in slices of temperature. The redshift is indicated at the lower left of each panel, and a identification letter is at the lower right. Each panel has a size of $7.7 \times 7.7\,\mathrm{Mpc}\,h^{-1}$ and is parallel to the *yz* plane. Temperature is color coded, with density contours overlayed. The highest temperature value of the scale (red) corresponds to $3.8 \times 10^7\,\mathrm{K}$

outwards, it interacts with the surrounding filaments and is broken in separate sections, as clearly visible in the bottom row of Fig. 2.

The propagation velocity of the shock is up to $1500\,\mathrm{km\,s^{-1}}$, corresponding to Mach numbers initially in the range from 2.5 to 7. The maximum temperature in the cluster region exceeds $10^8\,\mathrm{K}$ at $z = 0.2$, several times larger than the cluster virial temperature (about $3 \times 10^7\,\mathrm{K}$).

The estimated size of the post-shock region along the direction of shock propagation is of the order of $300\,\mathrm{kpc}\,h^{-1}$, with the velocity dispersion larger than $100\,\mathrm{km\,s^{-1}}$. The time scale for the shock propagation is about 2 Gyr. Within this time after the major merger, the ratio of turbulent to total pressure at the cluster center is larger than 10%, and still larger than that of a relaxed cluster simulated for comparison, until $z = 0$.

A very interesting similarity in morphology can be noticed between our simulations and some observed clusters with mergers ongoing, like for example the Mpc-scale giant arc seen in the radio observations of A3376 [39]. Besides the morphological similarities, the physical link between the radio emission and the turbulence injected in the post-shock region lies in the assumption of the coupling of the compressional modes resolved in the hydrodynamic turbulence and the Alfvénic turbulence, needed for particle acceleration according to the Fermi mechanism. The radio morphology matches the morphology of the simulated structure in temperature and, more importantly, in vorticity, which traces turbulence. Of course, a more detailed comparison should make use of more sophisticated tools for converting the information from the hydrodynamical simulations in synthetic observations; this is left for future work.

From the computational point of view, the choice of a refinement criterion and of its relevant thresholds and parameters is a very delicate task in AMR simulations. Cosmological simulations of galaxy clusters are favored by the clumped behavior of these objects: criteria based on baryon or dark matter overdensity, when properly set, are able to catch the relevant structures with a good compromise between accuracy and saving of computational resources [40, 41]. Unfortunately the simulations presented in this work are extremely challenging from this point of view, because we are mainly interested in the evolution of a shocked region, whose size at late times is comparable with the cluster size, and that for our purpose should be carefully refined.

A typical run, with root resolution of 64^3 grid cells and N-body particles, two nested static grids and the AMR to $l = 6$, consumes from 1000 to 2000 CPU-h on 128 CPUs of the SGI Altix 4700 *HLRB II*, depending on the merger evolution. Doubling the root grid resolution was not computationally convenient (see also the comment on the code issues at the end of Sect. 4), whereas a simulation with the same setup and a root resolution of 96^3, used for a resolution study, runs in about 10000 CPU-h on 128 CPUs.

4 FEARLESS Simulation of a Galaxy Cluster

The full *FEARLESS* approach, briefly presented in Sect. 2, is applied in [30] to cosmological simulations of the formation and evolution of a galaxy cluster. This work has been performed jointly with the computational resources granted to this project and to *h0972*.

Two simulations with and without the subgrid scale model are compared in detail. The simulations were done using a flat ΛCDM background cosmology with a dark energy density $\Omega_\Lambda = 0.7$, a total (including baryonic and dark matter) matter density $\Omega_\mathrm{m} = 0.3$, a baryonic matter density $\Omega_\mathrm{b} = 0.04$, the Hubble parameter set to $\mathrm{h} = 0.7$, the mass fluctuation amplitude $\sigma_8 = 0.9$, and the scalar spectral index $n = 1$. Both simulations were started with the same initial conditions at redshift $z_\mathrm{in} = 60$, using the [38] transfer function, and evolved to $z = 0$. The simulations are adiabatic with a heat capacity ratio $\gamma = 5/3$ assuming a fully ionized gas with a mean molecular weight $m_\mu = 0.6u$. Cooling physics, magnetic fields, feedback, and transport processes are neglected.

The simulation box has a comoving size of $128\,\mathrm{Mpc}\,h^{-1}$. It is resolved with a root grid (level $l = 0$) of 128^3 cells and 128^3 N-body particles. A static child grid

($l = 1$) is nested inside the root grid with a size of 64 Mpc h^{-1}, 128^3 cells and 128^3 N-body particles. The mass of each particle in this grid is $9 \times 10^9 M_\odot\ h^{-1}$. Inside this grid, in a volume of 38.4 Mpc h^{-1}, adaptive grid refinement from level $l = 2$ to $l = 7$ is enabled using the overdensity refinement criterion as described in [7] with an overdensity factor $f = 4.0$. The refinement factor between two levels was set to 2, allowing for an effective resolution of 7.8 kpc h^{-1}.

The static and dynamically refined grids were nested around the place of formation of a galaxy cluster, identified using the HOP algorithm [42]. The cluster has a virial mass of $M_{\rm vir} = 5.95 \times 10^{14} M_\odot\ h^{-1}$ and a virial radius of $R_{\rm vir} = 1.37$ Mpc h^{-1}.

In Fig. 3 a short time series of density and turbulent velocity slices is presented, and several merger events in the cluster outskirts can be identified (for example, in the rectangle in panel 3*a*). A considerable amount of turbulent energy is localized in front and in the wake of the merging clumps (panels 3*b* and *d*). From this point of view, the distribution of turbulent energy traces the local merging history of a galaxy cluster, until it is dissipated into internal energy completely. The morphological evolution of the cluster gives a clear sense of the markedly local behavior of the production and dissipation of turbulence.

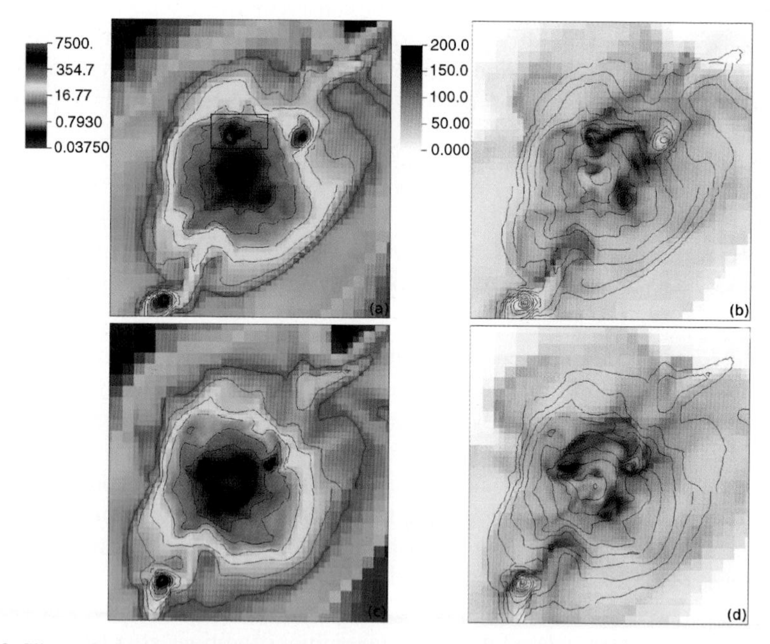

Fig. 3 Slices of baryon density (left-hand panels, *a* and *c*) and turbulent velocity $q = \sqrt{2e_t}$ (right-hand panels, *b* and *d*) at different redshifts z, for the cosmological simulation run with *FEARLESS*. The density is logarithmically color coded as overdensity with respect to the average baryon density in the colorbar on the left of panel *a*, whereas q is linearly coded in km s^{-1}, according to the colorbar on the left of panel *b*. The overlayed contours show density. The slices show a region of 6.4×6.4 Mpc h^{-1} around the center of the main cluster followed in the simulation. Panels *a* and *b* refer to $z = 0.05$, panels *c* and *d* to $z = 0$

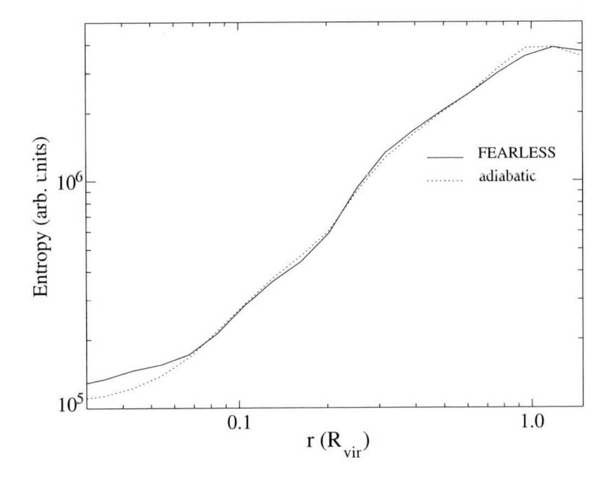

Fig. 4 Radial profiles of mass-weighted entropy (defined in the text) at $z = 0$. The dotted line refers to the simulation without SGS model, whereas the solid line is for the *FEARLESS* simulation

The merger events contribute therefore to the cluster energy budget, at the location where the turbulent dissipation modeled in *FEARLESS* is added to the numerical dissipation of the numerical scheme. Since the flow in the ICM is subsonic, the global turbulent energy contribution at the unresolved length scales is however smaller than 1% of the internal energy.

Another interesting result comes from the comparison between radial profiles in the *FEARLESS* simulation and in the standard adiabatic run, and is the change in the temperature and density profiles at the cluster center. In particular, in the *FEARLESS* run T is larger (3%) for central distances $r < 0.07\ R_{\rm vir}$, with respect to the standard run. Consequently the core in that run is less dense, so that the ICM remains in hydrostatic equilibrium. The local energy budget in the cluster core is therefore modified by the SGS model. This change is more clearly reflected on the entropy which is defined, as is customary in astrophysics, as

$$K = \frac{T}{\rho^{\gamma-1}} \tag{3}$$

with $\gamma = 5/3$ and ρ is the gas density. The entropy in the cluster core is higher in the *FEARLESS* run as compared to the standard run (Fig. 4). This result is consistent both with the locally increased dissipation of turbulent to internal energy provided by the SGS model and with the higher degree of mixing induced in the cluster core, as evaluated by the local velocity dispersion (cf. [30] for more details).

The level on entropy in cluster cores has been long debated, in particular for the discrepant results between SPH and grid codes [4, 43, 44]. Recently, [45] pointed out that the core temperature and entropy in grid-based codes are affected by a spurious increase, caused by the N-body noise in the gravitational force field. In our opinion, the higher entropy core value in the FEARLESS run suggests that the typical flat entropy core is a hydrodynamical feature which requires a better understanding of the numerics in mesh codes, and is at least not primarily caused by N-body noise. Further investigations are certainly needed on this issue.

From the computational viewpoint, the SGS model in *FEARLESS* does not increase significantly the use of CPU time ($\sim +20\%$). These simulations were run on the SGI Altix 4700 *HLRB II* on 128 CPUs, with a typical consumption time of 1100 and 1350 CPU-h for the adiabatic and *FEARLESS* simulations, respectively. A large part of the development and testing work was done with this resolution, although one could argue, from the rather small consumption of resources, that a higher root grid resolution could have been used for the production runs. Unfortunately, this is not the case, basically for known inefficiencies of the code ENZO, v.1.0, in managing the output and the memory. The implementation of structured AMR in the code [46] produces, at every data dump, one file for each grid patch produced, and the run performance decreases dramatically for more than about 10000 grids (about 2600 are produced in our runs). The new release (v.1.5) of the ENZO code overcomes this inefficiency by producing one output file for each CPU, so we expect to make a more efficient use of the available computational resources in our future work.

5 Conclusions and Outlook

Large Eddy Simulations (LES) are based on the notion of filtering the fluid dynamic equations at a specific length scale, thus performing a scale separation between the resolved and the unresolved flow. The latter is treated by means of a subgrid scale model, which in turn is coupled to the hydrodynamical equations governing the former. In principle, a single scale separation is not practical in simulations of clumped media, and furthermore it is not compatible with the concept of adaptive mesh refinement (AMR), often used to study astrophysical phenomena. The solution proposed to this problem is the development of a new numerical scheme that uses AMR and LES in combination, which was called *FEARLESS*.

The initial aim of the present project was to address the numerical problem of turbulence modeling in galaxy clusters, by focusing in particular on the AMR methods (Sect. 3). With the extension of our project, the work made a further step, applying the full *FEARLESS* scheme to simulations of galaxy clusters (Sect. 4). The novel tool appears suitable for modeling turbulent flows over a wide range of length scales, a key feature in the treatment of many astrophysical flows including the intracluster medium. The results give rise to several interesting implications with regard to the physics of galaxy clusters and to the numerical methods employed for their exploration in computational cosmology.

Future research with *FEARLESS* in the cluster physics will include a follow-up of [24] with the full numerical scheme; this study is particularly interesting because the turbulent energy injected by mayor mergers has a significant role in the energy budget, with respect to minor mergers. It will be also important to explore the turbulent flows not only in the ICM, but also in colder baryon phases like the Warm-Hot Intergalactic Medium ($10^5\,\mathrm{K} < T < 10^7\,\mathrm{K}$). The cosmic filaments and the cluster outer regions consist of baryons in this temperature range, and knowledge of the turbulent state of this gas is potentially important for shaping the emission and absorption lines associated with it (see [47] for a review).

Also the development work associated with *FEARLESS* is still in progress. The current version of the SGS model has to be considered as an intermediate solution to address some basic questions related to dynamics of thc turbulcnt intra-cluster medium. A more elaborate model that is able to handle the complexity of the flow (wide range of Mach numbers and large density gradients as well as pronounced inhomogeneities) in simulations of large scale structure evolution is under development (also see the contribution on *Numerical Models of Turbulence in Isothermal and Thermally Bistable Interstellar Gas* [48] in this volume). The infrastructure for the further development is the new release 1.5 of the ENZO code, which fixes many of the numerical inefficiencies we described. The first application reviewed here shows the promising perspectives for the use of an SGS model in combination with AMR, and its potential impact on many branches of numerical astrophysics.

Acknowledgements Thanks to the whole support team of the Leibniz Rechenzentrum for their invaluable help during our work.

References

1. J.P. Ostriker, ARA&A **31**, 689 (1993)
2. M.L. Norman, G.L. Bryan, LNP Vol. 530: The Radio Galaxy Messier 87 **530**, 106 (1999)
3. R.A. Sunyaev, M.L. Norman, G.L. Bryan, Astronomy Letters **29**, 783 (2003)
4. K. Dolag, F. Vazza, G. Brunetti, G. Tormen, *Mon. Not. Roy. Soc.* **364**, 753 (2005)
5. M. Brüggen, M. Hoeft, M. Ruszkowski, *Astrophys. J.* **628**, 153 (2005)
6. P. Rebusco, E. Churazov, R. Sunyaev, H. Böhringer, W. Forman, MNRAS **384**, 1511 (2008)
7. L. Iapichino, J.C. Niemeyer, *Mon. Not. Roy. Soc.* **388**, 1089 (2008)
8. K. Dolag, M. Bartelmann, H. Lesch, *Astron. Astrophys.* **387**, 383 (2002)
9. M. Brüggen, M. Ruszkowski, A. Simionescu, M. Hoeft, C. Dalla Vecchia, ApJ **631**, L21 (2005)
10. K. Subramanian, A. Shukurov, N.E.L. Haugen, *Mon. Not. Roy. Soc.* **366**, 1437 (2006)
11. A.R. Bell, *Mon. Not. Roy. Soc.* **353**, 550 (2004)
12. F. Miniati, T.W. Jones, H. Kang, D. Ryu, ApJ **562**, 233 (2001)
13. F. Miniati, D. Ryu, H. Kang, T.W. Jones, ApJ **559**, 59 (2001)
14. G. Brunetti, A. Lazarian, *Mon. Not. Roy. Soc.* **378**, 245 (2007)
15. M.L. Norman, ArXiv Astrophysics e-prints, astro-ph/0402230 (2004)
16. O. Agertz, B. Moore, J. Stadel, D. Potter, F. Miniati, J. Read, L. Mayer, A. Gawryszczak, A. Kravtsov, A. Nordlund, F. Pearce, V. Quilis, D. Rudd, V. Springel, J. Stone, E. Tasker, R. Teyssier, J. Wadsley, R. Walder, MNRAS **380**, 963 (2007)
17. N.L. Mitchell, I.G. McCarthy, R.G. Bower, T. Theuns, R.A. Crain, *Mon. Not. Roy. Soc.* **395**, 180 (2009)
18. B.W. O'Shea, G. Bryan, J. Bordner, M.L. Norman, T. Abel, R. Harkness, A. Kritsuk, in *Adaptive Mesh Refinement – Theory and Applications, ed. T. Plewa, T. Linde, V.G. Weirs (Berlin; New York: Springer), Lecture Notes in Computational Science and Engineering*, vol. 41 (2005), p. 341
19. P. Woodward, P. Colella, Journal of Computational Physics **54**, 115 (1984)
20. G.L. Bryan, M.L. Norman, J.M. Stone, R. Cen, J.P. Ostriker, Computer Physics Communications **89**, 149 (1995)

21. L. Iapichino, J.C. Niemeyer, J. Adamek, S. Paul, M. Scuderi, in *High Performance Computing in Science and Engineering, Garching/Munich 2007*, ed. by S. Wagner, M. Steinmetz, A. Bode, M. Brehm (2009), pp. 45–56
22. W. Schmidt, C. Federrath, M. Hupp, S. Kern, J.C. Niemeyer, *Astron. Astrophys.* **494**, 127 (2009)
23. L. Iapichino, J. Adamek, W. Schmidt, J.C. Niemeyer, *Mon. Not. Roy. Soc.* **388**, 1079 (2008)
24. S. Paul, L. Iapichino, F. Miniati, J. Bagchi, K. Mannheim, ArXiv e-prints (2010), 1001.1170, submitted to ApJ
25. W. Schmidt, J.C. Niemeyer, W. Hillebrandt, *Astron. Astrophys.* **450**, 265 (2006)
26. T.W. Jones, in *Extragalactic Jets: Theory and Observation from Radio to Gamma Ray, ASP Conference Series*, vol. 386, ed. by T. Rector, D. De Young (2007), p. 398, arXiv e-prints: 0708.2284
27. M. Lesieur, O. Metais, Annual Review of Fluid Mechanics **28**, 45 (1996)
28. M. Germano, Journal of Fluid Mechanics **238**, 325 (1992)
29. A. Favre, SIAM: Problems of hydrodynamics and continuum mechanics pp. 231–266 (1969)
30. A. Maier, L. Iapichino, W. Schmidt, J.C. Niemeyer, *Astrophys. J.* **707**, 40 (2009)
31. K. Roettiger, J. Burns, C. Loken, ApJ **407**, L53 (1993)
32. K. Roettiger, J.O. Burns, C. Loken, *Astrophys. J.* **473**, 651 (1996)
33. K. Roettiger, C. Loken, J.O. Burns, *Astrophys. J. Suppl.* **109**, 307 (1997)
34. P.M. Ricker, C.L. Sarazin, *Astrophys. J.* **561**, 621 (2001)
35. H. Mathis, G. Lavaux, J.M. Diego, J. Silk, *Mon. Not. Roy. Soc.* **357**, 801 (2005)
36. G.B. Poole, M.A. Fardal, A. Babul, I.G. McCarthy, T. Quinn, J. Wadsley, *Mon. Not. Roy. Soc.* **373**, 881 (2006)
37. G.B. Poole, A. Babul, I.G. McCarthy, M.A. Fardal, C.J. Bildfell, T. Quinn, A. Mahdavi, *Mon. Not. Roy. Soc.* **380**, 437 (2007)
38. D.J. Eisenstein, W. Hu, ApJ **511**, 5 (1999)
39. J. Bagchi, F. Durret, G.B.L. Neto, S. Paul, Science **314**(5800), 791 (2006)
40. B.W. O'Shea, K. Nagamine, V. Springel, L. Hernquist, M.L. Norman, *Astrophys. J. Suppl.* **160**, 1 (2005)
41. K. Heitmann, Z. Lukić, P. Fasel, S. Habib, M.S. Warren, M. White, J. Ahrens, L. Ankeny, R. Armstrong, B. O'Shea, P.M. Ricker, V. Springel, J. Stadel, H. Trac, Computational Science and Discovery **1**(1), 015003 (2008)
42. D.J. Eisenstein, P. Hut, *Astrophys. J.* **498**, 137 (1998)
43. J.W. Wadsley, G. Veeravalli, H.M.P. Couchman, *Mon. Not. Roy. Soc.* **387**, 427 (2008)
44. D. Kawata, T. Okamoto, R. Cen, B.K. Gibson, ArXiv e-prints (2009), 0902.4002
45. V. Springel, *Mon. Not. Roy. Soc.* **401**, 791 (2010)
46. M.J. Berger, P. Colella, Journal of Computational Physics **82**, 64 (1989)
47. S. Bertone, J. Schaye, K. Dolag, Space Science Reviews p. 28 (2008)
48. W. Schmidt, D. Seifried, M. Niklaus, J. Niemeyer, (2010). (in this volume)

Project h1021: Dynamics of Binary Black Hole Systems

Bernd Brügmann, Ulrich Sperhake, Doreen Müller, Roman Gold, Pablo Galaviz, Nobert Lages, and Marcus Thierfelder

Abstract The goal of this project is to compute the last phase of the inspiral and merger of two black holes in Einstein's theory of general relativity. Numerical simulations of binary black hole systems have only recently reached the point where for the first time several complete orbits can be computed. Such simulations pose on the one hand a significant theoretical and technical challenge since the time-dependent Einstein equations have to be solved in the regime of highly dynamic and non-linear gravitational fields. On the other hand, theoretical predictions for the gravitational waves generated in a binary black hole merger are eagerly awaited by a new generation of gravitational wave experiments. The plan is therefore to map out the parameter space of gravitational wave signals from binary black hole inspirals. This research is funded in part by the SFB/Transregio 7 on Gravitational Wave Astronomy, the BMBF/DLR grant for LISA Germany, and the Graduiertenkolleg on Quantum and Gravitational Fields.

1 Goals and Motivation

Only very recently (2007) it has become possible to compute up to ten orbits of two orbiting black holes in Einstein's theory of general relativity. The main goal of this project is a large scale and continued effort to move from the first proof of principle

B. Brügmann · D. Müller · R. Gold · P. Galaviz · N. Lages · M. Thierfelder
Friedrich-Schiller-Universität Jena, Theoretisch-Physikalisches Institut, Max-Wien-Platz 1, 07743 Jena, Germany
e-mail: bernd.bruegmann@uni-jena.de
e-mail: doreen.m@uni-jena.de

Ulrich Sperhake
California Institute of Technology, 1200 East California Boulevard, Pasadena CA 91125, USA
e-mail: sperhake@tapir.caltech.edu

S. Wagner et al. (eds.), *High Performance Computing in Science and Engineering, Garching/Munich 2009*, DOI 10.1007/978-3-642-13872-0_33,

simulations to large production runs that explore and map the entire parameter space of the binary black hole problem.

The two-body problem of general relativity in the strong field regime has a rich phenomenology which is still far from being satisfactorily understood, although there has been enormous progress in the last few years. The numerical solution of the full Einstein equations (in their standard form ten non-linear, coupled partial differential equations) is a very complex problem, and for two black holes there is the additional challenge to deal with the spacetime singularities that are encountered in the interior of black holes. Given initial data for configurations of two black holes, the time evolution of the system is computed. Currently, typical runs are limited by the achievable evolution time before the simulations become too inaccurate, or before the computer code becomes unstable and crashes at numerical infinities. State of the art are simulations that cover about 10 full orbits of two black holes shortly before their merger (details depending on the initial parameters of the binary).

Orbiting black holes generate gravitational waves (GW). Gravitational waves represent the analogue in Einstein's theory of general relativity to electromagnetic waves in the Maxwell theory of electro-magnetism, and in general any accelerated mass will produce gravitational radiation. Among the main sources of gravitational radiation are compact objects such as black holes and neutron stars. A correct modeling of these sources and their expected emission of gravitational waves can only be achieved by *numerical relativity* performing supercomputer simulations.

A key scientific objective of computer simulations of black hole binaries is to provide substantial support from the theoretical side to the ongoing effort to detect and measure gravitational waves. For this purpose, several laser-interferometric detectors have been constructed as large international science projects. These detectors, the American LIGO, the Anglo-German GEO600, the French-Italian VIRGO and the Japanese TAMA, use the latest advances in laser technology to facilitate the measurement of extremely small variations in the detector's arms' length produced by gravitational waves. Their indirect observation in astrophysical observations of the neutron star system $1913+16$ has lead to the 1993 Noble Prize awarded to Hulse and Taylor. The high accuracy of the detectors is made necessary by the low amplitude oscillations of 1 part in about 10^{20} caused by the passage of gravitational waves. This extremely low amplitude of the signal also makes it very important to gain as much theoretical knowledge about the expected signal patterns as possible in order to support the detection process.

2 Model and Methodology

There is large freedom in writing the Einstein equations as a system of partial differential equations, and much research has gone into finding optimal choices. In our work we employ the currently most popular choice, the Baumgarte-Shapiro-Shibata-Nakamura (BSSN) system [1–4], which so far is the only system for which results have been reported of long-time black-hole binary simulations that do not rely on black hole excision as has been used for example in [5, 6].

In order to represent black holes in the initial data, we use the so-called "puncture method" [7]. Puncture data employ the analytic Bowen-York solution for the initial values of the extrinsic curvature and provide a particularly convenient form of the remaining elliptic equation, the Hamiltonian constraint [8, 9]. The key initial parameters of the black holes, i.e. their initial linear and angular momenta are specified directly as free parameters of the analytic solution of the extrinsic curvature. Specific values, for example, for the linear momenta corresponding to a non-eccentric inspiral can be determined for example from an equilibrium condition at third post-Newtonian (PN) order as described in [10]. The elliptic constraint equations are solved in BAM with the pseudo-spectral collocation method described in [11] using adapted coordinates. Cartesian data are then obtained by barycentric interpolation, typically with eighth-order polynomials for both the fourth- and sixth-order finite differencing methods.

For time evolutions, we use the BSSN system together with the 1+log and gamma driver coordinate gauges [3, 12, 13] as described in [10]. These gauge conditions allow the "punctures" to move across the grid ("moving puncture" approach [14, 15]) and allow an effective softening of the singularity in the metric associated with an internal asymptotic region [16–18], which had been prohibited by the traditional "fixed punctures" approach. It is this approach of moving punctures that we and most groups follow. The BSSN system is based on a conformal decomposition of the spatial geometry, writing the physical spatial metric as $g_{ij} = \chi^{-1}\tilde{g}_{ij}$ (following [14]). The blowup of the metric at the punctures is absorbed into the conformal factor χ, which vanishes at the punctures.

3 Computational Infrastructure

For our numerical evolutions we mainly use the BAM code [19, 20], which is designed to solve partial differential equations on Cartesian meshes, in particular a coupled system of (typically hyperbolic) evolution equations and elliptic equations. Discretization in space is implemented by finite differences. Integration in time is performed by the method of lines using explicit Runge-Kutta algorithms. The main technical features provided by BAM are adaptive mesh refinement (AMR) and parallelization. AMR is based on the standard Berger-Oliger algorithm, which relies on interpolation in time and space to exchange data between the different levels. At the refinement boundaries we use additional buffer zones along the lines of [21, 22] as described in [10]. Parallelization is implemented with MPI [23].

The complexity of the equations is addressed by using a Mathematica package integrated into the code, which produces C-code from Mathematica expressions in tensor notation. Using such a system as we do in BAM, or as has been discussed in detail for the Cactus environment in [24] drastically simplifies the modification of complex codes for black-hole binary simulations, as was required to adapt codes from the "fixed puncture" to the "moving puncture" paradigm, or in the present case to implement the improved numerical algorithms discussed here. The BAM code is organized as a "framework", similar in spirit to the Cactus code [25, 26], but dropping much of its complexity. The structure of the BAM code has also made it particularly straightforward to implement higher order finite differencing methods.

The effectiveness of AMR is dependent on the different length scales in the problem. In the case of the black hole simulations in this project (which are performed without additional matter sources) we can point out four length scales. The horizon of a black hole of mass $1M$ has a radius of about $1M$ in the coordinates we employ (using geometric units in which the speed of light and Newton's constant of gravity are set equal to one, $c = G = 1$). The gravitational waves generated in a binary have a wave length of roughly $100M$ – $300M$. Since implementing proper outer boundary conditions for the Einstein equations becomes easier with distance from the sources, the outer boundary is placed at around $700M$, while requiring less resolution than the computation of the waves closer to the sources. The highest resolution is required inside the horizon of the black holes to provide sufficient resolution near the "puncture". This suggests a simple refinement strategy using nested rectangular boxes centered around the black holes, see e.g. [10, 19, 20]. In a typical run, the computational domain may be decomposed into nested rectangular boxes with resolution changing by factors of 2 from roughly $2^{-6}M$ near the punctures up to $2^{+4}M$ at the outer boundary.

While BAM is our main production code, there is also collaboration involving other codes. One of our collaborators (Dr. U. Sperhake, now at Caltech) has developed an independent code, the LEAN code [27], based on the Cactus Toolkit as infrastructure for parallelization and an independent set of physics modules. This has already offered several opportunities for fruitful comparisons, see for example [10, 28].

4 Performance and Scaling

In summer 2005 the Jena group was granted support by the LRZ in its porting and optimization initiative for the new Altix system. Over the years we performed several tests and optimizations in collaboration with I. Christadler at the LRZ.

In the beginning of 2007 a bottleneck in scaling was identified and removed for wave extraction and horizon finding due to inefficient scaling of interpolations to a coordinate sphere. In addition, the memory usage of the time stepping method, wave extraction and horizon finding was reduced by up to 40 % in some cases.

Following various code upgrades, we investigated the scaling properties once more in May 2008 using simulations of binary black hole spacetimes with adaptive mesh refinement. These simulations were performed using sixth order discretization combined with the grid setups as listed in Table 1. For these tests we kept the processor load approximately constant by increasing the grid resolution as we use more processors. The table shows the corresponding grid setups and the resulting

Table 1 Performance of the 5/2008 version of BAM following upgrades during the 2007/2008 grant period. Binary black hole spacetimes were evolved until $t = 1\,M$. The columns list the number of processors used, the total number of grid points in the simulation, the resolution on the finest and coarsest refinement level and the speed of the simulations. The speed is given as the number of iterations 10^6 points have been advanced per cpu hour (and normalized relative to that for 16 processors)

procs.	$N_{\text{pts}}/10^6$	h_{min}	h_{max}	iters.$\times 10^6$ pts./(cpu$\times$ h)	
16	12.12	0.00984	10.08	19.9	
32	22.86	0.00781	8.00	20.2	(101.5%)
64	43.32	0.00620	6.35	19.9	(100.0%)
80	52.60	0.00576	5.89	18.3	(92.0%)
100	64.86	0.00534	5.47	20.1	(101.0%)
128	80.63	0.00492	5.04	18.8	(94.5%)

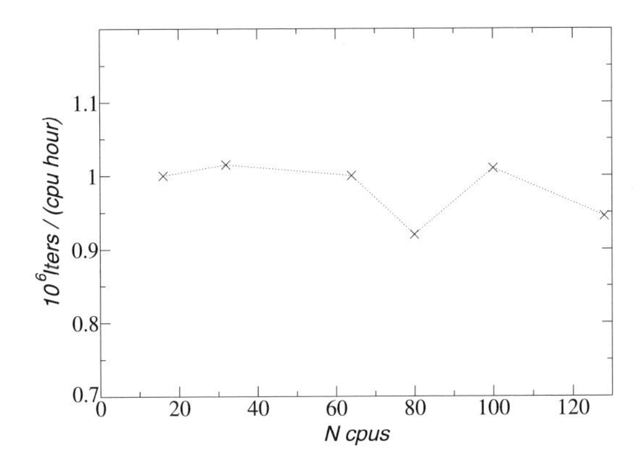

Fig. 1 Scaling properties of BAM after the upgrades in 2007 and early 2008 on the HLRB2. The points show the performance of the code *for short runs without heavy use of AMR regridding*. The data is normalized to the speed using 8 processors

total number of grid points for each simulation. All simulations lasted for $\Delta t = 1\,M$, where M is the total black hole mass. Because we work with a constant Courant factor, this implies a larger number of iterations as we increase the resolution. In order to calibrate the scaling properties of BAM we therefore measure the number of iterations performed for 10^6 grid points per cpu hour. The result is also shown in Fig. 1 and demonstrates good scaling in the range from 16 to 128 processors.

However, the current generation of production runs relies heavily on "moving boxes", i.e. nested cartesian grids of increasing resolution that track the motion of the black holes. These boxes move relatively to each other, change size, and sometimes merge and separate, which implies certain regridding and parallelization operations in the AMR algorithm. These operations were identified as the bottleneck to scaling in the 5/2008 version of BAM.

The optimized code with this bottleneck removed was ready for production runs in 06/2009. Specifically, the regridding was rearranged for maximal processor lo-

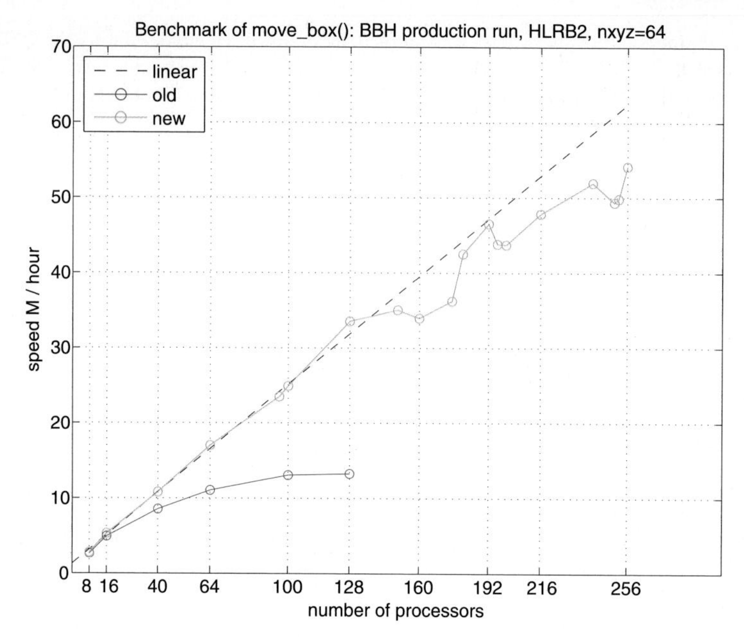

Fig. 2 Scaling properties of BAM after the most recent upgrade in 06/2009 on the HLRB2. The points show the performance of the code *for production runs with significant use of AMR regridding*. Shown is the computational speed in terms of physical time computed during one cpu hour versus the number of processors for a fixed global problem size, where the typical, smallest cartesian box consists of 64^3 points. The lower curve labeled "old" refers to the 5/2008 result, the upper curve labeled "new" to the 06/2009 version of the code. The old version of BAM showed degradation of scaling by about a factor of 2 when increasing the number of processors from 8 to 100, while the new version scales almost linearly up to 128 processors, and remains reasonably close to linear scaling up to 256 processors

cality. The grid transfer operators between different grid resolutions are now completely processor local using parallelization ghost zones if the boxes do not move by more than one or two grid points per time step (which usually they do not because the black holes do not move faster than the speed of light). Boxes that change by more than a couple of grid points, say when they merge and split, are handled by a newly optimized polynomial interpolator. To a large degree the new algorithm avoids communication of three-dimensional grid volumes between different refinement levels, since a typical regridding step now consists of processor local two-dimensional exchanges between refinement levels.

Figure 2 shows the improvement for 8 to 256 processors on the HLRB2 for a fixed global problem size, where the typical, smallest cartesian box consists of 64^3 points (and the global point count is constant while the processor number is changed). The test runs are representative for binary black hole production runs at medium resolution with significant use of moving box regridding. The lower curve labeled "old" refers to the 5/2008 result, the upper curve labeled "new" to the 06/2009 version of the code. Shown is the computational speed in terms of phys-

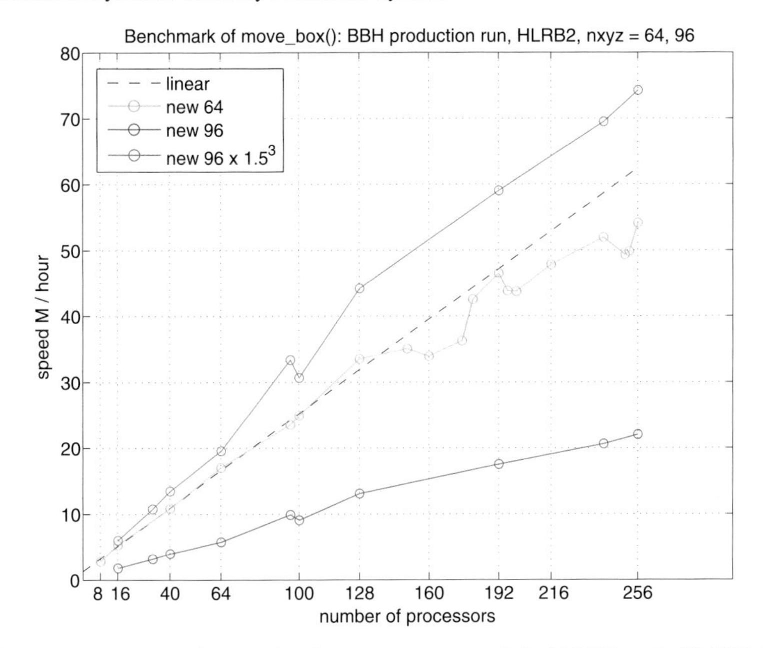

Fig. 3 Scaling properties of BAM after the most recent upgrade in 06/2009 on the HLRB2. Black hole example as in Fig. 2, but now comparing speeds for 64^3 and 96^3 nominal grid sizes after the optimizations. The computation proceeds more slowly for 96^3, but the larger box size results in less communication overhead. Indeed, just comparing box sizes, the high resolution run involves $1.5^3 \approx 3.4$ times more floating point operations than the medium resolution run, but scaling the resulting speed of the high resolution run accordingly (the uppermost curve) indicates better overall performance than the medium resolution run. Scaling improves with the number of grid points

ical time computed during one cpu hour. The dashed straight line indicates linear scaling.

The old version of BAM showed degradation of scaling by about a factor of 2 when increasing the number of processors from 8 to 100, while the new version scales almost linearly up to 128 processors, and remains reasonably close to linear scaling up to 256 processors. Some loss of scaling is to be expected since 64^3 points are distributed among e.g. $4 \times 4 \times 8 = 128$ processors in a 3d processor grid, which leaves comparatively few floating point operations per processor compared to the parallelization overhead.

Figure 3 shows results for a nominal grid size of 96^3 points corresponding to a high resolution version of the previous 64^3 point example. The computation proceeds more slowly, but the larger box size results in less communication overhead. Indeed, just comparing box sizes, the high resolution run involves $1.5^3 \approx 3.4$ times more floating point operations than the medium resolution run, but scaling the resulting speed of the high resolution run accordingly (the uppermost curve) indicates better overall performance than the medium resolution run.

In conclusion, the last round of optimizations has essentially removed all scaling problems for up to 256 processors. This implies that we can perform *longer* evolutions in a given wall-clock time using more processors efficiently, or alternatively, that we can perform *higher* resolution runs using more processors efficiently in a reasonable wall-clock time.

5 Resources Required for Typical Simulations

The exact computational requirements of a particular simulation depend on whether the system under study obeys simplifying symmetries such as equatorial symmetry across the $x-y$ plane. Such symmetries are not available for the majority of simulations using the most general parameters for spins and masses of the black holes and the following numbers are calculated for these general cases.

A typical production simulation requires about 10 refinement levels with about 100^3 points each. The evolution of the Einstein equations requires approximately 120 grid functions (including storage on previous timesteps) of double precision data type, i. e. 8 bytes per grid point and function. Together with additional variables not directly used in the main evolution of the Einstein equations, this corresponds to about 40 Gb of memory. In order to test convergence of certain simulations, evolutions with higher resolutions and thus larger memory usage are necessary, resulting in up to 250 Gb for a single run.

A key issue in finite difference simulations is to establish convergence with increasing resolution. To this end we have to perform a minimum of three runs at different resolution for a selected set of parameters. Even though in principle the convergence rate of the code is fully determined by the algorithms used, in practice this turns out to be a tricky issue. There are various parts of the method that converge differently, for example the derivatives in space, the time stepping method, the puncture region inside the black holes and the outer boundary, the refinement boundaries in space and time. Increasing the resolution by a factor of 2 increases the resource requirements by a factor of 16. We therefore perform runs at resolutions that are closer spaced, say $M/48$, $M/56$, $M/64$, $M/72$, also keeping in mind that we use 6th order finite differencing. The bottom line is that for consistency more than one run is required to firmly establish one data point.

The length of the computation is rather dependent on the initial black hole configuration and thus on the scientific target of the runs. In particular simulations for testing the post-Newtonian approximation require the use of large initial separations and, thus, long evolution times. In extreme cases, evolution times of up to a couple of weeks (using checkpointing) are possible, but the majority of simulations is likely to require significantly shorter times, of the order of several days. In terms of cpu hours required for one simulation, the range covered is about 1,000 for a low resolution, exploratory run up to 50,000 for the longest lasting simulation at highest resolution for convergence testing. An average value of 6,000 cpu hours can be assumed for a typical exploratory run and 24,000 cpu hours for a production run.

Of high importance is the total number of simulations necessary for the establishment of a satisfactory gravitational wave catalogue. This number depends on several factors. First, we note that a black hole configuration has at least seven key parameters, the mass ratio and three values each for the spin of either hole. Additionally, it is interesting to consider deviations from circular orbits which adds the eccentricity as one additional parameter. The parameter space thus has at least eight dimensions which makes it nearly impossible to cover the entire parameter space using brute force. The strategy of the community is instead to systematically isolate the dependency of the waveforms on a smaller subset of parameters, modeling these dependencies and finally validating the resulting models for more general classes of black hole configurations. A further key factor determining the total number of simulations required is directly related to gravitational wave data analysis, namely the question with what density the parameter space of numerical simulations needs to be filled. This question is subject to ongoing scientific work and the only statement one can make with certainty is that the current set of numerical simulations performed by the community is way below what will eventually be required in gravitational wave observations.

6 Status Report for h1021

The results obtained so far from our simulations performed at HLRB II can roughly be divided into two groups: results on the physical behavior of black hole spacetimes and their repercussions on astrophysical questions, and furthermore results pertaining to the use of gravitational waveforms in data analysis. We will briefly mention the main results of all the publications in the following. For more details, the reader is referred to the cited articles.

In one of our studies, we looked at the ultra-relativistic regime. This is of particular interest as the ultra-relativistic scattering of particles, for example in the Large Hadron Collider, should be well described by black hole scattering if the center of mass energy is beyond the Planck scale. In [36] the collision of two non rotating black holes with very high energy can lead to velocities of up to 94% of the speed of light. For the first time we were able to compare analytic calculations, black hole perturbation theory, and strong field, nonlinear numerical calculations for this problem.

In a second study on this topic, published in [44], we find a zoom-whirl behavior when we fine-tune the impact parameter. We further found that the remarkable amount of up to 35% of the center-of-mass energy can be radiated in form of gravitational waves, which is extremely high compared to black hole binary inspirals, where about ten times less energy is radiated. A further observation is, that such collisions can leave nearly extremely spinning black holes as a remnant. Figure 4 shows the radiated energy and final spin of the remnant black hole for varying impact factor.

While the ultra-relativistic regime is a highly interesting case to study with numerical relativity simulations, most of the black hole binaries in the universe are expected to inspiral on nearly circular orbits. A lot of our effort goes into studying

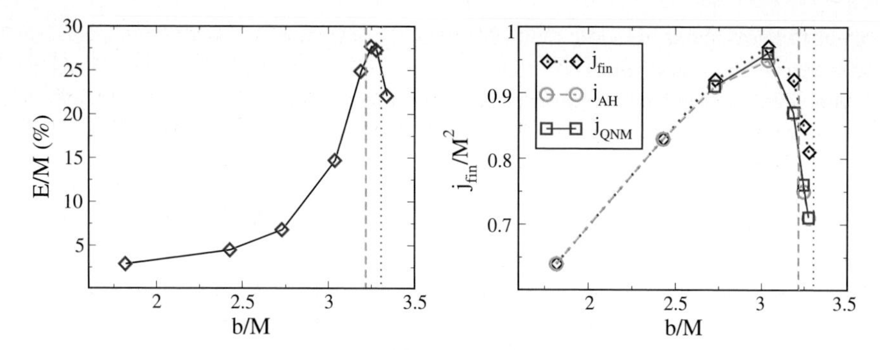

Fig. 4 Total energy radiated (left) and final BH spin (right) vs. impact parameter from one of the sequences studied in [44], the latter calculated using several methods. The vertical dashed green (dotted red) line is the estimated immediate merger threshold b (the scattering threshold b_{scat})

such binaries with different ratios of masses of both holes. One of the main questions is how to achieve stable simulations for mass ratios which are significantly larger than 1:1. The difficulties arise from the large asymmetry in the numerical grid when the masses differ by a huge amount and a mechanism which prevents grid points from falling into the larger black hole has to be found. In [38] we presented the first numerical simulations of an initially non-spinning black-hole binary with a mass ratio as large as 10:1 in full general relativity. With this study we were able to validate existing fit formulas for the kick and spin of the final black hole after merger and for the radiated energy. These formulas had been established using low mass ratios of 1:1 to 4:1 on one end and the extreme mass ratio (test particle) limit on the other one. Our results filled the gap between the two ends and showed the correctness of these formulas.

A separate question studied in [39] concerned the construction of initial data in binary black hole evolutions. As mentioned above, we often study black hole binaries in *quasi-circular* orbits. It is, however, not trivial to construct initial data resulting in such orbits and simultaneously fulfilling the constraints of full general relativity. We addressed this issue and computed post-Newtonian orbital parameters for quasi-spherical orbits using the method of Buonanno, Chen and Damour (2006) [45]. To quantify the quality of the data we examined the resulting eccentricity in numerical simulations of non-spinning binaries with mass ratios from 1:1 to 4:1 and of equal mass binaries with a selection of special spin configurations. The method was successful and provided us with initial data which produce orbits with eccentricities as low as or lower than found in previous studies.

Ensuring and improving the quality of our numerical waveforms is a constant effort. To this end, we constantly carry out convergence studies and error analyses. Such checks can give information about internal errors which can then be minimized. In order to verify the overall accuracy of our simulations it is also important to check the consistency of waveforms resulting from *different* numerical codes. Such an external comparison has been carried out in [40] where results from five numerical codes for the $(\ell = 2, |m| = 2)$ mode of the gravitational waves from an equal mass, non-spinning binary completing about six orbits before merger were found to be consistent within the numerical errors of the codes. This publication also established a connection to data analysis related issues and showed that the waveforms would be indistinguishable in all ground-based detectors if detected with a signal-to-noise ratio of less than ≈ 25 (which was proposed as a practical measure of the consistency of different results).

The above accuracy study shows that it is nowadays very important to establish and maintain a close contact between numerical relativity and data analysis groups. To foster closer collaboration between the numerical relativity and data analysis communities, the Numerical INJection Analysis (NINJA) project was founded and we took part in this effort by contributing waveforms (together with many other numerical relativity groups) which were used to generate a set of gravitational-wave signals. These signals were injected into a simulated data set, designed to mimic the response of the Initial LIGO and Virgo gravitational-wave detectors. Different data analysis groups searched for the original waveforms using different search and parameter-estimation pipelines in order to see if their particular algorithm is able to find various signals in noisy detector data. This collaboration marks the beginning of a new era in the business of gravitational wave detection and first results on the efficiency of the different search methods are described in Refs. [41, 42].

Finding GW signals in the data of various detectors is the first and basic step toward GW detection and a lot of work goes into modeling detector noise curves and building efficient search algorithms. Once a detection has been successful, we are highly interested in the parameters of the source. This includes its position in the universe and we were able to show in Ref. [37] how numerical relativity can drastically improve the accuracy of locating the position for coalescing Super Massive Black Hole (SMBH) binaries with the space-based GW detector LISA. Not only using the inspiral signal but also the merger and ring-down of the binary, position estimates can be improved by an order of magnitude. For calculating the late inspiral and merger waveforms, numerical relativity simulations are essential because approximation techniques are not valid in these strong field regimes. This study again shows how important the above mentioned use of numerical relativity simulations is for detection of gravitational waves and data analysis.

A further effort towards improving GW detection concerned the continued generation of gravitational wave templates. It is crucial for this purpose to generate long waveforms which cover approximately 5-10 orbits of the binary inspiral phase prior to merger. We contributed waveforms to the construction of a new class of templates in Ref. [43] which for the first time includes spins. It was found that using non-spinning templates for the detection of GW from black hole binaries with non-precessing spins, we could miss up to 50% of the signals compared to using the new spinning templates. These templates therefore have the potential to dramatically improve GW detection.

In summary, simulations at HLRB2 led to publications [29–44] during the last grant period from 07/2008 – 10/2009.

7 Conclusion

Binary black hole mergers reveal a wealth of physics we can only study using numerical simulations on computer clusters. Finding solutions of the Einstein Equations in this strong field regime is currently not possible otherwise. The resulting informations on e.g. the kick velocity or spin of the final black hole are very important for astrophysical studies and predictions. Moreover, numerical relativity waveforms are essential for the growing field of gravitational wave detection. As a significantly larger fraction of energy is radiated during merger compared to the inspiral of two black holes, numerically calculated waveforms are used in the construction of highly accurate templates employed in matched filtering of the detectors' output.

The simulations themselves are run using highly sophisticated codes. The main difficulty lies in handling singularities at the position of the puncture, which represents a black hole. In addition, several length scales have to be resolved properly whereby the run time should not exceed the limit of a few weeks. This is why adaptive mesh refinement has to be used. In order to get accurate results, we so far use 6th order finite differencing in the inner regions of the grid. With the BAM code, we can nowadays use up to 256 processors with nearly no loss of scaling properties.

References

1. T.W. Baumgarte, S.A. Hughes, L. Rezzolla, S.L. Shapiro, M. Shibata, (1999)
2. M. Shibata, T. Nakamura, Phys. Rev. D **52**, 5428 (1995)
3. M. Alcubierre, B. Brügmann, P. Diener, M. Koppitz, D. Pollney, E. Seidel, R. Takahashi, Phys. Rev. D **67**, 084023 (2003)
4. C. Gundlach, J.M. Martin-Garcia, Class. Quantum Grav. **23**, S387 (2006)
5. M.A. Scheel, H.P. Pfeiffer, L. Lindblom, L.E. Kidder, O. Rinne, S.A. Teukolsky, Phys. Rev. D **74**, 104006 (2006)
6. H.P. Pfeiffer, D.A. Brown, L.E. Kidder, L. Lindblom, G. Lovelace, M. Scheel, Class. Quant. Grav. **24**, S59 (2007). DOI 10.1088/0264-9381/24/12/S06
7. S. Brandt, B. Brügmann, Phys. Rev. Lett. **78**(19), 3606 (1997)
8. R. Beig, N. O'Murchadha, Class. Quantum Grav. **11**, 419 (1994)
9. S. Dain, H. Friedrich, Comm. Math. Phys. **222**, 569 (2001)
10. B. Brügmann, J.A. González, M. Hannam, S. Husa, U. Sperhake, W. Tichy, Phys. Rev. D **77**, 024027 (2008)
11. M. Ansorg, B. Brügmann, W. Tichy, Phys. Rev. D **70**, 064011 (2004)
12. M. Alcubierre, B. Brügmann, Phys. Rev. D **63**, 104006 (2001)
13. J. Baker, B. Brügmann, M. Campanelli, C.O. Lousto, R. Takahashi, Phys. Rev. Lett. **87**, 121103 (2001)
14. M. Campanelli, C.O. Lousto, P. Marronetti, Y. Zlochower, Phys. Rev. Lett. **96**, 111101 (2006)
15. J.G. Baker, J. Centrella, D.I. Choi, M. Koppitz, J. van Meter, Phys. Rev. Lett. **96**, 111102 (2006)
16. M. Hannam, S. Husa, D. Pollney, B. Brügmann, N. O'Murchadha, Phys. Rev. Lett. **99**, 241102 (2007)
17. M. Hannam, S. Husa, N. O'Murchadha, B. Brügmann, J.A. González, U. Sperhake, J. Phys. Conf. Ser. **66**, 012047 (2007). DOI 10.1088/1742-6596/66/1/012047
18. J.D. Brown, Phys. Rev. D **77**, 044018 (2008)

19. B. Brügmann, Int. J. Mod. Phys. D **8**, 85 (1999)
20. B. Brügmann, W. Tichy, N. Jansen, Phys. Rev. Lett. **92**, 211101 (2004)
21. E. Schnetter, S.H. Hawley, I. Hawke, Class. Quantum Grav. **21**(6), 1465 (2004)
22. L. Lehner, S.L. Liebling, O. Reula, (2005)
23. W. Gropp, E. Lusk, N. Doss, A. Skjellum, Parallel Computing **22**(6), 789 (1996)
24. S. Husa, I. Hinder, C. Lechner, Comput. Phys. Comm. **174**, 983 (2006)
25. G. Allen, T. Goodale, J. Massó, E. Seidel, in *Proceedings of Eighth IEEE International Symposium on High Performance Distributed Computing, HPDC-8, Redondo Beach, 1999* (IEEE Press, 1999)
26. C.C. Toolkit. `http://www.cactuscode.org`
27. U. Sperhake, Phys. Rev. D **76**, 104015 (2007). DOI 10.1103/PhysRevD.76.104015
28. J.A. González, M.D. Hannam, U. Sperhake, B. Brügmann, S. Husa, Phys. Rev. Lett. **98**(23), 231101 (2007)
29. U. Sperhake, E. Berti, V. Cardoso, J.A. Gonzalez, B. Brügmann, M. Ansorg, Phys. Rev. D **78**, 064069 (2008). DOI 10.1103/PhysRevD.78.064069
30. E. Berti, V. Cardoso, J.A. Gonzalez, U. Sperhake, B. Brügmann, Class. Quant. Grav. **25**, 114035 (2008). DOI 10.1088/0264-9381/25/11/114035
31. A. Gopakumar, M. Hannam, S. Husa, B. Brügmann, Phys. Rev. D **78**, 064026 (2008). DOI 10.1103/PhysRevD.78.064026
32. M. Hannam, S. Husa, B. Brügmann, A. Gopakumar, Phys. Rev. D **78**, 104007 (2008). DOI 10.1103/PhysRevD.78.104007
33. T. Damour, A. Nagar, M. Hannam, S. Husa, B. Brügmann, Phys. Rev. D **78**, 044039 (2008). DOI 10.1103/PhysRevD.78.044039
34. M. Hannam, S. Husa, F. Ohme, B. Brügmann, N. O'Murchadha, Phys. Rev. D **78**, 064020 (2008). DOI
35. L. Gualtieri, E. Berti, V. Cardoso, U. Sperhake, Phys. Rev. D **78**, 044024 (2008). DOI 10.1103/PhysRevD.78.044024
36. U. Sperhake, V. Cardoso, F. Pretorius, E. Berti, J.A. Gonzalez, Phys. Rev. Lett. **101**, 161101 (2008). DOI 10.1103/PhysRevLett.101.161101
37. S. Babak, M. Hannam, S. Husa, B. Schutz, arXiv: 0806.1591 [gr-qc] (2008)
38. J.A. González, U. Sperhake, B. Brügmann, Phys. Rev. D **79**, 024006 (2009)
39. B. Walther, B. Brügmann, D. Müller, Phys. Rev. D **79**, 124040 (2009). DOI 10.1103/PhysRevD.79.124040
40. M. Hannam, S. Husa, J.G. Baker, M. Boyle, B. Brügmann, T. Chu, N. Dorband, F. Herrmann, I. Hinder, B.J. Kelly, L.E. Kidder, P. Laguna, K.D. Matthews, J.R. van Meter, H.P. Pfeiffer, D. Pollney, C. Reisswig, M.A. Scheel, D. Shoemaker, Phys. Rev. D **79**, 084025 (2009)
41. B. Aylott, et al., Class. Quant. Grav. **26**, 165008 (2009)
42. B. Aylott, et al., Class. Quant. Grav. **26**, 114008 (2009). DOI 10.1088/0264-9381/26/11/114008
43. P. Ajith, M. Hannam, S. Husa, Y. Chen, B. Brügmann, N. Dorband, D. Müller, F. Ohme, D. Pollney, C. Reisswig, L. Santamaria, J. Seiler, submitted to Phys. Rev. Lett., arXiv: 0909.2867 [gr-qc] (2009)
44. U. Sperhake, V. Cardoso, F. Pretorius, E. Berti, T. Hinderer, N. Yunes, Phys. Rev. Lett. **103**, 131102 (2009)
45. A. Buonanno, Y. Chen, T. Damour, Phys. Rev. D **74**, 104005 (2006)

Sheared Magnetic Field and Kelvin-Helmholtz Instability

Bocchi Matteo

Abstract We studied the evolution of the Kelvin Helmholtz Instability (KHI) in the context of plasma flows and jets stability. We focused on the presence of a magnetic field reversal along the interface of the jet with the ambient medium. The aim of the work is to compare this case with the known results from the uniform magnetic field case. Indeed, we performed magnetohydrodynamic simulations of a single shear layer and of a jet in slab geometry. In the single shear layer simulations we found significant differences in the growth rates and the behavior of the instability. In particular, the KHI presented higher growth rates and a more unstable nature in the regime of high magnetic fields. In the slab jet simulations, the magnetic field is amplified around the jet even for supersonic Mach numbers, contrary to previous believe.

1 Introduction

Although astrophysical jets are observed to propagate for long distances keeping a high degree of collimation, thus showing remarkable stability properties, it is still up to debate how exactly jets survive magnetohydrodynamic (MHD) instabilities (see for example the reviews by Hardee [7]). A recent comprehensive review on jet stability, based on numerical investigation, has been published by Keppens et al. ([9]). The Kelvin Helmholtz Instability (KHI) in particular has been extensively studied (see for example [2, 6, 14] and subsequent papers). We decided to focus on a particular model of the interface between the jet and the ambient medium embedded in a magnetic field reversal. Some jet launching models like the X-wind ([13]) predict the presence of antiparallel magnetic field lines above the accretion disk, a scenario confirmed by numerical simulations of the star-disk interaction ([3]). This configuration is also relevant for the Earth magnetospheric boundary layer and is

Bocchi Matteo
Imperial College, Blackett Laboratory, London SW7 2BW - United Kingdom
e-mail: m.bocchi@imperial.ac.uk

S. Wagner et al. (eds.), *High Performance Computing in Science and Engineering, Garching/Munich 2009*, DOI 10.1007/978-3-642-13872-0_34,

well studied (see for example [11]). A previous study on this physical setup, with emphasis on the numerical aspects is present in the literature ([8]). A linear stability analysis for the problem has been already carried out in a different magnetic regime ([4]), and in the same regime but for different parameters ([5]). The aim of our work is to study the evolution of the KHI growing on the interface layer, and to compare the results with the well known literature on the case with uniform magnetic field. An upcoming paper will contain all the details.

2 Model and Numerical Methods

This section shortly describes the model and the numerical methods used to obtain the results.

2.1 The Model

As initial conditions, we impose a flow initially only along the x direction (in a cartesian coordinate system) following a simple hyperbolic tangent profile:

$$v_x(y) = \frac{V_0}{2} \tanh\left(\frac{y}{a}\right), \tag{1}$$

where a is the width of the shear layer. For the sake of simplification, the magnetic field is reversing with the same hyperbolic tangent profile:

$$B_x(y) = -B_0 \tanh\left(\frac{y}{a}\right). \tag{2}$$

In the above equations, V_0 and B_0 are parameters that vary from case to case, so they will be specified later in Sect. 3. Both v_y and B_y are initially set to zero. The layer is assumed to be in pressure equilibrium. The density profile is chosen in order to initially impose a constant temperature over the whole domain.

2.2 Numerical Methods

We implemented the initial conditions described above in the Pluto code. Pluto ([10]) is a finite volume, shock capturing fluid dynamical code designed to integrate the (R)(M)HD (special Relativistic)(Magneto)HydroDynamic equations in conservative form; several solvers and integration options are available, as well as non-ideal modules, like cooling and resistivity (http://plutocode.to.astro.it). To prepare the MHD simulations, I used the linear code Ledaflow ([12]) to compute the growth rates of the instability.

3 Results

This section provides the results obtained both from the numerical simulations of a single shear layer as described in Sect. 2, and from simulations of two layers to form the so called slab-jet. For the single layer simulations, we found significant differences from the uniform case in term of growth rates, saturation and disruption levels. The slab-jet simulations presented an interesting behavior characterized by an enhancement of the magnetic field at the interface, an effect previously believed to be present only in subsonic jets (see [15]).

3.1 Single Layer Simulations

In this set of simulations V_0 is conveniently chosen in order to set a Mach number equal to 1. The magnetic field strength is controlled by the Alfvén Mach number (M_A), which we varied between 2.5 and 100.

3.1.1 Linear Results

Our results from the Ledaflow code are summarised in Fig. 1. Two important differences are present from the uniform magnetic field case:

1. The wave vector k corresponding to the maximum growth rate does not depend on the magnetic field strength, but is constant for all the values of M_A. In the uniform case, low values of M_A have the maximum shifted towards smaller wave vectors.
2. The growth rates for M_A smaller than 5 are sensibly higher than the uniform case.

3.1.2 Non-Linear Results

The previous results suggest that the character of the instability is different from the uniform case. This is confirmed by the examination of the saturation levels and by

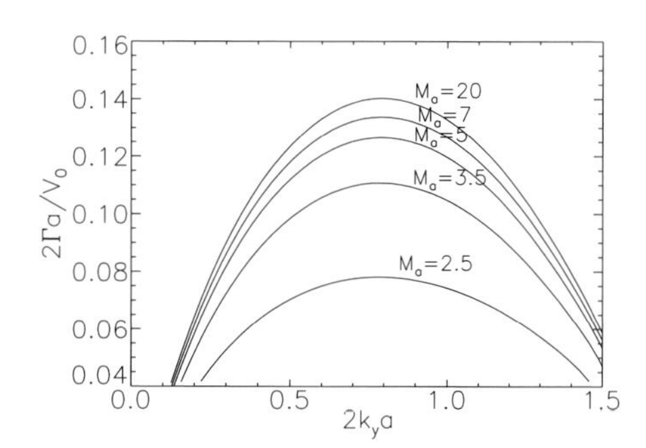

Fig. 1 Growth rates Γ as a function of the wave vector k for different values of the Alfvén Mach number

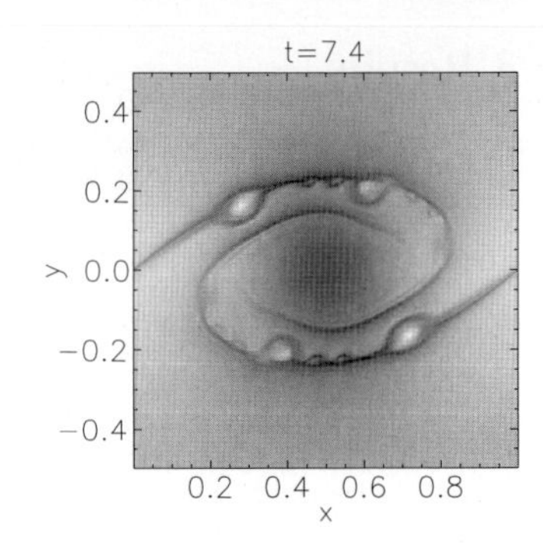

Fig. 2 Density plot of a single layer simulation at 7.4 time units. For this realization $M_A = 7$. The image is scaled with gray tones corresponding to the range $[0.7, 1.16]$

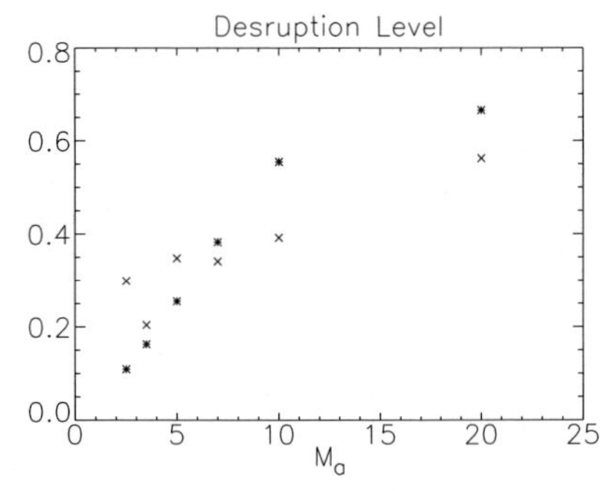

Fig. 3 Desruption levels of the KHI for different M_A. Stars represent the uniform magnetic field case, while crosses represent the reversed field case

the ranges of unstable M_A. Indeed, for uniform magnetic fields when $2 < M_A < 4$ the KHI is unstable, but stabilization occurs through non-linear processes ([1]). In the reversed field case, however, this effect is not present, and the shear is fully unstable for any value of $M_A > 2$. Fig. 2 represents the density structure at the time of saturation of the instability. Along the low density spiral structure typical of the KHI, a series of plasmoid structures can be noticed. These features proved to be magnetic islands driven by the KHI. The final state of the simulations is also a big magnetic island located at the centre of the domain, an outcome expected from the initially reversed magnetic field configuration.

In Fig. 3 we show the disruption level, measured as the quantity of kinetic energy that is taken away from the initial flow. For values of $M_A < 5$ the reversed field is more desruptive than the uniform case, but it is less desruptive for values of $M_A > 6$.

3.2 Slab-Jet Simulation

In the Slab-jet simulations, we set the Mach number to 3, while the Alfvén Mach number is set to 7. The simulations adopt a spatial approach to solve the problem. As a consequence we prescribed the plasma to flow into the domain from the left side of the x axis, and to flow out of the domain on the opposite side. We briefly summarise here our results:

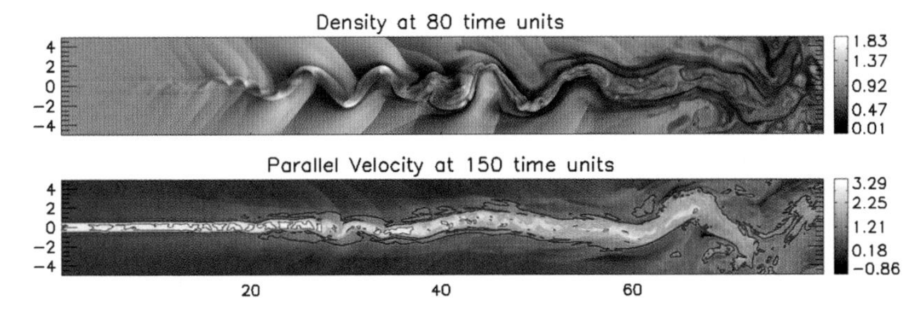

Fig. 4 **Upper panel**: Density plot at 80 time units. **Lower Panel**: Plot of the velocity parallel to the jet axis at 150 time units. The color contours represent the Mach number. Blue: $M = 1$, Green: $M = 2$, Red: $M = 3$

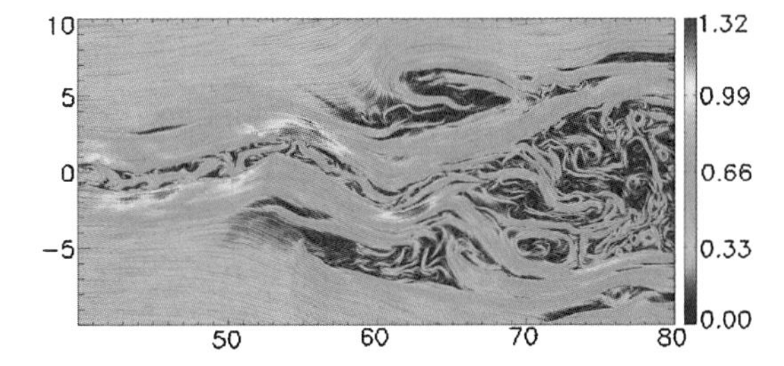

Fig. 5 Line integral convolution image of the magnetic field structure at 150 time units. The colors represent the magnitude of the magnetic field

1. Early in the instability evolution there is a transition from pinch to sinusoidal modes, and from short to long wavelengths (see Fig. 4, upper panel).
2. The jet is episodically disrupted by the instability, but the flow is later revived and keeps its coherence (see Fig. 4, lower panel).
3. The compression of the external medium due to the sinusoidal deformation of the jet causes a magnetic field enhancement propagated backwards towards the inflow boundary through Alfvén waves (see Fig. 5).
4. The inner part of the jet and the plasma left by the jet disruption display a turbulent magnetic structure (see Fig. 5).

These results point towards the conclusion that a mechanism of amplification of the magnetic field, generated by the instability itself, is operating on the beam and is responsible for the change of behavior of the instability (mode and wavelength). Such mechanism provides a partial stabilization of the flow, and is therefore an important piece of the puzzle in understanding the stability of astrophysical jets.

Acknowledgements The present work is supported by the European Communities Marie Curie Actions-Human Resources Mobility with the JETSET (Jet, Simulations, Experiments and Theory) Network under Contract MRTN-CT-2004-005592.

Part of the simulations for this work were performed on the supercomputers of the Leibnitz-Rechenzentrum (LRZ) in Muenchen, and on the JUMP computer of the John von Neumann Institute for Computing, Forschungszentrum Juelich.

References

1. Baty, H., Keppens, R., Comte, P. Physics of Plasmas. **10**, 4661 (2003)
2. Baty, H., Keppens, R., A&A **447**, 9 (2006)
3. Bessolaz, N., Zanni, C., Ferreira, J., Keppens, R., Bouvier, J., A&A. **478**, 155 (2008)
4. Dahlburg, R. B., Boncinelli, P., Einaudi, G., Phys. Plasmas **4**, 1213 (1997)
5. Dahlburg, R. B., Einaudi, G., Phys Plasmas **7**, 1356 (1999)
6. Frank, A., Jones, T. W., Ryu, D., Gaalaas, J. B., ApJ **460**, 777 (1996)
7. Hardee, P. E. ApJ. **664**, 26 (2007)
8. Keppens, R., Tóth, G., Westermann, R. H. J., Goedbloed, J. P. Jurnal of Plasma Physics. **61**, 1 (1999)
9. Keppens, R., Meliani, Z., Baty, H., van der Holst, B., Lecture Notes in Physics **791**, 179 (2009)
10. Mignone, A., Bodo, G., Massaglia, S., et al. ApJS. **170**, 228 (2007)
11. Nakamura, T. K. M., Fujimoto M., Otto, A. Geophys. Research Letters **33**, L14106 (2006)
12. Nijboer, R., van der Holst B., Poedts, S., Goedbloed, J. P. Comput. Phys. Commun. **101**, 39 (1999)
13. Shu, F., Najita, J., Ostriker, E., et al. ApJ. **429**, 781 (1994)
14. Trussoni, E., Protostellar Jets in Context, Astrophysics and Space Science Proceedings Series, 285 (2009)
15. Viallet, M., Baty, H. A&A. **473**, 1 (2007)

Solar Surface Flow Simulations at Ultra-High Resolution

Friedrich Kupka, Herbert J. Muthsam, Florian Zaussinger, Hannes Grimm-Strele, Natalie Happenhofer, Bernhard Löw-Baselli, Eva Mundprecht, and Christof Obertscheider

Abstract The dynamics of the surface layers of our Sun is of interest to solar and stellar astrophysics but also to the geophysical sciences. Since the solar surface dynamics operates on a vast range of temporal and spatial scales, numerical simulations of these processes require supercomputers for their study. This holds in particular, if shear driven turbulence generated by the vigorous convection present in the outer layers of our Sun is to be retrieved from within the calculation itself, rather than being modelled explicitly or implicitly. Our project SOLARSURF at HLRB-II aims at performing simulations of solar surface convection at ultra-high resolution to investigate the effects of shear driven turbulence and the generation of acoustic energy in the uppermost layers of the solar convection zone. The simulations are also performed as a reference for distinguishing hydrodynamical from magneto-hydrodynamical phenomena generally and particularly in their role in heating the solar chromosphere. We focus on magnetically quiescent solar regions with ordinary surface granulation. In this paper we compare some first results of our simulations with previous results obtained in 2D and at lower resolution in 3D and present a brief outlook on future project activities.

Friedrich Kupka
Universität Wien, Fakultät für Mathematik, Nordbergstr. 15, A-1090 Wien, Austria
(formerly at Max-Planck-Institute for Astrophysics, Garching)
e-mail: `Friedrich.Kupka@univie.ac.at`

Herbert J. Muthsam · Hannes Grimm-Strele · Natalie Happenhofer · Bernhard Löw-Baselli · Eva Mundprecht · Christof Obertscheider
Universität Wien, Fakultät für Mathematik, Nordbergstr. 15, A-1090 Wien, Austria
e-mail: `Herbert.Muthsam@univie.ac.at`

Florian Zaussinger
Max-Planck-Institute for Astrophysics, Karl-Schwarzschild Str. 1, D-85748 Garching bei München, Germany
e-mail: `fzaussinger@mpa-garching.mpg.de`

S. Wagner et al. (eds.), *High Performance Computing in Science and Engineering, Garching/Munich 2009*, DOI 10.1007/978-3-642-13872-0_35,

1 Introduction

Numerical simulations of solar surface convection at moderate resolution of 20 km to 50 km have reached a mature state (see, for instance, [14, 16, 17, 21, 22]). The goals of this previous research included the construction of realistic models of the solar surface layers assuming a hydrodynamical description (the fully compressible Navier-Stokes equations), a realistic treatment of radiative transfer, and realistic microphysics (equation of state, radiative absorption coefficients, etc.). These models require solving dynamical equations numerically on a grid covering a finite test volume over a finite amount of time. Typical volumes extend 3 to 4 Mm vertically and 3 to 12 Mm horizontally. Along lateral directions periodic boundary conditions are assumed. This is possible because the horizontal interaction length of the flow is small enough for the Sun (or similar stars) to fit within a box measuring just a view multiples of the length scales for which the kinetic energy has its maximum (i.e., the scales of the convectively driven up- and downflows observed at the solar surface). This "box-in-a-star" strategy allows studying the flow dynamics around those length scales and the astrophysical models based on it have passed important observational tests posed by detailed comparisons with spectroscopy and also helioseismology (see [12] for further details). With typical grid sizes of 100^3 to about 250^3 these simulations have become computationally quite affordable, at least for the solar case and moderate integration times of a few hours of solar time at previously mentioned resolutions.

More recent work at resolutions h of less than 10 km indicates [20] that such resolution or an even higher one is necessary to understand the dynamics of the magnetic field at the solar surface. An analysis of resolution requirements [7] easily demonstrates that for the case of solar surface convection $h \sim 5$ km or better is required to observe shear driven turbulence within the simulation. Even if one generally expects the large, energy carrying scales to appear laminar, as indeed found when looking at images of solar surface radiation intensity (see [7] for a discussion of the literature on this subject), the shear generated by the fast convective downflow channels has already been identified at moderate resolution to be a possible source of turbulence (see [18]). The consequences of this turbulence on the flow properties and also for the generation of acoustic energy and global solar oscillations can only be studied by performing numerical simulations of sufficiently small h. At $h \sim 10$ km we begin to see the counterparts of flow phenomena found in 2D simulations at ultra-high resolution ($h < 5$ km) [10]: shockfronts and pairs of vortex tubes and also features more akin to waves as well as local tornados. For the project SOLARSURF at HLRB-II we have thus suggested to use our modular Fortran90 code ANTARES [9, 10] and its capabilities of local grid refinement and high resolution numerics with minimized viscosity at a given grid scale to perform such ultra-high resolution simulations with $h \sim 5$ km in 3D. We focus on the case of ordinary granulation to see, if the same levels of turbulence can also be reached in a more quiescent region than in previous calculations [9] (we note here that typically about 95% of the solar surface are covered by ordinary granulation). In addition, we want to study the dependence of these results on boundary conditions to corroborate their robustness.

2 Current Status of the Simulations

Our ongoing computations have taken about 220 000 CPU hours thus far, with most runs using 196 CPU cores and up to 650 GB of RAM. Pure MPI parallelization based on spatial domain decomposition was used in these runs with a scaling efficiency of up to 80% (for purely hydrodynamical calculations it is linear, see [9]). These simulations can be regarded as the continuation and extension of work which we have commenced previously within the DEISA grant SOLEX. Earlier on only part of our calculations with the ANTARES code (see [9, 10]) have been non-grey. All the simulations performed at HLRB-II avoid the grey approximation for radiative transfer. Instead, frequency dependence of opacities is accounted for through a binning scheme where frequencies are grouped together for which emitted photons originate from the same layers ([8, 9, 11]). This greatly enhances the physical realism of the simulations and the possibility to eventually compare synthetic spectra computed from our simulations with high-resolution spectroscopic observations. At the same time, it makes the computations more expensive by a factor > 3. Let us describe now their setting more closely.

Into a simulation with a grid spacing of 22.2 km (horizontally) and 10.8 km (vertically) we plug a refined grid with spacings of 7.4 km and 5.4 km (again horizontally and vertically). Starting from a rather laminar state, grid refinement allows, in the course of time, a much more turbulent state to develop in the refined region. The effects are drastic as witnessed by Fig. 1 and 2 which correspond to the time at which grid refinement is first applied and to an instant of time later by 457 seconds, respectively. In both figures we view to the isosurface corresponding to a temperature of $T = 6000$ K (yellow surface). Very roughly, this corresponds to the depth in the Sun down to which one can peek when observing it in white light. The perspective for these figures, however, was chosen such that we look at this surface from below. Already the isosurface is considerably more rugged in Fig. 2. In particular, the developing true turbulence is evident considering the isolines of pressure in a plane somewhat below the solar surface (blue lines in both figures).

3 Results

A similar resolution has now been reached as previously only in 2D [10]. We have investigated here ordinary granulation as opposed to the case of an exploding granule discussed in [9]. Similarly, [20] have presented a high resolution 3D simulation for the *magnetic* case. They consider, however, a smaller domain not embedded into a larger volume and work at lower effective resolution.

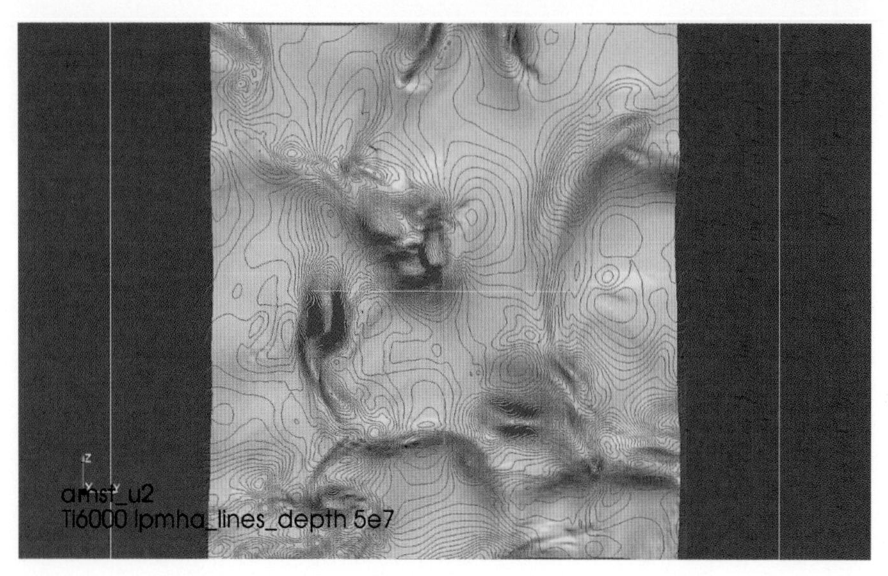

Fig. 1 Snapshot of the T = 6000 K isosurface of our ultra-high resolution simulation viewed from below at the initial time $t = 0$ sec. Only the grid-refined domain is shown. Blue isolines of pressure for a layer just underneath that isosurface indicate a rather smooth flow

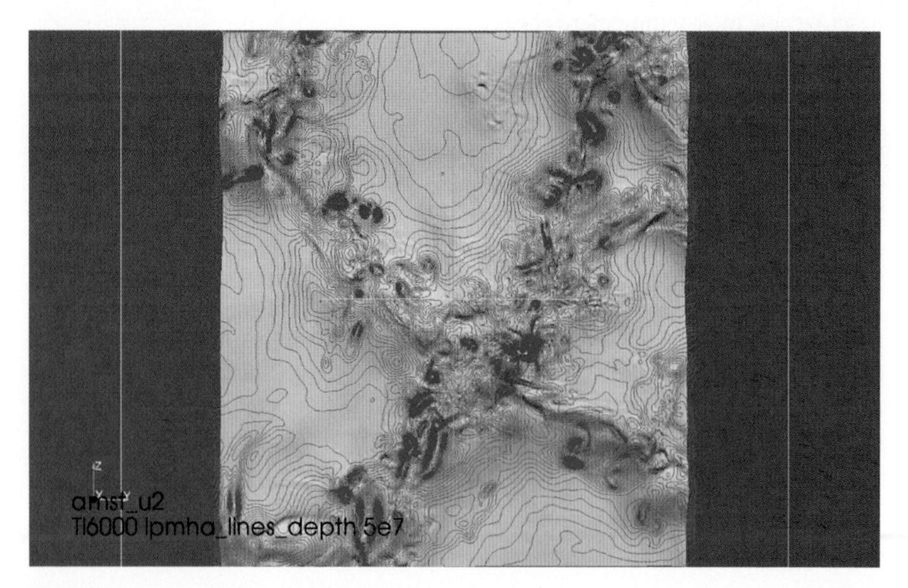

Fig. 2 The same quantities are shown here for the same simulation as in Fig. 1, but for a snapshot made $t = 457$ sec of solar time later. Note the considerably more rugged temperature isosurface and the much smaller structures visible in the isobaric lines compared to Fig. 1. This demonstrates the development of a fully turbulent flow (at least within the downflow regions)

Since ANTARES does not require a hyperviscosity to stabilize the simulation, optimum resolution can be obtained down to the grid scale [9]. As a consequence, we can observe the transition from a laminar looking flow into a fully turbulent flow, shrouded underneath the observable (solar) surface. Such a transition can be observed in 2D already at somewhat lower resolution (compare 2D simulations discussed in [9] at about 1/3 of the resolution used in the present calculations). This is due to the different dynamics of 2D and 3D simulations [7]. The present 3D simulations are sufficiently turbulent to make a comparison with the earlier results of [10] meaningful. They demonstrate that the vertical vortices present in 2D have a 3D counterpart appearing as vortex tubes in the simulations. These are already evident in simulations of slightly lower resolution (9.8 km horizontally and 7.1 km vertically) presented in [9] for the case of grey (frequency independent) radiative transfer and initial conditions containing an exploding granule.

In the present simulations, we have not only been able to infer the abundant occurrence of the vortex tubes even in normal granulation, but also the existence of a counterpart of the acoustic pulses we have seen in 2D earlier on [10]. In 2D, these pulses have clearly stood out and could not be missed once the effective resolution has been sufficient. In 3D, however, they have been difficult to identify previously [9], whereas now they can be seen as witnessed by Fig. 3–6. Note, however, that a movie makes it much easier to discern the pulses. For such a movie, see `http://www.univie.ac.at/acore/apuls3d.avi`.

There are substantial differences between the two- and the three-dimensional case regarding the acoustic pulses. In 2D, their fronts have been very sharp and their amplitude has been considerable. In 3D, the amplitudes are lower and the fronts less sharp. The latter fact may possibly be explained by a formation of the pulses not just at a point (as is largely true in 2D) but along a line or perhaps even a small surface. The issue is difficult to answer due to the difficulty of locating the spot of origin in 3D (see the chapter on visualization).

This does not necessarily mean that the acoustic pulses are in 3D less important than in 2D, or perhaps even completely unimportant. After all, although less conspicuous in 3D they are, at the same time, *much* more abundant than in the 2D case. In fact, on suitable spots in the atmosphere, for example, one can in the 3D calculation see new pulses moving up from deeper down arriving every few seconds. In 2D, they constitute more prominent but rarer events. Note furthermore that the energy such pulses carry has to spread only to a 1D manifold (a curve) in the 2D case, however to a 2D manifold (a surface) in 3D which makes for more dilution when moving away from the spot of origin.

4 Applications and Interpretation

When asking for the benefit of such ultra-high resolution simulations we have, on a general level, simply the basic physical description of the phenomena at hand as sketched in the previous paragraph.

Fig. 3 Snapshot of deviation of logarithmic pressure $\ln(p)$ (where p is the gas pressure in CGS units) from its horizontal average at a reference time t_0. Colour scale with values ranging from low (in black) to high (red / white). A black arrow in the lower right corner is used as a reference for an acoustic pulse moving downwards during the next few seconds ($t > t_0$)

Fig. 4 Same quantity as shown in Fig. 3, but at time $t = t_0 + 3.1$ sec. Note the acoustic pulse reaching the tip of the reference arrow (compare with Fig. 3)

Fig. 5 Snapshot of deviation of logarithmic pressure $\ln(p)$ (where p is the gas pressure in CGS units) from its horizontal average at time $t = t_0 + 6.2$ sec, but otherwise as in Fig. 3. Colour scale with values ranging from low (in black) to high (red / white). Note the acoustic pulse now underneath the reference arrow

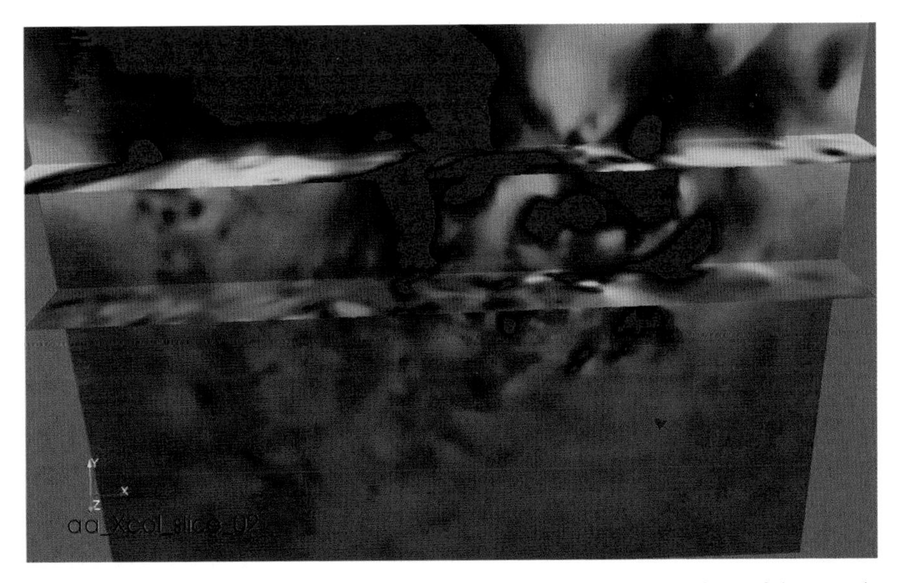

Fig. 6 Same quantity as shown in Fig. 5, but at time $t = t_0 + 9.3$ sec. The front of the acoustic pulse has now moved beyond the reference arrow

Let us turn to more specific questions, in particular to what the calculations performed within the LRZ project can be expected to contribute. One such issue is the excitation of p-modes in which the Sun is vibrating. These global oscillations of the Sun are, according to quite unanimous consensus, excited by convection near the surface, the subject of our computations [2, 5, 19]. Given that our truly turbulent convection is different in obvious aspects from the more laminar cases as we have seen, such models may prove essential in properly assessing details of the excitation mechanism. This holds true, more generally and beyond the present project, for asteroseismology [1, 3, 4, 15] which studies oscillations in other types of stars.

The region immediately above the photosphere (chromosphere) is heated by a mechanism still not uniquely identified. Either purely hydrodynamic waves (shocks, sound waves) or magnetohydrodynamic waves are being considered as candidates. Purely hydrodynamic waves are, in particular, considered in connection with the basal part of the chromospheric heating which is present in the quiet Sun and in comparable stars with little magnetic activity [6]. Ultra-high resolution simulations will help us to infer whether hydrodynamic waves are, from the standpoint of energetics, sufficient to account for the basal heating mechanism of the chromosphere. Models reaching higher into the solar atmosphere (which we intend to construct in the near future) will enable us to study the energetics of the outgoing hydrodynamic waves, above all the acoustic pulses which our calculations exhibit and which may possibly develop into shocks high in the atmosphere.

The ubiquity of the vortex tubes in the truly turbulent model hint to an intriguing possibility in connection with the solar magnetic field (even if ultra-high resolution simulations with magnetic field are outside the scope of the present project). There have long been suspicions that, besides the well-known solar dynamo operating near the base of the solar convection zone, there is a second dynamo situated near the surface (the solar surface dynamo) [13]. Recent simulations render such a dynamo viable from the modelling point of view [20]. Rapidly rotating plasma of high conductivity, as occurs in the present highly resolved runs, may quite conceivably play an essential role regarding the surface dynamo through its induction effects.

5 Visualization

Visualization and other types of interpretation of the data pose several challenges. First of all there is the sheer amount of data. Already the database for the high-resolution part of the simulations comprises $\sim$ 1TB as of now, and a multiple of that amount will be generated during the remaining time of the project. This is not so much a problem with respect to data transfer via the Internet from the LRZ to the home institutions (given the time the calculations themselves need to proceed), and also not regarding disk storage in the first place. The main challenge posed by the amount of data is rather related to the capabilities of the hardware and software which can be used for their analysis. While there are powerful tools readily available for this task, several problems remain which we briefly discuss below.

On the positive side, extensive software packages for 3D visualization and analysis can now be downloaded for free from the internet. We mention, in particular, Paraview (`http://www.paraview.org`), VAPOR (`http://www.vapor.ucar.edu`), and VISIT (see `https://wci.llnl.gov/codes/visit`). Each of these packages has its advantages.

While, in principle, it is feasible to work with several packages at a time, we are using mostly Paraview. One of its main advantages is that it allows addition of user defined filters in an easy manner. While Paraview comes along with a large number of basic functionalities as required in CFD, these are not sufficient for each application. This concerns already very simple operations such as subtracting the horizontal mean of some quantity and working with the outcome. After all, the horizontal mean is not likely to play a role in many technical applications, yet it is decisive in atmospheric sciences and in solar and stellar astrophysics. – A further advantage of Paraview is that it works increasingly reliably also in the parallel mode.

One problem when working with the current release (v3.6.1) and dealing with large data sets, at least if given in vtk-format, is that the whole data cube appears to be read in (which lasts dearly) even when working only with a subdomain, as is often the case. In this regard, a hierarchical data format as used in VAPOR would have its advantages. As a consequence of this and also of the time volume rendering takes on a modest machine with, say, 8 CPU-cores, interactive visualization has its limitations and one is often bound to rather produce and then inspect movies instead of working really interactively.

We are aware that the LRZ offers (among others) Paraview on a larger machine. If it should turn out to be feasible to have a user's version of Paraview compiled and installed there, using such dedicated visualization hardware would be a valuable option to perform fully interactive data analysis (we note that the inclusion of user defined filters requires Paraview to be compiled anew).

Even imagining the most perfect visualization system, some issues are likely to remain tough, though. We mention, for example, the task of locating the spot of formation and thus elucidating the cause of the acoustic pulses discussed in Sect. 3.

6 Next Project Stages and Conclusions

The next project stages emanate from open issues addressed within this paper.

First of all, we are extending the domain for the high-resolution region higher up into the (solar) atmosphere. This will allow the simulation data to be used for the computation of realistic absorption line spectra which can be compared to direct observations obtained for the Sun (these spectra can either be averaged over the whole visible surface patch or spectra mimicking a slit being placed over the region of interest, as also done by observers, can be computed for the study of individual structures, for example, the signatures of granular downflows). Furthermore, the energetics of acoustic pulses can then be investigated near the location where they actually are supposed to heat the chromosphere. A study of acoustic pulses deeper down may prove useful in the context of assessing their possible role in the excitation of solar p-modes.

We furthermore plan to insert secondary grid refinement at an interesting place within the primary refinement region. The resulting grid spacing will then be in the range of $3-4$ km, bringing us, in full three dimensions, close to the highest resolution ever achieved in 2D. Targets of investigation would include the vortex rolls. After all, as of now they measure only a few, at most $10-15$ grid points in diameter. For such a small number of grid points the flow is laminar (*on this small scale only!*). We need to know whether these tubes bend on smaller length scales than those resolved presently and perhaps excite new types of waves etc. once they are better resolved.

A resolution in the $3-4$ km range can be achieved only by using secondary grid refinement. Otherwise, the refined region would have to be unreasonably small or the computational demands would exceed present computing capacities. Proceeding hierarchically instead we expect to have only $\sim 50\%$ additional costs compared to the run we have described in the present paper (and thus, for instance, require ~ 1.1 TB of RAM for this new application), and yet attain such a much higher resolution in a domain of meaningful size.

In these calculations we will have to pay special attention to proper parallelization in order to avoid assigning too small volumes to MPI nodes. Already now we have, for such reasons, built an OpenMP capability into ANTARES such that in actual runs the two parallelization principles are used concurrently. For the simulations with double grid refinement extensions to the presently implemented OpenMP capabilities of ANTARES are necessary. We work on these and will then be able to fully exploit the opportunities provided by the LRZ machines also for those purposes.

Acknowledgements The numerical simulations described in this article have been performed at the HLRB-II of the LRZ in Munich. The discussion also benefits from earlier simulations performed as part of the DEISA project SOLEX. F. Kupka and H. Muthsam gratefully acknowledge support from the Austrian Science Foundation through projects P21742 and P20762, respectively.

References

1. Christensen-Dalsgaard, J.: Probing convection with helio- and asteroseismology. In: A. Giménez, E. Guinan, B. Montesinos (eds.) Theory and Tests of Convection in Stellar Structure, *ASP Conf. Ser.*, vol. 173, pp. 51–65. Astron. Soc. Pacific, San Francisco (1999)
2. Christensen-Dalsgaard, J.: Physics of Solar-Like Oscillations. Solar Physics **220**, 137–168 (2004)
3. Gautschy, A., Saio, H.: Stellar pulsations across the HR diagram: Part 1. Ann. Rev. Astron. Astrophys. **33**, 75–114 (1995)
4. Gautschy, A., Saio, H.: Stellar pulsations across the HR diagram: Part 2. Ann. Rev. Astron. Astrophys. **34**, 551–606 (1996)
5. Goldreich, P., Keeley, D.: Solar seismology. II - The stochastic excitation of the solar p-modes by turbulent convection. Astrophys. Jour. **212**, 243–251 (1977)

6. Judge, P., Carpenter, K.: On Chromospheric Heating Mechanisms of "Basal Flux" Stars. Astrophys. Jour. **494**, 828–839 (1998)
7. Kupka, F.: Turbulent Convection and Numerical Simulations in Solar and Stellar Astrophysics. In: W. Hillebrandt, F. Kupka (eds.) Interdisciplinary Aspects of Turbulence, *Lecture Notes in Physics*, vol. 756, pp. 49–105. Springer Verlag, Berlin (2009)
8. Ludwig, H.G., Jordan, S., Steffen, M.: Numerical simulations of convection at the surface of a ZZ Ceti white dwarf. Astron. & Astrophys. **284**, 105–117 (1994)
9. Muthsam, H., Kupka, F., Löw-Baselli, B., Obertscheider, C., Langer, M., Lenz, P.: ANTARES – A Numerical Tool for Astrophysical RESearch with applications to solar granulation. New Astronomy **15**, 460–475 (2010)
10. Muthsam, H., Löw-Baselli, B., Obertscheider, C., Langer, M., Lenz, P., Kupka, F.: High-resolution models of solar granulation: the two-dimensional case. Mon. Not. Roy. Astron. Soc. **380**, 1335–1340 (2007)
11. Nordlund, Å.: Numerical simulations of the solar granulation. I - Basic equations and methods. Astron. & Astrophys. **107**, 1–10 (1982)
12. Nordlund, Å., Stein, R., Asplund, M.: Solar Surface Convection. Living Rev. Solar Phys. **6** (2009)
13. Petrovay, K., Szakaly, G.: The origin of intranetwork fields: a small-scale solar dynamo. Astron. & Astrophys. **274**, 543–554 (1993)
14. Robinson, F., Demarque, P., Li, L., Sofia, S., Kim, Y.C., Chan, K., Guenther, D.: Three-dimensional convection simulations of the outer layers of the sun using realistic physics. Mon. Not. Roy. Astron. Soc. **340**, 923–936 (2003)
15. Samadi, R., Belkacem, K., Goupil, M.J., Dupret, M.A., Kupka, F.: Modeling the excitation of acoustic modes in α Centauri A. Astron. & Astrophys. **489**, 291–299 (2008)
16. Steffen, M.: Radiative hydrodynamics models of stellar convection. In: F. Kupka, I. Roxburgh, K. Chan (eds.) Convection in Astrophysics, *IAU Symposium*, vol. 239, pp. 36–43. Cambridge Univ. Press, Cambridge (2007)
17. Stein, R., Nordlund, Å.: Simulations of Solar Granulation. I. General Properties. Astrophys. Jour. **499**, 914–933 (1998)
18. Stein, R., Nordlund, Å.: Realistic Solar Convection Simulations. Solar Physics **192**, 91–108 (2000)
19. Stein, R., Nordlund, Å.: Solar Oscillations and Convection. II. Excitation of Radial Oscillations. Astrophys. Jour. **546**, 585–603 (2001)
20. Vögler, A., Schüssler, M.: A solar surface dynamo. Astron. & Astrophys. **465**, L43–L46 (2007)
21. Vögler, A., Shelyag, S., Schüssler, M., Cattaneo, F., Emonet, T., Linde, T.: Simulations of magneto-convection in the solar photosphere. Equations, methods, and results of the MURaM code. Astron. & Astrophys. **429**, 335–351 (2005)
22. Wedemeyer, S., Freytag, B., Steffen, M., Ludwig, H.G., Holweger, H.: Numerical simulation of the three-dimensional structure and dynamics of the non-magnetic solar chromosphere. Astron. & Astrophys. **414**, 1121–1137 (2004)

Part V
High-Energy Physics

Part V
High-Energy Physics

Lattice Investigation of Nucleon Structure: Towards the Physical Point

Y. Nakamura, G. Schierholz, H. Stüben, and J. Zanotti — QCDSF Collaboration
(HU Berlin, ZIB Berlin, DESY Hamburg and Zeuthen, U Edinburgh, U Leipzig, U Liverpool, UNAM Mexico City, IHEP Protvino, ITEP Moscow, TU Munich, U Regensburg)

Abstract QCD is the underlying quantum field theory describing the strong interactions, and lattice QCD is the technique to solve it. Large scale computing resources afford the opportunity to answer key questions regarding the structure and spectrum of hadrons and systems of hadrons. By considering new simulations at the physical quark masses, we review the recent progress that has been made in this exciting area by the QCDSF collaboration.

1 Introduction

Lattice QCD provides the only *ab initio* method for solving QCD, and for acquiring a quantitative description of the physics of hadrons and nuclear forces. Since the cost of full QCD computations grows with a large inverse power of the quark mass, initial calculations were restricted to relatively heavy quarks. However, in order to capture the physics of quarks and gluons in captivity, and reach the needed accuracy requested by the experiments, simulations at physical quark masses, on large volumes and at small lattice spacings are required.

Y. Nakamura
Institut für Theoretische Physik, Universität Regensburg, 93040 Regensburg, Germany
e-mail: `Yoshifumi.Nakamura@desy.de`

G. Schierholz
Deutsches Elektronen-Synchrotron DESY, 22603 Hamburg, Germany
e-mail: `Gerrit.Schierholz@desy.de`

H. Stüben
Konrad-Zuse-Zentrum für Informationstechnik Berlin, 14195 Berlin, Germany
e-mail: `stueben@zib.de`

J. Zanotti
School of Physics and Astronomy, University of Edinburgh, Edinburgh EH9 3JZ, UK
e-mail: `jzanotti@ph.ed.ac.uk`

S. Wagner et al. (eds.), *High Performance Computing in Science and Engineering, Garching/Munich 2009*, DOI 10.1007/978-3-642-13872-0_36,

Starting in 2001, we have performed extensive lattice QCD calculations at LRZ, first on the Hitachi SR8000 and more recently on the Altix 4700. Due to continuous improvements of the algorithm, and the advent of Teraflop/s-scale computing facilities, we are now able to perform simulations at the physical quark masses on moderately large lattices and cut-offs.

We do not have space to give a full account of what we have achieved so far [1–11]. Rather, we shall concentrate on results of some very recent simulations with pion masses reaching down to the physical value. A key feature of the new results is that we are now beginning to achieve quantitative control over the approach to the chiral limit.

2 The Simulation

The lattice QCD action, and QCDSF specific features, of the Hybrid Monte Carlo (HMC) algorithm have been reported in [12]. From the numerical point of view, the task is to solve the system of linear equations

$$b = Q^\dagger Q x \tag{1}$$

iteratively, where Q is a sparse $(12L^3T) \times (12L^3T)$ complex matrix, with L (T) being the number of lattice points in the spatial (temporal) direction. In a typical run more than 80% of the total time is spent in multiplication of vectors with the matrix Q. To improve the performance, this was implemented in assembler. A further means of improvement is provided by performing part of the calculations in single precision, and finally to implement such mixed-precision calculation in assembler. This can be done without loss of precision. The calculation requires the exchange of ghost cells of the input vector. In order to scale QCD programs to high numbers of processers, a communication network with high bandwidth and low latency is needed. Although the Altix 4700 has an excellent network, the loss due to communication is typically of the order of 50%.

Sustained performance results for a $32^3 \times 64$ lattice are plotted in Fig. 1 and given in Table 1. Remember that the sustained performance on a single core for a 4^4 sublattice was $\approx 75\%$ of the peak performance [12]. Figure 1 indicates that both

Table 1 Sustained performance of the conjugate gradient solver on the Altix 4700 for three different implementations: the Fortran code in double precision, the assembler in double precision, and the Fortran code in single precision. The numbers given refer to Mflop/s per core

Lattice	# cores	Fortran (double) [Mflop/s]	Assembler (double) [Mflop/s]	Fortran (single) [Mflop/s]
$32^3 \times 64$	512	1195	1619	1822
$32^3 \times 64$	1024	1395	1409	1817
$32^3 \times 64$	2048	996	841	1165

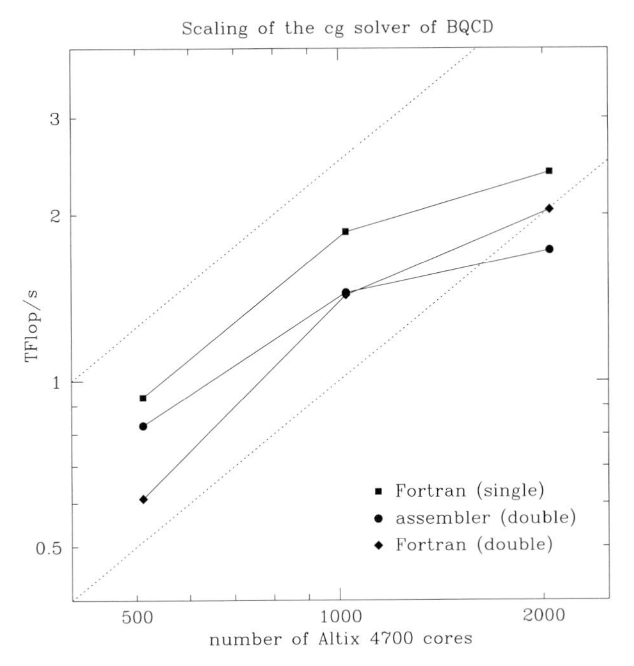

Fig. 1 The overall performance in Tflop/s as a function of cores for the Fortran code in double precision, the assembler in double precision, as well as the Fortran code in single precision

the Fortran code in double precision and the Fortran code in single precision scale almost linearly with the number of cores, when going from 512 to 2048 cores. Remarkably, the Fortran code in double precision outperforms the assembler in double precision on 2048 cores. This indicates that the MPI calls, that are handled in the assembler part, lead in this case to an inefficient communication pattern. The overall sustained performance of the Fortran code in single precision on the $32^3 \times 64$ lattice on 1024 nodes is 1.9 Tflop/s, which corresponds to 28 % of the peak performance. Currently, we are working on an assembler code for single precision. That should increase the sustained performance to $\approx 40\%$ for large volume lattice calculations.

3 Nucleon Structure at Light Quark Masses

The internal quark and gluon structure of the nucleon is the defining component of hadron physics. Hence, the goal of lattice QCD is precision calculation of fundamental quantities characterizing the nucleon. These include form factors and moments of parton densities, helicity and transversity distributions, and moments of structure functions.

Before we go into detail, let us look at the nucleon mass m_N, which enters all calculations reported here. In Fig. 2 we show m_N as a function of the pion mass. The

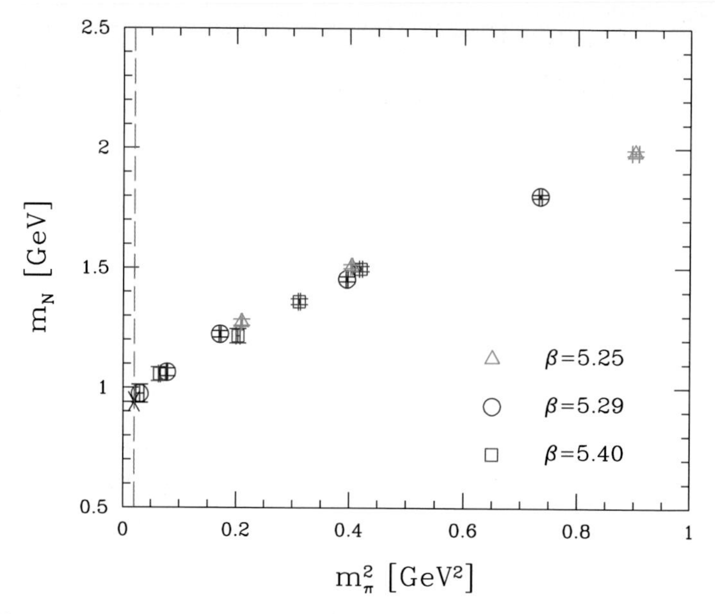

Fig. 2 The nucleon mass as a function of the pion mass squared for various couplings. The lattice spacings vary between $a = 0.079$ ($\beta = 5.25$) and $a = 0.067$ fm ($\beta = 5.40$). The star ($*$) indicates the physical nucleon mass

pion mass and the masses of the light quarks u and d are related by the Gell-Mann–Oakes–Renner relation, $m_\pi^2 = B(m_u + m_d)$. We may therefore express the dependence of m_N on the light quark masses in terms of its dependence on the corresponding pion mass. The new lattice data extend right to the physical point now, so that no unqualified extrapolation to the chiral limit is needed anymore. The functional dependence of m_N, and other hadron observables as well, on m_π is still of great interest though, because it involves a variety of low-energy constants of the underlying chiral effective theory. We use the nucleon mass to fix the scale. It is customary to express the scale in terms of the parameter r_0, which is the distance, at which the slope of the static interquark potential assumes a certain value. We find $r_0 = 0.467(15)$ fm. We do not see any cut-off dependence, which we expect to be of $O(a^2)$ at most for the improved fermion action used in the calculations. Between our lowest and highest couplings, $\beta = 5.25$ and 5.40, the lattice spacing a^2 varies by 40%.

3.1 Electromagnetic Form Factors

On the lattice we determine the form factors $F_1(q^2)$ and $F_2(q^2)$ by calculating the nucleon matrix element of the electromagnetic current j_μ:

$$\langle p', s' | j_\mu | p, s \rangle = \bar{u}(p', s') \left[\gamma_\mu F_1(q^2) + i\sigma_{\mu\nu} \frac{q_\nu}{2m_N} F_2(q^2) \right] u(p, s), \tag{2}$$

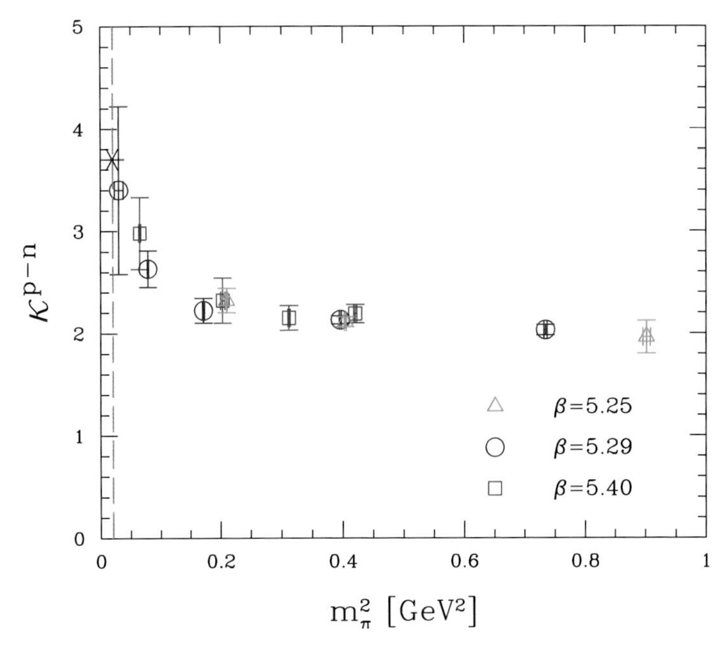

Fig. 3 The anomalous magnetic moment of proton – neutron as a function of the pion mass squared. The star ($*$) denotes the experimental result

where $u(p,s)$ is a Dirac spinor with momentum p and polarization s, $q = p' - p$ is the momentum transfer, and m_N the nucleon mass. Such a lattice calculation allows for the extraction of phenomenologically interesting quantities, including the anomalous magnetic moment $\kappa = F_2(0)$ and electric and magnetic charge radii r_1 and r_2, respectively, defined by the low-q^2 expansion

$$F_{1,2}(q^2) = 1 + r_{1,2}^2 q^2/6 + O(q^4), \quad q^2 \leq 0. \tag{3}$$

In Fig. 3 we show the anomalous magnetic moment of proton minus neutron $\kappa^{p\quad n} \equiv \kappa^p - \kappa^n$, *i.e.* the isovector anomalous magnetic moment, and in Fig. 3 we plot the isovector magnetic radius $\langle r_2^2 \rangle^{p-n}$, both as a function of the pion mass. The anomalous magnetic moment is seen to increase towards the experimental value in a relatively narrow interval of the pion mass, $m_\pi \lesssim 300\,\text{MeV}$. Such a behavior is predicted by chiral perturbation theory [13]. A similar effect is observed for $\langle r_2^2 \rangle^{p-n}$. Previous lattice studies of the Pauli form factor $F_2(q^2)$ found the radius $\langle r_2^2 \rangle^{p-n}$ to be a factor ≈ 3 smaller than experiment, and to show little variation as a function of m_π^2 for pion masses $m_\pi \gtrsim 300\,\text{MeV}$. The new results below $m_\pi \lesssim 300\,\text{MeV}$ are showing signs of upward curvature. The chiral effective theory predicts the magnetic radius $\langle r_2^2 \rangle^{p-n}$ to diverge like $\propto 1/m_\pi$ in the chiral limit [13], which appears to be supported by the lattice results. The solid curve in Fig. 4 shows a fit of chiral perturbation theory to the lattice data. Of course, at such light quark masses finite

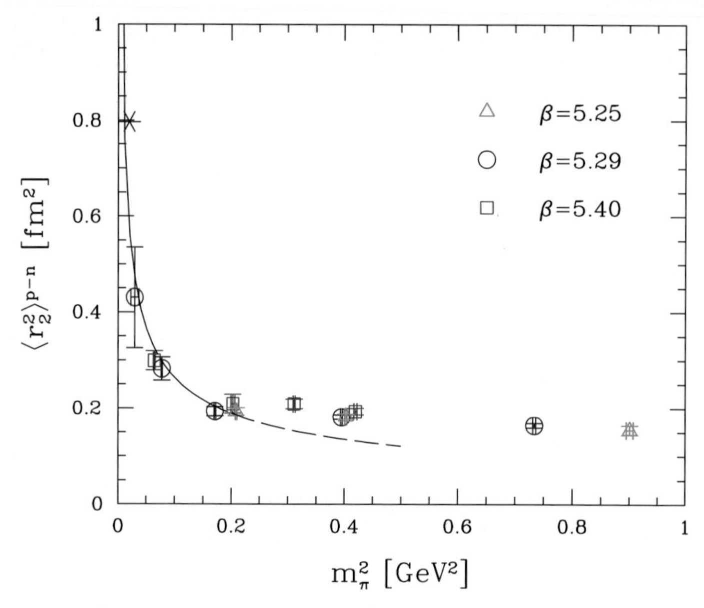

Fig. 4 The radius r_2^2 of the Pauli form factor $F_2(q^2)$ as a function of the pion mass squared. The solid line is a fit to the predictions of the chiral effective theory. The star ($*$) indicates the phenomenological radius

volume effects are expected to suppress the nucleon charge radii [14]. In the future calculations on larger volumes will have to be done to settle this issue.

3.2 Moments of Structure Functions

The axial coupling constant g_A of the nucleon is an important quantity, as it governs neutron β-decay and also provides a quantitative measure of spontaneous chiral symmetry breaking. It corresponds to the zeroth moment of the isovector helicity dependent quark distribution function,

$$g_A \equiv \Delta u - \Delta d = \int_0^1 dx \left[\Delta u(x) - \Delta d(x)\right], \tag{4}$$

and has been studied theoretically and experimentally for many years. Its value, $g_A = 1.2695(29)$, is known to very high accuracy. Hence it is an important quantity to study on the lattice, and serves as useful yardstick for lattice simulations of nucleon structure.

The axial charge is determined by the forward nucleon matrix element of the axial current A_μ:

$$\langle p', s' | A_\mu^{u-d} | p, s \rangle = 2 g_A s_\mu, \tag{5}$$

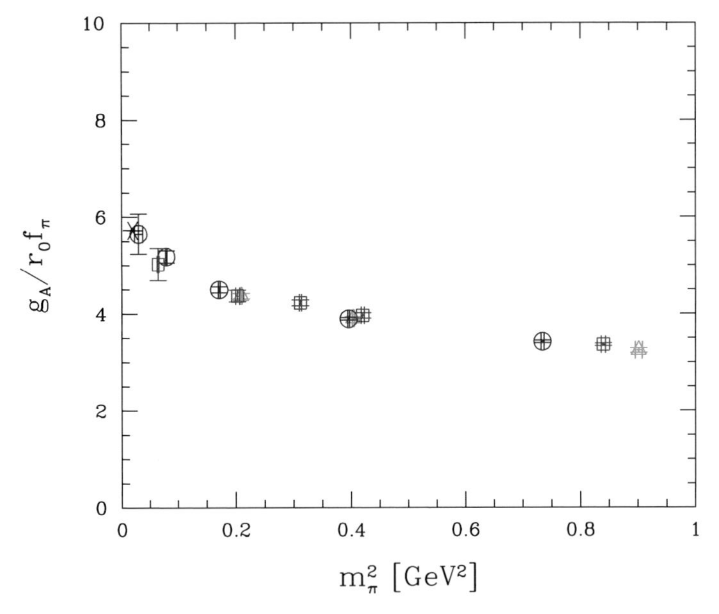

Fig. 5 The axial coupling g_A of the nucleon divided by the pion decay constant f_π in dimensionless units as a function of the pion mass squared. The star (∗) marks the experimental result

where s_μ is the spin vector with $s^2 = -m_N^2$. Unlike the electromagnetic current, the axial current has to be renormalized [15], because it is not conserved. In Fig. 5 we show $g_A/r_0 f_\pi$ as a function of the pion mass. The reason for plotting this ratio is that the renormalization constant of the axial current drops out and, furthermore, that the finite size corrections cancel to leading order [16, 17]. In fact, g_A has been studied in-depth for many years [16, 18, 19], and has been shown to suffer from large finite size effects. In Fig. 5 we have extended the earlier results from [16] to smaller pion masses. We find that our new data extrapolate nicely to the experimental value.

Quark distribution functions, unpolarized and polarized, are an important ingredient for LHC phenomenology. Lattice studies of the first moment of the unpolarized isovector quark distribution function

$$\langle x \rangle_{u-d} = \int_0^1 dx x \left[u(x) - d(x) \right] \tag{6}$$

are notorious in that all lattice results at heavy quark masses exhibit an almost constant behavior in quark mass towards the chiral limit, and are a factor of ≈ 1.5 larger than the phenomenological value $\langle x \rangle_{u-d}^{\overline{MS}} = 0.1754(41)$ [20]. Figure 6 shows the latest QCDSF results including new results below $m_\pi < 300\,\text{MeV}$. While the results above $m_\pi \gtrsim 300\,\text{MeV}$ show the same constant behavior seen in many lattice simulations, the new results are beginning to bend down to the phenomenological value. What is particularly interesting is the behavior at fixed pion mass $m_\pi \approx 300\,\text{MeV}$, but with three different lattice volumes, $L = 24, 32$ and 40. Here we observe that the

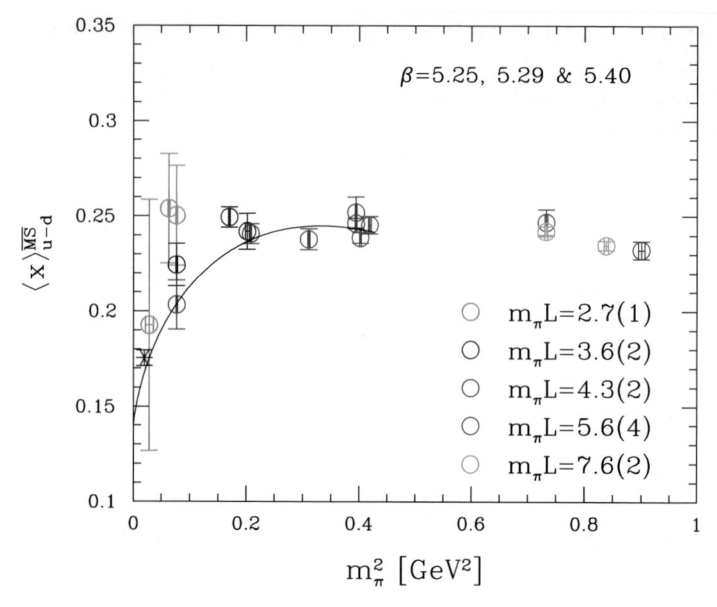

Fig. 6 The first moment of the isovector quark distribution function $\langle x \rangle^{u-d}$ in the $\overline{MS}$ scheme at the scale $\mu = 2\,\text{GeV}$ as a function of the pion mass squared. The star ($*$) marks the experimental result

result from the smallest volume continues the trend at heavier pion masses, while the results from the larger volumes lie significantly lower. This is not unexpected, as the dramatic downward curvature has been shown to be only achievable on larger volumes [21]. For this reason, we expect that even after a large increase in statistics, the current point at the physical pion mass with $L = 40$ will still lie somewhat higher than phenomenological determinations. A further simulation at this pion mass, but on $L = 64$ lattices, is currently under way to further test this behavior. The solid curve in Fig. 6 is a fit of the predictions of chiral perturbation theory [22] to the lattice data with $m_\pi L > 4$.

A similar behavior is seen in the first moment of the spin dependent quark distribution function

$$\langle x \rangle_{\Delta u - \Delta d} = \int_0^1 dx x \left[\Delta u(x) - \Delta d(x)\right], \tag{7}$$

but with $\langle x \rangle^{\overline{MS}}_{\Delta u - \Delta d} \approx 1.2 \, \langle x \rangle^{\overline{MS}}_{u-d}$.

4 Conclusion and Outlook

Lattice QCD, like other areas of science, is limited by computer performance. In the long run it will demand sustained speeds of Petaflop/s or more. This presents a greater challenge than the recent step to Teraflop/s.

We have presented some latest results on nucleon structure from the QCDSF collaboration. These calculations are now becoming available at physical pion masses, so direct comparison with experiment, *i.e.* without extrapolation to the physical point, will soon be possible. However, as we have seen, finite size effects are starting to become a serious issue. This is not surprising, given the fact that the nucleon is about $\approx 1.8\,\mathrm{fm}$ in diameter. As a result, we are now planning new simulations on volumes of $(5\,\mathrm{fm})^3$, in order to minimize finite volume corrections.

The long-term goal is to extend the calculations to lattice spacings $a \lesssim 0.05\,\mathrm{fm}$, which corresponds to a momentum cut-off of $\gtrsim 4\,\mathrm{GeV}$, so as to accommodate charmed quarks on the lattice. Charm physics is one of the main research topics at FAIR/GSI. To reach this goal, Petaflop/s-scale computing facilities are required.

The Altix 4700 so far has been very well suited for our tasks. The performance scales almost linearly with the number of cores, up to 2048 cores. Currently we have achieved a sustained performance of 28 % of the peak performance of the machine on 1024 cores. We hope to increase that number to 40%, once we have implemented the assembler code for single precision matrix times vector multiplication.

Acknowledgements The Hitachi SR8000 and the Altix 4700, as well as customer support of LRZ contributed considerably to the success of our calculations. We are grateful to LRZ for continuous and generous support of our projects.

References

1. M. Göckeler, P. Hägler, R. Horsley, Y. Nakamura, D. Pleiter, P.E.L. Rakow, A. Schäfer, G. Schierholz, H. Stüben and J.M. Zanotti, Phys. Rev. Lett. **98** (2007) 222001 [arXiv:hep-lat/0612032].
2. D. Brömmel, M. Diehl, M. Göckeler, P. Hägler, R. Horsley, Y. Nakamura, D. Pleiter, P.E.L. Rakow, A. Schäfer, G. Schierholz, H. Stüben and J.M. Zanotti, Phys. Rev. Lett. **101** (2008) 122001 [arXiv:0708.2249 [hep-lat]].
3. M. Göckeler, R. Horsley, T. Kaltenbrunner, Y. Nakamura, D. Pleiter, P.E.L. Rakow, A. Schäfer, G. Schierholz, H. Stüben, N. Warkentin and J.M. Zanotti, Phys. Rev. Lett. **101** (2008) 112002 [arXiv:0804.1877 [hep-lat]].
4. D. Brömmel, M. Göckeler, P. Hägler, R. Horsley, Y. Nakamura, M. Ohtani, D. Pleiter, P.E.L. Rakow, A. Schäfer, G. Schierholz, W. Schroers, H. Stüben and J.M. Zanotti, Eur. Phys. J. ST **162** (2008) 63 [arXiv:0804.4706 [hep-lat]].
5. M. Göckeler, R. Horsley, T. Kaltenbrunner, Y. Nakamura, D. Pleiter, P.E.L. Rakow, A. Schäfer, G. Schierholz, H. Stüben, N. Warkentin and J.M. Zanotti, Nucl. Phys. B **812** (2009) 205 [arXiv:0810.3762 [hep-lat]].
6. M. Göckeler, R. Horsley, Y. Nakamura, D. Pleiter, P. E. L. Rakow, G. Schierholz and J. Zanotti, arXiv:0810.5337 [hep-lat].
7. V.M. Braun, M. Göckeler, R. Horsley, T. Kaltenbrunner, Y. Nakamura, D. Pleiter, P.E.L. Rakow, A. Schäfer, G. Schierholz, H. Stüben, N. Warkentin and J.M. Zanotti, Phys. Rev. D **79** (2009) 034504 [arXiv:0811.2712 [hep-lat]].
8. N. Cundy, M. Göckeler, R. Horsley, T. Kaltenbrunner, A. D. Kennedy, Y. Nakamura, H. Perlt, D. Pleiter, P. E. L. Rakow, A. Schäfer, G. Schierholz, A. Schiller, H. Stüben and J. M. Zanotti, Phys. Rev. D **79** (2009) 094507 [arXiv:0901.3302 [hep-lat]].

9. V.M. Braun, M. Göckeler, R. Horsley, T. Kaltenbrunner, A. Lenz, Y. Nakamura, D. Pleiter, P.E.L. Rakow, J. Rohrwild, A. Schäfer, G. Schierholz, H. Stüben, N. Warkentin and J.M. Zanotti, Phys. Rev. Lett. **103** (2009) 072001 [arXiv:0902.3087 [hep-ph]].
10. W. Bietenholz, N. Cundy, M. Göckeler, R. Horsley, H. Perlt, D. Pleiter, P.E.L. Rakow, G. Schierholz, A. Schiller, T. Streuer and J.M. Zanotti, arXiv:0910.2437 [hep-lat].
11. W. Bietenholz, V. Bornyakov, N. Cundy, M. Göckeler, R. Horsley, A.D. Kennedy, Y. Nakamura, H. Perlt, D. Pleiter, P.E.L. Rakow, A. Schäfer, G. Schierholz, A. Schiller, H. Stüben and J.M. Zanotti, arXiv:0910.2963 [hep-lat].
12. Y. Nakamura, G. Schierholz, T. Streuer and H. Stüben, in *High Performance Computing in Science and Engineering, Transactions of the Third Joint HLRB and KONWIHR Status and Result Workshop, Dec. 3–4, 2007, Leibniz Supercomputing Centre, Garching, Germany*, p. 627, Springer (Berlin, Heidelberg, New York, 2009)
13. M. Göckeler, T. R. Hemmert, R. Horsley, D. Pleiter, P. E. L. Rakow, A. Schäfer and G. Schierholz, Phys. Rev. D **71** (2005) 034508 [arXiv:hep-lat/0303019].
14. T. R. Hemmert, PoS **LAT2009** (2009), 146.
15. G. Martinelli, C. Pittori, C. T. Sachrajda, M. Testa and A. Vladikas, Nucl. Phys. B **445** (1995) 81 [arXiv:hep-lat/9411010]; M. Göckeler, R. Horsley, H. Oelrich, H. Perlt, D. Petters, P.E.L. Rakow, A. Schäfer, G. Schierholz and A. Schiller, Nucl. Phys. B **544** (1999) 699 [arXiv:hep-lat/9807044].
16. A. Ali Khan, M. Göckeler, P. Hägler, T.R. Hemmert, R. Horsley, D. Pleiter, P.E.L. Rakow, A. Schäfer, G. Schierholz, T. Wollenweber and J.M. Zanotti, Phys. Rev. D **74** (2006) 094508 [arXiv:hep-lat/0603028].
17. G. Colangelo, S. Dürr and C. Haefeli, Nucl. Phys. B **721** (2005) 136 [arXiv:hep-lat/0503014].
18. R.G. Edwards, G.T. Fleming, P. Hägler, J.W. Negele, K. Orginos, A. Pochinsky, D.B. Renner, D.G. Richards and W. Schroers, Phys. Rev. Lett. **96** (2006) 052001 [arXiv:hep-lat/0510062].
19. T. Yamazaki, Y. Aoki, T. Blum, H. W. Lin, M.F. Lin, S. Ohta, S. Sasaki, R.J. Tweedie and J.M. Zanotti, Phys. Rev. Lett. **100** (2008) 171602 [arXiv:0801.4016 [hep-lat]].
20. S. Alekhin, J. Blümlein, S. Klein and S. Moch, arXiv:0908.2766 [hep-ph].
21. W. Detmold, W. Melnitchouk and A. W. Thomas, Mod. Phys. Lett. A **18** (2003) 2681 [arXiv:hep-lat/0310003]; W. Detmold and C. J. D. Lin, Phys. Rev. D **71** (2005) 054510 [arXiv:hep-lat/0501007].
22. M. Dorati, T.A. Gail and T.R. Hemmert, Nucl. Phys. A **798** (2008) 96 [arXiv: nucl-th/0703073].

Dynamical Lattice QCD with Ginsparg-Wilson-Type Fermions

T. Burch, C. Ehmann, G. Engel, C. Gattringer, M. Göckeler, C. Hagen, P. Hasenfratz, D. Hierl, C. B. Lang, M. Limmer, V. Maillart, T. Maurer, D. Mohler, F. Niedermayer, A. Schäfer, and S. Solbrig

Abstract Lattice Quantum Chromodynamics (LQCD) is the most versatile and powerful method to investigate non-perturbative effects in strong interaction physics. In view of the much improved statistical accuracy the control of systematic uncertainties became the most important issue in recent years. The most severe problem is the implementation of chiral symmetry. The BGR collaboration works with two different Dirac operators, both having good chiral properties but are suited best for the analysis of different aspects of QCD: Chirally Improved (CI) fermions are especially well suited for the investigation of excited hadrons while Fixed-Point (FP) fermions are ideal for studies in the so-called ε- and δ-regime.

T. Burch · C. Ehmann · M. Göckeler · C. Hagen · D. Hierl · T. Maurer · A. Schäfer · S. Solbrig
Institut für Theoretische Physik, Universität Regensburg, D-93040 Regensburg, Germany
e-mail: tommy.burch@physik.uni-regensburg.de
e-mail: christian.ehmann@physik.uni-regensburg.de
e-mail: meinulf.goeckeler@physik.uni-regensburg.de
e-mail: christian.hagen@physik.uni-regensburg.de
e-mail: dieter.hierl@physik.uni-regensburg.de
e-mail: thilo.maurer@physik.uni-regensburg.de
e-mail: andreas.schaefer@physik.uni-regensburg.de
e-mail: stefan.solbrig@physik.uni-regensburg.de

G. Engel · C. Gattringer · C. B. Lang · M. Limmer · D. Mohler
Institut für Physik, FB Theoretische Physik, Universität Graz, A-8010 Graz, Austria
e-mail: georg.engel@uni-graz.at
e-mail: christof.gattringer@uni-graz.at
e-mail: christian.lang@uni-graz.at
e-mail: markus.limmer@uni-graz.at
e-mail: daniel.mohler@uni-graz.at

P. Hasenfratz · V. Maillart · F. Niedermayer
Institut für Theoretische Physik, Universität Bern, CH-3012 Bern, Switzerland
e-mail: hasenfra@itp.unibe.ch
e-mail: vidushi@itp.unibe.ch
e-mail: niederma@itp.unibe.ch

S. Wagner et al. (eds.), *High Performance Computing in Science and Engineering, Garching/Munich 2009*, DOI 10.1007/978-3-642-13872-0_37,

1 The Research Field and Our Research Strategy

The fundamental elements of Quantum Chromodynamics (QCD) are gluons and quarks and their interaction. QCD is a relativistic, strongly coupled and highly non-linear theory, which results in an extremely rich phenomenology, much of which is only poorly understood. This fact is of far-reaching importance for the whole field of particle physics, as the uncertainties due to our lack of understanding of QCD effects result in uncertainties which severely constrain our ability to identify potential signs of New Physics as such. However, QCD and hadron physics are not only a necessary tool in our search for New Physics, but also fascinating research areas in their own right. To illustrate this statement let us mention that recent experiments have discovered whole new families of particles called X, Y, and Z resonances of which some seem to be predominantly four-quark bound-states rather than the usual mesons. The large future European research laboratory FAIR, under construction at GSI, Darmstadt will focus (among other aims) at the investigation of such and other novel hadronic states. Excited hadrons, which decay very fast into hadron ground states, are collectively called resonances. Understanding the properties of such resonances is closely related to the more general problem of understanding the mechanisms leading to confinement, i.e., the fact that quarks and gluons exist only in color-neutral bound states. These (and other) studies depend crucially on LQCD calculations.

Because of its fundamental importance LQCD became a very competitive field dominated by ever more and ever larger international collaborations with very significant computing resources (typically tens or even more than hundred TFlops). We followed this trend by the establishment of a Sonderforschungsbereich in the summer of 2008 (SFB/TRR-55, together with Wuppertal university, spokes university is Regensburg) which we continuously try to strengthen and extend. The BGR (Bern-Graz-Regensburg) collaboration is an important element of this SFB and a prime example for the kind of collaboration intended by the so-called D-A-CH-Agreement (D: Germany; A: Austria; CH-Switzerland). In addition a European Marie-Curie network (STRONGnet, spokes university Regensburg) was recently established, which strengthens the European perspective of our lattice work.

Physicswise BGR has specialized on two highly relevant research fields in which we are especially competitive, both being intimately related to chiral symmetry which reflects the fundamental left-right symmetry of QCD. While chiral symmetry is broken for hadronic ground states (with the pion being the associated Goldstone boson) it seems to get effectively restored both for highly excited resonances at zero temperature and for high temperature. Moreover, deconfinement and chiral symmetry restoration are intimately connected in a manner which is not understood. Onviously, in order to understand what is really going on, one has to perform lattice simulations in which chiral symmetry is not massively destroyed by lattice artefacts, i.e., one has to perform simulations with chiral quarks. Such simulations are, however, extremely costly and so far all large volume simulations done world-wide had to make rather significant approximations. The chirally improved (CI) fermions used by the Graz-Regensburg group are also based on an approximation, which allows us to reduce the computational costs very substantially. Domain-wall fermions, which approach exact chiral symmetry in the limit of infinite extension of an artificial additional time coordinate, and which are the choice of a large US-centered collaboration, are, e.g. at least an order of magnitude more expensive (but also a better approximation to chirality for small residual mass). In addition the approximation of chirality reached in practical domain-wall fermion simulations is open for dispute [1]. CI fermions are used primarily to investigate excited hadrons [2–5, 10, 11].

Bern is since decades an established center of chiral perturbation theory (ChPT), i.e., the low-energy limit of QCD. Therefore, it was natural for the Bern-Regensburg group to concentrate on the determination of the low energy constants (LECs) which parametrize ChPT. This can be done on very small volumes which are either hypercubic (ε-regime) or elongated in the time direction (δ-regime) but requires very good chiral symmetry, which we realize with Fixed-Point (FP) fermions, see [6–9, 12–19].

2 Investigations with Dynamical Chirally Improved Quarks

While our early investigations still used quarks in the quenched approximation (i.e., neglecting the creation of virtual quark-antiquark pairs) we simulated in the last year full $N_f = 2$ QCD.

2.1 Generation of Ensembles

The Hybrid Monte Carlo algorithm was implemented for the CI action, suitable for larger lattices and parallelization on the ALTIX. Several improvements (like the Hasenbusch preconditioning, single/double precision implementations of the conjugate gradient inverter, better memory organization, optimized inter-node transfer) led to a significant speed-up (by a factor of more than 2). The currently running program executes and performs nicely on 128 nodes for a lattice size $16^3 \times 32$.

In the first allocation period we generated three sets of configurations with three different pion mass values (see Table 1 and [21]). The parameters were chosen such that the lattice spacing (derived from the Sommer parameter [22] through fits to the lattice static potential) was close to 0.15 fm. This constitutes a mass dependent scheme and we plot most of our masses in units of the nucleon mass. Presently we perform studies with several different bare quark masses to be able to extrapolate to the chiral limit.

Table 1 gives the parameters for those configurations which we analysed so far. In our analysis, in addition to the fully dynamical results where valence and sea quark masses agree, we also obtained results for partially quenched cases, i.e., for differing valence and sea quark masses. These we then compare with the fully quenched results to verify the expected agreement with the heavy quark situation.

Table 1 Parameters of the used configurations for lattice size $16^3 \times 32$. The numbers given refer to configurations analyzed (i.e., every 5th of the actually determined configurations, since the autocorrelation time was below 5 for the quantities measured). The AWI (Axial Ward Identity) mass has been obtained as usual from the ratio of axial vector over pseudoscalar correlators. The values in MeV correspond to a Sommer parameter of 0.48 fm. Details of the analysis are given in [21]

set	β_{LW}	m_0	# analyzed conf.s	a [fm]	m_{AWI} [MeV]	m_π [MeV]	$am_\pi L$
A	4.70	-0.05	100	0.1507(17)	42.8(4)	526(7)	6.4
B	4.65	-0.06	200	0.1500(12)	34.1(2)	469(4)	5.8
C	4.58	-0.077	200	0.1440(12)	15.3(3)	318(5)	3.8

The gauge configurations generated with dynamical quarks on the ALTIX were transferred to Regensburg and Graz. There the quark propagators and subsequently the hadron propagators were computed on local clusters, and then analyzed. In constructing the quark and hadron source and sink operators we profited from our experience with the variational method and our quenched results.

2.2 Ground State and Excited Hadron Masses

In Fig. 1 we compare the nucleon mass (assuming a Sommer parameter of 0.48 fm; filled symbols) with the partially quenched results (open symbols). Fig. 2 shows the rho-meson mass. The used parameters and more results can be found, e.g., in [20, 21].

The configurations with light dynamical quarks were also used for a study of mesons and baryons with a heavy quark [23–25]. The heavy quark is approximated by a static propagator, appropriate for the b quark on our lattices ($1/a \sim 1-2$ GeV). We computed meson and baryon mass splittings and ratios of meson decay constants (e.g., f_{B_s}/f_B and $f_{B_s'}/f_{B_s}$).

In order to isolate the resonance states we study the cross correlation matrix of several lattice operators O_i with the correct quantum numbers and solve the generalized eigenvalue problem for the correlation matrix

$$\begin{aligned} C(t)_{ij} &= \langle O_i(t)\overline{O}_j(0)\rangle = \sum_n \langle 0|O_i|n\rangle\langle n|O_j^\dagger|0\rangle \mathrm{e}^{-tM_n}\,, \\ C(t)\,\mathbf{v}_i &= \lambda_i(t)\,C(t_0)\,\mathbf{v}_i\,, \\ \lambda_i(t) &\propto \mathrm{e}^{-tM_i}\left(1+\mathscr{O}\left(\mathrm{e}^{-t\Delta M_i}\right)\right)\,. \end{aligned} \tag{1}$$

giving the masses M_i and the eigenvectors $\mathbf{v}_i$ which allow for conclusions on the properties of the internal wave function.

In the last year we developed and used also derivative quark sources [26], which substantially improved the results for some resonances. Another fascinating topic being studied is the scale dependence of the composition of the vector meson ρ and its relationship to the ρ wave function [27, 28]. In generally the resulst for excited mesons are now really good [29], Fig. 3 and 4 just illustrate this for two cases.

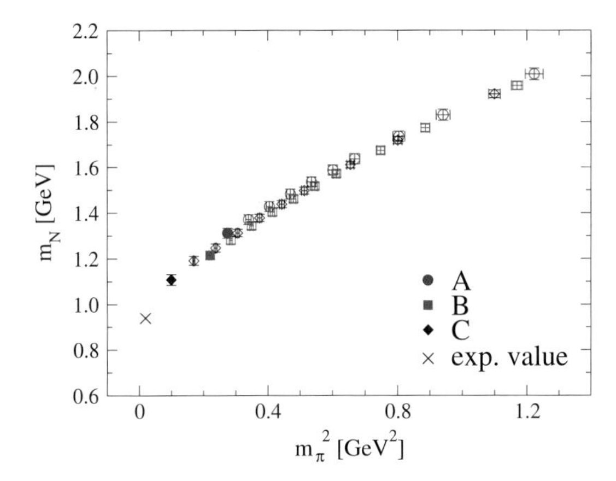

Fig. 1 Results for dynamical CI quarks: Nucleon (positive parity) [21]

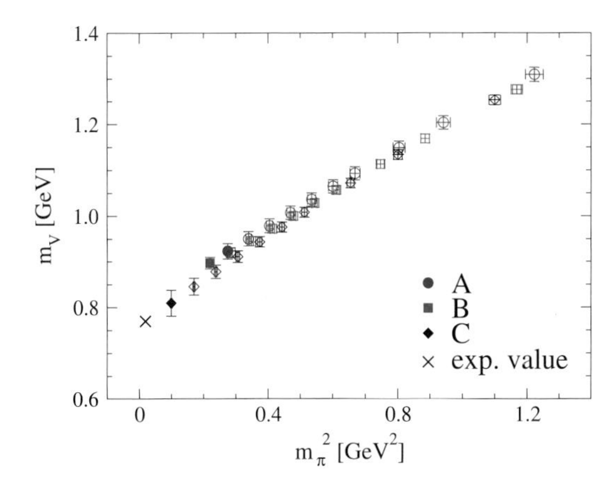

Fig. 2 Results for dynamical CI quarks: $\rho(772)$ meson [21]

We also searched for the much discussed 'tetraquarks', which are bound states of two quarks and two antiquarks. We got rather stable results [30] but did not find sign for a new state beyond the expected ground and continuum states.

For excited baryons we are still plagued by artefacts due to the small volume we simulate, a typical example is shown in Fig. 5. While the ground state Δ baryon comes out relatively well, the excited state lies much to high.

2.3 Filtering and Topological Charge Densities

QCD has non-trivial topological properties which seem to play a major role for confinement and chiral symmetry breaking, though nearly all aspects of this role are still hotly disputed. In principle LQCD could provide crucial hints, because

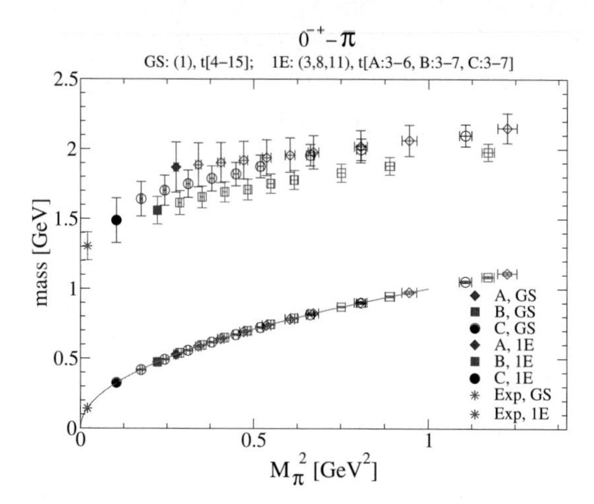

Fig. 3 The ground state pion (GS) and its first excitation (1E); open symbols denote partially quenched data, the three full symbols denote the dynamical data for the three ensembles A, B and C [29]

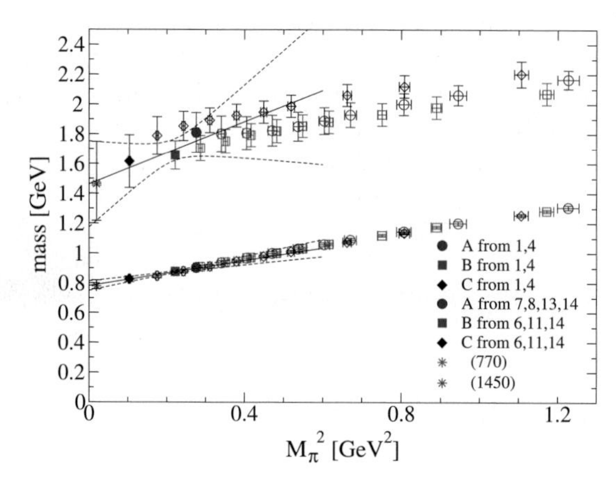

Fig. 4 Ground state and excited state in the 1^{--} channel. The open symbols detote partially quenched results [29]

it provides the microscopic densities of topological charge. However, topological properties have to be filtered out from the overwhelming noise due to quantum fluctuations and it seems that they are so delicate that the results depends very strongly on filtering and lattice artefacts. We have developed a special power-law analysis which allowed us to reduce these artefacts very substantially by combining several filtering techniques. In the last year we have applied this technique to the new, dynamical configurations. Fig. 6 shows some results [31]. All plots show the characteristic power-exponent as a function of the number of filtering steps performed, both for the quenched and dynamical CI configurations. If the widely discussed idea that the vacuum can be adequately represented by an instanton gas were correct the

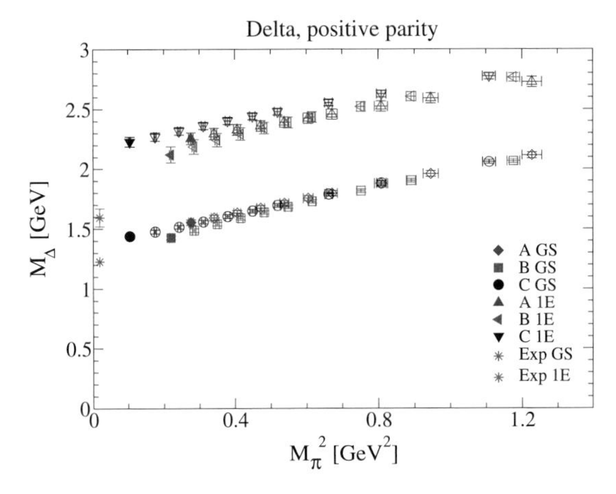

Fig. 5 L.h.s.: the Δ ground and excited state [29]

data should approach the two asymptotic values represented by the horizontal lines at the rigth side of the plot. The results show that both smearing and Laplace filtering alone would suggest that this hypothesis is wrong, while the combined filtering method shows that such a conclusion would be the consequence of artefacts and that in reality this hypotheses could be rather close to the truth.

3 2+1 Flavour QCD Results Obtained with the Fixed-Point Action

3.1 The ε-regime for Different Topological Sectors

In the ε-regime the Goldstone bosons feel the boundaries strongly. ChPT provides a systematic way to calculate these finite size corrections in terms of some low-energy constants, like the pion decay constant F, the quark masses and the chiral condensate Σ. Comparing the simulations with the analytic prediction offers a way to fix these important low-energy constants. A nice additional feature of the ε-regime is that Random Matrix Theory (RMT) [32] predicts additional microscopic observables. Among others, RMT connects the distribution of the low-lying eigenvalues of the Dirac operator to the chiral condensate.

In the ε-regime we use the so-called Fixed Point (FP) Dirac operator D_{FP} which should have excellent chiral properties, as has our parametrized fixed point action [33]. In [34] we devised a partially global update with three nested accept/reject steps which can reach small quark masses even on coarse lattices. All the three steps are preconditioned. Pieces of the quark determinant are switched on gradually in the order of their computational expenses. We reduce the ultraviolet fluctuations

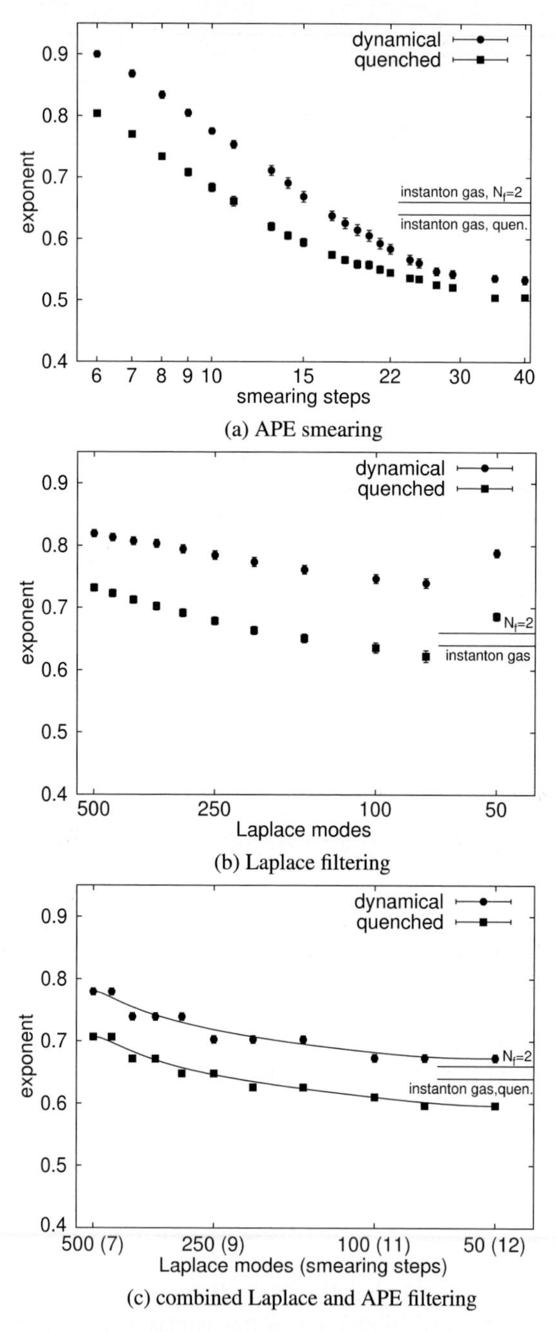

Fig. 6 The characteristic exponent ξ for the clusters found by APE-smearing (a), Laplace filtering (b) and for a matched filtering of both methods (c). The exponents predicted by the dilute instanton gas have been included in all plots [31]

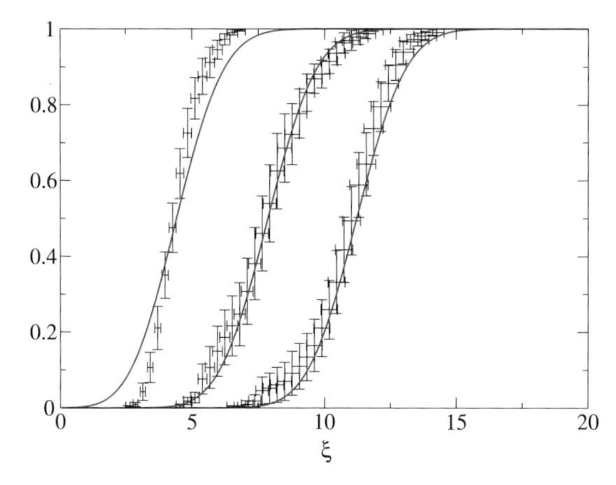

Fig. 7 Cumulative distribution of the first three eigenvalues in the zero topological sector. Here $\xi = \lambda \Sigma V$ where λ is the corresponding eigenvalue, Σ is the quark condensate and V is the lattice volume [39]

by calculating the trace of $D^n_{\rm FP}, n = 1,...,4$. The infrared part is tamed by calculating the ~ 100 lowest-lying modes and subtracting their contribution. The determinant of the reduced and subtracted $D_{\rm FP}$ is calculated stochastically.

In order to make the code more effective on the ALTIX, we replaced the polynomial expansion by a rational approximation in the stochastic estimator of the strange quark, which brought 30% performance gain in this part [35]. We made the parallelization significantly finer by dividing all the Euclidean directions in small pieces [36] and by rewriting the code on the gauge update significantly [37].

We fixed the lattice spacing from the Sommer parameter [22] $r_0 = 0.49\,\text{fm}$ and found $a = 0.129(5)\,\text{fm}$. Our box has approximately a size of $(1.6\,\text{fm})^4$. RMT predicts the probability distribution of the k-th low-lying eigenvalue of the Dirac operator in the different topological sectors. The only fitting parameter is the bare quark condensate Σ. The first three eigenvalue distributions in the zero topological sector are shown in Fig. 7. Correcting the result by the leading finite size effect (which is 12%) gives [B4,B5] $\Sigma_\infty^{1/3} = 0.255(3)(9)\,\text{GeV}$.

RMT makes some interesting predictions in the ε-regime which can not be easily approached in ChPT. On the other hand, for many important quantities, ChPT gives a systematic procedure for higher order calculations. These predictions contain additional low-energy constants which enter in many different ChPT results. Comparing the predictions with controlled numerical results gives a new possibility to fix these constants. In [40] we measured different two point correlation functions and compared them with the 1-loop ChPT results [15]. The covariant pseudoscalar (PP) correlator is proportional to Σ. Fig. 8 shows the ChPT fit to the measured curve predicting $\Sigma_\infty^{1/3} = 0.240(5)(9)\,\text{GeV}$.

The quark condensate Σ above is the bare value and needs renormalization. The non-perturbative RI/MOM technique is relatively simple and broadly used in the community. In [41] we have found, however, that the method has potentially large

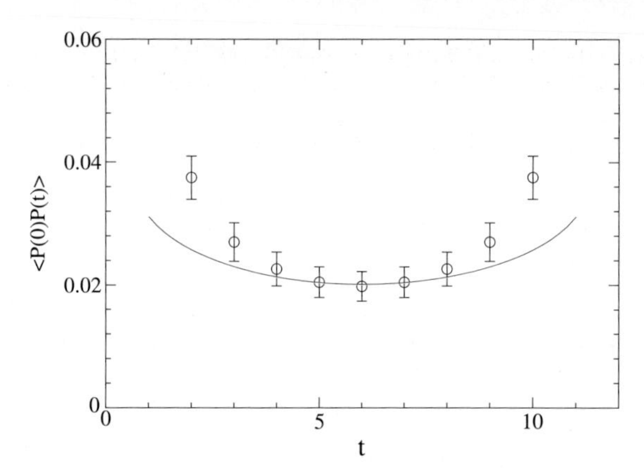

Fig. 8 The correlator of the covariant pseudoscalar density. Its height is proportional to Σ^2. Note the the points close to the boundaries have significant contributions from non-Goldstone excitations [39]

special cut-off effects. These cut-off effects can be (and must be) eliminated. Similar observations have been made in the literature before, however, it seems that this improvement practically has not been used in the published works.

Our numerical work on the δ regime has only started recently, as we were waiting the analytic result for the expected energy levels predicted by Next-to-Next-to-Leading order in chiral perturbation theory. This analytic work is described in [44].

Acknowledgements This work was supported by DFG, BMBF, the Fonds zur Förderung der Wissenschaftlichen Forschung in Österreich (FWF DK W1203-N08) and the Schweizerischer Nationalfonds.

References

1. K. Jansen, Plenary talk presented at the XXVI International Symposium on Lattice Field Theory, July 14-19, 2008, Williamsburg, arXiv: 0810.5634.
2. T. Burch, C. Gattringer, L. Ya. Glozman, R. Kleindl, C. B. Lang, A. Schäfer, Phys. Rev D **70**, 0154502, 2004.
3. T. Burch, C. Hagen, D. Hierl, A. Schäfer, C. Gattringer, L. Ya. Glozman, C. B. Lang, POS LAT2005, 097, 2006.
4. T. Burch, C. Gattringer, L. Ya. Glozman, C. Hagen, C. B. Lang, A. Schäfer, Phys. Rev D **73**, 094505, 2006.
5. T. Burch, C. Gattringer, L. Ya. Glozman, C. Hagen, C. B. Lang, A. Schäfer, Phys. Rev D **74**, 014504, 2006.
6. P. Hasenfratz, Nucl. Phys. Proc. Suppl. **63**, 53, 1998.
7. P. Hasenfratz, V. Laliena and F. Niedermayer, Phys. Lett. B **427**, 125, 1998.
8. P. Hasenfratz, Nucl. Phys. B **525**, 401, 1998.
9. F. Niedermayer, Nucl. Phys. Proc. Suppl. **73**, 105, 1999.
10. C. Gattringer, Phy. Rev. D **63**, 114501, 2001.

11. C. Gattringer, I. Hip and C. B. Lang, Nucl. Phys. B **597**, 451, 2001.
12. P. Hasenfratz, S. Hauswirth, K. Holland, T. Jörg, F. Niedermayer, U. Wenger, Nucl. Phys. B **643**, 280, 2002.
13. J. Gasser and H. Leutwyler, Phys. Lett. B **184**, 83, 1987, **188**, 477, 1987.
14. J. Gasser and H. Leutwyler, Nucl. Phys. B **307**, 763, 1988.
15. P. Hasenfratz and H. Leutwyler, Nucl. Phys. B **343**, 241, 1990.
16. F. C. Hansen, Nucl. Phys. B **345**, 685, 1990.
17. S. Weinberg, Physica A **96**, 327, 1979.
18. J. Gasser and H. Leutwyler, Phys. Lett. B **125**, 321, 325, 1983.
19. J. Gasser and H. Leutwyler, Ann. Phys. (N.Y.). **158**, 142, 1984.
20. C. Gattringer, C. B. Lang, M. Limmer, T. Maurer, D. Mohler and A. Schäfer, PoS (LATTICE 2009) (2009) 093; [arXiv:0809.4514 [hep-lat]].
21. C. Gattringer, C. Hagen, C. B. Lang, M. Limmer, D. Mohler and A. Schäfer Phys. Rev. D **79** (2009) 054501; arXiv:0812.1681 [hep-lat].
22. R. Sommer, Nucl. Phys. B **411**, 839, 1994.
23. T. Burch, D. Chakrabarti, C. Hagen, C. B. Lang, M. Limmer, T. Maurer and A. Schäfer PoS LATTICE 2007 (2008) 091; arXiv:0709.3708.
24. T. Burch, C. Hagen, C.B. Lang, M. Limmer, and A. Schäfer Phys. Rev. D **79** (2009) 014504; arXiv:0809.1103 [hep-lat].
25. T. Burch, C. Hagen, C. B. Lang, M. Limmer, and A. Schäfer PoS LATTICE 2008 (2009) 110, [arXiv:0809.3923 [hep-lat]].
26. C. Gattringer, L. Y. Glozman, C. B. Lang, D. Mohler and S. Prelovsek, Phys. Rev. D **78** (2008) 034501 [arXiv:0802.2020 [hep-lat]].
27. L. Y. Glozman, C. B. Lang and M. Limmer, Phys. Rev. Lett. **103** (2009) 121601 [arXiv:0905.0811 [hep-lat]].
28. L. Y. Glozman, C. B. Lang and M. Limmer, to appear in Few Body Systems [arXiv:0909.2939 [hep-lat]].
29. G. Engel, C. Gattringer, C. B. Lang, M. Limmer, D. Mohler and A. Schafer, arXiv:0910.2802 [hep-lat].
30. S. Prelovsek, T. Draper, C. B. Lang, M. Limmer, K. F. Liu, N. Mathur and D. Mohler, arXiv:0910.2749 [hep-lat].
31. F. Bruckmann, F. Gruber, and A. Schäfer, to be published.
32. E. V. Shuryak and J. J. M. Verbaarschot, Nucl. Phys. A **560**, 306, 1993 [arXiv:hep-th/9212088].
    J. J. M. Verbaarschot and T. Wettig, Ann. Rev. Nucl. Part. Sci **50**, 343, 2000 [arXiv:hep-ph/0003017].
    P. H. Damgaard and S. M. Nishigaki, Phys. Rev. **63**, 045012, 2001 [arXiv:hep-th/0006111].
33. P. Hasenfratz and F. Niedermayer, Nucl. Phys. B **414**, 785, 1994.
34. A. Hasenfratz, P. Hasenfratz and F. Niedermayer, Phys. Rev. D **72**, 114508, 2005.
35. 'Stochastic Estimator of the s Quark Determinant in Full QCD Simulation', M. Weingart diploma thesis, Uni Bern, 2007.
36. 'Lattice Quantum ChromoDynamics with approximately chiral fermions', D. Hierl, PhD thesis, Uni Regensburg, 2008.
37. 'Mass gap in NNL order in the δ-regime of QCD' C. Weiermann, PhD thesis, Uni Bern, 2009.
38. P. Hasenfratz, D. Hierl, V. Maillart, F. Niedermayer, A. Schäfer, C. Weiermann and M. Weingart arXiv:0707.0071 [hep-lat].
39. ε-regime P. Hasenfratz, D. Hierl, V. Maillart, F. Niedermayer, A. Schäfer, C. Weiermann and M. Weingart PoS **LAT2007** (2007) 077, arXiv:0710.0551 [hep-lat].
40. P. Hasenfratz, D. Hierl, V. Maillart, F. Niedermayer, A. Schäfer, C. Weiermann and M. Weingart, in progress.
41. V. Maillart, F. Niedermayer arXiv:0807.0003 [hep-lat].
42. C. Gattringer, M. Göckeler, P. Hasenfratz, S. Hauswirth, K. Holland, Th. Jörg, K. J. Juge, C. B. Lang, F. Niedermayer, P. E. L. Rakow, S. Schaefer and A. Schäfer, Nucl. Phys. B **677**, 3, 2004.

43. ‘NNL corrections to the δ-regime spectrum in the chiral limit using lattice regularization’, F. Niedermayer, Ch. Weiermann, in progress.
‘NNL chiral symmmetry breaking corrections to the δ-regime spectrum’ M. Weingart, in progress.
44. P. Hasenfratz, ‘The QCD rotator in the chiral limit’, arXiv:0909.3419 [hep-th].

Continuum-Limit Scaling of Chirally Symmetric Fermions as Valence Quarks

Krzysztof Cichy, Gregorio Herdoiza, and Karl Jansen

Abstract We present the results of mixed action lattice QCD simulations, employing dynamical twisted mass fermions in the sea sector and chirally symmetric overlap valence fermions. The aim is testing the continuum limit scaling of physical observables and by that, eventually address the question of the universality of such a mixed action approach. We focus on the study of the pion decay constant by using simulated pion masses of around 300 MeV. To render the computation practical, we impose a fixed volume corresponding to a lattice size $L \approx 1.3$ fm. We furthermore review the techniques used to deal with overlap fermions and discuss the computational details of our numerical studies.

1 Introduction

Chiral symmetry plays a central rôle in the theory of the strong interactions, Quantum Chromodynamics (QCD). This theory is invariant under the exchange of massless left-handed and right-handed quarks. There are strong indications that in nature this symmetry is spontaneously broken, giving rise to the development of a quark anti-quark condensate (the scalar condensate) and to the appearance of light pseudoscalar mesons, the observed pions.

Lattice QCD is considered to be the most effective way of studying the non-perturbative aspects of QCD. It is a regularization of QCD, which consists in discretising the relevant degrees of freedom by putting them on a four dimensional lattice with lattice spacing a, its inverse being the ultraviolet cutoff of the theory.

Krzysztof Cichy
Adam Mickiewicz University, Faculty of Physics, Umultowska 85, 61-614 Poznan, Poland
e-mail: kcichy@amu.edu.pl

Gregorio Herdoiza · Karl Jansen
NIC, DESY, Platanenallee 6, D-15738 Zeuthen, Germany
e-mail: Gregorio.Herdoiza@desy.de, Karl.Jansen@desy.de

S. Wagner et al. (eds.), *High Performance Computing in Science and Engineering, Garching/Munich 2009*, DOI 10.1007/978-3-642-13872-0_38,

With the present generation of supercomputers, large scale dynamical simulations are performed by a number of collaborations using different types of gauge and fermionic lattice actions. However, for some classes of actions, such as overlap fermions, which exactly preserve chiral symmetry, these simulations are extremely demanding[1]. A promising alternative for fully dynamical simulations with overlap fermions is the mixed action approach, which consists in using a computationally faster formulation for the fermions in the sea, such as Wilson twisted mass fermions, while using overlap fermions in the valence sector. In this way, one avoids the most computer-intensive part of a dynamical simulation with overlap fermions, but at the same time one profits from having such fermions in the valence sector. In the case of overlap fermions, its exact chiral symmetry gives the benefit of simplifying the operator mixing problem, something which is essential in several types of lattice computations, like the determination of the kaon bag parameter B_K which play an important rôle in the study of CP violation via CKM unitarity triangle analyses.

In this paper, we present the results of a mixed action approach with overlap fermions in the valence sector and Wilson twisted mass fermions in the sea sector[2]. In Section 2 we briefly review the overlap formulation and the techniques necessary to effectively apply it. Section 3 gives the description of our setup and in Section 4 we discuss the computational details. Section 5 presents the continuum-limit scaling test of overlap fermions and the discussion of the results. Section 6 concludes.

2 A Brief Review of Overlap Fermions

Since this contribution is written for a general audience, we provide here a very short review of overlap fermions.

2.1 The Need for Overlap Fermions

A very general problem of lattice field theory with fermions is the doubling problem. With a naive fermion discretisation, instead of one fermion in the continuum limit, one has 16 fermions in 4 dimensions of space-time. As was originally proposed by Wilson [4], the doubling problem can be solved by using the following Wilson-Dirac operator:

$$D_W(m) = \frac{1}{2}\left(\gamma_\mu(\nabla^*_\mu + \nabla_\mu) - ar\nabla^*_\mu\nabla_\mu\right) + m,$$

[1] For instance, the JLQCD and TWQCD collaborations are simulating dynamical overlap fermions, but in their case the global topological sector is kept fixed [1].

[2] For an account of earlier stages of the project, see [2, 3].

where ∇^*_μ (∇_μ) is the backward (forward) lattice derivative, r – Wilson parameter, m – bare quark mass. However, the Wilson term in the above Dirac operator explicitly breaks chiral symmetry, which is in accordance with a general no-go theorem proved by Nielsen and Ninomiya [5] back in 1981 that it is impossible to have at the same time for a Dirac operator D: locality, translational invariance, no doublers and chiral symmetry, i.e. $\{D(p),\gamma_5\} = 0$. Since Wilson's proposal, much of the effort went into finding a lattice theory without doublers which preserves the largest possible number of symmetries, and at the same time reaches the continuum limit as fast as possible (which practically means the absence of $\mathcal{O}(a)$ leading cutoff dependence on physical observables determined on the lattice).

In 1982 (i.e. only one year after establishing the Nielsen-Ninomiya theorem), it was shown by Ginsparg and Wilson [6] that a remnant of chiral symmetry is present on the lattice, without the doubler modes, if the corresponding Dirac operator D obeys an equation now called the Ginsparg-Wilson relation:

$$\gamma_5 D + D\gamma_5 = aD\gamma_5 D\,.$$

However, the solutions to this equation were not known for many years, until a particularly simple form of a Dirac operator obeying the Ginsparg-Wilson relation was given by Neuberger [7] in 1997. Neuberger's discretisation is now usually referred to as overlap fermions[3] and the (massless) overlap Dirac operator is of the form:

$$D_{ov}(0) = \frac{1}{a}\left(1 - A(A^\dagger A)^{-1/2}\right), \tag{1}$$

where:

$$A = 1 - aD_W(0)\,.$$

It was also found that the overlap Dirac operator is local under very general conditions [9]. Moreover, in 1998 Lüscher [10] found that the Ginsparg-Wilson relation leads to a non-standard realization of lattice chiral symmetry. The action is invariant under:

$$\psi \to e^{i\theta\gamma_5\left(1-\frac{aD}{2}\right)}\psi\,,$$
$$\bar{\psi} \to \bar{\psi}e^{i\theta\gamma_5\left(1-\frac{aD}{2}\right)}\,.$$

Although it is not the standard form of chiral symmetry, it still correctly reproduces the anomaly and protects the fermions from additive mass renormalization and from $\mathcal{O}(a)$ lattice artifacts. The non-standard realization of chiral symmetry means also that the conditions of the Nielsen-Ninomiya theorem do not apply and one can have chiral symmetry (which becomes standard chiral symmetry in the continuum limit) without the doublers.

---

[3] For a review of overlap fermions see e.g. [8].

In order to simulate massive overlap quarks, one uses the following form of the Dirac operator:

$$D_{ov}(m) = \left(1 - \frac{am}{2}\right) D_{ov}(0) + m,$$

where m is the bare overlap quark mass and $D_{ov}(0)$ the massless overlap Dirac operator of eq. (1).

2.2 Techniques to Effectively Deal with Overlap Fermions

The main disadvantage of overlap fermions is that they are much more costly to simulate – by a factor of 30-120 in comparison to maximally twisted mass fermions [11]. Moreover, this factor increases when reducing the simulated pseudoscalar mass down to the physical pion mass. Therefore, it was essential to develop techniques to effectively deal with overlap fermions. The aim of this subsection is to briefly review them.

Computation of the Overlap Operator. The first important thing is to effectively calculate the overlap Dirac operator itself. This is non-trivial, since it involves the inverse of the square root of a matrix. There are several ways to do this, including polynomial approximations, Lanczos based methods and partial fraction expansion. For overviews of these methods, see e.g. [12, 13]. The method that we have chosen is a Chebyshev polynomial approximation, which consists in constructing a polynomial $P_{n,\varepsilon}(x)$ of degree n, which has an exponential convergence rate in the interval $x \in [\varepsilon, 1]$. The advantages of using this approximation are the well-controlled exponential fit accuracy and the possibility of having numerically very stable recursion relations which allows for large degrees of the polynomial. To ensure that the Ginsparg-Wilson relation (for massless Dirac operator) is fulfilled to a very high precision, the degree of Chebyshev polynomial n has to satisfy the following condition:

$$||X - P_{n,\varepsilon}(A^\dagger A) A^\dagger A P_{n,\varepsilon}(A^\dagger A) X||^2 / ||X||^2 < \xi,$$

where ξ has to be a very small number, typically set to 10^{-16} to achieve a compromise between good quality of approximation and its cost. The degree of the polynomial depends on the condition number of the matrix $A^\dagger A$, i.e. the ratio of the highest to the lowest eigenvalue. The lowest eigenvalue can be a very small number and hence the condition number can be prohibitively large, if one constructs the approximation on the interval $[\varepsilon, 1]$, with ε being the lowest eigenvalue. Fortunately, one can do much better with the following method.

Eigenvalue Deflation. To achieve a considerably smaller degree of Chebyshev polynomial, one should calculate a certain number[4] of the lowest eigenvalues and eigen-

[4] The exact number has to be found empirically and tends to increase with volume.

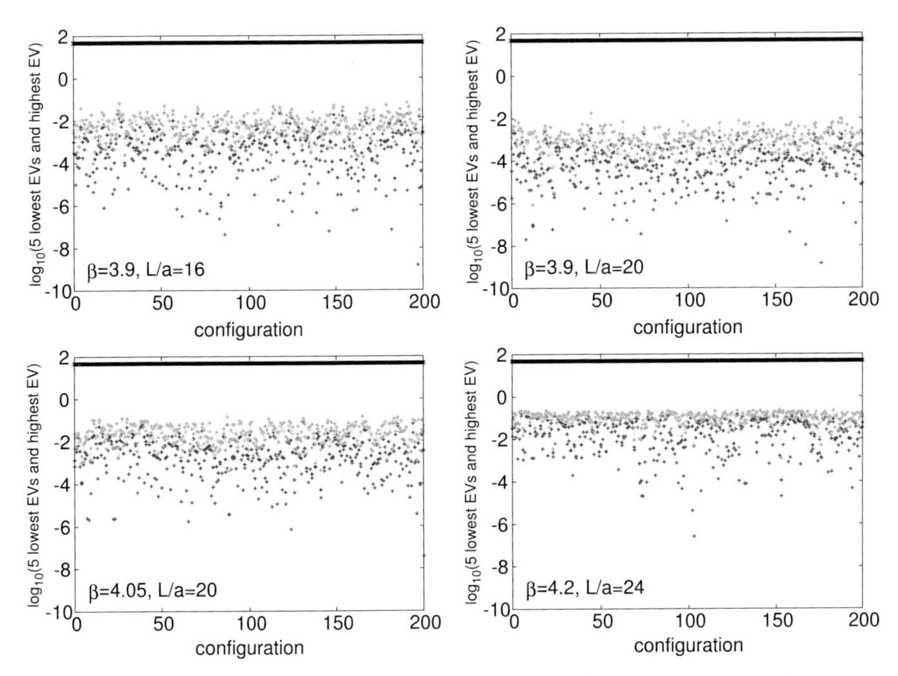

Fig. 1 Five lowest eigenvalues and the highest eigenvalue of $A^{\dagger}A$ for various gauge field ensembles. The lattice spacing is $a \approx 0.084$ fm for upper plots, $a \approx 0.066$ fm for bottom left and $a \approx 0.053$ fm for bottom right plot. The physical volume is kept fixed for the upper left plots and the bottom plots

vectors of $A^{\dagger}A$ and project them out of this matrix. In this way, the Chebyshev approximation is constructed on the interval $[\varepsilon, 1]$ with ε is now equal to the highest of the computed eigenvalues. To illustrate how low the lowest eigenvalues of the matrix $A^{\dagger}A$ can go, we plotted in Fig. 1 the five lowest eigenvalues and the highest eigenvalue for four gauge field ensembles, each with 200 configurations. The general dependence that can be deduced from these plots is that increasing the lattice spacing (decreasing β) moves the spectrum down (eigenvalues in lattice units tend to become smaller) and increases the probability of having very low eigenvalues.

HYP Smearing of Gauge Fields. Another way to lower the condition number of the matrix $A^{\dagger}A$ and thus the degree of the Chebyshev polynomial is to perform one iteration of HYP smearing on the gauge fields. This technique was introduced by A. Hasenfratz and F. Knechtli [14] and allows to achieve much better convergence of the solver due to improved chiral properties[5]. In comparison with other link fattening methods (e.g. APE smearing), HYP smearing is believed to preserve better

[5] Hasenfratz and Knechtli [14] remark that fat links lead to an order of magnitude improvement in convergence.

the short-distance quantities, because it mixes links from hypercubes attached only to the original link.

SUMR Solver with Adaptive Precision and Multiple Mass Capability. Having constructed the Dirac operator (with Chebyshev approximation), to find the propagator one has to solve the equation:

$$D_{ov}(m)\psi = \eta \,, \tag{2}$$

where ψ is the propagator and η is the source – a vector whose choice will be commented on below. To effectively solve this equation, one has to choose the most appropriate solver. Chiarappa et al. [11] found that in the case of (quenched) overlap and small lattice volume (12^4 and 16^4), the most effective solver is the chiral conjugate gradient algorithm, with the SUMR solver just behind[6]. However, the former algorithm can only be used for exact overlap operator, which means that the polynomial approximation would lead to some corrections that would have to be explicitly calculated. This makes the latter algorithm more attractive and we decided to use it. To improve its performance, we have also used adaptive precision and multiple masses. The former means that the degree of the Chebyshev polynomial is adapted to the accuracy that is actually needed in the present iteration step. In practice, this means that when the solver is moving towards the requested precision, the accuracy of the approximation can be substantially decreased[7], which saves a factor of around 2 in inversion time.

Multiple mass capability of an algorithm means that for the cost of one inversion for the smallest bare quark mass, one can obtain the solution also for heavier quark masses for practically no additional cost. Since the dependence on the quark mass is central in the present project, the use of a multiple mass variant of the SUMR solver was absolutely essential.

Stochastic Sources. Another important aspect of solving eq. (2) is the choice of the source η. The most obvious choice is the point source, which means that the vector η is chosen to be 1 at one space-time point and spin-color component and 0 otherwise. This leads to 12 inversions for each gauge configuration, one per spin-color component. However, one can do much better when it comes to statistical error on the pion mass and especially the pion decay constant, with the use of stochastic sources. To keep the signal-to-noise ratio high enough, one has to use time-slice sources, i.e. sources that are non-zero for all spatial points on a given time-slice and for a given spin. This requires 4 inversions per gauge configuration, one per spin component. Moreover, the noise can be further reduced by using the one-end trick, introduced in [15]. A reasonable choice of stochastic sources for mesonic correlators are the $\mathscr{Z}(2)$ sources – the random numbers are of the form $(\pm 1 \pm i)/\sqrt{2}$. Empirical observations show that this method reduces the statistical error on the pion decay constant by a factor of approximately two for the smallest quark masses that we

[6] SUMR = Shifted Unitary Minimal Residual.

[7] For example, if the degree of Chebyshev polynomial at the start of inversion is typically (for our parameters) of order 150-200, in the final iterations it can go down typically to 30-40 with adaptive precision.

consider, which means that the number of inversions to achieve the same statistical error can be even four times smaller than with point sources.

3 Simulation Setup

To perform the scaling test of overlap fermions we fix the volume to $L \approx 1.3\,\mathrm{fm}$ and use the following ensembles of dynamical $N_f = 2$ maximally twisted mass fermions[8] [17, 18], generated by the European Twisted Mass Collaboration (ETMC) [19, 20]:

- $\beta = 3.90,\ a \approx 0.084\ \mathrm{fm},\ V/a^4 = 16^3 \times 32,\ a\mu = 0.004,\ \kappa = 0.160856,$
- $\beta = 4.05,\ a \approx 0.066\ \mathrm{fm},\ V/a^4 = 20^3 \times 40,\ a\mu = 0.003,\ \kappa = 0.157010,$
- $\beta = 4.20,\ a \approx 0.053\ \mathrm{fm},\ V/a^4 = 24^3 \times 48,\ a\mu = 0.002,\ \kappa = 0.154073.$

The valence quarks are overlap fermions and the overlap quark mass was chosen to vary from the unitary light quark mass up to the physical strange quark mass. In total we have 20 quark masses, which allows for a precise determination of the quark mass dependence of the pion mass and decay constant.

In addition, we also perform a tree-level scaling test, for which we fix $Nm = 0.5$ (which is the equivalent of fixing volume), where N is the number of lattice points in spatial directions and go from $N = 4$ to $N = 64$. Thus, the change in N introduces the scaling towards the continuum limit, which corresponds to $N \rightarrow \infty$. The temporal extent was set to be 64 times larger than the spatial extent for each value of N, which makes it possible to extract the relevant quantities without any contamination from the excited states. For the details of this test and expressions for the tree-level quark propagators and correlation functions, see [21, 22].

4 Computational Details

4.1 Parallelisation

Paralellisation in lattice QCD simulations consists in dividing the lattice in blocks and handling each block by a single node. For the current project we have used a 3-dimensional parallelisation. For the respective lattice sizes we use the following number of processors:

- $V/a^4 = 16^3 \times 32$ – $8 \times 8 \times 2$ processors, block size = $2 \times 2 \times 8 \times 32$,
- $V/a^4 = 20^3 \times 40$ – $10 \times 10 \times 2$ processors, block size = $2 \times 2 \times 10 \times 40$,
- $V/a^4 = 24^3 \times 48$ – $8 \times 8 \times 2$ processors, block size = $3 \times 3 \times 12 \times 48$.

[8] The twisted mass (TM) Dirac operator is defined by: $D_{TM} = D_W(m)\mathbb{1}_f + i\mu\gamma_5\tau^3$, where: m – untwisted quark mass, μ – twisted quark mass, $\mathbb{1}_f$ and τ^3 act in flavour space. For a review of twisted mass fermions, see [16].

Table 1 Average and total timings for the three analyzed ensembles

Ensemble β	Avg. time per conf. [CPU hours]	Time per site [CPU seconds]	Nr. of confs.	Total time used [CPU hours]
3.90	188.7	5.18	544	102653
4.05	511.2	5.75	300	153360
4.20	1024.3	5.56	400	409720

Notation: *Time per site* = average time per configuration divided by the number of lattice points, *Nr. of confs* = number of independent gauge field configurations used in the analysis.

4.2 Timings

Table 1 presents the average and total times achieved on the HLRB system. We present the average times for the inversion of one gauge field configuration[9], but since the lattice sizes vary, we also give this time divided by the number of lattice points to have an estimate of the efficiency. As can be seen from Table 1, the efficiency for different lattice sizes is comparable, with slightly smaller efficiency for the case where a bigger number of processors was used ($V = 20^3 \times 40$, $\beta = 4.05$), which is due to communication between the increased number of processors.

5 Results

5.1 Tree-Level Test

We investigated the relative (with reference to the continuum-limit value) cut-off effects of three observables[10]: pseudoscalar (PS) correlation function $N^3 C_{PS}$ at a fixed physical distance $t/N = 4$, pseudoscalar meson mass Nm_{PS} and pseudoscalar decay constant Nf_{PS}. The results are presented in Fig. 2. As expected, all observables show only $\mathcal{O}(a^2)$ scaling violations, since overlap fermions are $\mathcal{O}(a)$-improved. We also observe that the smallest lattice artifacts are observed for the pseudoscalar mass and the largest for the correlation function itself.

5.2 The Interacting Case – Matching the Pion Mass

In the interacting case, we are interested in the continuum limit scaling of the pion decay constant for three reference values of the pion mass[11] – the sea quark pion

[9] This is the combined time of the computation of a certain number of eigenvalues/eigenvectors and the inversion of the Dirac operator. The time of HYP smearing and of the generation of stochastic sources is not taken into account and is negligible in comparison with the time of eigenvalue computation and inversion.

[10] All of the observables are multiplied by an appropriate power of N to make them dimensionless.

[11] The pion mass and decay constant were obtained from the pseudoscalar correlation function. Thus, pion refers here the light pseudoscalar meson.

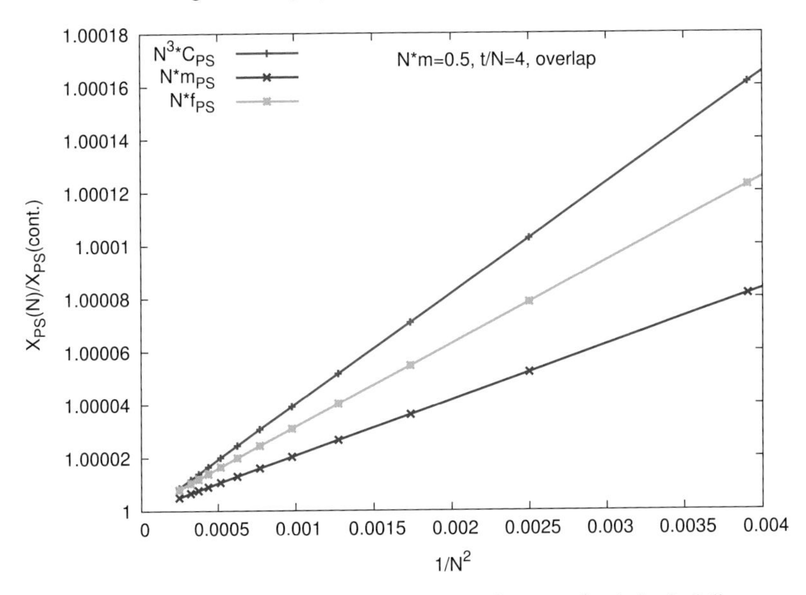

Fig. 2 The relative cutoff effects of the pseudoscalar correlator at a fixed physical distance, pseudoscalar mass and decay constant

mass, an intermediate mass $r_0 m_\pi \approx 1$ and a heavy mass (around the strange quark region) $r_0 m_\pi \approx 1.5$.[12]

The matching quark mass $\hat{m}$ is defined, for each ensemble, by the condition:

$$m_\pi^{ov}(\hat{m}) = m_\pi^{TM}(\mu),$$

where m_π^{ov} and m_π^{TM} are the mixed action and unitary pion masses, respectively.

Fig. 3 shows the procedure of matching for the ensemble at $\beta = 3.9$. The "overlap" curve shows the dependence of the pion mass as a function of the bare overlap quark mass. As the plot indicates, when the overlap quark mass equals 0.007 (with an uncertainty of approx. 0.001), the TM pion and overlap pion have equal masses, corresponding in infinite volume to ca. 300 MeV. An analogous procedure was applied for the other ensembles.

One should add here a word of caution. For the situation of rather small lattice extent $L \approx 1.3$ fm, the chiral zero-modes of the overlap operator can play an important role. Note that since we use here a mixed action approach, these zero-modes are not compensated for by the fermion determinant. The influence of the interplay between the finite box size, the quark mass and the mixed action on the effects of the zero-modes for physical observables will be discussed elsewhere.

[12] These two higher reference masses correspond to the partially quenched setup.

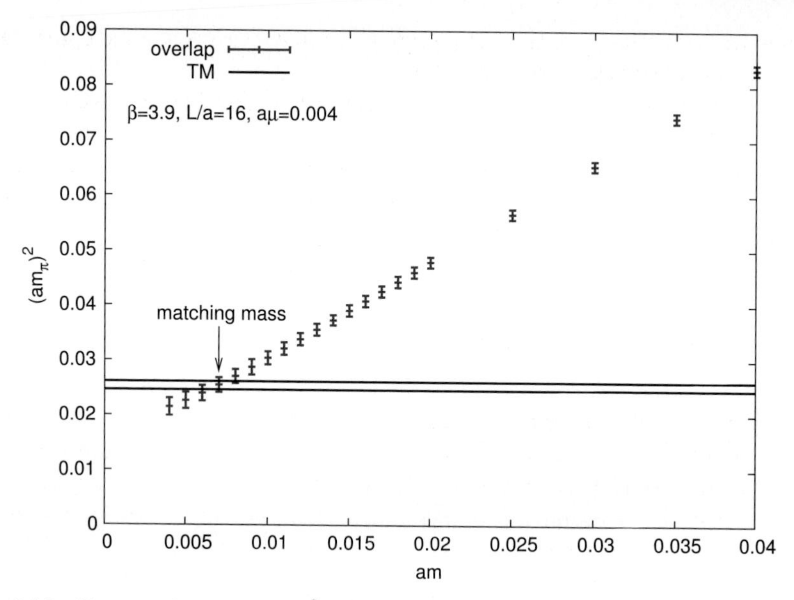

Fig. 3 Matching mass for ensemble $16^3 \times 32$, $a \approx 0.084$ fm ($\beta = 3.9$), $a\mu = 0.004$, $\kappa = 0.160856$. The matching procedure gives $a\hat{m} = 0.007(1)$

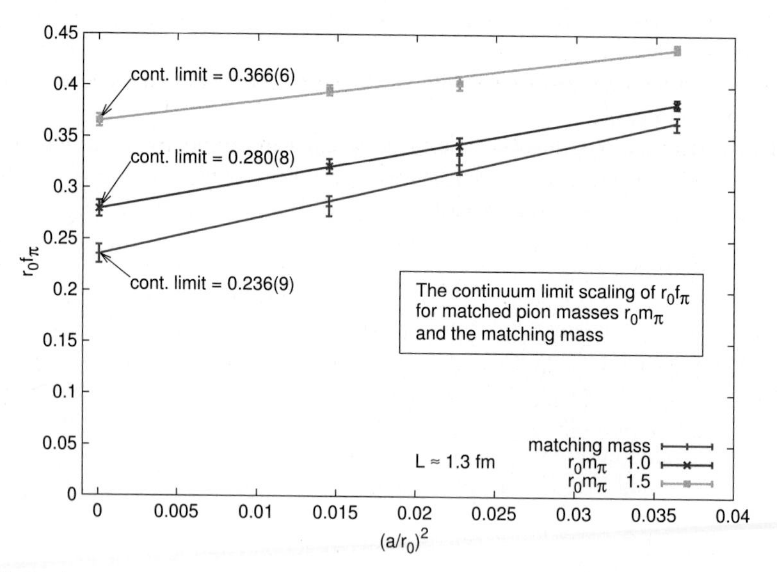

Fig. 4 Continuum limit scaling of the (overlap) pion decay constant for fixed volume $L \approx 1.3$ fm

5.3 *Continuum-Limit Scaling of the Pion Decay Constant*

Fig. 4 shows the continuum limit scaling of $r_0 f_\pi$ for three choices of the pion mass. The lower curve corresponds to the sea quark pion mass, i.e. the bare overlap quark mass is set to the relevant $\hat{m}$ for each ensemble. For the intermediate and upper curve we fix the pion mass $r_0 m_\pi$ to 1.0 and 1.5, respectively.

We expect $\mathscr{O}(a^2)$ scaling violations and hence we plot the decay constant against the lattice spacing squared.

Indeed, for all three cases, we observe good scaling behaviour towards the continuum limit. The fact that the unitarity violations, proper to a mixed action approach, do not spoil the scaling of the pion decay constant is reassuring.

6 Conclusion and Prospects

In this paper, we briefly reviewed the overlap discretisation of the fermionic action and the techniques we have used to deal with overlap fermions. We presented the continuum limit scaling test of the pion decay constant at tree-level and in the interacting case, using three values of the lattice spacing coming from dynamical twisted mass configurations at roughly matched physical box length of $L \approx 1.3$ fm and with three reference values of the pion mass.

We observe very good continuum limit scaling properties for all cases, i.e. at tree-level and in the interacting case for pion masses corresponding to light (with overlap quark mass leading to the same pion mass as the sea quark), intermediate and somewhat heavier quark masses.

In the next stage of the project, we plan to extend the scaling test to other observables (e.g. the nucleon mass).

In order to make predictions for physically relevant observables it is required to increase the lattice size, with respect to the one mentioned in this work, in order to have finite size effects under control. From our first simulations at larger physical volumes, we can estimate that performing this continuum-limit scaling study with a lattice size $L \approx 2.5$ fm, would imply around 10 Million CPUh on the HLRB II system for a pseudoscalar meson mass of around 300 MeV. In addition to setting the ground for a test of the universality of the mixed action approach at large volume, such a physical situation would allow to properly study observables for which chiral symmetry is crucial and to address the question of the applicability of chiral perturbation theory in this setup.

Acknowledgements The computer time for this project was made available to us by the Leibniz computer centre in Munich on the HLRB II system. We thank this computer centre and its staff for all technical advice and help.

References

1. Aoki S. et al., Phys. Rev. D78 (2008), 014508; arXiv:0803.3197.
2. Bär O., Jansen K., Schaefer S., Scorzato L., Shindler A., PoSLAT2006:199,2006; arXiv: hep-lat/0609039.

3. Garron N., Scorzato L., PoSLAT2007:083,2007; arXiv: 0710.1582 (hep-lat).
4. Wilson K.G., New Phenomena In Subnuclear Physics. Part A. Proceedings of the First Half of the 1975 International School of Subnuclear Physics, Erice, Sicily, July 11 - August 1, 1975, ed. A. Zichichi, Plenum Press, New York, 1977, p. 69.
5. Nielsen N.B., Ninomiya M., Phys. Lett. B105 (1981), 211.
6. Ginsparg P., Wilson K., Phys. Rev. D25 (1982), 2649.
7. Neuberger H., Phys. Lett. B 417 (1998), 141; arXiv: hep-lat/9707022.
8. Niedermayer F., Nucl. Phys. B (Proc. Suppl.) 73 (1999), 105; arXiv:hep-lat/9810026.
9. Hernandez P., Jansen K., Lüscher M., Nucl. Phys. B552 (1999) 363-378; arXiv: hep-lat/9808010.
10. Lüscher M., Phys. Lett. B428 (1998), 342; arXiv: hep-lat/9802011.
11. Chiarappa T. et al.; arXiv: hep-lat/0609023.
12. Van den Eshof J. et al., Comput.Phys.Commun. 146 (2002) 203-224; arXiv: hep-lat/0202025.
13. Frommer A., Lippert T., Medeke B., Schilling K. (eds.), Numerical Challenges in Lattice Quantum Chromodynamics, Lecture Notes in Computational Science and Engineering 15, Heidelberg 2000.
14. Hasenfratz A., Knechtli F., Phys.Rev. D64 (2001), 034504; arXiv:hep-lat/0103029.
15. Foster M., Michael C., Phys.Rev. D59 (1999), 074503; arXiv: hep-lat/9810021.
16. Shindler A., Phys. Rept. 461 (2008), 37; arXiv: 0707.4093 (hep-lat).
17. Frezzotti R., Grassi P.A., Sint S., Weisz P., JHEP 0108 (2001) 058; arXiv: hep-lat/0101001.
18. Frezzotti R., Rossi G.C., JHEP 0408 (2004) 007; arXiv: hep-lat/0306014.
19. Boucaud Ph. et al., Phys. Lett. B650 (2007), 304; arXiv: hep-lat/0701012.
20. Boucaud Ph. et al., Comput. Phys. Commun. 179 (2008) 695-715; arXiv: 0803.0224 (hep-lat).
21. Cichy K., Gonzalez Lopez J., Jansen K., Kujawa A., Shindler A., Nucl.Phys. B800 (2008), 94; arXiv: 0802.3637 (hep-lat).
22. Cichy K., Gonzalez Lopez J., Kujawa A., Acta Phys. Pol. B, Vol. 39 (2008), No. 12, 3463-3472; arXiv: 0811.0572 (hep-lat).

Quantum Boltzmann Equations in the Early Universe

Mathias Garny and Markus Michael Müller

Abstract Physics of the early universe unavoidably falls into the domain of quantum dynamics far from thermal equilibrium. Since properties of specific theories, like leptogenesis or preheating, are to be tested experimentally in the near future, it is important to assess the validity of (semi-)classical or Boltzmann-type approximations underlying their predictions. A complete quantum-mechanical description of nonequilibrium processes is furnished by so-called Kadanoff-Baym equations, which are quantum mechanical generalizations of classical Boltzmann equations. Kadanoff-Baym equations take memory effects into account, which is reflected by their non-Markovian structure.

1 Introduction

The combination of astronomical precision observations on the one hand with results from high-energy physics experiments on the other hand provides a unique possibility to answer fundamental questions about the structure and the evolution of our universe. A prime example for the interplay of astro- and particle physics is the explanation of the matter-antimatter asymmetry in our cosmos. Within the Big Bang paradigm of cosmology, such an asymmetry can be dynamically generated in the very early universe. This so-called *baryogenesis* requires a tiny mismatch of the fundamental forces between particles and anti-particles. Conversely, the search for such a mismatch is one of the motivations for building high-energy collider ex-

Mathias Garny
Physik-Department, Technische Universität München, James-Franck-Straße, 85748 Garching bei München, Germany
e-mail: mgarny@ph.tum.de

Markus Michael Müller
Leibniz Supercomputing Centre, Boltzmannstrasse 1, 85748 Garching bei München, Germany
e-mail: Markus.Michael.Mueller@lrz.de

S. Wagner et al. (eds.), *High Performance Computing in Science and Engineering, Garching/Munich 2009*, DOI 10.1007/978-3-642-13872-0_39,

periments such as LHC as well as high-precision neutrino oscillation experiments. Another example is the search for hypothetical, extremely weakly interacting *dark matter* particles, whose existence is inferred from numerous cosmological observations including the large-scale structure of the universe.

An important building block for relating physics at smallest and largest length-scales is the dynamical description of processes taking place in the expanding hot and dense universe shortly after the Big Bang. These include reheating after inflation, baryogenesis, and dark-matter freeze-out. Typically, the kinematics of these processes are calculated based on (approximations to) Boltzmann equations. However, Boltzmann equations are only a classical approximation of the quantum mechanical processes under consideration. In particular, Boltzmann equations rely on a classical particle concept, requiring well-localized quasi-stable particle excitations. Furthermore, they rely on the assumption of instantaneous interactions of uncorrelated particles. These conditions may be violated in the early universe, due to the extremely high temperatures and the rapid cosmological expansion. On the other hand, Kadanoff-Baym equations provide a means to describe such quantum mechanical processes from first principles.

This raises the question how reliable Boltzmann equations are as approximations to the complete Kadanoff-Baym equations. Therefore, we performed detailed comparisons of Boltzmann and Kadanoff-Baym equations in the framework of relativistic quantum field theories in 3+1 space-time dimensions [7, 8]. Furthermore, we derived extended Kadanoff-Baym equations, that can describe non-Gaussian initial states [6]. This is a requirement for obtaining quantitative results independent of the discretization.

After recalling some basic facts about Boltzmann and Kadanoff-Baym equations in section 2, we present numerical results obtained for extended Kadanoff-Baym equations in section 3. The numerical methods and the parallelization strategy are discussed in section 4, and section 5 contains conclusions and an outlook.

2 Kadanoff-Baym Equations

The success of the Standard Model of particle physics suggests that nature can be described by quantum field theory at very high energy scales. However, there are a number of reasons why the Standard Model cannot be complete, and possible extensions will be tested for example by the proton-proton collisions observed at LHC experiments. An important guideline for the search for new physics is provided by cosmological observations, such as dark matter. In order to compare theory with experiment, it is generally necessary to relate fundamental parameters of quantum field theory on the one hand with experimentally measurable quantities, such as scattering cross-sections, on the other hand.

Within astro-particle physics, an additional complication arises. Namely, the information gained by cosmological observations is only sensitive to the outcome of the complex many-particle dynamics in the very early universe. Therefore, it is crucial to obtain a reliable description of the non-equilibrium time-evolution of such many-particle systems. Typically, the dynamics is parameterized in terms of a few macroscopic parameters, such as energy densities or effective chemical potentials. In order to relate their dynamics with parameters of the fundamental theory, it is necessary to employ a suitable microscopic picture of the underlying processes. Such a microscopic description is one of the main properties of Boltzmann equations, which are usually used for this task.

Boltzmann equations provide a kinematic description of elastic and inelastic scattering processes of a system of quasi-classical particles. Their basic degree of freedom is the one-particle distribution function $f_i(t,\mathbf{x},\mathbf{k})$. It describes the classical phase-space density of particles of type i at position $\mathbf{x}$ with momentum $\mathbf{k}$, for each instant of time t. The underlying fundamental theory enters the Boltzmann equations via so-called collision integrals, which require scattering probabilities as basic input.

The classical particle concept underlying Boltzmann equations naturally leads to certain limitations in their range of applicability, especially for processes occurring in the very early universe. In order to obtain a microscopic description of the non-equilibrium dynamics within quantum field theory, it is necessary to replace the one-particle distribution function by a suitable quantum-mechanical object. A particularly useful quantity is the two-point correlation function. For a quantum field $\Phi(x)$, the so-called statistical propagator

$$G_F(x,y) = \frac{1}{2}\langle \Phi(x)\Phi(y) + \Phi(y)\Phi(x)\rangle \ ,$$

encodes information about the state of the system. Additionally, the spectral function

$$G_\rho(x,y) = i\langle \Phi(x)\Phi(y) - \Phi(y)\Phi(x)\rangle \ ,$$

encodes information about the particle spectrum. Here, $x = (x^0, \boldsymbol{x})$ and $y = (y^0, \boldsymbol{y})$ denote two space-time points, and the expectation value $\langle\cdot\rangle \equiv \mathrm{Tr}(\rho\,\cdot)$ is taken with respect to an ensemble characterized by the density matrix ρ.

The semi-classical limit, which corresponds to the Boltzmann picture, requires a spectral function that is sharply peaked on the mass-shell of a quasi-classical particle excitation,

$$G_\rho^{Boltzmann}(X,k) = 2\pi \mathrm{sgn}\left(k^0\right)\ \delta\left(k^2 - m^2\right) \ .$$

Here $X = (x+y)/2$ denotes the central coordinate, and the momentum k is the Fourier mode with respect to the relative coordinate $s = x - y$. Furthermore, in the Boltzmann picture, the statistical propagator is forced to obey a fluctuation-dissipation relation,

$$G_F^{Boltzmann}(X,k) = \left(f(X,\mathbf{k}) + \frac{1}{2}\right) G_\rho^{Boltzmann}(X,k) \ ,$$

where $f(X,\mathbf{k})$ corresponds to the one-particle distribution function.

In the hot and expanding early universe, we expect that the spectral function has a non-zero width, and peaks at a time-dependent effective mass. If multiple species are present, their mixing can make the spectral function even more complex. The time-evolution of a completely general system, possibly far from equilibrium, can be described based on the quantum-mechanical equations of motion of the spectral function and the statistical propagator, the so-called Kadanoff-Baym equations. For scalars they have the form [7]

$$\begin{aligned}&\left(-\partial_{x^\mu}\partial_{x_\mu}+M^2(x)\right)G_F(x,y)\\&=\int_0^{y^0}d^4z\,\Pi_F(x,z)\,G_\rho(z,y)-\int_0^{x^0}d^4z\,\Pi_\rho(x,z)\,G_F(z,y)\end{aligned}\tag{1}$$

and

$$\left(-\partial_{x^\mu}\partial_{x_\mu}+M^2(x)\right)G_\rho(x,y)=-\int_{y^0}^{x^0}d^4z\,\Pi_\rho(x,z)\,G_\rho(z,y)\;.\tag{2}$$

In the *setting-sun* approximation, the *self-energies* M^2, Π_F, and Π_ρ read:

$$M^2(x)=m_B^2+\frac{\lambda}{2}G_F(x,x)\;,\tag{3}$$

$$\Pi_F(x,y)=-\frac{\lambda^2}{6}\Big(G_F(x,y)\,G_F(x,y)\,G_F(x,y)-\frac{3}{4}G_\rho(x,y)\,G_\rho(x,y)\,G_F(x,y)\Big)\;,\tag{4}$$

$$\Pi_\rho(x,y)=-\frac{\lambda^2}{6}\Big(3G_F(x,y)\,G_F(x,y)\,G_\rho(x,y)-\frac{1}{4}G_\rho(x,y)\,G_\rho(x,y)\,G_\rho(x,y)\Big)\;.\tag{5}$$

In contrast to Boltzmann equations, Kadanoff-Baym equations determine both the spectrum and the state of the system self-consistently. Furthermore, they describe on- and off-shell processes in a consistent way, thereby avoiding double-counting problems occurring in the Boltzmann picture. The memory integrals on the right-hand side take the complete time-history of the system into account. They describe the effect of correlations on the time-evolution, which is neglected in the probabilistic Boltzmann approach.

3 Numerical Results

We performed detailed comparisons of numerical solutions of Boltzmann and Kadanoff-Baym equations in the framework of relativistic quantum field theories in 3+1 space-time dimensions. In a first step, for simplicity, we considered a real scalar Φ^4 quantum field theory [7] and in a second step we generalized our results to a chirally invariant Yukawa-type quantum field theory including fermions [8]. The obtained numerical solutions reveal significant discrepancies in the results predicted by both types of equations. Apart from quantitative differences, on a qualitative level the late-time universality respected by Kadanoff-Baym equations is severely

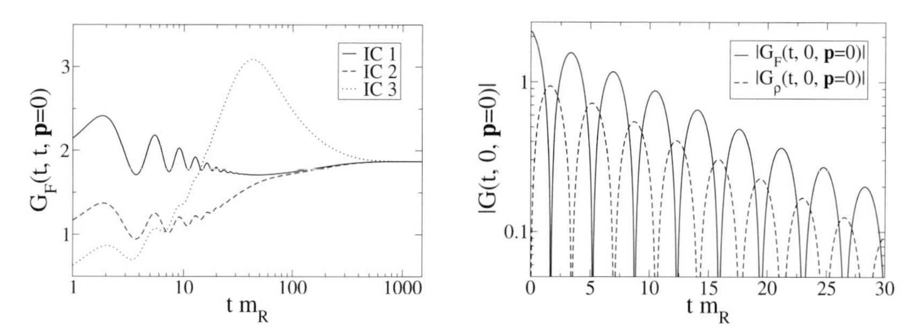

Fig. 1 Left: Equal-time propagator as function of time for a fixed momentum mode. The three lines correspond to three different initial conditions with identical average energy density. At late times, all solutions approach a universal equilibrium value. Right: Unequal-time propagator as function of time for a fixed momentum mode. Correlations between earlier and later times are exponentially damped, which allows to introduce a history cut-off

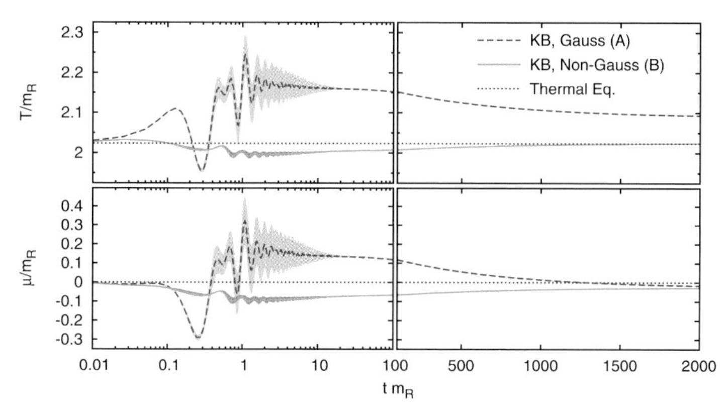

Fig. 2 Time evolution of the effective temperature and effective chemical potential obtained from Kadanoff-Baym equations [6] with thermal initial 2-point correlation function (initial state (A), red dashed lines) as well as thermal initial 2- and 4-point correlation functions (initial state (B), green solid lines). For the Gaussian initial state (A), the solution is considerably driven away from equilibrium for times $tm_R \sim 1$ due to correlation build-up. The shaded areas illustrate qualitatively the deviation from kinetic equilibrium, which becomes negligibly small for $tm_R \gg 10$. Chemical equilibration occurs for times $tm_R \gg 100$, when the effective chemical potential approaches zero

restricted in the case of Boltzmann equations (see Fig. 1). Furthermore, Kadanoff-Baym equations strongly separate the time scales between kinetic and chemical equilibration (see Fig. 2). In contrast to this, standard Boltzmann equations cannot describe the process of quantum-chemical equilibration, and consequently also cannot feature the above separation of time scales.

In order to perform a *quantitative* comparison of Boltzmann and Kadanoff-Baym equations, it is important to answer some non-trivial technical questions related to renormalization out of thermal equilibrium. Recently, some theoretical progress on this issue has been achieved [4]. In particular, we derived extended Kadanoff-Baym equations, which can describe the time-evolution starting from non-Gaussian ini-

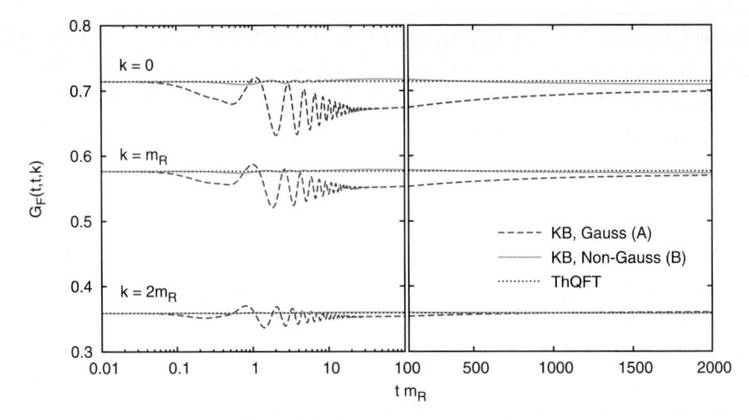

Fig. 3 Time evolution of the equal-time propagator $G_F(t,t,\boldsymbol{k})$ obtained from Kadanoff-Baym equations with thermal initial 2-point correlation function (initial state (A), red dashed lines) as well as thermal initial 2- and 4-point correlation functions (initial state (B), green solid lines), for three momentum modes, respectively. The dotted horizontal lines show the renormalized thermal propagator $G_{th}(0,0,\boldsymbol{k})$, which serves as initial condition at $t = 0$

tial states [6]. This is important, since physical states of the interacting theory are non-Gaussian. In particular, the vacuum state and thermal equilibrium states feature relevant non-Gaussian correlations. The inclusion of these correlations is non-trivial due to the non-perturbative nature of Kadanoff-Baym equations.

In Fig. 3 we compare solutions obtained for two sets of initial states. Initial state (A) is Gaussian, while initial state (B) takes the leading non-Gaussian correction into account. The difference is most relevant for initial states close to equilibrium. Therefore, we chose thermal initial correlations. The truncation of higher correlations drives the solution (A) away from equilibrium at short time-scales, while thermalization occurs on a longer time-scale. In contrast to this, solution (B) always remains close to equilibrium. This property is a condition for studying solutions for initial states that fall into the domain of validity of Boltzmann equations.

Fig. 2 reveals a qualitative difference between the Gaussian and non-Gaussian cases. For the former, the effective temperature of the initial and final state differ from each other. The reason is that a Gaussian initial state captures only kinetic, but no correlation energy. In particular, we note that this offset diverges when the continuum limit is approached. This short-coming is completely remedied already by taking the leading non-Gaussian correction into account.

4 Numerical Methods

Even more than for Boltzmann equations the numerical solution of Kadanoff-Baym equations lies far beyond the scope of any integrated computer software package such as Mathematica or Maple. Consequently, we developed our own Kadanoff-Baym solver from scratch.

The code is written in the C programming language, supplemented by appropriate libraries, such as the Fastest Fourier Transform in the West and the Message Passing Interface. Our solver has a modular layout, which separates the algorithm from the particular physical model. Up to now, we implemented two algorithms, namely a single-host algorithm and an MPI-based distributed-memory algorithm, as well as a hand-full of physical models. The set of physical models comprises e.g. the real scalar Φ^4 theory or the linear σ model on various levels of physical approximations. The Kadanoff-Baym equations for any model may be solved with either algorithm, depending on discretization parameters and available computational resources. Exploiting the interoperability with C offered by Fortran 2003, we also set up an interface, which allows to implement certain computationally expensive model routines in Fortran.

Due to its modular design our code can easily be extended to comprise new theories. The code has been highly optimized, but still portability has been retained. OpenMP threads may be used on a model basis, thus offering the possibility for pure shared-memory as well as hybrid programming.

4.1 Single-Host Algorithm

4.1.1 Memory Layout

For the numerical solution of the Kadanoff-Baym equations we follow exactly the lines of Refs. [1–3, 9], i.e. for the spatial coordinates we employ a standard discretization on a three-dimensional lattice with lattice spacing a_s and N_s lattice sites in each direction. Thus, the lattice momenta are given by

$$\hat{p}_{n_j} = \frac{2}{a_s}\sin\left(\frac{\pi n_j}{N_s}\right) ,$$

where n_j, $j \in \{1,2,3\}$, enumerates the momentum modes in the j-th dimension. As we consider a spatially homogeneous and isotropic system, for given times (x^0, y^0) we only need to store the propagator for momentum modes with $\frac{N_s}{2} \geq n_1 \geq n_2 \geq n_3 \geq 0$. This reduces the required amount of computer memory by a factor of 48. For the practical implementation, it is convenient to map this three-dimensional structure of the independent momentum modes onto a one-dimensional array.

As we will see in the next subsubsection, for a given time x^0 we have to compute the memory integrals for all times (x^0, y^0) with $y^0 \leq x^0$. This requires that we know the propagator for all times $(t_1, t_2) \in [0, x^0]^2$. Accordingly, the discretization in time leads to a history matrix $H = \{0, a_t, 2a_t, \ldots\}^2$, where a_t is the time-step size. Exploiting the symmetry of the statistical scalar propagator G_F with respect to the interchange of its time arguments, we only need to store the values of the statistical scalar propagator for all (x^0, y^0) with $x^0 \geq y^0$. In very much the same way we can exploit the corresponding anti-symmetry of the scalar spectral-function G_ρ. This reduces the memory consumption by another factor of 2. A convenient way to implement this structure in a computer program is to use a two-dimensional array of pointers, where each entry aims to a one-dimensional array consecutively containing the values of all propagator components in all momemtum modes for times (x^0, y^0). The afore mentioned symmetry of the statistical propagator can be realized by having the pointers for the times (x^0, y^0) and (y^0, x^0) aim to the same propagator-components array. Of course, when computing the memory integrals, one has to guarantee that the spectral function is accompanied by an additional minus sign where necessary.

4.1.2 Time Stepping

In order to discretize the second derivative with respect to time, which appears in the Kadanoff-Baym equations (1) and (2), we use forward and backward derivatives [9]:

$$\begin{aligned} \partial_{x^0}^2 G\left(x^0, y^0, k\right) \quad &\rightarrow \quad \Delta_0^b \Delta_0^f G\left(x^0, y^0, \hat{\boldsymbol{k}}\right) \\ &= \frac{G\left(x^0 + a_t, y^0, \hat{\boldsymbol{k}}\right) - 2G\left(x^0, y^0, \hat{\boldsymbol{k}}\right) + G\left(x^0 - a_t, y^0, \hat{\boldsymbol{k}}\right)}{a_t^2} . \end{aligned}$$

Solving the discretized Kadanoff-Baym equations for $G(x^0 + a_t, y^0, \hat{\boldsymbol{k}})$, we obtain

$$\begin{aligned} G\left(x^0 + a_t, y^0, \hat{\boldsymbol{k}}\right) &= 2G\left(x^0, y^0, \hat{\boldsymbol{k}}\right) - G\left(x^0 - a_t, y^0, \hat{\boldsymbol{k}}\right) \\ &+ a_t^2 \left[MEMINT\left(x^0, y^0, \hat{\boldsymbol{k}}\right) - \left(\hat{\boldsymbol{k}}^2 + M^2\left(x^0\right)\right) G\left(x^0, y^0, \hat{\boldsymbol{k}}\right)\right] . \end{aligned} \tag{6}$$

Suppose, for a given x^0 we know the propagator throughout the history matrix $H(x^0) = \{0, \ldots, x^0\}^2$, which is represented by the dark-gray square in the left illustration of Fig. 4. The time-stepping algorithm then proceeds as follows: First of all, we compute the statistical and spectral self energies (4) and (5) for all $y^0 \leq x^0$ and all momentum modes $\hat{\boldsymbol{k}}$. Above all, here it is crucial to compute the emerging convolutions using a Fast Fourier Transform algorithm [5]. Then we can compute the memory integrals and apply Eq. (6) for all (x^0, y^0) with $y^0 \leq x^0$ and all momentum modes $\hat{\boldsymbol{k}}$ as indicated on the left hand side in Fig. 4 by steps 1, 2 and 3. Now, we know the propagators for all times $(x^0 + a_t, y^0)$ with $y^0 \leq x^0$. At this point, it is important to take the (anti-)symmetry of the various propagator components into account, due to which we also know the propagators for $(y^0, x^0 + a_t)$. As a matter of fact, the upper boundary terms of all memory integrals vanish due to the anti-symmetry of the spectral function and the spectral self-energy. Hence, we do not need to compute any self energy for times $(x^0, x^0 + a_t)$, and we immediately can compute the memory integrals and apply Eq. (6) (step 4 on the left hand side in Fig. 4). Thereby, we have completed the history matrix $H(x^0 + a_t)$, which is represented by the light-gray box in the left drawing of Fig. 4.

In order to compute the memory integrals one can employ a simple trapezoidal rule. However, at this point there may be room for improvements. Maybe, numerical

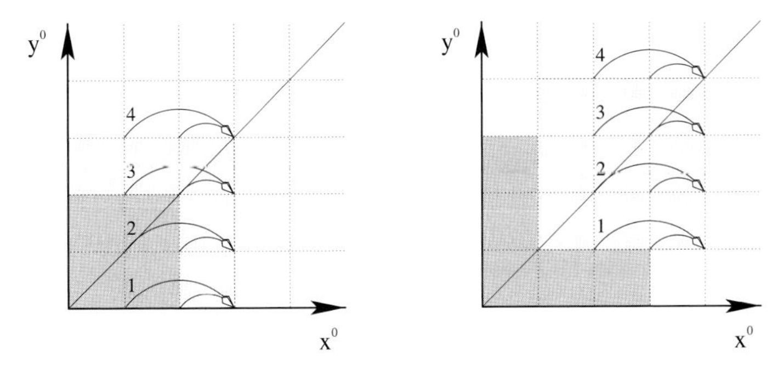

Fig. 4 In order to evolve the Kadanoff-Baym equations one step from x^0 to $x^0 + a_t$, we have to advance the propagators for all (x^0, y^0) with $y^0 \leq x^0 + a_t$ in the order specified in the left figure. The right figure illustrates the shifting of the history matrix along the bisecting line of x^0-y^0-plane

errata can be reduced by a more sophisticated integration scheme. So far, however, this has not been investigated.

4.1.3 History Cut-Off

Obviously, the memory layout described in previous subsubsections calls for an enormous amount of computer memory. Therefore it is impossible to evolve the Kadanoff-Baym equations up to reasonable times keeping the complete history matrix of all propagator components in memory. Fortunately, according to Fig. 1 correlations between earlier and later times are damped exponentially, which allows us to introduce a history cut-off. A convenient way to implement this history cut-off is to use a history matrix of fixed size

$$H = \{0, a_t, 2a_t, \ldots, (N_t - 1)\, a_t\}^2 ,$$

and to store the propagator components at indices $(x^0 \bmod N_t, y^0 \bmod N_t)$. The time-stepping algorithm and the computation of the memory integrals described in the previous subsubsection can easily be adjusted to account for this history cut-off. After the history matrix has been filled completely, at each time x^0 we advance the Kadanoff-Baym equations only for times $y^0 \in \{x^0 - N_t + a_t, \ldots, x^0 + a_t\}$ in the order specified on the right hand side in Fig. 4. This history cut-off amounts to shifting the history matrix along the bisecting line of the x^0-y^0-plane as indicated by the gray squares.

4.2 *Parallel Distributed-Memory Algorithm*

Our distributed-memory algorithm has been realized using the Message Passing Interface. In this subsection, we describe how one can distribute the data and the computations described in the previous section on an arbitrary number of processes on

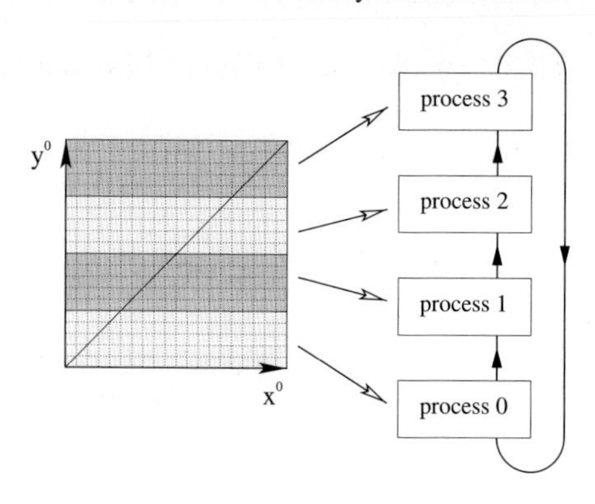

Fig. 5 The history matrix is divided into stripes and distributed among a number of processes, which communicate their results

various computer nodes, and in particular how these processes can communicate their results.

4.2.1 Memory Layout

The sequential algorithm described in the previous subsection was designed to run on a single computer. Therefore memory consumption was a critical issue and we had to exploit the (anti-)symmetry of the various propagator components in order to reduce the required amount of memory by a factor of 2. For the parallel algorithm described in this subsection, however, it is convenient to abstain from this reduction of memory. Instead we will allocate memory for the complete history matrix. The increasing requirement of computer memory can easily be compensated by an increasing number of participating computer nodes. Suppose, we started P processes on P identical computers[1]. Then we divide the history matrix in P equally sized stripes parallel to the x^0-axis and each of the P processes allocates the memory needed for one of those stripes. Fig. 5 adumbrates the situation for the case $P = 4$.

4.2.2 Time Stepping

The integration of the Kadanoff-Baym equations according to steps 1, 2 and 3 of both illustrations in Fig. 4 can be performed by all processes simultaneously. However, not all processes are working on this task from the very beginning. Rather, they start working one after another during the filling of the history matrix. Next, taking their (anti-)symmetry properties into account the obtained propagator components $G(x^0 + a_t, y^0, \hat{\boldsymbol{k}})$ are sent to the process maintaining the history-matrix stripe which contains the time $y^0 = (x^0 + a_t) \bmod N_t$. This is most easily accomplished with appropriate derived MPI-datatypes. Performing step 4 of Fig. 4 this latter process then completes the time-stepping procedure.

[1] For simplicity of the algorithm and in order to achieve an optimal work-load balance P must be a divisor of the size of the history matrix N_t.

4.2.3 Self Energies, Memory Integrals, and History Cut-Off

Each process computes the self energies (4) and (5) for all times, which belong to its history-matrix stripe. The self energies are then distributed to all processes by a repeated round-robin shift operation as indicated by the arrows in Fig. 5. After the self energies have been distributed among all processes, the computation of the memory integrals may proceed exactly along the lines of the single-host algorithm described in the previous subsection. Eventually, the history cut-off also can be adapted to the parallel distributed-memory algorithm.

4.3 Memory Requirements

In order to give an impression on the consumption of computer memory, it might be illustrative to put together some numbers: In the purely scalar case considered in Ref. [7] there are 2 propagator components, the statistical propagator and the spectral function. Our numerical solutions were achieved on a single Pentium4 desktop PC with $N_t = 500$ and $N_s = 32$. Due to isotropy, for $N_s = 32$ there are 969 independent momentum modes. Thus using double precision requires

$$2 \times \frac{500 \times 501}{2} \times 969 \times 8 \text{ Bytes} = 1.9 \text{ GBytes} .$$

This number is doubled in the case of the linear σ-model, where one has to cope with 6 propagator components. However, due to the leap-frog algorithm the 4 fermion-propagator components can be stored in an alternating fashion, such that they only require as much computer memory as the 2 scalar propagator components.

In contrast to this, studies of tachyonic preheating including fermions and scalars cannot be performed on a single desktop PC anymore. In this model, we have to cope with 6 *effective* propagator components. Thus, using the distributed-memory algorithm with double precision, $N_t = 2000$, and $N_s = 32$, we need

$$6 \times 2000 \times 2000 \times 969 \times 8 \text{ Bytes} = 187.1 \text{ GBytes} .$$

5 Conclusions and Outlook

Studying the quantum dynamics of nonequilibrium transport phenomena in the early universe is important in order to compare theoretical predictions with cosmological observations as well as particle physics experiments. Beyond that, it also includes important aspects for adjacent research fields, such as relativistic heavy ion collisions. We argued that the numerical solution of the emerging equations requires sophisticated HPC techniques.

The techniques developed so far are important in order to assess the validity of the (semi-)classical treatment of the out-of-equilibrium decay process of heavy Majorana neutrinos, which can explain the observed overabundance of matter over antimatter in the universe via the leptogenesis mechanism. The latter provides a lever arm to new physics at extremely high energies by relating parameters observable at low-energy neutrino experiments with those of the heavy Majorana neutrinos via the seesaw mechanism. However, this relation relies on the correct calculation of the produced baryon asymmetry based on Boltzmann equations. Quantum corrections are expected to be important here, especially in the case of resonant leptogenesis. Therefore, a treatment based on Kadanoff-Baym equations which incorporate off-shell effects is desirable.

References

1. Berges, J.: Controlled nonperturbative dynamics of quantum fields out of equilibrium. Nucl. Phys. **A699**, 847–886 (2002)
2. Berges, J.: Introduction to nonequilibrium quantum field theory. AIP Conf. Proc. **739**, 3–62 (2005)
3. Berges, J., Borsányi, S., Serreau, J.: Thermalization of fermionic quantum fields. Nucl. Phys. **B660**, 51–80 (2003)
4. Borsanyi, S., Reinosa, U.: Renormalised nonequilibrium quantum field theory: scalar fields (2008)
5. Frigo, M., Johnson, S.G.: The Design and Implementation of FFTW3. Proceedings of the IEEE **93**(2), 216–231 (2005). Special issue on "Program Generation, Optimization, and Platform Adaptation"
6. Garny, M., Müller, M.M.: Kadanoff-Baym Equations with Non-Gaussian Initial Conditions: The Equilibrium Limit. Phys. Rev. **D80**, 085011 (2009). 10.1103/PhysRevD.80.085011
7. Lindner, M., Müller, M.M.: Comparison of Boltzmann equations with quantum dynamics for scalar fields. Phys. Rev. **D73**, 125002 (2006)
8. Lindner, M., Müller, M.M.: Comparison of Boltzmann Kinetics with Quantum Dynamics for a Chiral Yukawa Model Far From Equilibrium. Phys. Rev. **D77**, 025027 (2008)
9. Montvay, I., Münster, G.: Quantum fields on a lattice. Cambridge University Press (1994)

Topological Structure of the QCD Vacuum Revealed by Overlap Fermions

Ernst-Michael Ilgenfritz, Karl Koller, Yoshiaki Koma, Gerrit Schierholz, and Volker Weinberg

Abstract Overlap fermions preserve a remnant of chiral symmetry on the lattice. They are a powerful tool to investigate the topological structure of the vacuum of Yang-Mills theory and full QCD. Recent results concerning the localization of topological charge and the localization and local chirality of the overlap eigenmodes are reported. The charge distribution is radically different, if a spectral cut-off for the Dirac eigenmodes is applied. The density $q(x)$ is changing from the scale-a charge density (with full lattice resolution) to the ultraviolet filtered charge density. The scale-a density, computed on the Linux cluster of LRZ, has a singular, sign-coherent global structure of co-dimension 1 first described by the Kentucky group. We stress, however, the cluster properties of the UV filtered topological density resembling the instanton picture. The spectral cut-off can be mapped to a bosonic smearing procedure. The UV filtered field strength reveals a high degree of (anti)selfduality at "hot spots" of the action. The fermionic eigenmodes show a high degree of local chirality. The lowest modes are seen to be localized in low-dimensional space-time regions.

Ernst-Michael Ilgenfritz
Institut für Physik, Humboldt-Universität zu Berlin, 12489 Berlin, Germany
e-mail: ilgenfri@physik.hu-berlin.de

Karl Koller
Fakultät für Physik, Ludwig-Maximilians-Universität München, 80333 München, Germany
e-mail: Karl.Koller@lrz.uni-muenchen.de

Yoshiaki Koma
Numazu College of Technology, 3600 Ooka, Numazu-shi, Shizuoka 410-8501, Japan
e-mail: koma@numazu-ct.ac.jp

Gerrit Schierholz
Institut für Theoretische Physik, Universität Regensburg, 93040 Regensburg, Germany
Deutsches Elektronen-Synchrotron DESY, 22603 Hamburg, Germany
e-mail: gsch@mail.desy.de

Volker Weinberg
Leibniz-Rechenzentrum der Bayerischen Akademie der Wissenschaften, 85748 Garching b. München, Germany
e-mail: Volker.Weinberg@lrz.de

S. Wagner et al. (eds.), *High Performance Computing in Science and Engineering, Garching/Munich 2009*, DOI 10.1007/978-3-642-13872-0_40,

1 Introduction: Overlap Fermions and Topological Charge

Quantum chromodynamics (QCD) is the theory of strong interactions. It is formulated in terms of quarks and gluons. The task is twofold: to describe the composite structure and the high-energy interactions of strongly interacting *hadrons*, in both cases taking the substructure in terms of quarks and gluons into account. One problem for theorists and experimentalists is that quarks are permanently *confined* inside hadrons. Apart from details, the spectrum and symmetries of hadrons are dictated by an approximate *chiral symmetry* and its spontaneous breaking by the interaction via gluons. These properties are unique for QCD as part of the standard model and result from the vacuum fluctuations of gluons. At a temperature of about 160 MeV the hadronic world experiences a transition to a quark-gluon plasma phase with rather unusual properties. Then, the vacuum structure has changed.

Simulations on a space-time lattice are the only *ab initio* approach to these phenomena. This approach has the virtue that also structural information on the vacuum fluctuations is accessible. On the other hand, there are models attempting to give a qualitative understanding. Some of them, like the instanton model, were partially successful. The instanton model became challenged more recently. A study of how the *topological charge density* is distributed in space-time and how the gluon field localizes the quarks (in analogy to the Anderson effect) is crucial to understand the microscopic mechanisms.

Overlap fermions [1, 2] possess an exact chiral symmetry on the lattice [3] and realize the Atiyah-Singer index theorem at a finite lattice spacing a [4]. This is possible because they allow a clear distinction between chiral zero modes and non-chiral non-zero modes. Depending on their chirality, counting of n_+ or n_- zero modes determines the topological charge of the gauge field as $Q = n_- - n_+$. Furthermore, they give rise to a *local* definition of the topological charge density [5]. Altogether, this makes overlap fermions an attractive tool for investigating the chiral and topological structure of the QCD vacuum.

The topological structure attracts attention because of the old hope that local excitations contributing to the winding number might not only realize the breaking of $\mathrm{U_A}(1)$ symmetry but simultaneously provide a mechanism for confinement and chiral symmetry breaking. The instanton liquid model (ILM) does not fulfill this expectation: it can account for chiral symmetry breaking but fails to give an explanation of confinement.

When the QCDSF collaboration had started to analyze ensembles of lattice configurations with overlap fermions for quenched QCD [6, 7] and later for QCD with dynamical quarks [8] by diagonalizing the massless overlap Dirac operator and to store an – until then unprecedent – number of eigenmodes ($\mathscr{O}(150)$ per configuration), the way was open for a serious investigation of the structure of topological charge [9–11]. Not long before we started this investigation, the instanton model had been challenged [12–14] by the observation that the topological charge, rather than appearing in $4d$ clusters, possesses a global, sign-coherent $3d$ membrane-like structure [15–17]. We shall discuss here what remains from the instanton picture.

The overlap Dirac operator D has to fulfill the Ginsparg-Wilson [18] equation,

$$\gamma_5 D^{-1} + D^{-1} \gamma_5 = a 2 R \gamma_5 , \tag{1}$$

with a local operator R. This is what maximally can remain of chiral symmetry on the lattice. A possible solution – for any input Dirac operator, i.e. for the Wilson-Dirac operator D_W as well – is the following zero mass overlap Dirac operator

$$D(m=0) = \frac{\rho}{a} \left(1 + \frac{D_W}{\sqrt{D_W^\dagger D_W}} \right) = \frac{\rho}{a} \left(1 + \mathrm{sgn}(D_W) \right) , \tag{2}$$

with $D_W = M - \frac{\rho}{a}$ where M is the Wilson hopping term and $\frac{\rho}{a}$ a negative mass term to be optimized. This operator is diagonalized using a variant of the Arnoldi algorithm. The operator can be improved by projecting the Ginsparg-Wilson circle on the imaginary axis, $\lambda \rightarrow \lambda_{\mathrm{imp}}$. The topological density can be defined (with maximal resolution a) as follows:

$$q(x) = -\mathrm{tr} \left[\gamma_5 \left(1 - \frac{a}{2} D(m=0;x,x) \right) \right] . \tag{3}$$

Using the spectral representation of (3) in terms of the eigenmodes $\psi_\lambda(x)$ with eigenvalue λ, an UV smoothed form of the density can be defined by filtering,

$$q_{\lambda_{\mathrm{sm}}}(x) = - \sum_{|\lambda| < \lambda_{\mathrm{sm}}} \left(1 - \frac{\lambda}{2} \right) \sum_c \left(\psi_\lambda^c(x) , \gamma_5 \psi_\lambda^c(x) \right) , \tag{4}$$

summed over color c and with λ_{sm} as an UV cut-off. Our intention was to study the topological charge density at *all scales*, in particular the spectral representation in terms of overlap modes defining $q_{\lambda_{\mathrm{sm}}}(x)$. Also, we had the opportunity to evaluate the topological charge density without filtering (i.e. at the resolution scale a) on the Linux cluster of the LRZ in Munich. We call this the scale-a topological density $q(x)$ in contrast to $q_{\lambda_{\mathrm{sm}}}(x)$ defined with UV smoothing. (For details see Ref. [11].)

2 Topological Density with Different Resolution

Figure 1 shows the topological density calculated with four different smoothing scales λ_{sm} given in units of $1/a$. The presentation is in the form of isosurfaces at a fixed value of $|q(x)| = q_{\mathrm{cut}}$. With increasing λ_{sm}, the number and shape of the visible clusters changes. The given configuration has $Q = 1$.

Therefore exactly one zero mode exists. In Fig. 2 we show two extreme cases, the highly localized contribution of the single zero mode alone (left) and the scale-a topological density (right). For the same value of q_{cut} the isosurfaces of the latter fill the whole volume, separating essentially a single positive from a single negative cluster. The net charge is naively expected to reside in an unpaired instanton (located at the zero mode). Actually, however, the positive and negative charge (differing by one) is globally distributed over the two clusters.

3 Cluster Analysis

More quantitative information can be obtained from a cluster analysis. The value q_{cut}, that has characterized the isosurfaces in Figs. 1 and 2, is now treated as a running parameter. The region characterized by $|q(x)| > q_{\mathrm{cut}}$ (i.e. the interior of the isosurfaces) consists of some number of mutually disconnected clusters. We have classified them according to their size and other properties, for example their fractal dimension. Results of the cluster analysis are presented in Fig. 3. In Fig. 3a we show the number of clusters as function of $q_{\mathrm{cut}}/q_{\mathrm{max}}$ for three lattices characterized by different coarseness. Except for the coarsest lattice with $\beta = 8.1$ ($a = 0.142$ fm), the behavior is similar. Around $q_{\mathrm{cut}}/q_{\mathrm{max}} = 0.2 \ldots 0.25$ the multiplicity of clusters reaches a maximum with a cluster density of $\mathcal{O}(75\ \mathrm{fm}^{-4})$ for $\beta = 8.45$ ($a = 0.105$ fm). The density is even higher for the finer lattice with $\beta = 8.60$ ($a = 0.096$ fm). In the limit $q_{\mathrm{cut}}/q_{\mathrm{max}} \to 0$ the number of clusters reduces to the two oppositely charged global clusters. Figure 3b shows the volume fraction of the biggest cluster relative to all clusters. Except for the coarsest ($\beta = 8.1$) lattice, above $q_{\mathrm{cut}}/q_{\mathrm{max}} = 0.2$ where the number of clusters is still growing towards the maximum, this fraction is small (< 10 %). It changes rapidly below $q_{\mathrm{cut}}/q_{\mathrm{max}} = 0.2$ and approaches 50 % for $q_{\mathrm{cut}}/q_{\mathrm{max}} < 0.1$ and remains so in the limit when only the two biggest clusters remain. The distance between the two largest clusters is shown in Fig. 3c. The "distance" is defined as the *maximum* taken over all points in one cluster

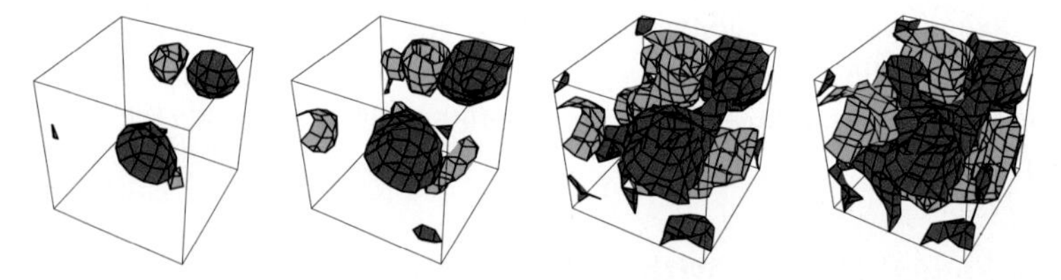

Fig. 1 Distribution of the topological charge density in a given time slice of a $12^3 \times 24$ lattice at $\beta = 8.10$: the spectral cut-off $a\,\lambda_{\mathrm{sm}} = 0.14, 0.28, 0.42, 0.56$ (from left to right). The isosurfaces are shown for $q(x) = \pm 0.0005$. Colors red/green denote the sign of topological charge. Fig. from [9]

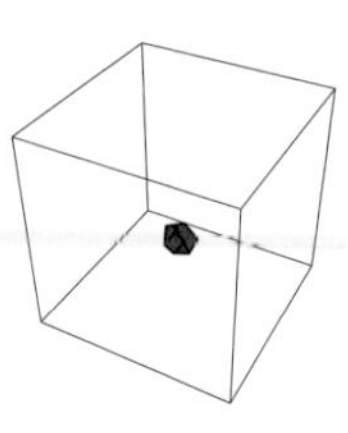

Fig. 2 The same plot as in Fig. 1 for the zero mode contribution (left) and the unfiltered density (right). Fig. from [9]

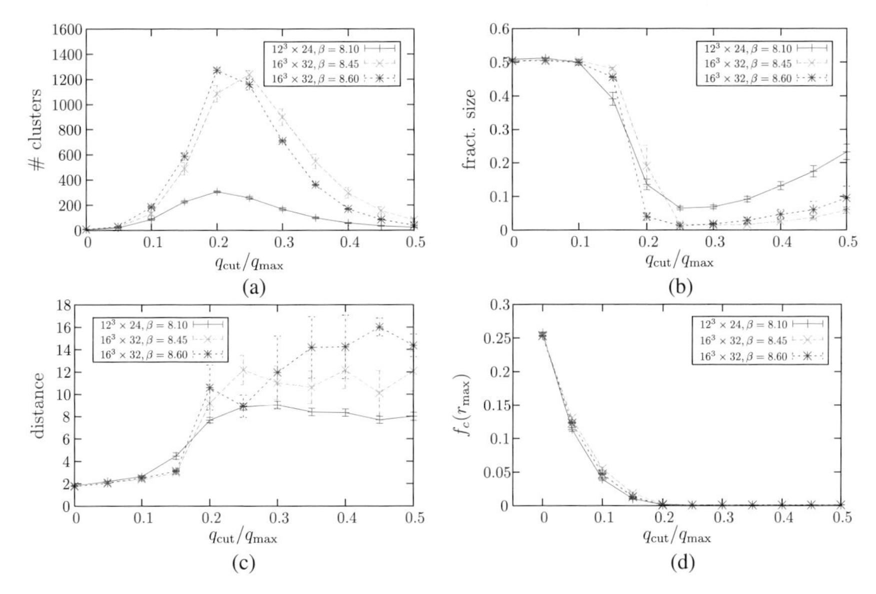

Fig. 3 Cluster analysis of the unfiltered topological density: the q_{cut}-dependence is shown (a) of the number of separate clusters, (b) of the volume of the largest cluster relative to all clusters, (c) of the distance between the two largest clusters in lattice units and (d) of the connectivity describing cluster percolation [11]. The data is plotted for the $12^3 \times 24$ lattice configurations at $\beta = 8.10$ (red), for the $16^3 \times 32$ lattice configurations at $\beta = 8.45$ (green) and for the $16^3 \times 32$ lattice configurations at $\beta = 8.60$ (blue). Fig. from [11]

of the *minimal distance* to any point of the other cluster. It approaches $2a$ for $q_{cut}/q_{max} < 0.1$, meaning that the two clusters are closely entangling each other. There are no points deep in the interior, such that the clusters must be considered to be of lower dimension than $4d$. Figure 3d shows the connectivity (explained in [11]) of the biggest cluster, describing the ability to span the whole lattice. The connectivity signals the onset of percolation which begins below $q_{cut}/q_{max} = 0.2$, too. All this confirms the description of "melting instantons" given earlier by the Kentucky group [12–14].

The meaning of the cluster analysis is geometrically visualized in Fig. 4, again contrasting the scale-a density with the filtered density. Both types of clusters are presented as a function of q_{cut}. The upper row shows, say for $q_{cut}/q_{max} \approx 0.2$, that more and more irregular, spiky clusters of the unfiltered density are popping up before percolation sets in. The lower row demonstrates that the number of clusters of the filtered density remains small and practically independent of q_{cut}. The density is $\mathcal{O}(1\ \text{fm}^{-4})$, that means comparable with what has been estimated for the instanton density. Of course, with lowering q_{cut}, the clusters slightly grow before they eventually merge. Coalescence happens finally below $q_{cut}/q_{max} = 0.1$. We see that the UV smoothed density exhibits cluster properties similar to the instanton model [19] or what cooling studies on the lattice have shown earlier [20, 21].

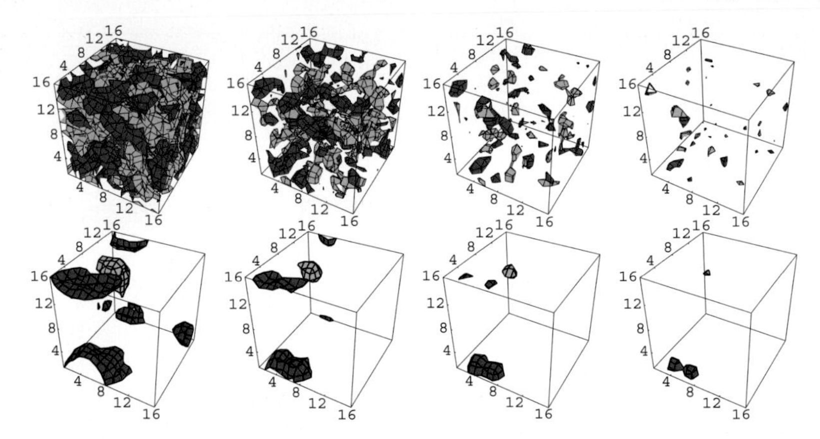

Fig. 4 Isosurfaces of topological charge density with $|q(x)|/q_{\max} = 0.1, 0.2, 0.3, 0.4$, in one timeslice of a single $16^3 \times 32$ configuration at $\beta = 8.45$. The upper pictures are based on the scale-a density, the lower pictures on the eigenmode-truncated density with $a\lambda_{\text{sm}} = 0.076$. Color encodes the sign of the charge enclosed by the isosurface. Fig. from [10]

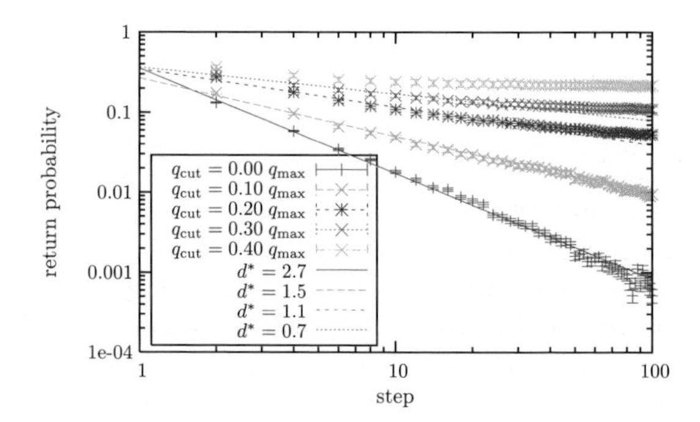

Fig. 5 The return probability of random walks in the biggest cluster of the scale-a topological density as a function of the step number, and the fractal dimensions inferred from that for various lower cut-off values q_{cut}. Fig. from [11]

4 Fractal Dimensions

The visualization in three dimensions does not immediately expose the (fractal) dimension of the clusters of the scale-a topological density. We have measured the (fractal) dimension of these clusters by a random walk method. While the cut-off q_{cut} defines the extension of the clusters as discussed above, a random walk is arranged inside the biggest cluster that is unable to penetrate the (fractal) border. A rough estimate of the dimensionality of the cluster at the respective level of q_{cut} can be inferred from the return probability to the starting point of the random walk. In d^* dimensions it behaves as a function of the number of steps like $P(\mathbf{0},\tau) = 1/(2\pi\tau)^{d^*/2}$. Figure 5 shows the observed power like behavior for different $q_{\text{cut}}/q_{\max}$. The dimensions d^* extracted from this study are shown in the legend. This analysis reveals a continuous change of the fractal dimension from $d^* < 1$ (characterizing the irregular spikes) to $d^* \approx 3$ (characterizing the $3D$ membranes) with a lowering cut-off $q_{\text{cut}}/q_{\max} = 0.4\ldots0.0$.

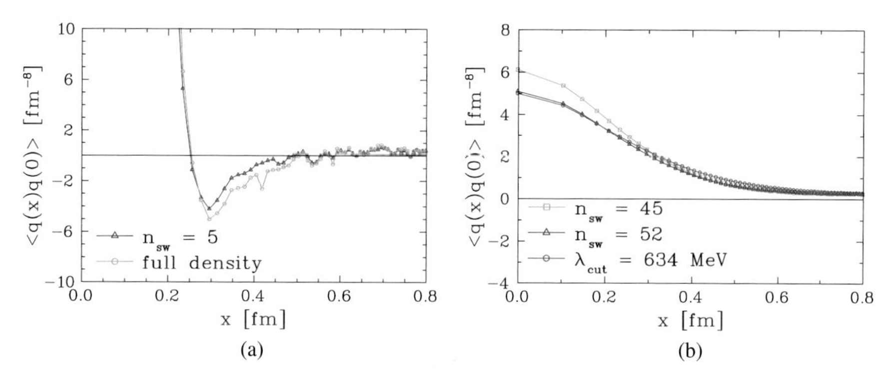

Fig. 6 Left: the two-point function of the fermionic topological density at scale a (green) compared with the bosonic definition after 5 steps of stout-link smearing (blue). Right: the two-point function of the fermionic topological density with an UV cut-off at $\lambda_{\rm sm} = 634$ MeV (red) compared with the bosonic definition after 52 steps of stout-link smearing (blue) when the correlator is fitted best. The best point-by-point matching of the density $q_{\lambda_{\rm sm}}$ with $\lambda_{\rm sm} = 634$ MeV is achieved with 45 steps. For this case of less smearing the correlator is steeper. Fig. from [23]

5 Smearing vs. Filtering

The similarity of the filtered topological density to clusters obtained earlier by smearing methods can be studied more in detail by direct comparison. First let us recall that the overlap operator is *not ultralocal*. Even the scale-a topological density given by overlap fermions represents some inherent non-locality of the overlap operator [22]. This is seen by considering the two-point function of the topological density. Theoretically, one expects that it is negative for all non-zero distances, with an infinite, δ-like contact term at vanishing distances. The two-point function actually measured is shown as the green curve in Fig. 6a. One needs to go to larger β (a finer lattice) in order to see the positive core slowly shrinking and the minimum becoming deeper. To compare with smearing, the alternative measurement of the topological density (and of the two-point function) uses an improved bosonic definition of the topological charge density, essentially

$$q(x) = \frac{1}{16\pi^2} \mathrm{tr}\left(\mathbf{E}(x) \cdot \mathbf{B}(x)\right) , \tag{5}$$

in terms of an highly improved electric and magnetic field strength [24], that includes plaquettes and bigger loops. To apply it to a generic lattice configuration requires a few iterations of smearing (actually, of stout link smearing with respect to an overimproved action [25]) before integer-valued topological charges Q are measured with good precision. Figure 6a shows that the overlap definition and the bosonic definition of the topological density agree best to reproduce a similar two-point function at 5 smearing steps. Figure 6b shows the gluonic two-point function after 45 and 52 smearing steps compared with the overlap definition of the two-point

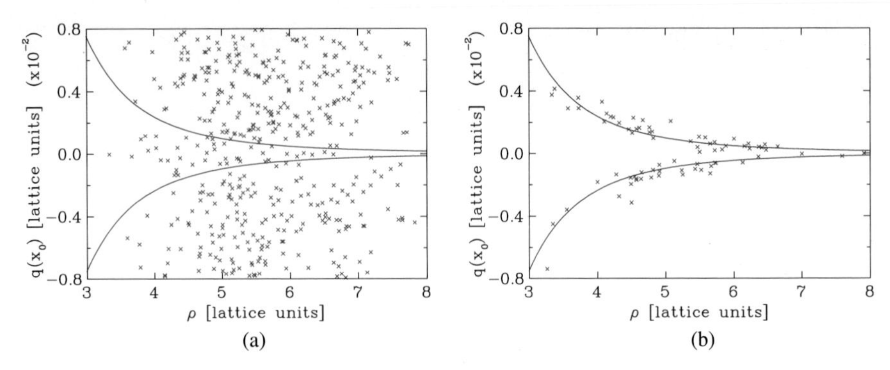

Fig. 7 The cluster content after 5 and 40 smearing steps in the q_{max}-ρ_{inst} plane. Fig. from [23]

function with $\lambda_{sm} = 634$ MeV. At this level of smearing, the negativity is already washed out and extended clusters are visible. A point-by-point matching of the topological density and a matching of the two-point function is achieved with almost the same number of smearing iterations.

There are two parameters, q_{max} and the size ρ_{inst} (historically the instanton radius) suitable to characterize each cluster. Now the size parameter ρ_{inst} is estimated by the curvature of $q(x)$ close to the maxima x_0 where $q(x_0) = q_{max}$. In the form of a scatter plot in the plane spanned by these two parameters one can describe and compare the cluster composition of the gauge field. In Fig. 7 this is shown after 5 and 40 stout-link smearing steps. After 40 smearing steps the cluster multiplicity is essentially reduced, while an instanton-like relation (the curves drawn in Fig. 7)

$$q_{max} = \frac{6}{\pi^2 \rho_{inst}^4}, \tag{6}$$

is enforced within some tolerance. At the same time the clusters are far from being ideal O(4) symmetric instanton solutions. The bosonic and the fermionic view of the cluster structure is shown in the left and right panel of Fig. 8. They show a remarkable similarity.

6 Selfduality

This leads to the question to what extent the UV smoothed *fieldstrength tensor* is *locally* selfdual or antiselfdual, or even a semiclassical (instanton) solution. Gattringer was the first to ask this question in Ref. [26]. We have applied his technique using the eigenmodes of the overlap Dirac operator. Truncating the sum over eigenvalues, an UV filtered form of the field strength $F_{\mu\nu}$ can be obtained that allows to evaluate the respective infrared (IR), UV smoothed topological charge density,

$$q_{IR}(x) \propto \mathrm{Tr}\left(F_{\mu\nu}(x)\tilde{F}_{\mu\nu}(x)\right), \tag{7}$$

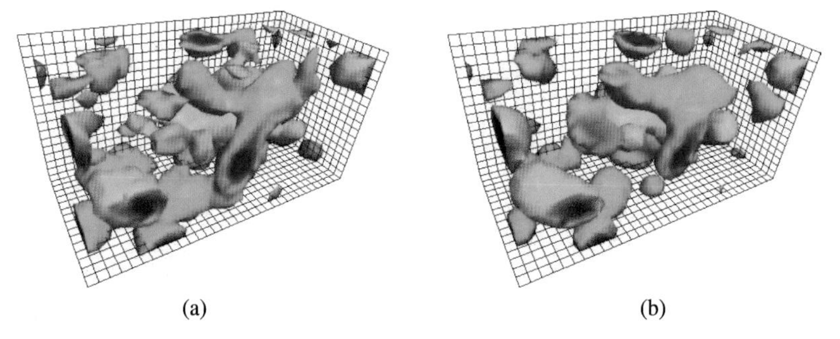

Fig. 8 The fermionic topological charge density of a $Q = 0$ configuration with $\lambda_{\rm sm} = 634$ MeV (left) compared with the bosonic one after 48 sweeps of stout-link smearing (right). Negative: blue/green, positive: red/yellow. Fig. from [23]

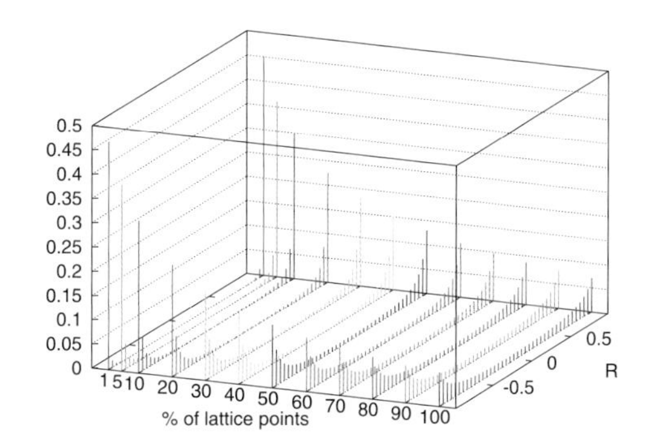

Fig. 9 Normalized histogram with respect to the UV smoothed (anti)selfduality R for different subsamples of lattice points (1%, 5%, 10% etc.) ordered with respect to the UV smoothed action. 20 overlap modes have been included in the smoothing. Fig. from [11]

and action density

$$s_{IR}(x) \propto \mathrm{Tr}\left(F_{\mu\nu}(x)F_{\mu\nu}(x)\right) . \tag{8}$$

Analogously to the local chirality of the non-zero modes one can define the ratio

$$r(x) = \frac{s_{IR}(x) - q_{IR}(x)}{s_{IR}(x) + q_{IR}(x)} , \tag{9}$$

which can be converted to

$$R(x) = \frac{4}{\pi} \arctan\left(\sqrt{r(x)}\right) - 1 \in [-1, +1] . \tag{10}$$

Regions with $R(x) \approx -1$ or $R(x) \approx +1$ are characterized by the field strength being locally selfdual (antiselfdual). Figure 9 shows histograms with respect to R for the whole lattice and for subsets of lattice points selected by ranking according to the action density. With higher action density (say, for $< 30\%$ of the lattice points forming "hot spots" of filtered action S_{IR}) the parameter R is distributed close to ± 1 (i.e. $|R| > 0.7$). Clusters with respect to $|R| > 0.98$ are overlapping with clusters of a suitably filtered topological density.

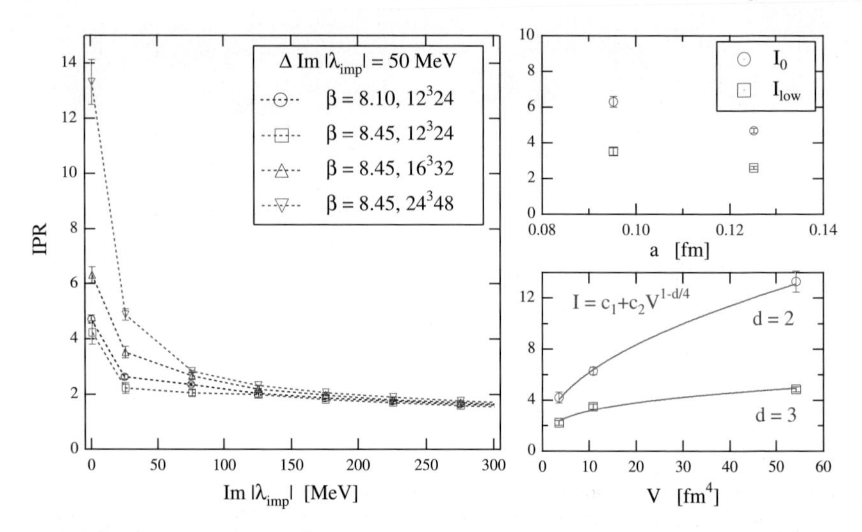

Fig. 10 IPR: the dependence on λ (left), the a-dependence (right upper), and the V-dependence (right lower). Fig. from [9]

7 Localization and Local Chirality of Overlap Eigenmodes

We have also analyzed the localization behavior of individual overlap eigenmodes. The interest in the localization properties of Dirac and scalar eigenmodes has been discussed by de Forcrand [27]. In the left panel of Fig. 10 we show the average inverse participation ratio (IPR) typical for zero modes and for bins of eigenvalues in dependence on coarseness and lattice size. The localization grows approaching the continuum limit and with increasing physical volume. The right panel shows, separately for zero modes and the first bin ($|\lambda| < 50$ MeV), the change of the IPR with a and with V. From fits we have concluded that zero modes are typically 2-dimensionally extended, while the lowest non-zero modes are 3-dimensional objects.

Similar to (9) the local chirality $X(x)$ of the non-zero modes can be defined. This quantity is highly correlated with the UV smoothed topological density. For $\lambda_{\mathrm{sm}} = 200$ MeV the correlation function between $X(x)$ and $q_{\lambda_{\mathrm{sm}}}(y)$ is shown in Fig. 11. The correlator is positive for the lowest 120 eigenmodes within a range of distance $R < \mathcal{O}(1 \text{ fm})$.

8 Technical Details

To compute the low lying eigenvalues of the overlap Dirac operator (2) we use an adaption of the implicitly restarted Arnoldi method which is used in the ARPACK [28] library. The advantage of this method is that the action of a matrix on a vector can be computed freely without the need to express the input matrix explicitly.

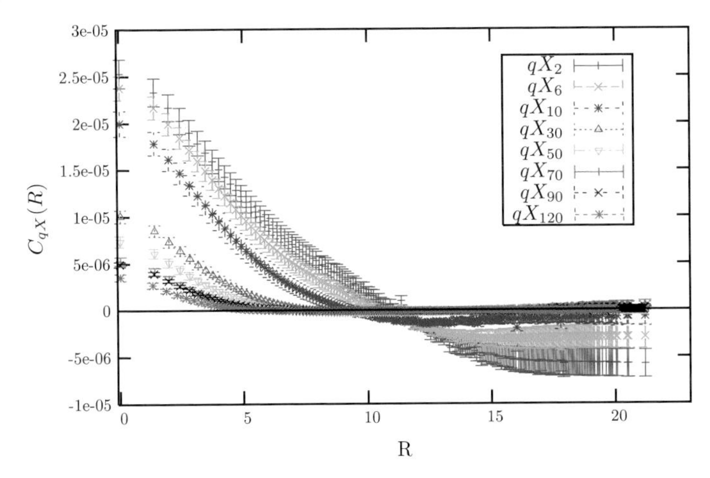

Fig. 11 The correlation function between the local chirality $X(x)$ of selected eigenmodes and the filtered topological density $q_\lambda(y)$ in a configuration with topological charge $Q = 0$. Fig. from [11]

The main computational challenge using the overlap Dirac operator as an input matrix is the computation of the sign function $\text{sgn}(D_W)$ in (2). We project out the lowest $\mathcal{O}(50)$ eigenvalues of D_W and treat them exactly. The rest acting in the orthogonal subspace is approximated using minmax polynomials [29]. The application of the Wilson-Dirac operator D_W on a vector, the main computational kernel of almost every lattice QCD code, is implemented in assembler and highly optimized using SSE2 instructions. The data is aligned on full cache lines and stored in a special way to make optimal use of the 128-bit XMM registers. Prefetch instructions are issued to move the data to the cache accurately timed. The SSE2 version reaches approx. $1/3$ of peak performance on a $16^3 \times 32$ lattice on the Linux cluster of the LRZ.

The computation of the scale-a topological density $q(x)$ requires to evaluate the local trace over Dirac and color indices of the overlap Dirac operator with the γ_5 matrix included according to (3). This can be trivially parallelized, but is numerically extremely expensive, since the overlap operator has to be applied on $12\ V_{\text{lat}}$ unit vectors. To compute $q(x)$ on one single configuration on a $16^3 \times 32$ lattice approx. $1.3 \cdot 10^4$ CPU hours for totally 1.6 million applications of the overlap Dirac operator were required.

9 Conclusions

We have confirmed that the scale-a topological density is highly singular and the topological charge of either sign is globally filling the two extended $3d$ percolating structures. A similar low-dimensionality is found for chiral zero-modes (two-dimensional) and the lowest non-chiral non-zero modes (three-dimensional). The microscopic topological density is correlated with vortices and monopoles [30].

The UV filtered topological density shows cluster properties reminiscent of instantons. We found a highly correlated behavior of a huge number of lowest modes resembling the presence of a semiclassical background. The suitably UV smoothed field strength tensor becomes selfdual (antiselfdual) at "hot spots" where the corresponding action becomes maximal. The reason for this self-organization between the fermionic modes is still unclear.

References

1. Neuberger, H.: Exactly massless quarks on the lattice. Phys. Lett. **B417**, 141–144 (1998). DOI 10.1016/S0370-2693(97)01368-3
2. Neuberger, H.: More about exactly massless quarks on the lattice. Phys. Lett. **B427**, 353–355 (1998). DOI 10.1016/S0370-2693(98)00355-4
3. Lüscher, M.: Exact chiral symmetry on the lattice and the Ginsparg-Wilson relation. Phys. Lett. **B428**, 342–345 (1998). DOI 10.1016/S0370-2693(98)00423-7
4. Hasenfratz, P., Laliena, V., Niedermayer, F.: The index theorem in QCD with a finite cut-off. Phys. Lett. **B427**, 125–131 (1998). DOI 10.1016/S0370-2693(98)00315-3
5. Niedermayer, F.: Exact chiral symmetry, topological charge and related topics. Nucl. Phys. Proc. Suppl. **73**, 105–119 (1999). DOI 10.1016/S0920-5632(99)85011-7
6. Galletly, D., et al.: Quark spectra and light hadron phenomenology from overlap fermions with improved gauge field action. Nucl. Phys. Proc. Suppl. **129**, 453–455 (2004). DOI 10.1016/S0920-5632(03)02611-2
7. Galletly, D., et al.: Hadron spectrum, quark masses and decay constants from light overlap fermions on large lattices. Phys. Rev. **D75**, 073015 (2007). DOI 10.1103/PhysRevD.75.073015
8. Göckeler, M., et al.: Simulating at realistic quark masses: light quark masses. PoS **LAT2006**, 160 (2006)
9. Koma, Y., et al.: Localization properties of the topological charge density and the low lying eigenmodes of overlap fermions. PoS **LAT2005**, 300 (2006)
10. Ilgenfritz, E.M., et al.: Probing the topological structure of the QCD vacuum with overlap fermions. Nucl. Phys. Proc. Suppl. **153**, 328–335 (2006). DOI 10.1016/j.nuclphysbps.2006.01.028
11. Ilgenfritz, E.M., et al.: Exploring the structure of the quenched QCD vacuum with overlap fermions. Phys. Rev. **D76**, 034506 (2007). DOI 10.1103/PhysRevD.76.034506
12. Horváth, I., et al.: On the local structure of topological charge fluctuations in QCD. Phys. Rev. **D67**, 011501 (2003). DOI 10.1103/PhysRevD.67.011501
13. Horváth, I., et al.: Low-dimensional long-range topological charge structure in the QCD vacuum. Phys. Rev. **D68**, 114505 (2003). DOI 10.1103/PhysRevD.68.114505
14. Horváth, I., et al.: Inherently global nature of topological charge fluctuations in QCD. Phys. Lett. **B612**, 21–28 (2005). DOI 10.1016/j.physletb.2005.03.004
15. Thacker, H.B.: D-branes, Wilson bags, and coherent topological charge structure in QCD. PoS **LAT2006**, 025 (2006)
16. Thacker, H.B.: Topological charge and the laminar structure of the QCD vacuum. AIP Conf. Proc. **892**, 223–226 (2007). DOI 10.1063/1.2714378
17. Thacker, H.B.: Melting instantons, domain walls, and large N. PoS **LAT2008**, 260 (2008)
18. Ginsparg, P.H., Wilson, K.G.: A remnant of chiral symmetry on the lattice. Phys. Rev. **D25**, 2649 (1982). DOI 10.1103/PhysRevD.25.2649
19. Schäfer, T., Shuryak, E.V.: Instantons in QCD. Rev. Mod. Phys. **70**, 323–426 (1998). DOI 10.1103/RevModPhys.70.323

20. Negele, J.W.: Instantons, the QCD vacuum, and hadronic physics. Nucl. Phys. Proc. Suppl. **73**, 92–104 (1999). DOI 10.1016/S0920-5632(99)85010-5
21. García Pérez, M.: QCD vacuum structure. Nucl. Phys. Proc. Suppl. **94**, 27–34 (2001). DOI 10.1016/S0920-5632(01)00924-0
22. Horváth, I., et al.: The negativity of the overlap-based topological charge density correlator in pure-glue QCD and the non-integrable nature of its contact part. Phys. Lett. **B617**, 49–59 (2005). DOI 10.1016/j.physletb.2005.04.076
23. Ilgenfritz, E.M., et al.: Vacuum structure revealed by over-improved stout-link smearing compared with the overlap analysis for quenched QCD. Phys. Rev. **D77**, 074502 (2008). DOI 10.1103/PhysRevD.77.074502
24. Bilson-Thompson, S.O., Leinweber, D.B., Williams, A.G.: Highly-improved lattice field-strength tensor. Ann. Phys. **304**, 1–21 (2003). DOI 10.1016/S0003-4916(03)00009-5
25. Moran, P.J., Leinweber, D.B.: Over-improved stout-link smearing. Phys. Rev. **D77**, 094501 (2008). DOI 10.1103/PhysRevD.77.094501
26. Gattringer, C.: Testing the self-duality of topological lumps in SU(3) lattice gauge theory. Phys. Rev. Lett. **88**, 221601 (2002). DOI 10.1103/PhysRevLett.88.221601
27. de Forcrand, P.: Localization properties of fermions and bosons. AIP Conf. Proc. **892**, 29–35 (2007). DOI 10.1063/1.2714343
28. `http://www.caam.rice.edu/software/ARPACK/`
29. Giusti, L., Hoelbling, C., Lüscher, M., Wittig, H.: Numerical techniques for lattice QCD in the epsilon-regime. Comput. Phys. Commun. **153**, 31–51 (2003). DOI 10.1016/S0010-4655(02)00874-3
30. Ilgenfritz, E.M., et al.: Localization of overlap modes and topological charge, vortices and monopoles in SU(3) LGT. PoS **LAT2007**, 311 (2007)

Part VI
Condensed Matter Physics

Gyrokinetic Turbulence Investigations Involving Ion and Electron Scales

T. Görler, F. Jenko, M.J. Pueschel, D. Told, and H. Lesch

Abstract Plasma microinstabilities are one of the key physics problems on the way to efficient power plants based on nuclear fusion. They cause anomalous heat and particle transport which significantly degrades the plasma confinement quality, thus preventing self-sustaining plasma burning in present-day experiments. Hence, extensive experimental studies are dedicated to understanding and predicting turbulence features. They are accompanied by numerical simulations which are typically based on the gyrokinetic theory. While experimental diagnostics are about to address the role of fine-scale turbulence within a bath of large-scale turbulence, nonlinear gyrokinetic codes are already able to investigate turbulent transport at a wide range of wave numbers simultaneously. However, such simulations covering several space and time scales self-consistently are computationally extremely demanding and thus need to be massively parallelized.

1 Introduction

In the present section, some basic information on plasma turbulence in the context of magnetic confinement fusion will be provided. Furthermore, the underlying theory for theoretical investigations and the requirements for massively parallel computations will be discussed briefly.

1.1 Magnetic Confinement Fusion and Plasma Turbulence

In order to manage an increasing world energy consumption, an important approach is to explore new energy sources whose resources are not running out, or which suf-

T. Görler · F. Jenko · M.J. Pueschel · D. Told · H. Lesch
Max-Planck-Institut für Plasmaphysik, EURATOM Association, Boltzmannstr. 2, 85748 Garching, Germany
e-mail: fsj@ipp.mpg.de

S. Wagner et al. (eds.), *High Performance Computing in Science and Engineering, Garching/Munich 2009*, DOI 10.1007/978-3-642-13872-0_41,

fer from side-effects, such as global warming. A well working example is provided by our star which converts energy mostly by fusing protons to helium. However, efficiency dictates that an artificial, terrestrial sun has to rely on a different but similar reaction where the nuclear fusion involves the hydrogen isotopes deuterium and tritium. During this process, the Coulomb potential barrier has to be overcome which requires temperatures of more than 100 million degrees. At those extreme conditions matter is found to be an (almost) fully ionized gas – a so-called plasma. The most promising way of isolating the hot plasma from the walls of a containing vessel is realized by the application of correspondingly shaped magnetic fields which are often reminiscent of doughnout or floating tire forms. Although motion perpendicular to magnetic field lines is significantly reduced by Lorentz forces, one nevertheless observes outward heat and particle fluxes which degrade the energy confinement in current devices in a way that self-sustained plasma burning cannot be achieved. At present, it is widely accepted that small scale instabilities, i.e. instabilities on scales of the order of the Larmor radius, are responsible for the so-called anomalous transport. Driven by the unavoidably steep density and temperature gradients occurring in fusion devices, they initially grow exponentially in time until the fluctuation amplitudes become sufficiently large for nonlinear effects to come into play. Due to the according redistribution of free energy to stabilizing modes, the system eventually reaches a quasi-stationary state far from thermal equilibrium in which significant core-to-edge transport levels are observed. A general understanding of anomalous transport is therefore crucial to predict and control fusion scenarios. In this context, it is noteworthy that various turbulence types exist which may be distinguished by means of their characteristic wave numbers and frequencies. Three prominent examples are the ion temperature gradient (ITG) driven mode, the trapped electron mode (TEM), and the electron temperature gradient (ETG) mode. While the first two types are typically found on space-time scales associated with to the ion dynamics, the ETG mode resides on electron scales. Obviously, the question arises whether both turbulence scales can be treated independently which is often done due to the enormous computational effort linked to resolving multiple space and time scales. Furthermore, the significance of ETG modes for heat transport fluxes is currently a controversial issue. Namely, mixing length estimates predicting negligible contributions have recently been challenged by new theoretical and experimental findings. A clarification along these line is desperately needed since future fusion devices will exhibit strong electron heating caused by fusion-born α particles.

1.2 Plasma Turbulence Investigations Using Gyrokinetic Theory

It is impossible to solve the equations of motion for each and every plasma particle separately. Additionally, due to the high temperatures and low densities, fusion plasmas are only very weakly collisional which limits the use of fluid models like those known from hydrodynamic turbulence investigations. Instead, a kinetic description is called for. Here, a six-dimensional Vlasov equation for each particle species needs to be solved. Furthermore, they are all coupled via Maxwell's equations. The resulting integro-differential system of equations is extremely difficult to evaluate numerically. However, in strongly magnetized plasmas, the gyromotion around the field line is decoupled from the typical turbulence time scales. The corresponding description can thus be reduced to the dynamics of a charged ring using sophisticated perturbation theories [3]. The most popular and successful such model is the so-called modern gyrokinetic theory [1, 11, 15, 16] which forms the basis for almost all ab-initio simulations of plasma microturbulence in fusion devices. A prominent example of a corresponding implementation is the GENE code which will be further discussed in the next section. With such software, many problems crucial for the next generation fusion devices can be targeted. For instance, turbulent transport contributions originating from scales much smaller than the ion gyroradius can be investigated.

2 The Plasma Turbulence Code GENE

All simulation results which will be presented in the following sections are generated using the nonlinear gyrokinetic GENE code. A brief software introduction shall therefore be given at this point.

Initially developed by F. Jenko [18], the GENE (Gyrokinetic Electromagnetic Numerical Experiment) code has now been maintained and extended at the Max-Planck-Institut für Plasmaphysik and the Garching Computing Centre for about a decade. Some of the most important milestones along this way are reported in [4, 5, 14, 22–25]. Since 2007, regular public releases have been distributed [21], and several significant code extensions have been performed in the context of international cooperations, particularly with the Centre de Recherches en Physique en Plasma at the École Polytechnique Fédérale de Lausanne. By now, GENE is considered to be Open Source, thus encouraging collaborators around the world to extend the physical comprehensiveness.

The discretized nonlinear gyrokinetic equations are solved on a fixed, five-dimensional (three spatial and two velocity space dimensions) grid which is aligned with the magnetic field lines in order to reduce the computational requirements describing highly anisotropic plasma turbulence structures. Furthermore, a so-called δf splitting technique consistent with the ordering used in the derivation of the gyrokinetic theory is employed so that only the fluctuating parts of each distribution function are propagated in time. The phase space and time operators are treated separately, following the so-called method of lines. While the time stepping is typically done with a fourth-order explicit Runge-Kutta scheme in the initial value solver mode, the numerical schemes in the spatial directions depend on the chosen type of operation. Large fusion devices where the gyroradius is much smaller than the machine size may be simulated in the local approximation. Here, all profiles are evaluated at a single radial position which allows for the application of pseudo-spectral methods in both perpendicular space directions. For smaller devices, GENE has been recently enhanced to consider the full radial profile information. Here, finite-element and finite-difference techniques have to be applied in the radial coordinate. The third spatial direction (parallel to the background magnetic field) and the velocity space directions are always discretized via finite difference methods. For an efficient use of high-performance architectures, GENE is hybrid-parallelized, i.e. all core parts can be run in parallel using the OpenMP and/or the MPI paradigm. The latter is achieved through domain composition in the species, in the velocity space directions and in two of the three spatial directions (all three in the nonlocal code, respectively). Very good scaling properties have been demonstrated on up to 32,768 processors [23], and the GENE software package has successfully been ported and used efficiently on various massively parallel systems, for instance IBM BlueGene/P, IBM Power5/6, Cray XT4/5, and of course SGI Altix 4700.

3 Nonlinear Gyrokinetic Simulations Covering Multiple Spatio-Temporal Scales

This section is dedicated to the presentation of one of the first efforts to self-consistently simulate spatio-temporally separated turbulence modes, an attempt which has been made possible by the current and a previous LRZ (DEISA) project. More detailed information can be found in Refs. [12–14].

3.1 Introduction and Context

A large variety of modes involving a wide range of space and time scales may potentially contribute to the heat and particle transport in magnetically confined fusion plasmas. However, based on simple mixing-length estimates, it was often assumed that sub-ion-gyroradius scale turbulent fluctuations do not contribute significantly, e.g., to the heat fluxes. Furthermore, simulations covering all scales involved turn out to be on the verge of today's supercomputing resources. As a remedy, turbulence modes on different scales are often assumed to be decoupled so that investigations of modes e.g. on the ion gyroradius scale become feasible.

In fact, such single-scale simulations [10, 17, 19] provided first evidence for significant electron-scale driven transport. These fluctuations which are spatio-temporally separated from ion-scale turbulence types – for instance, ion temperature gradient (ITG) driven or trapped electron modes (TEMs) – by the square root of the ion-to-electron mass ratio are driven by electron temperature gradients (ETG) and are linearly isomorphic to ITG modes. However, the nonlinear saturation mechanisms turn out to be different. While ITG modes are often dominated by zonal (shear) flows which suppress long, radially elongated structures, ETG modes develop streamer-like vortices which boost the radial heat transport.

In realistic cases where different turbulence types are excited simultaneously, the question immediately arises whether, and if so, how cross-scale coupling potentially alters those findings. Hence, multiscale simulations covering ion and electron space and time scales self-consistently are urgently called for. Corresponding examples will be shown in the next subsection.

3.2 Simulation Details

The physical and numerical parameters chosen in these investigations are detailed in the present section. It is important to avoid including too many effects which have an impact on the turbulent systems under investigation, for that would needlessly complicate a subsequent interpretation and cause significantly more computational effort. Thus, magnetic field fluctuations and collisions are neglected in the following study, even though GENE is able to include them. Furthermore, all of the below simulations were performed in a simplified, so-called $\hat{s}$-α flux tube geometry with vanishing Shafranov shift $\alpha = 0$, which is consistent with the electrostatic limit $\beta \ll 1$ that allows for employing a relatively moderate number of grid points in the parallel direction. Most physical parameters correspond to the so-called Cyclone Base Case (CBC) [9], i.e., the safety factor $q_0 = 1.4$, the magnetic shear $\hat{s} = 0.8$, the inverse aspect ratio $\varepsilon = r/R_0 = 0.18$, as well as equal densities $n_{0i} = n_{0e} \equiv n_0$ and temperatures $T_{0i} = T_{0e} \equiv T_0$ are employed. The density and temperature gradients, $R_0/L_n \equiv R_0 \nabla \ln n$ and $R_0/L_{Tj} \equiv R_0 \nabla \ln T_j$ $(j = e, i)$, being normalized to the major tokamak radius R_0 are varied on a case-to-case basis.

Addressing the numerical parameters, the perpendicular box size is chosen to be $(L_x, L_y) = (64, 64)$ in units of the ion gyroradius ρ_s, and $768 \times 384 \times 16$ real space grid points are used in the radial (x), binormal (y), and parallel (z) direction, respectively, complemented by 32×8 grid points in $(v_\parallel, \mu)$ space. Here, $v_\parallel$ is the parallel velocity space direction and μ the magnetic moment. With these settings, one simulation consumes about 100-200 kCPUh. However, at this point, we note that a reduced mass ratio of $m_i/m_e = 400$ is considered. Otherwise, a single multiscale simulation would have exhausted the whole project budget since the computational time roughly scales as $T_{\mathrm{CPU}} \sim (m_i/m_e)^{3/2}$.

Typically, those simulations have been run in a pure MPI mode on 384 cores with an average performance of about 16% of the theoretical maximum value. The memory consumption has been on the order of 100 GB which fits very well within the hardware specifications. To ensure a reasonably well resolved resolved quasi-stationary state of the nonlinear simulations, on the order of 10^5 time steps had to be calculated. At regular intervals which have been adapted to the data size, files containing 1D, 3D, and 6D fields have been written for post-processing reasons.

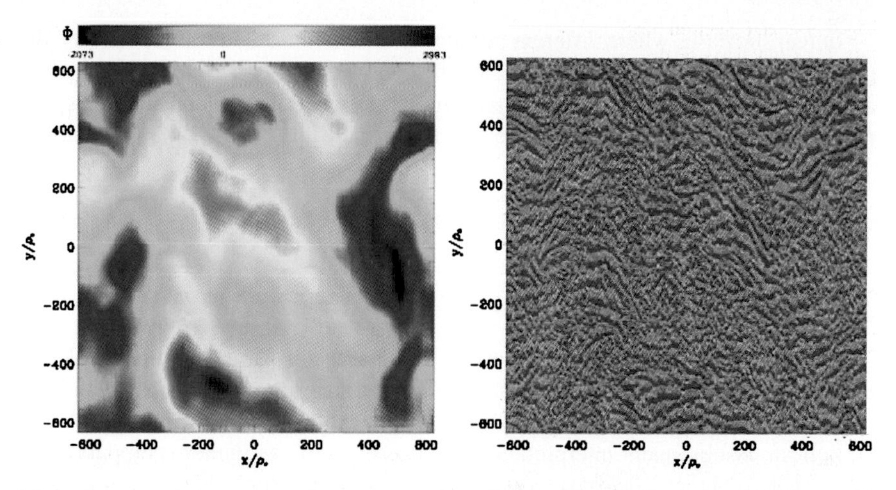

Fig. 1 Snapshot of the electrostatic potential at the outboard mid-plane for $R_0/L_{Ti} = R_0/L_{Te} = 6.9$ and $R_0/L_n = 2.2$, showing a dominance of large-scale, ITG vortices; and the same data with all $k_y\rho_s < 2$ modes filtered out, demonstrating the existence of small-scale ETG streamers which are subject to vortex stretching. Source: [12]

3.3 Simulation Results

The linear driving terms in the gyrokinetic Vlasov equation strongly depend on the temperature and density gradients. In order to be close to an experimentally relevant regime, many simulations thus employ the CBC nominal values, $R_0/L_{Ti} = 6.9$ and $R_0/L_n = 2.2$, which are based on a specific DIII-D tokamak discharge. A snapshot of a corresponding multiscale simulation with $R_0/L_{Te} = R_0/L_{Ti}$ is shown in Fig. 1. In this case, the electrostatic potential is dominated by large-scale vortices which show weak zonal flow behaviour (prevalent orientation in y direction). However, a high-pass filter (in terms of wave numbers) reveals the presence of small-scale streamers whose radial extension seems to be limited by large-scale vortex shearing. The high-k ($k_y\rho_s > 1$) fraction of the electron heat flux is hence quite small (about 10%), see Fig. 2. Both findings imply that single-scale simulations assuming isolated subsystems can, in general, not be applied for transport predicitions. The significance of ETG driven modes, however, depends on the specific parameter set, as will be seen in the following.

Comparing the simulated heat fluxes with the actual experimental values reveals an overestimation of the ion heat flux by almost two orders of magnitude. A likely key reason for this dramatic difference is that the normalized ion temperature gradient R_0/L_{Ti} – on which ITG turbulence depends very strongly but whose extraction from experimental temperature profile data is usually difficult – has been chosen somewhat too large. Therefore, several simulations with smaller values have been performed and analyzed. The ion heat flux is indeed decreased, as can be seen in Fig. 2. But additionally, the high-k contribution to the electron heat flux becomes more and more pronounced with decreasing R_0/L_{Ti} and eventually drives about

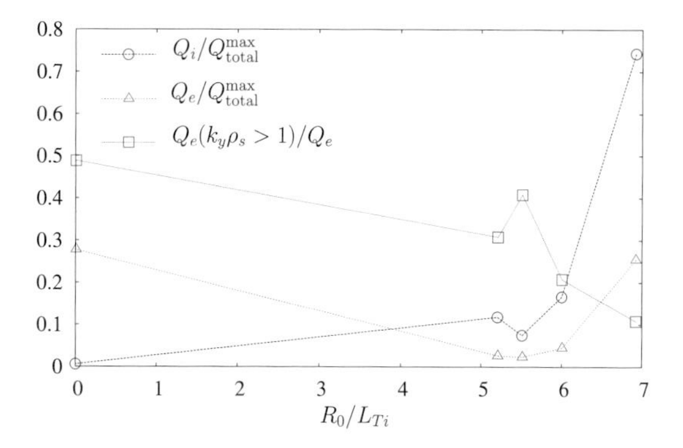

Fig. 2 Ion and electron heat flux Q as functions of the ion temperature gradient, normalized to the maximum total heat transport which is reached for $R_0/L_{Ti} = R_0/L_{Te} = 6.9$ and $R_0/L_n = 2.2$. In addition, the high-k fractions of the electron heat flux are displayed. The default choices for the remaining gradients are $R_0/L_{Te} = 6.9$ and $R_0/L_n = 0.0$. Source: [14]

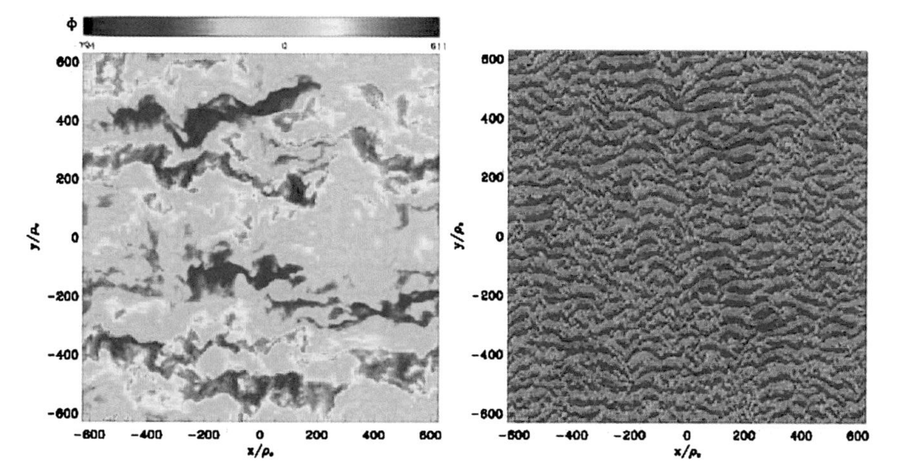

Fig. 3 Electrostatic potential contour at the low-field side for $R_0/L_{Ti} = 0$, $R_0/L_{Te} = 6.9$, $R_0/L_n = 0$, and the same contour neglecting all modes $k_y\rho_s < 2$. Source: [12]

50% when ITG modes are stable. One possible reason can be deduced from Fig. 3 where a contour of the electrostatic potential is shown for the $R_0/L_{Ti} = 0$ case. The character of the remaining large-scale turbulence mode – here, temperature gradient driven TEM – resembles the general behaviour of the small-scale ETG modes. Both seem to develop radially elongated structures so that thin ETG mode streamers may evolve much more easily within thick TEM vortices than in the ITG dominated case.

While heat and particle fluxes and their spectra are typically the quantities of interest in plasma turbulence research, the latter turn out to be hardly measurable in experiments. For spectral comparisons, one thus has to fall back on related quantities as, for instance, densities. Figure 4 shows corresponding examples for different scenarios. Compared to pure ITG, TEM, and ETG mode driven turbulence cases, mixtures exhibit a tendency to flatten density spectra in the $k_y\rho_e \sim 0.1$ region. For experimentalists, such a behaviour may thus serve as a signature for strong ETG activity (note that for a realistic mass ratio of $m_i/m_e = 1836$ or $m_i/m_e = 3675$, the wave numer region corresponds to $k_y\rho_s \sim 4$ and $k_y\rho_s \sim 6$, respectively).

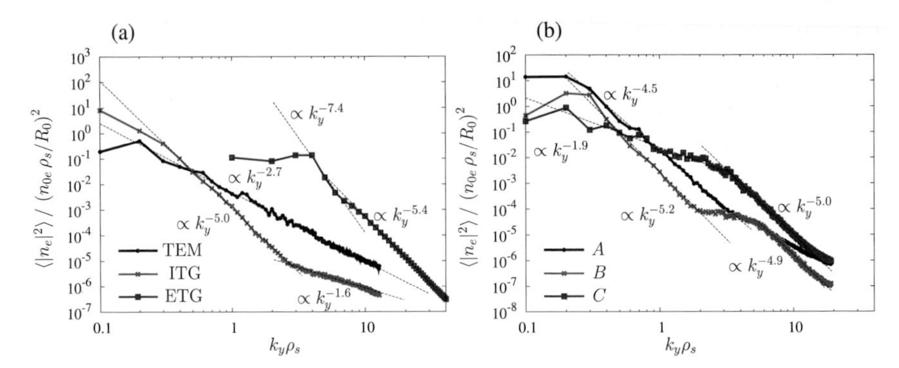

Fig. 4 Squared electron density fluctuations evaluated at $k_x\rho_s = 0$ and averaged over the parallel direction and time for (a) pure turbulence cases and (b) turbulence mixtures. Source: [13]

Summarizing, we have shown that massively parallized multiscale simulations predict a tendency towards a scale separation between ion and electron heat transport. In contrast to its ion counterpart which is only driven by large scales, the electron heat channel may thus exhibit substantial or even dominant high wave number contributions carried by ETG modes and short-wavelength TEMs. Therefore, these investigations might help to understand residual electron heat fluxes in cases where the low-k drive becomes small compared to the ETG drive, as for instance in discharges with dominant electron heating, high β, or transport barriers. Special focus on the latter will be put in the next section.

4 ETG Turbulence in Edge Transport Barriers

Having shown that small-scale ETG turbulence can contribute significantly to the heat transport observed in today's fusion experiments, we now turn to another study which examines ETG turbulence in an edge transport barrier. The formation of such a barrier has been observed for the first time almost three decades ago [26] and has since been found in many other experiments. The physics of its formation, however, is still not fully understood. Experimental measurements show that there are strong shear flows in the plasma edge, which are usually thought to suppress large-scale turbulence. On the other hand, it has been unclear whether ETG turbulence is affected by these flows and what sets the residual transport that is found in the barrier region.

To illuminate this issue, we performed nonlinear GENE simulations with ASDEX Upgrade edge parameters [20], restricting the simulation domain to electron scales and assuming ion turbulence to be suppressed. Under such conditions, it was found that ETG turbulence indeed produces enough heat flux to match the values inferred by transport modeling codes. The simulations furthermore show that the electron heat flux peaks at very small scales of $k_y\rho_s \approx 50$ which corresponds to a physical

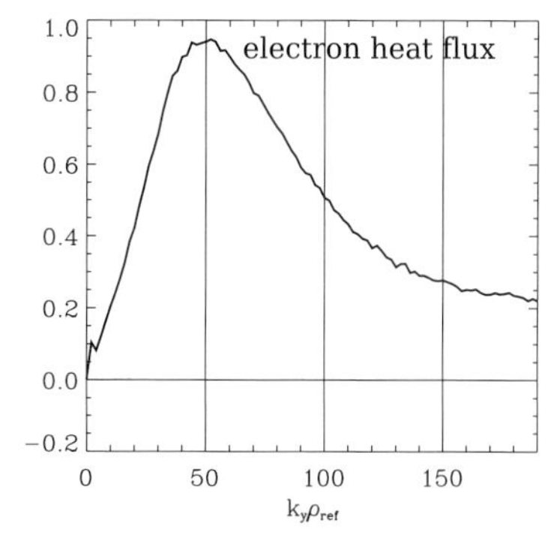

Fig. 5 k_y spectrum of the electron heat flux. As can be seen, most of the transport is produced at wavenumbers around $k_y\rho_s \approx 50$

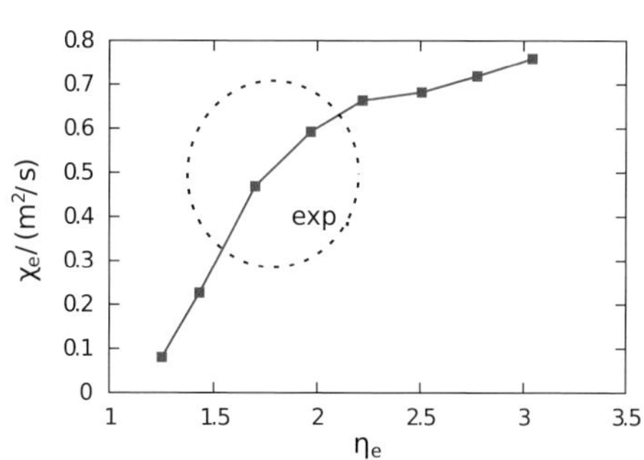

Fig. 6 Electron heat diffusivity at different electron temperature gradients. The obtained diffusivities are of the same order of magnitude as the experimental values

wavenumber of $k_\perp\rho_s \approx 15$ (see Fig. 5). In addition, linear simulations show that the ETG modes examined here are unstable for $\eta_e > 1.2$, where $\eta_e = n_e/T_e \cdot \nabla T_e/\nabla n_e$. This finding, which is confirmed by a nonlinear gradient scan (see Fig. 6), implies that ETG modes are unstable in the edge of ASDEX Upgrade H-mode and even L-mode (low confinement) discharges, since η_e is often around a value of 2. Therefore, ETG turbulence is a chief candidate for the process governing the relation between the electron temperature and density profiles that are found in the experiment.

5 High-β Simulations and Microturbulence in Astrophysics

Most gyrokinetic simulations have been performed either in the electrostatic limit or at low values of the normalized plasma pressure, $\beta = \beta_e = n_eT_e/(8\pi B^2) \ll 1$. At typical experimental values $\beta \sim 0.01$, however, electromagnetic effects become important and can even change the turbulence type completely. Corresponding studies were published in Refs. [6, 25]. There, it was found that for CBC parameters, raising β continuously from zero to values beyond the ballooning threshold had a fundamental impact on the nonlinear transport levels. More specifically, at the experimental value, the transport was suppressed by a factor of ~ 10 compared with the electrostatic limit. Such studies require high parallel resolutions which also has an adverse impact on the time step, resulting in significant computational requirements.

Apart from their importance in understanding tokamak core turbulence, electromagnetic effects are also key to using the GENE code to investigate anomalous transport in magnetized space plasmas. In some astrophysical contexts, density and temperature gradients can occur on scales small enough to drive plasma microinstabilities. Even when the system scale is too big to allow for any such instability to develop, MHD turbulence may create filaments and small structures, and microturbulence is able to enhance the transport by taking over at the end of the MHD cascade (see, e.g., Ref. [7]). Preliminary results regarding turbulence in evaporating clouds around Active Galactic Nuclei (AGN) have been published in Ref. [25]. There, dense, cold gas is immersed in a hot, dilute medium. Since unhindered evaporation would lead to cloud lifetimes much shorter than observations report, mechanisms have to exist which protect the clouds. Primarily, magnetic fields have been considered [8], preventing heat exchange perpendicularly to the field lines. However, microturbulent transport may be strong enough in some of these objects to have a significant impact on the clouds by providing an additional heat transport channel.

6 Conclusions

Employing massively parallel simulations on the HLRB2 machine with the nonlinear gyrokinetic code GENE, we were able to address several key questions concerning plasma microturbulence. First, we investigated whether earlier predictions on the role of sub-ion scale turbulence hold true if such small-scale modes are considered self-consistently with ion-gyroradius-scale turbulence types. It has been found for the first time that ion and electron heat fluxes tend to exhibit a scale separation where the former is only driven by large-scale turbulence whereas the latter may have significant or even dominant electron-gyroradius-scale contributions. Future investigations involving a realistic mass ratio and more physics are thus currently planned. In a second study, we concentrated on pure ETG modes in the edge regime of the Garching experiment ASDEX-Upgrade. It could be shown that such small-scale modes can indeed provide an explanation for the residual transport in edge barriers. Finally, first results of applications to astrophysical plasmas have been discussed which are going to be extended in the near future.

References

1. A. Brizard, J. Plasma Phys. **41**, 541 (1989)
2. A. Brizard, Phys. Fluids B **1**, 1381 (1989)
3. J. R. Cary, R. G. Littlejohn, Ann. Phys. **151**, 1 (1983)
4. T. Dannert, F. Jenko, Phys. Plasmas **7**, 072309 (2005)
5. T. Dannert, *Gyrokinetische Simulation von Plasmaturbulenz mit gefangenen Teilchen und elektromagnetischen Effekten*, PhD thesis, Technische Universität München, 2005

6. M. J. Pueschel, M. Kammerer, F. Jenko, Phys. Plasmas **15**, 102310 (2008)
7. G. G. Howes *et al.*, Phys. Rev. Lett. **100**, 065004 (2008)
8. Z. Kuncic, E. G. Blackman, M. J. Rees, Mon. Not. R. Astron. Soc. **283**, 1322 (1996)
9. A. M. Dimits *et al.*, Phys. Plasmas **7**, 969 (2000)
10. W. Dorland, F. Jenko, M. Kotschenreuther, B. N. Rogers, Phys. Rev. Lett. **85**, 5579 (2000)
11. D. H. E. Dubin, J. A. Krommes, C. Oberman, W. W. Lee, Phys. Fluids **26**, 3524 (1983)
12. T. Görler, F. Jenko, Phys. Rev. Lett. **100**, 185002 (2008)
13. T. Görler, F. Jenko, Phys. Plasmas **15**, 102508 (2008)
14. T. Görler, *Multiscale effects in plasma microturbulence*, PhD thesis, Universität Ulm, 2009
15. T. S. Hahm, W. W. Lee, A. Brizard, Phys. Fluids **31**, 1940 (1988)
16. T. S. Hahm, Phys. Fluids **31**, 2670 (1988)
17. F. Jenko, W. Dorland, M. Kotschenreuther, B. N. Rogers, Phys. Plasmas **7**, 1904 (2000)
18. F. Jenko, Comput. Phys. Commun. **125**, 196 (2000)
19. F. Jenko, W. Dorland, Phys. Rev. Lett. **89**, 225001 (2002)
20. F. Jenko, D. Told, P. Xanthopoulos, F. Merz, L. D. Horton, Phys. Plasmas **16**, 055901 (2009)
21. F. Jenko and The GENE development team, The GENE code, `http://www.ipp.mpg.de/~fsj/gene`. Cited 15 Oct 2009
22. M. Kammerer, F. Merz, F. Jenko, Phys. Plasmas **15**, 052102 (2008)
23. H. Lederer, R. Tisma, R. Hatzky, A. Bottino, F. Jenko, *Application Enabling in DEISA: Petascaling of Plasma Turbulence Codes, Advances in Parallel Computing, Vol. 15* (IOS Press, Amsterdam, 2008)
24. F. Merz, *Gyrokinetic Simulation of Multimode Plasma Turbulence*, PhD thesis, Westfälische Wilhelms-Universität Münster, 2008
25. M. J. Pueschel, *Electromagnetic Effects in Gyrokinetic Simulations of Plasma Turbulence*, PhD thesis, Westfälische Wilhelms-Universität Münster, 2009
26. F. Wagner et al., Phys. Rev. Lett. **49**, 1408 (1982)

Quantum Monte Carlo Studies of Strongly Correlated Electron Systems

Thomas C. Lang, Martin Bercx, David Luitz, Gang Li, Fakher F. Assaad, and Werner Hanke

Abstract Electronic correlations are at the heart of modern solid state physics. The interest lies in emergent collective phenomena which appear at low energy scales and which often originate from competing interactions. In this article, we summarize three research subjects where the effects of correlations dominate and can be elucidated with the combined use of supercomputers and state-of-the-art stochastic algorithms.
i) We show that the semimetallic state of the two-dimensional honeycomb lattice with a point-like Fermi surface is unstable towards a canted antiferromagnetic insulator upon application of an in-plane magnetic field. The magnetic field shifts the up- and the down-spin cones in opposite directions thereby generating a finite density of states at the Fermi surface which triggers a nesting instability leading to antiferromagnetic insulating state. Our conclusions, based on mean-field arguments, are confirmed by large scale auxiliary field projective quantum Monte Carlo methods.
ii) Unbiased weak-coupling continuous time quantum Monte Carlo (CTQMC) is used to study the transition between the singlet and doublet (local moment) states of a single magnetic impurity coupled to s-wave superconducting leads, focusing on the Josephson current with 0 to π phase shift and the crossing of the Andreev bound states in the single particle spectral function. Extended to dynamical mean-field theory (DMFT), this impurity problem provides a link to the periodic Anderson model with superconducting conduction electrons (BCS-PAM). We compute the spectral functions which signal the transition from a coherent superposition of Andreev bound states to incoherent quasiparticle excitations.
iii) Dynamical quantum-cluster approaches, such as different cluster extensions of the DMFT (cluster DMFT) or the variational cluster approximation (VCA), combined with efficient cluster solvers, such as CTQMC provide controlled approximations of the single-particle Green's function for lattice models of strongly correlated

Thomas C. Lang · Martin Bercx · David Luitz · Gang Li · Fakher F. Assaad · Werner Hanke
Institut für Theoretische Physik & Astrophysik, Universität Würzburg, Am Hubland 97074 Würzburg, Germany
e-mail: lang@physik.uni-wuerzburg.de

S. Wagner et al. (eds.), *High Performance Computing in Science and Engineering, Garching/Munich 2009*, DOI 10.1007/978-3-642-13872-0_42,

electrons. To access the thermodynamics, however, a thermodynamical potential is needed. We compute the numerically exact cluster grand potential within VCA using CTQMC in combination with a quantum Wang-Landau technique to reweight the coefficients in the expansion of the partition function of the two-dimensional Hubbard model at finite temperatures.

1 Magnetic Field Induced Semimetal-to-Canted-Antiferromagnet Transition on the Honeycomb Lattice

Graphene, or the physics of electrons on the honeycomb lattice, has recently received tremendous attention due to its semimetallic nature with low-energy quasiparticles behaving as massless Dirac spinors; for a recent review see Ref. [20]. A crucial point, is the stability of this semimetallic phase to particle-hole pairing. In particular, research activities have been devoted to the investigation of magnetic-field-induced transitions as a function of magnetic fields [1, 16, 17] and electronic correlations [14, 21, 25].

In this project we have shown that the semimetallic state of the two-dimensional honeycomb lattice with a point-like Fermi surface is unstable towards a canted antiferromagnetic insulator upon application of an in-plane magnetic field [8]. The magnetic field shifts the up and down spin cones in opposite directions thereby generating a finite density of states at the Fermi surface and perfect nesting between the up and down spin Fermi sheets. This triggers a canted antiferromagnetic insulating state. The mechanism behind this instability can be understood already at the mean-field level [6, 19]. We have shown that those mean-field arguments indeed capture the correct physics, since exact quantum Monte-Carlo simulations on the honeycomb lattice with up to 12×12 unit cells compare favorably with those mean-field results.

Our starting point is the Hubbard model on the honeycomb lattice (cf. Fig. 1) $H = H_0 + H_U + H_B$, where

$$
\begin{aligned}
H_0 &= -t \sum_{\mathbf{i},\mathbf{r},\sigma} \left(\hat{a}^{\dagger}_{\mathbf{i},\sigma} \hat{b}_{\mathbf{i}+\mathbf{r},\sigma} + \hat{b}^{\dagger}_{\mathbf{i}+\mathbf{r},\sigma} \hat{a}_{\mathbf{i},\sigma} \right) \\
H_U &= U \sum_{l=a,b} \sum_{\mathbf{i}} \left(\hat{n}_{\mathbf{i},l,\uparrow} - 1/2 \right) \left(\hat{n}_{\mathbf{i},l,\downarrow} - 1/2 \right) \\
H_B &= \frac{g}{2} \mu_{\mathrm{B}} B \sum_{l=a,b} \sum_{\mathbf{i},\sigma} p_{\sigma} \hat{n}_{\mathbf{i},l,\sigma} \, .
\end{aligned}
\tag{1}
$$

The electron operator $\hat{a}^{\dagger}_{\mathbf{i},\sigma}$ ($\hat{b}^{\dagger}_{\mathbf{i},\sigma}$) creates an electron on the orbital a (b) in the unit cell $\mathbf{i}$ and the associated electron density operator is $\hat{n}^{l}_{\mathbf{i},\sigma} = \hat{a}^{\dagger}_{\mathbf{i},\sigma} \hat{a}_{\mathbf{i},\sigma}$ ($\hat{b}^{\dagger}_{\mathbf{i},\sigma} \hat{b}_{\mathbf{i},\sigma}$), for $l = a(b)$. Owing to the bipartite nature of the lattice, hopping with matrix element t, occurs only between the a- and b-orbitals of unit cells related by lattice vector $\mathbf{r} = \{\mathbf{0}, \mathbf{a}_1 - \mathbf{a}_2, -\mathbf{a}_2\}$. The on-site electron-electron repulsion is denoted by $U > 0$

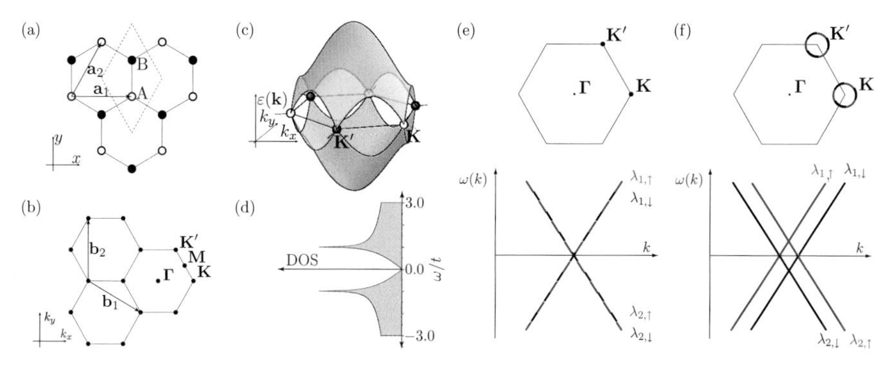

Fig. 1 Real space lattice of the honeycomb (a) with $\mathbf{a}_1 = \sqrt{3}a_0\,(1,0)$, $\mathbf{a}_2 = \sqrt{3}a_0\,(1/2,\sqrt{3}/2)$ and a_0 being the lattice constant. The unit cell with the orbitals A and B is indicated by the dashed diamond. The free dispersion π and π^* band touch each other in the two inequivalent K-points located at the corners of the hexagonal Brillouin zone (b) which leads to a linearly vanishing density of states at the Fermi level at half-filling. The nesting of spin-up and spin-down Fermi surface for (c) $B = 0$, where the spin bands collapse onto each other whereas for (d) $B > 0$ the bands are shifted by virtue of the magnetic field

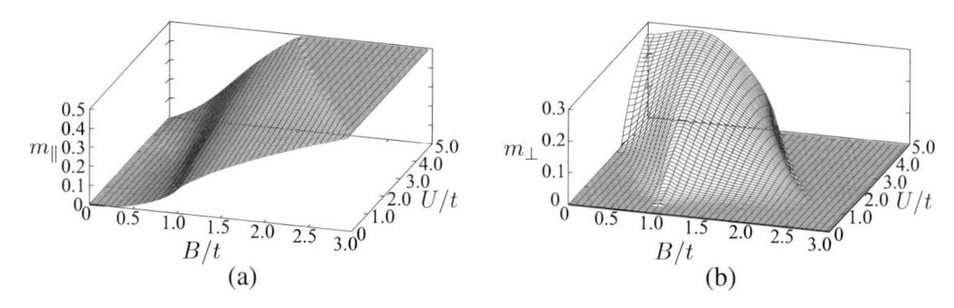

Fig. 2 Parallel magnetization $m_{\parallel}$ (a) and staggered magnetization $m_{\perp}$ (b) vs. U and B

and $p_\sigma = \pm 1$ for $\sigma = \uparrow, \downarrow$. In the following, we set $(g/2)\mu_{\mathrm{B}} \equiv 1$. For comparison with experiments, we only consider setups with magnetic field orientations parallel to the lattice plane. The Hamiltonian H_0 gives rise to two bands, $\lambda_n(\mathbf{k}) = p_n \left| t \sum_{\mathbf{r}} e^{-i\mathbf{k}\cdot\mathbf{r}} \right|$, with $\lambda_{n,\sigma}(\mathbf{k}) = \lambda_n(\mathbf{k}) + p_\sigma B$ and $p_n = \pm 1$ for $n = 1,2$ respectively, the latter being single particle states of $H_0 + H_B$. At half-band filling the Fermi surface consists of two points, K, K' in Fig. 1. At zero magnetic field the nesting instability is cut off by the vanishing density of sates (cf. Fig. 1d,e). At $B > 0$ the spin degeneracy of is lifted and the bands shift. The low energy density of states becomes finite leading to the nesting relation $\lambda_{1,\uparrow}(\mathbf{k}) = -\lambda_{2,\downarrow}(\mathbf{k})$ (cf. Fig. 1f).

Given the above instability, a mean-field ansatz is derived by assuming the magnetization $\mathbf{m}$ to be alternating on the sublattices: $\mathbf{m}_l = \left(m_{\parallel},\, m_{\perp}(-1)^l,\, 0\right)$, with the index $l = 0,1$ referring to the orbitals in the unit cell. That is, the magnetization $\mathbf{m}$ has a constant component $m_{\parallel}$ parallel to the field axis and a staggered component $m_{\perp}$ in the plane perpendicular to the field. Neglecting fluctuations we can derive the gap-equations and solve them self-consistently.

Our mean-field results are plotted in Figs. 2, 4(a), (c) and (e). At zero magnetic field, we observed as a function of U/t the expected transition from the semimetal-

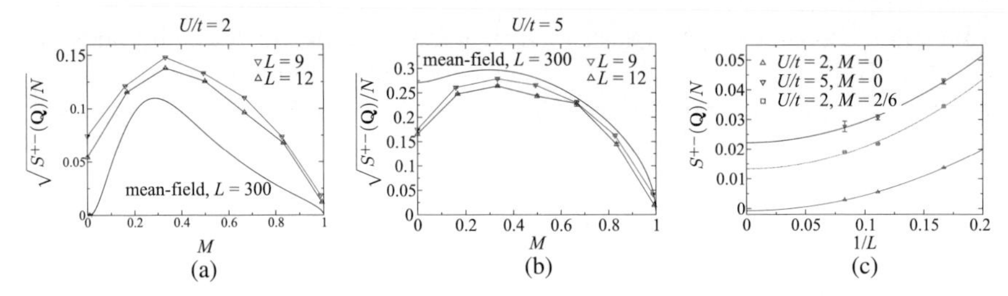

Fig. 3 Staggered magnetization $\sqrt{S^{+-}(\mathbf{Q})/N}$ below (a) and above (b) the critical interaction strength. (c) Finite size extrapolation of $S^{+-}(\mathbf{Q})/N$. In the semimetallic phase, $U/t = 2$, the data is consitent with the onset of transeverse staggered order at finite magnetization M. For comparison, we have plotted the $U/t = 5$ data in the absence of a magnetic field. This value of the Hubbard interaction places us in the antiferromagnetic Mott insulating state

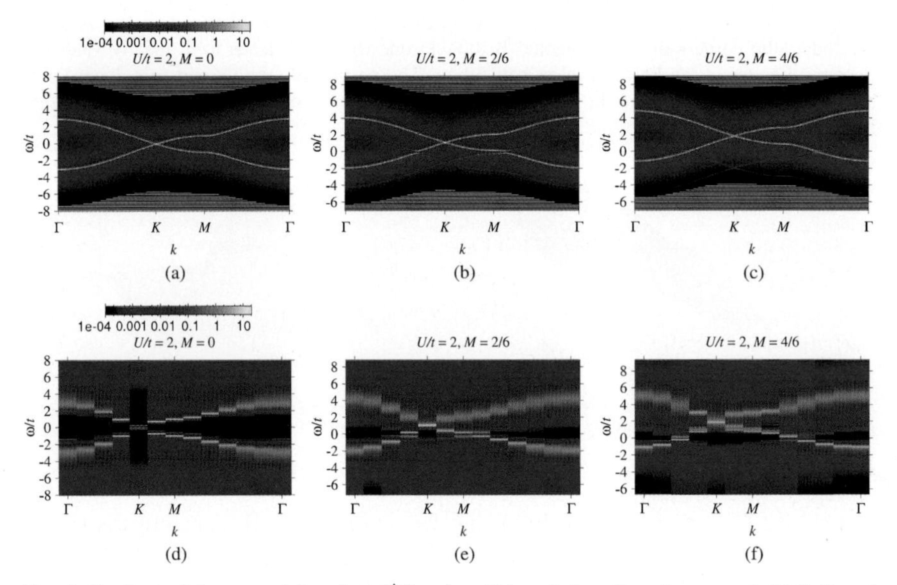

Fig. 4 Single-particle spectral function $A^{\uparrow}(\mathbf{k},\omega)$ at $U/t = 2$, based on the mean-field (left) and the QMC calculations (right), respectively. The magnetization M takes the values 0, $2/6$, and $4/6$ (from top to bottom). For the QMC calculations, the lattice size was set to 12×12 unit cells

lic state ($\mathbf{m} = 0$) at $U < U_c$ to the antiferromagnetic Slater insulator characterized by $|m_\perp| > 0$. The semimetallic state at $B = 0$ is characterized by the spin degenerate dispersion relation, $\lambda_{n,\sigma}(\mathbf{k})$, as shown in Fig. 4. Ramping up the magnetic field lifts this degeneracy thereby producing nested Fermi sheets of opposite spin indices. Hence, and irrespectiv of the magnitude of $U < U_c$, energy can be gained by ordering the spins in a canted antiferromagnet. The dispersion relation of this canted antiferromagnetic state is plotted in Fig. 4(c). To compare at best with the QMC simulations we consider the quantity

$$A^{\uparrow}(\boldsymbol{k},\omega) = -\frac{1}{\pi}\mathrm{Im}\left(G^{\uparrow}_{aa}(\boldsymbol{k},\omega) + G^{\uparrow}_{bb}(\boldsymbol{k},\omega)\right) \tag{2}$$

with a finite broadening. As apparent, the features with dominant weight follow the dispersion relation $\lambda_{1,\uparrow}(\mathbf{k})$ and $\lambda_{2,\uparrow}(\mathbf{k})$ and a gap at the Fermi level is apparent. Due to the transverse staggered moment, mixing between the up and down dispersion relations occurs thereby generating shadow features following the dispersion relations of $\lambda_{1,\downarrow}(\mathbf{k})$ and $\lambda_{2,\downarrow}(\mathbf{k})$. The intensity of the shadow features tracks $m_\perp$. As apparent from Fig. 2 the growth of $m_\perp$ as a function of the magnetic field is countered by the polarization of the spins along the magnetic field. It is interesting to note that irrespectively of U/t the maximal value of $m_\perp$ and hence of the magnetic field induced gap is at $B = 1$ corresponding to an energy scale matching the position of the Van-Hove singularity in the non-interacting density of states. At this point a maximal amount of energy can be gained by the opening of the gap.

To confirm our mean-field results, we have carried out projector auxiliary field QMC calculations. This projective QMC (PQMC) algorithm is based on the equation:

$$\frac{\langle \Psi_0 | A | \Psi_0 \rangle}{\langle \Psi_0 | \Psi_0 \rangle} = \lim_{\theta \to \infty} \frac{\langle \Psi_T | e^{-\theta H} A e^{-\theta H} | \Psi_T \rangle}{\langle \Psi_T | e^{-2\theta H} | \Psi_T \rangle} . \tag{3}$$

The trial wave function $|\psi_T\rangle$ has to be non-orthogonal to the ground state wave function, $\langle \psi_T | \psi_0 \rangle \neq 0$ and the ground state is assumed to be non-degenerate. For the details of the formulation of this approach, we refer the reader to [2]. In this canonical approach, we fix the magnetization $M = \frac{N_\uparrow - N_\downarrow}{N_\uparrow + N_\downarrow}$ rather than the magnetic field. N_σ corresponds to the total number of electrons in the spin sector σ. Furthermore, due to the particle-hole symmetry which locks in the signs of the fermionic determinants in both spin sectors one can avoid the so-called negative sign problem irrespectively of the choice of the magnetization. In practice, for each finite system, we choose a value of the projection parameter, θ, large enough so as to guarantee convergence within statistical uncertainty.

To detect transverse staggered magnetic order under an applied magnetic field, we have computed the spin-spin correlation functions

$$S^{+-}(\mathbf{q}) = \frac{1}{N} \sum_{\mathbf{i},\mathbf{j}} e^{-\imath \mathbf{q}(\mathbf{i}-\mathbf{j})} \langle S_a^+(\mathbf{i}) S_a^-(\mathbf{j}) - S_a^+(\mathbf{i}) S_a^-(\mathbf{j}) , \tag{4}$$

with $\lim_{N \to \infty} S^{+-}(\mathbf{Q})/N = (m_\perp^{\mathrm{QMC}})^2$. We have computed this quantity on 6×6, 9×9 and 12×12 lattices, and our results are plotted in Fig. 3 both for $U < U_c$ and $U > U_c$. At $U/t = 2 < U_c/t$ and zero magnetization, $M = 0$, our results are consistent with $m_\perp^{\mathrm{QMC}} = 0$ whereas at $M = 2/6$, $m_\perp^{\mathrm{QMC}}$ takes a finite value. Although we cannot reproduce the essential singularity of the mean-field calculation at $U < U_c$, the overall form the transverse staggered magnetization compares favorably with the mean-field results (see Fig. 3(a)(b)) both at $U < U_c$ and $U > U_c$.

Within the PQMC, the zero temperature single particle Green function along the imaginary time axis can be computed efficiently with methods introduced in [11]. From this quantity, we can obtain the spectral function of Eq. (2) with the use of a stochastic formulation of the Maximum Entropy method [5, 24]. The so obtained results for $A^\uparrow(\mathbf{k}, \omega)$ are plotted in Fig. 4(b),(d) and (f). As apparent the features in the QMC calculation which are associated with substantial spectral weight are well reproduced by the mean-field calculation. The particle-hole transformation, $\hat{a}^{\dagger}_{\mathbf{i},\sigma} \rightarrow \hat{a}_{\mathbf{i},-\sigma}$ and $\hat{b}^{\dagger}_{\mathbf{i},\sigma} \rightarrow -\hat{b}_{\mathbf{i},-\sigma}$, leads to the relation $A^{\uparrow}(\mathbf{k},\omega) = A^{\downarrow}(\mathbf{k},-\omega)$. At finite magnetic fields or equivalently at finite magnetizations, the staggered transverse order leads to a gapless Goldstone mode of which the quasiparticle can spin-flip scatter. As a consequence, and as already observed in the mean-field calculation, the features of the down spectral function should be visible in $A^{\uparrow}(\mathbf{k},\omega)$. Upon close inspection of Fig. 4 one will observe that for each dominant low energy peak at $\omega(\mathbf{k})$ in $A^{\uparrow}(\mathbf{k},\omega)$ a shadow feature at $-\omega(\mathbf{k})$ is present.

Experimentally, such a transition could be observed by magneto resistance measurements. The transition to the canted antiferromagnet breaks a U(1) symmetry and hence occurs at finite temperatures, T_{KT}, in terms of a Kosterlitz-Thouless transition. Below T_{KT} the power-law decay of the transverse spin-spin correlation function should suffice to produce a visible pseudo-gap in the charge sector and hence an increase of the resistivity as a function of decreasing temperature. The primary issue to observe the transition is the magnitude of the required magnetic field so as to obtain a visible gap. With $t \approx 2.5$eV and $U \approx 5-12$eV [13] one can readily see that very large magnetic fields will be required to obtain charge gaps in the meV region. In particular, we can predict from our calculations that values of the order of $B \propto 10^2 - 10^3$T are required to obtain an acceptable gap.

In conclusion, we have carried out mean-field calculations and projective quantum Monte Carlo simulations for the Hubbard model with the Zeeman spin coupling on the honeycomb lattice in a magnetic field oriented parallel to the lattice plane [8]. Our results show the inherent instability of the semimetallic state to a canted antiferromagnet upon application of the magnetic field.

2 CTQMC Study of the Single Impurity and Periodic Anderson Models with s-Wave Superconducting Baths

2.1 Quantum Dot with Two Superconducting Baths

Recent advances in manufacturing carbon nanotube quantum dots coupled to superconducting leads [9, 10, 15] have renewed the interest in the problem of a magnetic impurity embedded in a superconducting host. Such a system shows the interesting phenomenon of the Josephson current, an equilibrium supercurrent driven by the phase difference of the superconducting order parameters of a right and a left lead coupled to the quantum dot as well as a quantum phase transition from a singlet regime of the quantum dot to a doublet regime.

We studied this quantum phase transition in detail, calculating not only the Josephson current $I_j(\phi)$ as a function of the phase difference of the superconducting order parameter Δ of the right and left lead, but also various other static quantities as well as the single particle spectral function and the dynamical spin structure factor using the weak coupling continuous time quantum Monte Carlo method (CTQMC) [3, 23].

We model the carbon nanotube quantum dot coupled to two superconducting leads (L=left, R=right) by a single impurity Anderson model with two baths described by the BCS reduced Hamiltonian:

$$H = \sum_{k,\sigma} \xi_k \tilde{c}^{\dagger}_{k,\sigma,\alpha} \tilde{c}_{k,\sigma,\alpha} - \sum_{k} \left(\Delta e^{i\phi_\alpha} \tilde{c}^{\dagger}_{k,\uparrow,\alpha} \tilde{c}^{\dagger}_{-k,\downarrow,\alpha} + \text{h.c.} \right) + \sum_{\sigma} \xi_d \tilde{d}^{\dagger}_{\sigma} \tilde{d}_{\sigma} +$$
$$U \left(\tilde{d}^{\dagger}_{\uparrow} \tilde{d}_{\uparrow} - \frac{1}{2} \right) \left(\tilde{d}^{\dagger}_{\downarrow} \tilde{d}_{\downarrow} - \frac{1}{2} \right) - \frac{V}{\sqrt{N}} \sum_{\alpha=L}^{R} \sum_{\sigma,k} \left(\tilde{c}^{\dagger}_{k,\sigma,\alpha} \tilde{d}_{\sigma} + \tilde{d}^{\dagger}_{\sigma} \tilde{c}_{k,\sigma,\alpha} \right). \quad (5)$$

Here, $\tilde{c}^{\dagger}_{k,\sigma,\alpha}$ is the creation operator for electrons with momentum k and z-component of spin σ in lead α and $\tilde{d}^{\dagger}_{\sigma}$ is a creation operator for electrons on the quantum dot. In our calculations, we have assumed a one dimensional cosine band $\xi_k = -2t\cos(k) - \mu$ and we chose to fix the chemical potential μ and ξ_d at $\mu = \xi_d = 0$. Further, we express all quantities in units of the hopping matrix element $t = 1$. We also fix the hybridisation matrix element V at $V = 0.5$ and only vary the modulus Δ and the phase ϕ_α of the superconducting order parameter in the baths as well as the Coulomb blockade U.

For the CTQMC simulation, it is convenient to eliminate non-normal Greenfunctions, therefore, we employ a canonical particle-hole transformation on the spin down sector as detailed in Ref. [18]. This enabled us to calculate the Josephson current through the quantum dot as a function of the phase difference ϕ of the superconducting order parameters $\Delta e^{i\phi}$ in the two leads for different values of Δ at finite temperature $1/\beta$.

In the CTQMC method, the Josephson current can be easily calculated, as it is nothing but the equal time Green's function $I_j = \langle j_\alpha \rangle = i\frac{V}{\sqrt{N}} \sum_{k,\sigma} \langle \tilde{c}^{\dagger}_{k,\sigma,\alpha} \tilde{d}_{\sigma} - \tilde{d}^{\dagger}_{\sigma} \tilde{c}_{k,\sigma,\alpha} \rangle$. In Fig. 5, we show the change of the current-phase relation of the Josephson current for different values of the superconducting order parameter in the leads Δ. For small values of Δ, the current phase relation is sinusoidal. As Δ increases,

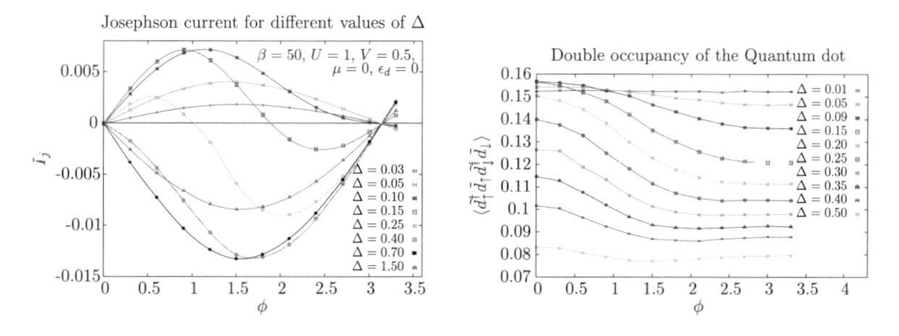

Fig. 5 Left panel: Josephson current I_j as a function of the phase difference ϕ. Right panel: Double occupancy $\langle \tilde{d}^{\dagger}_{\uparrow} \tilde{d}_{\uparrow} \tilde{d}^{\dagger}_{\downarrow} \tilde{d}_{\downarrow} \rangle$ of the quantum dot as a function of the phase difference for the same parameter set as in the left panel. Lines are guides to the eye only

we observe a change of the current-phase relation, as for larger ϕ, the sign of the Josephson current changes. This results ultimately in a complete phase-shift by π of the current phase relation for $\Delta \gtrsim 0.35$. For even larger Δ, the maximal Josephson current begins to decrease again and vanishes in the limit of infinite Δ.

We have already seen, that the π shift of the Josephson current seems to occur first for larger values of the phase difference ϕ. This can be confirmed by studying the double occupancy $\langle \tilde{d}_{\uparrow}^{\dagger}\tilde{d}_{\uparrow}\tilde{d}_{\downarrow}^{\dagger}\tilde{d}_{\downarrow}\rangle$. This quantity is also well accessible within CTQMC by the use of Wick's theorem for every Monte Carlo configuration [18] and is proportional to the derivative $\frac{\partial \Omega}{\partial U}$ of the grand potential Ω. In Ref. [18], we showed that the double occupancy shows a jump as a function of U which is a clear sign for a first order quantum phase transition. Using this knowledge, we can now look at the double occupancy as a function of the phase difference ϕ for different values of the order parameter Δ depicted in figure 5.

The double occupancy shows a very interesting behaviour in the parameter region, in which the π-shift of the Josephson current occurs. Figure 5 is best understood by studying simultaneously the current phase relation of the Josephson current in Fig. 5. For small values of $\Delta \lesssim 0.5$, the current-phase relation is sinusoidal and the double occupancy is a constant function of ϕ. For larger values of Δ, the double occupancy develops a jump at a certain critical phase which is smeared out by the finite temperature. The jump is accompanied by a current-phase relation which deviates significantly from the sinusoidal form. This intermediate regime is referred to as the $0'$ or π' regime, depending on the global minimum of the grand potential Ω, which can be derived from the current-phase relation of the Josephson current, because of $I_j(\phi) \propto \frac{\partial \Omega}{\partial \phi}$ [7, 18, 22].

In order to characterize the nature of the quantum phase transition from the 0 to the π Junction regime of the Josephson current, it is useful to study dynamical quantities, such as the dynamical spin structure factor $S(\omega)$ and the single particle spectral function $A(\omega)$. These quantities cannot be calculated directly within the CTQMC method and have to be extracted from imaginary time correlation functions such as the spin spin correlation function $C_s(\tau) = \langle \tilde{S}_z(\tau)\tilde{S}_z\rangle$ or the single particle Green's function $G_{\sigma\sigma'}(\tau) = \langle T\tilde{d}_{\sigma}^{\dagger}(\tau)\tilde{d}_{\sigma'}\rangle$ using stochastic analytic continuation [6, 24].

First, we look at the result for the dynamical spin structure factor $S(\omega) = \frac{\pi}{Z}\sum_{n,m} \mathrm{e}^{-\beta E_n}\left|\langle n|\tilde{S}_{+}|m\rangle\right|^2 \delta(\omega + E_n - E_m)$ depicted in Fig. 6. From this quantity, we can read off the energy scale required to flip the spin on the quantum dot. We observe, that for small values of Δ, $S(\omega)$ shows a suppressed spectral weight at $\omega = 0$, which corresponds to the Kondo effect. The energy scale of the peak in $S(\omega)$ is associated with the energy needed to break the Kondo singlet. This gives us a rough estimate of the Kondo temperature of $T_K \approx 0.06$ from Fig. 6. For this parameter region, the Josephson current is well in the 0-Junction regime. With increasing Δ, a sharp peak develops at $\omega = 0$ which dominates and has fully developped in a δ function for $\Delta \gtrsim 0.35$. This peak is a clear sign for a local magnetic moment, as no energy is required to flip the spin on the quantum dot.

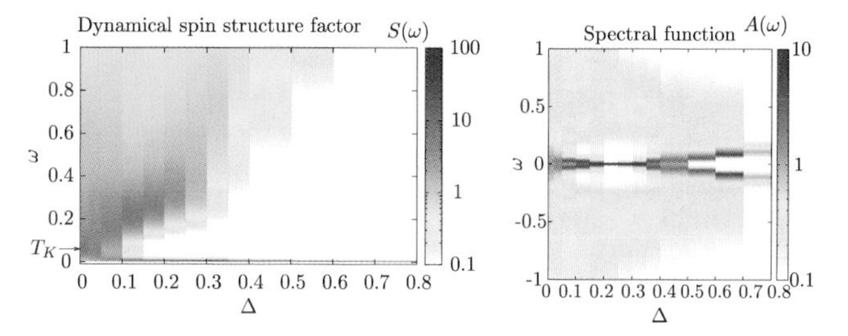

Fig. 6 Left panel: Dynamical spin structure factor $S(\omega)$. Right panel: Spectral function $A(\omega)$. The parameters in both plots are given by $U = 1$, $\beta = 100$, $V = 0.5$ and $\xi_d = \mu = 0$

Finally, we study the single particle spectral function $A(\omega)$ of the quantum dot. Using an effective model valid in the limit Δ much larger than the bandwidth, it can be seen that the lowest lying excitation in the spectral function is linked to the energy difference between the ground state of the system and the first excited state (see reference [18]). These two states differ in their magnetic properties corresponding to a singlet and a local moment state. In Fig. 6, we can see, that the lowest lying excitation crosses $\omega = 0$ at $\Delta \approx 0.2$, which corresponds to a crossing of the energies of the ground state and the first excited state of the system. This means, that the 0-Junction to π-Junction transition is in fact a first order quantum phase transition.

2.2 Periodic Anderson Model with Superconducting Conduction Band

A very interesting extension of the impurity problem studied in Sec. 2.1 is the periodic Anderson model with superconducting conduction band (BCS-PAM). This problem is linked to the impurity problem within the dynamical mean-field theory (DMFT) and we were particularly interested in how the first order phase transition of the impurity problem is reflected in the BCS-PAM.

The Hamiltonian of the BCS-PAM is given by

$$H = \sum_{k,\sigma} \xi(k) \tilde{c}^{\dagger}_{k,\sigma} \tilde{c}_{k,\sigma} - \Delta \sum_{k} \left(\tilde{c}^{\dagger}_{k,\uparrow} \tilde{c}^{\dagger}_{-k\downarrow} + \text{h.c.} \right) + \sum_{k,\sigma} \xi_f \tilde{f}^{\dagger}_{k,\sigma} \tilde{f}_{k,\sigma} + \\ U \sum_{i_f} \left(\tilde{n}_{i_f,\uparrow} - \frac{1}{2} \right) \left(\tilde{n}_{i_f,\downarrow} - \frac{1}{2} \right) - V \sum_{k,\sigma} \left(\tilde{c}^{\dagger}_{k,\sigma} \tilde{f}_{k,\sigma} + \text{h.c.} \right). \tag{6}$$

The operators $\tilde{c}^{\dagger}_{k,\sigma}$ and $\tilde{f}^{\dagger}_{k,\sigma}$ are creation operators for electrons of momentum k and z-component of the spin σ. As in the impurity problem, we concentrate on the particle-hole symmetric point $\mu = \xi_f = 0$. We study the case of a cosine band in two dimensions with $\xi(k) = -2t(\cos(k_x) + \cos(k_y)) - \mu$. Δ is the superconducting order parameter in the conduction band, which we chose to be real in this case, V is

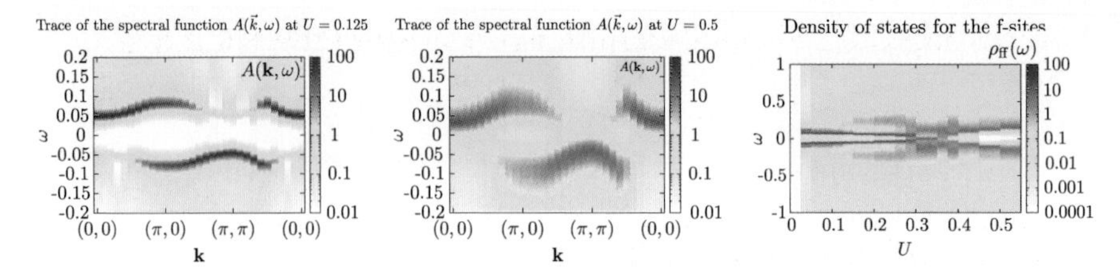

Fig. 7 Left panels: **k**-resolved spectral function for the singlet (left) and the magnetic moment (middle) regime. In the singlet regime, the bands are sharp and the spectrum is very similar to the $U = 0$ case. If U exceeds the critical value for the quantum phase transition, the bands become incoherent and the spectrum corresponds to finite lifetime quasiparticles. Right panel: Local density of states of the f-sites as a function of U for $V = 0.5$, $\Delta = 2$ and $\beta = 100$

the hybridisation matrix element and U is the strength of the Coulomb interaction between f electrons on the same site.

We solve the BCS-PAM within the framework of DMFT, which can be easily extended to superconducting baths as in the present case (see Ref. [12]). As for the impurity problem, we use again the CTQMC method as an impurity solver and we use stochastic analytical continuation in order to obtain the dynamical quantities, such as the **k**-resolved spectral function and the local density of states.

The quantity corresponding to the single particle spectral function $A(\omega)$ is the local density of states of the f-sites ρ_{ff}. Figure 7 depicts ρ_{ff} and we can again see the crossing of the lowest lying excitations if the Coulomb interaction U on the f-sites is increased. This behaviour can be compared directly with the impurity case in Fig. 6, because Bauer et al. were able to observe the same crossing of the lowest lying excitations for varying U instead of Δ in Ref. [4]. Together with further results detailed in Ref. [18], we argue, that this is a sign for a first order quantum phase transition in the BCS-PAM at least within the framework of the DMFT. As in the impurity case, we observe a transition from a singlet regime to a regime with local magnetic moments on the f-sites. In order to gain a better understanding of the nature of the magnetic moment regime, we calculated the **k**-resolved basis independent spectral function $A(\mathbf{k},\omega) = \text{Tr}\mathbf{A}(\mathbf{k},\omega)$ from the self energy obtained in the DMFT and from the bare lattice Green's function of the noninteracting BCS-PAM using Dyson's equation and stochastic analytic continuation.

Figure 7 compares $A(\mathbf{k},\omega)$ for the singlet and the local moment regimes. The spectral function for the singlet regime is continuously linked to the noninteracting case in the sense, that the bands remain sharp and well defined and the overall band structure is very similar to the noninteracting case. In the local moment regime, the situation becomes more complicated. Here, the bands are broadened and correspond to a finite lifetime of the quasiparticles. From the study of the local spin-structure factor of the BCS-PAM (see Ref. [18]), we know, that each f-site acts like a local magnetic moment and therefore, we propose the interpretation of randomly distributed magnetic moments throughout the lattice. The disorder of the moments is then responsible for the destruction of the infinite lifetime of the quasiparticles and the system goes from a coherent to an incoherent state. This fits very well with a mean-field calculation including the disorder.

3 Accessing the Thermodynamic Properties in the Hubbard Model

The thermodynamics of strongly correlated fermion systems is governed by a thermodynamical potential, which is related to the cluster Green's function and self-energy in a dynamical cluster embedding approach via:

$$\Omega = \Omega' + \mathrm{Tr}\ln\mathbf{G} - \mathrm{Tr}\ln\mathbf{G}'\,. \tag{7}$$

Here Ω' is the grand potential of the cluster reference system, and $\mathbf{G}$ the approximate Green's function of the original lattice model in the thermodynamic limit which is obtained from the lattice Dyson equation using the cluster self-energy. The trace refers to both, spatial and temporal lattice degrees of freedom, i.e. involves sums over lattice sites and Matsubara frequencies. From the grand potential on obtains the complete thermodynamics such as the free energy, entropy, specific heat, etc. properties one can directly compare with experiments.

3.1 Method and Numerical Aspects

As we know, Exact Diagonalization (ED) method is very convenient for the evaluation of the thermodynamics which may then be directly calculated from the cluster many-body eigenenergies. The limitation of the application of ED lies in the smaller cluster size which can be accessed. In this work, we will use the QMC method, including both the weak and strong coupling version, to evaluate the system grand potential through a cluster-embedding method. The basic difficulties involved in our calculations include

- The accurate determination of the cluster grand potential Ω'.
- The infinite Matsubara frequency summation, where the cluster Green's function and self energy function are only known for a finite number of frequency points.

The straightforward application of QMC is not suitable for the calculation of the grand potential, since the importance sampling used in QMC introduces an unknown factor to it. In order to determine this factor, we used the quantum Wang-Landau approach [26, 27] to generate approximately the same number of configurations in the QMC. In weak-coupling (continuous-time) CT-QMC, the partition function is expanded as

$$\frac{\mathscr{Z}}{\mathscr{Z}_0} = \sum_{k=0}^{\infty} U^k w(k) = 1 + \sum_{k=1}^{\infty} U^k w(k)\,, \tag{8}$$

where $\mathscr{Z}_0$ is the partition function of the non-interacting system. The coefficient of the k-th order is given by

$$w(k) = \sum_{C_k} w(k, C_k)\,, \tag{9}$$

Note that $w(k)$ is not normalized to unity (Eq. (8) yields $\sum_{k=0}^{\infty} w(k) = \mathscr{Z}_{U=1}/\mathscr{Z}_0$). We introduced a Wang-Landau factor $g(k)$ to re-define the weights $w(k, C_k)$:

$$w(k,C_k) \rightarrow \widetilde{w}(k,C_k) = w(k,C_k)/g(k)\,. \tag{10}$$

This Wang-Laudau factor $g(k)$ is chosen to make the new histogram flat, i.e. $\tilde{p}(k) = U^k \tilde{w}(k) = const$, for $k < k_c$.

$$\frac{\mathscr{Z}}{\mathscr{Z}_0} = \sum_{k=0}^{\infty} U^k w(k) = \sum_{k=0}^{\infty} U^k \tilde{w}(k) g(k) = \tilde{p}(0) \sum_{k=0}^{\infty} g(k)\,. \tag{11}$$

With $p(0) = 1$ we find $\mathscr{Z}/\mathscr{Z}_0 = \sum_{k=0}^{\infty} g(k)/g(0)$. Note that in practice, the Wang-Landau re-weighting is performed up to certain order k_c only. In any practical simulation $\tilde{p}(k) = const$ is approximate, but can be obtained to arbitrary precision, in principle. Then, the cluster grand potential can be calculated as $\Omega' = -T \ln \mathscr{Z}$.

The second problem listed above can be solved by introducing the high-frequency corrections in the trace in Eq. (7). This is done by approximating the second and the third terms in Eq. (7) with their high frequency asymptotic behavior for which the infinite Matsubara frequency summation can be analytically determined. Then the final expression of Eq. (7) under the high frequency corrections becomes

$$\begin{aligned}\Omega = \Omega' \, - \, 2T \sum_{\tilde{\boldsymbol{k}}} \frac{N_c}{N} \sum_{n=-\Lambda}^{\Lambda} \ln \frac{\det\left[1 - \boldsymbol{V}_{\tilde{k}} \boldsymbol{G}_{\boldsymbol{t}',U}(i\omega_n)\right]}{\det\left[1 - \boldsymbol{V}_{\tilde{k}}/i\omega_n\right]} \\ + \, 2TN_c \ln 2 - 2T \sum_{a,\tilde{\boldsymbol{k}}} \frac{N_c}{N} \ln(1 + e^{-\tilde{\boldsymbol{V}}_{\tilde{k}}^{aa}/T})\end{aligned} \tag{12}$$

This equation is now ready for numerical evaluation. Fig. 8 demonstrates that our technique can in fact be used for the determination of variational parameters within the framework of the self-energy-functional theory for the two-dimensionaly Hubbard model with on-site interaction U and the nearest-neighbor hoppting t. The figure shows the lattice grand potential Ω as a function of the strength of the Weiss

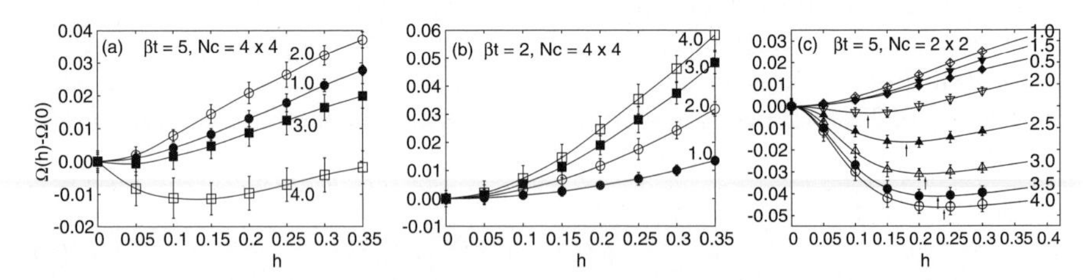

Fig. 8 Lattice grand potential Ω as a function of the variational parameter h as obtained by embedding a 4×4 cluster (a,b) and a 2×2 cluster (c). Results for different interaction strengths U/t as indicated in the figure. Note that the difference $\Omega(h) - \Omega(0)$ is plotted. (a) $\beta t = 5.0$, $N_c = 4 \times 4$, (b) $\beta t = 2.0$, $N_c = 4 \times 4$, (c) $\beta t = 5.0$, $N_c = 2 \times 2$. Symbols with statistical errors represent the numerical data. Lines are obtained by a spline interpolation and serve as a guide to the eye only in panel (a) and (b). Lines in panel (c) are from full ED calculations for comparison

field h for different values of U and T. Let us concentrate on the results obtained for the 4×4 cluster first. For $U/t = 1$, $U/t = 2$ and $\beta t = 5.0$ (panel a) we find a single stationary point at $h = 0$. This is indicative of the paramagnetic phase at high temperatures and weak interaction. For $U/t = 4$ (panel a), the SFT grand potential clearly displays a minimum around $h = 0.15$. This corresponds to antiferromagnetic order. Here we also get a non-zero value for the order parameter. Note that the variation of Ω with h is small and comparable to the statistical error. This shows that $\beta t = 5$ is close to the Neel temperature for $U/t = 4$. As the trend of $\Omega(h)$ is symmetric with respect to h, the point $h = 0$ represents another stationary point corresponding to the paramagnetic phase. The latter is metastable as $\Omega(h = 0)$ is higher than $\Omega(h = 0.15)$.

Fig. 8, panel (b) displays results for a higher temperature ($\beta t = 2$). Here we are again left with the paramagnetic phase for all U/t only. Obviously, the variation of the grand potential with h is most pronounced for $U/t = 4$ while is becomes more and more flat with decreasing interaction strength. This is due to the fact that the h dependence enters the self-energy functional via the self-energy only, i.e. $\Omega(h) = \Omega[\Sigma(h)]$, and $\Omega[\Sigma] \equiv 0$ in the non-interacting limit. For $\beta t = 5$ (panel a), this is different. Here, the trend of $\Omega(h)$ first becomes stronger (comparing $U/t = 1$ with $U/t = 2$) as explained above. For stronger interactions, however, this mechanism has to compete with the tendency to form a minimum at a finite h. This explains the non-monotonic trend in U visible in panel a.

References

1. Aleiner, I.L., Kharzeev, D.E., Tsvelik, A.M.: Spontaneous symmetry breaking in graphene subjected to an in-plane magnetic field. Phys. Rev. B **76**(19), 195415 (2007)
2. Assaad, F.F., Evertz, H.G.: World-line and determinental quantum Monte Carlo methods. In: H. Fehske, R. Schneider, A. Weisse (eds.) Computational Many-Particle Physics. Springer, Berlin (2008)
3. Assaad, F.F., Lang, T.C.: Diagrammatic determinantal quantum Monte Carlo methods: Projective schemes and applications to the Hubbard-Holstein model. Phys. Rev. B **76**(3), 035116 (2007)
4. Bauer, J., Oguri, A., Hewson, A.C.: Spectral properties of locally correlated electrons in a Bardeen-Cooper-Schrieffer superconductor. Journal of Physics: Condensed Matter **19**(48), 486211 (2007)
5. Beach, K.S.D.: Identifying the maximum entropy method as a special limit of stochastic analytic continuation. arXiv:cond-mat/0403055 (2004)
6. Beach, K.S.D., Lee, P.A., Monthoux, P.: Field-induced antiferromagnetism in the Kondo insulator. Phys. Rev. Lett. **92**, 026401 (2004)
7. Benjamin, C., Jonckheere, T., Zazunov, A., Martin, T.: Controllable π junction in a Josephson quantum-dot device with molecular spin. Eur. Phys. J. B **57**, 279–289 (2007)
8. Bercx, M., Lang, T.C., Assaad, F.F.: Magnetic field induced semimetal-to-canted-antiferromagnet transition on the honeycomb lattice. Phys. Rev. B **80**, 045412 (2009)
9. Cleuziou, J.P., Wernsdorfer, W., Bouchiat, V., Ondarçuhu, T., Monthioux, M.: Carbon nanotube superconducting quantum interference device. Nature Nanotechnology **1** (2006)

10. Eichler, A., Weiss, M., Oberholzer, S., Schönenberger, C., Yeyati, A.L., Cuevas, J.C., Martín-Rodero, A.: Even-odd effect in Andreev transport through a carbon nanotube quantum dot. Phys. Rev. Lett. **99**(12), 126602 (2007)
11. Feldbacher, M., Assaad, F.F.: Efficient calculation of imaginary-time-displaced correlation functions in the projector auxiliary-field quantum Monte Carlo algorithm. Phys. Rev. B **63**(7), 073105 (2001)
12. Georges, A., Kotliar, G., Krauth, W., Rozenberg, M.J.: Dynamical mean-field theory of strongly correlated fermion systems and the limit of infinite dimensions. Rev. Mod. Phys. **68**(1), 13 (1996)
13. Gloor, T.A., Mila, F.: Strain induced correlation gaps in carbon nanotubes. Eur. Phys. J. B **38**, 9–12 (2004)
14. Herbut, I.F.: Interactions and phase transitions on graphene's honeycomb lattice. Phys. Rev. Lett. **97**(14), 146401 (2006)
15. Jørgensen, H.I., Novotný, T., Grove-Rasmussen, K., Flensberg, K., Lindelof, P.E.: Critical 0-π transition in designed Josephson quantum dot junctions. Nano Letters **7**(8), 2441–2445 (2007)
16. Kempa, H., Semmelhack, H.C., Esquinazi, P., Kopelevich, Y.: Absence of metal-insulator transition and coherent interlayer transport in oriented graphite in parallel magnetic fields. Solid State Communications **125**(1), 1 (2003)
17. Kharzeev, D.E., Reyes, S.A., Tsvelik, A.M.: Spin density wave formulation in graphene facilitated by the in-plane magnetic field. arXiv:cond-mat/0611251 (2006)
18. Luitz, D.J., Assaad, F.F.: A weak coupling ctqmc study of the single impurity and periodic Anderson models with s-wave superconducting baths. arXiv:0909.2656v2 (2009)
19. Milat, I., Assaad, F., Sigrist, M.: Field induced magnetic ordering transition in Kondo insulators. Eur. Phys. J. B **38**, 571–580 (2004)
20. Neto, A.H.C., Guinea, F., Peres, N.M.R., Novoselov, K.S., Geim, A.K.: The electronic properties of graphene. Reviews of Modern Physics **81**(1), 109 (2009)
21. Paiva, T., Scalettar, R.T., Zheng, W., Singh, R.R.P., Oitmaa, J.: Ground-state and finite-temperature signatures of quantum phase transitions in the half-filled Hubbard model on a honeycomb lattice. Phys. Rev. B **72**, 085123 (2005)
22. Rozhkov, A.V., Arovas, D.P.: Josephson coupling through a magnetic impurity. Phys. Rev. Lett. **82**(13), 2788–2791 (1999)
23. Rubtsov, A.N., Savkin, V.V., Lichtenstein, A.I.: Continuous-time quantum Monte Carlo method for fermions. Phys. Rev. B **72**(3), 035122 (2005)
24. Sandvik, A.: Stochastic method for analytic continuation of quantum Monte Carlo data. Phys. Rev. B **57**, 10287 (1998)
25. Sorella, S., Tosatti, E.: Semi-metal-insulator transition of the Hubbard model in the honeycomb lattice. Europhys. Lett. **19**, 699 (1992)
26. Wang, F., Landau, D.P.: Determining the density of states for classical statistical models: A random walk algorithm to produce a flat histogram. Phys. Rev. E **64**, 056101 (2001)
27. Wang, F., Landau, D.P.: Efficient, multiple-range random walk algorithm to calculate the densiy of states. Phys. Rev. Lett. **86**, 2050 (2001)

Deacon Process over RuO_2 and TiO_2-Supported RuO_2

Ari P. Seitsonen, Jan Philipp Hofmann, and Herbert Over

Abstract We use density functional theory (DFT) calculations to study a novel Deacon-like process over pure $RuO_2(110)$ and TiO_2-supported $RuO_2(110)$, a commercial process which was recently introduced by Sumitomo Chemical. During the HCl oxidation reaction the surface becomes chlorinated, which makes this catalyst stable under the harsh reaction conditions. The reaction mechanism for the chlorination of unsupported $RuO_2(110)$ proceeds via water formation in the bridging position and the final replacement of this water species by chlorine. The actual HCl oxidation reaction takes place via a one-dimensional Langmuir-Hinshelwood mechanism. The recombination of on-top chlorine atoms to form the desired product Cl_2 is identified as the rate-determining step with an activation energy E_a of 114 kJ/mol. DFT calculations predict that one monolayer of $RuO_2(110)$ supported on $TiO_2(110)$ reveals catalytic activity which is almost as high as that on the unsupported $RuO_2(110)$ (E_a = 124 kJ/mol). Further reduction of the Ru loading to half of a monolayer still keeps the $TiO_2(110)$-$RuO_2(110)$ as a good catalyst for the Deacon process, with an activation barrier of the rate-determining step of 29 kJ/mol higher than at pure $RuO_2(110)$.

Ari P. Seitsonen
IMPMC, CNRS & Université Pierre et Marie Curie, 4 place Jussieu, case 115, F-75252 Paris Cedex 05, France
Department of Applied Physics, Helsinki University of Technology, P.O. Box 1100, FI-02015 TKK, Espoo, Finland
Physikalisch-Chemisches Institut der Universität Zürich, Winterthurerstr. 190, CH-8057 Zürich, Switzerland
e-mail: Ari.P.Seitsonen@iki.fi

Jan Philipp Hofmann · Herbert Over
Physikalisch-Chemisches Institut, Justus Liebig Universität Giessen, Heinrich Buff Ring 58, D-35392 Gießen, Germany
e-mail: Herbert.Over@Phys.Chemie.Uni-Giessen.de

S. Wagner et al. (eds.), *High Performance Computing in Science and Engineering, Garching/Munich 2009*, DOI 10.1007/978-3-642-13872-0_43,

1 Introduction

The original Deacon process is known since more than 140 years [1]. However, it was only recently that Sumitomo Chemical succeeded in developing an efficient and stable Deacon-like process, which is catalysed by ultra thin RuO_2 layers supported on TiO_2 [2]. The Sumitomo process is considered as a true breakthrough in recent catalysis research since chlorine can be recycled from waste HCl with low energy cost and high conversion yields of 95 %. The energy consumption of the Sumitomo process is only 15 % of that required by the recently developed Bayer & UhdeNora electrolysis method [3]. The stability and the catalytic activity of RuO_2 have recently been elucidated by a combined experiment/theory approach on $RuO_2(110)$ model catalyst [4, 5] and on RuO_2 powder catalyst [6]. It was shown that the replacement of bridging O by chlorine is responsible for the extraordinary stability of RuO_2 in the HCl oxidation [4]. Yet the actual reaction mechanism for the replacement reaction is not (fully) understood, calling for the ab-initio calculations presented in Chapter 3. The reaction mechanism of the HCl oxidation over the chlorinated $RuO_2(110)$ is presented in Chapter 4, summarising the results of Ref. [5].

One serious problem encountered with the Sumitomo process is, however, related to the limited abundance of ruthenium. With a worldwide production of only 8 t/a it is obvious that availability of ruthenium may impose a bottleneck for a broader commercialisation of the Sumitomo process. Detailed studies performed by Sumitomo Chemical [2] indicated that the actual Sumitomo process takes place over ultra thin RuO_2 films (about 1 or 2 monolayers) supported on rutile-TiO_2, which is preferentially oriented along the (110) direction. Experimental data of well-defined model $RuO_2/TiO_2(110)$ catalysts are not available at the moment since the growth of such model systems is time consuming. Here the predictive power of Density Functional Theory (DFT) calculations can be efficiently utilised to capture detailed information about the reactivity of such layered systems as a function of the RuO_2 film thickness. These DFT calculations are presented in Chapter 5, which is the main focus of the present report. In Chapter 2 calculational details are presented about the applied DFT program package and how this program performs on the HLRB system. Chapter 6 concludes with a short summary and poses some still open questions.

2 Calculational Details

In the DFT calculations the electronic structure of an atomistic system is solved explicitly in contrast to methods where the interactions between the atoms are parametrised. This approach causes a vast increase in computing time required, but leads to a *first-principles* approach with no adjustable parametres, thus lending credibility to the results. In practise the Kohn-Sham method [7] is employed; there the solution of the ;quantum-mechanical many-body Schrödinger equation is replaced with the solution of just single-particle equations for the Kohn-Sham orbitals ψ_i

with an adjoint effective, local Kohn-Sham potential V_{KS}, that intrinsically carries the many-body character of the complicated interactions:

$$\begin{aligned}\left\{-\frac{1}{2}\nabla^2+V_{KS}(\mathbf{r})\right\}\psi_i(\mathbf{r})&=\varepsilon_i\psi_i(\mathbf{r})\\ n(\mathbf{r})&=\sum_i|\psi_i(\mathbf{r})|^2\\ V_{KS}(\mathbf{r})&=V_{ext}(\mathbf{r})+V_H(\mathbf{r})+V_{xc}(\mathbf{r})\end{aligned} \tag{1}$$

These equations have to be solved self-consistently, since the Kohn-Sham potential depends explicitly on the electron density n. All the terms in V_{KS} are exactly known — up to the exchange-correlation potential V_{xc} that has to be approximated in practise. A reasonable compromise between accuracy and computational effort are the Generalised Gradient Approximations (GGA); in particular we use here the PBE-GGA [8].

For the solution of the Kohn-Sham equations the wave functions are expanded in a basis set. A popular choice for the basis are the plane waves; they combine a straight-forward implementation with no positional bias, thus avoiding the Pulay forces in the expression for the gradients on the atomic positions, and the quality of the basis set and thus the accuracy of the results can be adjusted with a single parameter. The large number of plane waves and the enforcement of using either pseudo potentials or the projected-augmented wave method [9] complicate the plane wave codes, but they still remain widely used and are developed further.

An efficient practical scheme of using plane waves in the Kohn-Sham equations was pioneered by Roberto Car and Michele Parrinello [10]: Fast Fourier Transforms (FFT) are used for switching between the real and reciprocal space, so as to work in the space best suited for each term in the algorithms. The parallelisation for large-scale computers follows by either dividing the FFT's into different processor groups, or splitting the three-dimensional space onto the processors. The parallelisation of the 3D-FFT's is thus the major task in adjusting the DFT codes to run efficiently on parallel computers; the remaining operations are mainly global sums over small number of elements, and thus less critical for high degree of parallellisation.

We have used the computer code `VASP` [11] for the calculations. The code is efficiently parallel up to a modest number of parallel tasks, as demonstrated in Figure 1. Some calculations also necessitate the use of nudged elastic band algorithm [12] for the evaluation of reaction paths and energy barriers; in this method several replicas are calculated simultaneously along the reaction path, and as minimal amount of communication is required between them, large number of processors, typically 16 replicas $\times$ 32 processors/replica = 512 processors can be employed for the parallel computation.

Calculated adsorption energies of Cl, O, H_20 are given with respect to half a molecule of Cl_2 and O_2 and a molecular H_2O in the gas phase, respectively. Positive energy values are used for exothermic adsorption processes.

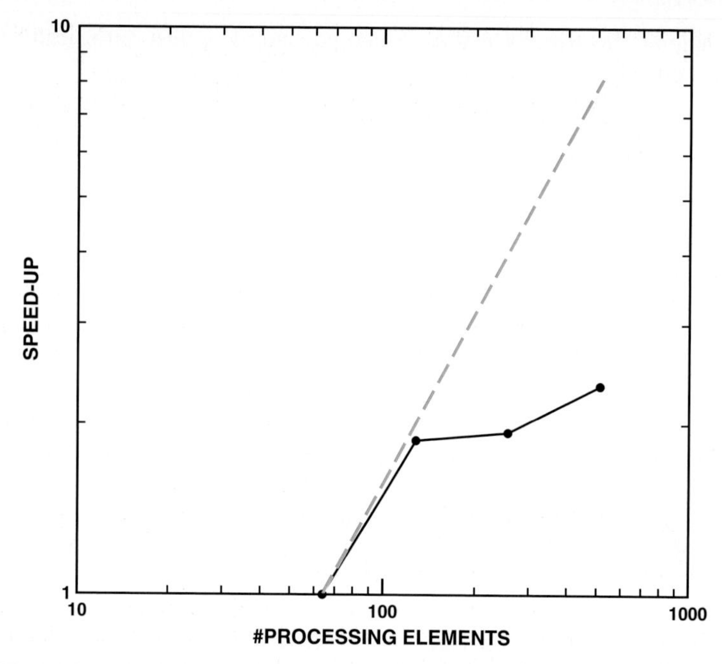

Fig. 1 The scaling of the execution time in a system of 340 atoms of Ru@TiO_2(110) surface as a functions of the number of tasks used on the Altix HLRB-II system. The time at 64 tasks is used as the reference

3 Reaction Mechanism for the Chlorination of RuO_2(110)

As shown in our recent publication [4] the actual chlorination process of RuO_2(110) takes place by replacing the bridging O atoms with chlorine. In this chapter we focus on mechanistic aspects of this replacement reaction [13]. First of all simple exposure of molecular chlorine is not able to chlorinate the RuO_2(110) surface. Molecular chlorine dissociates readily, however, the exchange process $Cl_{ot}+O_{br} \rightarrow Cl_{br}+O_{ot}$ is activated by 221 kJ/mol according to our DFT calculations. Therefore hydrogen interaction is an important ingredient in the chlorination mechanism which is automatically provided by HCl exposure. The hydrogen of adsorbed HCl is directly transferred to the neighbouring bridging O atom without any noticeable activation barrier. However, the next step for the chlorination, *ie.* the replacement of the bridging hydroxyl group by on-top Cl, Cl_{ot} + O_{br}-H $\rightarrow$ Cl_{br} + O_{ot}-H is again strongly activated by more than 300 kJ/mol as determined by DFT calculations. Therefore the chlorination process requires the formation of bridging water as a reaction intermediate to substitute bridging O by chlorine.

There are two possible reaction mechanisms conceivable which both have been studied by DFT calculations (mechanisms A and B) and which will be discussed in some detail (*cf.* Figure 2) in the following:

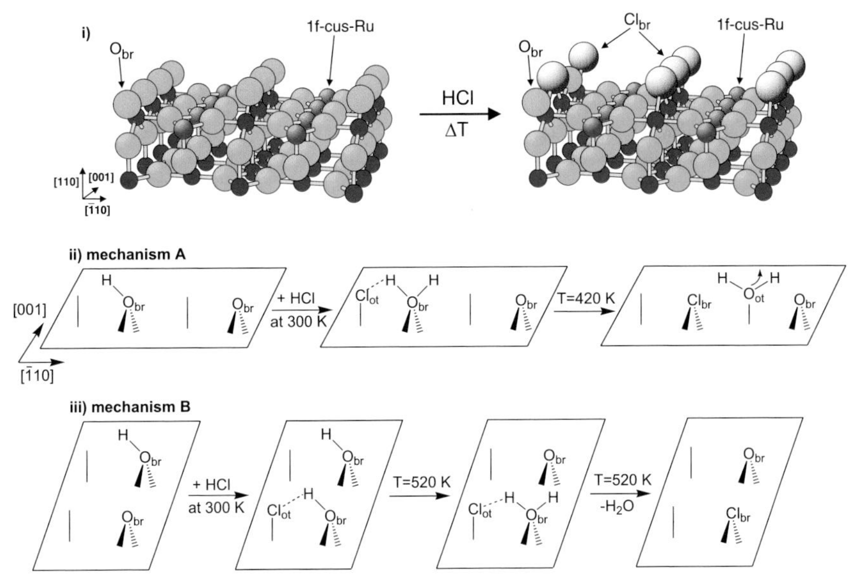

Fig. 2 Schematic representation of the chlorination mechanism of $RuO_2(110)$. i) Ball-and-stick model of bulk-truncated $RuO_2(110)$ revealing undercoordinated surface atoms: bridging O atoms (O_{br}) and one fold-coordinatively unsaturated Ru-sites (Ru 1f-cus), which are called on-top sites. Upon HCl exposure at higher temperatures the stoichiometric surface transforms into a chlorinated surface where the bridging O atoms are replaced by bridging chlorine (Cl_{br}) atoms (shown in grey color). ii) and iii) Simplified representation of the chlorination reaction on $RuO_2(110)$: Solid upright lines refer to empty cus-sites and double wedges indicate bridge positions. ii) Mechanism A: Bridging water H-O_{br}-H is formed by hydrogen transfer from adsorbed HCl to an adjacent O_{br}-H group. Water shifts subsequently to a 1f-cus site from which it desorbs. The vacancies are in turn filled by Cl_{ot} to form Cl_{br}. iii) Mechanism B: Bridging water is produced by the recombination of two neighbouring O_{br}-H groups, assisted by Cl_{ot}. As soon as bridging water is formed, it shifts to an on-top site from which it desorbs immediately and leaves a vacancy. These vacancies are in turn filled by Cl_{ot} according to mechanism A. In both mechanisms A and B the replacement of H-O_{br}-H with Cl_{br} may take place in a concerted way according to preliminary NEB calculations

Mechanism A (Figure 2ii): If the bridging oxygen atoms are selectively converted into O_{br}-H group, a non-activated hydrogen transfer from HCl to a neighbouring O_{br}-H group forms a hydrogen-bonded $Cl_{ot} \cdots$ H-O_{br}-H complex. The direct desorption of water from this complex is activated by −203 kJ/mol and therefore not feasible. However, the complex can release water by shifting the bridging water H-O_{br}-H first to an on-top site, a process which is activated by only 134 kJ/mol and which is followed by the desorption of $(H_2O)_{ot}$ at temperatures above 420 K (with an activation energy of E_{des} = 84 kJ/mol). The still high activation barrier for shifting the water is partially due to a stabilisation of the bridging water by the Cl_{ot} atom. Without the stabilising effect of Cl_{ot} the bridging water species shifts readily to an on-top position, with an activation energy of only 83 kJ/mol. Therefore a removal of the Cl_{ot} from the vicinity of the bridging water by diffusion along the [001] direction may open another promising reaction pathway, since the diffusion barrier of

Cl_{ot} along a 1f-cus row amounts to 89 kJ/mol. Regardless of how the water removal from the bridging position creates a vacancy, the final reaction step is that Cl_{ot} slides into the bridging vacancy either directly and practically without an activation barrier (9 kJ/mol) when Cl_{ot} is directly facing the vacancy, or Cl_{ot} diffuses along the 1f-cus row to reach the vacancy, a process which is activated by 63 kJ/mol.

Mechanism B (Figure 2iii): A second way to produce bridging water is the recombination of two neighbouring O_{br}-H groups. The rest of the reaction is similar to mechanism A. This process is activated by 106 kJ/mol according to our preliminary NEB calculations.

For both mechanisms it is also possible that the replacement of bridging water by chlorine proceeds in a concerted step as suggested by preliminary NEB calculations. Both mechanisms (A and B) of bridging water formation depend crucially on the presence of hydrogen on the surface. For stoichiometry reasons HCl exposure alone is not sufficient to provide enough hydrogen to replace all bridging O atoms: $2HCl_{ot} + O_{br} \rightarrow H_2O + Cl_{ot} + Cl_{br}$. Therefore, the chlorination process may be facilitated by pre-exposing hydrogen to the surface so that all the O_{br} sites are already transformed into $O_{br}H$ groups. Previouslym it was shown that hydrogen exposure indeed transforms O_br into O_brH [14]. Subsequent HCl adsorption and an additional H transfer from HCl to O_{br}-H can lead one-to-one to H_2O and Cl_{br}: $HCl_{ot} + O_{br} \rightarrow H_2O + Cl_{br}$. Recent HRCLS experiments [13] have supported and substantiated the DFT-motivated chlorination mechanisms.

4 Reaction Mechanism of the HCl Oxidation over $RuO_2(110)$

In a recent publication [5] we have shown that the kinetics of the HCl oxidation reaction over chlorinated $RuO_2(110)$, where all bridging O atoms are replaced by chlorine, is purely determined by surface thermodynamics, *ie.* the adsorption energies of the reaction intermediates rather than by true kinetic barriers. A very similar conclusion has been drawn by Lopez *et al* in the case of pure $RuO_2(110)$ powder [6]. The microscopic steps of the HCl oxidation are summarised in Figure 3. Adsorption of O_2 is non-activated, forming on-top O (O_{ot}) on the 1f-cus Ru sites. HCl adsorbs on 1f-cus sites directly transfering the H atom to the on-top O species. Without on-top O, HCl may not adsorb on the chlorinated $RuO_2(110)$ surface. According to our DFT calculations the dehydrogenation of HCl_{ot} via O_{ot} proceeds without any noticeable activation barrier. The final production of adsorbed water (H_2O_{ot}) via H transfer [14, 15] between two neighbouring $O_{ot}H$ groups is activated by 29 kJ/mol, an energy barrier that is easily surmounted at typical reaction temperatures. The recombination of two on-top Cl to form the desired product Cl_2 constitutes the rate determining step with an activation barrier of 114 kJ/mol.

Since hydrogen cannot be accepted by the bridging Cl atoms (activation barrier is as high as 250 kJ/mol), an efficient communication between neighbouring 1f-cus Ru rows is efficiently suppressed. Therefore the chlorinated $RuO_{2-x}Cl_x(110)$ catalyst can be envisioned as a one-dimensional catalyst offering isolated rows of

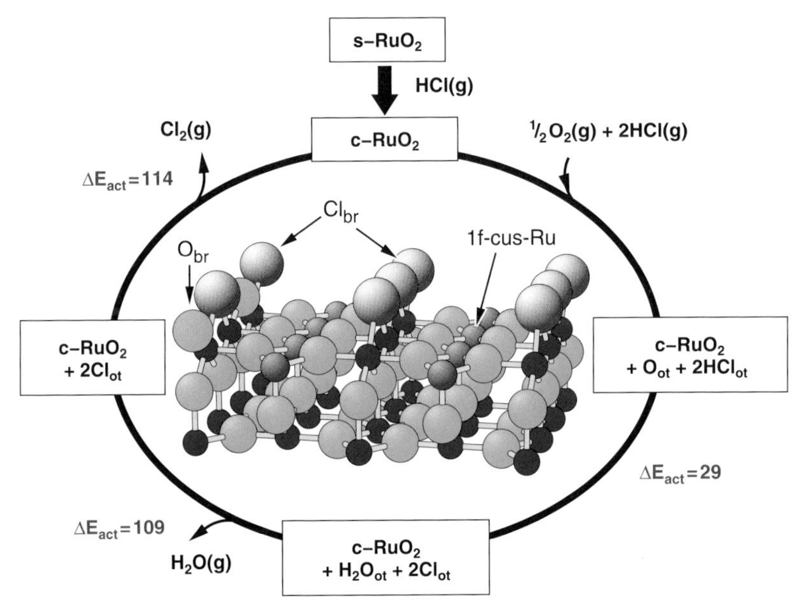

Fig. 3 The catalytic cycle of the HCl oxidation over $RuO_2(110)$, starting with a selective and self-limiting replacement of the bridging O atoms by bridging chlorine forming $RuO_{2-x}Cl_x(110)$. The reactant molecules O_2 and HCl both adsorb first on the 1f-cus Ru sites. O_2 dissociates to form adsorbed O and HCl dehydrogenates via a hydrogen transfer to form Cl and OH species in on-top positions. H-transfer among the OH species leads to water formation which is released from the surface around 420 K. Neighbouring on-top adsorbed Cl atoms recombine to form Cl_2 which is immediately liberated into the gas phase. The activation energies ΔE_{act} are determined by DFT calculations. The rate determining step is constituted of the association of two neighbouring Cl_{ot} atoms to form Cl_2. From Ref. [5]

1f-cus Ru sites where a LH-type dehydrogenation reaction between HCl and O_2 takes place. As shown in Figure 3 only the hydrogen transfer from adsorbed HCl to adsorbed OH is kinetically activated by 29 kJ/mol.

All the other activation barriers are determined by the adsorption energies (equals the acitivation energies for desorption) of reaction intermediates such as water (ΔE_{act} = 109 kJ/mol) and on-top Cl (ΔE_{act} = 114 kJ/mol). Under typical reaction conditions in excess of oxygen and temperatures between 500 and 600 K, the chlorinated $RuO_2(110)$ surface is mainly covered with on-top Cl and O atoms. The released hydrogen is removed from the catalyst surface via water formation and subsequent desorption above 400 K (*cf.* Figure 3). The remaining Cl_{ot} atoms on the surface have to diffuse along the 1f-cus Ru rows to meet a second Cl_{ot} to react with it. This diffusion process is activated by 63 kJ/mol (DFT) and is therefore not rate determining at reaction temperatures between 500 and 600 K.

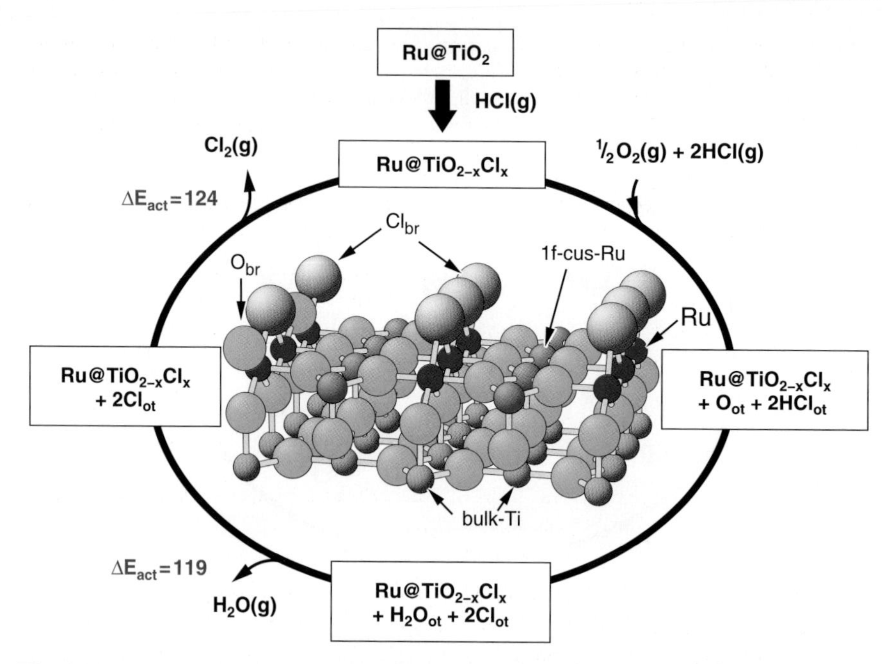

Fig. 4 The catalytic cycle of the HCl oxidation over $TiO_2(110)$-supported $RuO_2(110)$, starting with a selective and self-limiting replacement of the bridging O atoms by bridging chlorine forming $(RuO_{2-x}Cl_x)@(TiO_2)$. The activation energies ΔE_{act} are determined by DFT calculations. The rate determining step is constituted of the association of two neighbouring Cl_{ot} atoms to form Cl_2

5 Reaction Mechanism of the HCl Oxidation over $RuO_2(110)$ Supported on $TiO_2(110)$: DFT Predictions

In this chapter we report on extensive state-of-the art DFT calculations exploring the catalytic cycle (*cf.* Figure 4) of the oxidation of HCl on $TiO_2(110)$-supported $RuO_2(110)$ model catalyst of various thicknesses of the RuO_2 film. These calculations may help to elucidate the role of the TiO_2 support on the $RuO_2(110)$ catalysed Sumitomo-process without relying on a tedious preparation of well-defined RuO_2-$TiO_2(110)$ model catalysts. Without Ru the pure, stoichiometric s-$TiO_2(110)$ slab is completely inactive towards the HCl oxidation due to the endothermic adsorption process of molecular oxygen [16] and also compare 1. However, as soon as the 1f-cus Ti sites are replaced by Ru dissociative adsorption of oxygen is facile, making TiO_2 a reasonably good Deacon catalyst. Since the kinetics of the HCl oxidation process is mainly determined by the adsorption/desorption energies of the reaction intermediates (*cf.* Chapter 4), we calculated first the adsorption energies of all relevant reaction intermediates on the ultra-thin RuO_2 films supported on $TiO_2(110)$ in the case where all bridging O atoms have been replaced by chlorine. These DFT calculations are summarised in Table 1.

Table 1 Adsorption energies of various reaction intermediates on the fully chlorinated, mixed Ru-TiO_2(110) layers, where all bridging surface O atoms are replaced by chlorine. All energies are given in kJ/mol. The energy reference is the fully chlorinated surface. RuO_2 = 7 trilayers of RuO_2 with native lattice constant of RuO_2. $(RuO_2)_7@(TiO_2)_0$ = 7 trilayers of RuO_2 with lattice constant of TiO_2. $(RuO_2)_2@(TiO_2)_5$ = 2 trilayers RuO_2 + 5 trilayers of TiO_2, lattice constant of TiO_2. $(RuO_2)_1@(TiO_2)_6$ = 1 trilayer of RuO_2 + 6 trilayers of TiO_2, lattice constant of TiO_2. $(RuO_2)_{0.5}@(TiO_2)_{6.5}$ = 7 trilayers of TiO_2 where all 1f-cus Ti sites are replaced by Ru. Notice that exothermic interactions are indicated by positive energies, while endothermic steps are indicated by negative energies

System	Reaction Adsorbates +1/2 O_2 O_{ot}	+HCl OH_{ot} + Cl_{ot}	+HCl $OH_{2,ot}$ + $2Cl_{ot}$	$-H_2O$ $2Cl_{ot}$	(Cl_{ot})
RuO_2	100	226	404	295	114
$(RuO_2)_2@(TiO_2)_5$	122	231	412	310	122
$(RuO_2)_1@(TiO_2)_6$	118	247	432	313	124
$(RuO_2)_{0.5}@(TiO_2)_{6.5}$	144	286	478	352	143
s-TiO_2	−219	−101	31	−54	−60

The reference energy is chosen to be that of the respective chlorinated surface of the model catalyst, where all bridging O atoms are replaced by chlorine and no other adsorbate is present. Positive energy values indicate exothermic steps. The values in the second column indicate the energy changes when 1/2 O_2 is added to the catalyst, forming oxygen (O_{ot}) on-top of a 1f-cus position. On this O_{ot}-precovered surface HCl is accommodated, thereby forming $O_{ot}H$ and Clot; the energies are provided in the third column. On this surface another HCl is adsorbed, resulting in $O_{ot}H_2 + 2Cl_{o}t$ with energies given in the fourth column. Lastly, the water desorbs from the surface, reducing the energies (due to the endothermic desorption step) as indicated in the fifth column. In order to be able to determine the reaction energies, the adsorption energies of on-top chlorine are compiled in the last column. The difference of energies in the fifth column and those of twice the adsorption energy of Clot results in the reaction energy of the net reaction 2 HCl + 1/2 $O_2 \rightarrow H_2O$ + Cl_2, which turned out to be 66 kJ/mol and therefore being only slightly higher than the experimental value of 59 kJ/mol.

When going to a single trilayer of RuO_2(110) which is pseudomorphically grown on TiO_2(110) [$(RuO_2)_1@(TiO_2)_6$], the adsorption energies of on-top chlorine decreases to 124 kJ/mol so that the activation energy for association of on-top chlorine (the rate determining step) is only 9 kJ/mol higher than that of native RuO_2 (110) bulk-like slab. Therefore we expect from these calculations that the activity of 1 ML RuO_2(110) on TiO_2(110) is practically identical to that of pure RuO_2(110). Increasing the RuO_2 film thickness to two monolayers does not change the adsorption energies.

When further decreasing the amount of Ru to 1/2 ML, *ie.* in the 7 trilayers of TiO_2(110) only the 1f-cus Ti sites are replaced by Ru [$(RuO_2)_{0.5}@(TiO_2)_{6.5}$], then the adsorption energy of on-top chlorine increases by 29 kJ/mol with respect to RuO_2(110), *ie.* the activity is expected to be significantly reduced. But we have to bear in mind that merely 1/2 ML of Ru is already able to switch the catalytically dead s-$TiO_2(110)$ surface into a reasonably active catalyst for the HCl oxidation.

From Table 1 we can also recognise that the oxygen activation on s-$TiO_2(110)$ is the true rate determining step in the HCl oxidation reaction, since dissociative adsorption of oxygen molecules is endothermic by −219 kJ/mol. Our DFT calculations indicate that with small amounts of Ru in the outermost surface layer, this deficiency of TiO_2 can be overcome. Therefore Ru may be envisioned to promote the activation of oxygen on TiO_2.

The microscopic steps of the HCl oxidation on a fully chlorinated 1 ML-$RuO_2(110)/TiO_2(110)$ model catalyst are summarised in Figure 4. The adsorption of O_2 is non-activated, forming on-top O (O_{ot}) and HCl_{ot} on vacant 1f-cus Ru sites. According to our DFT calculations the dehydrogenation of HCl_{ot} via O_{ot} proceeds without any noticeable activation barrier. The final production of adsorbed water (H_2O_{ot}) via H transfer [14] between two neighbouring $O_{ot}H$ groups is activated by about 15 kJ/mol according to our preliminary calculations. Water desorbs with an activation energy of 119 kJ/mol. The rate determining step is still identified with the association of two neighbouring Cl_{ot} atoms to form Cl_2. This reaction step is activated by 124 kJ/mol, the adsorption energy of on-top Cl. An additional barrier due to kinetics has not been identified with DFT calculations. Most of these energies are very close to the case of fully surface-chlorinated bulk-$RuO_2(110)$ [5], rendering $RuO_2(110)$ indeed a proper model catalyst for the practical Sumitomo process.

6 Concluding Remarks

The stability of $RuO_2(110)$ in the HCl oxidation reaction is traced back to the replacement of bridging O by chlorine. DFT calculations have shown that the chlorination of the $RuO_2(110)$ surface is mediated by the formation of a bridging water species. Bridging H_2O is produced either by H-transfer from adsorbed HCl at a 1f-cus site to a neighbouring O_{br}-H group (Figure 2ii, low temperature mechanism A) or by the recombination of two neighbouring O_{br}-H groups to produce water, O_{br} and a vacancy V_{br} (Figure 2iii, high temperature mechanism B). The final step in the chlorination process constitutes the replacement of the bridging water species by on-top chlorine during which the bridging water shifts to an on-top position and desorbs immediately above 420 K. On the basis of preliminary NEB calculations we presume that this final replacement process occurs in a concerted way.

Density functional theory calculations reveal that the oxidation of HCl with oxygen producing Cl_2 and water proceeds on $TiO_2(110)$-supported- and pure-$RuO_2(110)$ surfaces via a one-dimensional Langmuir-Hinshelwood mechanism. The recombination of two adjacent chlorine atoms on the surface of the catalyst constitutes the rate determining step in this novel Deacon-like process. For Ru coverage of 1 ML, where all Ti atoms in the topmost double layer are replaced by Ru, the activation barriers for rate determining step of Cl association is within 9 kJ/mol identical to that of a pure $RuO_2(110)$ surface. If only the 1f-cus Ti atoms are replaced by Ru then the 1/2 ML Ru-$TiO_2(110)$ catalyst is still active, however with an activation barrier which is 29 kJ/mol higher than at $RuO_2(110)$.

Very important for industrial applications is the finding that we can save Ru resources in the preparation of the Sumitomo catalyst as already 1 ML of $RuO_2(110)$ supported on $TiO_2(110)$ is sufficient to maintain practically the full activity of RuO_2 in the HCl oxidation reaction. However, we have to bear in mind that Ru and Ti can easily form mixed oxides above 600-700 K [17]. This means that once prepared, 1 ML of $RuO_2(110)$ on $TiO_2(110)$ is thermodynamically not stable under UHV conditions but rather forms a mixed Ti-Ru dioxide with unknown catalytic activity. Since the affinity of RuO_2 to adsorb oxygen is substantially higher than for $TiO_2(110)$ one may expect that under reaction conditions with oxygen excess in the gas ($HCl+O_2$) feed ruthenium is driven out of the mixed Ti-Ru dioxide towards the surface for thermodynamic reasons. Future ab-initio thermodynamic calculations will clarify this point.

Acknowledgements We would like to thank Leibniz Rechenzentrum in Munich for providing us massive parallel super-computing time, and the Deutsche Forschungsgemeinschaft and Fond der Chemischen Industrie for financial support.

References

1. Deacon, H.: U.S. Patent 165,802,1875
2. Iwanaga, K., Seki, K., Hibi, T., Issoh, K., Suzuta, T., Nakada, M., Mori, Y., Abe, T.: The Development of Improved Hydrogen Chloride Oxidation Process", Sumitomo Kagaku 2004-I, 1-11
3. Gestermann, F., Ottavini, A.: Chlorine Production with Oxygen-depolarised Cathodes on an Industrial Scale. In: Moorhouse J. (Ed.): Modern Chlor-Alkali Technology. **8**, 49–56 (2001)
4. Crihan, D., Knapp, M., Zweidinger, S., Lundgren, E., Weststrate, C.J., Andersen, J.N., Seitsonen, A.P., Over, H.: Stable Deacon Process for HCl Oxidation over RuO_2. Angew. Chemie Int. Ed. **47**, 2131-2134 (2008); doi: 10.1002/anie.200705124
5. Zweidinger, S., Crihan, D., Knapp, M., Hofmann, J.P., Seitsonen, A.P., Westrate, C.J., Lundgren, E., Andersen, J.N., Over, H.: J. Phys. Chem. C **112**, 9966-9969 (2008); doi: 10.1021/jp803346q
6. López, N., Gómez-Segura, J., Marín, R.P., Pérez-Ramírez, J.: Mechanism of HCl oxidation (Deacon process) over RuO_2. J. Catal. **255**, 29-39 (2008); doi: 10.1016/j.jcat.2008.01.020
7. Kohn, W., Sham, L.J.: Self-Consistent Equations Including Exchange and Correlation Effects. Phys. Rev. **140** A1133-A1138 (1965); doi: 10.1103/PhysRev.140.A1133
8. Perdew, J.P., Burke, K., Ernzerhof, M.: Generalised Gradient Approximation Made Simple. Phys. Rev. Lett. **77**, 3865-3868 (1996); doi: 10.1103/PhysRevLett.77.3865
9. Blöchl, P.E.: Projector augmented-wave method. Phys. Rev. B 50, 17953-17979 (1994); doi: 10.1103/PhysRevB.50.17953
10. Car, R., Parrinello, M.: Unified approach for molecular dynamics and density functional theory. Phys. Rev. Lett. **55**, 2471-2474 (1985); doi: 10.1103/PhysRevLett.55.2471
11. Kresse, G., Furthmüller, J.: Efficiency of ab-initio total energy calculations for metals and semiconductors using a plane-wave basis set. Comput. Mat. Sci. **6**, 15-50 (1996); doi: 10.1016/0927-0256(96)00008-0

12. Jónsson, H., Mills, G., Jacobsen, K.W.: Nudged Elastic Band Method for Finding Minimum Energy Paths of Transitions. In: Berne, B.J., Ciccotti, G., Coker, D.F. (Eds.): Classical and Quantum Dynamics in Condensed Phase Simulations, pp. 385. World Scientific (1998)
13. Hofmann, J.P., Zweidinger, S., Knapp, M., Seitsonen, A.P., Schulte, K., Andersen, J.N., Lundgren, E., Over, H.: in preparation
14. Knapp, M., Crihan, D., Seitsonen, A.P., Lundgren, E., Resta, A., Andersen, J.N., Over, H.: Complex Interaction of Hydrogen with the $RuO_2(110)$ Surface. J. Phys. Chem. C **111**, 5363-5373 (2007); doi: 10.1021/jp0667339
15. Knapp, M., Crihan, D., Seitsonen, A.P., Over, H.: Hydrogen Transfer Reaction on the Surface of an Oxide Catalyst. J. Am. Chem. Soc. **127**, 3236 (2005); doi: 10.1021/ja043355h
16. Rasmussen, M.D., Molina, L.M., Hammer, B.: Adsorption, diffusion, and dissociation of molecular oxygen at defected $TiO_2(110)$: A density functional theory study. J. Chem. Phys. **112**, 988-997 (2004); doi: 10.1063/1.1631922
17. Chambers, S.A., Epitaxial Growth and Properties of thin film oxides. Surf. Sci. Rep. **39**, 105-180 (2000); doi: 10.1016/S0167-5729(00)00005-4

Charge-Carrier Transport Through Guanine Crystals and Stacks

Frank Ortmann, Lars Matthes, Björn Oetzel, Friedhelm Bechstedt, and Karsten Hannewald

Abstract We study small-polaron motion through guanine-based materials. The temperature dependence and anisotropy of charge carrier (hole) transport in crystalline guanine is investigated by employing the Kubo formalism and calculating the hole mobilities with ab initio DFT material parameters. We discuss our findings in relation to transport pathways in DNA-based structures like guanine quadruplexes and ribbons which are considered to play a major role in DNA-based nanoelectronics. The mobility results are interpreted by help of a novel visualization method for transport channels which we derive from overlapping wavefunctions. An analysis of coherent and incoherent contributions to the mobility shows that even in materials with high purity and long-range order like crystals only incoherent phonon-assisted hopping occurs at room temperature.

1 Introduction

Miniaturization in nanotechnology has already reached length scales of a few tens of nanometers. Further progress is expected from a bottom-up approach of self-assembling building blocks arranging in supramolecular clusters of nanometer size. Thereby, π electron rich small molecules such as the DNA bases are promising materials for electronic and optoelectronic devices. In particular, a conducting DNA wire for use in nanotechnology is a very intriguing vision. It combines the conductivity of a wire with the self-assembly properties of DNA and the enormous knowledge of synthesizing DNA gathered over the past decades. Besides studies using the DNA as a template for a metallic wire, [5] the inner core of the double helix itself, the base stack, is considered to be a (semi-) conducting transport channel. [11] Thereby it is assumed that carrier transport is confined to the DNA bases

Frank Ortmann · Lars Matthes · Björn Oetzel · Friedhelm Bechstedt · Karsten Hannewald
Institut für Festkörpertheorie und -optik, Friedrich-Schiller-Universität Jena and European Theoretical Spectroscopy Facility (ETSF), Max-Wien-Platz 1, 07743 Jena, Germany

S. Wagner et al. (eds.), *High Performance Computing in Science and Engineering, Garching/Munich 2009*, DOI 10.1007/978-3-642-13872-0_44,

[25] and that this channel is built up by π electrons along the stack of the helix. [12] Many experimental studies have been carried out focusing on various aspects of charge transport through DNA on surfaces [34], in solution, [19] or suspended DNA molecules [7]. A recent review has sorted the various experimental situations and parameters (e.g., sequence, length, etc.) and the different results for transport properties. [12]

However a severe drawback of the above-mentioned studies is that the conformation of the DNA is not known in any of these experiments. Although in some cases at least the base pair sequence was fixed to a well defined guanine-cytosine sequence (poly(dG)-poly(dC)), the geometries are affected by thermal and/or static disorder. [39] In particular the temperature becomes an important parameter whose strong influence has been demonstrated experimentally. [37, 38] Also theoretically, temperature induced dynamic disorder is one of the great unknowns in charge transport through DNA and DNA-related systems.

In contrast to DNA where the charge transport occurs along the stack through intrastrand or interstrand pathways, Ref. [27] investigated the charge transport along a ribbon structure of guanine derivatives within the molecular planes in a transistor geometry. Transport in this direction does not fit into the simple overlap-of-π-orbitals picture. In order to better understand the anisotropy arising from the relative orientation of the guanine molecules, it is instructive to consider three-dimensional guanine crystals as a model system. [33] Such a crystal is displayed in Fig. 1. It is a very advantageous model system because it allows to study different transport pathways through the material with their different characteristics. The green arrows in Fig. 1 indicate a direct path through the stack in vertical direction as in intrastrand transport in DNA, whereas the yellow arrows indicate interstrand pathways. The charge transport along the molecular planes like in the ribbon structure used in the transistor experiment of Ref. [27] is indicated by the blue arrow.

From the theoretical side, several first principles studies on DNA have focused on assemblies of its charge-carrying building blocks (e.g. stacked guanine molecules) to mimic basic electronic properties of the DNA but remain tractable for ab initio methods. [3, 10] For more complex assemblies like quartet structures [9] or the 3D guanine crystals of Fig. 1 [31] it has been shown that modifications of the electronic structure of the stack are induced by the influence of the electronic coupling perpendicular to the stacking direction. Electronic coupling within such a single layer is essential to understand the current along the ribbons in the transistor experiment as well as base pair coupling in DNA which is a prerequisite for interstrand transport pathways (see Fig. 1). Therefore, the influence of the relative orientation of guanine molecules deserves high attention in exploring this material as an advanced one. For that reason, we address the important aspect of anisotropy effects in guanine based materials and charge carrier mobilities along various directions and planes in the 3D guanine crystals.

The complexity of the charge transport problem, however, does not only originate from the various possible pathways for the charge carriers and the relative orientations of the molecules. It also arises from the ambient temperature and its influence on the carriers. [18] While previous first principles studies have picked out single

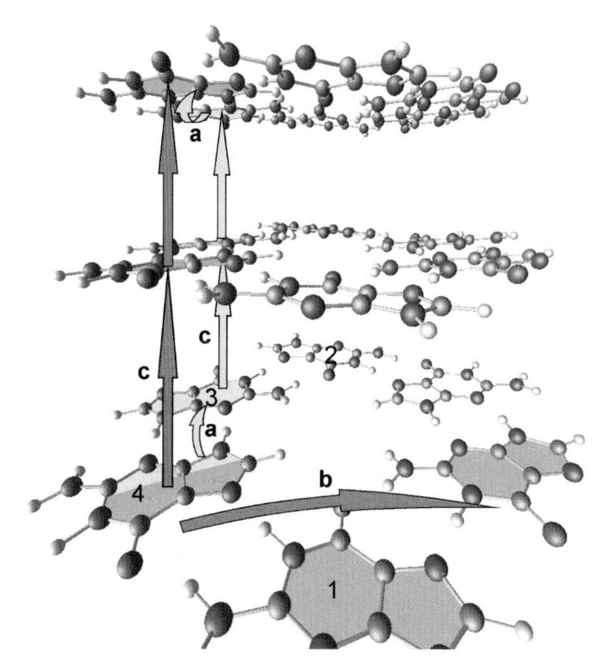

Fig. 1 Guanine crystal as viewed from the side (three molecular planes are shown). For clarity, every second molecular plane in the vertical stacking direction has been removed. The four molecules in a single unit cell are labeled from 1 to 4 each of which represents also a vertical stack. Transport pathways are indicated by arrows (c: along the stack, a and b: between the stacks)

conformations (fixed geometry at $T = 0$K), the experiments (mostly at room temperature) average over many of them. However, the number of possible configurations at a given temperature is huge and can only be crudely represented by a single geometry at $T = 0$K. Therefore, in the present paper we go beyond the $T = 0$ scenario by taking into consideration vibrational degrees of freedom and the influence of their interaction with the charge carriers. This interaction is important because the electronic properties (e.g. the electronic coupling between neighboring molecules) are strongly affected by variations of nuclei positions. [4, 12, 14, 36] Hence, neglecting them in theoretical approaches prevents such studies from being comparable to experiment. The dependence of the electronic properties on the geometry can be rationalized with the polaron concept [20] based on the electron-phonon interaction. By means of the polaron concept we take into account the change in the geometry with the moving charge and how this is influenced by temperature.

After describing the computational procedure, we start the discussion with the anisotropy of the electronic coupling followed by a description of the polaron formation and the influence of the temperature. Applying linear response theory, we calculate the hole mobilities as a function of temperature and direction. Finally, the transport channels are visualized to get an impression of the anisotropy of the carrier mobility.

2 Theory, Computational Method, and Performance

In order to describe temperature-induced effects in the electronic coupling, we additionally take into account vibrational degrees of freedom [with frequencies $\omega_{\mathbf{Q}} \equiv \omega_\lambda(\mathbf{q})$] and the coupling between the charges and the phonons. We use the Hamiltonian

$$H = \sum_{MN} \left[\varepsilon_{MN} + \sum_{\mathbf{Q}} \hbar\omega_{\mathbf{Q}} g_{MN}^{\mathbf{Q}} \left(b_{\mathbf{Q}}^\dagger + b_{-\mathbf{Q}} \right) \right] a_M^\dagger a_N + \sum_{\mathbf{Q}} \hbar\omega_{\mathbf{Q}} \left(b_{\mathbf{Q}}^\dagger b_{\mathbf{Q}} + \frac{1}{2} \right) \quad (1)$$

which, besides the electron transfer integrals ε_{MN}, explicitly contains coupling constants $g_{MN}^{\mathbf{Q}}$ between particles and phonons. For a moving particle these dimensionless parameters describe the strength of scattering by an individual phonon mode. By means of the Holstein-Peierls Hamiltonian (1) one can simulate the effect of a quasiparticle (polaron) which is heavier than the bare charge-carrier (hole) due to the accompanying phonon cloud that has to be carried through the system as well. This increased effective mass is also reflected in the reduced transfer integrals for the polarons

$$\widetilde{\varepsilon}_{MN} = (\varepsilon_{MN} - \Delta_{MN})\, e^{-\sum_\lambda G_\lambda (1+2N_\lambda)}, \quad (2)$$

where zero point vibration effects and temperature effects give rise to a damping factor depending on the phonon occupation number $N_\lambda = \left[\exp(\frac{\hbar\omega_\lambda}{k_B T}) - 1\right]^{-1}$. [17] The quantity $G_\lambda = \left(g_{MM}^\lambda\right)^2 + \frac{1}{2}\sum_{K\neq M}\left(g_{MK}^\lambda\right)^2$ measures the total scattering strength of the mode λ. Additional polaron binding terms like Δ_{MN} have been discussed elsewhere. [1, 16] Here, we focus on the temperature dependence of the coupling between different sites $\widetilde{\varepsilon}_{MN}$ that determine the transfer rates and hence the mobility of the charge carriers.

The basis for the derivation of the charge-carrier mobility is the Kubo formula for conductivity [26] which describes the linear response to an applied electric field. The final result of the evaluation of the Kubo formula including a non-perturbative treatment of electron-phonon coupling is derived elsewhere. [16] The mobility along the Cartesian axis α takes the form

$$\begin{aligned} \mu_\alpha = {} & \frac{e_0}{2k_B T\hbar^2} \sum_{M,N} (R_{M\alpha} - R_{N\alpha})^2 \\ & \times \int_{-\infty}^{\infty} dt F_{MN} \exp\left\{ -2\sum_\lambda G_\lambda \left[1 + 2N_\lambda - \Phi_\lambda(t)\right] \right\} e^{-\Gamma^2 t^2} \quad (3) \\ F_{MN} = {} & (\varepsilon_{MN} - \Delta_{MN})^2 + \frac{1}{2}\sum_\lambda \left(\hbar\omega g_{MN}^\lambda\right)^2 \Phi_\lambda(t). \end{aligned}$$

The damping term $e^{-\sum_\lambda G_\lambda(1+2N_\lambda)}$ has already been discussed in Eq. (2) as polaron band narrowing and appears here as well, however, with the counteracting amplification factor $e^{+\sum_\lambda G_\lambda \Phi_\lambda}$. The function $\Phi_\lambda(t) = (1+N_\lambda)e^{-i\omega_\lambda t} + N_\lambda e^{i\omega_\lambda t}$ takes additional incoherent phonon-assisted hopping contributions into account. This function represents phonon emission ($\propto 1+N_\lambda$) and absorption ($\propto N_\lambda$) processes and its appearance in the exponential factor means that all orders of electron-phonon interaction contribute. The broadening parameter Γ takes scattering mechanisms into account that are not explicitly treated in this formalism (e.g. static disorder due to locally varying on-site energies ε_{MM}). A previous approach calculating mobilities in a stack of guanine molecules was based on the band conduction regime (Boltzmann transport equation) treating the electron-phonon interaction as a perturbation. [2] Our approach is more general. It does not assume coherence *a priori* but covers both coherent (band-like) and incoherent (thermally activated hopping) transport including processes of all orders in the electron-phonon interaction. Note that is has been shown previously that the theory at high T contains the activation law known from Marcus theory. [16, 29]

The material parameters are determined from first principles DFT the details of which are described below. The Kohn-Sham band structure of the HOMO bands is mapped onto a set of parameters using a tight-binding expression with effective transfer integrals ε_{MN}. The expression reads $\varepsilon_{\mathbf{k}} = \pi_{\mathbf{k}} \pm \sqrt{\sigma_{\mathbf{k}} \pm \sqrt{\tau_{\mathbf{k}}}}$ where the four different combinations of signs lead to the four bands derived from the four molecules present in the unit cell (cf. Figure 1). If there were no interactions between molecules of the same cell or inequivalent molecules of neighboring unit cells ($i \neq j$), the four branches would be degenerate and adopt the mean value $\pi_{\mathbf{k}}$. In general, this is not the case and the bands can be characterized by the resulting Davydov splitting for each $\mathbf{k}$. The mean of the four branches, $\pi_{\mathbf{k}} \equiv \sum_{\mathbf{R}} \varepsilon_{ii}^{\mathbf{R}} e^{i\mathbf{kR}}$, is used to fit the mean of the four DFT bands throughout the Brillouin zone leading to the transfer integrals $\varepsilon_{ii}^{\mathbf{R}}$ (for each molecule $i \in \{1,2,3,4\}$). Our desription includes terms according to $\mathbf{R} = 0, \pm\mathbf{a}, \pm\mathbf{b}, \pm\mathbf{c}, \pm(\mathbf{b}\pm\mathbf{c}), \pm 2\mathbf{c}$. Already these transfer integrals give rise to the main feature in the band structure, namely the strong dispersion in c direction due to the large value for $\varepsilon_{11}^{\mathbf{c}}$ (cf. Table 1). Finally, the transfer integrals between inequivalent molecules of neighboring cells are determined using the difference $\varepsilon_{\mathbf{k}} - \pi_{\mathbf{k}}$. Thereby we include $\varepsilon_{ij}^{\mathbf{R}}$ with $(ij) \in \{(12),(21),(34),(43)\}$ at $\mathbf{R} = 0, \pm\mathbf{c}$ and $\varepsilon_{ij}^{\mathbf{R}}$ with $(ij) \in \{(14),(41),(23),(32)\}$ at $\mathbf{R} = 0, \mathbf{a}, \mathbf{b}, \mathbf{c}, \mathbf{a}+\mathbf{b}, \mathbf{a}+\mathbf{c}, \mathbf{b}+\mathbf{c}, \mathbf{a}+\mathbf{b}+\mathbf{c}$. The transfer integrals for inequivalent molecules $i \neq j$ give rise to the splitting of the four bands. We neglect terms with $(ij) \in \{(13),(31),(24),(42)\}$ since they make the expression untreatable but have no additional value over the previous ones. Symmetry allows to reduce the parameters further.

In order to determine the carrier-phonon coupling constants we perform DFT calculations with distorted lattices according to the phonon eigenvectors and obtain the values g_{MN}^{λ} from linear fits of resulting changes in the band structure with the phonon amplitude. This method is has been sucessfully applied to calculate mobilities in other organic materials before. [15, 30]

The calculations are performed using the Vienna Ab Initio Simulation Package (VASP) implementation [22] of DFT. We use the PW91 functional [35] to model electron exchange and correlation. The electron-ion interaction is described by the projector-augmented wave (PAW) method [23], which allows for an accurate treatment of the first-row elements. The Brillouin zone integration is performed on a regular $2\times3\times10$ grid and an energy cutoff of 500 eV has been used.

The VASP package has been demonstrated to show very good performance on the SGI Altix 4700. For the particular system described here, parallelization over 256 cores has been demonstrated to work sufficiently fast, while an optimum between speed-up and total consumption of ressources was found at 128 cores. A detailed analysis has been discussed previously. [33]

3 Results and Discussion

The electronic coupling (the transfer integrals ε_{MN}) between highest occupied molecular orbitals (HOMOs) of neighboring molecules (sites M and N) are compiled in Table 1. The anisotropy in the electronic coupling is evident, since there is one dominating entry, $\varepsilon_{11}^{\mathbf{c}} = 170.4$ meV. This matrix element belongs to a pair of molecules along the stacking (c) direction (see Fig. 1). Its value is on the same order of magnitude as in quasi-1D guanine stacks [28] but, importantly, it is one order of magnitude larger than the largest contribution for any other direction. We note that this pair of molecules is not perfectly cofacially aligned like in the ordinary quasi-1D stacks but slightly shifted parallel to the molecular plane similar to adjacent bases in DNA. The displacement results in a somewhat reduced transfer integral compared to the quasi-1D stack [28]. This is clear from the nodal structure of the frontier orbitals and has been described recently for organic semiconductors. [8, 21] Besides the dominating transfer integral in stacking direction, the electronic coupling across a hydrogen bridge to a neighboring stack in either a or b direction is smaller but non-zero, e.g., $\varepsilon_{12}^{\mathbf{c}} = 48.2$ meV or $\varepsilon_{14}^{\mathbf{c}} = 12.0$ meV. In terms of transport pathways, this makes a stack change during transport less favorable but still possible.

For the guanine crystal we obtain effective coupling values of G_λ =0.25, 0.09, 0.34 for the three symmetrical rotations at $\hbar\omega_\lambda = 10.6, 15.2, 17.1$ meV according to the method described above. The strong influence of the electron-phonon coupling is shown in Fig. 2 where we have plotted the effective hole polaron mass and an effective transfer integral as obtained from Eq. (2). First we recognize that already at zero Kelvin the polaron mass is somewhat larger than the bare hole mass due to zero point vibrational effects. In other words the coupling seen by the electrons (ε_{MN}) is reduced for the polarons ($\widetilde{\varepsilon}_{MN}$). The inclusion of finite temperatures gives rise to an even stronger increasing mass by an order of magnitude since more phonons are excited that scatter the charge carriers. The same effect leads to a strong decrease of the electronic coupling with rising T and is known as polaron band narrowing.

Table 1 Transfer integrals that best fit to the crystal band structure. Values are given in meV. For denotation see Fig. 1

$\varepsilon_{11}^{\mathbf{a}}$	0.0	ε_{12}^{0}	-10.2	ε_{14}^{0}	1.2
$\varepsilon_{11}^{\mathbf{b}}$	4.1	$\varepsilon_{12}^{\mathbf{c}}$	48.2	$\varepsilon_{14}^{\mathbf{a}}$	8.0
$\varepsilon_{11}^{\mathbf{c}}$	170.4	$\varepsilon_{12}^{-\mathbf{c}}$	5.5	$\varepsilon_{14}^{\mathbf{c}}$	12.0
$\varepsilon_{11}^{\mathbf{2c}}$	11.7			$\varepsilon_{14}^{\mathbf{ac}}$	-5.7
$\varepsilon_{11}^{\mathbf{bc}}$	-1.2				

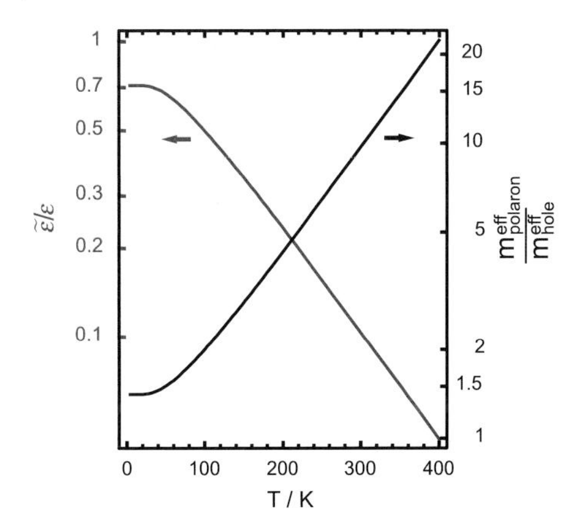

Fig. 2 Effective mass and coupling parameter (transfer integral) of hole polarons as a function of temperature. The value at T =0K is due to zero-point vibrational effects. The mass increases with temperature since more phonons are available which couple to the charge carriers

[17, 26] With the quantities $\widetilde{\varepsilon}_{MN}$ we have replaced the bare electronic structure of our model system (including bare hole bandwidth and anisotropy) by a polaronic energy structure whose strong temperature dependence has been visualized in Fig. 2 in terms of an increasing effective mass.

Now, the natural question arises what can be learned from such a modified electronic structure with respect to the resulting charge transport? This is a non-trivial problem because discussing solely the electronic/polaronic band structure as a measure for transport properties tends to imply that the carriers are transported only by a coherent (band transport) mechanism. For guanine-based materials this is a very unlikely scenario (see below) even though it has been considered previously for the DNA [13] and for guanine stacks. [2] Therefore, in order to investigate the temperature dependent motion of the polarons, we go beyond the study of their effective mass and band narrowing as in the previous paragraph. Instead, we are aiming for a measurable transport quantity and calculate directly the mobility of the carriers.

The mobility from Eq. (3) is calculated using the above-discussed material parameters. The results for hole transport as a function of temperature are displayed in Fig. 3 (a) giving an overview over the temperature dependence for different directions. With increasing T the mobilities are strongly reduced underlining the effect of electron-phonon interaction and polaron formation. Decreasing mobilities with rising T as seen in Fig. 3 (a) are known from band transport. In contrast to the simple band picture, here we find that at room temperature actually only a small fraction of the mobility is really due to coherent transport even for this model system of high purity (we assumed a broadening of $\hbar\Gamma$=0.1 meV). This is plotted in Fig. 3 (b) which shows the composition of the total mobility (in stacking direction) with respect to incoherent and coherent contributions. The coherent mobility $\mu^{(\mathrm{inc})}$ is obtained by setting $\Phi_\lambda = 0$ in Eq. (3). In the low-T limit, coherent processes contribute most since the number of scatterers (N_λ) is small. The incoherent processes obtained by $\mu^{(\mathrm{inc})} = \mu - \mu^{(\mathrm{coh})}$ are negligible in this limit but become more important with rising T. The break even point depends on the purity of the sample and shifts to lower temperatures if the purity/crystallinity is reduced, i.e., the lower

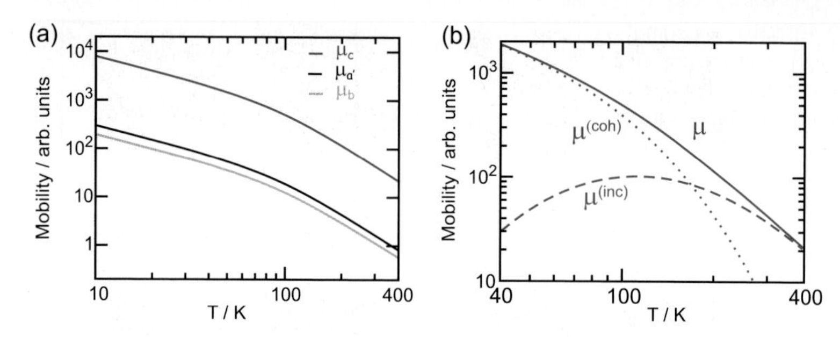

Fig. 3 Temperature dependence of hole mobilities in guanine crystals. (a) The mobility along the stacking direction is given by μ_c, whereas $\mu_{a'}$ and μ_b measure the mobilities in perpendicular directions. (b) Coherent ($\mu^{(\mathrm{coh})}$) and incoherent ($\mu^{(\mathrm{inc})}$) contributions to the total mobility μ

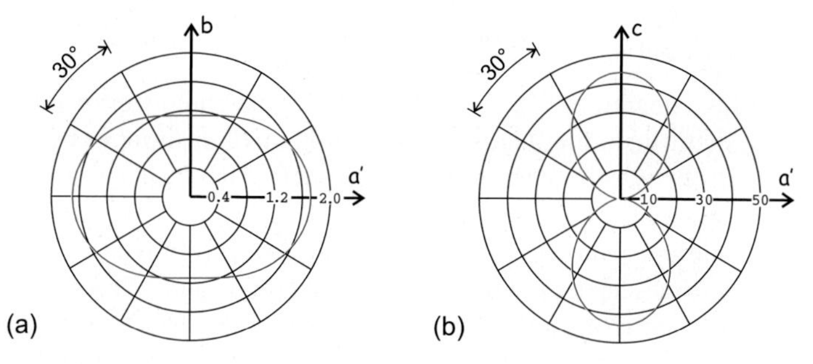

Fig. 4 Hole mobility anisotropy in guanine crystals

the crystallinity the lesser is the contribution from band transport. The total mobility in Fig. 3 (a) represents the limit of ultrapure crystals. It is also interesting to consider the limit of strong disorder. For example, the case of stacked molecules in DNA-like structures can be regarded as a limit with high impurities/low crystallinity compared to our model guanine crystal. In such a system the coherent contribution vanishes (in DNA coherent tunneling is only observed experimentally across very few bases [24]) and the total mobility is then given by $\mu^{(\mathrm{inc})}$ in Fig. 3 (b). However, for high-purity crystals as assumed here it should be possible to achieve band-like conduction at least for low T.

We now proceed with the analysis of the anisotropy of charge transport. From Fig. 3 it is obvious that the mobility anisotropy does not change with temperature and we restrict our discussion to room temperature. In order to gain a more comprehensive impression of the direction dependence, we present polar plots in Fig. 4. Low mobilities are found in the plane perpendicular to the stacking direction ($a'b$ plane) in contrast to much ($\approx 30\times$) higher mobilities along the stack (c direction). This quasi one-dimensional characteristic is a direct consequence of the electronic coupling implied by the stacking motif (Fig. 1). Although the molecules within the guanine planes are hydrogen-bridge bonded, the electronic coupling within such planes is small. In contrast, electronic coupling across the planes is strong and leads to high mobilities. A small off-angle from the low-mobility $a'b$-plane gives strong contributions from the principal (c) direction.

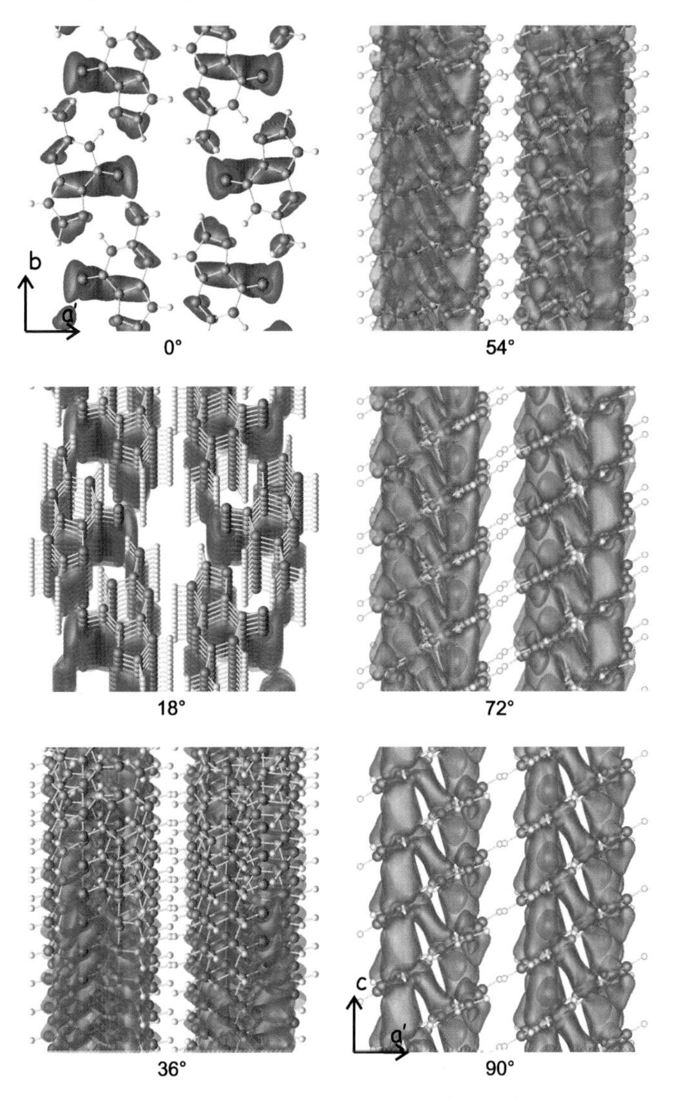

Fig. 5 Visualization of the hole channels in guanine crystals, for details see text

The discussed direction-dependence of the mobility is a direct consequence of the anisotropy of the bare electronic coupling. The reason are the relatively small non-local coupling constants for electron-phonon interaction and hence small non-local currents. [16] A highly significant measure of the electronic coupling is the overlap of the involved molecular orbitals Ψ_i. [6, 30] Therefore one can directly access the origin of the mobility anisotropy by the anisotropy of the overlap of contributing electronic wavefunctions. In Fig. 5, we have plotted the absolute value of the overlap density which is a total of all pairwise overlaps according to $\sum'_{ij} \psi_i \psi_j$ where the sums run over all molecules but excluding $i = j$ as indicated by the prime. The various snapshots in Fig. 5 show different angular views to illustrate the anisotropy of the overlap density. High density between the planes (see picture taken at 90°) indicates a strong overlap of the π-orbitals between adjacent molecules in stacking direction. The resulting vertical columns represent hole transport channels. On the other hand, these vertical channels are very weakly connected to each other. More precisely, the overlap density between molecules of adjacent stacks is small. A view along the stacking direction (see picture taken at 0°) reveals the weak coupling within each stacking plane. As a result, charge carriers can hardly overcome the gaps resulting in low carrier mobility in a'- and b-direction. This is in accordance to the anisotropy factor of 30 in the mobilities seen in Fig. 3. For a three-dimensional view to the transport channels the reader is referred to the movie published with the original paper. [32]

4 Conclusions

The most important findings of the present paper can be summarized as follows: (i) The temperature dependence of the hole transport in guanine-derived systems can only be understood within the polaron concept. (ii) At room temperature only a small contribution to the mobility in guanine-based materials comes from coherent transport. Due to dynamical disorder, the dominating contribution originates from incoherent phonon-assisted hopping. (iii) As a consequence, no contribution to the hole mobility along a poly(dG) sequence in DNA-like structures should be due to coherent transport since the required crystallinity cannot be achieved. This, however, does not necessarily prevent high mobilities. (iv) The analysis of the anisotropy of the mobility shows that transport along the stacks is strongly preferred over motion across the hydrogen bonds between the stacks. (v) It is, however, possible to have also interstrand pathways in DNA-like structures although the hopping rate between adjacent guanines is reduced by a factor of 30. (vi) The quasi one-dimensional transport channels made visible by a novel technique are aligned along the stack in striking accordance to the calculated mobility anisotropy. (vii) The anisotropy of the bare electronic coupling is a strong indicator for the anisotropy in the carrier mobilities, even in the presence of strong polaronic effects.

Acknowledgements We would like to thank the Deutsche Forschungsgemeinschaft for financial support (Projects HA 2900/3-2,3). One of the authors (B.O.) would also like to thank the Carl-Zeiss-Stiftung for scholarship funding. This work was further supported by the European Community within the framework of the Network of Excellence NANOQUANTA (Contract No. NMP4-CT-2004-500198) and the European Theoretical Spectroscopy Facility ETSF (GA No. 211956). The calculations were carried out using grants of computer time from the Leibniz-Rechenzentrum Garching.

References

1. Alexandre, S.S., Artacho, E., Soler, J.M., Chacham, H.: Small Polarons in Dry DNA. Phys. Rev. Lett. **91**, 108105 (2003)
2. Beleznay, F.B., Bogár, F., Ladik, J.: Charge carrier mobility in quasi-one-dimensional systems: Application to a guanine stack. J. Chem. Phys. **119**, 5690 (2003)
3. Bogár, F., Ladik, J.: Correlation corrected energy bands of nucleotide base stacks. Chem. Phys. **237**, 273 (1998)
4. Bongiorno, A.: Energy landscape of an electron hole in hydrated dna. J. Phys. Chem. B **112**, 13,945 (2008)
5. Braun, E., Eichen, Y., Sivan, U., Ben-Yoseph, G.: DNA-templated assembly and electrode attachment of a conducting silver wire. Nature **391**, 775 (1998)
6. Brédas, J.L., Calbert, J.P., da Silva Filho, D.A., Cornil, J.: Organic semiconductors: A theoretical characterization of the basic parameters governing charge transport. Proc. Natl. Acad. Sci. USA **99**, 5804 (2002)
7. Cohen, H., Nogues, C., Naaman, R., Porath, D.: Direct measurement of electrical transport through single DNA molecules of complex sequence. Proc. Nat. Acad. Sci. **102**, 11,589 (2005)
8. Coropceanu, V., Cornil, J., da Silva Filho, D.A., Olivier, Y., Silbey, R., Brédas, J.L.: Charge transport in organic semiconductors. Chem. Rev. **107**, 926 (2007)
9. Di Felice, R., Calzolari, A., Garbesi, A., Alexandre, S.S., Soler, J.M.: Strain-dependence of electronic properties in periodic quadrupel helical g4-wires. J. Phys. Chem. B **109**, 22,301 (2005)
10. Di Felice, R., Calzolari, A., Molinari, E., Garbesi, A.: Ab initio study of model guanine assemblies: The role of π-π coupling and band transport. Phys. Rev. B **65**, 045104 (2001)
11. Eley, D.D., Spivey, D.I.: Semiconductivity of organic substances. Trans. Faraday Soc. **58**, 411 (1962)
12. Endres, R.G., Cox, D.L., Singh, R.R.P.: Colloquium: The quest for high-conductance DNA. Rev. Mod. Phys. **76**, 195 (2004)
13. Grozema, F.C., Berlin, Y.A., Siebbeles, L.D.A.: Mechanism of charge migration through dna: Molecular wire behavior, single-step tunneling or hopping? J. Am. Chem. Soc. **122**, 10,903 (2000)
14. Grozema, F.C., Tonzani, S., Berlin, Y.A., Schatz, G.C., Siebbeles, L.D.A., Ratner, M.: Effect of structural dynamics on charge transfer in dna hairpins. J. Am. Chem. Soc. **130**, 5157 (2008)
15. Hannewald, K., Bobbert, P.A.: Ab initio theory of charge-carrier transport in ultrapure organic crystals. Appl. Phys. Lett. **85**, 1535 (2004)
16. Hannewald, K., Bobbert, P.A.: Anisotropy effects in phonon-assisted charge-carrier transport in organic molecular crystals. Phys. Rev. B **69**, 075212 (2004)
17. Hannewald, K., Stojanović, V.M., Schellekens, J.M.T., Bobbert, P.A., Kresse, G., Hafner, J.: Theory of polaron band width narrowing in organic molecular crystals. Phys. Rev. B **69**, 075211 (2004)
18. Hatcher, E., Balaeff, A., Keinan, S., Venkatramani, R., Beratan, D.N.: PNA versus DNA: Effects of Structural Fluctuations on Electronic Structure and Hole-Transport Mechanisms. J. Am. Chem. Soc. **130**, 11,752 (2008)
19. Henderson, P.T., Jones, D., Hampikian, G., Kan, Y., Schuster, G.B.: Long-distance charge transport in duplex dna: The phonon-assisted polaron-like hopping mechanism. Proc. Natl. Acad. Sci. USA **96**, 8353 (1999)
20. Holstein, T.: Studies of Polaron Motion, Part I. Ann. Phys. **8**, 325 (1959)
21. Kim, E.G., Coropceanu, V., Gruhn, N.E., Sánchez-Carrera, R.S., Snoeberger, R., Matzger, A.J., Brédas, J.L.: Charge transport parameters of the pentathienoacene crystal. J. Am. Chem. Soc. **129**, 13,072 (2007)
22. Kresse, G., Furthmüller, J.: Efficiency of *ab-initio* total energy calculations for metals and semiconductors using a plane-wave basis set. Comput. Mater. Sci. **6**, 15 (1996)

23. Kresse, G., Joubert, D.: From ultrasoft pseudopotentials to the projector augmented-wave method. Phys. Rev. B **59**, 1758 (1998)
24. Lewis, F.D., Wu, T., Zhang, Y., Letsinger, R.L., Greenfield, S.R., Wasielewski, M.R.: Distance-Dependent Electron Transfer in DNA Hairpins. Science **277**, 673 (1997)
25. Liu, T., Barton, J.K.: DNA Electrochemistry through the Base Pairs Not the Sugar-Phosphate Backbone. J. Am. Chem. Soc. **127**, 10,160 (2005)
26. Mahan, G.D.: Many-Particle Physics. Kluwer Academic Publishers, New York (2000)
27. Maruccio, G., Visconti, P., Arima, V., D'Amico, S., Biasco, A., D'Amone, E., Cingolani, R., Rinaldi, R., Masiero, S., Giorgi, T., Gottarelli, G.: Field Effect Transistor Based on a Modified DNA Base. Nano Lett. **3**, 479 (2003)
28. Matthes, L., Ortmann, F., Oetzel, B., Bechstedt, F., Hannewald, K.: (unpublished)
29. Ortmann, F., Bechstedt, F., Hannewald, K.: Theory of charge transport in organic crystals: Beyond holstein's small-polaron model. Phys. Rev. B **79**, 235206 (2009). DOI 10.1103/PhysRevB.79.235206
30. Ortmann, F., Hannewald, K., Bechstedt, F.: Ab initio description and visualization of charge transport in durene crystals. Appl. Phys. Lett. **93**, 222105 (2008). DOI 10.1063/1.3033830
31. Ortmann, F., Hannewald, K., Bechstedt, F.: Guanine crystals: A first principles study. J. Phys. Chem. B **112**, 1540 (2008). DOI 10.1021/jp076455t
32. Ortmann, F., Hannewald, K., Bechstedt, F.: Charge Transport in Guanine-Based Materials. J. Phys. Chem. B **113**, 7367 (2009). DOI 10.1021/jp901029t
33. Ortmann, F., Preuss, M., Oetzel, B., Hannewald, K., Bechstedt, F.: Charge Transport through Guanine Crystals, p. 687. Springer-Verlag (2009)
34. de Pablo, P.J., Moreno-Herrero, F., Colchero, J., Gómez Herrero, J., Herrero, P., Baró, A.M., Ordejón, P., Soler, J.M., Artacho, E.: Absence of dc-Conductivity in λ-DNA. Phys. Rev. Lett. **85**, 4992
35. Perdew, J.P.: p. 11. Akademie-Verlag, Berlin (1991)
36. Porath, D., Bezryadin, A., de Vries, S., Dekker, C.: Direct measurement of electrical transport through DNA molecules. Nature **403**, 635 (2000)
37. Tran, P., Alavi, B., Gruner, G.: Charge Transport along the l-DNA Double Helix. Phys. Rev. Lett. **85**, 1564 (2000)
38. Yoo, K.H., Ha, D.H., Lee, J.O., Park, J.W., Kim, J., Kim, J.J., Lee, H.Y., Kawai, T., Choi, H.Y.: Electrical Conduction through Poly(dA)-Poly(dT) and Poly(dG)-Poly(dC) DNA Molecules. Phys. Rev. Lett. **87**, 198102 (2001)
39. Young, M.A., Ravishanker, G., Beveridge, D.L.: A 5-Nanosecond Molecular Dynamics Trajectory for B-DNA: Analysis of Structure, Motions, and Solvation. Biophys. J. **73**, 2312 (1997)

Nanomagnetism in Transition Metal Doped Si Nanocrystals

Christian Panse, Roman Leitsmann, and Friedhelm Bechstedt

Abstract We study the influence of transition-metal doping on silicon nanocrystals by means of *ab-initio* first-principles calculations and a supercell approach. As transition-metals Manganese and Iron are investigated on either substitutional or interstitial doping sites. For description of the localized d-levels we use the GGA+U approach with an Hubbard-like Coulomb repulsion of about 3 eV. We focus on the energetic stability of the different doping sites and on electronic and magnetic properties of the most stable configurations. We found multiple energy-minima for the doping site right beneath the surface, induced by some deformation processes. In addition, a self-purification effect, which hampers an incorporation of transition-metal atoms in small nanocrystals, is observed.

1 Introduction

In recent years there has been a great interest in diluted magnetic semiconductors (DMS). They are supposed to be the materials for new devices, which could be used in spintronics. For industrial applications spintronics must conquer silicon. One way could be alloying or doping silicon with transition-metal (TM) atoms such as Mn or Fe. Additionally the discovery of an increasing number of spin-related effects shows that the spin of electrons offers unique posibilities for novel mechanisms of information processing like "spin effect transistors" or "spin qubits" [1–3].

C. Panse · F. Bechstedt
Institut für Festkörpertheorie und -optik, Friedrich-Schiller-Universität Jena, Max-Wien-Platz 1, 07743 Jena, Germany
e-mail: `christian.panse@uni-jena.de`

R. Leitsmann
GWT-TUD GmbH, Abteilung Material Calculations, Annabergerstr. 240, 09125 Chemnitz, Germany
e-mail: `leitsmann@matcalc.de`

S. Wagner et al. (eds.), *High Performance Computing in Science and Engineering, Garching/Munich 2009*, DOI 10.1007/978-3-642-13872-0_45,

On the other hand, quantum confinement of electrons alters various properties of materials. Especially for semiconductor nanocrystals this could be demonstrated under several aspects [4]. For spin and magnetic properties the unusual increase of spin-dephasing time is of special interest [1]. Current theoretical and experimental studies focus on Mn-doped II-VI and III-V semiconductors but also Ge nanocrystals [4–9]. As recently shown for Si nanowires [10], TM doped Si nanostructures becomes interesting. Under certain conditions those quasi-one-dimensional structures exhibit an half-metallic character. Room-temperature ferromagnetism seems to be discovered. A detailed understanding of ferromagnetism due to TM incorporation on the nanometer scale requires a deeper analysis of the TM doping of silicon for various nanostructure sizes, dopants, and doping sites. Their formation energy as a function of the impurity position and the defect-induced electronic structure has to be studied, as well as the influence of the electronic correlation especially within the localized semicore TM $3d$ states on the defect-level positions, their occupation, and the accompanying magnetic moments [11].

In the present paper we investigate the effects of TM doping in hydrogen-passivated Si nanocrystals (NC). The incorporation of Mn and Fe atoms is investigated in the framework of the spin-polarized density functional theory. Special attention is paid on the stability of the NC structures with respect to the different doping sites and NC diameters. We study possible self-purification effects and the influence of the TM $3d$-derived impurity states on the electronic and magnetic properties. Additionally strong electron correlation effects are taken into account via a Hubbard-like Coloumb repulsion U [12].

2 Computational Methods

2.1 DFT Framework

In order to investigate clean and doped silicon nanocrystals by first-principles techniques we use the spin-polarized density functional theory (DFT) as implemented in the Vienna *ab-initio* simulation package (VASP) [13]. The pseudopotentials are generated by means of the projector-augmented-wave (PAW) method [14] that allows for the accurate treatment of the TM $3d$, TM $4s$, Si $3s$, and Si $3p$ valence electrons. In the region between the atomic cores the wave functions are expanded in plane waves up to a cutoff energy of 200 eV. To approximate the exchange and correlation (XC) functional we use the generalized gradient approximation including an additional Coulomb repulsion U (GGA+U) [15]. For intermediate spin polarizations we apply the interpolation proposed by von Barth and Hedin [16]. Within the supercell approach we build up unit cells containing one nanoparticle and a certain amount of vacuum. The supercells have to be choosen big enough so that the NCs do not interact with each other through periodic boundary conditions. Since these used unit cells are large, the k-point mesh in the reciprocal space can be restricted to the center of the Brillouin zone (Γ point). The structures of the NCs are optimized by means of the quasi- Newton (variable-metric) or conjugate gradient algorithms. The atomic forces are minimized to 25 meV/Å or less. The approach has been tested for Mn doping of bulk Si. Good agreement with other recent first-principles calculations [17–19] was found, at least with respect to the atomic geometries and magnetic properties.

2.2 Numerical Simulations and Performance

The DFT ground state calculations are based on an iterative and self-consistent solution of the Fourier transformed Kohn-Sham equation using fictitious electron dynamics. The application of the plane wave expansion and the PAW pseudopotential method to the Kohn-Sham equation results in a generalized eigenvalue problem that can be efficiently solved with iterative algorithms. The resulting diagonalization procedures can be straightforwardly parallelized over the bands ("inter-band distribution") and, if enough nodes are available, within bands ("intra-band distribution"). Therefore each node must be endowed with enough memory to store the chargedensity and the calculated wave functions. Since the orthogonalization of the eigenstates requires the redistribution of the wave functions between the nodes, a fast and low-latency network is a prerequisite for a fast execution of the algorithms.

The algorithms implemented in VASP use different iterative matrix diagonalization schemes: such as the conjugate gradient scheme, the block Davidson scheme, and the residual minimization scheme - direct inversion in the iterative subspace (RMM-DIIS). For the mixing of the charge density an efficient Broyden/Pulay [20] mixing scheme or the Vosko-Wilk-Nusair [21] interpolation is used. Within each selfconsistent loop the charge density is used to set up the Hamiltonian, then the wave functions are optimized iteratively so that they get closer to the exact wave functions of this Hamiltonian. From the optimized wave functions a new charge density is calculated, which is then mixed with the old input charge density. The parallelization in VASP is done using the messagepassing paradigm MPI. The communication overhead is reduced by using a two-dimensional Cartesian topology in which the bands are distributed among a group of nodes in a round-robin fashion. The code is parallelized with respect to the plane-wave coefficients and with respect to the eigenstates of the Kohn-Sham equation.

However, a optimal scaling on several hundreds of CPUs could be limited by the accessible bandwith and communication overhead. We have tested the code and looked at the scaling behavior on our server-computer Cray T3E and the supercomputer SGI Altix 4700. In the Figure 1 we plot the scaling for a typical mid-size system including 76 atoms in the unit cell with 168 bands and 15399 plane-waves up to 256 cores. It is obvious that already the single processor performance of the SGI Altix 4700 is much better than our local machine. The scaling behavior of our code on the SGI Altix 4700 is almost ideal up to a number of 128 CPUs. For larger CPU numbers the administration cost rises faster than the gain of computer time. How-

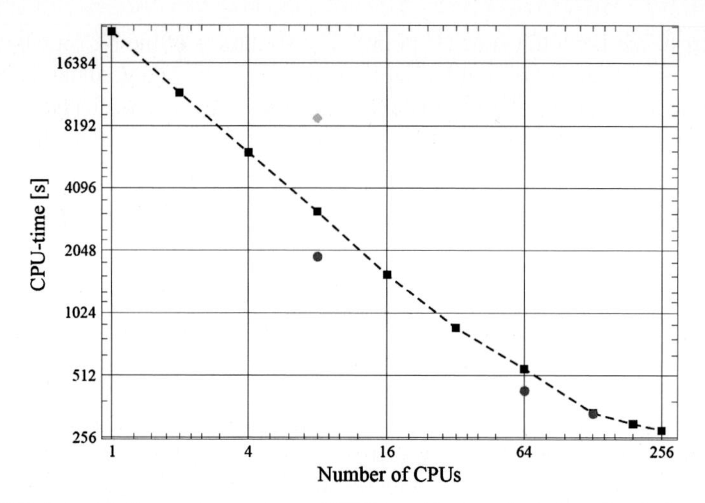

Fig. 1 CPU-time regarding to the numbers of CPUs used in the calculations. In all cases a system of 76 atoms with 168 bands and 15399 plane-waves is calculated including three ionic relaxation steps. The green diamond represents our Cray T3E, the red circle marks the NEC-SX-8/512 in Stuttgart, the black dotted line pictures the SGI Altix 4700 in Munich

ever, most of our systems will even scale much better, because they contain more atoms so that the administration problem probably begins far beyond 256 involved CPUs.

To achieve our results for the single doped nanocrystals approximately 300 calculations were done at the HLRB, mostly using 32 or 64 CPUs. The main part (200 jobs) were groundstate and dipole-corrected calculations, which in general take only up to 6 h. About 70 jobs were needed to perform the time consuming relaxations. Using 64 CPUs the relaxations take an averaged time of 18 h (up to 44 h in case of the $Si_{147}H_{148}$ nanocrystal).

3 Modeling and Methods

3.1 Nanocrystal Construction

The Si NCs are modeled using the supercell method [22] in a simple cubic arrangement. The chosen edge lengths guarantee a distance larger than 1 nm between the surfaces of the NCs. In addition, we restrict the studies to single doped NCs, i.e. only one incorporated atom. The atomic positions are determined by successively adding Si atoms in a local tetrahedron coordination and a subsequent structural optimization (shell-by-shell construction). For the passivation of the surface dangling bonds hydrogen atoms are used. This procedure leads to NCs with varying total number of shells $N = 2, 3, 4, 5$, and 6 (cf. Fig. 2 and Table 1). Together with the hydro-

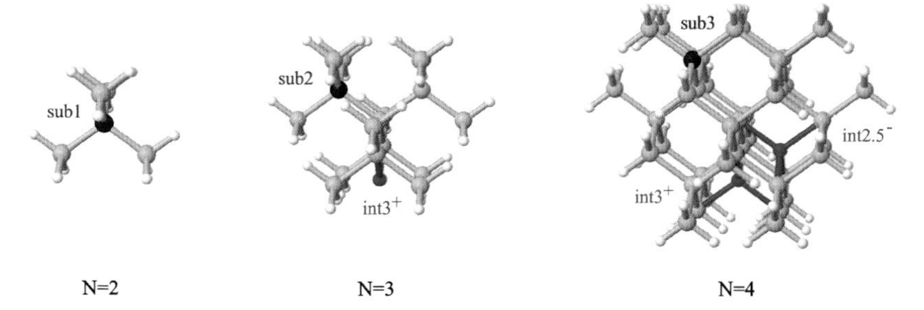

Fig. 2 Schematic representation of the 2-4 shell Si NCs. The positions of the subsurface substitutional and different interstitial doping sites are indicated

Table 1 Doping concentration, NC diameter of the investigated Si NC structures and the average distance d_{Si-TM} of the TM atom to its four Si neighbors are listed

number of shells N		2	3	4	5	6
diameter [Å]		6.3	10.4	14.3	18.2	21.8
dopant	sub	25	6.25	2.50	1.22	0.68
concentration [%]	int	-	5.88	2.44	1.20	0.68
d_{Si-Mn} [Å]	sub (N-1)	2.34	2.39	2.46	2.46	2.49
	int $(N\text{-}1)^+$	-	-	2.61	2.61	2.62
d_{Si-Fe} [Å]	sub (N-1)	2.26	2.33	2.42	2.43	2.46
	int $(N\text{-}1)^+$	-	-	2.55	2.55	2.60

gen passivation the molecular formulars of our NCs are Si_5H_{12}, $Si_{17}H_{36}$, $Si_{41}H_{60}$, $Si_{83}H_{108}$, and $Si_{147}H_{148}$. A detailed description of the construction of pure Si NCs can be found elsewhere [23, 24].

The possible doping positions are marked in Fig. 2 either with red (substitutional) or with blue (interstitial) atoms. To distinguish the different doping positions we introduce the following notation: $(type\ n)^i$ with $i \in \{+,-\}$. The *type* can be either "sub" or "int" for a substitutional or an interstitial position, respectively. The shell number $n \leq N$ determines the radial position, where $n = 1$ indicates the position in the NC center. If the dopant is inside a Si shell this number is an integer. In the case the dopant occupies a site between two Si shells, "n" is replaced by the averaged shell number, i.e., by a half-integer value. The index i denotes the position of the dopant with respect to the type of the surface facet. Dopants under a rectangular facet are indicated by "+", and dopants under triangular facets by "-".

Since all NCs are doped with one TM atom, different dot sizes lead to different doping concentrations (see Tab. 1). The point group of a TM-doped Si crystal is only a subgroup of the T_d symmetry of bulk Si and undoped Si NCs. Nevertheless, if possible we will keep the T_d-symmetry based notation of the impurity states in terms of e and t_2 orbital symmetries in order to visualize their origin.

3.2 Approaches Beyond GGA

As recently discussed in the literature [11, 15, 25] TM atoms experience a strong on-site Coulomb repulsion U among their $3d$ electrons due to the narrow d bandwidth, which is not correctly described in a spin-polarized DFT treatment. Within the DFT+U method this error can be corrected, combining the DFT and a Hubbard-like description of the Coulomb repulsion in the Hamiltonian. Because the strength of the on-site repulsion is proportional to the localization of the corresponding wave-functions, it has to be considered only in the case of strongly localized TM $3d$ states. For the present calculations we use the DFT+U version proposed by Dudarev *et al.* [15]. The spin-dependent interaction leads to a difference in the band energies between majority- and minority-spin channels

$$\varepsilon_{3d,\uparrow} - \varepsilon_{3d,\downarrow} = \Delta = U(n_{3d,\downarrow} - n_{3d,\uparrow}), \tag{1}$$

where we have introduced the exchange splitting Δ and the partial occupation numbers of the TM $3d$ states, $n_{3d,\uparrow}$ and $n_{3d,\downarrow}$. As proposed in [26] we limited the Hubbard-like Coulomb repulsion U to a value of 3 eV to obtain the correct band ordering. This value leads to the correct energetic ordering of the valence and conduction bands with e and t_2 symmetry, at least for antiferromagnetic MnO. More sophisticated methods like the spatially non-local hybrid functional HSE03 yield qualitatively the same results as a GGA+U treatment, apart from a splitting of partially filled impurity states. A detailed description about the influence of XC-functionals on the magnetic properties of TM-doped Si-nanocrystals can be found in [11].

4 Results and Discussion

4.1 Stability and Geometry

To examine different preparation conditions Planck's grand canonical thermodynamic potential has to be taken into account for a given NC with $N_{\mathrm{Si:TM}}$ Si atoms and N_{TM} TM atoms. The preparation conditions are determined by the actual chemical potentials μ_{Si} and μ_{TM} of the reservoirs of Si and TM atoms. More precisely, we calculate the formation energy of a doped NC

$$\Omega_f = \gamma_f - \mu_{\mathrm{TM}} N_{\mathrm{TM}} \tag{2}$$

with

$$\gamma_f = E_{tot}^{\mathrm{Si:TM}} - E_{tot}^{\mathrm{Si}} - \mu_{\mathrm{Si}}(N_{\mathrm{Si:TM}} - N_{\mathrm{Si}}), \tag{3}$$

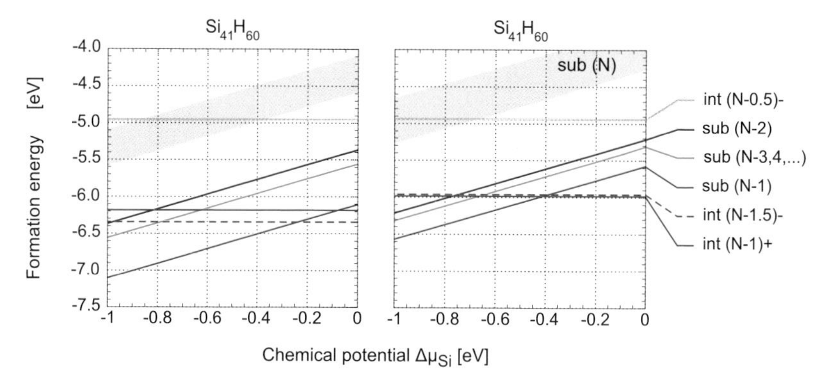

Fig. 3 Relative formation energy of the four shell NC with incorporation of Mn (left panel) or Fe (right panel) versus the variation of the Si chemical potential $\Delta\mu_{\mathrm{Si}}$ with respect to its bulk value. The corresponding doping sites are indicated

where $E_{tot}^{\mathrm{Si:TM}}$ is the total energy of the doped NC, E_{tot}^{Si} the total energy of the undoped NC with N_{Si} Si atoms, and μ_X ($X = \mathrm{Si,TM}$) are the chemical potentials of the two species. With the chemical potential of bulk silicon $\mu_{\mathrm{Si}}^{\mathrm{bulk}}$ the relative formation energy γ_f (2) can be studied as a function of the variation of the Si chemical potential $\Delta\mu_{\mathrm{Si}} = \mu_{\mathrm{Si}} - \mu_{\mathrm{Si}}^{\mathrm{bulk}}$. The relative formation energy γ_f is sufficient because we limit our calculations to a fixed number $N_{\mathrm{TM}} = 1$ of TM dopants. With the variation of $\Delta\mu_{\mathrm{Si}}$ we can compare Si-rich ($\Delta\mu_{\mathrm{Si}} = 0$) and Si-poor ($\Delta\mu_{\mathrm{Si}} \to -\infty$) preparation conditions. The relative formation energy γ_f is, therefore, appropriate to discuss the relative stability of different doping positions and different numbers $N_{\mathrm{Si:TM}}$ in a NC with a certain size.

The calculated formation energies are represented in Fig. 3. For interstitial positions the γ_f curves remain constant, while for substitutional positions they vary linearly with $\Delta\mu_{\mathrm{Si}}$. As expected from bulk results [17–19], the interstitial TM positions are preferred for Si-rich conditions, while the substitutional TM positions are more stable under Si-poor conditions. This holds for all NCs with a diameter larger than 7 Å. In the case of small NCs with $N = 2$ a definition of interstitial doping sites is difficult. In addition, the range of the chemical potential, for which substitutional or interstitial sites are preferred, depends on the radial distance from the NC center.

Beside this general trend two different effects are visible in Fig. 3. Firstly, under Si-poor conditions we find that the sub-surface doping site (sub N-1) directly beneath the NC edges is the most stable one. Interestingly in this case we found three different structural configurations very close in energy. They are indicated by the subscripts {s,m,i}. The first one (sub N-1)$_{\mathrm{s}}$ is accompanied by only small lattice distortions. In the second case (sub N-1)$_{\mathrm{m}}$ the TM atom together with the Si corner-atom moves inside a neighboring cage, i.e., the cage spanned by eight Si atoms within the diamond structure, towards an interstitial site. In contrast to the third case (sub N-1)$_{\mathrm{i}}$, where the TM atom reached the interstitial position and the Si corner-atom closes the cage, the (sub N-1)$_{\mathrm{m}}$ TM atom retains at an intermediate po-

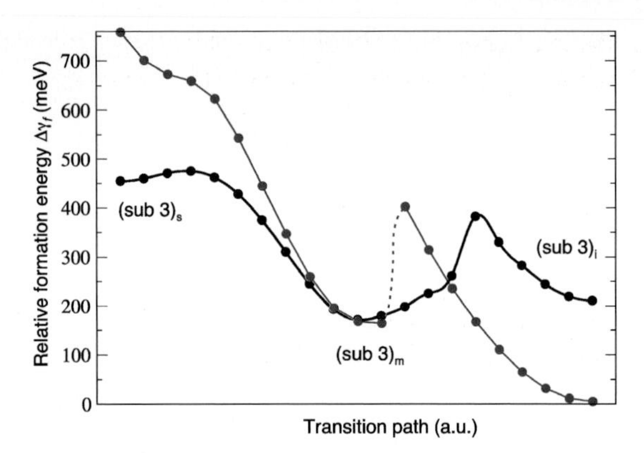

Fig. 4 Energy barriers between the three subsurface doping positions in a $Si_{40}TMH_{60}$ NC: red for Fe and black for Mn incorporation

sition. The configurations (sub N-1)$_m$ and (sub N-1)$_i$ are the most stable ones with formation energy differences < 40 meV (Mn) or < 140 meV (Fe). As can be seen in Fig. 4 they are separated by an energy barrier of about 200 meV. Therefore, in room-temperature experiments both configuration types may appear. The first configuration (sub N-1)$_s$ is only stable in the case of Mn doping. For Fe dopants no energy barrier to (sub N-1)$_m$ exists.

The second point is that under Si-rich conditions the interstitial doping sites directly beneath the rectangular $\{001\}$ NC-surface facets (int N-1)$^+$ are prefered. The combination of two simple geometric effects could explain the observed results. Due to a symmetry reduction we find an increase of the formation energy from the NC centre to the surface of the NC. And on the other hand, directly beneath the NC surface the distortion of the (incomplete) tetrahedron bonding configuration in order to incorporate the TM atom gets easier.

The computed total magnetic moments μ_{tot} of the most stable doping sites are shown in Table 2. For all Mn or Fe-doped NCs on subsurface doping sites, we observe nonvanishing total magnetic moments. In both cases, Mn- and Fe-doped Si NCs, the energetic positions of the impurity states with respect to the Fermi level are more or less independent of the NC size, which is a result of the strong localization of the TM-3d-derived states at the impurity sites. This results in an total magnetization independent of the NC size. We achieve for Mn a high-spin phase with $\mu_{tot} = 3\mu_B$. The six shell Mn interstitial, which relax into a low-spin phase, has not to be taken into account, because its occurrence is only a surface effect, due to

Table 2 Total magnetic moment μ_{tot} in units of μ_B for differently sized single doped Si NCs. Only the energetically most favored substitutional or interstitial sites are listed

number of shells N		3	4	5	6
Mn	sub $(N-1)$	3	3	3	3
	int $(N-1)^+$	-	3	3	1
Fe	sub $(N-1)$	2	2	2	2
	int $(N-1)^+$	-	2	2	2

the interaction between Mn and the hydrogen passivation. For Fe the mediate-spin phase, i.e., a magnetic moment of $2\mu_B$ is preferred.

4.2 Self-purification Effect

To study the influence of the NC size on the TM defect formation we introduce the relative formation energy $\Delta\gamma_f$ of TM incorporation with respect to one of the corresponding bulk material [6]

$$\Delta\gamma_f = \gamma_f - \gamma_f^{\text{bulk}}. \tag{4}$$

Here γ_f is the relative formation energy of TM defects in Si nanocrystals as defined in (3), while γ_f^{bulk} refers to the relative formation energy in the bulk material. We set γ_f^{bulk} to a value of a Si crystal with a doping concentration of about 0.5%. Details about these calculations can be found elsewhere [18]. The results are plotted in Fig. 5 for different NC sizes and the most stable doping sites. With decreasing NC size the relative formation energy $\Delta\gamma_f$ increases, so that the incorporation of TM atoms is hampered (at least for preparation conditions close to the thermodynamic equilibrium). This is a consequence of the increased doping concentration in small NCs. However, the change of the relative formation energy with increasing doping concentration in bulk Si is much smaller than the effect shown here. Therefore, together with the most stable sites for TM incorporation in subsurface positions, we interprete Fig. 5 as a clear evidence for the so called self-purification effect proposed for several nanostructures in the literature [4–6, 27].

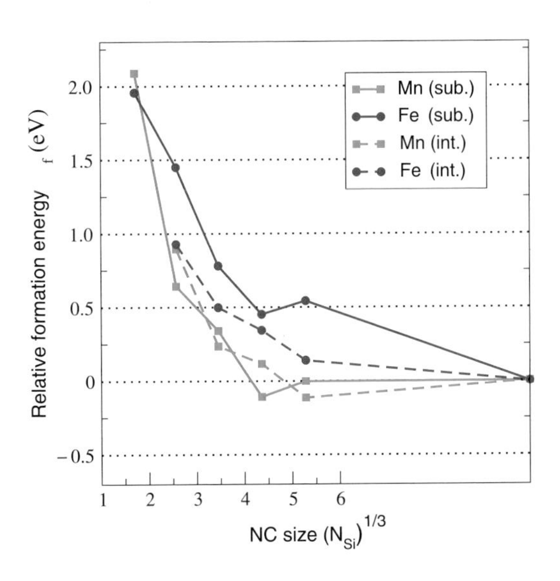

Fig. 5 Relative formation energy $\Delta\gamma_f$ for Mn (green) and Fe (red) incorporation at the most stable substitutional (solid lines) and interstitial (dashed lines) doping sites

4.3 Electronic and Magnetic Properties of Subsurface Doping Sites

An selection of the calculated electronic structures is shown for Mn doping and the influence of different subsurface doping positions in Fig. 6. The vacuum level is chosen as energy zero. The total magnetic moment of the TM-doped Si NCs can easily be obtained by counting the electrons on occupied levels within the two spin channels. A detailed view over electronic and magnetic properties (dependence on size, radial doping position and XC-functional) could be found in our paper [28].

Since for the most stable substitutional subsurface doping sites three different geometries (sub 3)$_s$, (sub 3)$_m$, and (sub 3)$_i$ appear, their electronic properties are plottet in Fig. 6. While the doping sites (sub 3)$_s$ and (sub 3)$_m$ exhibit an high-spin (Mn, $\mu_{\text{tot}} = 3\mu_B$) or mediate-spin (Fe, $\mu_{\text{tot}} = 2\mu_B$) character, the (sub 3)$_i$ geometry corresponds to a low-spin phase (Mn: $\mu_{\text{tot}} = 0\mu_B$, Fe: $\mu_{\text{tot}} = 1\mu_B$). This spin-phase transition is the result of the structural rearrangement which is reflected in the electronic structure. The properties for the doping site (sub 3)$_i$ considerably differ from those for the doping sites (sub 3)$_s$ and (sub 3)$_m$, which have qualitatively similar electronic properties. In detail, the occupied Si sp^3-like state below the Fermi level in the majority-spin channel is shifted to higher energies which results in a spin-flip and leads to the observed spin-phase transition. Therefore, one has to state a significant sensibility of the electronic and magnetic properties of the NCs with subsurface doping sites on the actual geometry.

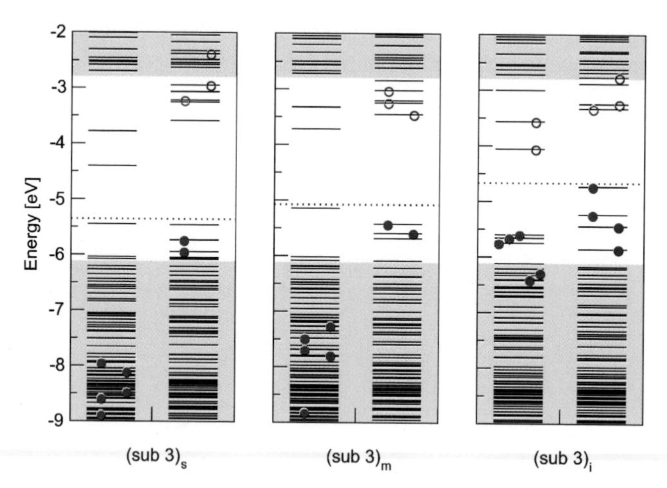

Fig. 6 Energy level diagrams for the three substitutional subsurface doping sites for a $Si_{40}MnH_{60}$ NC. The black lines represent the energy levels. The degeneracy of the TM d-states (e - blue, t_2 - red) is illustrated by open (empty) and filled (occupied) circles. The grey regions correspond to conduction and valence states of the undoped NCs. The horizontal dotted line defines the Fermi level. Majority-spin (left) and minority-spin (right) channels are shown

5 Summary

The doping of single Mn and Fe atoms into silicon nanocrystals at substitutional and interstitial sites has been studied within the framework of density functional theory including a Hubbard-like on-site Coloumb repulsion U to describe the more localized $3d$ states. The most stable doping sites strongly depends on the preparation conditions, i.e., under Si-rich conditions the interstitial doping sites below the rectangular $\{001\}$ facets are preferred and under Si-poor conditions subsurface doping sites are energetically favored. In addition, we have shown a clear evidence of self-purification effects in small TM-doped Si NCs.

The resulting electronic properties depend on the size of the Si NCs, the impurity site, and the chemical nature of TM atoms. There are several impurity-induced levels in the fundamental gap mainly derived from TM $3d$ states with e and t_2 symmetry. For impurity sites outside the NC centre the T_d symmetry is disturbed and the degeneracy of the e and t_2 levels is partly or fully lifted. The total magnetic moment of the TM-doped Si NCs can easily be obtained by counting the electrons on occupied levels within the two spin channels. Therefore, the total magnetic moment μ_{tot} of the doped Si NC depend on the number and the relative position of the TM impurity-induced states in the minority-spin and majority-spin channels. The most favoured doping sites are high-spin states ($3\mu_B$) for Mn doping and mediate-spin states ($2\mu_B$) in the case of Fe doping.

References

1. D. Awschalom, J. Kikkawa, Nature **397**, 139 (1999)
2. D. Awschalom, J. Kikkawa, Physics Today **52**, 33 (1999)
3. G. Burkard, D. Loss, D. D. Vincenzo, Phys. Rev. B **59**, 2070 (1999)
4. S. C. Erwin, L. Zu, M. I. Haftel, A. L. Efros, T. A. Kennedy, D. J. Norris, Nature **436**, 91 (2005)
5. J. T. Arantes, G. M. Dalpian, A. Fazzio, Phys. Rev. B **78**, 045402 (2008)
6. G. M. Dalpian, J. Chelikowsky, Phys. Rev. Lett. **96**, 226802 (2006)
7. X. Huang, A. Makmal, J. R. Chelikowsky, L. Kronik, Phys. Rev. Lett. **94**, 236801 (2005)
8. Y. Leger, L. Besomber, L. Maingautt, J. Fernandez-Rossior, D. Ferrand, H. Mariette, Phys. Stat. Sol. (B) **243**, 3912 (2006)
9. A. Nag, D. Sarna, J. Phys. Chem. C **111**, 13641 (2007)
10. E. Durgun, D. Cakir, N. Akman, S. Ciraci, Phys. Rev. Lett. **98**, 117202 (2001)
11. F. Kuewen, R. Leitsmann, C. Roedl, C. Panse, F. Bechstedt, J. Chem. Theory Comput. **submitted** (2009)
12. V. Anisimov, J. Zaanen, O. Andersen, Phys. Rev. B **44**, 943 (1991)
13. G. Kresse, J. Furthmüller, Comp. Mat. Sci. **6**, 15 (1996)
14. G. Kresse, D. Joubert, Phys. Rev. B **59**, 1758 (1999)
15. S. L. Dudarev, G. A. Bolton, S. Y. Savrasov, C. J. Humphreys, A. P. Sutton, Phys. Rev. B **57**, 1505 (1998)
16. U. von Barth, L. Hedin, J. Phys. C: Solid State Phys. **5**, 1629 (1972)
17. I. Appelbaum, B. Huang, D. Monsma, Nature **447**, 295 (2007)
18. F. Kuewen, R. Leitsmann, F. Bechstedt, Phys. Rev. B **80**, 045203 (2009)

19. A. Stroppa, S. Picozzi, A. Contineza, A. J. Freeman, Phys. Rev. B **68**, 155203 (2003)
20. P. Pulay, Chem. Phys. Lett. **73**, 393–398 (1980)
21. S. H. Vosko, L. Wilk, M. Nusair, Can. J. Phys. **58**, 1200 (1980)
22. M. C. Payne, M. P. Teter, D. C. Allan, T. A. Arias, J. D. Joannopoulos, Rev. Mod. Phys. **64**, 1045 (1992)
23. L. E. Ramos, J. Furthmüller, F. Bechstedt, Phys. Rev. B **72**, 045351 (2005)
24. L. E. Ramos, H.-C. Weissker, J. Furthmüller, F. Bechstedt, Phys. Stat. Sol. (B) **15**, 3053 (2005)
25. V. Anisimov, O. Gunnarsson, Phys. Rev. B **43**, 7570 (1991)
26. C. Rödl, F. Fuchs, J. Furthmüller, F. Bechstedt, Phys. Rev. B **77**, 184408 (2008)
27. G. Cantele, E. Degoli, E. Luppi, R. Magri, D. Ninno, G. Iadonisi, S. Ossicini, Phys. Rev. B **72**, 113303 (2005)
28. R. Leitsmann, C. Panse, F. Kuewen, F. Bechstedt, Phys. Rev. B **80**, 104412 (2009)

High Performance Computing for the Simulation of Thin-Film Solar Cells

C. Jandl, K. Hertel, W. Dewald, and C. Pflaum

Abstract To optimize the optical efficiency of silicon thin-film solar cells, the absorption and reflection of sunlight in these solar cells has to be simulated. Since the thickness of the layers of thin-film solar cells is of the size of the wavelength, a rigorous simulation by solving Maxwell's equations is important. However, large geometries of the cells described by atomic force microscope (AFM) data lead to a large computational domain and a large number of grid points in the resulting discretization. To meet the computational amount of such simulations, high performance computing (HPC) is needed. In this paper, we compare different high performance implementations of a software for solving Maxwell's equations on different HPC machines. Simulation results for calculating the optical efficiency of thin-film solar cells are presented.

1 Introduction

Silicon thin-film solar cells are an innovative, inexpensive, and environmentally friendly technique to produce renewable energy. Increasing the efficiency of these cells is currently an important research topic in the field of photovoltaic. To this end, the optical path in the solar cell has to be improved to enhance absorption. Therefore e.g. the light coupling into the cell can be adjusted by structured front

C. Jandl · K. Hertel · C. Pflaum
Erlangen Graduate School in Advanced Optical Technologies (SAOT) and Department of Computer Science 10, University Erlangen-Nuremberg, Cauerstr. 6, 91058 Erlangen, Germany
e-mail: christine.jandl@informatik.uni-erlangen.de
e-mail: kai.hertel@informatik.uni-erlangen.de
e-mail: pflaum@informatik.uni-erlangen.de

W. Dewald
Fraunhofer Institute for Surface Engineering and Thin Films IST, Bienroder Weg 54 E, 38108 Braunschweig, Germany
e-mail: wilma.dewald@ist.fraunhofer.de

S. Wagner et al. (eds.), *High Performance Computing in Science and Engineering, Garching/Munich 2009*, DOI 10.1007/978-3-642-13872-0_46,

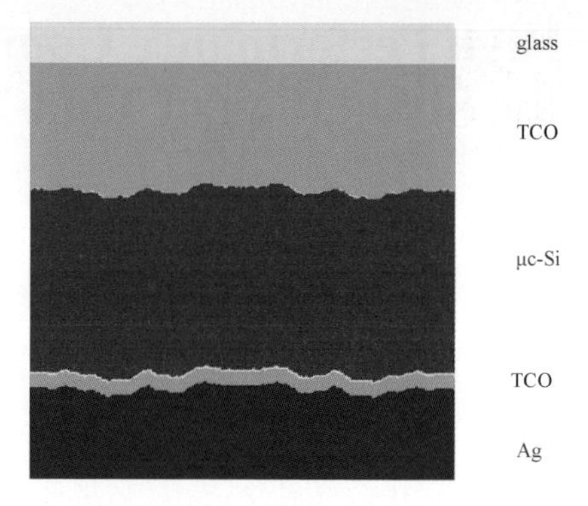

Fig. 1 Cross-section of a thin-film solar cell interface described by AFM scan

contacts. Since the rough morphology of the front TCO is passed to all following films of the cell, light trapping in the cell can be achieved. In order to design such a solar cell, numerical simulations of Maxwell's equations are needed (see [1] and [2]). Widely used methods for solving Maxwell's equations for high frequencies include the finite difference time domain method (FDTD) and the finite integration technique (FIT). The computational effort required by these methods is extremely high in case of thin-film solar cell simulations. Therefore, the use of HPC is needed. In this paper, we study different implementations of the FDTD iteration method.

To show the practicability of our simulation tool, we present results of the simulation of a solar cell structure based on AFM scan data. In this simulation, we study a thin-film solar cell consisting of a layer of glass, a transparent conductive oxide (TCO) layer, a layer of microcrystalline silicon (μc-Si:H), a second layer of TCO, and a layer of silver (Ag) (see Fig. 1). The texture of the first TCO layer influences the texture of all the other layers. There exist several publications analyzing the optical efficiency of a thin-film solar cell based on this kind of model structure (see [1, 2, 4–6], and [8]). However, the real texture of TCO in a thin-film solar cell is a random texture. Therefore, simulations based on model structures can give a hint on what an optimal texture might look like. In order to compare different TCO textures though, it is important that simulations based on AFM scans are performed on large computational domains to gain statistical information. Results gained from these simulations greatly help in optimizing the optical properties of the TCO layer without the need to manufacture a complete thin-film solar cell.

2 Simulation of Thin-Film Solar Cells

To calculate the short-circuit current density of a solar cell, it is necessary to take into account the intensity of light with respect to the wavelength. Fig. 2 depicts the corresponding solar spectral irradiance AM1.5. Using these data, the short-circuit current density can be calculated by

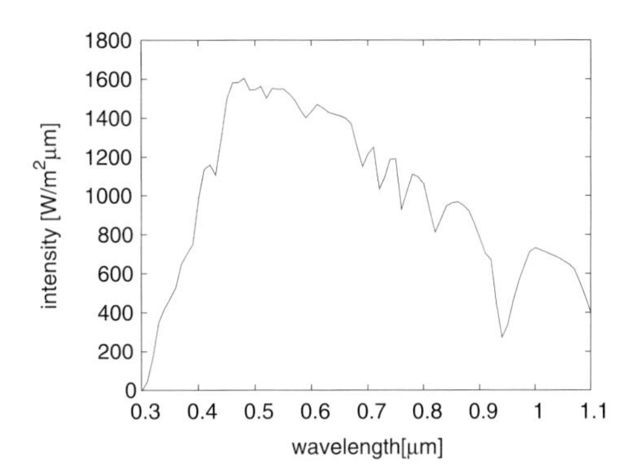

Fig. 2 Intensity of the solar spectrum AM1.5 which ranges from $\lambda = 0.3\,\mu\text{m}$ to $\lambda = 1.1\,\mu\text{m}$

$$J_{SC} = \sum_{\lambda} QE(\lambda) P_{\lambda}^{\mathrm{AM1.5}} \Delta\lambda \frac{e\lambda}{hc} , \tag{1}$$

where $P_{\lambda}^{\mathrm{AM1.5}}$ is the intensity of light with linewidth $\Delta\lambda$, e is the elementary charge, h the Planck constant, and c the velocity of light in vacuum. Furthermore, the quantum efficiency is defined by the quotient of the absorbed power in silicon over input power:

$$QE(\lambda) := \frac{P_{\mathrm{abs},\lambda}}{P_{\mathrm{in},\lambda}} . \tag{2}$$

The main difficulty is to calculate the absorbed power $P_{\mathrm{abs},\lambda}$ for an incoming wave of power $P_{\mathrm{in},\lambda}$. Since the relevant structures of the thin-film solar cell are of the order of magnitude of one single wavelength, it is important that Maxwell's equations are solved by an accurate simulation method. A common approach is to apply the finite integration technique.

2.1 Modeling Maxwell's Equations

Maxwell's equations describe the propagation of electromagnetic waves in a very broad manner. Based on some model assumptions about the materials involved in the production of solar cells (see [3, 9]), we can reduce them to the following set of equations:

$$\mu\,\partial_t H = -\nabla \times E - \sigma^{\star} H , \tag{3}$$

$$\varepsilon\,\partial_t E = \nabla \times H - \sigma E , \tag{4}$$

$$\div (\varepsilon E) = 0 , \tag{5}$$

$$\div H = 0 . \tag{6}$$

In order to be able to solve these equations numerically, a discretization scheme has to be employed. In the following, finite differences will be shown to be a practicable method for this kind of problem. Our implementation relies on rectangular staggered grids for spatial discretization. It is well known that such a staggered grid discretization satisfies (5) and (6) in a discrete sense, if these conditions hold for the initial value of the iteration. So, with two out of four equations being dealt with implicitly, the finite difference discretization allows for Maxwell's equations to be solved by a relatively simple system of equations. In order to be able to model interfaces of curvilinear geometry more accurately, the finite integration technique is applied. Resulting stencils in the time-harmonic case are equivalent to the ones produced by the FDTD method, while integrals are hidden in the coefficients. So, for reasons of simplicity we limit the following discussion to the FDTD and derived finite difference frequency domain (FDFD) cases, which turns out to be equivalent to the FIT method where coefficients are constant. Due to interdependencies between Eqs. (3) and (4), temporal decoupling is an important step towards developing an iterative solver. It turns out that a leap frog scheme can be used to this end. This leads to the following update steps for the time evolution:

$$\left(1+\rho\frac{\tau\sigma^\star}{\mu}\right)H_C^{t+\frac{\tau}{2}} = H_C^{t-\frac{\tau}{2}} - \frac{\tau}{\mu}\left(\operatorname*{grad}_h \times E_C^t\right) - \frac{\tau\sigma^\star}{\mu}(1-\rho)H_C^{t-\frac{\tau}{2}} \text{ and} \tag{7}$$

$$\left(1+\rho\frac{\tau\sigma}{\varepsilon}\right)E_C^{t+\tau} = E_C^t + \frac{\tau}{\varepsilon}\left(\operatorname*{grad}_h \times H_C^{t+\frac{\tau}{2}}\right) - \frac{\tau\sigma}{\varepsilon}(1-\rho)E_C^t\,, \tag{8}$$

where $\rho \in [0;1]$ is a steering parameter for interpolating the given quantities as needed. Note that the subscript C simply identifies the node in the mesh relative to the adjacent ones while iterating over the vector field (refer to Fig. 3 for further details). In order to allow for a linear iterative solver to converge towards a solution of this evolution scheme, we exploit the fact that electromagnetic waves expose periodic behavior, and thus solutions can also be expected to be periodic. This leads to the related stationary problem in the frequency domain:

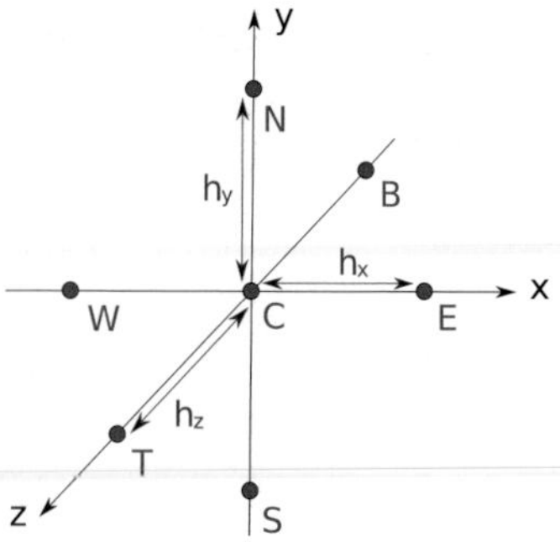

Fig. 3 A node in the finite difference mesh and its neighbors

$$\left(1+\rho\frac{\tau\sigma^\star}{\varepsilon}\right)H_C^{n+\frac{1}{2}}e^{i\omega\tau} = H_C^{n-\frac{1}{2}} - \frac{\tau}{\mu}\left(\text{grad}_h \times E_C^n\right)e^{i\omega\frac{\tau}{2}} - \frac{\tau\sigma^\star}{\mu}(1-\rho)H_C^{n-\frac{1}{2}} \text{ and} \tag{9}$$

$$\left(1+\rho\frac{\tau\sigma}{\varepsilon}\right)E_C^{n+1}e^{i\omega\tau} = E_C^n + \frac{\tau}{\varepsilon}\left(\text{grad}_h \times H_C^{n+\frac{1}{2}}\right)e^{i\omega\frac{\tau}{2}} - \frac{\tau\sigma}{\varepsilon}(1-\rho)E_C^n, \tag{10}$$

where t and τ have been dropped in favor of n as a label for iterative updates to the E and H fields, since this scheme does not represent a time evolution in the strict sense anymore. In order to meet stability requirements for materials of negative permittivity, especially silver in the case of thin-film solar cells, this iterative scheme can be slightly modified as has been shown in [7]. To gain better control over spurious reflections at the computational boundaries, perfectly matching layers are deployed. To this end, a split field approach is helpful, effectively turning the above two equations into four. Take, for example, the x component of the electric field. In its original form, the equation reads:

$$\left(1+\rho\frac{\tau\sigma}{\varepsilon}\right)E_{C_x}^{n+1}e^{i\omega\tau} = E_{C_x}^n + \frac{\tau}{\varepsilon}\left(h_y^{-1}\left(H_{N_z}^{n+\frac{1}{2}} - H_{S_z}^{n+\frac{1}{2}}\right) - h_z^{-1}\left(H_{T_y}^{n+\frac{1}{2}} - H_{B_y}^{n+\frac{1}{2}}\right)\right)e^{i\omega\frac{\tau}{2}}$$

after resolving the discrete curl operator in the finite difference formulation. Note that h denotes the spatial granularity of the mesh. After splitting up this equation subject to the constraint of $E_{C_x} = E_{C_{xy}} + E_{C_{xz}}$ being satisfied for all steps n of the iteration, we get:

$$\left(1+\rho\frac{\tau\sigma}{\varepsilon}\right)E_{C_{xy}}^{n+1}e^{i\omega\tau} = E_{C_{xy}}^n - \frac{\tau}{\varepsilon}h_z^{-1}\left(H_{T_{yx}}^{n+\frac{1}{2}} + H_{T_{yz}}^{n+\frac{1}{2}} - H_{B_{yx}}^{n+\frac{1}{2}} - H_{B_{yz}}^{n+\frac{1}{2}}\right)e^{i\omega\frac{\tau}{2}} \text{ and}$$

$$\left(1+\rho\frac{\tau\sigma}{\varepsilon}\right)E_{C_{xz}}^{n+1}e^{i\omega\tau} = E_{C_{xz}}^n + \frac{\tau}{\varepsilon}h_y^{-1}\left(H_{N_{zx}}^{n+\frac{1}{2}} + H_{N_{zy}}^{n+\frac{1}{2}} - H_{S_{zx}}^{n+\frac{1}{2}} - H_{S_{zy}}^{n+\frac{1}{2}}\right)e^{i\omega\frac{\tau}{2}}.$$

For reasons of reducing computational complexity, coefficients can be absorbed into separate vectors, as they are not subject to change after the beginning of the iterative process:

$$\hat{c}_{E_{xy}} = \frac{1}{1+\rho\frac{\tau\sigma}{\varepsilon}}e^{-i\omega\tau},$$

$$c_{E_{xy}} = -\frac{1}{1+\rho\frac{\tau\sigma}{\varepsilon}}e^{-\frac{1}{2}i\omega\tau}\frac{\tau}{\varepsilon}h_z^{-1},$$

$$\hat{c}_{E_{xz}} = \frac{1}{1+\rho\frac{\tau\sigma}{\varepsilon}}e^{-i\omega\tau},$$

$$c_{E_{xz}} = +\frac{1}{1+\rho\frac{\tau\sigma}{\varepsilon}}e^{-\frac{1}{2}i\omega\tau}\frac{\tau}{\varepsilon}h_y^{-1}.$$

Note that ε and σ generally are not constant throughout the computational domain, and as a consequence neither are the coefficients above. This gives rise to the following formulation of the split-field approach:

$$E_{C_{xy}}^{n+1} = \hat{c}_{E_{C_{xy}}} E_{C_{xy}}^{n} + c_{E_{C_{xy}}} \left(H_{T_{yx}}^{n+\frac{1}{2}} + H_{T_{yz}}^{n+\frac{1}{2}} - H_{B_{yx}}^{n+\frac{1}{2}} - H_{B_{yz}}^{n+\frac{1}{2}} \right), \tag{11}$$

$$E_{C_{xz}}^{n+1} = \hat{c}_{E_{C_{xz}}} E_{C_{xz}}^{n} + c_{E_{C_{xy}}} \left(H_{N_{zx}}^{n+\frac{1}{2}} + H_{N_{zy}}^{n+\frac{1}{2}} - H_{S_{zx}}^{n+\frac{1}{2}} - H_{S_{zy}}^{n+\frac{1}{2}} \right). \tag{12}$$

So, the resulting iteration scheme effectively solves for a vector of approximately 6 components times the number of grid points in the computational domain.

2.2 Solar Cell Model

In order to demonstrate the efficiency of our simulation technique, let us study a thin-film solar cell model with one layer of microcrystalline silicon. Fig. 1 depicts the cross-section of a simulated solar cell with the interface being described by an AFM scan. The size of the complete simulated solar cell is 2.4 μm in x and y directions, and 2.38 μm in z direction. The thickness of the glass layer is 0.2 μm, the size of the upper TCO layer is 0.6 μm, and that of the lower TCO layer is 0.08 μm. The height of the microcrystalline silicon is 1 μm, and the silver layer has a thickness of 0.5 μm. A real thin-film solar cell consists of a glass layer of a thickness larger than 0.5 cm. Simulating such a thick layer of glass by using the FDTD method is computationally extremely intensive. However, light waves approximately behave like plane waves inside the glass layer. Therefore it is sufficient to start with an incoming plane wave at the top of a thin layer of glass (thickness of 0.2 μm). The AFM scans of the TCO layer are courtesy of Fraunhofer IST Braunschweig.

To calculate the short-current density, we apply the finite integration technique. The mesh size of the grid is calculated as follows. We compute the local wavelength in every grid point p, which is defined by $\lambda(p) := \frac{\lambda}{n_r}$, where n_r is the real part of the refractive index of the local material (see Fig. 4). Then this local wavelength is used for discretizing the domain with a fixed number of grid points per wavelength. For the AFM based simulations presented in this paper, we choose 20 grid points per wavelength. This calculation leads to grids of varying mesh size and a fairly large number of discrete grid points. We calculate the quantum efficiency for the most relevant part of the spectrum, ranging from 0.55 μm to 0.995 μm with a linewidth $\Delta\lambda = 5$nm. The number of grid points varies between 6 and 46 millions depending on the wavelength of the incoming wave.

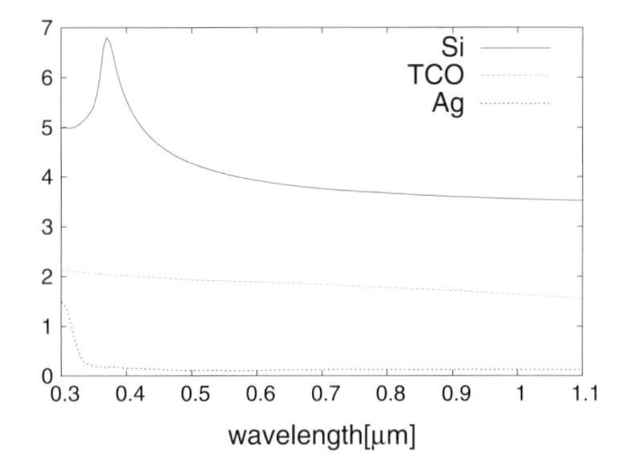

Fig. 4 The real part of the refractive indices of silicon, TCO, and silver dependent on the wavelength

2.3 Integration of AFM Scans

By standard simulation programs for solving Maxwell's equations, it is difficult to simulate the real roughness of the solar cell layers. Hence, imitated structures (e.g. a pyramidal structure) are used to calculate the short circuit current density of a thin film solar cell. To overcome this restriction, we developed a simulation tool which is able to integrate AFM scan data. As an example we present simulation results with AFM scan data provided by Fraunhofer IST Braunschweig (see Fig. 5 and Table 1). We need periodic boundary conditions in x and y directions to compute the efficiency of the scanned thin-film solar cells. Hence, we have to mirror the scan. This leads to a domain of size $10\,\mu\text{m} \times 10\,\mu\text{m}$. If we wanted to compute the effi-

Table 1 Samples done by Fraunhofer IST Braunschweig

	Sample A	Sample B
rms-roughness	32 nm	53 nm
O2-partial pressure	6 mPa	12 mPa
specific resistance	315 μΩcm	410 μΩcm
layer thickness	1.08 μm	1 μm

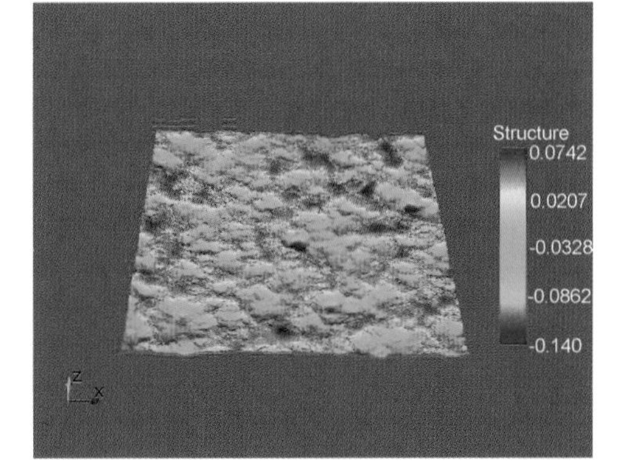

Fig. 5 AFM scan "Sample A" from Fraunhofer IST Braunschweig sized $5\,\mu\text{m} \times 5\,\mu\text{m}$ with a resolution of 512×512 points

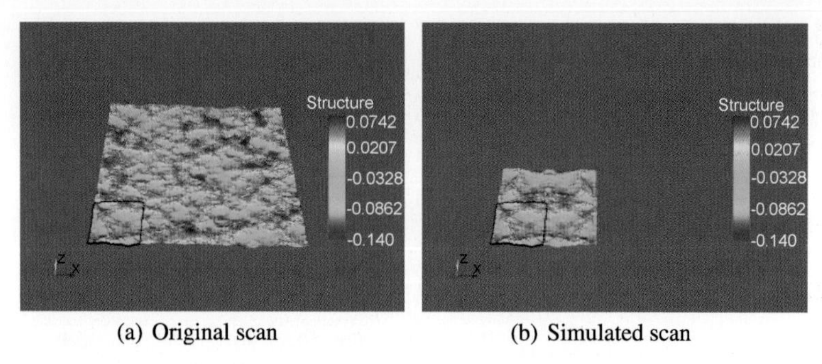

(a) Original scan (b) Simulated scan

Fig. 6 AFM Scan: Black framed part is the part of the original scan

ciency for a region of this size, we would need more than 10^9 grid points for the resulting discretization. In order to decrease the size of the computational domain and, with it, the number of grid points, we cut the AFM domain into smaller pieces (see Fig. 6). There are two ways of picking the part of the scan to be used. One is to choose a fixed part of the AFM scan, which is then simulated for every wavelength. Another is to choose a different part of the AFM scan for each wavelength. In this case, the chosen part is computed reproducibly from pseudo random numbers. Numerical simulations have shown that choosing different random pieces for each wavelength lead to more accurate results. Another aspect when dealing with AFM scans in simulations is that the spatial resolution of the scan generally does not match the one of the computational domain, so the positions of the resulting grid points do not necessarily coincide. To resolve this issue, AFM data are linearly interpolated.

2.4 Parallelization

Since the spatial resolution of the computational domain needs to be relatively high in order to obtain accurate results, parallelization and the use of high performance compute clusters are essential. This is especially true where simulations are based on AFM data, since a minimum of 20 nodes per wavelength are required on the grid in order to obtain useful results that can yield information about the efficiency of applied nanostructures. Therefore, the implementation effort has strongly focused on the solver being MPI-capable throughout the development process. The scaling behavior turns out to be near-linear as long as the ratio problem size vs. number of nodes is considered to be same (see Fig. 7, 8).

For parallel runs, a box-shaped computational domain is decomposed into rectangular sub-boxes. This is done in such a way that a maximum of available compute nodes is assigned a computational partition, while at the same time interfaces between partitions are kept to a minimum.

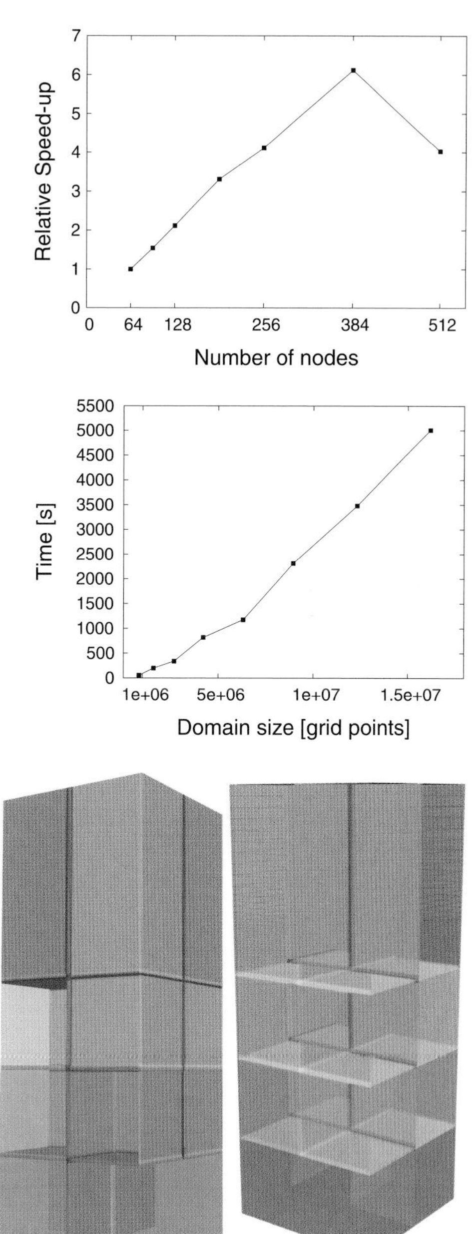

Fig. 7 Speed-up of the solver relative to the 64 compute node configuration. Simulation of a synthetic 3-dimensional moth structure consisting of roughly 12.28 million grid points. Comparative run on 64, 96, 128, 192, 256, 384 and 512 nodes of the LRZ Itanium2 Cluster

Fig. 8 Runtime against the resolution of the computational domain on a fixed number of 64 compute nodes. Simulation of the same synthetic 3-dimensional moth structure with varying numbers of grid points

Fig. 9 Rectangular partitioning scheme balancing the computational load between compute nodes

Generally, parallelization is a relatively straight-forward matter where rectangular grids are concerned. Communication across local boundaries need to be performed between iteration steps in order to ensure that neighboring stencils are correctly fed current data. The only detail that needs special attention when exchanging

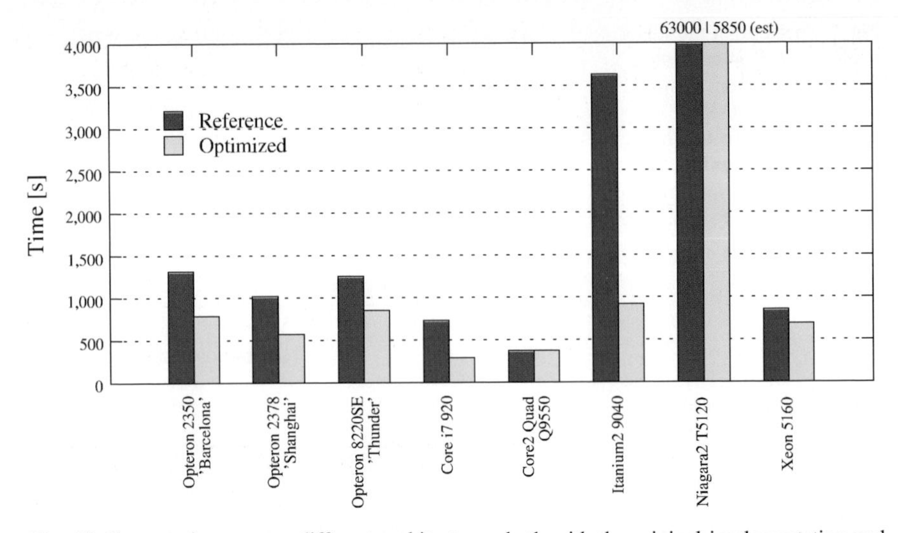

Fig. 10 Comparative runs on different architectures, both with the original implementation and current progress in optimization. Serial simulation runs of a synthetic 3-dimensional moth structure. Fixed number of 1000 iteration steps, 1.5 million grid points

data between compute nodes is the fact that dual grids are in place for different aspects of the electromagnetic field, leading to a slight index shift.

Besides parallelization, a number of optimizations steps are currently being explored in order to allow for the Maxwell code to run faster and, as a consequence, reduce workloads and expensive runtime on high performance machines. For once, a relatively simple cache-blocking technique has been implemented lately. As optimization is highly dependent on the architectures involved, results seem rather ambivalent at the time and are currently under investigation. Also, we are currently pursuing to reduce application complexity in critical sections and introduce vectorization where available (see Fig. 10 for a preliminary analysis of the current state of affairs).

3 Simulation Results

The aim of our simulation in regard to physics is to compare quantum efficiencies and short-circuit current densities dependent on the geometry and the simulated parts of AFM scans.

To this end, we compare quantum efficiencies of two different AFM scans "Sample A" and "Sample B" provided by Fraunhaufer IST Braunschweig. The geometry of "Sample A" is depicted in Fig. 5. "Sample B" is of the same size and spatial resolution as "Sample A", resulting in the same amount of discretized grid points, but it has a rougher structure. The scan domains are chosen by pseudo random numbers. Fig. 11 shows the quantum efficiency (QE) of both solar cells computed by Eq. (2).

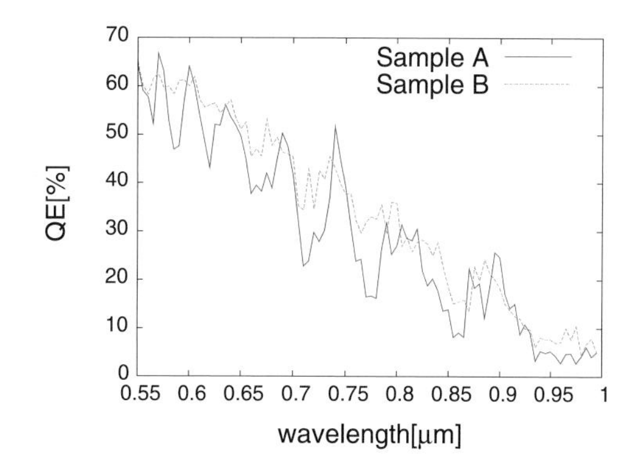

Fig. 11 Quantum efficiency of thin-film solar cells with different rough layers

Table 2 Short-circuit current density

	Sample A	Sample B
J_{SC} $\left[\mathrm{mA/cm^2}\right]$	8.9821	10.2116

The short-circuit current density J_{SC} is obtained by using Eq. (1) (see Table 2). It can be seen that rougher surface structures lead to an increase in short-circuit current density.

The simulations are performed on the HLRBII of LRZ Munich with up to 512 processors. Depending on the speed of convergence the runtime varies between 4½ and 7 hours per solar cell simulation. The number of discretization points varies from 6 million for a wavelength of 0.95 µm to 46 millions for 0.55 µm.

The quantum efficiency and short-circuit current density were computed for each solar cell structure with three different sets of pseudo random numbers. A maximum deviation of 2.4% of the short-circuit current density is observed for "Sample A" and 1.1% for "Sample B" respectively. Therefore it is necessary to expand the computational domain to obtain more accurate results.

4 Conclusion

Computationally intensive simulations allow a detailed analysis of light trapping in thin-film silicon solar cells. In this paper, we present first simulaton results for a complete thin film solar cell structure using AFM scan data. Our aim is to extend these simulations to a tandem solar cell which consists of amorphous (a-Si:H) and microcrystalline (μc-Si:H) silicon and has a thickness of 3.3 µm. To gain more accurate results a larger part of the AFM scan has to be simulated. This will increase the number of discretization points and computing time. In addition, simulations for oblique incident waves have to be done, because real outdoor conditions often lead to oblique incident sunlight.

Acknowledgements The authors gratefully acknowledge funding from the Erlangen Graduate School of Advanced Optical Technologies (SAOT) by the German Research Foundation (DFG) in the framework of the German Excellence Initiative, the Bavarian Competence Network for Technical and Scientific High Performance Computing (KONWIHR) and funding in the frame of the joint project LiMa by the Federal Ministry for the Environment, Nature Conservation and Nuclear Safety BMU under contract No. 0327693A.

References

1. Haase, C. and Stiebig, H.: Optical Properties of Thin-film Solar Cells with Grating Couplers. Progress in Photovoltaics: Research of Applications 14(7), 629–641 (2006)
2. Haase, C. and Stiebig, H.: Thin-film Silicon Solar Cells with Efficient Periodic Light Trapping Texture. Applied Physics Letters, 92 (2008)
3. Hecht, E.: Optics. Addison Wesley (2001)
4. Krč, J., Zeman, M., Smole, F., Topič, M.: Study of Enhanced Light Scattering in Microcrystalline Silicon Solar Cells. Journal of Non-Crystalline Solids 338–340, 673–676 (2004)
5. Krč, J., Smole, F., Topič, M.: Optical Modelling of Thin-film Silicon Solar Cells Deposited on Textured Substrates. Thin Solid Films 451–452, 298–302 (2004)
6. Müller, J., Rech, B., Springer, J., Vanecek, M.: TCO and Light Trapping in Silicon Thin Film Solar Cells. Solar Engergy 77, 917–930 (2004)
7. Pflaum, C. and Rahimi, Z.: An Iterative Solver for the Finite-Difference Frequency-Domain (FDFD) Method for Simulation of Materials with Negative Permittivity. Submitted to Numerical Linear Algebra with Applications (2009)
8. Sittinger, V., Dewald, W., Szyszka, B.: Large Area Deposition of Al-doped ZnO for Si-based Thin Film Solar Cells by Magnetron Sputtering. Proceedings of Glass Peformance Day 2009. Solar and Glass Technology – Market and Applications, 1–5 (2009)
9. Taflove, A. and Hagness, S.: Computational Electrodynamics — The Finite-Difference Time-Domain Method. Artech House, Boston, London (2000)

Origin of Interface Magnetism in $Fe_2O_3/FeTiO_3$ Heterostructures

Hasan Sadat Nabi and Rossitza Pentcheva

Abstract Nanoscale exsolutions of the canted antiferromagnet hematite (α-Fe_2O_3) and the room temperature paramagnet ilmenite ($FeTiO_3$) show a surprisingly stable room temperature remanent magnetization, making the material interesting for spintronics applications. To understand the nature of this phenomenon at the atomic scale, density functional theory calculations with an on-site Coulomb repulsion parameter were performed on $Fe_{2-x}Ti_xO_3$, varying the concentration, distribution and charge state of cations. We find that the polar discontinuity at the interface is accommodated by the formation of a mixed Fe^{2+}/Fe^{3+} layer. The uncompensated interface moments give rise to ferrimagnetism in this system. We also explore the effect of strain, showing that it can be used to tune the electronic properties (e.g. band gap, position of impurity levels). Furthermore, we find that epitaxial growth on an $Al_2O_3(0001)$ substrate is energetically unfavorable compared to substrates with a larger lateral lattice parameter, providing thereby a guideline for an optimal choice of the substrate in growth experiments.

1 Introduction

The unexpected properties evolving at oxide interfaces have raised a lot of interest and excitement in the scientific community. The best known example is undoubtedly the heterointerface between the two band insulators $LaAlO_3$ and $SrTiO_3$, showing conductivity [1], superconductivity [2] and magnetic scattering [3]. For a review on the theoretical work on these systems the reader is referred to Ref. [4] and references therein.

Hasan Sadat Nabi · Rossitza Pentcheva
Department of Earth and Environmental Sciences, University of Munich, Theresienstr. 41, 80333 Munich, Germany
e-mail: hasan.sadat@lrz.uni-muenchen.de
e-mail: pentcheva@lrz.uni-muenchen.de

S. Wagner et al. (eds.), *High Performance Computing in Science and Engineering, Garching/Munich 2009*, DOI 10.1007/978-3-642-13872-0_47,

Such intriguing properties are by far not limited to the perovskite structure. Already in the 50s Akimoto and Ishikawa [5] reported a remanent magnetization up to 900 K in intermediate members of the Fe_2O_3-$FeTiO_3$ join, although the end members are a canted antiferromagnet (CAF) ($T_N = 948$ K) and a room temperature paramagnet ($T_N = 56$ K), respectively, both with a corundum(-derived) structure. The complex phase diagram of this system is determined by the interplay of cation and magnetic ordering and exsolution processes [6]. Recently, this system has moved in the focus of the paleomagnetic community as a possible source of magnetic anomalies in the Earth's crust [6–8]. Moreover, Ti doped hematite is one unconventional example for a diluted magnetic semiconductor that is promising for spintronics applications [9–11]. Both the resistivity and the type of conductivity (*n*- or *p*-type) can be tuned, depending on the Ti concentration, suggesting use in electronics devices [12].

As mentioned above the end members crystallize in the corundum structure. In α-Fe_2O_3 (space group $R\bar{3}c$) layers of $2Fe^{3+}$ and $3O^{2-}$ alternate with antiparallel alignment between the adjacent Fe-layers with a small spin-canting. In $FeTiO_3$ ($R\bar{3}$) there is a stacking of $2Fe^{2+}/3O^{2-}/2Ti^{4+}$ with a slight preference for antiferromagnetic (AFM) coupling between the Fe-layers. Hence, a polar discontinuity arises at an interface or in a solid solution of these oxides if all ions preserved their charge state in the respective end members. To explain the origin of magnetism in this system, Robinson et al. [7] recently proposed that a mixed Fe^{3+}/Fe^{2+} contact layer forms at the interface. We note that the compensation on the iron sublattice is only one out of several possibilities, because also the Ti-ion can change its charge state and thus accommodate the valence discontinuity.

The so called *lamellar magnetism hypothesis* [7] is based on bond valence arguments and Monte Carlo simulations using empirical chemical and magnetic interaction parameters. In order to obtain a material-specific atomistic understanding of the underlying mechanisms, we have performed systematic density functional theory (DFT) calculations both for the end members and for intermediate compositions [13]. We have thereby varied not only the concentration of Ti in a hematite host and of Fe in ilmenite, but also the distribution of the impurities and their charge state and magnetic order. Moreover, we have taken into account electronic correlations which play an important role in these transition metal oxides by using a Coulomb corrected term within the LDA+U method [14]. The goal of this study is (*i*) to identify the compensation mechanism and its influence on the electronic and magnetic properties, (*ii*) to determine the relative stability of layered configurations versus solid solutions as well as (*iii*) to extract accurate magnetic interaction parameters for the interface that are not accessible from experiment.

A further important aspect in this system is the effect of lattice strain. We note that the volume of the end member ilmenite ($a = 5.18$ Å, $c = 14.27$ Å) is 9.7% larger than that of hematite ($a = 5.04$ Å, $c = 13.75$ Å). Even more severe is the problem for epitaxial $Fe_{2-x}Ti_xO_3$ films which are typically grown on $Al_2O_3(0001)$. The smaller lateral constant of Al_2O_3 ($a = 4.76$ Å, $c = 12.99$ Å) introduces a substantial compressive strain of 5.8% and 8.8% compared to Fe_2O_3 and $FeTiO_3$. Therefore we have also investigated the effect of strain on the energetic stability and the electronic properties of the system [15].

2 Method and Details of the Calculations

The DFT calculations are performed with the full-potential linearized augmented plane wave (FP-LAPW) method as implemented in the WIEN2k-code [16]. For the exchange correlation potential we have used the generalized gradient approximation [17]. Electronic correlations beyond the local density approximation are considered within the LDA/GGA+U method [14]. To describe correctly the end member ilmenite $U = 8$ eV and $J = 1$ eV is needed. Therefore, these parameters have been applied on the $3d$ states of both Fe and Ti throughout the calculations.

The systems are modeled in a hexagonal setup with 30 and 60 atoms in the unit cell. The most time consuming part of the calculation (around 70-90%) is the set up and diagonalization of the Hamiltonian (`lapw1`). The cutoff parameter for the plane wave basis set ($E_{cut} = 19$ Ry) corresponds to matrix dimensions of 3200×3200 and 6800×6800 for the 30 and 60 atom systems, respectively. For these intermediate matrix sizes we have used mainly the k-point parallelization on HLRB II. The typical running times of `lapw1` are 210 s and 618 s with 24 and 15 k-points in the irreducible part of the Brillouin zone for the 30 and 60-atom u.c., respectively. A single iteration of the spin-polarized calculations takes 608/1477s for 30/60 atom u.c., respectively. The convergence for a single geometry requires approximately 30-80 iterations. On the average 10 geometries are needed to optimize the structure. Keeping in mind that more than 100 different systems have been considered, these calculations would not have been possible without using the high performance computing resources at the Leibniz Rechenzentrum. Since our last report the hardware has not changed, therefore we refer the reader to our previous detailed performance benchmarks on HLRB II and other platforms in Refs. [18, 19].

A major improvement beyond the previous version of WIEN2k is that now the forces can be derived also for orbitally dependent potentials [20]. This allows the structural optimization within LDA/GGA+U. As shown in heterostructures of $LaAlO_3$ and $SrTiO_3$ [21], the optimization of the internal parameters within GGA and GGA+U can lead to significantly different electronic behavior.

3 Results and Discussion

The question that we address here is how the local charge imbalance is compensated at the interface of hematite and ilmenite. As mentioned above, besides the *lamellar magnetism hypothesis* proposed by Robinson et al. [7], there are further possibili-

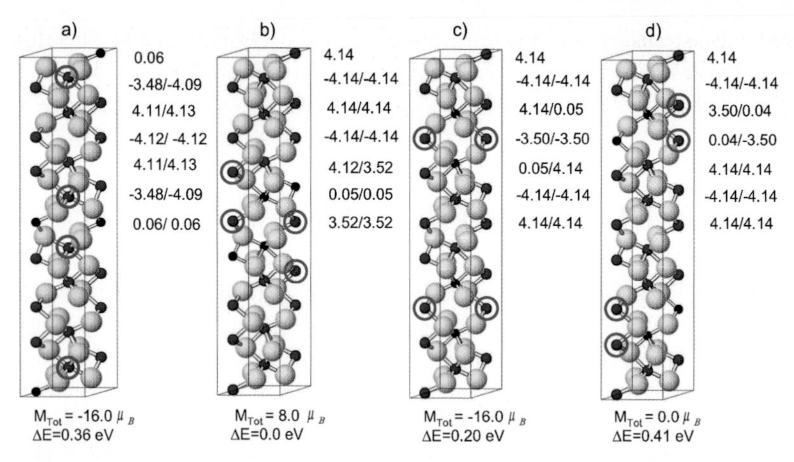

Fig. 1 Selected structures of Ilm_{33} (33% Ti doped Fe_2O_3) modeled in a 60-atom unit cell: a) single Ti-layer incorporated in Fe_2O_3; b) an ilmenite block in Fe_2O_3; two more homogeneous distributions (solid solutions) with Ti incorporated in c) the same and d) different spin sublattices of the host. Also shown are the local magnetic moments of the cations (at the side) and the total magnetic moments and relative energy (in eV/60-atom unit cell) with respect to the most stable configuration (beneath each structure). Fe, Ti and O are shown in red, black and grey spheres, respectively. Positions of the Fe^{2+} are marked by pink circles, while the rest of the iron cations are Fe^{3+}

ties, e.g. a disproportionation through Ti^{3+}/Ti^{4+} or Fe^{3+}/Ti^{3+}. To identify the most favorable compensation mechanisms we have considered all these possibilities and have varied systematically the concentration, distribution and the charge state of Ti in a Fe_2O_3-host and Fe in a $FeTiO_3$-host. In particular we have investigated ilmenite concentrations $x = 17\%$, 33%, 50%, 66% and 83%.

3.1 Structural Relaxation

A full structural optimization of the internal parameters has been performed. Table 1 lists the cation-anion distances of first and second nearest neighbors in Fe_2O_3, $FeTiO_3$ and three different concentrations of Ti doped hematite (33%, 50% and 66%). An interesting trend is observed in the shortest cation-oxygen bond lengths, which tend to relax towards the values in the respective end member: While $d_{Fe^{3+}-O}$ remains close to the value in bulk hematite (1.96Å), the bond lengths of the Ti-impurity ($d_{Ti^{4+}-O}$) and the neighboring Fe^{2+} ($d_{Fe^{2+}-O}$) relax towards the values in bulk ilmenite (1.94 and 2.11Å, respectively). As we will see in the following this has an important effect on the compensation mechanism.

Table 1 Cation-anion distances up to 2nd nearest neighbor in Fe_2O_3, $FeTiO_3$ and for three different concentrations of Ti in hematite. Subscript IF refers to the cations at the interface. Δd_1 and Δd_2 are the deviations of the two shortest bond lengths in each system with respect to the values in the corresponding end member

Systems	$d_{Cat.-Oxy.}$(Å)	1st	$\Delta d_1(\%)$	2nd	$\Delta d_2(\%)$
Fe_2O_3	$d_{Fe^{3+}-O}$	1.96		2.09	
$FeTiO_3$	$d_{Fe^{2+}-O}$	2.11		2.25	
	$d_{Ti^{4+}-O}$	1.94		2.06	
33%	$d_{Fe^{2+}_{IF}-O}$	2.08	-1.42	2.12	-5.78
	$d_{Fe^{3+}_{IF}-O}$	1.90	-3.06	2.18	4.31
	$d_{Fe^{2+}-O}$	2.04	-3.32	2.16	-4.00
	$d_{Ti^{4+}-O}$	1.89	-2.58	2.03	-1.46
	$d_{Fe^{3+}-O}$	1.95	-0.51	2.08	-0.48
50%	$d_{Fe^{2+}_{IF}-O}$	2.08	-1.42	2.11	-6.22
	$d_{Fe^{3+}_{IF}-O}$	1.90	-3.06	2.18	4.31
	$d_{Fe^{2+}-O}$	2.04	-3.32	2.17	-3.56
	$d_{Ti^{4+}-O}$	1.90	-2.06	2.02	-1.94
	$d_{Fe^{3+}-O}$	1.95	-0.51	2.09	0.00
66%	$d_{Fe^{2+}_{IF}-O}$	2.03	-3.79	2.42	7.56
	$d_{Fe^{3+}_{IF}-O}$	2.05	4.59	2.09	0.00
	$d_{Fe^{2+}-O}$	2.13	0.95	2.27	0.89
	$d_{Ti^{4+}-O}$	1.94	0.00	2.06	0.00
	$d_{Fe^{3+}-O}$	2.00	2.04	2.15	2.87

3.2 Energetic Stability

Selected configurations for an ilmenite concentration of 33% which corresponds to four Ti-ions out of 24 cations in the 60-atoms cell are shown in Fig. 1. These include a single Ti-layer incorporated in the hematite host (Fig. 1a) and an ilmenite block containing an Fe layer sandwiched between two Ti layers (Fig. 1b). Furthermore we have considered some more homogeneous distributions by exchanging 50% of an Fe-layer by Ti (Fig. 1c-d). As we will see in the following, the magnetic properties of the system depend sensitively on whether Ti is incorporated into the same spin-sublattice (Fig. 1c) or different spin-sublattices (Fig. 1d).

We find that the formation of a compact ilmenite-like block with a Fe-layer sandwiched between two Ti-layers (Fig. 1b), is by 0.36 eV more favorable than incorporation of a single Ti layer in the hematite host (Fig. 1a). Compensation mechanisms involving Ti^{3+} are always less favorable, especially after structural relaxation. Concerning the charge state, iron in the central layer turns into Fe^{2+}. Substitution of Ti^{4+} in hematite leads to change in the oxidation state of neighboring iron from Fe^{3+} to Fe^{2+}. We find that, the octahedron surrounding Ti^{4+} shares faces with the one of Fe^{3+}, while Fe^{2+} in the contact layer is face-sharing with Fe^{3+} from the next hematite layer. The GGA+U results thus support the *lamellar magnetism hypothesis* [7].

The Fe^{2+}-layer sandwiched between two Ti-layers in Fig. 1b is only weakly coupled to the next Fe-layer (parallel and antiparallel orientation of the magnetic moments is nearly degenerate as in the ilmenite end member). Therefore, at temperatures above the Néel temperature of ilmenite, such layers will not contribute to the total magnetization. In contrast, Fe^{2+} in the contact layer shows a strong antiferromagnetic coupling to the neighboring Fe-layer of the hematite host. [22] These defect interface moments are responsible for the ferrimagnetic behavior of the system, e.g. for the structure in Fig. 1b the total magnetic moment is 8.0 μ_B.

Each Ti ion adds a magnetic moment of $4\mu_B$ independent of whether the extra electron is localized at Ti (Ti^{3+}) or at a neighboring Fe (Fe^{2+}). However, the total magnetization of the system depends on the distribution of Ti in the Fe_2O_3-lattice. We find that the formation of layered arrangements (Fig. 1b) is 0.20 eV and 0.41 eV more stable compared to more random distributions with 50% substituted cation layers shown in Fig. 1c) and Fig. 1d), respectively. Furthermore, incorporation of Ti in the same spin sublattice (Fig. 1c), which maximizes the magnetic moment of the system, is favored compared to incorporation in different spin-sublattices (Fig. 1d) by 0.21 eV. Similar results were obtained by Velev [23] for single Ti impurities in α-Fe_2O_3. This feature promotes ferrimagnetic behavior in the system.

The trend towards layered arrangements is retained for 66%, while for 83% the ordered and disordered phases are nearly degenerate in agreement with thermodynamic data of the system (see e.g. [6] and references therein). When substituting Fe for Ti in $FeTiO_3$, iron is Fe^{3+} and additionally one of the Fe in the next layer becomes Fe^{3+} to compensate the charge, resulting again in a mixed Fe^{2+}, Fe^{3+} contact layer. The substituted Fe shows a strong tendency to couple antiparallel to the neighboring Fe layer.

3.3 Electronic Properties

The density of states (DOS) for different concentrations of Ti doped in hematite is shown in Fig. 2. The formation of Fe^{2+} in the contact layer leads to an impurity state in the band gap that is pinned at the Fermi level and reduces the band gap of α-Fe_2O_3 from ~ 2.1 eV to 1.65, 1.68, 1.73, 1.64 for $x = 17\%$, 33%, 50%, and 66%, respectively. These values refer to systems with a layered configuration. For more random distribution of Ti in the hematite host, the reduction of the band gap is even more pronounced as the impurity level moves close to the conduction band. A further interesting feature is that Fe^{3+} at the interface has a much narrower lower Hubbard band than Fe^{3+} in hematite.

3.4 Influence of Epitaxial Strain

As mentioned above, the lateral lattice parameters of Fe_2O_3, $FeTiO_3$ and Al_2O_3 differ significantly, prompting us to investigate the effect of strain on the energetic stability and electronic properties of the Fe_2O_3-$FeTiO_3$ system. To this end DFT calculations were performed for systems strained laterally at $a_{Fe_2O_3}$, a_{FeTiO_3} and $a_{Al_2O_3}$, relaxing in each case the c/a ratio [15]. The energetic stability with respect to the end members is described by the formation energy $E_f = E_{Fe_{2-x}Ti_xO_3} -$

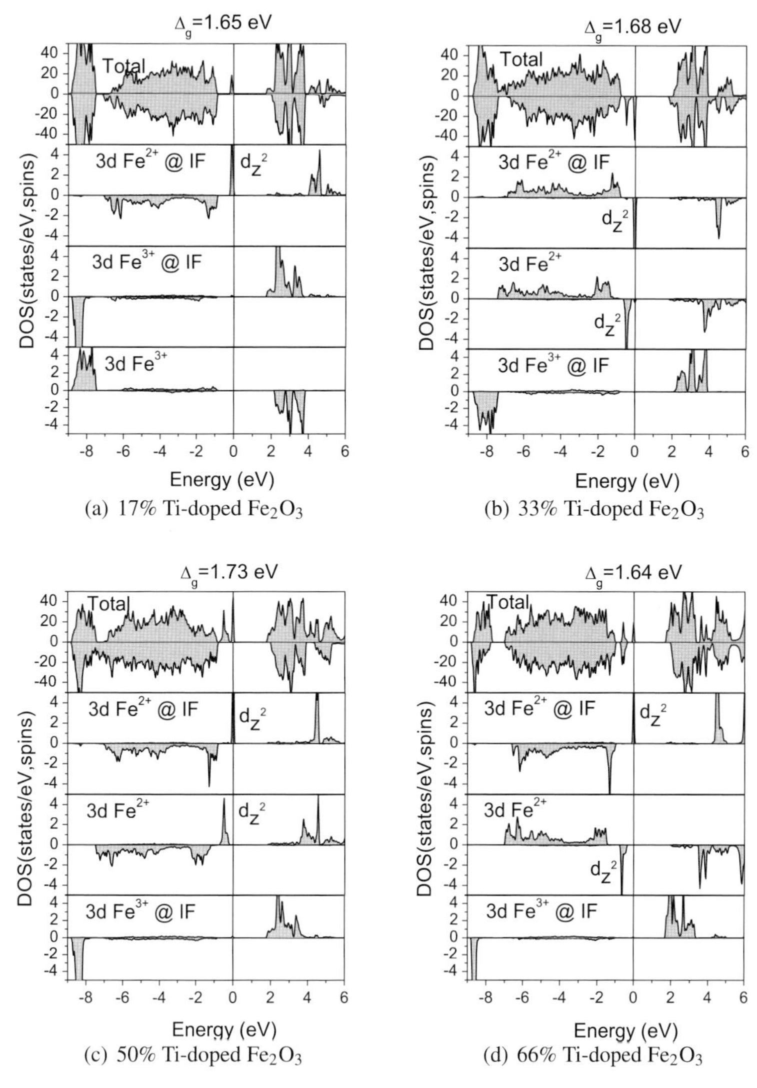

Fig. 2 Density of states (DOS) for layered configurations of $Fe_{2-x}Ti_xO_3$ for concentration x of a) 17%, b) 33%, c) 50% and d) 66%: The top panel shows the total DOS and the lower panels the projected DOS of Fe^{3+} and Fe^{2+} in the interface layer. For comparison also the DOS of cations further away from the interface are shown

$(1-x)E_{Fe_2O_3} - xE_{FeTiO_3}$, where $E_{Fe_{2-x}Ti_xO_3}$, $E_{Fe_2O_3}$ and E_{FeTiO_3} are the total energies of the system with a concentration x of ilmenite and the two end members, respectively. The formation energy as a function of x is plotted in Fig. 3. We find that the compensation mechanism through Ti^{4+} and disproportionation in Fe^{2+}, Fe^{3+} (circles) is not affected by strain. Furthermore, E_f shows a linear dependence on x for all lateral parameters. Systems strained laterally at a_{FeTiO_3} are more stable than the

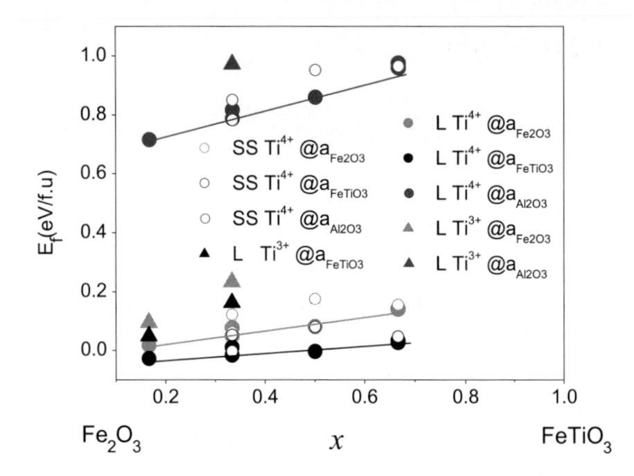

Fig. 3 The formation energy (eV/f.u) versus ilmenite concentration x for $Fe_{2-x}Ti_xO_3$ strained at the Al_2O_3 (red/dark grey), Fe_2O_3 (grey) and $FeTiO_3$ (black) lateral lattice constants. Circles (triangles) denote compensation involving Ti^{4+}(Ti^{3+}). Open/filled symbols refer to solid solutions (SS)/ layered configurations (L) [15]

ones on $a_{Fe_2O_3}$. In contrast, the formation energy of films strained at $a_{Al_2O_3}$ increases by 0.7 eV as compared to films on $a_{Fe_2O_3}$. This implies that the strong compressive strain is energetically unfavorable and gives a possible explanation why a lateral strain relaxation occurs in $Fe_{2-x}Ti_xO_3$ films grown on Al_2O_3(0001) [24, 25]. While for systems strained on hematite and ilmenite substrates layered arrangements (full symbols) are more favorable than homogeneous distributions (open symbols), the trend is reversed for $x = 0.33$ and $x = 0.66$ on an Al_2O_3(0001)-substrate.

The influence of strain on the electronic properties of $Fe_{2-x}Ti_xO_3$ films with $x = 50\%$ and $x = 66\%$ is presented in Fig. 4 a and b, respectively. The main effect is the broadening/reduction of bands for $a_{Al_2O_3}/a_{FeTiO_3}$, compared to those at $a_{Fe_2O_3}$. Furthermore, the position and degree of localization of the impurity state can be tuned by strain. For compressive strain the impurity level is broadened and shifted to lower energies. Generally, we observe a reduction of the band gap between the impurity state and the bottom of the conduction band, e.g. for $x = 66\%$ from 1.64 (a_{FeTiO_3}) to 1.46 ($a_{Fe_2O_3}$) and 0.78 eV ($a_{Al_2O_3}$).

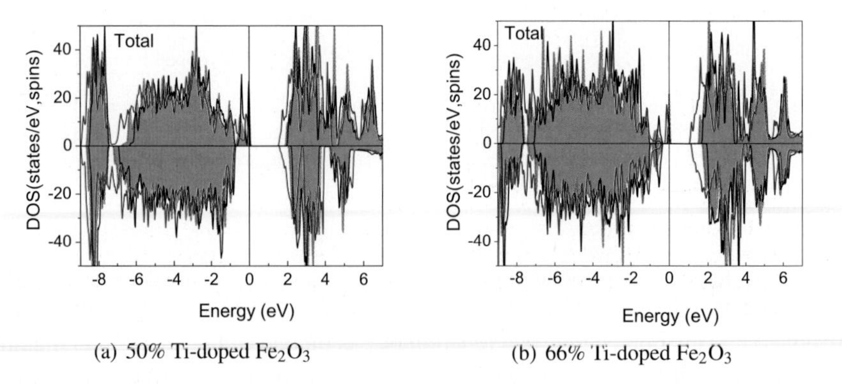

(a) 50% Ti-doped Fe_2O_3 (b) 66% Ti-doped Fe_2O_3

Fig. 4 Total density of states for concentrations of 50% and 66% Ti doped in hematite strained laterally to $a_{Fe_2O_3}$ (grey area), a_{FeTiO_3} (black) and $a_{Al_2O_3}$ (red)

4 Summary

In summary we have performed a comprehensive study of the energetic stability, electronic and magnetic properties in the Fe_2O_3-$FeTiO_3$ system. The GGA+U results show that the most favorable compensation mechanism is through a disproportionation into Fe^{2+}, Fe^{3+} in the contact layer [13], giving first theoretical evidence for the *lamellar magnetism hypothesis* [7]. The uncompensated moments generated at the interface due to the charge mismatch show a strong tendency to couple antiferromagnetically to the next hematite layer thus leading to ferrimagnetism in the system. The formation of Fe^{2+} at the interface leads to impurity levels in the band gap that are pinned at the Fermi level. Their position with respect to the conduction band depends on the distribution of the impurities and strain. One of the main findings is that ordered configurations such as hematite/ilmenite heterostructures are more stable than solid solutions, except for very high Ti concentrations. This trend can be reversed e.g. by using Al_2O_3(0001) as a substrate. We note, however, that the strong compressive strain generated by this substrate is energetically unfavorable compared to substrates with a larger lateral lattice constant [15]. The interface magnetism in this system is a further example how the polar discontinuity at oxide interfaces can lead to novel functional properties, thereby opening possibilities for electronics and spintronics applications.

Acknowledgements We acknowledge stimulating discussions with M. Winklhofer, W. Moritz, M. Lübbe, R. Harrison, S. McEnroe and R. Robinson, as well as support by the German Science Foundation (Pe883/4-1) and the European Science Foundation within the EuroMinSci programme. The calculations were performed at the supercomputer HLRBII at the Leibniz Rechenzentrum within project h0721.

References

1. A. Ohtomo, D. A. Muller, J. L. Grazul and H. Y. Hwang, Nature **419**, 378 (2002).
2. N. Reyren, S. Thiel, A. D. Caviglia, L. Fitting Kourkoutis, G. Hammerl, C. Richter, C. W. Schneider, T. Kopp, A.-S. Rüetschi, D. Jaccard, M. Gabay, D. A. Muller, J.-M. Triscone and J. Mannhart, Science **317**, 1196 (2007).
3. A. Brinkman, M. Huijben, M. van Zalk, J. Huijben, U. Zeitler, J. C. Maan, W. G. van der Wiel, G. Rijnders, D. H. A. Blank and H. Hilgenkamp, Nature Mater. **6**, 493 (2007).
4. R. Pentcheva and W. E. Pickett, J. Phys.: Cond. Mat. **22**, 043001 (2010).
5. Y. Ishikawa and S. Akimoto, J. Phys. Soc. Jpn. **12**, 1083 (1957).
6. S. A. McEnroe, K. Fabian, P. Robinson, C. Gaina, L. L. Brown, Elements **5**, 241 (2009).
7. P. Robinson, R. J. Harrison, S. A. McEnroe, and R. B. Hargraves, Nature **418**, 517 (2002).
8. T. Kasama, S. A. McEnroe, N. Ozaki, T. Kogure and A. Putnis, Earth Planet. Sci. Lett. **224**, 461 (2004).
9. S. A. Chambers, T. C. Droubay, C. M. Wang, K. M. Rosso, S. M. Heald, D. A. Schwartz, K. R. Kittilstved, and D. R. Gamelin, Materials Today **9**, 28 (2006).
10. Y. Takada, M. Nakanishi, T. Fujii and J. Takada, J. Magn. Magn. Matter. **310**, 2108 (2007).
11. H. Hojo, K. Fujita, K. Tanaka, and K. Hirao, Appl. Phys. Lett. **89**, 142503 (2006).
12. F. Zhou, S. Kotru, and R. K. Pandey, Thin Solid Films **408**, 33 (2002).

13. R. Pentcheva and H. Sadat Nabi, Phys. Rev. B **77**, 172405 (2008).
14. V. I. Anisimov, I. V. Solovyev, M. A. Korotin, M. T. Czyżyk, and G. A. Sawatzky, Phys. Rev. B **48**, 16929 (1993).
15. H. Sadat Nabi and R. Pentcheva, J. Appl. Phys. **105**, 053905 (2009).
16. P. Blaha, K. Schwarz, G. K. H. Madsen, D. Kvasnicka, and J. Luitz, WIEN2k: An Augmented Plane Wave+Local Orbitals Program for Calculating Crystal Properties (Techn. Universität Wien, Austria, 2001), ISBN 3-9501031-1-2.
17. J. P. Perdew, K. Burke, and M. Ernzerhof, Phys. Rev. Lett. **77**, 3865 (1996).
18. R. Pentcheva, F. Wagner, W. Moritz, and M. Scheffler, Structure, Energetics and Properties of $Fe_3O_4(001)$ from First Principles, *High Performance Computing in Science and Engineering*, p. 375 Springer Verlag, 2004.
19. R. Pentcheva, N. Mulakaluri, W. E. Pickett, H.-G. Kleinhenz, W. Moritz and M. Scheffler, Compensation Mechanisms and Functionality of Transition Metal Oxide Surfaces and Interfaces: A Density Functional Theory Study, *High Performance Computing in Science and Engineering*, p. 709 Springer Verlag, 2007.
20. F. Tran, J. Kunes, P. Novak, P. Blaha, and L. D. Marks, Comp. Phys. Comm. **179**, 784 (2008).
21. R. Pentcheva and W.E. Pickett, Phys. Rev. B **78**, 205106 (2008).
22. H. Sadat Nabi, R. Harrison and R. Pentcheva, Phys. Rev. B, submitted.
23. J. Velev, A. Bandyopadhyay, W. H. Butler and S. Sarker, Phys. Rev. B **71**, 205208 (2005).
24. Y. Takada, M. Nakanishi, T. Fujii, J. Takada, and Y. Muraoka, J. Appl. Phys. **104**, 033713 (2008).
25. E. Popova, H. Ndilimabaka, B. Warot-Fonrose, M. Bibes, N. Keller, B. Berini, F. Jomard, K. Bouzehouane, and Y. Dumont, Appl. Phys. A **93**, 669 (2008).

Evaluation of Magnetic Spectra Using the Irreducible Tensor Operator Approach

Jürgen Schnack and Roman Schnalle

Abstract In the current project magnetic properties of special magnetic molecules are described by quantum mechanical models. The aim is to determine all energy eigenvalues which enables us to evaluate all thermodynamic properties of these materials. This is achieved by a group theoretical decomposition of the Hilbert space and a subsequent numerical diagonalization of the remaining matrices using self-written programs with openMP parallelization. Besides our scientific results, which in part are already published, we also obtained results about the scaling behavior of our program code which will serve as a guide for future improvements.

1 Introduction

This contribution reports about a project on quantum magnetism. It deals with the theoretical description of physical properties of magnetic molecules as well as low-dimensional spin systems. A central aspect consists in solving the stationary Schrödinger equation which from a mathematical point of view constitutes an eigenvalue problem. The new aspect of our calculations is that we are able to combine the $SU(2)$ symmetry of the Heisenberg model with point group symmetries of the investigated molecules. This enables us to model spin systems with huge and unprecedented Hilbert space sizes since we are able to decompose the Hamilton matrix into many smaller matrices according to irreducible representations of the employed point group symmetries.

Jürgen Schnack
Universität Bielefeld, Fakultät für Physik, Postfach 100131, 33501 Bielefeld, Germany
e-mail: jschnack@uni-bielefeld.de

Roman Schnalle
Universität Osnabrück, Fachbereich Physik, 49069 Osnabrück, Germany
e-mail: rschnall@uos.de

S. Wagner et al. (eds.), *High Performance Computing in Science and Engineering, Garching/Munich 2009*, DOI 10.1007/978-3-642-13872-0_48,

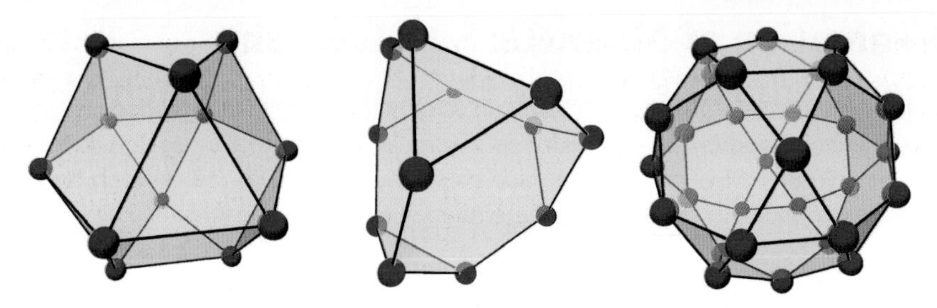

Fig. 1 Structure of the cuboctahedron (left), the truncated tetrahedron (middle), and the icosidodecahedron (right)

Regarding the magnetism this led us to a deeper understanding of the magnetic properties of frustrated antiferromagnetic spin systems as exemplary given by certain molecules [1–4]. Although our project runs for about a year only until now, first results could already be published [5–7], and a dissertation [8] could be finished.

Motivated by very recent specific heat measurements of a molecule-based triangular spin tube – a very attractive spin system [9, 10] – we currently evaluate the energy spectrum of this system for various sizes. Another system for which investigations have started recently is the molecular nanomagnet Mn_6Cr [11].

2 Goals of the Project

Molecular Magnetism is an interesting and ever growing new field in condensed matter physics. Topics of interest are for instance the behavior of Single Molecule Magnets (SMM) or the characteristics of frustrated antiferromagnetic clusters that can show extraordinary magnetization behavior or magnetocaloric properties.

The energy eigenvalue spectrum of many interesting molecular magnets can rather accurately be modeled by the Heisenberg model. Nevertheless, it is theoretically often inaccessible due to the prohibitive size of the underlying Hilbert space. Experimentally the system of interest might very well be accessible for instance by neutron scattering, EPR, specific heat or magnetization measurements. The interpretation of the experimental data thus suffers from numerical restrictions. A prominent example for this problem is given by the giant Keplerate molecules $\{Mo_{72}Fe_{30}\}$ [3, 4], $\{Mo_{72}Cr_{30}\}$ [12], and $\{Mo_{72}V_{30}\}$ [13], in which 30 paramagnetic ions occupy the vertices of an icosidodecahedron and interact by nearest-neighbor antiferromagnetic exchange, see r.h.s. of Fig. 1. These molecules are rather similar to the Kagome lattice antiferromagnet. Thus by studying them one can also gain further insight into the properties of the Kagome lattice antiferromagnet.

In this project we employ state of the art irreducible tensor operator techniques (ITO) together with point group symmetries in order to obtain an as small as possible decomposition of the Hilbert space into orthogonal subspaces. The possibility to use both kinds of symmetries simultaneously was recently developed in our group [8]. The related matrices are composed and diagonalized using self-written openMP programs. The obtained complete energy spectra allow an evaluation of all thermodynamic observables as functions both of temperature and applied magnetic field. This enables us to investigate interesting features especially of so-called frustrated antiferromagnetic molecules as there are: many low-lying singlet states below the first triplet state, magnetization plateaus, unusual magnetocaloric properties, or zero-temperature phase transitions [14–17].

3 Scientific Results from the HLRB Computations

3.1 Basic Background

The physics of many magnetic molecules can be well understood with the help of the isotropic Heisenberg model with nearest-neighbor coupling. The action of an external magnetic field is accounted for by an additional Zeeman term

$$\underset{\sim}{H}(\mathbf{B}) = -\sum_{i,j} J_{ij} \underset{\sim}{\mathbf{s}}(i) \cdot \underset{\sim}{\mathbf{s}}(j) + g\mu_B \underset{\sim}{\mathbf{S}} \cdot \mathbf{B} \,. \tag{1}$$

The sum reflects the exchange interaction between single spins given by spin operators $\underset{\sim}{\mathbf{s}}$ at sites i and j. A negative value of J_{ij} refers to an antiferromagnetic coupling. For the sake of simplicity we assume a common isotropic g–tensor. Then the Zeeman term couples the total spin operator $\underset{\sim}{\mathbf{S}} = \sum_{i=1}^{N} \underset{\sim}{\mathbf{s}}(i)$ to the external magnetic field $\mathbf{B}$. Since the commutation relations

$$\left[\underset{\sim}{H}(0), \underset{\sim}{\mathbf{S}}\right] = 0 \tag{2}$$

hold it is possible to find a common eigenbasis $\{|\nu\rangle\}$ of $\underset{\sim}{H}$, $\underset{\sim}{\mathbf{S}}^2$ and $\underset{\sim}{S}_z$. We denote the corresponding eigenvalues as E_ν, S_ν and M_ν.

Calculating the eigenvalues here corresponds to finding a matrix representation of the Hamiltonian (1) and diagonalizing it numerically. A very efficient and elegant way to evaluate the matrix elements is based on the use of irreducible tensor operators and the application of the Wigner-Eckart-theorem [18, 19]. Apart from its elegance it drastically reduces the dimensionality of the problem because it becomes possible to work directly within the subspace $\mathscr{H}(S, M = S)$ of the total Hilbert space $\mathscr{H}$ characterized by the total spin quantum number S and the total magnetic quantum number $M = S$.

3.2 General Point-Group Symmetries

The use of irreducible tensor operators for the calculation of the matrix elements of the Hamiltonian and as a result also of the energy spectrum is essential for the treatment of magnetic molecules containing many interacting paramagnetic ions. Nevertheless, it is sometimes necessary to further reduce the dimensionality of the problem, either because computational resources are limited or a labeling of certain energy levels becomes advantageous, e.g. for spectroscopic classification. Such a reduction can be done if the Hamiltonian remains invariant under certain permutations of spin centers. Often the spin-permutational symmetry of the Hamiltonian coincides with spatial symmetries of the molecule, i.e. point-group symmetries. Therefore the term point-group symmetry is used while one refers to the invariance of the Hamiltonian under permutations of spins.

Using point-group symmetries of the system results in a decomposition of the Hamilton matrix $\langle \alpha S M | \underset{\sim}{H} | \alpha' S M \rangle$ into irreducible representations $\Gamma^{(n)}(\mathscr{G})$ of a group $\mathscr{G}$ whose elements $\underset{\sim}{G}(R)$, i.e. the operators corresponding to the symmetry operations R, do commute with $\underset{\sim}{H}$. The symmetrized basis functions which span the irreducible representations n are found by the application of the projection operator $\mathscr{P}^{(n)}$ to the full set of basis vectors $|\alpha S M\rangle$ and subsequent orthonormalization. The over-complete set of basis states $\{|\alpha S M \Gamma^{(n)}\rangle\}$ spanning the n-th irreducible representation $\Gamma^{(n)}(\mathscr{G})$ is generated by [20]

$$\mathscr{P}^{(n)} |\alpha S M\rangle = \left(\frac{l_n}{h} \sum_R \left(\chi^{(n)}(R) \right)^* \underset{\sim}{G}(R) \right) |\alpha S M\rangle \,, \tag{3}$$

where l_n is the dimension of the irreducible representation $\Gamma^{(n)}$, h denotes the order of $\mathscr{G}$, and $\chi^{(n)}(R)$ is the character of the n-th irreducible representation of the symmetry operation R.

Eq. (3) contains the main challenge while creating symmetrized basis states. The action of the operators $\underset{\sim}{G}(R)$ on basis states of the form $|\alpha S M\rangle$ has to be known. Following Ref. [21] the action of $\underset{\sim}{G}(R)$ on states $|\alpha S M\rangle$ can directly be evaluated without expanding it into product states. Suppose there is a certain coupling scheme a in which spin operators $\underset{\sim}{\mathbf{s}}(i)$ are coupled to yield the total spin operator $\underset{\sim}{\mathbf{S}}$. Generally the action of operators $\underset{\sim}{G}(R)$ on states $|\alpha S M\rangle$ leads to a different coupling scheme b. Now those states which belong to the coupling scheme b have to be reconverted into a linear combination of states belonging to a. This is technically a rather involved calculation. To the best of our knowledge it has never been noted or even used that the conversion from any arbitrary (!) coupling scheme b into the desired coupling scheme a can be well automatized. Supposing that there is a state $|\alpha S M\rangle_a$ belonging to the coupling scheme a, the action of an arbitrary group element $\underset{\sim}{G}(R)$ on this state results in a state $|\alpha S M\rangle_b$ belonging to a different coupling scheme b. Then the above mentioned re-expression takes the following form

$$\underset{\sim}{G}(R) |\alpha S M\rangle_a = \sum_{\alpha'} |\alpha' S M\rangle_a \, {}_a\langle \alpha' S M | \alpha S M\rangle_b \,, \tag{4}$$

where a term like ${}_a\langle\alpha' \, SM|\beta \, SM\rangle_b$ is known as *general re-coupling coefficient*. The calculation of general re-coupling coefficients and the evaluation of Eq. (4) can be performed with the help of graph theoretical methods [22, 23]. An implementation of these methods within a computer program is a straightforward task (follow directions given in Refs. [22, 23]).

3.3 Result I: Complete Energy Spectra of Frustrated Magnetic Molecules

In the following two subsections we report on some very recent scientific results that could be obtained in 2009.

Frustrated antiferromagnetic spin systems are – simply speaking – characterized by a classical ground state where neighboring interacting spins cannot align their moments in an antiparallel way due to competing interactions with other spins. These systems possess rather unusual properties which they share with extended lattices like the Kagome or Pyrochlore antiferromagnets. Therefore, such finite size antiferromagnets offer the possibility to discover and understand properties that are shared by the infinitely extended lattices. An example is the discovery of localized independent magnons [14], which explain the unusual magnetization jump at the saturation field. Also the plateau at $1/3$ of the saturation magnetization that appears in systems built of corner sharing triangles could be more deeply investigated by looking at the cuboctahedron and the icosidodecahedron [24].

In the following we like to present numerical studies for two realistic spin systems that can be treated using irreducible tensor operator techniques and point-group symmetries, but not otherwise [5, 6, 8]. Both systems – cuboctahedron and truncated tetrahedron – consist of $N = 12$ spins of spin quantum number $s = 3/2$ (Hilbert space dimension 16,777,216) and are realized as antiferromagnetic molecules [1, 2], see Fig. 1 for the structures. Both systems as well as all frustrated magnetic systems in general cannot be treated by means of Quantum Monte Carlo, especially at low temperatures.

The l.h.s. of Fig. 2 shows the energy spectrum of the antiferromagnetic cuboctahedron with $s = 3/2$. This spectrum was obtained using only D_2 point-group symmetry which is already sufficient in order to obtain sufficiently small Hamilton matrices. In addition the r.h.s. of Fig. 2 demonstrates for the subspaces of total spin $S = 0, 1, 2, 3$ that a representation in the full O_h group can be achieved which yields level assignments according to the irreducible representations of this group.

A complete energy spectrum allows to calculate thermodynamic properties as functions of both temperature T and magnetic field B. We like to discuss this aspect for the truncated tetrahedron which was synthesized quite recently [2]. In principle this geometry permits two different exchange constants, one inside the triangles (J_1) and the other one between the triangles (J_2), compare Fig. 1. A practical sym-

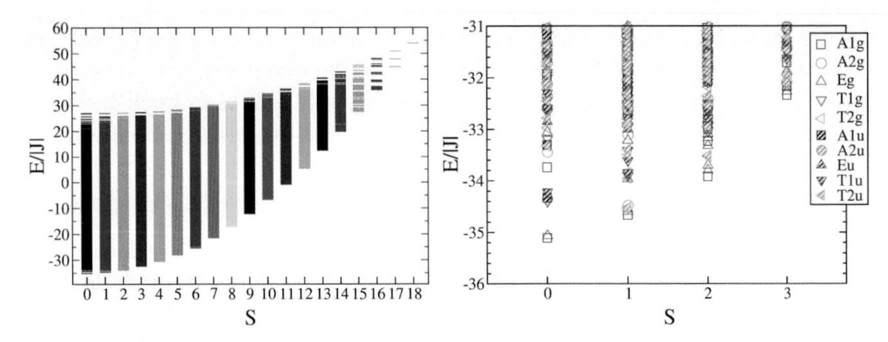

Fig. 2 L.h.s.: Complete energy spectrum of the antiferromagnetic cuboctahedron with $s = 3/2$. R.h.s.: Low-lying energy spectrum in subspaces of $S = 0, 1, 2, 3$. The symbols denote the irreducible representations of the O_h group

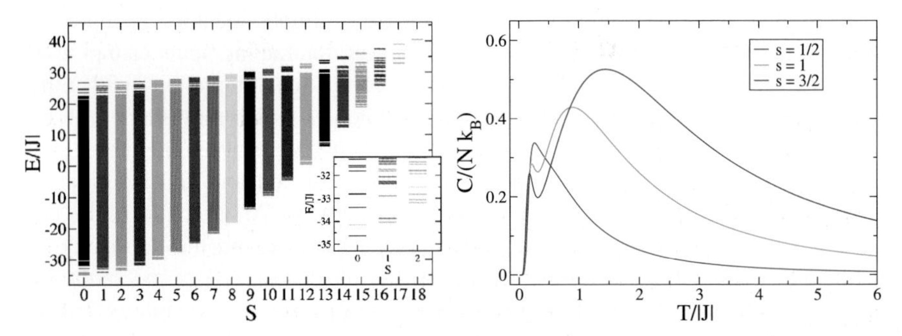

Fig. 3 L.h.s.: Complete energy spectrum of the antiferromagnetic truncated tetrahedron with $J_1 = J_2 = J$. The inset shows low-lying levels in subspaces with $S = 0, 1, 2$. R.h.s.: Specific heat $c(T, B = 0)$

metry for this molecule is for instance C_{2v}, whereas the full symmetry is T_d. The l.h.s. of Fig. 3 displays the complete energy spectrum for the case $J_1 = J_2 = J$. The inset magnifies the low-energy sector. As in the case of many other frustrated antiferromagnetic systems the spectrum exhibits more than one singlet below the first triplet [15].

On the r.h.s. of Fig. 3 we show the zero-field specific heat. The fine structure of the specific heat, which is especially pronounced for $s = 3/2$, results from the low-energy gap structure. The sharp peak is an outcome of the gap between the lowest singlet and the group of levels around the second singlet and the first two triplets, the latter being highly degenerate (both nine-fold including M-degeneracy). This unusual degeneracy of the lowest triplets is also the origin of the quick rise and subsequent flat behavior of the susceptibility in the case of $s = 3/2$.

In connection with the truncated tetrahedron it might be interesting to realize the technical progress. The truncated tetrahedron with $s = 1/2$ was investigated in 1992 [25]. The dimension of its Hilbert space is 4,096, whereas the dimension for $s = 3/2$ is 16,777,216.

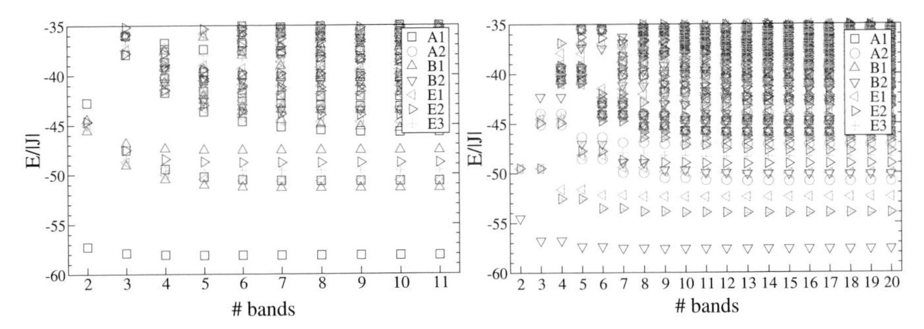

Fig. 4 Energy levels of an a antiferromagnetic spin ring $N = 8$ with $s = 5/2$ as a function of the number of occupied rotational-bands used for diagonalization in subspaces $S = 0$ (l.h.s.) and $S = 1$ (r.h.s.). The states are labeled according to irreducible representations of D_8

3.4 Result II: Approximate Energy Spectra of Magnetic Molecules

The previous subsection demonstrates that numerical exact diagonalization in connection with irreducible tensor operators is a powerful tool to investigate thermodynamic properties of large magnetic molecules. Nevertheless, sometimes the use of total-spin and point-group symmetries is not sufficient to obtain small enough matrices. For such cases we suggest an approximate diagonalization. The approximation is partially based on perturbation theory arguments. First ideas along this line were already suggested in Ref. [26]. We generalized and largely extended this idea [5, 7, 8].

The approximation consists in the diagonalization of the full Hamiltonian within a reduced set of basis states. Due to the observation that the low-lying eigenstates of bipartite antiferromagnetic spin systems can be well approximated by so-called rotational band states [27] these states are chosen for the approximate diagonalization.

In the following we discuss the properties of the proposed approximate diagonalization for the example of an antiferromagnetic spin ring of $N = 8$ spins with $s = 5/2$. Fig. 4 shows the convergence of the energy levels. In order to label the levels the full symmetry group D_8 of an octagon was used. One clearly sees that the convergence within the $S = 0$ subspace is fast and smooth (looking almost exponential). In subspaces of $S = 1$ and $S = 2$ (not shown) the convergence is also fast, but when only few bands are incorporated sharp steps can be observed.

As expected the approximate diagonalization scheme based on rotational bands yields good results for bipartite, i.e. unfrustrated antiferromagnetic spin systems. We now want to investigate how robust the approximate diagonalization is against the introduction of frustration. To this end we study a spin ring with $N = 8$ and $s = 5/2$ with antiferromagnetic nearest-neighbor coupling $J = J_{nn}$ and an additional antiferromagnetic next-nearest-neighbor-coupling J_{nnn} which acts frustrating. In a corresponding classical system the Néel state (up-down-up-down- ...) would no longer be the ground state, instead canting can occur. One can qualitatively say that with increasing J_{nnn}/J_{nn} also the frustration increases.

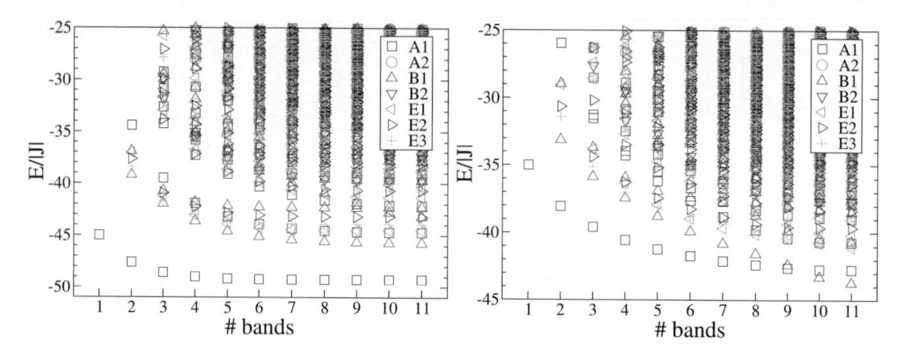

Fig. 5 Energy levels of an a antiferromagnetic spin ring $N = 8$ with $s = 5/2$ and additional next-nearest-neighbor coupling J_{nnn} as a function of the number of occupied rotational-bands used for diagonalization in subspaces $S = 0$ with $J_{nnn}/J_{nn} = 0.2$ (l.h.s.) and $J_{nnn}/J_{nn} = 0.4$ (r.h.s.). The states are labeled according to irreducible representations of D_8

Fig. 5 displays the effect of J_{nnn} in the subspace $\mathscr{H}(S = 0, M = S)$ for the same system that is discussed in Fig. 4 for $J_{nnn} = 0$. The energy gap between the ground state and the first excited state decreases with increasing frustration. Moreover, the convergence of the ground state as well as of excited states becomes slower. With $J_{nnn}/J_{nn} = 0.4$ the convergence is rather poor and the quantum mechanical ground state now belongs to the irreducible representation B_1 of the symmetry group D_8. This means that the true ground state is not the result of an adiabatic continuation from the ground state of the rotational-band model, which belongs to A_1. We just like to mention for the interested reader, that this change of the character of the ground state constitutes a so-called Quantum Phase Transition; in this case for the antiferromagnetic chain with next-nearest-neighbor exchange. Summarizing, if frustration is only small the approximate diagonalization still yields good results.

In a subsequent investigation we demonstrated that in contrast to bipartite systems for frustrated spin systems the rotational band states of all non-trivially different families of coplanar classical ground states, i.e. sublattice colorings, have to be taken into account in order to achieve a reliable convergence of energy levels [7, 8]. Numerically this is very demanding since a huge number of re-coupling coefficients has to be evaluated. Without parallel treatment this would simply be impossible.

As a single example we like to show a result again for the cuboctahedron. In Fig. 6 the energy difference between selected low-lying energy levels and the ground state of the system is displayed for a cuboctahedron $s = 1$ within $S = 0$ subspace. The difference is shown in dependence on the number of incorporated rotational bands. The last column refers in both sub-figures to the exact values, taken from a complete diagonalization. The dotted red line indicates the energy difference to the E_g-state.

In the first case (l.h.s.) only rotational-band states of one coloring are taken into account for the approximate diagonalization. In the second case (r.h.s.) linear combinations of states of all four colorings are used. The first noticeable difference is that when taking basis states of all colorings into account the low-lying levels start and remain in close proximity to their final (true) values. This is especially obvious

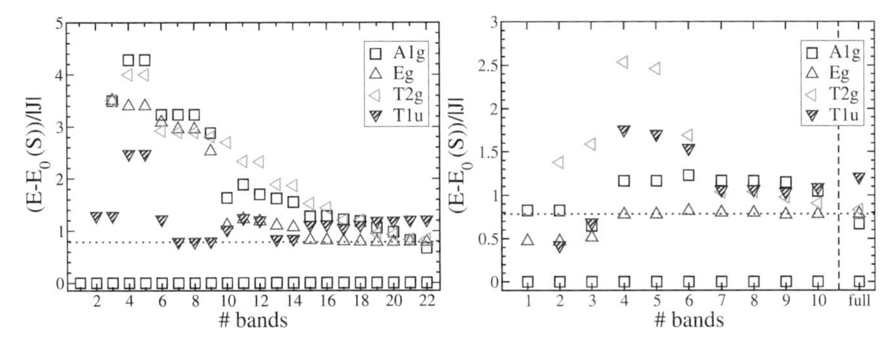

Fig. 6 Energy difference to the ground state of selected low-lying energy levels of the cuboctahedron $s = 1$ within $S = 0$ subspace in dependence on the number of incorporated rotational-bands. The reduced set of basis states is taken from rotational-band states of the $\Gamma(q = 0)$-family only (left) and from linear combinations of states of all four colorings (right). The classification is according to O_h point-group symmetry. The dotted red line refers to the exact difference to the E_g-state

when comparing the convergence of the lowest E_g-state in both sub-figures. The second difference is that when using linear combinations of states of all four colorings the convergence is smoother and more rapidly. This can be traced back to the fact that the different colorings contribute to low-lying states of different irreducible representations which in the other families would only be available as high-lying states. Therefore, it is advantageous to use fewer bands but basis states of all classical colorings for the approximate diagonalization.

As a summary one can state that although the necessary calculations are rather involved, especially for large frustrated systems, the approximate diagonalization can be a valuable method. Numerically the evaluation of re-coupling coefficients constitutes the strongest challenge, i.e. calculations are rather limited by runtime than by available memory. However, recent developments show that highly parallelized program code or public resource computing can help to overcome this barrier.

4 Computations Run on the HLRB II

The computations were performed on the SGI Altix 4700.

4.1 Technical and Algorithmic Methods

As discussed above the evaluation of energy eigenvalue spectra of magnetic molecules uses decompositions of the underlying Hilbert space into mutually orthogonal subspaces.

In addition we use Lanczos methods [28, 29] to evaluate low-lying energy levels for certain very large magnetic systems, see [16] for an earlier example.

4.2 Programming Techniques

The program using the ITO algorithm is written in C. The evaluation of the elements of the Hamilton matrix is parallelized using openMP directives. The diagonalization of the complete matrix is performed with the aid of MKL routines.

The Lanczos code is written in FORTRAN again using openMP directives.

5 Technical Results from the HLRB Computations

5.1 Scaling of Calculations Using the Irreducible Tensor Operator Technique

The calculations using the irreducible tensor operator technique can be divided into two completely independent parts. On the one hand a matrix representation of the Hamiltonian is set up with the help of the irreducible tensor operator approach. On the other hand this matrix or independent blocks of it are diagonalized, i.e. the eigenvalues and eigenvectors are determined numerically.

In the irreducible tensor operator approach the matrix elements are calculated using the Wigner-Eckart theorem and the decoupling procedure [8]. Since the decoupling procedure splits up reduced matrix elements of compound tensor operators, calculations are most easily performed by a recursive implementation of this particular step. In every single decoupling process a Wigner-9J symbol appears. Due to calculational reasons the Wigner-9J symbols are expressed in terms of Wigner-6J symbols. As a result, a large amount of Wigner symbols has to be calculated in order to set up the Hamilton matrix.

If point-group symmetries are used, a very decisive role concerning the computation time is played by the construction of symmetrized basis states. These functions appear as linear combinations of the initial basis states. The weight of the states that are included in these linear combinations is determined by general re-coupling coefficients. If a coupling scheme can be found that is invariant under all point-group operations, the number of states that contribute to a linear combination representing a symmetrized basis state is minimized.

The program has been parallelized by using openMP directives. There are three regions that are executed in parallel: the construction of symmetrized basis states, the calculation of the Hamilton matrix, and its diagonalization. The diagonalization has been performed using MKL routines by INTEL. These routines are already parallelized and cannot be further optimized. It turned out that on HLRB II the diagonalization routines only scale well up to the use of four cores. For this reason, the matrix diagonalization always contributes as a constant to the execution times in the following considerations since it has always been performed using four cores. The remaining part of the source code has shown a much better scaling behavior. Mostly, a considerable speed up has been achieved by calculating single matrix elements in

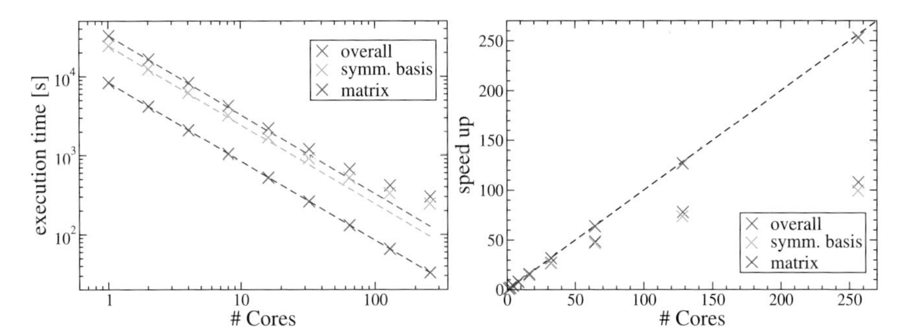

Fig. 7 Overall scaling (red crosses) and scaling of openMP parallelized regions (green and blue crosses) of exemplary calculations on the SGI Altix 4700 platform of HLRB II (l.h.s.). Speed up of the calculations (r.h.s.). The dashed lines refer to optimal scaling behavior

parallel, because the matrix elements only depend on the quantum numbers of the involved states and are completely independent from each other. In addition to the parallel set up of the matrices several steps within the construction of symmetrized basis states have also been parallelized.

Fig. 7 shows the performance of the parallelized program for the determination of the energy eigenvalues of a cuboctahedral system with $s = 3/2$ and applied O_h point-group symmetry in the subspace $\mathscr{H}(S = 0, M = 0, E_g)$.

One clearly sees that the overall performance is dominated by contributions from the construction of symmetrized basis states. Using an increasing number of cores, the speed up of the construction of symmetrized basis states already starts deviating from the optimal behavior if more than 16 cores are used. This deviation is not surprising since the construction of the symmetrized basis states can only be partially parallelized. In contrast, the scaling behavior within the region that is concerned with the calculation of the matrix elements is optimal up to the use of 256 cores. Resulting from the contributions of the parallelized and the remaining sequential regions, the overall speed up seems to be limited to a value of about 120. The above considerations show that the construction of symmetrized basis states plays an important role for extending the limits of numerical exact diagonalization.

5.2 *Scaling of our Lanczos Diagonalization Routine*

The scaling of our Lanczos diagonalization routine is very good [29]. The underlying reason is that all necessary matrix elements are evaluated *on the fly*. This has become possible since an analytical coding scheme could be developed for many-spin basis states [29].

Thanks to a cooperation with *ScaleMP Inc.* we could also compare the performance of the SGI Altix with up to date Xeon-based multiprocessor machines, in this case DELL M600 (Xeon X5460) and M610 (Xeon X5570) [30]. As can be deduced

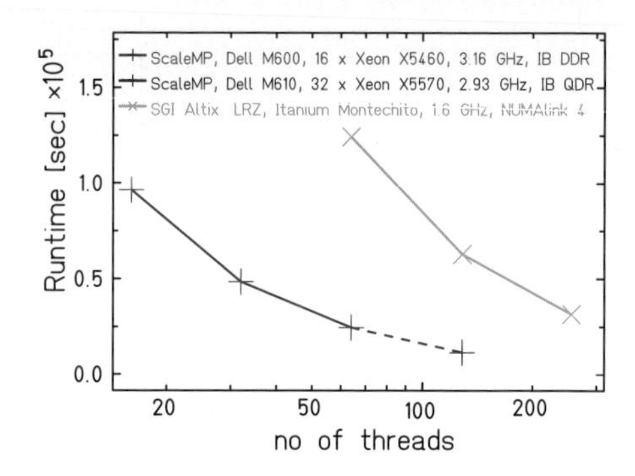

Fig. 8 Scaling of openMP parallelized Lanczos code: the SGI Altix is compared to new DELL machines using ScaleMP [30]. The dimension of the underlying Hilbert space, i.e. the length of the used Lanczos vectors is 601,080,390

from Fig. 8 the code scales very well on the SGI and on the DELL. The speed difference is merely due to progress in processor technology. It is rather interesting to note that the DELL M600 and M610 actually used commodity servers and interconnect, along with software from ScaleMP, rather than hard-wired electronics, to form the shared memory machine.

6 Technical and Numerical Issues for the Future

A very important technical issue for the future is the problem of complete matrix diagonalization. The routines provided by LAPACK or MKL for this purpose scale horribly. A usage of more than 8 cores turns out to be unfavorable and actually wasting CPU time. Since matrix diagonalization is at the core of quantum mechanics we would encourage the LRZ to support any effort to improve these methods. A promising starting point is given by investigations of Andreas Honecker (University of Göttingen) [31, 32].

Acknowledgements We thank the Leibniz-Rechenzentrum for providing the CPU time and technical support especially by Dr. Momme Allalen. We are looking forward to continuing a very fruitful relationship. This work was also supported within a Ph.D. program of the State of Lower Saxony in Osnabrück as well as by the Deutsche Forschungsgemeinschaft (FOR 945). We thank Christian Karlewski for producing Fig. 1 for us.

References

1. A. J. Blake, R. O. Gould, C. M. Grant, P. E. Y. Milne, S. Parsons, R. E. P. Winpenny, *Reactions of copper pyridonate complexes with hydrated lanthanoid nitrates*, J. Chem. Soc. Dalton Trans. (1997) 485
2. C. P. Pradeep, D.-L. Long, P. Kögerler, L. Cronin, *Controlled assembly and solution observation of a 2.6 nm polyoxometalate 'super' tetrahedron cluster:* $[KFe_{12}(OH)_{18}(\alpha\text{-}1,2,3\text{-}P_2W_{15}O_{56})_4]^{29-}$, Chem. Commun. (2007) 4254

3. A. Müller, S. Sarkar, S. Q. N. Shah, H. Bögge, M. Schmidtmann, S. Sarkar, P. Kögerler, B. Hauptfleisch, A. Trautwein, V. Schünemann, *Archimedean synthesis and magic numbers: "Sizing" giant molybdenum-oxide-based molecular spheres of the Keplerate type*, Angew. Chem., Int. Ed. **38** (1999) 3238
4. A. Müller, M. Luban, C. Schröder, R. Modler, P. Kögerler, M. Axenovich, J. Schnack, P. C. Canfield, S. Bud'ko, N. Harrison, *Classical and quantum magnetism in giant Keplerate magnetic molecules*, Chem. Phys. Chem. **2** (2001) 517
5. R. Schnalle, J. Schnack, *Numerically exact and approximate determination of energy eigenvalues for antiferromagnetic molecules using irreducible tensor operators and general point-group symmetries*, Phys. Rev. B **79** (2009) 104419
6. J. Schnack, R. Schnalle, *Frustration effects in antiferromagnetic molecules: The cuboctahedron*, Polyhedron **28** (2009) 1620
7. R. Schnalle, A. Läuchli, J. Schnack, *Approximate eigenvalue determination of geometrically frustrated magnetic molecules*, Condens. Matter Phys. **12** (2009) 331
8. R. Schnalle, *Symmetry assisted exact and approximate determination of the energy spectra of magnetic molecules using irreducible tensor operators*, Ph.D. thesis, Osnabrück University (2009)
9. J. Schnack, H. Nojiri, P. Kögerler, G. J. T. Cooper, L. Cronin, *Magnetic characterization of the frustrated three-leg ladder compound* $[(CuCl_2tachH)_3Cl]Cl_2$, Phys. Rev. B **70** (2004) 174420
10. A. Lüscher, R. M. Noack, G. Misguich, V. N. Kotov, F. Mila, *Soliton binding and low-lying singlets in frustrated odd-legged* $S = 1/2$ *spin tubes*, Phys. Rev. B **70** (2004) 060405(R)
11. T. Glaser, M. Heidemeier, T. Weyhermüller, R. D. Hoffmann, H. Rupp, P. Müller, *Property-oriented rational design of single-molecule magnets: A* C_3*-symmetric* Mn_6Cr *complex based on three molecular building blocks with a spin ground state of* S_t*=21/2*, Angew. Chem.-Int. Edit. **45** (2006) 6033
12. A. M. Todea, A. Merca, H. Bögge, J. van Slageren, M. Dressel, L. Engelhardt, M. Luban, T. Glaser, M. Henry, A. Müller, *Extending the* $\{(Mo)Mo_5\}_{12}M_{30}$ *Capsule Keplerate Sequence: A* $\{Cr_{30}\}$*–Cluster of* $S = 3/2$ *Metal Centers with a* $\{Na(H_2O)_{12}\}$ *Encapsulate*, Angew. Chem. Int. Ed. **46** (2007) 6106
13. A. Müller, A. M. Todea, J. van Slageren, M. Dressel, H. Bögge, M. Schmidtmann, M. Luban, L. Engelhardt, M. Rusu, *Triangular Geometrical and Magnetic Motifs Uniquely Linked on a Spherical Capsule Surface*, Angew. Chem., Int. Ed. **44** (2005) 3857
14. J. Schulenburg, A. Honecker, J. Schnack, J. Richter, H.-J. Schmidt, *Macroscopic magnetization jumps due to independent magnons in frustrated quantum spin lattices*, Phys. Rev. Lett. **88** (2002) 167207
15. R. Schmidt, J. Schnack, J. Richter, *Frustration effects in magnetic molecules*, J. Magn. Magn. Mater. **295** (2005) 164
16. C. Schröder, H.-J. Schmidt, J. Schnack, M. Luban, *Metamagnetic Phase Transition of the Antiferromagnetic Heisenberg Icosahedron*, Phys. Rev. Lett. **94** (2005) 207203
17. J. Schnack, R. Schmidt, J. Richter, *Enhanced magnetocaloric effect in frustrated magnetic molecules with icosahedral symmetry*, Phys. Rev. B **76** (2007) 054413
18. A. Bencini, D. Gatteschi, *Electron paramagnetic resonance of exchange coupled systems*, Springer, Berlin, Heidelberg (1990)
19. B. S. Tsukerblat, *Group theory in chemistry and spectroscopy: a simple guide to advanced usage*, Dover Publications, Mineola, New York, second edition (2006)
20. M. Tinkham, *Group theory and quantum mechanics*, Dover Publications, New York (2003)
21. O. Waldmann, *Symmetry and energy spectrum of high-nuclearity spin clusters*, Phys. Rev. B **61** (2000) 6138
22. V. Fack, S. N. Pitre, J. van der Jeugt, *Calculation of a general recoupling coefficient using graphical methods*, Comp. Phys. Comm. **101** (1997) 155
23. V. Fack, S. N. Pitre, J. van der Jeugt, *New efficient programs to calculate general recoupling coefficients, Part II: Evaluation of a summation formula*, Comp. Phys. Comm. **86** (1995) 105
24. C. Schröder, H. Nojiri, J. Schnack, P. Hage, M. Luban, P. Kögerler, *Competing Spin Phases in Geometrically Frustrated Magnetic Molecules*, Phys. Rev. Lett. **94** (2005) 017205

25. D. Coffey, S. A. Trugman, *Correlations for the $S = 1/2$ antiferromagnet on a truncated tetrahedron*, Phys. Rev. B **46** (1992) 12717
26. O. Waldmann, *E-band excitations in the magnetic Keplerate molecule Fe_{30}*, Phys. Rev. B **75** (2007) 012415
27. J. Schnack, M. Luban, *Rotational modes in molecular magnets with antiferromagnetic Heisenberg exchange*, Phys. Rev. B **63** (2000) 014418
28. C. Lanczos, *An iteration method for the solution of the eigenvalue problem of linear differential and integral operators*, J. Res. Nat. Bur. Stand. **45** (1950) 255
29. J. Schnack, P. Hage, H.-J. Schmidt, *Efficient implementation of the Lanczos method for magnetic systems*, J. Comput. Phys. **227** (2008) 4512
30. B. Galili, ScaleMP, Inc., private communication
31. A. Honecker, J. Schüle, in C. Bischof, M. Bücker, P. Gibbon, G. Joubert, T. Lippert, B. Mohr, F. Peters, editors, *Advances in Parallel Computing*, volume 15, IOS Press (2008)
32. A. Honecker, J. Schüle, in C. Bischof, M. Bücker, P. Gibbon, G. Joubert, T. Lippert, B. Mohr, F. Peters, editors, *Parallel Computing: Architectures, Algorithms and Applications*, volume 38 of *NIC Series*, Jülich (2007) pages 271–278, pages 271–278

Simulating Strongly Coupled Plasmas on High-Performance Computers

M. Bussmann, U. Schramm, P. Thirolf, and D. Habs

Abstract Simulating strongly coupled plasmas is a demanding computational task. When a plasma is strongly coupled, the mutual Coulomb energy between the plasma particles is much stronger than their kinetic energy. Such a system can undergo a phase transition into a state in which long-range ordering of the plasma constituents can be observed. In a realistic simulation of the plasma dynamics one has to compute the total mutual interaction of each particle with each other particle for particle numbers up to hundred thousand particles. To study the microscopic and macroscopic dynamics of the plasma on a long time scale one thus has to rely on the computational power which is only available at supercomputing centers such as the Leibniz Rechenzentrum.

1 Introduction to the Physics of Strongly Coupled Plasmas

In this paper we will introduce the reader to the basic physics of the interaction of charged particles in a non-relativistic plasma, where the interaction between the particles can be modeled as a sequence of static states determined by the Coulomb interaction. Based on the physical properties of the system we will show that at strong coupling all techniques presently available which avoid the quadratic scaling of the problem with respect to the particle number simulated must fail. We will finally present results on the stopping of highly charged ions in a strongly coupled plasma, which introduces a fast, efficient cooling scheme that can deliver highly charged ions for fundamental high precision experiments.

M. Bussmann · U. Schramm
Forschungszentrum Dresden-Rossendorf e.V., Bautzner Landstrasse 400, D-01328 Dresden, Germany
e-mail: m.bussmann@fzd.de

P. Thirolf · D. Habs
Ludwig-Maximilians-University, Am Coulombwall 1, D-85748 Garching, Germany

S. Wagner et al. (eds.), *High Performance Computing in Science and Engineering, Garching/Munich 2009*, DOI 10.1007/978-3-642-13872-0_49,

In the following we will concentrate on a one-component plasma (OCP), in which all plasma constituents are of the same particle species, mass and charge. Such plasmas are frequently studied in particle traps and storage rings [16]. In such machines the mutual Coulomb repulsion is counteracted by external confining forces, usually in such a way that the restoring confinement forces are harmonic [17]. Thus, the space charge of the plasma is compensated by the three-dimensional harmonic confinement.

In a strongly coupled one-component plasma of density n the mutual Coulomb interaction between particles of charge Q in a is given by

$$E_{\text{Coulomb}} = \frac{Q^2 e^2}{4\pi\varepsilon_0 a_{\text{WS}}} \tag{1}$$

where e is the elementary charge, ε_0 the dielectric constant in vacuum and the so-called Wigner-Seitz radius

$$a_{\text{WS}} = \left(\frac{4\pi n}{3}\right)^{-1/3} \tag{2}$$

denotes the average inter-particle distance [9]. The kinetic energy can be described by the plasma temperature T by

$$E_{\text{kin}} = k_{\text{B}} T \tag{3}$$

where k_{B} is Boltzmann's constant. When the mutual Coulomb interaction overcomes the kinetic energy of the particles, the repulsion between two particles due to their like-sign charge is so strong that the thermal velocity of the particles is too small to allow for two particles to come close to each other. This situation can be described by the dimensionless plasma parameter [20]

$$\Gamma = \frac{E_{\text{Coulomb}}}{E_{\text{kin}}} = \frac{Q^2 e^2}{4\pi\varepsilon_0 a_{\text{WS}} k_{\text{B}} T} \tag{4}$$

to be greater than unity. In a plasma where $\Gamma \gg 1$, the maximum impact parameter, given by the screening length [21]

$$\lambda_{\text{scr}} = \frac{1}{\omega_{\text{p}}} \sqrt{\frac{3k_{\text{B}}T}{m}} \tag{5}$$

can become much smaller than the Wigner-Seitz-Radius a_{WS}, the inter-particle distance, which means that close encounters of particles are greatly suppressed. The dynamics of these plasmas cannot be modeled analytically [21], since long-range correlations have to be taken into account, see the discussion in the following part. It is thus necessary to use simulations to analyze the plasma dynamics at large plasma parameters. However, resolving the different length scales involved is a very challenging demand, as discussed next.

2 Resolving Different Length Scales in the Simulation of Strongly Coupled OCPs

A realistic simulation of the particle dynamics must resolve length scales much smaller than the inter-particle distance to correctly model the spatial particle distribution. This means that the dynamics cannot be modeled using particle distribution functions as used in particle-in-cell methods [13] or hydrodynamic approaches which average over the the inter-particle distance, describing only the ensemble dynamics.

In a spatially confined strongly coupled plasma, the particles can undergo a phase transition from an unordered, fluid-like state to an ordered, crystal-like state [6]. In the special case of a three-dimensional, harmonically confined OCP the plasma will take an ellipsoidal form and long-range ordering of particles is observed. The particles will be ordered in an fcc lattice on concentric ellipsoidal shells [4].

This long-range order cannot be resolved if a mean-field approach such as the multipole method [3] is used to calculate the long-range interaction, it would instead wash out the long range order. Fast electrostatic methods such as Ewald summation [12] cannot be applied either, since they require periodic boundary conditions which are not present in the system of interest. Thus, one has to rely on the calculation of the complete mutual Coulomb interaction to realistically resolve the different length scales that govern the dynamics of the system.

3 Resolving Different Time Scales in the Simulation of Strongly Coupled OCPs

The simulation must resolve the particle dynamics not only on very different length-scales, but also on very different time scales. One time scale to be resolved is given by the plasma frequency

$$\omega_{\mathrm{p}} = \sqrt{\frac{Q^2 e^2 n}{\varepsilon_0 m}}. \tag{6}$$

When the plasma is locally perturbed, it will oscillate with the plasma frequency. Stable integration of the particle trajectories is only assured if collisions at the smallest possible impact parameter [21]

$$b_{\min} = \frac{Q^2 e^2}{4\pi\varepsilon_0 m\, v_{\mathrm{rel}}^2} \tag{7}$$

are well resolved. Here, v_{rel}^2 denotes the highest relative velocity observed in the simulation. In a cold, strongly coupled plasma this velocity can be approximated by the thermal velocity

$$v_{\mathrm{th}} = \sqrt{\frac{3k_{\mathrm{B}}T}{m}} \tag{8}$$

without sacrificing much in accuracy, because impacts at high relative velocity are scarce. The simulation time step must thus satisfy the criterion [13]

$$\Delta\tau_{\text{step}} \leq \sqrt{\frac{16\pi^3\varepsilon_0 m b_{\text{min}}^3}{Q^2 e^2}}. \tag{9}$$

There is however a much longer time scale that is related to the collective motion of the plasma. The OCP can oscillate in the confining harmonic potential. In experiment, the axial confinement Ψ_{axial} of the plasma is usually two to three orders of magnitude smaller than the radial confinement Ψ_{radial}. While the plasma density and thus the plasma frequency is mainly governed by the strong radial confinement, the plasma can oscillate as a whole in the weak axial harmonic potential. In extreme but realistic cases, the axial oscillation frequency can be on the order of a few hundred Hertz to few kHz, while the radial harmonic oscillation will be on the order of MHz and on the order of the plasma frequency. Thus, very different time-scales must be resolved in the simulation.

4 Numerical Techniques used in Integrating the Equation of Motion

The simulation thus computes the complete mutual Coulomb interaction of all particles with all other particles as well as the plasma oscillation in the confining potentials. With $\mathbf{r}_i = (x_i, y_i, z_i)$ denoting the coordinate vector of particle i, the equation of motion

$$m_i \frac{{}^2\mathbf{r}_i}{{}^2 t} = \left(\sum_{j\neq i}^{N} \frac{Q_i Q_j e^2}{4\pi\varepsilon_0 |\mathbf{r}_i - \mathbf{r}_j|^3} (\mathbf{r}_i - \mathbf{r}_j)\right) - Q_i e(\Psi_{\text{radial}}\, x_i, \Psi_{\text{radial}}\, y_i, \Psi_{\text{long}}\, z_i) \tag{10}$$

is integrated at each time step using a simple `Velocity Verlet` integration scheme [10]. This choice is governed by the need to maintain energy conservation over a large number of time steps. The time-reversible structure of the Velocity Verlet algorithm ensures for this. It however requires a small time step and thus is computationally expensive.

The most demanding part of the integration is the calculation of the mutual Coulomb force, which for N particles scales as N^2, since we have to take into account the interaction of each particle with each other particle. In our `C++` code implementation, we fully unroll any loop over particles, using the `valarray` library provided by the C++ standard [18]. This library makes writing the integration algorithm easy since it provides a natural syntax that allows to operate directly on large arrays instead on single array elements. For future implementations we will certainly look into expression templates [2] to optimize these calculations even further.

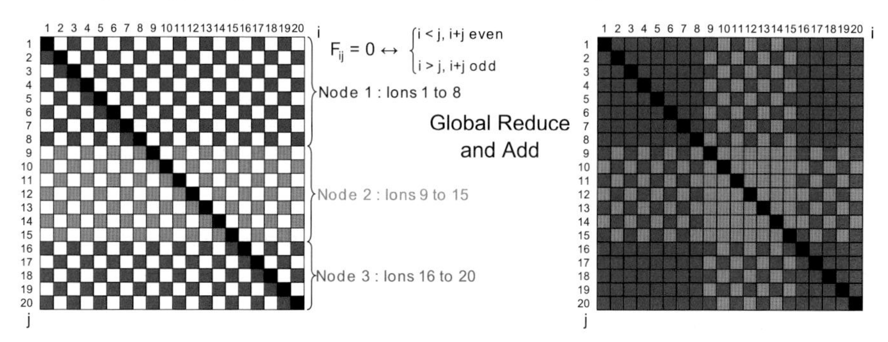

Fig. 1 The checkerboard algorithm is a force decomposition algorithm. On each node a number of particles is stored, denoted in the $N = 20$ example above by the colors blue, green and red, according to the computing power of the node. The force between the particles in this local subset and all particles in the simulation is computed at each time step. On each node the force between two particles denoted by their global particle indexes i and j is only computed if $i > j$ and $i + j$ even or $i < j$ and $i + j$ odd. Otherwise, the force equals zero. The computational steps computed on each node are proportional to the number of locally stored particles, thus load balancing is easy. On a homogeneous compute cluster the number of particles stored on each node is the same and each node computes the same number of instructions. After each node has finished calculating the entries in the force matrix, a checkerboard pattern denotes the non-zero elements of the force matrix as shown on the left side. With Newtons third law, the computation of all force terms is then simply done by an all to all global reduction operation, summing up the individual force contributions

Another bottleneck is found in the inter-node communication. We use the Message-Passing-Interface `MPI` [1] to distribute after each time step the new particle positions between computing nodes. On each note a subset of particles is stored and new particle positions must be interchanged after each time step to compute the interaction of the particles stored on one node with those on all other nodes. For the all to all communication of the particle positions after each time step we use the `checkerboard algorithm` [19], a force decomposition algorithm that makes use of Newtons third law, which states that the mutual actions of two particles upon each other are always equal. This effectively reduces the force calculation to a global reduce operation performing one addition. Load balancing for this algorithm is very simple. In the case of equal nodes we simply assign to each node the same number of particles. With the checkerboard algorithm each node will then process the same amount of data and load balancing is assured.

For large particle numbers N we observe that the time spent in calculating the force contributions scales worse than the all to all communication even on a standard Linux cluster. Therefore an increase in particle number usually increases the overall computational load of the cluster nodes to values up to 90 %. For our algorithm it is thus wise to use multi-core nodes with large local memory to allow for larger particle subsets to be computed locally on a single node and thus to reduce the ratio of data communication over computation time.

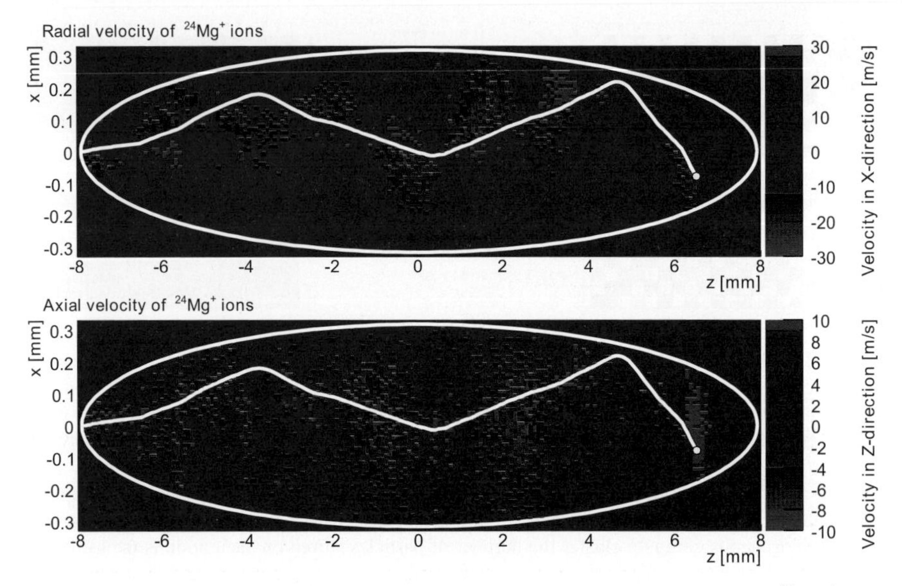

Fig. 2 Color-coded radial (top) and axial (bottom) velocities of the initially ultra-cold $^{24}Mg^{+}$ ions after the HCI has passed through the plasma and 18 μs after it has entered the plasma. The initial time step was chosen to be one nanosecond. The boundaries of the plasma ellipsoid are marked by a white ellipse, while the trajectory of the HCI going from left to right is marked by a white line with the HCI at the end depicted as a white circle. The velocities are integrated in y-direction, where blue shows positive velocities and red negative. A clear oscillation pattern is visible in both cases. The OCP reacts collectively to the HCI and oscillates in both the axial and radial direction

5 Recent Results on Stopping Highly Charged Ions in a Strongly Coupled Ion Plasma

Highly charged ions (HCI) can be used for precision mass measurements [5]. If the precision is high enough, fundamental physical experiments such as a precise determination of the fine structure constant or tests of quantum electrodynamics.

This however requires efficient cooling of the ions of interest to mK temperatures and below. Very efficient and fast cooling can be provided by laser cooling [15]. However, it requires an atomic cooling transition in the ion of interest and a laser system that can address this transition. A more general approach is to store the HCI together with another species of ions such as $^{24}Mg^{+}$ that can be directly laser cooled in a single trap [8]. In this approach, energy transfer from the HCI to the $^{24}Mg^{+}$ ions via close Coulomb collisions reduces the kinetic energy of the HCI until it thermalizes with the ultra-cold $^{24}Mg^{+}$ ions [7].

We have performed simulations of the stopping of a single HCI in a harmonically confined OCP of laser cooled $N = 10^5$ $^{24}Mg^{+}$ ions, which requires the computation of $N_{\text{int}} = N \times N/2 = 5 \times 10^9$ interactions per time step. The time step was chosen to be 1 ns, satisfying Eq. 9. To follow the energy transfer from the HCI to the OCP, cooling is switched off before the HCI enters the plasma.

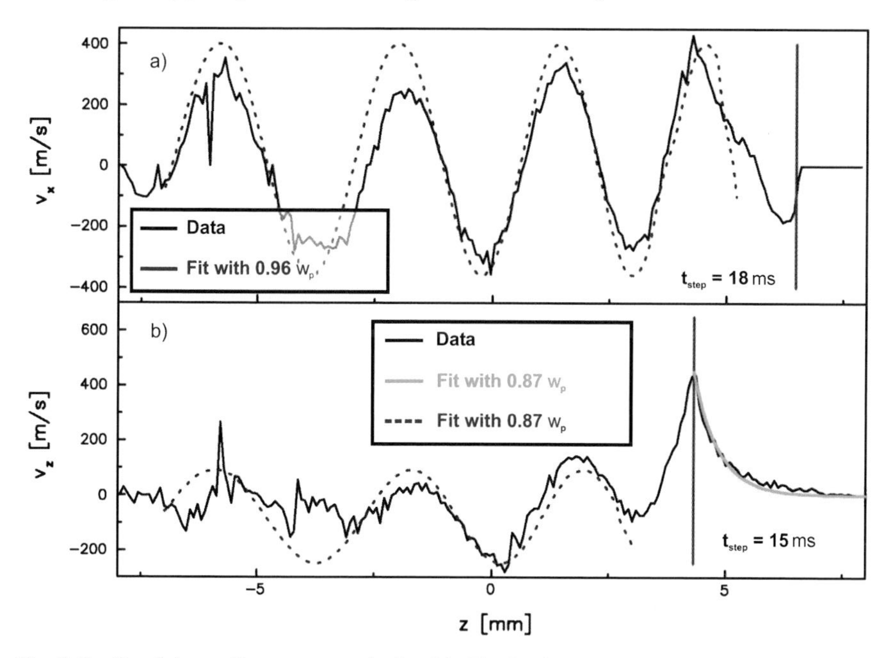

Fig. 3 Profile of the oscillatory pattern depicted in Fig. 3. The velocities of all particles below the HCI trajectory are summed. One observes a clear oscillatory pattern in both the radial (top) and axial (bottom) direction. The position of the HCI after 18 μs and 15 μs, respectively, after entering the plasma is marked by a red line. Note that in the case of the axial oscillation the plasma is also perturbed in front of the HCI, showing an exponential decrease in velocity. This decrease is fitted by a green line, while the oscillation is fitted by the blue curves. The fitting curves agree well with the simulation data

We present here results on the stopping dynamics of the HCI that show the collective response of the OCP to the passage of the HCI through the plasma. The OCP is characterized by its density of $n = 3.02 \times 10^{13}$ m^{-3} and its temperature of $T = 1$ mK, which gives a plasma parameter $\Gamma = 814$, an inter-particle spacing of $a_{\mathrm{WS}} = 20\ \mu$m and a plasma frequency of $\omega_{\mathrm{p}} = 2\pi \times 1.482$ MHz. The highly charged ion has a charge state of $Q_{\mathrm{HCI}} = 20$ and an initial kinetic energy of $E_{\mathrm{kin}} = 400$ meV, with its initial velocity vector pointing in the axial direction towards the center of the plasma ellipsoid. The confinement was chosen to be $\Psi_{\mathrm{radial}} = 2.5 \times 10^5$ Vm^{-2} and $\Psi_{\mathrm{axial}} = 2.5 \times 10^3$ Vm^{-2} respectively.

The collective response of the plasma to the passage of the OCP is depicted in Fig. 2. One observes both an axial and radial oscillation of the plasma trailing the HCI. The oscillation follows the HCI through the plasma, forming a wave. When summing those parts of the oscillation pattern that are located below the HCI trajectory along the x-direction, one gets a profile of the radial oscillation as seen in the top part of Fig. 3, while summing all axial velocities over the x-direction gives the axial oscillation pattern in the bottom part.

With the plasma density n determined from the simulation one can compute the plasma frequency ω_p following Eq. 6. This is the frequency with which the ultracold plasma locally oscillates when disturbed by the HCI [11]. This oscillation can be well resolved for individual particles, and the individual particle oscillation frequency is very well approximated by the plasma frequency derived from the ensemble density.

The radial oscillation pattern is caused by transverse oscillations of the plasma due to the charge of the HCI. The HCI pushes aside the ions in its vicinity due to the Coulomb repulsion. The plasma ions are moved out of their rest position and start to oscillate with the plasma frequency. The axial extend of those particles which have followed only one half period of their oscillation is related to the the velocity v_{HCI} of the HCI relative to the plasma ions. The faster v_{HCI}, the longer the wavelength of the resulting oscillation and one finds

$$\lambda_{\mathrm{radial}} = \frac{v_{\mathrm{HCI}}}{\omega_p} \equiv \lambda_{\mathrm{ad}}. \tag{11}$$

This is by definition equivalent to the adiabatic screening length [21] the HCI sees when it passes through the plasma. This length defines the distance up to which the plasma is locally perturbed by the HCI before a local changes in the polarization of the plasma screen the charge of the HCI. Using a simple fit

$$V_{\mathrm{fit,radial}}(z) = V_0 \sin\left(\frac{z+z_0}{\lambda_{\mathrm{radial}}(z)}\right) + V_{\mathrm{off}} \tag{12}$$

with the fit parameters V_0 as the amplitude, z_0 as the axial displacement and V_{off} as a constant velocity offset we find very good agreement by choosing $0.96\omega_p$ for the oscillation frequency of the particles.

In a similar fashion we can analyze the axial oscillation pattern. Here, the plasma in front of the HCI shows an exponentially decaying perturbation that has the length of the adiabatic screening length. When fitting

$$V_{\mathrm{fit,axial}}(z) = V_{0,\mathrm{axial}}\, e^{-\frac{z-z_{\mathrm{HCI}}}{\lambda_{\mathrm{ad}}}}, \tag{13}$$

to the exponential decay, using z_{HCI} as the axial position of the HCI given by the simulation and choosing $V_{0,\mathrm{axial}}$ as the only free fitting parameter, the decay is very well reproduced by the green fitting curve, this time choosing $0.87\omega_p$ as the oscillatory frequency. The axial oscillation trailing the HCI was fitted in the same manner as indicated in Eq. 12, again using $0.87\omega_p$.

The small discrepancy between the plasma frequency extracted from the simulation and that used in the fits can be reduced by calculating the plasma frequency only from the confining forces, which then results in fits using $1.00\omega_p$ for the radial oscillation and $0.91\omega_p$ for thc axial oscillation.

6 Conclusion and Outlook

We have presented simulation results showing the passage of a highly charged ion through an ultracold one-component plasma of ions. The collective response of the plasma is well described by local plasma oscillations following the trajectory of the highly-charged ion.

This result is very important not only for the understanding of the basic physics of strongly coupled plasmas. It can be directly applied to the cooling of HCI for precision mass measurements in Penning traps such as the MLLTrap setup in Munich [14]. Here, it is planned that the HCI is cooled by injecting it into a plasma of laser-cooled $^{24}Mg^{+}$ which are confined in a linear radio-frequency Paul trap. This setup has the advantage of fast and efficient cooling at virtually no degradation of the HCI charge due to charge exchange with the plasma ions [8].

The results presented here show that a great part of the energy transferred from the HCI to the plasma disperses over the complete bulk of the plasma. This means that the plasma stays intact and no local heating or disruption of the plasma is observed. From this we follow that it will be possible to constantly cool the plasma, keeping its temperature close to the mK temperatures required for precision measurements.

For future applications we want to include heating effects due to the alternating confining fields of the Paul trap and study the ejection of HCI from the plasma after they have thermalized with the plasma ions. Extraction of the HCI will be difficult, because at low temperatures the HCI will become strongly coupled to the plasma ions and external fields cannot access the HCI anymore.

On the simulation side we want to optimize the communication load testing various communication schemes suitable for all-to-all communication and optimize the numerics by using expression templates.

Acknowledgements We would like to thank the Linux-Cluster team at Leibniz Rechenzentrum for their invaluable support, without them this work would have not been possible. This work was partially funded by the LRZ project `pr26ka`.

References

1. http://www-unix.mcs.anl.gov/mpi/
2. http://www.oonumerics.org/blitz/
3. Barnes, J., Hut, P.: A hierarchical o(n log n) force-calculation algorithm. Nature **324**(6096), 446–449 (1986). DOI 10.1038/324446a0
4. Birkl, G., Kassner, S., Walther, H.: Multiple-shell structures of laser-cooled mg^{+} ions in a quadrupole storage ring. Nature **357**, 310–313 (1992). DOI 10.1038/357310a0
5. Blaum, K.: High-accuracy mass spectrometry with stored ions. Physics Reports **425**(1), 1–78 (2006). DOI 10.1016/j.physrep.2005.10.011
6. Bollinger, J.J., Wineland, D.J.: Strongly coupled nonneutral ion plasma. Physical Review Letters **53**(4), 348–351 (1984). DOI 10.1103/PhysRevLett.53.348

7. Bowe, P., Hornekær, L., Brodersen, C., Drewsen, M., Hangst, J.S., Schiffer, J.P.: Sympathetic crystallization of trapped ions. Physical Review Letters **82**(10), 2071–2074 (1999). DOI 10.1103/PhysRevLett.82.2071
8. Bussmann, M., Schramm, U., Habs, D., Kolhinen, V., Szerypo, J.: Stopping highly charged ions in a laser-cooled one component plasma of ions. International Journal of Mass Spectrometry **251**(2-3), 179–189 (2006). DOI 10.1016/j.ijms.2006.01.042
9. Bussmann, M., Schramm, U., Schätz, T., Habs, D.: Structural changes in bunched crystalline ion beams. Journal of Physics A **36**(22), 6119–6127 (2003). DOI 10.1088/0305-4470/36/22/339
10. Butcher, J.: Numerical Methods for Ordinary Differential Equations. John Wiley and Sons, Ltd. (2003)
11. Dubin, D., O'Neil, T.M.: Trapped nonneutral plasmas, liquids, and crystals (the thermal equilibrium states). Reviews of Modern Physics **71**(1), 87 (1999). DOI 10.1103/RevModPhys.71.87
12. Ewald, P.P.: Die berechnung optischer und elektrostatischer gitterpotentiale. Annalen der Physik **369**(3), 253–287 (1921). DOI 10.1002/andp.19213690304
13. Hockney, R., Eastwood, J.: Computer Simulation using Particles. Adam Hilger, Bristol and Philadelphia (1988)
14. Kolhinen, V.S., Bussmann, M., Gartzke, E., Habs, D., Neumayr, J.B., Schurmann, C., Szerypo, J., Thirolf, P.G.: Commissioning of the double penning trap system mlltrap. Nuclear Instruments and Methods in Physics Research Section A: Accelerators, Spectrometers, Detectors and Associated Equipment **600**(2), 391–397 (2009). DOI 10.1016/j.nima.2008.12.004
15. Metcalf, H., van der Straten, P.: Laser Cooling and Trapping. Springer Berlin Heidelberg New York (2002)
16. Schramm, U., Habs, D.: Crystalline ion beams. Progress in Particle and Nuclear Physics **53**(2), 583–677 (2004). DOI 10.1016/j.ppnp.2004.03.002
17. Schramm, U., Schätz, T., Habs, D.: Bunched crystalline ion beams. Physical Review Letters **87**(18), 184801 (2001). DOI 10.1103/PhysRevLett.87.184801
18. Stroustrup, B.: The C++ Programming Language (Special 3rd Edition). Addison-Wesley Professional (2000). URL `http://www.amazon.com/exec/obidos/redirect?tag=citeulike07-20&path=ASIN/0201700735`
19. Sutmann, G.: Classical molecular dynamics. NIC Series **10**, 1–46 (2002). URL `http://www.fz-juelich.de/nic-series/volume10/volume10.html`
20. Walther, H.: From a single ion to a mesoscopic system - crystallization of ions in paul traps. Physica Scripta **1995**(T59), 360–368 (1995). DOI 10.1088/0031-8949/1995/T59/049
21. Zwicknagel, G., Toepffer, C., Reinhard, P.G.: Stopping of heavy ions in plasmas at strong coupling. Physics Reports **309**(3), 117–208 (1999). DOI 10.1016/S0370-1573(98)00056-8

Material-Specific Investigations of Correlated Electron Systems

Arno P. Kampf, Marcus Kollar, Jan Kuneš, Michael Sentef, and Dieter Vollhardt

Abstract We present the results of numerical studies for selected materials with strongly correlated electrons using a combination of the local-density approximation and dynamical mean-field theory (DMFT). For the solution of the DMFT equations a continuous-time quantum Monte-Carlo algorithm was employed. All simulations were performed on the supercomputer HLRB II at the Leibniz Rechenzentrum in Munich. Specifically we have analyzed the pressure induced metal-insulator transitions in Fe_2O_3 and NiS_2, the charge susceptibility of the fluctuating-valence elemental metal Yb, and the spectral properties of a covalent band-insulator model which includes local electronic correlations.

1 Introduction

The basic concepts of solid-state physics explain the physical properties of numerous materials such as simple metals, semiconductors, and insulators. For materials with open d and f shells, where electrons occupy narrow orbitals, the additional understanding of the role of electron-electron interactions is crucial. In transition metals such as vanadium or iron, and also in their their oxides, electrons experience a strong Coulomb repulsion because of the spatial confinement in their respective orbitals. Such strongly interacting or "correlated" electrons cannot be described as embedded in a static mean field generated by the other electrons [14, 17, 22]. The d and f electrons have internal degrees of freedom (spin, charge, orbital moment) whose interplay leads to many remarkable ordering phenomena at low temperatures. As a consequence of the competition between different ordering phenomena, strongly correlated electron systems are very sensitive to small changes in their con-

Arno P. Kampf · Marcus Kollar · Jan Kuneš · Michael Sentef · Dieter Vollhardt
Theoretical Physics III, Center for Electronic Correlations and Magnetism, Institute of Physics, University of Augsburg, 86135 Augsburg, Germany
e-mail: arno.kampf@physik.uni-augsburg.de

S. Wagner et al. (eds.), *High Performance Computing in Science and Engineering, Garching/Munich 2009*, DOI 10.1007/978-3-642-13872-0_50,

trol parameters (temperature, pressure, doping, etc.), resulting in strongly nonlinear responses, and tendencies to phase separate or to form complex patterns in chemically inhomogeneous situations.

Understanding the metal-insulator transition (MIT) in strongly correlated electron systems has been one of the key topics in condensed matter theory since the 1930's. It was realized early on that the local Coulomb repulsion plays a decisive role in the physics of an important class of materials known as Mott insulators, examples of which are many transition-metal oxides. While the role of correlations due to the on-site Coulomb repulsion has been appreciated for over half a century only quite recently the theory progressed to the point where they can be treated quantitatively.

In the last two decades, a new approach to electronic lattice models, the dynamical mean-field theory (DMFT), has led to new analytical and numerical techniques to study correlated electronic systems [14, 22]. This theory – initiated by Metzner and Vollhardt in 1989 – is exact in the limit of infinite dimensions ($d = \infty$) [31]. In this limit, the lattice problem reduces to a single-impurity Anderson model with a self-consistency condition [13, 18]. After the initial studies of conceptually simple models, DMFT has advanced to models for real materials and only quite recently to purely numerical Hamiltonians obtained from bandstructure calculations, an approach which goes by the name LDA+DMFT.

2 Computational Method

2.1 The LDA+DMFT Approach

In the LDA+DMFT approach [16] the LDA band structure, represented by a one-particle Hamiltonian H^0_{LDA}, is supplemented with the local Coulomb repulsion, while the correction for double-counting the effect of the local interaction is absorbed in $\hat{H}^0_{\mathrm{LDA}}$,

$$\hat{H} = \hat{H}^0_{\mathrm{LDA}} + \sum_{m,m'} \sum_i U_{mm'} \hat{n}_{im} \hat{n}_{im'} . \tag{1}$$

Here, i denotes a lattice site; m and m' enumerate different orbitals on the same lattice site. In the present implementation we use an approximate form of the local interaction consisting of products of the occupation operators $\hat{n}_{im}$.

During the last ten years, DMFT has proved to be a successful approach for investigating Hamiltonians with local interactions as in Eq. (1) [14]. DMFT treats the local dynamics exactly while neglecting the non-local correlations. In this non-perturbative approach the lattice problem is mapped onto an effective single-site problem which has to be solved self-consistently together with the momentum-integrated Dyson equation connecting the self energy Σ and the Green function G at frequency ω:

$$G_{mm'}(\omega) = \frac{1}{V_B} \int d^3k \big[(\omega+\mu)\mathbf{1} - H^0_{\mathrm{LDA}}(\boldsymbol{k}) - \boldsymbol{\Sigma}(\omega) \big]^{-1}_{mm'} . \tag{2}$$

Here, $\mathbf{1}$ is the unit matrix, μ the chemical potential, and $\boldsymbol{\Sigma}(\omega)$ denotes the self-energy matrix which is nonzero only between the interacting orbitals. $[\ldots]^{-1}$ implies a matrix inversion in the space with orbital indices m, m'. The integration extends over the Brillouin zone with volume V_B.

2.2 QMC Method

The single-site impurity problem can be fomulated without explicit construction of the fermionic bath using the effective action formalism [13, 18]. The effect of the bath is represented by the frequency-dependent hybridization function $\Delta(\omega)$ given implicitly by

$$G_m^{-1}(\omega) = \omega + \mu - \varepsilon_m - \Delta_m(\omega) - \Sigma_m(\omega), \tag{3}$$

leading to the action

$$\begin{aligned} S[\psi^*, \psi\ , \Delta] = & \int_0^\beta d\tau \sum_m \psi_m^*(\tau) \left(\frac{\partial}{\partial \tau} - \mu + \varepsilon_m \right) \psi_m(\tau) \\ & + \sum_{m,m'} U_{mm'} \psi_m^*(\tau) \psi_m(\tau) \psi_{m'}^*(\tau) \psi_{m'}(\tau) \\ & + \int_0^\beta d\tau \int_0^\beta d\tau' \sum_m \psi_m^*(\tau) \Delta_m(\tau - \tau') \psi_m(\tau'), \end{aligned} \tag{4}$$

and the local Green function is obtained by a functional integral over the Grassmann variables ψ and ψ^*

$$G_m(\tau - \tau') = -\frac{1}{\mathscr{Z}} \int \mathscr{D}[\psi] \mathscr{D}[\psi^*] \psi_m(\tau) \psi_m^*(\tau') e^{S[\psi, \psi^*, \Delta]}. \tag{5}$$

Here, $\mathscr{Z} = \int \mathscr{D}[\psi]\mathscr{D}[\psi^*] \exp(S[\psi, \psi^*, \Delta])$ is the partition function. Such a fomulation in terms of an effective action in imaginary time is well suited for quantum Monte-Carlo (QMC) methods. In the present work we have used predominantly the continuous-time hybridization expansion QMC algorithm [43], which consists in expanding the exponential in (5) in powers of the hybridization function Δ and sampling the contributions using a QMC random walk, on which almost all of the required CPU time is spent upon during the iterative solution of the DMFT equations.

The computational effort of the QMC algorithm based on the expansion of the effective impurity action in the impurity-bath hybridization scales as the cube of the matrix size [15], which is determined by the mean value of the order of the hybridization expansion. The required order of the expansion scales linearly with inverse temperature, and even decreases upon an increase in the electron-electron interaction strength and is thus an excellent method for the regime of strong correlations. Concerning the cluster extension of DMFT, however, the effort of the full matrix code grows exponentially with the cluster size. Therefore the hybridization expansion solver is efficient for small clusters, but calculations for clusters larger than four sites are prohibitively costly.

From a numerical perspective the continuous-time QMC is perfectly suited for runs on the HLRB II. It is almost parallel by definition, and delayed update optimizations [2] are irrelevant to the code since rank-one updates, i.e. single spin flips, are not required. Most of the CPU time is spent on the calculation of a trace, for which the multiplication of small matrices is needed. These matrices fit in the cache and thus the code scales basically linearly with the number of CPUs used.

The linear scaling in theory is slightly affected in practice by a small overhead due to the thermalization sweeps, which need to be performed on each CPU separately, and more importantly by the data collection procedure of the ALPS tools [42], if more than around 300 CPUs are involved. Our typical jobs, however, are rather small scale jobs with 32 to 256 CPUs and CPU times of 10 to 100 CPU hours per DMFT iteration, with a typical number of 20 DMFT iterations. The performance measured for multiorbital single-site calculations is around 100 to 200 MFlops/core, while it fluctuates between 100 and 2000 MFlops/core in simulations of small clusters using the full matrix code. In a typical scaling test for a 2-site cluster DMFT the total CPU time varies between 27 CPU-h on 32 CPUs and 31 CPU-h on 128 CPUs. At the same time the performance slightly decreases from 128 MFlops/core on 32 CPUs to 120 MFlops/core on 128 CPUs. Thus for a typical job, the run time and the performance per core scale almost linearly with the number of CPUs, with a slight decrease of around ten percent upon increasing the number of CPUs by a factor of four.

The computational requirements of the continuous-time QMC are rather simple. It basically needs fast processors in parallel. Not a lot of memory is needed for the QMC and, since the random walks on different CPUs are independent of each other, fast network communications are not required either.

3 Results and Discussion

3.1 Pressure-Driven Metal-Insulator Transition in Hematite

An important example for of the MIT is the pressure driven transition seen in MnO [46], $BiFeO_3$ [12] or Fe_2O_3 [24], which is accompanied by a change of the local spin state (high spin (HS) to low spin (LS) transition). Understanding the pressure-driven HS-LS transition and its relationship to the MIT and structural and/or volume changes is in fact relevant to a broader class of oxides, often with geophysical implications.

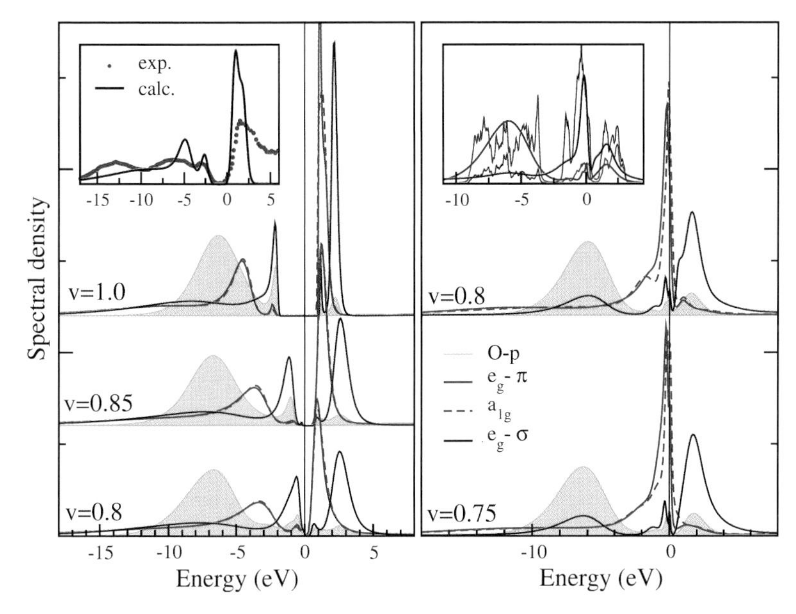

Fig. 1 Pressure evolution of the paramagnetic single-particle spectra (T=580 K). The high-spin solutions are shown in the left and the low-spin solutions in the right panel. All curves are normalized to unity, and v denotes the volume relative to the volume at ambient pressure. Left inset: comparison of experimental PES [27] and IPS [7] data (points) to the d-spectra (solid line) with 0.6 eV Gaussian broadening which is comparable to the experimental resolution. Right inset: comparison of the v=0.8 DMFT (smooth curves) and LDA spectra (jagged curves): total d-electron spectra (black), the total oxygen p-electron spectra (blue). Figure taken from Ref. [24]

We have studied the spin transition and the MIT in hematite (α-Fe_2O_3) under pressure using the LDA+DMFT approach [16] including the effects of temperature and magnetic long-range order (LRO). At ambient conditions, hematite has the corundum structure and is an antiferromagnetic (AFM) insulator with a Néel temperature T_N=956 K [38]. The iron ions have a formal Fe^{3+} valence with five d electrons giving rise to a local HS state. Photoemission spectroscopy (PES) classified hematite as a charge-transfer insulator [10, 27]. A charge gap of 2.0-2.7 eV was inferred from dc conductivity data [33]. Under pressure, a first-order phase transition is observed at approximately 50 GPa at which the specific volume decreases by almost 10% and the crystal symmetry is reduced (to the Rh_2O_3-II structure) [28, 35, 36]. The high-pressure phase is characterized by metallic conductivity and the absence of both magnetic LRO and the HS local moment [35]. Badro *et al.* showed that the structural transition actually precedes the electronic transition, which is, nevertheless, accompanied by a sizable reduction of the bond lengths [3].

The calculations started with the construction of Wannier based Hamiltonians from non-magnetic LDA bandstructures for various specfic volumes spanning the range of experimentally accessible pressures. Given the nominal d^5 configuration of the Fe-d shell in hematite, a transition between the high-spin $S = 5/2$ state and the low-spin $S = 1/2$ state is expected. Our LDA+DMFT calculations were first per-

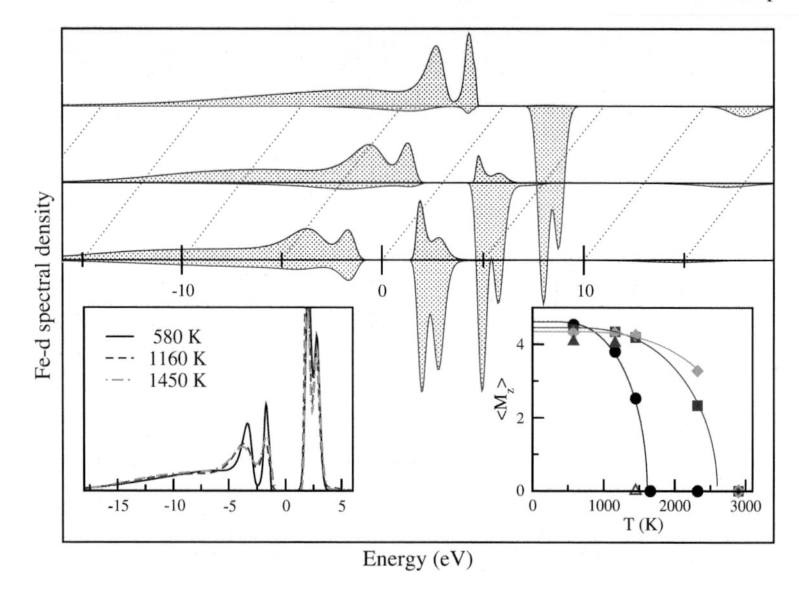

Fig. 2 Spin-polarized Fe-*d* spectra at ambient pressure for 580 K, 1160 K, and 1450 K (top to bottom). Left inset: the same spectra averaged over spin. Right inset: staggered magnetization vs. temperature for v=1.0 (circles; black), v=0.9 (squares; red), v=0.85 (diamonds; green), and v=0.8 (triangles; blue), where *v* denotes the volume relative to the volume at ambient pressure. For empty symbols only the low-spin solution exists. The lines represent mean-field fits. Figure taken from Ref. [24]

fomed in the paramagnetic phase and exhibited indeed such a HS-LS transition observed as a discontinuous drop of the expectation value $\langle S_z^2 \rangle$. The HS-LS transition is accompanied by the disappearance of the charge gap and a substantial change in the single-particle spectrum (Fig. 1). The comparison to iso-electronic MnO, which exhibits a similar HS-LS/insulator-metal transition, reveals a striking difference between the two materials. While the transition in MnO sets on before the charge gap is closed due to the pressure-induced increase in bandwidth and crystal-field splitting, the transition in hematite is instead characterized by a continuous closing of the gap followed by an abrupt spin transition. Comparing these two materials we have identified the microscopic mechanism behind the observed transitions [24].

A further issue addressed the question of long-range magnetic order. The occurence of magnetic order in strongly correlated oxides is important also in the broader context of first-principles electronic structure methods such as LDA. In several such materials a gap appears in the anti-ferromagnetic LDA solutions, while the non-magnetic solutions are metallic. This is sometimes interpreted as antiferromagnetism causing the opening of the gap. Our results explicitly showed that magnetic order has in fact only a marginal effect on the single-particle spectrum and the size of the charge gap does not change between the paramagnetic and the antiferromagnetic phase (see Fig. 2). We also clarified that the presence of antiferromagentic order has only a marginal effect on the HS-LS/insulator-metal transition [24].

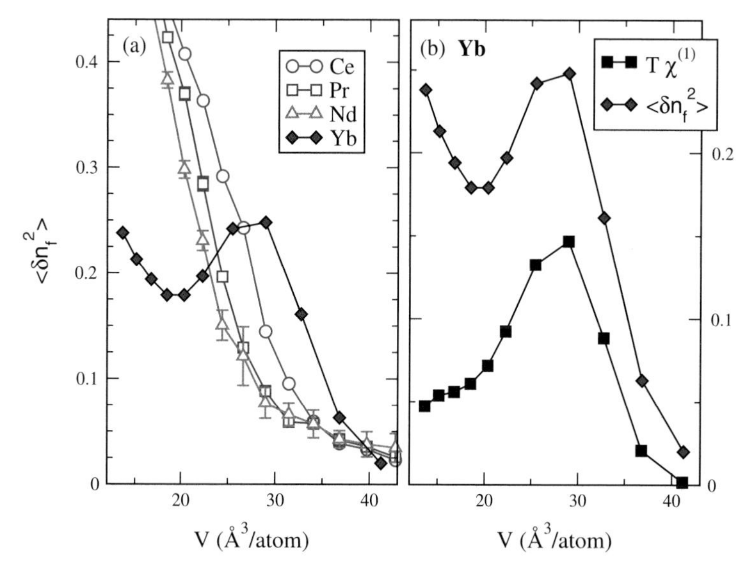

Fig. 3 (a) Charge fluctuations quantified by the mean-square deviation of the f occuancy in Yb, Ce, Pr, and Nd at 630 K. If not shown error bars are smaller than the symbol size. (b) Equal-time charge fluctuations (diamonds, the same as in left panel) compared to their imaginary-time average given by the product of temperature and the static susceptibility $T\chi^{(1)}$ (squares). Figure taken from Ref. [45]

3.2 Fluctuating Valence and Valence Transition of Yb Under Pressure

The valence state of rare-earth atoms in lanthanide compounds also affects their physical properties. The determination of the lanthanide valence from first principles and the description of the $4f$ electrons have been long-standing challenges due to the duality between their atomic character, with a strong local electron-electron interaction, and their itinerant character due to the lattice periodicity. Theories based on a two-species picture, which treat part of the $4f$ electrons as atomic and the rest as itinerant, succeeded in reproducing the trends across the lanthanide series for compounds with integer valence [39]. Nevertheless, besides being conceptually unsatisfactory, the two-species picture cannot describe transitions between different valence states as well as the heavy-fermion behavior of the charge carriers.

We have studied the valence transition in the elemental Yb metal under applied pressure. Yb and Eu in their elemental form behave quite differently from the other lanthanides. If we define the valence as the number of electrons participating in bonding, the majority of the lanthanide series is trivalent, however for Yb and Eu the 3+ and 2+ valence states are close to degeneracy with the 2+ state being more stable at ambient conditions [39]. This results in a number of anomalous properties, such as a larger molar volume compared to the general trend in the lanthanide series, and a lower bulk modulus [40]. The thermal-expansion coefficient of Yb is three times larger than for most other lanthanides [4].

In accord with the experimental data our calculations reveal a continuous decrease of the f-shell occupancy, which corresponds to a crossover from the f^{14} to the f^{13} local charge state as was verified by calculating the single-particle spectra [45]. The change of the f occupancy is connected to a transfer of electrons to spd-bands. The interesting question arises what the observable differences are between such self-doping in Yb and the rare-earth materials without valence-state degeneracy. The answer is shown in Fig. 3 where the mean-square deviations of the f-occupancy and the static local charge susceptibilites as a function of specific volume are compared for Ce, Nd, Pr, and Yb. The former three exhibit a monotonous increase of the charge fluctuations with pressure and a small charge susceptibility. The origin of the increasing charge fluctuations is the growing frequency of hopping processes involving the f electrons. However, these processes are energetically costly due to the on-site Coulomb repulsion and thus very short lived, which leads to the small charge susceptibility. The behavior of Yb is strikingly different. Besides the increase of charge fluctuations at the smallest volumes, which has essentially the same origin as in the other materials, the charge susceptibility vs. volume exhibits a local maximum approximately in the middle of the valence-transition region [45]. This feature is directly related to the quasi-degeneracy of the f^{13} and the f^{14} states. As a consequence, the charge fluctuations are energetically favorable and thus lead to a large susceptibility. The observation of the large charge susceptibility provides an explanation for the reported softness of Yb in the valence transition region.

We conclude that the observed behavior is a common and distiguishing feature of fluctuating valence systems characterized by the quasi-degeneracy of local charge states.

3.3 Metal-Insulator Transition in $NiS_{2-x}Se_x$

The $NiS_{2-x}Se_x$ series presents an important model system in which a metal-insulator transition can be controlled either by varying the Se content x, temperature T, or pressure P [26, 30, 32, 41, 44]. Despite a vast amount of available experimental data a satisfactory material-specific theory for the microscopic origin of the MIT in $NiS_{2-x}Se_x$ is still missing. Using the LDA+DMFT approach we have discovered an unexpected mechanism which controls the opening of the charge gap in $NiS_{2-x}Se_x$.

NiX_2 (X = S, Se) can be viewed as NiO with the O atom replaced by an X_2 dimer (Fig. 4), which accommodates two holes in its p_σ^* anti-bonding orbitals leading to an X_2^{-2} valence state. It was previously asserted that the empty p_σ^* bands do not play an active role in the physics of NiX_2. If true the analogy to NiO becomes complete in the sense that a charge-transfer gap forms between the ligand p-band and the upper Ni-d Hubbard band, whose position is sensitive to the strength of the on-site Coulomb repulsion U. Within such a scenario a stronger screening leading to smaller U in $NiSe_2$ can explain why NiS_2 is an insulator while $NiSe_2$ is a metal.

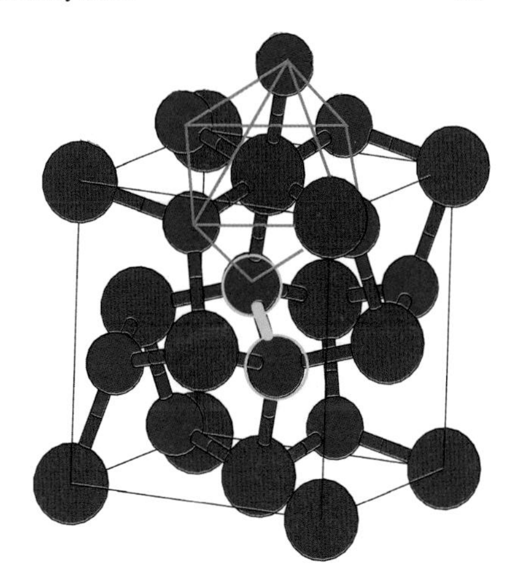

Fig. 4 The crystal structure of NiS_2: Ni atoms red (bright), S atoms blue (dark). The S-S dimer is highlighted in the center of the figure

However, our calculations revealed only a moderate difference in the Coulomb repulsion between NiS_2 and $NiSe_2$, and their respective groundstates turned out to be unaffected by variations of U. This is readily understood by the observation that it is not the upper Ni-d Hubbard band, but rather the X-p_σ^* band, which forms the bottom

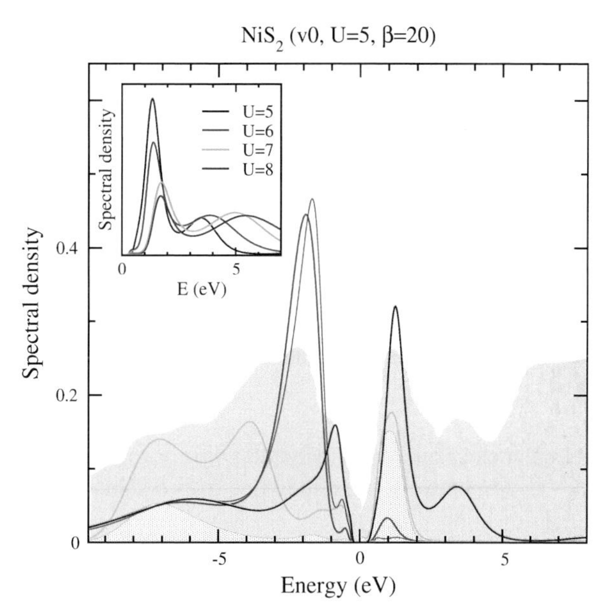

Fig. 5 The calculated (U=5 eV, T=580 K) orbitally resolved spectra of NiS_2 [Ni-d:e_g^σ (black) and t_{2g}=e_g^π (thick red) + a_{1g} (thin red); S-p: total (blue) and p_σ (blue-shaded)] compared to the experimental XPS+BIS data (gray, shaded) [9]. The inset shows the conduction band e_g^σ spectra for U=5, 6, 7, and 8 eV. Figure taken from Ref. [25]

of the conduction band (see Fig. 5). In the meantime this finding was corroborated by X-ray emission and absorption measurements at the K-edge of sulfur [25].

Our calculations provide the following picture. The different groundstates of NiS_2 and $NiSe_2$ are consequence of a larger bonding–anti-bonding splitting within the S-S dimer. The splitting in the Se-Se dimer is too small to open a gap. The closing of the gap in NiS_2 by the application of pressure is due to a broadening of the bands while the S-S bond-length related bonding–anti-bonding splitting remains unchanged since the S-S dimer behaves as a rigid object in a soft matrix.

3.4 Interaction Driven Insulator-to-Insulator Transition

While the Hubbard model has become a paradigm for the description of electronic correlations in metals and the metal-insulator transition [17], much less attention has so far been paid to electronic correlations in band insulators. E.g., modeling of Kondo insulators is far from trivial, and the only recently achieved progress in the topological classification of band insulators [21] demonstrates that our understanding of the insulating state is still incomplete.

Motivated by several investigations of the ionic Hubbard model [5, 6, 8, 11, 19, 20, 29, 34] we have analyzed a covalent insulator as a complementary example of a band insulator. As a covalent insulator we denote a band insulator with partially filled local orbitals. This definition implies that the band gap is a hybridization gap arising from a particular pattern of hopping integrals. It has been proposed that similar characteristics apply to materials such as FeSi, $FeSb_2$ or CoTiSb [23], some of which exhibit temperature dependent magnetic and transport properties reminiscent of Kondo insulators.

In our model study we use a simple particle-hole symmetric model at half-filling described by the Hamiltonian

$$H = \sum_{\boldsymbol{k}\sigma} \left(a^{\dagger}_{\boldsymbol{k}\sigma}, b^{\dagger}_{\boldsymbol{k}\sigma} \right) \mathbf{H}(\boldsymbol{k}) \begin{pmatrix} a_{\boldsymbol{k}\sigma} \\ b_{\boldsymbol{k}\sigma} \end{pmatrix} + U \sum_{i\alpha} n_{i\uparrow\alpha} n_{i\downarrow\alpha}, \quad (6)$$

$$\mathbf{H}(\boldsymbol{k}) = \begin{pmatrix} \varepsilon_{\boldsymbol{k}} & V \\ V & -\varepsilon_{\boldsymbol{k}} \end{pmatrix}, \quad (7)$$

with two semi-circular electronic bands of widths $4t$ ($t = 1$ in the following) and dispersions ε_k and $-\varepsilon_k$, respectively, corresponding to two sublattices coupled by the $\boldsymbol{k}$-independent hybridization V and a local electron-electron interaction of strength U. Here $n_{i\sigma\alpha} = \alpha^{\dagger}_{i\sigma}\alpha_{i\sigma}$ measures the number of electrons with spin $\sigma = \uparrow, \downarrow$ on site i of sublattice $\alpha = a, b$.

We use DMFT in conjunction with the recently developed continuous-time QMC [43] to study the evolution from the band insulator at small to the Mott insulator at large Coulomb interaction strength. The insulator-insulator transition is discontinuous at finite but low temperatures, with a region of interaction strengths where hysteretic behavior with two solutions of the DMFT equations is observed (see Fig. 6).

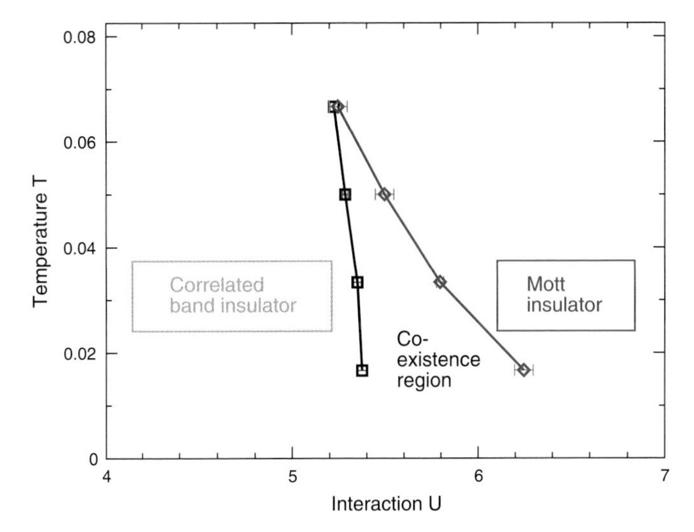

Fig. 6 The T-U phase diagram at fixed $V = 0.5$ shows the band and Mott insulating phases with a first-order phase transition upon increasing the interaction strength U below a critical temperature. In the coexistence region both band and Mott insulating solutions of the DMFT equations are found depending on the initial guess for the self-energy. For temperatures above the critical end point of the coexistence region, there is a regime where the spectral function has a single peak at the Fermi energy accompanied by broad Hubbard bands. All energies are given in units of $t = 1$. Figure taken from [37]

The behavior of charge and spin gaps upon increasing the interaction strength is shown in Fig. 7 [37]. Surprisingly we find that both gaps shrink with increasing Coulomb repulsion. This behavior is in contrast to the correlation-induced Mott insulator where the charge gap increases with increasing interaction strength. Furthermore, in the correlated insulator charge and spin gaps deviate from each other, and the spin gap is smaller than the charge gap. From the self-energy we extract a renormalization factor Z defined in the same way as the quasiparticle weight in a Fermi liquid. In the band insulator the Z factor describes the renormalization of the charge gap at moderate interaction strengths (see Fig. 7). Thus we obtain the remarkable finding that a concept from Fermi liquid theory can be applied to quantify correlation effects in a band insulator!

4 Conclusions

We have presented several examples for the application of dynamical mean-field theory to investigate quantitatively the properties of selected electronically correlated materials. As the main numerical tool a continuous-time quantum Monte-Carlo algorithm was used to solve the auxiliary impurity problem in the DMFT self-consistency cycle. The present method proved to be a powerful tool to calculate

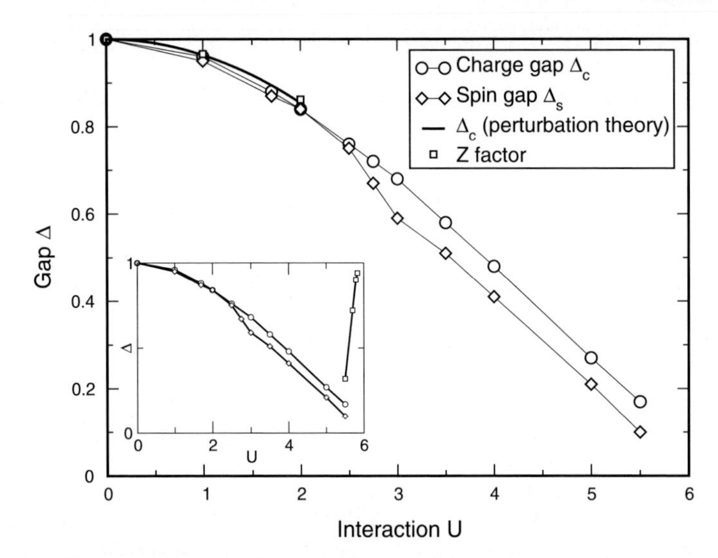

Fig. 7 Spin and charge gaps in the correlated band insulator as a function of U for $V = 0.5$ as determined from the spectral function and the spin susceptibility at $T = 1/30$, respectively. The thick solid line shows the charge gap obtained from second order perturbation theory. The squares represent the Z factor. Inset: Discontinuous change of spin and charge gaps at the band insulator to Mott insulator transition. In the Mott insulator the spin gap $\Delta_s = 0$. All energies are given in units of $t = 1$. Figure taken from [37]

spectral properties of materials with strong electronic correlations, as well as phase transitions near and above room temperature and at arbitrary pressure. The presented model analysis for a covalent band insulator with local Hubbard-type interaction provided new insights into the previously unexplored effects of correlations in band insulators.

Acknowledgements This work was conducted in project pr28je "Dynamical Mean-Field Theory for Electronically Correlated Materials" on the supercomputer HLRB II at the Leibniz-Rechenzentrum in Munich. We thank Philipp Werner for sharing his expertise regarding the continuous-time QMC algorithm. M.S. acknowledges support by the Studienstiftung des Deutschen Volkes. The research was performed within the Sonderforschungsbereich 484 funded by the Deutsch Forschungsgemeinschaft. The calculations made use of the ALPS library [1, 42].

References

1. Alet, F., Dayal, P., Grzesik, A., Honecker, A., Körner, M., Läuchli, A., Manmana, S.R., McCulloch, I.P., Michel, F., Noack, R.M., Schmid, G., Schollwöck, U., Stöckli, F., Todo, S., Trebst, S., Troyer, M., Werner, P., Wessel, S.: J. Phys. Soc. Jpn. Suppl. **74**, 30 (2005)
2. Alvarez, G., Summers, M.S., Maxwell, D.E., Eisenbach, M., Meredith, J.S., Larkin, J.M., Levesque, J., Maier, T.A., Kent, P.R.C., D'Azevedo, E.F., Schulthess, T.C.: New algorithm to enable 400+ tflop/s sustained performance in simulations of disorder effects in high-tc super-

conductors. Article No. 61 in *Proceedings of the 2008 ACM/IEEE conference on Supercomputing* pp. 1–10 (2008)
3. Badro, J., Fiquet, G., Struzhkin, V., Somayazulu, M., Mao, H.K., Shen, G., Bihan, T.L.: Phys. Rev. Lett. **89**, 205504 (2002)
4. Barson, F., Legvold, S., Spedding, F.H.: Thermal expansion of rare earth metals. Phys. Rev. **105**, 418–424 (1957)
5. Batista, C.D., Aligia, A.A.: Phys. Rev. Lett. **92**, 246405 (2004)
6. Byczuk, K., Sekania, M., Hofstetter, W., Kampf, A.P.: Phys. Rev. B **79**, 121103(R) (2009)
7. Ciccacci, F., Braicovich, L., Puppin, E., Vescovo, E.: Phys. Rev. B **44**, 10444 (1991)
8. Craco, L., Lombardo, P., Hayn, R., Japaridze, G.I., Müller-Hartmann, E.: Phys. Rev. B **78**, 075121 (2008)
9. Folkerts, W., Sawatzky, G.A., Haas, C., de Groot, R.A., Hillebrecht, F.U.: J. Phys. C **20**, 4135 (1987)
10. Fujimori, A., Saeki, M., Kimizuka, N., Taniguchi, M., Suga, S.: Phys. Rev. B **34**, 7318 (1986)
11. Garg, A., Krishnamurthy, H.R., Randeria, M.: Phys. Rev. Lett. **97**, 046403 (2006)
12. Gavriliuk, A.G., Struzhkin, V.V., Lyubutin, I.S., Ovchinnikov, S.G., Hu, M.Y., Chow, P.: Phys. Rev. B **77**, 155112 (2008)
13. Georges, A., Kotliar, G.: Phys. Rev. B **45**, 6479 (1992)
14. Georges, A., Kotliar, G., Krauth, W., Rozenberg, M.J.: Rev. Mod. Phys. **68**, 13 (1996)
15. Gull, E., Werner, P., Millis, A.J., Troyer, M.: Phys. Rev. B **76**, 235123 (2007)
16. Held, K., Nekrasov, I.A., Keller, G., Eyert, V., Blümer, N., McMahan, A.K., Scalettar, R.T., Pruschke, T., Anisimov, V.I., Vollhardt, D.: Phys. Status Solidi B **243**, 2599 (2006)
17. Imada, M., Fujimori, A., Tokura, Y.: Rev. Mod. Phys. **70**, 1039 (1998)
18. Jarrell, M.: Phys. Rev. Lett. **69**, 3410 (1992)
19. Kampf, A.P., Sekania, M., Japaridze, G.I., Brune, P.: J. Phys. Condens. Matter **15**, 5895 (2003)
20. Kancharla, S.S., Dagotto, E.: Phys. Rev. Lett. **98**, 016402 (2007)
21. Kane, C.L., Mele, E.J.: Phys. Rev. Lett. **95**, 146802 (2005)
22. Kotliar, G., Vollhardt, D.: Phys. Today **57**(3), 53 (2004)
23. Kuneš, J., Anisimov, V.I.: Phys. Rev. B **78**, 033109 (2008)
24. Kuneš, J., Korotin, D.M., Korotin, M.A., Anisimov, V.I., Werner, P.: Phys. Rev. Lett. **102**, 146402 (2009)
25. Kuneš, J. *et al.*: To be published
26. Kwizera, P., Dresselhaus, M.S., Adler, D.: Phys. Rev. B **21**, 2328 (1980)
27. Lad, R.J., Henrich, V.E.: Phys. Rev. B **39**, 13478 (1989)
28. Liu, H., Caldwell, W.A., Benedetti, L.R., Panero, W., Jeanloz, R.: Phys. Chem. Miner. **30**, 582 (2003)
29. Manmana, S.R., Meden, V., Noack, R.M., Schönhammer, K.: Phys. Rev. B **70**, 155115 (2004)
30. Matsuura, M., Hiraka, H., Yamada, K., Endoh, Y.: J. Phys. Soc. Jpn. **69**, 1503 (2000)
31. Metzner, W., Vollhardt, D.: Phys. Rev. Lett. **62**, 324 (1989)
32. Miyasaka, S., Takagi, H., Sekine, Y., Takahashi, H., Mori, N., Cava, R.J.: J. Phys. Soc. Jpn. **69**, 3166 (2000)
33. Mochizuki, S.: Phys. Status Solidi A **41**, 591 (1977)
34. Paris, N., Bouadim, K., Hebert, F., Batrouni, G.G., Scalettar, R.T.: Phys. Rev. Lett. **98**, 046403 (2007)
35. Pasternak, M., Rozenberg, G., Machavariani, G., Naaman, O., Taylor, R., Jeanloz, R.: Phys. Rev. Lett. **82**, 4663 (1999)
36. Rozenberg, G., Dubrovinsky, L., Pasternak, M., Naaman, O., Bihan, T.L., Ahuja, R.: Phys. Rev. B **65**, 064112 (2002)
37. Sentef, M., Kuneš, J., Werner, P., Kampf, A.P.: Correlations in a band insulator. Phys. Rev. B **80**, 155116 (2009)
38. Shull, C.G., Strauser, W.A., Wollan, E.O.: Phys. Rev. **83**, 333 (1951)
39. Strange, P., Svane, A., Temmerman, W.M., Szotek, Z., Winter, H.: Nature **399**, 756 (1999)
40. Takemura, K., Syassen, K.: Pressure-volume relations and polymorphism of europium and ytterbium to 30 gpa. J. Phys. F **15**, 543 (1985)

41. Takeshita, N., Takashima, S., Terakura, C., Nishikubo, H., Miyasaka, S., Nohara, M., Tokura, Y., Takagi, H.: ArXiv:0704.0591
42. Troyer, M., Ammon, B., Heeb, E.: Lect. Notes Comput. Sci. **1505**, 191 (1998)
43. Werner, P., Comanac, A., de' Medici, L., Troyer, M., Millis, A.J.: Phys. Rev. Lett. **97**, 076405 (2006)
44. Yao, X., Honig, J.M., Hogan, T., Kannewurf, C., Spałek, J.: Phys. Rev. B **54**, 17469 (1996)
45. Ylvisaker, E.R., Kuneš, J., McMahan, A.K., Pickett, W.E.: Phys. Rev. Lett. **102**, 246401 (2009)
46. Yoo, C.S., Maddox, B., Klepeis, J.H.P., Iota, V., Evans, W., McMahan, A., Hu, M.Y., Chow, P., Somayazulu, M., Häusermann, D., Pickett, W.E., Scalettar, R.T.: Phys. Rev. Lett. **94**, 115502 (2005)

Theoretical Study of Electron Transfer and Electron Transport Processes in Molecular Systems at Metal Substrates

Óscar Rubio-Pons, Rainer Härtle, Jingrui Li, and Michael Thoss

Abstract In this paper we present a study of electron transfer and electron transport processes in molecular systems at metal substrates. In particular, photoinduced electron transfer in molecules at metal surfaces and voltage driven electron transport in single-molecule junctions are investigated. The methodology is based on a combination of first-principle electronic structure methods to characterize the systems and dynamical basis-set as well as nonequilibrium Green's function methods to study electron transfer dynamics and transport properties, respectively. The results show the ultrafast character of electron transfer at molecule-metal interfaces and demonstrate the importance of electronic-vibrational coupling in single-molecule junctions. Furthermore, a mechanism for photoinduced switching of molecular junctions based on hydrogen translocation is discussed.

1 Introduction

Electron transfer (ET) processes in condensed phases are of fundamental importance in many areas of physics, chemistry and biology [3, 24, 29]. Important examples include ET processes in biological systems, such as photosynthetic reaction centers of bacteria and plants [29], as well as surface and interfacial ET reactions, such as injection of electrons from an excited state of a molecular chromophore

Óscar Rubio-Pons · Jingrui Li
Lehrstuhl für Theoretische Chemie, Technische Universität München, Lichtenbergstraße 4, 85747 Garching, Germany
e-mail: oscar.pons@ch.tum.de
e-mail: jingrui.li@ch.tum.de

Rainer Härtle · Michael Thoss
Institut für Theoretische Physik, Universität Erlangen-Nürnberg, Staudtstrasse 7 /B2, 91058 Erlangen, Germany
e-mail: Rainer.Haertle@physik.uni-erlangen.de
e-mail: Michael.Thoss@physik.uni-erlangen.de

S. Wagner et al. (eds.), *High Performance Computing in Science and Engineering, Garching/Munich 2009*, DOI 10.1007/978-3-642-13872-0_51,

into a metal or semiconductor substrate [11, 12, 23, 32, 36, 37]. This process at molecule-semiconductor interfaces is a key point in photonic energy conversion in nanocrystalline solar cells [17, 18]. ET processes at molecule-metal interfaces represent the fundamental step in the conductance of single-molecule junctions in the field of molecular electronics [9]. A detailed understanding of the basic mechanisms of ET in these systems is thus important also from the perspective of future technological applications.

In this paper, we present a theoretical study of ET in different molecule-metal systems. Specifically, we consider ultrafast photoinduced ET in molecular adsorbates at metal substrates and electron transport in molecular junctions, i.e. molecules which are chemically bound to metal electrodes. From the point of view of theory, electron transfer or electron transport at molecule-metal interfaces is particularly challenging and interesting because it requires a methodology that is able to describe properly a finite system - the molecular adsorbate- as well as an extended surface. Furthermore, purely static characterizations are not sufficient, because both photoinduced ET reactions and voltage driven electron transport processes typically involve significant nonequilibrium effects.

The paper is organized as follows. The two different processes are presented in Sects. 2 and 3, respectively. In each section we give a brief account on the theoretical methodology, the practical implementation and a discussion of results for representative systems. In addition, some computational details and information on the required computer resources are provided. Sec. 4 summarises and gives a conclusion.

2 Photoinduced Electron Transfer of Molecules at Surfaces

In this section, we consider photoinduced electron dynamics in molecules adsorbed at metal surfaces. After a review of the theory, we show results for benzonitrilethiolate adsorbed on the gold (Au) 111 surface depicted in Fig. 1 This system serves as a model for adsorbate-substrate interactions of aromatic molecules attached to noble metal surfaces.

2.1 *Electron-Transfer Hamiltonian*

To study ET dynamics in molecules adsorbed at metal substrates, we use an *ab-initio* based model for heterogeneous ET reactions, which is motivated by the Anderson-Newns model of chemisorption [7]. Within this model the Hamiltonian is represented in a basis of the following localized electronic states which are relevant for the photoreaction: the donor state of the ET process $|\psi_d\rangle$ (which, in the limit of vanishing coupling between chromophore and substrate, corresponds to the product of an electronically excited state of the chromophore and the equilibrium state of the metal substrate) and the (quasi)continuum of acceptor states of the ET reaction

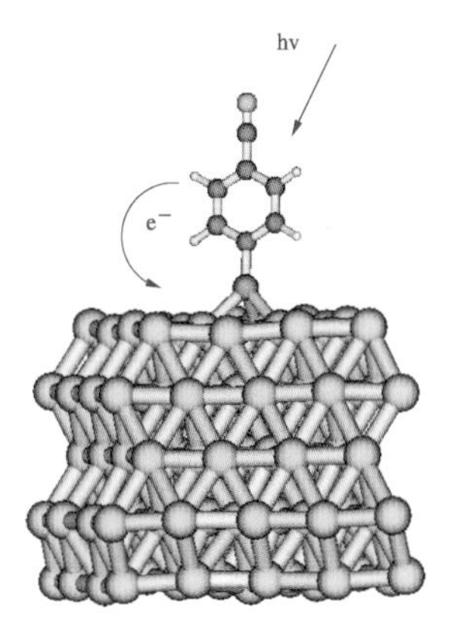

Fig. 1 Photoinduced electron transfer from a molecule into the Au(111) surface. After photoexcitation, an electron is injected into the conduction band of the metal

$|\psi_k\rangle$ (corresponding in the zero coupling limit to the product of the cationic state of the chromophore and an additional electron in the conduction band of the metal substrate). For simplicity, in the work presented here, we consider a system at low temperature and focus on the unoccupied states of the metal substrate. Thus, the Hamiltonian reads (we use atomic units, where $\hbar = 1$)

$$H = |\psi_d\rangle E_d \langle\psi_d| + \sum_k |\psi_k\rangle E_k \langle\psi_k| + \sum_k \left(|\psi_d\rangle V_{dk} \langle\psi_k| + |\psi_k\rangle V_{kd} \langle\psi_d|\right). \quad (1)$$

Here, E_d and E_k denote the energies of the electronic donor and acceptor states, respectively, while the non-diagonal elements, V_{dk}, characterize the donor-acceptor ET coupling. In this section we neglect the dependence of the energies and matrix elements on the nuclear degrees of freedom and consider the purely electronic dynamics.

The electronic injection dynamics is described by the time-dependent population of the donor state. Assuming a photoexcitation of the donor state by an ultrashort laser pulse, it is given by

$$P_d(t) = |\langle\psi_d|e^{-iHt}|\psi_d\rangle|^2.$$

2.2 Determination of Model Parameters

The parameters of the Hamiltonian (1) are determined from first-principle electronic structure calculations employing a partitioning method (for details see Ref. [25]). Briefly, the method employs a partitioning scheme based on density functional theory (DFT) calculations for the dye molecule under consideration with a finite gold cluster on the Au(111) surface.

The scheme for defining the localized donor and acceptor states $|\psi_d\rangle$ and $|\psi_k\rangle$ in the Hamiltonian (1) is based on three steps: (i) a partitioning of the Hilbert space in a donor and acceptor group using a localized basis, (ii) a partitioning of the Hamiltonian according to the donor-acceptor separation, and (iii) a separate diagonalization of the donor and acceptor blocks of the partitioned Hamiltonian. In the present paper, we work within the mean-field single-electron picture. Thus, we identify the effective Hamiltonian with the Fock (or Kohn-Sham) matrix and use the orbitals and orbital energies in the partitioning method.

The partitioning method discussed above is not limited to dye-metal systems with a finite metal cluster but can, in principle, be applied to a dye molecule adsorbed on an extended surface. One possibility is to employ a slab model and electronic structure calculations with periodic boundary conditions. Alternatively, the effect of an infinite cluster can also be described using surface Green's function techniques [38]. Within this method, the effect of the infinite substrate enters via the self-energy. In the application considered below, we have used a simpler approximate version of this method to mimic the effect of an extended surface. Thereby a constant imaginary part (σ = 1 eV) is added to the atomic orbital energies at the outermost layer of atoms of the gold cluster.

2.3 Electronic Structure Calculations

The electronic structure calculations performed to obtain the model parameters for the system benzonitrilethiolate adsorbed on the gold (111) surface can be summarized as follows.

To obtain the structure of benzonitrilethiolate on the Au(111) surface shown in Figure 1 we have proceeded in the following way: The Au(111) surface was modeled by a repeated slab geometry that consists of three or five atomic Au layers. We have used an in-plane 2×3 supercell. The adsorption position was determined to be at the hollow position according to previous work[14, 30]. The atomic positions of the last two layers from the cluster with five layers and only the last layer for the cluster of with three layers, have been frozen to keep the bulk structure of Au(111) during the structural relaxations. To avoid artificial interaction in the slab model, we imposed a distance of 60 Å between the slabs and a separation of 8.5 Å between the molecules. A representation of the cluster with five layers is depicted in Fig. 1.

All optimizations have been performed with the VASP program [1] employing DFT with the PBE functional and the projected augmented wave (PAW) [26] method. Results for surface structure relaxations have been checked for convergence with respect to the energy cutoff of the plane wave basis set with 415 eV and 600 eV employing a $6 \times 6 \times 1$ k-point mesh. All structural relaxations, except for calculation of the bulk parameters, have been performed at constant volume with an energy cutoff of 415 eV.

Based on the optimized surface structure of the extended molecule-adsorbate system, a finite cluster was selected to determine the parameters of the model Hamiltonian [25] as described in Sec.2.2. All electronic structure calculations for the finite cluster were performed with the TURBOMOLE package [2] employing DFT with the B3LYP functional [4] and SV(P) basis sets.

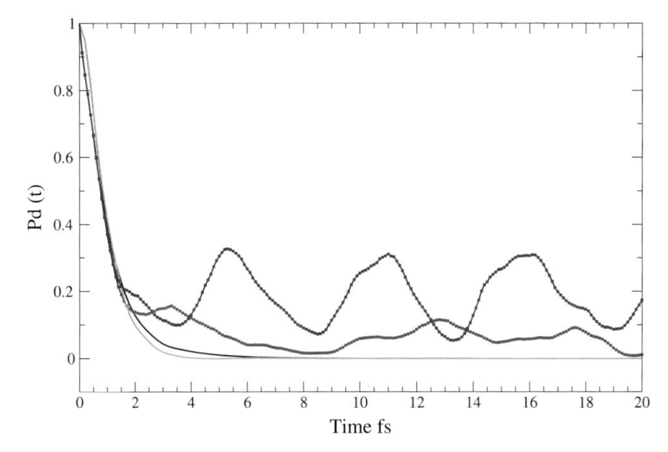

Fig. 2 Electron transfer dynamics of the system benzonitrilethiolate/Au(111). Shown are results for the population of the donor state after photoexcitation for a model with three (blue and green lines) and five (red and black lines) gold layers. The red and blue lines have been obtained for a finite gold cluster while the green and the black lines correspond to an infinite system obtained by including absorbing boundary conditions

2.4 Electron Dynamics in Benzonitrilethiolate at Au(111)

Figure 2 shows results of simulations for the ET dynamics of the system benzonitrilethiolate/Au(111). The different lines depict results for two different slab sizes (three and five layers). In addition to the results which have been obtained for a finite gold cluster (blue and red lines), results for an infinite cluster (green and black lines) modeled by absorbing boundary conditions ($\sigma = 1$ eV) are shown.

Overall, the dynamics of the population of the donor state exhibits an ultrafast decay corresponding to an electron injection time of about 2 fs. The results obtained for the finite gold clusters show oscillations and recurrences for longer times. These are caused by reflections of the wavepackets due to the finite size of the cluster and are quenched if absorbing boundary conditions, mimicking an extended system, are included. It is also interesting to note that already the smaller cluster with only three layers gives a rather good description of the ultrafast ET dynamics.

2.5 Computational Details

To conclude this section, we mention some computational details of the calculations. All electronic structure calculations were carried out at the HLRB-II employing parallelized versions of the program packages VASP [1] and TURBOMOLE [2]. The TURBOMOLE calculations were performed on 16 cores and typically took between 8 hours and one day on 16 cores, depending of the specific input parameters. The numerical performance of the parallel modules of TURBOMOLE has been discussed elsewhere [31, 33]

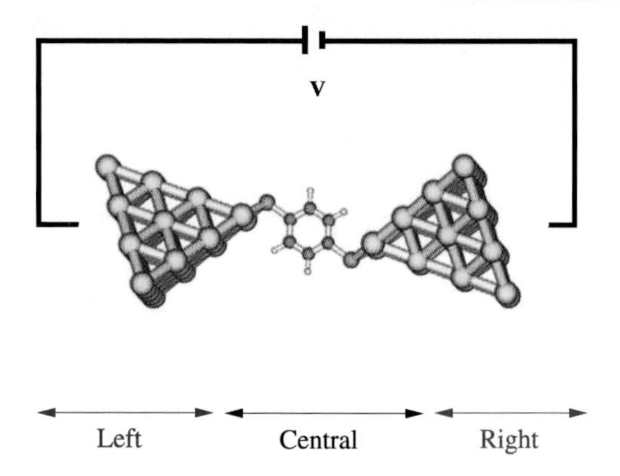

Fig. 3 Electron transport in a single-molecule junction. The central molecule is covalently bound via thiol groups to two gold electrodes, which are represented by gold clusters. In the theoretical treatment, the overall system is partitioned into left and right lead and the central molecule

The VASP calculations were performed on parallel nodes with 16, 32 and 64 cores, depending of the input parameters. The VASP code uses the discrete Fourier Transform and subroutines from Fastest Fourier Transform in The West (FFTW) libraries [22] together with Message Passing Interface (MPI) libraries [13] to optimize the performance. For our systems, calculations with VASP at HLRB-II show linear scaling up to 32 processors. With 64 processors VASP slightly deviates from the linear scaling. A typical VASP single point calculation took about 38 hours on 32 CPUs.

3 Electron Transport in Single-Molecule Junctions

As a second example of ET at metal surfaces, we consider electron transport in single-molecule junctions, *i.e.* molecules that are chemically bound to metal electrodes. The process is depicted schematically in Fig. 3. Specifically, we study electron transport in benzenedibutanethiolate (BDBT) and heptrienbutadienamine (HB). In the first system, we analyze the influence of the electron-vibrational (vibronic) coupling on the conductance properties of the junction. The second system is a prototype for a molecular switch.

3.1 Electron Transport Theory

To study vibrationally coupled electron transport in molecular junctions, we use a parametrized model Hamiltonian, which comprises an electronic and a vibrational part

$$H = H_{\mathrm{el}} + H_{\mathrm{vib}}.$$

The electronic part of the Hamiltonian reads:

$$H_{\mathrm{el}} = \sum_{i\in \mathrm{M}} \varepsilon_i c_i^\dagger c_i + \sum_{k\in\{\mathrm{L,R}\}} \varepsilon_k c_k^\dagger c_k + \sum_{k,i} (V_{ki} c_k^\dagger c_i + h.c.)$$

and describes the coupling of the electronic states of the molecular bridge, with energies ε_i, to electronic states in the leads, with energies ε_k, by interaction matrix elements V_{ki}. The vibrational degrees of freedom of the molecular bridge are described within the harmonic approximation by

$$H_{\mathrm{vib}} = \sum_{\alpha} \Omega_\alpha a_\alpha^\dagger a_\alpha + \sum_{\alpha, i\in \mathrm{M}} \lambda_{i\alpha}(a_\alpha + a_\alpha^\dagger) c_i^\dagger c_i.$$

Here, a_α is the annihilation operator for a vibration with frequency Ω_α, and $\lambda_{i\alpha}$ denote the corresponding vibronic coupling constants.

Transport properties, in particular the current-voltage characteristic $I(\Phi)$ of a molecular junction, are obtained employing a nonequilibrium Green's function (NEGF) method [15, 19, 20]. This method is based on the small polaron transformation $H \rightarrow \overline{H}$ [16] and provides a nonperturbative description of electron-vibrational coupling in molecular conductance. The transformed Hamiltonian $\overline{H} = \overline{H}_0 + \overline{V}$ comprises a part $\overline{H}_0$, which is diagonal in all degrees of freedom. The electronic-vibrational interaction is encoded in the renormalized molecule-lead coupling term $\overline{V} = \sum_{k,i}(V_{ki} X_i c_k^\dagger c_i + h.c.)$, which involves the shift operator

$$X_i = \exp\Big(\sum_{\alpha} (\lambda_{i\alpha}/\Omega_\alpha)(a_\alpha - a_\alpha^\dagger)\Big).$$

It is noted that for models with multiple electronic states $\overline{H}_0$ includes Hubbard-like terms $\sim \lambda_{i\alpha}\lambda_{j\alpha} c_i^\dagger c_i c_j^\dagger c_j$, $i \neq j$, which describe vibrationally mediated electron-electron coupling.

The calculation of the current, using the general expression

$$I = \frac{2e}{\hbar} \int \frac{\mathrm{d}E}{2\pi}\, tr\{\Sigma_{\mathrm{L}}^{<} \mathbf{G}^{>} - \Sigma_{\mathrm{L}}^{>} \mathbf{G}^{<}\},$$

requires the electronic Green's function matrix $\mathbf{G}(E)$ and the corresponding self-energy matrices $\Sigma_{\mathrm{L/R}}(E)$ that reflect the coupling of the molecule to the left (L) and the right (R) lead. Evaluating a Dyson equation, the electronic Green's function matrix $\mathbf{G}$ is determined by the self energy matrices $\Sigma_{\mathrm{L/R}}(E)$, with coefficients

$$\Sigma_{\mathrm{L/R},ij}(\tau,\tau') = \sum_{k\in\{\mathrm{L/R}\}} V_{ki}^{*} V_{kj} g_k(\tau,\tau') \langle \mathrm{T}_c X_j(\tau') X_i^\dagger(\tau)\rangle.$$

Here, g_k denotes the free Green's function of lead state k and T_c denotes time-ordering on the Keldysh contour. The calculation of the correlation functions $\langle \mathrm{T}_c X_j(\tau') X_i^\dagger(\tau)\rangle$ is based on a cumulant expansion in the dimensionless coupling parameters $\lambda_{i\alpha}/\Omega_\alpha$, which in turn involves the elements of the electronic Green's function matrix $\mathbf{G}$. Thus, $\mathbf{G}$ and $\Sigma_{\mathrm{L/R}}$ are obtained from a self-consistent scheme [15, 19]. Since the coupling of the molecule to the leads is treated within (self-consistent) second order perturbation theory, this NEGF approach comprises higher order processes, e.g. co-tunneling processes, which result in a broadening of the molecular levels. As a result, this scheme becomes exact for vanishing vibronic coupling.

3.2 Determination of Model Parameters

To determine the parameters in the Hamiltonian we employ first-principle electronic structure calculations based on a partitioning procedure similar to that described in Sec. 2. Thereby, the overall system was partitioned into a finite part that is treated explicitly in terms of electronic structure methods and an infinite part that is treated implicitly in terms of surface self-energies. The finite part comprises the central molecule and a left and a right cluster of gold atoms, representing the contacts of the molecule to the respective electrodes (cf. Fig. 3). A detailed description of the strategy used in the electronic structure calculations can be found in Ref. [5]. Briefly, a finite cluster of 30 gold atoms on each side of the molecule with a tip-like geometry was used to model the contacts (cf. Fig. 3). A partial geometry optimization of the molecule and the first layer of the gold clusters was employed to determine a realistic molecule-lead binding geometry.

The nuclear degrees of freedom of the molecular bridge were characterized based on a normal mode analysis of an extended molecule that includes seven gold atoms on each side of the molecule. The electronic-nuclear coupling constants $\lambda_{i\alpha}$ were obtained from the gradients of the electronic energies with respect to the normal coordinates.

All electronic structure calculations were performed with TURBOMOLE [2] employing DFT with the B3LYP functional and an SV(P) basis set (including ECP-60-MWB on the gold atoms).

3.3 Vibrational Nonequilibrium Effects in Electron Transport Through Benzenedibutanethiolate

As a first example, we consider electron transport through a benzenedibutanethiolate (BDBT) molecular junction [5, 20]. The butane-bridge in this system acts as an insulator between the delocalized electron systems of gold and benzene. As a result, the electronic coupling of the central phenyl ring to the leads is small and therefore electronic-vibrational coupling is significant.

First-principle electronic structure calculations for this system have shown that the transport properties of this junction can be well represented by a model that includes two electronic states, localized on the molecular bridge, and four vibrational modes [5]. The mechanisms of electron transport have been studied in detail recently [21]. As a representative example, Fig. 4 shows the current-voltage char-

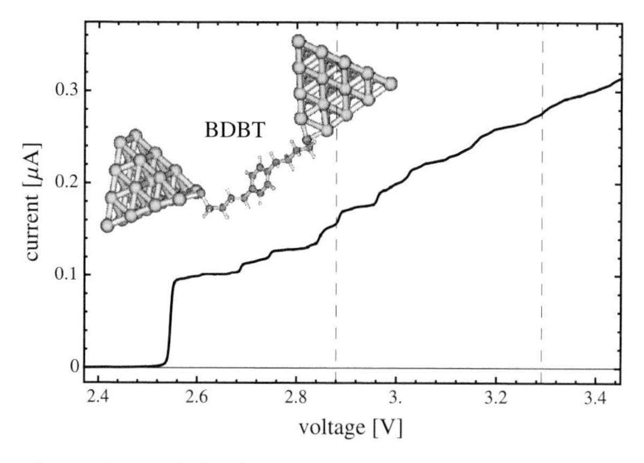

Fig. 4 Current-voltage characteristic of a benzenedibutanethiolate molecular junction (adapted from Ref. [20]). For details see text

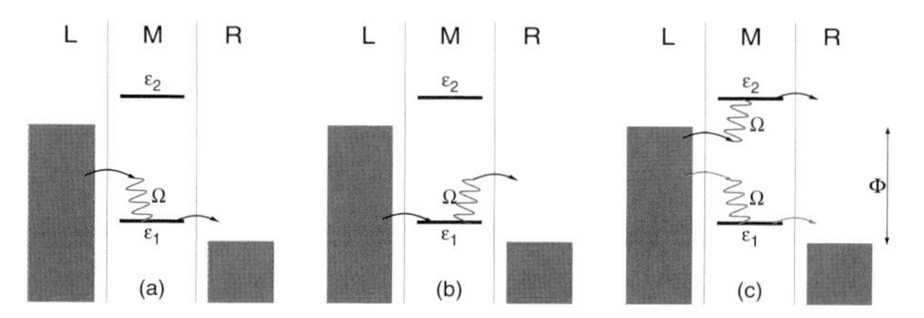

Fig. 5 Schematic representation of vibronic transport processes in a molecular junction involving emission (a) and absorption (b) of vibrational quanta upon transmission of an electron through a single electronic state, as well as emission and subsequent absorption associated with sequential electron transmission via two different electronic states (c)

acteristic for this molecular junction. The current-voltage characteristic exhibits a distinct resonance at $\Phi \approx 2.5$V, where the first electronic state enters the conductance window. For larger voltages, a multitude of weak resonances appear. These resonances correspond to inelastic electron transmission channels and indicate current induced excitation of the vibrational modes of the molecular bridge. The basic processes which are caused by electron-vibrational coupling have been analyzed in Ref. [20] and are depicted in Fig. 4.

The interaction of a transmitted electron with the vibrational degrees of freedom of the central molecule may result in excitation of vibrational quanta (emission, Fig. 5 (a)), or deexcitation (absorption, Fig. 5 (b)). In the stationary state, these processes can result in a pronounced heating of the molecule.

For molecules with multiple electronic states, as studied here, a number of additional vibronic processes have to be considered. These include, in particular, resonant absorption processes associated with the sequential transmission of two electrons (Fig. 5 (c)) as well as vibrationally induced effective electron-electron interaction. The latter results in a splitting of the electronic resonance associated with the second electronic state into two resonances (highlighted by vertical dashed lines in

Fig. 4). Resonant absorption processes, on the other hand, cause a wealth of additional vibronic structures in the current-voltage in Fig. 4 and facilitate vibrational cooling and thus, represent an important stabilization mechanism of molecular junctions. Since polyatomic molecules typically include numerous active vibrational modes and often exhibit multiple closely lying electronic states, these findings are expected to be of importance for most molecular junctions.

3.4 *Photoinduced Switching of a Molecular Junction via Hydrogen Transfer*

An important element for the design of molecular memory or logic devices is a molecular switch. A molecular junction may be used as a nanoswitch, if the molecule can exist in two or more differently conducting states that are sufficiently stable and can be reversibly transferred into each other. A variety of different optical mechanisms have been proposed to achieve reversible switching of molecular nanojunctions between different conductance states. Most mechanisms for optical switches considered so far are based on light-induced conformational changes, in particular isomerization reactions [8, 28], or ring-opening reactions [10, 27] of the molecular bridge. We have recently proposed an alternative mechanism for optical switching of a molecular junction between different conductance states based on photoinduced excited state hydrogen transfer. [6]

The basic idea of the mechanism is demonstrated in the prototypical reaction scheme depicted in Fig. 6. The molecule may exist in two different tautomers of planar structure, 3-hydroxy-2,4,6-heptatrien-butadienimine (in the following referred to as HB-enol) and 3-oxo-1,4,7-heptatrien-butadienamine (HB-keto) (for R=H). Both tautomers are connected via the translocation of a hydrogen atom, *i.e.* a hydrogen transfer reaction.

Fig. 7 shows the calculated current-voltage characteristic of the two tautomers chemically bound via thiol groups to gold electrodes. Although the two tautomers only differ by the position of a single hydrogen atom, their conductance properties are drastically different. The keto form exhibits, in particular for positive voltages, a rather low current. The enol form, on the other hand, shows a significant current and an almost Ohmic current-voltage characteristics. Thus, for a given voltage, the two

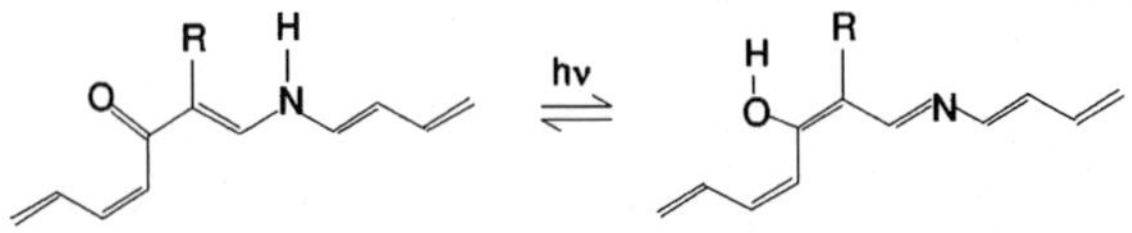

R H O N hν ⇌ H O R N

Fig. 6 Reaction scheme for photoinduced hydrogen transfer. The molecule may exist in two different tautomers, the hydroxylic (enol) form and the ketonic (keto) form. The side group *R* may be used as a molecular crane

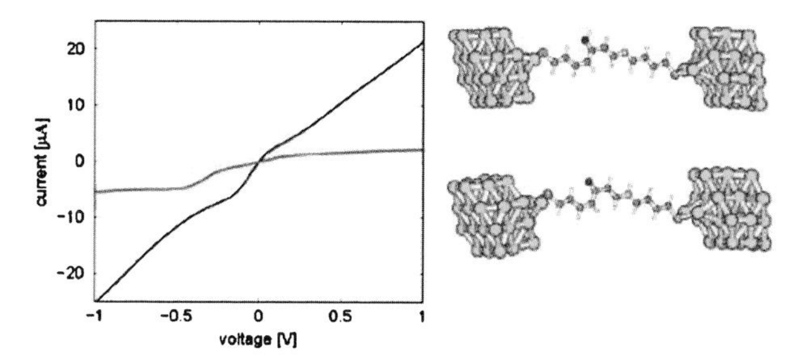

Fig. 7 Current-voltage characteristics for the heptrienbutadienamine molecular junction. The black line corresponds to the enol form and the red line depicts the result for the keto form. (Adapted from Ref. [6])

tautomers realize different conductance states of the molecular bridge corresponding to 'on' (enol form) and 'off' (keto form).

The very different conductance properties of the two tautomeric forms can be rationalized by the valence bond structures in Fig. 6. The enol tautomer corresponds to a fully conjugated system, while in the keto form the conjugation is broken at two sites (oxo and amino moieties). Typically, conjugated systems have a delocalized π-electron density that facilitates conduction.

The results presented above provide a proof of principle that tautomeric forms of a molecule, which are related to each other by the translocation of a hydrogen atom, may exhibit very different conductance behavior and thus realize different conductance states of the molecular bridge. A reversible change between the two tautomers would thus realize a molecular nanoswitch. The most promising mechanism to achieve efficient translocation of the hydrogen atom is photoinduced excited state hydrogen transfer [34, 35]. To this end, the molecule has to be functionalized by a hydrogen transferring unit (a molecular 'crane') at site R (cf. Fig. 6). The simulation of photoinduced switching in such molecular junctions is currently under study.

3.5 Computational Details

All electronic structure calculations have been performed with the TURBOMOLE package [2]. The TURBOMOLE calculations have been shown to scale almost linearly up to 16 processors for these systems with DFT methods.

A typical DFT calculation for the largest system of the molecular junction studied here took about 1 day and 20 hours to obtain the density converged to the required threshold on a 16 CPU machine at the HLRB-II supercomputer. The NEGF method has been used for the computation of the transport characteristics of a molecular junction.

The NEGF calculation requires to allocate a large quantity of memory to get converged results. The results presented in Sects. 3.3 and 3.4 are obtained with the following procedure: Parts of the transport calculation have been done at the Linux-

Cluster and lasted for two or three weeks of CPU time depending on the input parameters. This first step is needed to determine the important electronic degrees of freedom and all the necessary numerical parameters. However to obtain an accurate and a full description (including vibrational and electronic degrees of freedom) the HLRB-II machine is used for fully converged current-voltage characteristics free of numerical noise. We currently estimate a need of at least 32 GB per node to perform these calculations for the systems we are investigating here.

4 Concluding Remarks

In this paper we have presented a study of electron transfer in molecular systems adsorbed at metal substrates. Specifically, we have considered ultrafast photoinduced electron transfer in molecules at metal surfaces and voltage driven electron transport in molecular junctions, *i.e.* molecules which are chemically bound to metal electrodes.

In both cases the systems were characterized by first-principle electronic structure (DFT) calculations resulting in a parametrized model Hamiltonian. Based on the thus obtained model, the dynamics of the photoinduced ET process is described by the solution of the time-dependent Schrödinger equation employing basis-set methods. Transport properties of molecular junctions, on the other hand, are described using nonequilibrium Green's function methods.

The specific example considered in the first part of the paper shows the ultrafast timescale of electron transfer processes at metal surfaces. The systems considered in the second part demonstrate the importance of electronic-vibrational coupling in molecular junctions, which may influence the current-voltage characteristic significantly and result in substantial current-induced vibrational heating. Furthermore, we have discussed a model for a molecular switch based on photoinduced hydrogen translocation.

From the computational point of view, simulations for both type of surface ET processes investigated in this paper are challenging and require substantial computer resources as provided by the HLRB-II.

Future work will include the study of the influence of different anchor and bridge groups on ET dynamics, the simulation of electron transport in molecular junctions including the coupling to a laser field, as well as electron transport in molecular junctions with semiconductor and carbon-based (e.g. graphene) electrodes. Furthemore, strategies to characterize the systems beyond the single-particle level using correlated methods will be developed.

Acknowledgements Generous allocation of computing time by the Leibniz Computing Center (LRZ) is gratefully acknowledged. This work has been supported by the Deutsche Forschungsgemeinschaft (DFG) through the DFG-Cluster of Excellence Munich Center for Advanced Photonics and a research grant, the Fonds der Chemischen Industrie, the Federal Ministry of Education and Research (BMBF) and the German-Israel Science Foundation (GIF).

References

1. Vienna ab-intio simulation package (vasp), version 4.6/ http://cms.mpi.univie.ac.at/cmspage/main/ (2006)
2. Arnim, M.V., Ahlrichs, R.: J. Comput. Chem. **19**, 1746 (1998)
3. Barbara, P.F., Meyer, T.J., Ratner, M.A.: J. Phys. Chem. **100**, 13148 (1996)
4. Becke, A.D.: **98**, 5648 (1993)
5. Benesch, C., Cizek, M., Klimes, J., Kondov, I., Thoss, M., Domcke, W.: J. Phys. Chem. C **112**, 9880 (2008)
6. Benesch, C., Rode, M., Cizek, M., Hartle, R., Rubio-Pons, O., Thoss, M., Sobolewski, A.: J. Phys. Chem. C **113**, 10315 (2009)
7. Boroda, Y.G., Calhoun, A., Voth, G.A.: J. Chem. Phys. **107**, 8940 (1997)
8. Choi, B.Y., Kahng, S.J., Kim, S., Kim, H., Kim, H., Song, Y., Ihm, J., Kuk, Y.: Phys. Rev. Lett. **96**, 156106 (2006)
9. Cuniberti, G., Fagas, G., Richter, K.: Introducing Molecular Electronics. Springer, Heidelberg (2005)
10. Dulic, D., van der Molen, S., Kudernac, T., Jonkman, H., de Jong, J., Bowden, T., van Esch, J., Feringa, B., van Wees, B.: Phys. Rev. Lett. **91**, 207402 (2003)
11. Ernstorfer, R., Gundlach, L., Felber, S., Storck, W., Eichberger, R., Willig, F.: J. Phys. Chem. B **110**, 25383 (2006)
12. Fölisch, A., Feulner, P., Fink, A., Menzel, D., Sanchez-Portal, D., Echenique, P., W. Wurth: Nature (London) **436**, 373 (2005)
13. http://www.mpi-forum.org/
14. Frey, S., Shaporenko, A., Zharnikov, M., Harder, P., Allara, D.L.: J. Phys. Chem. B **107**, 7716 (2005)
15. Galperin, M., Ratner, M., Nitzan, A.: Phys. Rev. B **73**, 045314 (2006)
16. Galperin, M., Ratner, M.A., Nitzan, A.: Molecular transport junctions: vibrational effects. J. Phys.: Condens. Matter **19**, 103201 (2007)
17. Grätzel, M.: Nature **414**, 338 (2001)
18. Hagfeldt, A., Grätzel, M.: Chem. Rev. **95**, 49 (1995)
19. Härtle, R., Benesch, C., Thoss, M.: Multimode vibrational effects in single-molecule conductance: A nonequilibrium green's function approach. Phys. Rev. B **77**, 205314 (2008)
20. Härtle, R., Benesch, C., Thoss, M.: Phys. Rev. Lett. **102**, 146801 (2009)
21. Härtle, R., Benesch, C., Thoss, M.: Vibrational nonequilibrium effects in the conductance of single molecules with multiple electronic states. Phys. Rev. Lett. **102**, 146801 (2009)
22. http://www.fftw.org/
23. Huber, R., Moser, J.E., Grätzel, M., Wachtveitl, J.: Chem. Phys. Lett. **285**, 39 (2002)
24. Jortner, J., Bixon, M.: Electron Transfer: From Isolated Molecules to Biomolecules, Dynamics and Spectroscopy Adv. Chem. Phys. Vols. 106-107. Wiley, New York (1999)
25. Kondov, I., Cizek, M., Benesch, C., Wang, H., Thoss, M.: J. Phys. Chem. C **111**, 11970 (2007)
26. Kresse, G., Joubert, J.: Phys. Rev. B **59**, 1758
27. Li, J., Speyer, G., Sankey, O.: Phys. Rev. Lett. **93**, 248302 (2004)
28. M. del Valle, Gutierrez, R., Tejedor, C., Cuniberti, G.: Nature Nanotech. **2**, 176 (2007)
29. Michel-Beyerle, M.E.: The Reaction Center of Photosynthetic Bacteria: Structure and Dynamics. Springer, Berlin (1996)
30. Ragan, S., Gallet, J.J., Bournel, F., Kubsky, S., Guen, K.L., Dugour, G., Rochet, F., Sirotti, F., Carninato, S., Ilakovac, V.: Phys. Rev. B **71**, 165318 (2005)
31. Rubio-Pons, O., Hartle, R., Thoss, M.: HLRBII project (pr28lo) (2009)
32. Schnadt, J., Brühwiler, P.A., Patthey, L., O'Shea, J.N., Södergreen, S., Odellus, M., Ahuja, R., Karis, O., Bässler, M., Persson, P., Siegbahn, H., Lunell, S., Martenson, N.: Nature **418**, 620 (2002)
33. Shemesh, D., Domcke, W.: HLRBII project (h0833) (2008)
34. Sobolewski, A.: Phys. Chem. Chem. Phys. **10**, 1243 (2008)

35. Sobolewski, A., Domcke, W., Hättig, C.: J. Phys. Chem. A **110**, 6301 (2006)
36. Tominaga, K., Kliner, D.A.V., Johnson, A.E., Levinger, N.E., Barbara, P.F.: J. Chem. Phys. **98**, 1228 (1993)
37. Wang, C., Mohney, B.K., Williams, R.D., Petrov, V., Hupp, J.T., Walker, G.C.: J. Am. Chem. Soc. **120**, 5848 (1998)
38. Xue, Y., Ratner, M.: Phys. Rev. B **68**, 115406 (2001)

Fluctuations in the Photoionization Cross Sections of Highly Doubly Excited Two-Electron Atoms

Johannes Eiglsperger and Javier Madroñero

Abstract We report on the spectral properties of highly doubly excited states of planar helium up to energies around the 25th ionization threshold. For that purpose we have developed an approach which combines an exact representation of the Hamiltonian of the three-body Coulomb problem in terms of creation and annihilation operators with complex dilation and with a parallelized implementation of the Lanczos algorithm for the diagonalization of the rather large associated complex symmetric generalized eigenvalue problem. We corroborate the existence of an approximate quantum number $F = N - K$ and the dominance of the series of states associated to low values of F in the photoionization cross sections of triplet P states of planar helium. As the energy increases the dominant role of a single series as sole contributor is apparently lost as new series start to contribute significantly to the cross sections. This would result in an earlier onset of Ericson fluctuations than in the picture of a single dominant series, where the onset is expected around I_{34}.

1 Introduction

Despite the seemingly simple problem of three charged particles with known interactions, there exists no comprehensive understanding of highly doubly excited states of two-electron atoms. The electron-electron interaction term in the Hamiltonian of the unperturbed helium atom – which otherwise is just the sum of two hydrogen Hamiltonians with amended nuclear charge – renders the two-electron classical dynamics in general irregular or chaotic, with only rather small domains of the classical phase space occupied by regular motion. This loss of integrability caused the failure of early quantization attempts on the basis of Bohr's quantum pos-

J. Eiglsperger · J. Madroñero
Physik Department, Technische Universität München, 85747 Garching bei München, Germany
e-mail: Johannes.Eiglsperger@ph.tum.de
e-mail: Javier.Madronero@ph.tum.de

S. Wagner et al. (eds.), *High Performance Computing in Science and Engineering, Garching/Munich 2009*, DOI 10.1007/978-3-642-13872-0_52,

tulates [1]. Only with the development of the modern semiclassical theory [2, 3] and the subsequent semiclassical quantization of helium [4, 5] was the nonintegrability of the quantum system understood as the direct counterpart of the corresponding classical mixed regular-chaotic dynamics [6]. Indeed, due to the scaling invariance of the Hamiltonian, highly doubly excited helium can be described semiclassically. As a consequence, the quantum spectrum of highly doubly excited states should be influenced by the underlying classical chaotic dynamics and typical signatures of quantum chaos, such as a Wigner distribution of the energy spacings between nearest-neighbor resonances [7], semiclassical scaling laws for the fluctuations in the spectrum close to the double ionization threshold [8] or Ericson fluctuations [9–11], are expected to become observable [12].

The understanding of these issues requires an accurate description of doubly excited states. This indeed posses a formidable theoretical and numerical challenge: Doubly excited states of two-electron atoms are organized in series converging towards the single ionization thresholds I_N of $\mathrm{He}(N)^+$ states. Starting from the 4th single ionization threshold members of higher lying series interfere with lower series. Therefore, as the energy approaches the total break-up regime the density of states increases dramatically. In addition, doubly excited states cannot in general be described by a simple model based on independent-particle angular momentum quantum numbers $NL\nu\lambda$, as was first realized in 1963 through the famous experiment by Madden and Codling [13]. The inherent strongly correlated nature of the doubly excited states couples almost all degrees of freedom and leaves only the total angular momentum and the parity as conserved quantities. Therefore, the description of the high energy regime of two-electron atoms requires a basis representation which leads to rather huge eigenvalue problems.

In recent years an improvement of measurement techniques has allowed a detailed examination of the doubly excited states converging up to the $N = 16$ threshold of He [14–16]. From the theoretical side, currently available full 3D approaches are able to describe the spectrum up to the $N = 17$ threshold [15]. The analysis of the theoretical and experimental results up to $I_{N=17}$ in [15] reveals a clear dominance of principal Rydberg series in the total photoionization cross section. This together with the existence of an approximate quantum number for a large fraction of the states indicates that Ericson fluctuations are absent in this regime and no transition to full chaos is observed. Simplified one dimensional (1D) models or the s^2-model of the three-dimensional atom significantly reduce the calculational difficulties. However, the former models may underestimate the decay rates of the resonances by orders of magnitude [17] and the latter does not resolve all resonances that are important for Ericson fluctuations. While 1D models predict the Ericson regime to start around I_{34}, studies within the s^2-model find Ericson fluctuations in the partial inelastic cross sections between electrons and He^+ already around I_{16} [18].

In the present work we describe an energy regime far beyond the state-of-the-art to date and shed some light on the understanding of the regime close to the double ionization threshold. For that purpose we have developed an approach which combines the representation of the atomic Hamiltonian in a suitably chosen basis set, which allows for fully algebraic expressions of the matrix elements, and complex rotation to access the autoionization decay rates. The non-unitary complex dilation transformation finally leaves us with a large, generalized, *complex symmetric* eigenvalue problem, which has to be diagonalized on the most powerful parallel machines currently available. An efficient parallel implementation of the Lanczos diagonalization routine allows the extraction of a small number of complex eigenvalues. The description of the whole spectrum thus requires several executions of the code.

Note that the speed-up of program execution as one of the prominent advantages of a large parallel machine is vital for our project, since it accelerates our progresses tremendously. However, the availability of large storage space for the matrix to be diagonalized is a necessary condition.

Yet, due to the above-mentioned rapid increase of the Hilbert space dimension (and hence, on the numerical level, of the required storage capacities), we still restrict our problem to planar configurations of the two electrons and the nucleus. Whilst this certainly does restrict the generality of our model, semiclassical scaling arguments suggest that the unperturbed three body dynamics is essentially planar at high electronic excitations and small total angular momenta [19, 20].

2 Theory and Numerical Implementation

The computation and analysis of fluctuations in photoionization cross sections of highly doubly excited states demands a description of two-electron atoms with a minimum of approximations. The electrons are subject to the combined potentials of the nucleus and of the interelectronic interaction which gives rise to a very rich resonance structure close to the total break-up regime.

Therefore, our theoretical approach has to account for the following: (1) The singularities of the Coulomb potentials; (2) The spectrum of the helium atom consisting of bound states, and of resonances embedded into the atomic continua; (3) The photon induced transitions between initial and final states which provide the cross sections; (4) Adequate classification schemes of the resonances reflecting the underlying classical mixed regular chaotic dynamics.

Let us start with the Hamiltonian describing our problem, in atomic units (which will be used throughout this paper),

$$H = \frac{\mathbf{p}_1^2 + \mathbf{p}_2^2}{2} - \frac{Z}{r_1} - \frac{Z}{r_2} + \frac{1}{r_{12}}, \tag{1}$$

where $\mathbf{p}_i$ and r_i, $i = 1, 2$, denote the respective momenta and positions of both electrons, r_{12} represents the interelectronic distance, the nucleus (with infinite mass) is fixed at the origin, and Z is the nucleus charge. From now on we restrict our problem to two dimensions, i.e., the electrons and the nucleus are confined to a plane. In this case, two subsequent, parabolic coordinate transformations and a suitable rotation completely regularize all singularities in this Hamiltonian and finally allow us to identify the eigenvalue problem generated by (1) with an eigenvalue problem describing *four* coupled harmonic oscillators [19, 21, 22]. Consequently, (1) can be represented in a basis set defined by the tensor product

$$|n_1 n_2 n_3 n_4\rangle = |n_1\rangle \otimes |n_2\rangle \otimes |n_3\rangle \otimes |n_4\rangle \tag{2}$$

of Fock states of the individual harmonic oscillators, and has a purely algebraic representation in the associated annihilation and creation operators that define the four oscillator algebras. The final eigenvalue problem involves polynomials of maximal degree 12 in the creation and annihilation operators, with alltogether 1511 monomial terms (generated by a home made Mathematica code [22]), and thus allows for a purely analytical calculation of all matrix elements defining our eigenvalue problem [22]. The Hamiltonian (1) is invariant under the particle exchange symmetry and the reflection symmetry Π_x with respect to the x-axis. It also commutes with the total angular momentum l. This leads to a reduction of the eigenvalue problem.

The resolution of the resonances in the complex plane is achieved with the help of the complex rotation method [23–27], consisting of a rotation of the coordinates by a suitable angle θ. In addition to this we also introduce a dilation of the coordinates by a real parameter α. The complex dilation is thus achieved through the transformations $\mathbf{r} \to \alpha \mathbf{r} e^{i\theta}$ and $\mathbf{p} \to \mathbf{p} e^{-i\theta}/\alpha$. Though the dilation by α is a unitary transformation, the whole transformation leads to a complex symmetric matrix representation of the generalized eigenvalue problem. Its spectrum is complex in general with the following properties: (a) The bound eigenvalues of the Hamiltonian (1) are invariant under complex rotation; (b) There are isolated complex eigenvalues $E_{i,\theta} = E_i - i\Gamma_i/2$ in the lower half plane, corresponding to resonance states. These are stationary under changes of θ, provided the dilation angle is large enough to uncover their positions on the Riemannian sheets of the associated resolvent [28, 29]. The associated resonance eigenfunctions are square integrable [30], in contrast to the resonance eigenfunctions of the unrotated Hamiltonian; (c) The continuum states are located on half lines, rotated by an angle -2θ around the ionization thresholds of the unrotated Hamiltonian into the lower half of the complex plane.

These complex symmetric matrices have a sparse banded structure with 159 coupling matrix elements in the band. The eigenvalue problem is diagonalized with the help of an efficient parallel implementation of the Lanczos algorithm with a preceding Choleski decomposition. The required communication between the processing elements is achieved by MPI, while most of the local calculations are performed using BLAS-2 and BLAS-3, for matrix-vector and matrix-matrix operations [22, 31].

For the calculation of the photoionization cross sections the dipole operator is expressed in terms of the creation and annihilation operators [32]. The nonzero matrix elements (up to 20 per row) of its matrix representation are stored in each processor. Further insight in the spectral structure and important features of the underlying classical dynamics are obtained through the expectation value, for each resonance state, of the cosine of the angle θ_{12} between the electron positions $\mathbf{r}_1$ and $\mathbf{r}_2$ ($\langle\cos(\theta_{12})\rangle$). The operator $\cos(\theta_{12})$ also has an exact polynomial representation in terms of the creation and annihilation operators. Therefore, its representation in the basis (2) leads to sparse matrices with up to 171 nonzero matrix elements per row which is also stored in each processor. The calculation of both, the photoionization cross sections and the expectation value $\langle\cos(\theta_{12})\rangle$, finally require only matrix-vector multiplications of this matrix with the eigenvectors of the system.

In a typical production run, for the description of doubly excited helium above the 15th ionization threshold and up to the 25th ionization threshold, the matrix dimension reaches values of $4 \times 10^5 \ldots 6 \times 10^5$, with a bandwidth of $1.9 \times 10^4 \ldots 2.4 \times 10^4$. This corresponds to storage requirements between 175 GB and 340 GB. Our calculations up to the 20th ionization threshold were executed on the Linux Cluster at LRZ [33], whilst the larger ones were performed on the National Supercomputer HLRB-II: SGI Altix 4700 at LRZ [34]. Our parallel Lanczos code, which is composed of a Cholesky decomposition of the Hamiltonian matrix and a Lanczos iteration [22, 31], performs very well on both machines, with typical monoprocessor performances of approx. 1400–1900 MFlops on both, the Linux Cluster and the HLRB-II. The peak performance we observed (2375 MFlops) was obtained with 128 processors. On both machines, the code scales excellently with the number of processors (which varies between 64 and 128 on the HLRB II).

3 Results

In the following we present some of our results obtained up to the 25th single ionization threshold of planar helium. More precisely, we investigate the photoionization cross section for dipole transitions from the lowest-lying triplet bound state, with angular momentum $l = 0$ and $\Pi_x = +1$, of planar helium. The dipole operator couples this state with $|l| = 1$ triplet states of symmetry $\Pi_x = +1$. Therefore, our calculations require the diagonalization of the rotated Hamiltonian associated with this symmetry. The spectrum contains θ-dependent continuum states and resonance states, which are invariant under complex rotation. The continuum states are responsible for the smooth background of the cross sections and do not affect their fluctuations. Thus, only the resonances contribute to the fluctuating part of the spectrum $\sigma_{fl}(E)$. For the numerical calculation of $\sigma_{fl}(E)$ only converged resonances have been taken into account. To guarantee the convergence of our results we have checked the stability of the resonances, of the expectation value $\langle\cos(\theta_{12})\rangle$, of the oscillator strengths, and of the fluctuations as function of the parameters θ and α. Fig. 1 depicts $\sigma_{fl}(E)$.

The spectral structure of low-lying doubly excited states is rather well known. Above the first ionization threshold all bound states are embedded into the continuum of lower series, i.e. they are resonance states with finite width [29]. These states are organized in Rydberg series converging to single ionization thresholds. As the energy increases, series of states converging to the different single ionization thresholds will overlap. In 2D helium the 5th Rydberg series overlaps with the 6th series [19] in contrast to 3D helium where the overlapping of series starts with the 4th series. For moderate excitations – up to the 7th ionization threshold – the spectrum consist of well isolated resonances in the sense that the widths of individual

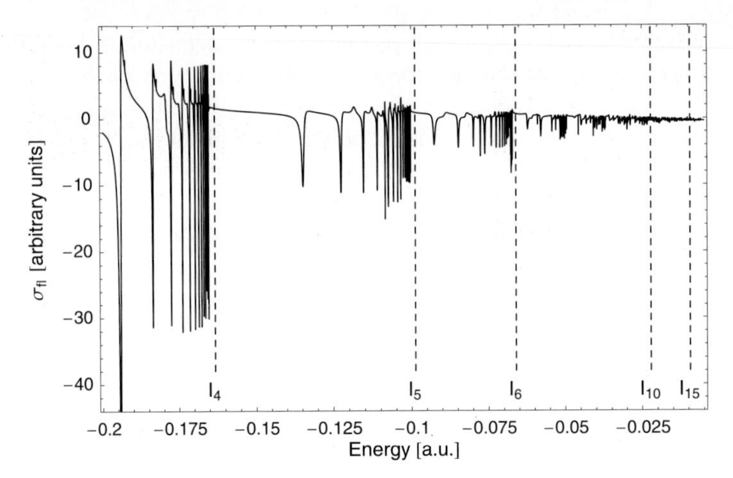

Fig. 1 Calculated fluctuations of the photoionization cross section of triplet planar helium from below I_4 to I_{20}

resonances are smaller than the separation from their nearest-neighbor resonances. Furthermore, up to the 9th ionization threshold the photoionization cross sections are the result of a superposition of Fano profiles as observed in the low-energy region of Fig. 1.

As the energy approaches the double ionization threshold the cross sections lose their regularity as a consequence of the dramatic increase in the density of states. Nevertheless, a systematic decrease of the fluctuations in magnitude is observed. Indeed, semiclassical studies based on closed orbit theory [8] predict an algebraic scaling law,

$$\sigma_{\mathrm{fl}}(E) \propto |E|^{\mu} \qquad \text{for } E \to 0^{-}\,,$$
$$\mu = \frac{1}{4}\left[\sqrt{\frac{100Z-9}{4Z-1}} + 2\sqrt{\frac{4Z-9}{4Z-1}}\right], \tag{3}$$

for the fluctuations in the cross section below the double ionization threshold. This scaling law has been corroborated for 1D helium restricted to the *eZe* configuration [8] and experimental evidence has been found below the 17th ionization threshold [35]. The scaled fluctuations $\sigma_{\mathrm{fl}}^{\mathrm{scaled}}(E) = |E|^{-\mu}\sigma_{\mathrm{fl}}(E)$ depicted in Fig. 2 exhibits rather small variations along the energy regime from I_4 to I_{20}. They thus provide further evidence for the scaling law.

The first term of μ in the square bracket of Eq. (3) results from the stability exponents of the collinear *eZe* configuration, while the second term can be associated with the stability exponents of the Wannier ridge. The contribution of the latter does not affect the fluctuations amplitude in the case of helium ($Z = 2$). Therefore, the total cross sections are expected to be dominated by the low-dimensional *eZe* dynamics. States associated with such configurations can be identified through the expectation value $\langle\cos(\theta_{12})\rangle$. Fig. 3 shows a plot of the calculated expectation values

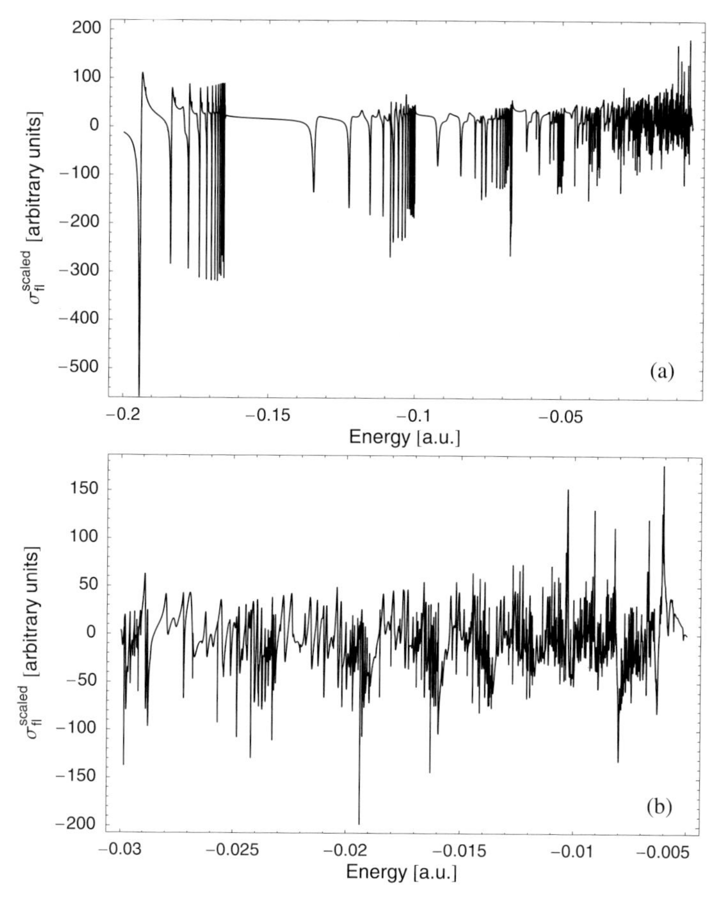

Fig. 2 Scaled fluctuations $\sigma_{\rm fl}^{\rm scaled}$ of the cross sections of triplet planar helium: (a) $\sigma_{\rm fl}^{\rm scaled}$ between I_3 and I_{20}; (b) A close-up of the region from below I_9 to below I_{20}. The amplitude of the fluctuations remains approximately constant as a function of the energy

of $\langle\cos(\theta_{12})\rangle$ as a function of $\sqrt{|\mathrm{Re}(E_0)|}$ for all converged resonances up to I_{25}. The resonances for which $\langle\cos(\theta_{12})\rangle$ is close to 1 are associated to the stable collinear classical frozen planet configuration (*Zee*) [36]. These states play a fundamental role in the formation of two-electron nondispersive wave packets [19, 37]. The region $|\langle\cos(\theta_{12})\rangle| < 0.5$ is characterized by strong mixing between the numbers N, K, and F, in agreement with the instability of the classical motion of the associated configurations. The resonances which satisfy $\langle\cos(\theta_{12})\rangle \lesssim -0.5$ are arranged in series along straight lines converging to -1 at $E = 0$. This observation corroborates the results obtained by Jiang *et al.* for the full 3D helium atom up to I_{16} [15]. In that case such series correspond to (small) constant values of $F = N - K$, where N and K are approximate quantum numbers from Herrick's algebraic classification [38, 39]. In addition, as a consequence of the relation

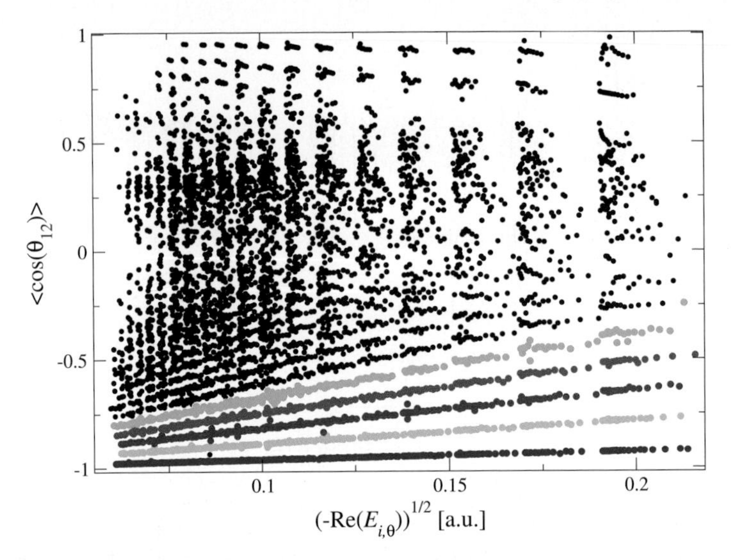

Fig. 3 Expectation values $\langle\cos(\theta_{12})\rangle$ as a function of square root of the resonance energy E below the 25th threshold. Each point represents a particular triplet state resonance with $\Pi_x = +1$ and $|l| = 1$. The resonances are displayed in color according to their allocation to Rydberg series: • (1st), • (2nd), • (3rd), • (4th), • (5th), and • for resonances not identified with any of these series

$$\langle\cos(\theta_{12})\rangle \overset{n\to\infty}{\longrightarrow} -\frac{K}{N}, \tag{4}$$

the *eZe* configuration can be identified with low values of $F = N - K$. Herrick's algebraic classification is also valid for planar helium [17]. Therefore, the approximate quantum number F also labels the series of 2D helium related to the *eZe* configuration. This approximate classification allows us to study separately the contributions of different series to the photoionization cross sections. The minimal differences between $\sigma_{\mathrm{fl}}(E)$ and the fluctuations from the resonances of the series $F = 1$ depicted in the left plot of Fig. 4 show that this series provides the most important contribution to the cross sections. Indeed, for triplet states the resonances which yield major contributions are characterized by rather small odd values of F, while series with even F, large value of F, and all resonances that can not be characterized by F – e.g. those resonances for which $\langle\cos(\theta_{12})\rangle \gtrsim 0.5$ –, result in almost no contribution. The contributions of the series associated with $F = 1$, 3 and 5 are illustrated in the right panel of Fig. 4. As the energy increases the weight of the contributions of the $F = 3$ and $F = 5$ series increases in comparison with the weight of the contribution of the $F = 1$ series. This suggest a breakdown of the dominant role of a single series.

The dominance of a subset of resonances has important consequences for the discussion of Ericson fluctuations in the cross sections. If one considers *all* resonances, the Ericson regime would be already reached around the 9th ionization threshold, since an important part of the resonance widths are larger than the mean level spacing. Furthermore, at energies as high as I_{20} more than 90% of the resonances overlap [32]. From this one might expect that single peaks in the cross

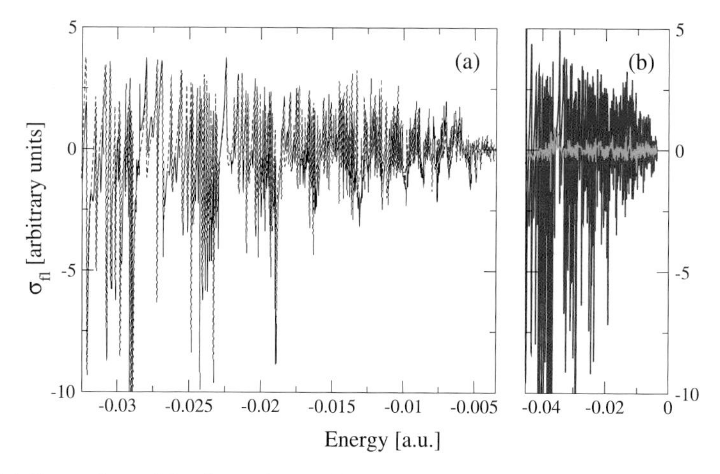

Fig. 4 (a) Comparison of the fluctuations of the photoionization cross sections from I_9 up to I_{25} including all resonances (solid black line) and resonances with $F = 1$ only (dashed blue line). The dominance of the $F = 1$ series is evident. (b) Separated contributions of the $F = 1$ (blue), $F = 3$ (red) and $F = 5$ (green) series to $\sigma_{\rm fl}$ from I_7 up to I_{25}. As the energy increases the relative weight of the $F = 3$ and $F = 5$ series increases and the $F = 1$ series starts to lose its dominant role

sections observed in Fig. 2 have random character and are not the result of individual resonances. However, if one looks only at the resonances of the dominant series $F = 1$, a very small fraction of the resonances overlap. The expectation value of $\cos(\theta_{12})$ for the resonances belonging to the $F = 1$ series is close to -1 even at low energies and approaches -1 rather fast as the energy increases. Therefore, practically all of these resonances can be associated to the collinear *eZe* configuration. Provided the picture of a dominant series remains valid for high enough energies, Ericson fluctuations in helium are expected around I_{34} [15]. However, the apparent breakdown of the dominance of a single series would lead to an earlier onset of Ericson fluctuations than expected for the 1D helium picture: the closer one gets to the double ionization threshold the larger is the number of series with odd F that one might have to consider. Nevertheless, preliminary results indicates that most of the peaks in the cross sections can be identified with single resonances even at energies around the 25th ionization threshold.

4 Summary and Outlook

We have presented a comprehensive study of the fluctuations in the photoionization cross sections of triplet P states of planar helium for energies as high as the 25th ionization threshold, far beyond the regimes any other non 1-dimensional approach has reached so far. The description of this regime is only possible after the diagonalization of large matrices. For that purpose we have used an efficient parallel implementation of the Lanczos algorithm. Even at these high energies, there exists an approximate quantum number $F = N - K$ for a large fraction of states. This imposes a hierarchy on the contributions to the total photoionization cross sections, where the subset of resonances with $F = 1$ gives the dominant contribution. $F = 3$ and $F = 5$ states contribute less significantly at low energies, but they start to compete with the dominant series at high energies. The existence of a dominant series implies that Ericson fluctuations are absent in the photoionization cross sections. The breakdown of the dominance of a single series suggests an earlier onset of Ericson fluctuations than expected in the picture of a single dominant series, which however has not been observed yet. The calculated fluctuations in the cross sections exhibit a systematic decay in agreement with the semiclassical scaling law (3). In the case of helium only contributions of the *eZe* configuration are observed. For other two-electron atomic species, e.g. Li^+, the Wannier ridge is also expected to contribute. Further calculations for singlet He and for Li^+ are being carried out at present.

Apart from the spectral properties of the field-free three-body Coulomb problem, we also plan for the near future to use our diagonalization code for the investigation of highly doubly excited states in external fields. For that purpose we will use our planar approach with some modifications [19] and a recently developed full 3D approach [40]. In particular we are interested in the possibility of creating two-electron nondispersive wave packets.

Acknowledgements We are indebted to Harald Friedrich for enlightening discussions and for carefully reading this manuscript. Access to the computing facilities of the Leibniz-Rechenzentrum der Bayerischen Akademie der Wissenschaft, as well as financial support by Deutsche Forschungsgemeinschaft under the Contract No. FR 591/16-1 are gratefully acknowledged.

References

1. A. Einstein, Verh. Dtsch. Phys. Ges. **19**, 82 (1917).
2. M. C. Gutzwiller, J. Math. Phys. **8**, 1979 (1967).
3. M. C. Gutzwiller, J. Math. Phys. **12**, 343 (1971).
4. G. S. Ezra, K. Richter, G. Tanner, and D. Wintgen, J. Phys. B **24**, L413 (1991).
5. D. Wintgen, K. Richter, and G. Tanner, Chaos **2**, 19 (1992).
6. G. Tanner, K. Richter, and J. M. Rost, Rev. Mod. Phys. **72**, 497 (2000).
7. O. Bohigas, M. J. Giannoni, and C. Schmit, Phys. Rev. Lett. **52**, 1 (1984).
8. C. W. Byun, N. N. Choi, M.-H. Lee, and G. Tanner, Phys. Rev. Lett. **98**, 113001 (2007).
9. T. Ericson, Phys. Rev. Lett. **5**, 430 (1960).
10. R. Blümel and U. Smilansky, Phys. Rev. Lett. **60**, 477 (1988).
11. J. Madroñero and A. Buchleitner, Phys. Rev. Lett. **95**, 263601 (2005).
12. R. Blümel, Phys. Rev. A **54**, 5420 (1996).
13. R. P. Madden and K. Codling, Phys. Rev. Lett. **10**, 516 (1963).
14. M. Domke, K. Schulz, G. Remmers, G. Kaindl, and D. Wintgen, Phys. Rev. A **53**, 1424 (1996).
15. Y. H. Jiang, R. Püttner, D. Delande, M. Martins, and G. Kaindl, Phys. Rev. A **78**, 021401(R) (2008).
16. A. Czasch, et al., Phys. Rev. Lett. **95**, 243003 (2005).

17. J. Madroñero, P. Schlagheck, L. Hilico, B. Grémaud, D. Delande, and A. Buchleitner, Europhys. Lett. **70**, 183 (2005).
18. J. Xu, A.-T. Le, T. Morishita, and C. D. Lin, Phys. Rev. A **78**, 012701 (2008).
19. J. Madroñero and A. Buchleitner, Phys. Rev. A **77**, 053402 (2008).
20. K. Sacha and B. Eckhardt, Phys. Rev. A **63**, 043414 (2001).
21. L. Hilico, B. Grémaud, T. Jonckheere, N. Billy, and D. Delande, Phys. Rev. A **66**, 022101 (2002).
22. J. Madroñero, Dissertation, Ludwig-Maximilians-Universität München (2004), http://edoc.ub.uni-muenchen.de/archive/00002187/.
23. J. Aguilar and J. M. Combes, Commun. Math. Phys. **22**, 269 (1971).
24. E. Balslev and J. M. Combes, Commun. Math. Phys. **22**, 280 (1971).
25. B. Simon, Ann. Math. **97**, 247 (1973).
26. W. P. Reinhardt, An. Rev. Phys. Chem. **33**, 223 (1982).
27. S. Graffi, V. Grecchi, and H. J. Silverstone, Ann. Inst. Henry Poincaré **42**, 215 (1985).
28. M. Pont and R. Shakeshaft, Phys. Rev. A **43**, 3764 (1991).
29. M. Reed and B. Simon, *Methods of modern mathematical physics IV. Analysis of operators*, Acad. Pr. New York (1978).
30. Y. K. Ho, Phys. Rep. **99**, 1 (1983).
31. A. Krug, Dissertation, Ludwig-Maximilians-Universität München (2001), URL http://edoc.ub.uni-muenchen.de/archive/00000336/.
32. J. Eiglsperger and J. Madroñero, Phys. Rev. A **80**, 022512 (2009); J. Eiglsperger and J. Madroñero, Phys. Rev. A **80**, 033902(E) (2009).
33. http://www.lrz-muenchen.de/services/compute/linux-cluster/.
34. http://www.lrz-muenchen.de/services/compute/hlrb/index.html.
35. G. Tanner, N. N. Choi, M.-H. Lee, A. Czasch, and R. Dörner, J. Phys. B **40**, F157 (2007).
36. K. Richter and D. Wintgen, Phys. Rev. Lett. **65**, 1965 (1990).
37. P. Schlagheck and A. Buchleitner, J. Phys. B **31**, L489 (1998).
38. D. R. Herrick and O. Sinanoğlu, Phys. Rev. A **11**, 97 (1975).
39. O. Sinanoğlu and D. Herrick, J. Chem. Phys. **62**, 886 (1975).
40. J. Eiglsperger, B. Piraux, and J. Madroñero, Phys. Rev. A **80**, 022511 (2009).

Part VII
Chemistry

Part VII
Chemistry

Photophysics of the Trp-Gly Dipeptide: Role of Electron and Proton Transfer Processes for Efficient Excited-State Deactivation

Dorit Shemesh and Wolfgang Domcke

Abstract The excited-state potential-energy surfaces of the lowest-energy conformer of the tryptophan-glycine dipeptide have been investigated with ab initio electronic-structure methods. The calculations reveal a potentially very efficient excited-state deactivation mechanism via conical intersections of the excited states of the indole chromophore with locally-excited and charge-transfer states of the peptide backbone. These findings suggest that the excited-state lifetime of the lowest-energy conformer of the Trp-Gly dipeptide may be too short to allow the detection of a resonant two-photon ionization signal. It is proposed that the efficient excited-state deactivation enhances the photostability of the dipeptide.

1 Introduction

Amino acids are the elementary building blocks of peptides and proteins. Electrostatic interactions, hydrogen bonds and salt bridges stabilize the three-dimensional structure of proteins. In the last decade, a rapidly increasing number of experiments and theoretical calculations have been devoted to isolated small peptides, eg. dipeptides and tripeptides, aiming at the understanding of the local structural preferences of particular amino acids. [1–13] These experiments provide detailed information on the structure and energetics of different conformers for small peptides and give insight into the driving forces for the formation of the secondary structure of peptides and proteins. Hydrogen bonds play an important role in stabilizing the secondary structure of peptides. The most common hydrogen bonds found in proteins are hydrogen bonds between the N-H group on one peptide bond and the O=C group of

Dorit Shemesh · Wolfgang Domcke
Lehrstuhl für Theoretische Chemie, Technische Universität München, 85747 Garching bei München, Germany
e-mail: dorit.shemesh@ch.tum.de
e-mail: wolfgang.domcke@ch.tum.de

S. Wagner et al. (eds.), *High Performance Computing in Science and Engineering, Garching/Munich 2009*, DOI 10.1007/978-3-642-13872-0_53,

another peptide bond. Several structural motifs are found in proteins: The α-helix is built from a peptide strand where the ith residue is connected with the (i+4)th residue via a hydrogen bond between the N-H group and the C=O group. The β-sheet, on the other hand, is built from two strands of peptides aligned either parallel or antiparallel and stabilized by hydrogen-bond interactions between the N-H group of one strand and the C=O group of the other strand (and vice versa). Other important structural motifs are the β-turn and the γ-turn. The β-turn (γ-turn) is built from a hydrogen bond where the donor and acceptor residues are separated by 2 residues (3 residues). The H-bonding interaction is between the N-H group of one residue and the C=O group of the other residue. Natural peptides terminate on one end with a free carboxylic group and on the other end with a free amino group. These groups can interact via hydrogen bonds with the peptide backbone. In this work, we have investigated the dipeptide tryptophan-glycine (Trp-Gly). The resonant multiphoton ionization (REMPI) spectrum of Trp-Gly and Trp-Gly-Gly was recorded recently. [12] This spectroscopic work was supported by a thorough theoretical exploration of the conformational space. [5] Unexpectedly, some of the energetically most stable conformers were not found in the REMPI spectra. The most stable conformers exhibit a strong hydrogen bond which connects the free carboxylic group on the C-terminus with the carbonyl group of the peptide bond, see the encircled region in Fig. 1. This hydrogen bond can be formed by every C-terminus in peptides. We concentrate here on the photophysics of the proton transfer along this hydrogen bond. We suggest that this strong hydrogen bond gives rise to an efficient excited-state deactivation mechanism for this particular conformer of Trp-Gly. Our results suggest that this efficient radiationless decay pathway is responsible for the absence of this conformer in the REMPI spectrum.

2 Computational Methods

The ground-state equilibrium geometry of the conformer wg01[g+.0.70] (see Ref. [5]) of Trp-Gly has been determined with the second-order Møller-Plesset (MP2) method. Excitation energies have been calculated with the ADC(2) (algebraic diagrammatic construction through second order) method. [14, 15] For comparison, the vertical excitation energies have also been calculated with the CC2 (approximate second-order coupled cluster) method. [16, 17] The performance of ADC(2) compared with CC2 and related methods has thoroughly been investigated by Hättig. [14] The advantage of the ADC(2) method is that the excitation energies are calculated as the eigenvalues of a Hermitian secular matrix, while in coupled cluster response theory they are obtained from a non-Hermitian Jacobi matrix. The ADC(2) method is therefore able to describe conical intersections between excited states physically correctly, which need not to be the case for coupled-cluster theory. [18] The equilibrium geometries and reactions paths of the excited singlet states have been determined at the ADC(2) level, making use of analytic ADC(2) gradients. [14] A relaxed scan for the transfer of the proton from the hydroxyl group to

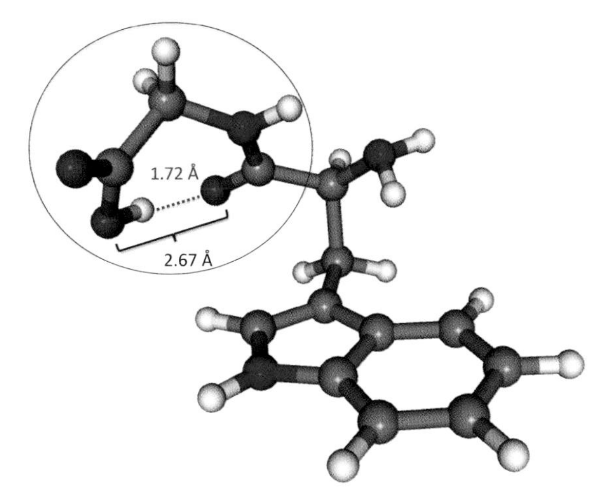

Fig. 1 MP2 ground-state equilibrium geometry of the conformer wg[g+.0.70] of Trp-Gly. The ellipse encircles a hydrogen-bond motif which is commonly found in the backbone of uncapped peptides

the carbonyl group has been computed: The OH bond length R_{OH} of the carboxyl group involved in the hydrogen bonding has been chosen as the driving coordinate of the reaction path; all other internal coordinates have been optimized for a given value of R_{OH} in the respective electronic state. All calculations were carried out with the TURBOMOLE program suite, [19] making use of the resolution-of-the-identity (RI) approximation for the evaluation of the electron-repulsion integrals. [20] The cc-pVDZ basis set has been employed throughout.

3 Scientific Results

3.1 Ground-State Equilibrium Structure

The structure of the conformer wg01[g+.0.70] of the Trp-Gly dipeptide is shown in Fig. 1.

This conformer has been identified as the lowest enthalpy conformer by Valdes et al [5]. This structure exhibits a hydrogen bond between the OH of the carboxyl group and the O=C of the nearest peptide bond ($O\text{-}H_{carb} \cdots O{=}C_{pep}$), see Fig. 1. This hydrogen bond also exists in other structures like wg02[t.140.-70] and wg03[g+.-20.70], see Ref. [5]. All of these conformers could not be identified in the REMPI spectrum. [12] The distance between the carboxylic hydrogen and the carbonyl oxygen is 1.72 Å. The distance between the two oxygens is 2.67 Å. According to established empirical rules, this distance corresponds to a strong hydrogen bond. [21, 22]

Table 1 Vertical excitation energies (ΔE, in eV), oscillator strengths (f) and dipole moments (μ, in Debye) of Trp-Gly (conformer wg01[g+.0.70]), calculated with the ADC(2)/cc-pVDZ and CC2/cc-pVDZ methods at the MP2/cc-pVDZ equilibrium geometry

CC2	ADC(2)			
State	ΔE	ΔE	f	μ
$\pi\pi$ (^1L_b)	4.87	4.85	0.038	4.49
$\pi\pi$ (^1L_a)	4.93 (3.66)^a	4.94 (3.63)^a	0.082	4.18

^{a} adiabatic exitation energy

3.2 Excitation Energies and Molecular Orbitals

The vertical excitation energies, dipole moments and oscillator strengths of the lowest two excited singlet states **of the chromophore** of conformer wg01[g+.0.70] of Trp-Gly are summarized in Table 1.

The lowest two excited singlet states are the well-known ^1L_b ($\pi\pi^*$) and ^1L_a ($\pi\pi^*$) states of the indole chromophore. The ADC(2) vertical excitation energies are 4.85 eV and 4.94 eV. CC2 gives very similar results: 4.87 eV and 4.93 eV. The lowest $^1\pi\pi^*$ excited state has an oscillator strength of 0.038, whereas the oscillator strength of the second state is 0.082. The dipole moment of the lowest excited state is 4.49 Debye, that of the second state is 4.18 Debye. The first excited state thus possesses the higher dipole moment, while the second excited state possesses the higher oscillator strength. On the basis of these properties, we tentatively assign the first excited state as the ^1L_b state and the second state as the ^1L_a state **at the ground-state equilibrium geometry**. Optimization of the lowest excited state of the chromophore yields an adiabatic excitation energy of about 3.63 (3.66) eV with ADC(2) (CC2). The oscillator strength of the lowest chromophore excited state at its optimized geometry is 0.073, compared to 0.0098 for the second state. The corresponding dipole moments are 5.73 Debye and 3.70 Debye. These properties clearly indicate that the lowest excited state **at its equilibrium geometry** is of ^1L_a character. The energetic ordering of the ^1L_b and ^1L_a states is thus reversed upon geometry optimization. Further details on the state reversal in the Franck-Condon region can be found in ref. [23] The MOs associated with the next two excited singlet states are n-type or σ-type orbitals mainly located on the backbone of the dipeptide.

3.3 Proton-Transfer Reaction Path

The energy profiles of the proton-transfer reaction path along the O-H$\cdots$O=C hydrogen bond are depicted in Fig 2.

The proton is transferred from the hydroxyl group of the carboxyl group to the carbonyl group along the hydrogen bond (see Fig. 1). All the symbols in the graph refer to optimized structures (with the constraint of a fixed OH distance); the lines are interpolations of these points. The $\pi\pi^*$ state (green) is denoted by top-down

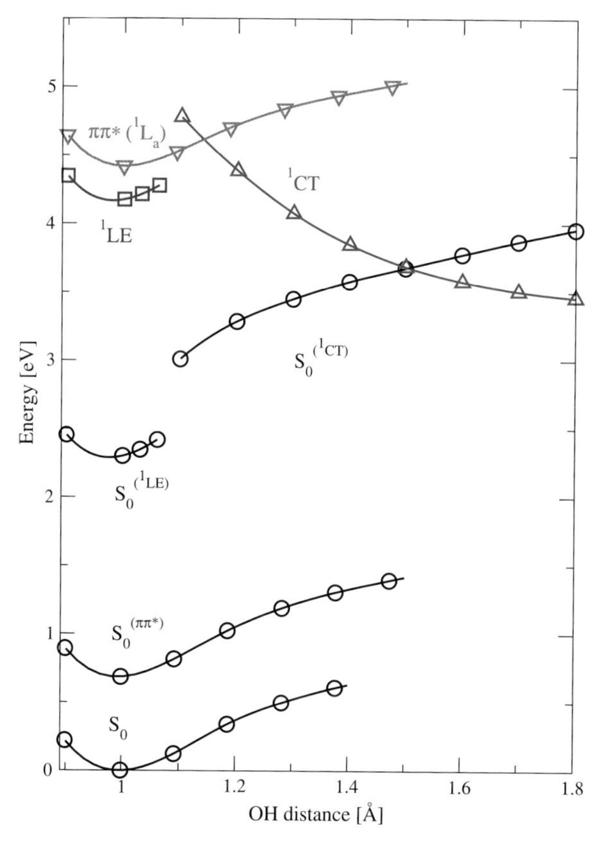

Fig. 2 Minimum-energy profiles (relaxed scans) of the ^1L_a state (top-down triangles), the locally excited state, (¹LE, squares) and the charge-transfer state, (¹CT, top-up triangles) of Trp-Gly along the proton-transfer coordinate. The ^1L_a, ¹LE and ¹CT energies have been determined at the minimum-energy geometries of the respective state. The ground-state energies designated as S_0 have been calculated at the minimum-energy geometries of the S_0 state, while the S_0 energies designated as $S_0^{(CT)}$ have been determined at the minimum-energy geometries of the ¹CT state

triangles, the locally excited state (^1LE, blue) by squares, the charge-transfer state (^1CT, red) by top-up triangles and the ground state (black) by circles. $S_0^{(\pi\pi^*)}$ represents the ground-state energy calculated at the geometry of the $^1\pi\pi^*$ state, etc. As discussed above, the lowest vertically excited state in the Franck-Condon region is the ^1L_b $(\pi\pi^*)$ state. Geometry optimization of the lowest $\pi\pi^*$ excited state yields the ^1L_a $(\pi\pi^*)$ state, see previous section. However, the excited state of Trp-Gly with the lowest adiabatic excitation energy is not the ^1L_a state, but a locally excited state **of the peptide group**. We refer to this state as ^1LE state. This electronic excitation is completely localized on the peptide backbone and the terminal groups. The geometric structure of Trp-Gly at the equilibrium geometry of the ^1LE state is very different from the ground-state equilibrium structure, as can be seen from the increase of the ground-state energy by more than 2 eV (see $S_0^{(LE)}$ curve in Fig. 2). Upon stretching the OH distance, a charge-transfer state (^1CT) is strongly stabilized, see Fig. 2. It is seen that in the ^1CT state one electronic charge unit is transferred from the terminal carboxyl group to the NH and CO groups of the backbone. Upon geometry optimization of the ^1CT state, the proton of the carboxyl group follows the electronic charge along the O-H$\cdots$O hydrogen bond, resulting in a stabilization of the ^1CT state by several electron volts (see Fig. 2). The result of this electron-driven proton transfer (EDPT) process [24, 25] is a biradical. The optimized geometric structure of the biradical is very different from the equilibrium structure of the ground state, which raises the ground-state energy by 3-4 eV (see the $S_0^{(CT)}$ curve in Fig. 2). It should be noted that the crossing of the ^1L_a and ^1CT curves in Fig. 2 is an apparent crossing (since these energy profiles correspond to separately optimized reaction paths), while the crossing of the ^1CT and $S_0^{(CT)}$ curves is a real crossing (conical intersection).

4 Discussion of Reaction Mechanisms

Fig. 2 provides the quantitative basis for the following qualitative mechanistic picture: Assuming absorption of a UV photon near the absorption threshold of Trp-Gly, the system is vertically excited to the ^1L_b state. This state is strongly vibronically coupled with the ^1L_a state. The ^1L_b and ^1L_a states are nearly degenerate in the Franck-Condon region and the oscillator strengths and dipole moments indicate a profound mixing of the ^1L_a and ^1L_b states. Relaxation to the minimum of the ^1L_a surface yields an enthalpy gain of 0.45 eV see ref [23]. From the ^1L_a state, a radiationless transition to the ^1LE state is possible via a low barrier (< 0.2 eV relative to the ^1L_b vertical excitation energy, see ref [23]) From the ^1LE state, another radiationless transition via a low barrier (< 0.15 eV relative to the ^1L_b vertical excitation energy, see ref [23]) can lead to the ^1CT state. The ^1CT state in turn leads in a barrierless manner to a low-lying conical intersection with the electronic ground state, see Fig. 2. A direct nonadiabatic transition from the ^1L_a state to the ^1CT state appears less likely, since a much higher barrier has to be overcome, see Ref [23]. When the system is back in the S_0 state, a considerable amount of energy is released by the back-transfer of the proton to the carboxyl group. By this sequence of radiationless transitions, the energy deposited by the UV photon is converted into vibrational energy in the electronically stable S_0 state. This mechanism enhances the photostability of the dipeptide, since potentially deleterious photochemical reactions are quenched by the ultrafast internal conversion to the ground state.

5 Conclusions

We have investigated possible mechanisms for ultrafast excited-state deactivation of the wg01[g+.0.70] conformer of the Trp-Gly dipeptide. The strong hydrogen bond involving the terminal carboxyl group and the CO group of the backbone plays an essential role in the radiationless decay process. The proposed mechanism provides an explanation for the short excited-state lifetime of this specific conformer and thus the lack of a REMPI signal. The rather detailed mechanistic picture is the following: The chromophore is vertically excited to the strongly mixed ^1L_b state and ^1L_a states. Low-energy curve crossings of the ^1L_a state with the lowest locally excited state of the backbone (^1LE) and of the latter state with a charge-transfer state (^1CT) of the backbone provide access to a conical intersection of the ^1CT with the electronic ground state. By this sequence of nonadiabatic transitions, the electronic excitation energy is converted into vibrational energy. This mechanism explains why certain conformers, namely those with a strong intramolecular hydrogen bond, have a very short excited state lifetime and therefore cannot be detected in the REMPI spectrum. The present findings, together with recent results for in the Gly-Phe-Ala tripeptide [26], indicate that ultrafast excited-state deactivation via photoinduced electron- and proton transfer along intramolecular hydrogen bonds is a generic process in peptides and proteins. The highly efficient radiationless deactivation of excited states may play an essential role for the photostability of these biological molecules.

6 Technical Results

Here we discuss the performance of TURBOMOLE on the HLRB. Calculations were done on an amino acid derivative (namely N-acetyl tryptophan methyl amide (NATMA), 36 atoms) and on a tripeptide (namely Trp-Gly-Gly, 41 atoms), to be published. Runtimes of the system investigated here (Trp-Gly, 34 atoms) are comparable to the runtimes for the systems above. The calculations on NATMA and Trp-Gly-Gly were performed on the HLRB. These calculations cannot be done on a standard computer, since the methods here used are very demanding and the systems investigated here are rather large for such accurate calculations. These calculations have shown, that ab initio calculations on aromatic amino acids and peptides are feasible on this high-performance machine, as can be seen from the results. Here are some details about the performance of TURBOMOLE on this machine.

- Vertical excitation energy calculations of the first 10 excited states of NATMA: 2 days 17 hours (serial). We tested, this calculation cannot be done on a standard machine, because of two reasons: 1) The program is aborted after 6 days and cannot be efficiently restarted. 2) The calculation itself is too time-consuming.
- Vertical excitation energy calculations of the first 5 excited states of Trp-Gly-Gly: about three days (serial). This calculation can also not be done on a regular computer.
- Optimization of the geometry of Trp-Gly-Gly in the first excited state at the CC2 level: On the Linux-Cluster, the calculation of two points at the CC2 level took about 6 days. A reasonable number of points that needs to be calculated until the optimal geometry is found is 50 points. This would take about 150 days to

accomplish. On the HLRB with the parallel implementation of TURBOMOLE with 4 processors about 12 points can be calculated in three days. This is a reasonable time for the calculations.

- Vertical excitation energy calculations of the first 5 excited states of Trp-Gly-Gly: Effect of the number of processors on the CPU time

Number of Processors	Walltime (hours)	Total CPU time (hours)
1	72	72
4	20	73
16	6	85

It is seen that parallelization is very effective up to 16 processors.

Acknowledgements This work has been supported by the Deutsche Forschungsgemeinschaft through the DFG cluster of excellence "Munich-Centre for Advanced Photonics". D.S. acknowledges support by a research fellowship of the Alexander von Humboldt foundation. A substantial amount of computing time on the Höchstleistungsrechner in Bayern (HLRB) has been provided by the Leibniz-Rechenzentrum der Bayerischen Akademie der Wissenschaften.

References

1. Masman, M.F., Lovas, S., Murphy, R.F., Enriz, R.D., Rodriguez, A.M. J Phys Chem A **111**, 10682 (2007)
2. Doslic, N., Kovacevic, G., Ljubic, I. J Phys Chem A **111** 8650 (2007)
3. Brenner, V., Piuzzi, F., Dimicoli, I., Tardivel, B., Mons, M. Angew Chem Int Edit **46** 2463 (2007)
4. Chin, W., Piuzzi, F., Dimicoli, I., Mons, M. Phys Chem Chem Phys **8** 1033 (2006)
5. Valdes, H., Reha, D., Hobza, P. J Phys Chem B **110** 6385 (2006)
6. Compagnon, I., Oomens, J., Meijer, G., von Helden, G. J Am Chem Soc **128** 3592 (2006)
7. Chin, W., Dognon, J.P., Piuzzi, F., Tardivel, B., Dimicoli, I., Mons, M. J Am Chem Soc **127** 707 (2005)
8. Chin, W., Mons, M., Dognon, J.P., Mirasol, R., Chass, G.A., Dimicoli, I., Piuzzi, F., Butz, P., Tardivel, B., Compagnon, I., von Helden, G., Meijer, G. J Phys Chem A **109** 5281 (2005)
9. Bakker, J.M., Plutzer, C., Hünig, I., Haber, T., Compagnon, I., von Helden, G., Meijer, G., Kleinermanns, K. Chemphyschem **6** 120 (2005)
10. Chin, W., Dognon, J.P., Canuel, C., Piuzzi, F., Dimicoli, I., Mons, M., Compagnon, I., von Helden, G., Meijer, G. J Chem Phys **122** 054317 (2005)
11. Reha, D., Valdes, H., Vondrasek, J., Hobza, P., Abo-Riziq, A.G., Crews, B.O., de Vries, M.S. Chem-Eur J **11** 6803 (2005)
12. Hünig, I., Kleinermanns, K. Phys Chem Chem Phys **6** 2650 (2004)
13. Gerhards, M., Unterberg, C., Gerlach, A., Jansen, A. Phys Chem Chem Phys **6** 2682 (2004)
14. Hättig, C. Adv Quantum Chem **50** 37 (2005)
15. Schirmer, J. Phys. Rev. A **26** 2395 (1982)
16. Christiansen, O., Koch, H., Jorgensen, P. Chem Phys Lett **243** 409 (1995)
17. Hättig, C., Weigend, F. J Chem Phys **113** 5154 (2000)
18. Köhn, A., Tajti, A. J Chem Phys **127** 044105 (2007)
19. Ahlrichs, R., Bär, M., Häser, M., Horn, H., Kölmel, C. Chem Phys Lett **162** 165 (1989)
20. Weigend, F., Häser, M. Theor Chem Acc **97** 331 (1997)
21. Perrin, C.L., Nielson, J.B. Annu Rev Phys Chem **48** 511 (1997)

22. Sobczyk, L., Grabowski, S.J., Krygowski, T.M. Chem Rev **105** 3513 (2005)
23. Shemesh, D., Hättig, C., Domcke, W. Chem Phys Lett **482** 38-43 (2009)
24. Sobolewski, A.L., Domcke, W. Chemphyschem **7** 561 (2006)
25. Sobolewski, A.L., Domcke, W. J Phys Chem A **111** 11725 (2007)
26. Shemesh, D., Sobolewski, A.L., Domcke, W. J Am Chem Soc **132** 1374 (2009)

Grid Workflows for Molecular Simulations in Chemical Industry

Ekaterina Elts, Ioan Lucian Muntean, and Hans-Joachim Bungartz

Abstract Today, molecular simulation is a very important technique not only for fundamental research in the science, but also for chemical engineering and chemical industry. Sometimes it is the most efficient or even the only way to obtain useful estimates for parameters and behaviour needed to do traditional chemical engineering process development and design. In order to elaborate appropriate molecular models, one has to carry out extensive parameter studies for workflows, consisting of multiple different operations. This paper discusses a number of aspects of using grid computing methods in support of molecular simulations, with examples drawn from the vapour-liquid equilibria simulations, concerns the GridSFEA software tools development, designed to facilitate the execution, monitoring, and management of such simulations in computational grids, and presents an integrated scientific workflow solution aiming at automating parameter studies for the fast elaboration of molecular models.

1 Introduction

For many technologically relevant tasks in chemical and thermal process engineering, such as gas treating, distillation, condensation, crystallisation, catalyst development and many others, simulations on the molecular level have turned out to be

Dr. Ekaterina Elts · Prof. Dr. Hans-Joachim Bungartz
Technische Universität München, Dept. of Informatics, Chair of Informatics V - Scientific Computing, Boltzmannstr. 3, 85748 Garching bei München, Germany
e-mail: elts@in.tum.de
e-mail: bungartz@in.tum.de

Dr. Ioan Lucian Muntean
Technical University of Cluj-Napoca, Dept. of Computer Science, Baritiu Str. 26-28, Cluj-Napoca, Romania
e-mail: ioan.lucian.muntean@cs.utcluj.ro

S. Wagner et al. (eds.), *High Performance Computing in Science and Engineering, Garching/Munich 2009*, DOI 10.1007/978-3-642-13872-0_54,

necessary. Computer simulation in this field is very important for both quantitative understanding of matter at the nanoscale and describing newly observed phenomena and prediction of reliable physical and chemical properties of the materials components *prior* to synthesis and experiment. Such work implies both the elaboration of models for relevant real fluids for reliable simulations and, due to the tremendous computational resources needed, the development of methods efficient for high capacity molecular simulations.

With the advent of Grid technologies, the complexity of systems that scientists, engineers, and practitioners wish to understand and model ever increases and application scenarios require means for composing and execution of complex scientific workflows and for running these workflows with many different parameter sets. The term *Workflow* is used here to define a series of analytical steps, which describe the process of computational experiments. In chemistry as well, it can be regarded as a big problem divided into a number of subproblems, where each subproblem represents an activity to solve itself. Such workflows make it easy to define the analysis, execute the necessary computations on distributed resources, collect information about the derived data products, and if necessary to repeat the analysis. On the other hand, it is a matter of fact that embedding HPC into such industrial workflows is mandatory for increasing HPC's acceptance and usage there.

Last decade many efforts have been made towards the development of scientific workflow management system (WMS), designed specially to compose and execute series of computational or data manipulation steps or scientific workflows. Among them are such famous systems like Taverna, Triana, Kepler, Pegasus, ASKALON and others. Such WMS are necessary to help scientists to monitor and control research tasks in a single user-friendly GUI, hide complicated distributed, heterogeneous platform issues from scientists, allow them to deal with multiple simulation experiments. Despite the attractiveness of such systems, they still remain underutilised and underexploited by the computational science engineering (CSE) community. Some of the most important reasons for that are:

- The system is not open source.
- Too complicated system, it is not easy to use for the non-IT specialists.
- Very restricted GUI functionality.
- Grid knowledge is necessary for a user to execute a grid workflow.
- The system does not satisfy the user's requirements (very specific application field, no support for parameter study, difficulties executing legacy code).
- Absence of portal-based access: even if workflow tools are made available to domain scientists they often cannot and will not download and install them [2].

It is well known that computational scientists and engineers are typically reluctant to approach the grid due to the usage overhead this one brings. The GridSFEA framework (Grid-based Simulation Framework for Engineering Applications) [9] was designed to bridge the gap between CSE applications and the grid and to provide an *easy-to-use* mechanism for the end user. It allows application-independent parameter studies and the checkpoint-based migration of long-running simulations in the grid. It also has a portal to provide custom interfaces for a broad community of users and was already successfully used for scenarios from molecular dynamics, astrophysics, computational fluid dynamics and others [8].

This ongoing work is devoted to the extension of GridSFEA with the workflow functionality. It would allow the CSE community in general, and the users of GridSFEA in particular, to benefit from the joint advantages of both. New application scenarios, such as grid workflows with parameter study and migration support, would then emerge. For that, we select an appropriate WMS from the existing ones and integrate it into the GridSFEA framework. We report here the first experiences and results.

The remainder of this paper is organised as follows. Sect. 2 presents the aspects of scientific workflow composition with the example drawn from the development of molecular models for vapour-liquid equilibria (VLE) simulation. The GridSFEA framework is briefly introduced in Sect. 3. Sect. 4 describes our work on the integration of GridSFEA with WS-VLAM. Sect. 5 presents current results and status of the work. In Sect. 6 as a conclusion we draw our further plans.

2 Workflows for VLE Simulations

Knowledge on vapour-liquid equilibria is essential for many problems (e.g., gas treating, distillation) in chemical and thermal process engineering. Among the different ways to study VLE, such as experimental and theoretical approaches, molecular simulation plays an important role [11, 13, 14], especially for multicomponent mixtures, where experimental data are available only for a limited number of mixtures and established theoretical methods often lack of predictive power due to the insufficient physical basis.

The search for an appropriate interaction model for a given fluid is usually a time consuming process and always involves extensive parameter studies. In general, the focus lies on *one pure fluid*, where the optimisation of the potential model is done by a number of simulations with subsequent variation of model parameters. Recently, a new method to develop interaction models was proposed by J. Stoll et al. [11] that allows fast adjustments of model parameters to experimental data for a *given class of pure fluids* and considerably reduced the time required for the development of new molecular models. Following this new route, in our article [8] we considered a model elaboration technique for the example of the two-centre Lennard-Jones plus point quadrupole model (2CLJQ). The model is composed of two identical Lennard-Jones sites, situated at a distance L from each other, plus a point quadrupole of momentum Q placed in the geometric centre of the molecule. The idea was to study the thermodynamic properties of the 2CLJQ model fluid systematically and in detail over a wide range of model parameters with the help of GridSFEA. Using reduced coordinates, for the 2CLJQ fluid with fixed angle ϑ, only 2 parameters have to be varied: the LJ centre-centre distance[1] L^* and the point quadrupole strength Q^{*2}.

[1] All values with * are reduced values (transformed to a dimensionless form).

Based on the results from the parameter study, it was straightforward to adjust the molecular interaction parameters of the 2CLJQ fluid to experimental data of real quadrupolar fluids. The systematic investigation of the vapour-liquid equilibria of two-centre Lennard-Jones plus point quadrupole model fluid was performed for a range of reduced quadrupolar momentum $0 \leq Q^{*2} \leq 4$ and of reduced elongation $0 \leq L^{*} \leq 0.8$, with steps 1 and 0.2, respectively. Combining these values led to a set of 25 individual 2CLJQ fluids with different L^{*} and Q^{*2} studied there. Temperatures investigated ranged from 55% to 95% of the critical temperature of fluid. Totally, 125 input files were generated with GridSFEA.

To obtain the VLE data the Grand Equilibrium method [13] was used. The simulation was carried out using the ms2[2] simulation tool, featuring molecular dynamics (MD) and Monte-Carlo (MC) simulation in classical ensembles. For the liquid phase, a MD simulation was performed in NPT ensemble. The data obtained from liquid simulation were further used as phase equilibrium conditions to set the chemical potential, vapour pressure and vapour composition for the vapour phase. This was then simulated in the pseudo μVT ensemble, using the Monte-Carlo method. It means that in order to obtain VLE results we had to execute at least six different operations: staging input data to the computer, MD simulation for liquid phase, generation of an input file for vapour phase MC Simulation from these MD results, MC simulation of the vapour phase, extraction of VLE data from the simulation results, and the graphical result analysis. While it would be easy to execute all these operations one after another manually starting from a single input file, it becomes very difficult if we need a parameter study and have hundreds of different input files. In this case we would need GridSFEA to provide automatic execution not only the first operation in workflow, but also the whole workflow for the all generated input files.

3 The GridSFEA Framework

GridSFEA is a tool for enabling computing applications to efficiently run on grid [8, 9]. This framework is focused on tasks specific to numerical simulation. Its mission is to simplify the act of gridifying an application and to improve the usability of computing grids by means of ready-to-use functionality. The main ones are: the migration of simulation tasks on heterogeneous grids, the remote preview of simulation results, and the enhanced support for parameter studies. Therefore, various computing scenarios can make use of GridSFEA, such as long simulation tasks and parameter studies.

GridSFEA acts rather as a grid application than as a middleware component. From the perspective of the user – a computational scientist or engineer, both end user or application developer – it is a central interaction point with the grid. No significant grid knowledge is required to operate this tool. Through its portal and stand-

[2] The ms2 simulation tool was developed in Inst. of Thermodynamics and Thermal Process Engineering, University of Stuttgart in cooperation with the chair of Informatics V - Scientific Computing, Dept. of Informatics, Technische Universität München.

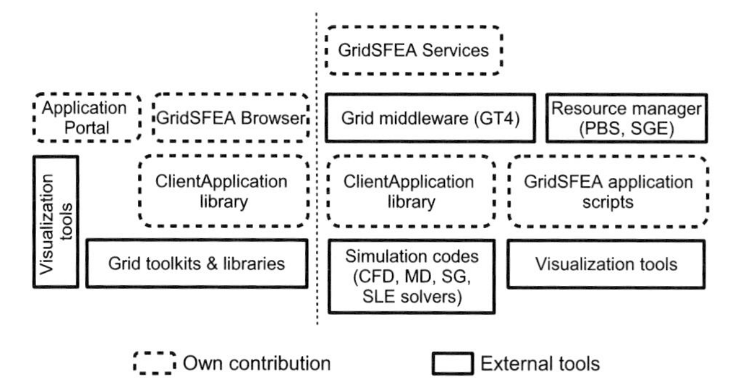

Fig. 1 Modules and components of GridSFEA and external tools that interact directly with the framework

alone tools, GridSFEA hides the specifics and complexity of the grid middleware from its users. From the middleware perspective, this framework is a lightweight grid application.

In Fig. 1 the different components and tools comprised in GridSFEA so far can be seen. They act at different grid levels, ranging from the computing resources (supercomputers) and the grid servers (Fig. 1, right), to the end-user machine (Fig. 1, left). The scientist interacts with the simulation code by means of framework tools.

Some of the framework functionalities can be used as key points in application integration scenarios. From this perspective we shortly present two such functionalities: migration of simulation tasks and enhanced parameter studies.

The first operation allows a simulation task to be migrated in a heterogeneous (both at the hardware and at the middleware level) grid environment. The migration is based on application checkpoints that are managed entirely by GridSFEA. The framework application scripts (see Fig. 1, left) compose the Checkpoint Migration Tool (CMT). CMT contains an extensible hierarchy of wrappers used by GridSFEA to control the interaction of a simulation task with the grid. New simulation codes can be integrated with the framework by simply adding further wrappers. This way they will benefit automatically from the rest of the features.

The second functionality with high integration potential is the support for parameter studies. Parameter definition files can be created using the framework's XML Schema definition. Several generic parameter types are already provided. Further custom types can be easily implemented based on the existing ones. Any program can be used with this feature, as long as it runs in a non-interactive way and it is parameterized.

These two features of GridSFEA could be highly valuable when employed in workflow application scenarios. They can be used to automate the parameter studies and to bring completely new functionalities such as the checkpoint-based migration of long running tasks.

4 Integration of GridSFEA with Workflow Management System

The Grid workflow sector is relatively crowded, with a number of different frameworks, languages, and representations for similar concepts, there is also much overlap in the functionality of the different scientific workflow systems. At the same time all these systems are based on different technologies, vary in terms of their intended scientific scope, their openness to incorporating new services, whether or not they are open source, and sometimes it is not so easy to choose which system to use, because many of them lack functionality.

Here we focus on the extension of GridSFEA with the workflow functionalities from the most appropriate WMS. Sect. 4.1 outlines the criteria we used for choosing the WMS and gives a short review of the present systems. In Sect. 4.2 we talk about the selected system. Sect. 4.3 devoted to the extension of GridSFEA.

4.1 Workflow Management Systems Review

To our mind and accordingly to many other authors and reviews [2, 3, 6, 12, 15, 16] the most important criteria for the evaluating scientific WMS are: 18pt=18pt

1. Open source software – to modify, reuse and improve the existing system.
2. Unrestricted scientific application domain.
3. Support for standard middleware, Web/Grid Services.
4. Legacy code[3] support.
5. User-friendly GUI for easy-to-use, intuitive workflow composition.
6. Rich library of workflow components, support for new component development.
7. Long-running and data-intensive workflow support:
   a. the possibility to detach and reattach the GUI client to the Workflow engine while performing a long running experiment [2];
   b. hierarchical workflows, where workflow components are no longer atomic units – to reuse simple workflows as components of complex workflows;
   c. support for parallel processes and jobs.
8. Parameter studies support – a workflow tool without it is not really usable [6].
9. Fault tolerance – for reliable execution in presence of concurrency and failures.
10. Workflow per-step execution monitoring – to simplify debugging.
11. Interoperability with other WMS.
12. Good documentation (for both user and developer).

The following popular WMS were considered and evaluated on the basis of the above criteria: Taverna, Triana, Kepler, WS-PGrade, ASKALON, ICENI, Pegasus, K-Wf Grid GWES, WS-VLAM, Karajan, gEclipse, UNICORE. In the Table 1 you

[3] The software is considered as legacy code if it cannot be modified.

Table 1 Workflow management systems review. The signs used to show the feature fulfilling grade: (++) – very good, system has all desired functionality; (+) – good, system provides this feature, but has some minor usability or functionality issues; (o) – adequate, system provides this feature, but with some severe restrictions; (-) – bad, system provides only rudimentary support for this feature; (--) – very bad, system does not provide any adequate support

	1	2	3	4	5	6	7 (a/ b/ c)	8	9	10	11	12
Taverna	+	-	-	+	+	++	+ / + / +	+	+	+	-	+/ o
Triana	+	+	+	+	++	++	++/ +/ +	+	o	+	+	+/ o
WS-PGrade	-	+	+	+	+	+	+/ +/ +	++	+	+	+	+/ --
WS-VLAM	+	+	+	++	++	++	++/ +/ +	+	-	+	+	+/ +
Kepler	+	+	-	+	++	++	+/ +/ +	+	++	o	o	+/ o
ASKALON	-	+	+	+	o	o	+/ +/ +	+	++	+	-	+/ --
ICENI	+	+	-	+	-	o	-/ --/ +	+	+	o	-	o/ o
Pegasus	+	+	+	o	-	o	+/ +/ +	o	++	+	o	o/ -
GWES	o	+	+	+	-	o	o/ --/ +	o	+	+	+	+/ +
Karajan	+	+	+	+	-	o	-/ +/ +	+	+	+	-	+/ o
g-Eclipse	+	+	-	+	+	+	?/ --/ +	--	?	+	+	+/ o
UNICORE	+	+	-	+	+	+	++/ --/ +	+	+	+	-	+/ +

can find the results of our review on these criteria. For the WMS Integration with GridSFEA it was important to assure that the WMS is capable to work with GT4 (Globus Toolkit 4), because GridSFEA services are GT4-based. That's why in the third column of the table only the support of GT4 was considered.

As we can see, there is no system that would satisfy all the criteria above. Some of them are not open source, which makes the integration with GridSFEA impossible, some of them are overly complex and have many dependencies from the different libraries (Taverna, Triana, ASKALON), which is also not optimal for the integration into GridSFEA, and in some of them grid execution support is still not handled transparently (Kepler). Other systems still have different problems with GUI or do not support GT4.

The WS-VLAM system more or less satisfies all criteria, but the fault tolerance issues are almost not addressed and there is no possibility for workflow checkpointing. However it has advantage that WS-VLAM engine was designed as WSRF service and, like GridSFEA services, can be easy deployed to GT4 container. It makes the integration with it more intuitive and effective, that is why this system has been selected for the integration with GridSFEA. In addition, there is a hope that it would be possible to implement workflow checkpointing using CMT tool of GridSFEA in GridSFEA framework extended with workflow feature.

4.2 WS-VLAM Organisation and Properties

WS-VLAM (Virtual Laboratory AMsterdam) [7] is a compact GT4-based data driven workflow management system that has been developed at the Institute for Informatics, University of Amsterdam as a part of the Virtual Laboratory for e-Science

(VL-e) project. The system was already successfully used for scientific applications from many different domains, among them medical diagnosis and imaging, biodiversity or bioinformatics [5, 7].

WS-VLAM consists of two core components: a graphical user interface (WS-VLAM client) and a WSRF compliant workflow engine for executing workflows. The GUI is based on an open source JGraph library and provides the visual workflow design using the drag and drop technique. For the workflow creation the user can browse and select proper workflow components from a repository and assemble the selected components by connecting their input and output ports. The topology of the workflow can be stored in XML format. Workflow components are connected through a high performance GSI enabled communication channels, which allow moving data streams between these components directly.

The workflow engine of the WS-VLAM is implemented as a WSRF compliant Web service making it one of the first workflow engines following the Execution Management Services described in the OGF documents. The WS-VLAM engine uses some of the services provided by the grid middleware such as the delegation service, the grid resource manager and the notification mechanisms. All communications between the WS-VLAM client and the WS-VLAM engine use a standard mechanism, based on the End Point Reference to identify and monitor running workflows. This enables an easy detaching of the WS-VLAM client from the engine while performing long running workflows.

WS-VLAM system is one of the few which allow dynamic runtime user interaction with scientific workflow during execution phase, e.g. content modification of a workflow at runtime is supported. The results, obtained while workflow executing, can be also visualized at runtime – the system provides the graphical X11 based workflow component monitoring over the GSI capable private VNC. To enable scalability of the system and the workflows running in the WS-VLAM environment a number of special algorithms for load distribution and resource management on the heterogeneous resources have been developed [7].

4.3 Integration of GridSFEA with WS-VLAM

It is well known that the most successful utilization of grids is done by parameter study applications, where the same application is executed several times with a large number of different input sets. Such parameter study applications are easy to implement on grids since the different execution runs of the applications are independent from each other so the instances can be managed separately [6]. The WS-VLAM system provides only the basic parameter study support (list and range options). In order to use it, the user will either need to provide a number of pre-generated input files (for the list parameter study) or to create a special module with all necessary parameters for each program and then to define a parameter study for each parameter separately. Our task here was to integrate the WS-VLAM System with GridSFEA in order to provide the possibility for the user to perform universal "GridSFEA-like" parameter study for the whole workflow.

GridSFEA uses a flexible and extensible data model for the parameter definition, based on XML schema. A special XML-based description language has been developed there for describing the parameter studies. It allows the description of all possible variations for parameters from the input file in a single XML file. Possible options for the parameter variation are value lists, enumerations, intervals with fixed step, and any combinations of these. This allows for a complete parameter space coverage in a single XML file. This XML file is then processed by the GridSFEA parser and all possible datasets are generated, to be submitted to a job. The job is then launched for each input dataset automatically. Thus it is possible to start thousands of jobs by giving a single XML file and the initial input file template to GridSFEA.

In order to provide the possibility to perform universal GridSFEA parameter study for the whole workflow a WS-VLAM module based on GridSFEA parser has been created. It enables the generation of multiple files from an input file template for a simulation program and an xml-based parameter variation descriptor. These multiple files can then be used as input files for the next module with the simulation program. Thus, the user will not need to create a special module with all necessary parameters for each program, but to use this new "FileGenerator" module. The rest of workflow will be then executed with all generated input files.

So, it is already possible to create and execute the parameter studies based on GridSFEA parameter descriptions for the whole workflow. During the implementation the parameter study execution possibilities have been extended even further. If the program requires more than one input file, which are hierarchically arranged, it is now possible to vary both of these, and not only the topmost one, like it was with GridSFEA.

A series of workflows for simple molecular simulations with the ms2 program were already performed in the grid environment of our chair, see, e.g., Fig. 2.

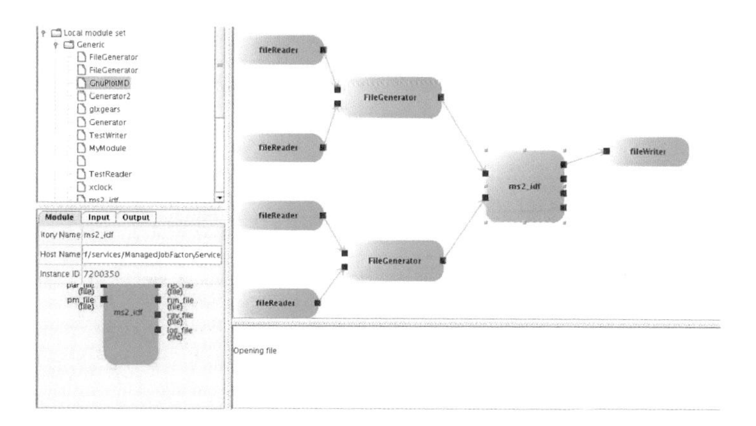

Fig. 2 Creation of scientific workflow for molecular dynamics simulation

Fig. 2 shows a workflow for ms2 program module execution. This program requires two input files, one with the parameters for potential molecular model (coordinates and descriptions of parameters for Lennard-Jones sites, charges, dipoles, quadrupoles), and one with the parameters for the simulation (ensemble, temperature, density, pressure and so on). The parameters from each file can be varied, if necessary, by providing an optional XML parameter variation descriptor file. These are passed by the file readers to a file generator. Each file generator module, based on GridSFEA parser, creates an appropriate set of input files, and each combination is then submitted for the simulation execution. This simple workflow can easily be extended with other modules, for example, for VLE simulations described in Sect. 2.

5 Results

The molecular simulation can be beneficial in reducing process development time and production costs in the chemical industry, but unfortunately it is still underutilised by it [4]. The main reason for that is a lack of reliable molecular models and of effective and easy-to-use tools that would provide fast elaboration of such models. In this work we are trying to solve this problem and to provide such a system.

The ms2 program, specially developed for VLE simulations was successfully tested in this field. Recently, it received a first place[4] in the International Simulation Challenge, organised by the Industrial Fluid Properties Simulation Collective [1]. The program is parallelised using the message passing interface (MPI). Plimpton's force decomposition method [10] was implemented for parallel execution, which scales reasonably well up to 16 processors. Fig. 3 shows how execution time decreases as number of processors increases for the simulations carried out on HLRB II. Usually such simulations are computationally expensive, we need about 5 hours of CPU time to simulate 10 thousand steps molecular dynamics for a system, consisting of 864 particles (see Fig. 3). Only the first step of VLE workflow, described in Sect. 2, would need 10-12 times more steps. The same workflow should be executed more than one hundred times with different sets of parameters. It is clear, that the parallel execution on HLRB II strongly reduces the time needed to provide the results (already 3 times while using 4 processors), while it would be much better in the case of parameter study, when hundreds of workflows will be executed simultaneously on the multiple cores each. In this case we achieve a two-level parallelism. Assuming a simulation for a single dataset taking 5 hours, and a hundred possible dataset variations, the simulation would need 500 hours in a usual case. Simultaneous execution on a HPC, like HLRB II, would allow the entire simulation to complete in 5 hours. Thus the performance increment is of the same order of magnitude as the number of dataset variations.

[4] http://fluidproperties.org/4th-challenge-results

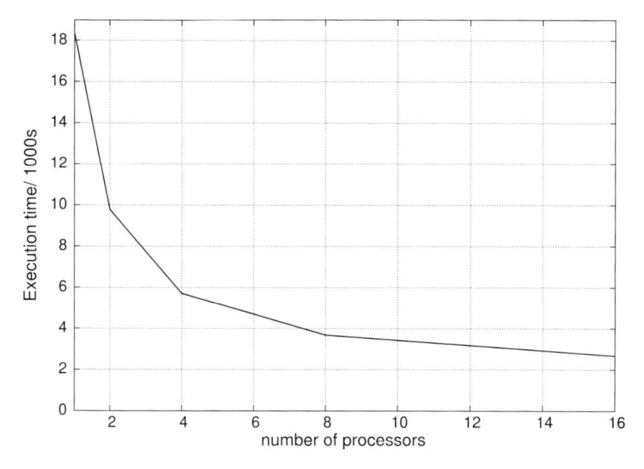

Fig. 3 Execution time vs. number of processors on HLRB II for ms2 program

The GridSFEA framework, with its support for parameter space investigations and for long running simulations, is currently being extended with the workflow functionality, based on WS-VLAM workflow system. It would make it even better suited not only for the VLE case study discussed here, but also for most complex CSE scenarios specific to capacity computing.

WS-VLAM integration with GridSFEA is already implemented to a large extent. The WS-VLAM module based on GridSFEA parser is already created. It makes a realisation of arbitrary parameter studies for the whole workflow easy, the user only needs to describe parameters that should be varied in an xml file. During the implementation the parameter study execution possibilities have been extended even further, than it was in GridSFEA. If the program requires more than one input file, which are hierarchically arranged, it is now possible to vary all of these, and not only the topmost one, like it was before. So, it is already possible to create and execute the parameter studies based on GridSFEA parameter descriptions for the whole workflow. So far a feasibility test was successfully completed on the grid environment of our chair.

6 Concluding Remarks

Here we presented first results and experiences of our ongoing work of extending GridSFEA with the workflow functionality. As a first step, we tackled the parameter study integration. Further, we intend to enable workflow operations with check-pointing. This way it would be possible to resume the workflow execution from an arbitrary point and from an arbitrary location in the grid. Having such complex and convenient functionalities available, we plan to perform a series of simulations using the new system on the HLRB II.

Acknowledgements We would like to thank the Bayerische Forschungsstiftung for financial support of one of the authors. We are grateful to the WS-VLAM team, and especially to Dr. V. Korkhov and Dr. A.S.Z. Belloum at the Institute of Informatics, University of Amsterdam for their collaboration. Furthermore, we acknowledge the collaboration on the VLE topic with Prof. Dr.-Ing. habil. J. Vrabec at the Chair of Thermodynamics and energy technologies, Department of Engineering, Universität Paderborn.

References

1. Industrial simulation collective. URL http://www.fluidproperties.org
2. Barker, A., van Hemert, J.: Scientific Workflow: A Survey and Research Directions. Lecture Notes in Computer Science **4967**, 746–753 (2008). Springer
3. Gil, Y., Deelman, E., Ellisman, M., Fahringer, T., Fox, G., Gannon, D., Goble, C., Livny, M., Moreau, L., Myers, J.: Examining the Challenges of Scientific Workflows. IEEE Computer **40**(12), 24–32 (2007)
4. Gupta, S., Olson, J.: Industrial needs in physical properties. Industrial & Engineering Chemistry Research **42**(25), 6359–6374 (2003)
5. Inda, M.A., Belloum, A.S.Z., Roos, M., Vasunin, D., de Laat, C., Hertzberger, L.O., Breit, T.M.: Interactive Workflows in a Virtual Laboratory for e-Bioscience: The SigWin-Detector Tool for Gene Expression Analysis. In: Proc. Second IEEE International Conference on e-Science and Grid Computing (e-Science'06), pp. 19–26 (2006)
6. Kacsuk, P., Karoczkai, K., Hermann, G., Sipos, G., Kovacs, J.: WS-PGRADE: Supporting parameter sweep applications in workflows. In: Proc. Third Workshop on Workflows in Support of Large-Scale Science WORKS 2008, pp. 1–10 (2008)
7. Korkhov, V., Vasunin, D., Wibisono, A., Guevara-Masis, V., A.Belloum, de Laat, C., Adriaans, P., L.O. Hertzberger: WS-VLAM: Towards a scalable workflow system on the Grid. In: 16th IEEE International Symposium on High Performance Distributed Computing, pp. 63–68. Monterey Bay, California, USA (2007)
8. Muntean, I., Elts, E., Buchholz, M., Bungartz, H.J.: Grid-supported simulation of vapour-liquid equilibria with GridSFEA. In: Proceedings of 8th International Conference Computational Science - ICCS 2008, vol. Part I, pp. 45–55. Springer-Verlag, Krakow, Poland (2008)
9. Muntean, I.L.: Efficient Distributed Numerical Simulation on the Grid. Ph.D. thesis, Institut für Informatik, Technische Universität München, Verlag Dr. Hut, München, ISBN 978-3-89963-870-7 (2008)
10. Plimpton, S.: Fast parallel algorithm for short-range molecular dynamics. Journal of Computational Physics **117**, 1–38 (1993)
11. Stoll, J., Vrabec, J., Hasse, H., Fischer, J.: Comprehensive study of the vapour-liquid equilibria of the two-centre Lennard-Jones plus point quadrupole fluid. Fluid Phase Equilibria **179**, 339–362 (2001)
12. Taylor, I.J., Deelman, E., Gannon, D., Schields, M. (eds.): Workflow for e-Science: Scientific Workflows for Grids. Springer-Verlag (2007)
13. Vrabec, J., Hasse, H.: Grand Equilibrium: vapour-liquid equilibria by a new molecular simulation method. Molecular Physics **100**(21), 3375–3383 (2002)
14. Vrabec, J., Stoll, J., Hasse, H.: A set of molecular models for symmetric quadrupole fluids. Journal of Physical Chemistry B **105**, 12126–12133 (2001)
15. Yu, J., Buyya, R.: A taxonomy of scientific workflow systems for grid computing. Journal of Grid Computing **3**(3), 171–200 (2005)
16. Zhao, Y., Raicu, I., Foster, I.: Scientific Workflow Systems for 21st Century, New Bottle or New Wine? In: Proc. IEEE Congress on Services - Part I SERVICES '08, pp. 467–471 (2008)

Global Chemistry-Climate Modelling with EMAC

R. Sausen, R. Deckert, P. Jöckel, V. Aquila, S. Brinkop, U. Burkhardt, I. Cionni, M. Dall'Amico, M. Dameris, S. Dietmüller, V. Eyring, K. Gottschaldt, V. Grewe, J. Hendricks, M. Ponater, and M. Righi

Abstract The Institute of Atmospheric Physics of the German Aerospace Center (DLR) uses the numerical model system ECHAM/MESSy Atmospheric Chemistry (EMAC). The model has a flexible modular structure and allows for coupled chemistry-climate simulations. Typical fields of application are related to questions regarding Earth's climate, atmospheric chemical composition, and aerosol characteristics. In its current setup, the performance of EMAC on LRZ/ALTIX allows for multi-decadal simulations with climatologically significant results. The good performance demonstrates the multi-purpose capabilities of LRZ/ALTIX because EMAC involves various different numerical concepts and implementations of parallel decomposition. Our EMAC activities on LRZ/ALTIX are devoted to both model development and production simulations. The former comprise a new upper-boundary representation, a chemistry-transport mode, the inclusion of a mixed-layer ocean, and full-Lagrangian transport and dynamics. The latter tackle, for instance, questions related to the environmental impact of anthropogenic aerosol and gaseous substances.

1 Introduction

The ECHAM/MESSy Atmospheric Chemistry (EMAC) model is a numerical chemistry and climate simulation system that includes sub-models describing tropospheric and middle atmosphere processes and their interaction with oceans, land

R. Sausen · R. Deckert · P. Jöckel · V. Aquila · S. Brinkop · U. Burkhardt · I. Cionni · M. Dall'Amico · M. Dameris · S. Dietmüller · V. Eyring · K. Gottschaldt · V. Grewe · J. Hendricks · M. Ponater · M. Righi
Deutsches Zentrum für Luft- und Raumfahrt (DLR), Institut für Physik der Atmosphäre, Oberpfaffenhofen, 82234 Wessling, Germany
e-mail: robert.sausen@dlr.de

S. Wagner et al. (eds.), *High Performance Computing in Science and Engineering, Garching/Munich 2009*, DOI 10.1007/978-3-642-13872-0_55,

and human influences [8]. It uses the first version of the Modular Earth Sub-model System (MESSy1) to link multi-institutional computer codes. The core atmospheric model is the 5th generation European Centre Hamburg general circulation model ECHAM5 [15]. More detailed information, including references, about the model system is available from `http://www.messy-interface.org`. Currently 32 different sub-models are available.

The modular structure of EMAC allows its application for a wide variety of scientific tasks. It can be applied in different configurations; for instance as pure atmospheric general circulation model, as atmospheric-chemistry transport model, the latter also as chemistry-climate model with full feedback between atmospheric chemistry and dynamics via the radiation calculations. Optionally, in order to represent the observed meteorological patterns, the model dynamics can be relaxed towards analysis and re-analysis data of the European Centre for Medium-Range Weather Forecasts (ECMWF). This is particularly useful for the evaluation of the model in comparison to observations.

Besides the flexibility in model configuration and setup the resolution is likewise flexible. The core model, ECHAM5, uses a spectral transform technique, the so-called T-value indicating the degree of triangular spectral truncation. T42, for instance, corresponds to a Gaussian grid of approximately 2.8° by 2.8° in longitude and latitude. Vertically, the model resolves the troposphere and lower stratosphere, implemented as model setups L41DLR and L31ECMWF. The top of the model layers are centered at 5 hPa and 10 hPa, respectively. In further versions, L39MA and L90M, the stratosphere and the lower mesosphere are better resolved, with the uppermost model layer centered at hPa and hPa, respectively. The L-value indicates the number of discrete hybrid-pressure levels. MA refers to "middle atmosphere", and DLR and ECMWF denote specific vertical-layer structures associated with these institutions.

EMAC is easily portable to various architectures because it follows some important standards: Fortran95 standard, ISO/IEC-1539-1, Message Passing Interface standard for parallel computation, MPI-2, and Network Common Data Form data format, NetCDF.

This article demonstrates our modelling activities on LRZ/ALTIX with EMAC. Our activities encompass a wide range of tasks in the field of atmospheric science with some being focused on model development. This makes is hard to combine our activities into a coherent text structure. Rather, we start with a brief overview on computational methods and performance of EMAC in standard configuration (Section 2) which forms the basis of our modelling work. Some more specific EMAC sub-modules are presented later in Section 3 and 4. While the former Section presents activities related to model development, the latter presents results from production-oriented simulations. Finally, we draw conclusions in Section 5.

2 Technical Information

The core model of EMAC, ECHAM5, solves the atmospheric primitive equations horizontally via a spectral transform technique. In the vertical it uses the method of finite differences and for time integration a semi-implicit leap-frog scheme with time filter.

Most of the EMAC sub-models act on a Gaussian transform grid, i.e. in the physical-space domain. Examples are parameterised physics, advection of atmospheric constituents, and atmospheric chemistry. Depending on the complexity of the chemistry setup, the chemical calculations usually consume the largest portion of the total CPU time. Chemistry is dealt with by the sub-model MECCA1 [17], which is based on the kinetic pre-processor software [1] and considers the combined chemical reactions as a stiff system of ordinary differential equations (ODE). The system is solved via a Rosenbrock sparse-matrix technique with adaptive time step.

The parallelisation is implemented based on the distributed memory concept with three different approaches for the spectral and the grid-point representation, and a specific representation for the tracer advection algorithm, respectively. In spectral representation variables are decomposed into their spherical harmonics, in Gaussian representation the global domain is decomposed into blocks in latitudinal and longitudinal direction, for the advection an additional vertical decomposition is applied. For each variable in Gaussian representation two blocks in North-South direction are combined for an improved load balancing with respect to the distribution of day and night grid-boxes. These combined blocks are further re-arranged into vectors of selectable length to optimally exploit vector registers on vector architectures and cache sizes on scalar architectures.

Table 1 provides information on the performance of EMAC on LRZ/ALTIX for a typical setup, including chemistry, in two different resolutions. The wall-clock time per model year is in the range of a few days and the corresponding CPU time in the range of 5–12 kilohours. This benchmark allows for multi-decadal simulations, particularly in T42L41DLR, the default resolution applied for production simulations at our institution. In case of T42L41DLR the performance gain decreases significantly from 64 to 128 CPUs and in case of T42L90MA it does so from 128 to 256 CPUs (not shown). Presumably, these breaks in scalability result from two different effects: first, from overhead enhancement associated with communication

Table 1 Performance of EMAC on LRZ/ALTIX. The model configuration includes detailed atmospheric physics and complex atmospheric chemistry but neglects detailed aerosol effects

resolution	number of CPUs	CPU time (10^3 h/model year)	wall-time (days/model year)
T42L41DLR	64	5.4	3.5
T42L41DLR	128	7.9	2.6
T42L90MA	64	12.0	7.9
T42L90MA	128	12.1	4.0

between the nodes on LRZ/ALTIX, second, from load-imbalance associated with photo-chemistry as the stiffness of the chemical ODE system varies throughout a simulation day by orders of magnitude. Further tests on this are required and will be conducted in near future. The load imbalance issue, which is architecture independent, is already being tackled so that future EMAC versions will have improved scalability characteristics.

3 Model Development

Much of our modelling activity on LRZ/ALTIX is related to indispensable model development which represents a prerequisite for many of our planned production simulations. At the same time, these activities are absolutely necessary to keep pace with scientific questions becoming more complex and to push the state of the art of global chemistry-climate modelling. Our activities are focused on

- establishing a new upper-boundary parameterisation for our 41-layer setup of EMAC (Section 3.1),
- creating a chemistry-transport mode for EMAC where dynamics and chemistry are completely decoupled (Section 3.2),
- coupling a mixed layer ocean to EMAC atmospheric dynamics and chemistry (Section 3.3),
- creating a Lagrangian base model for EMAC as an alternative to ECHAM5 (Section 3.4).

3.1 Upper-Boundary Parameterisation

A high spatial resolution of our global chemistry-climate model is desirable to resolve all relevant physical and chemical processes. It is limited however, by the available resources: CPU time, memory, and storage. Our standard spectral horizontal resolution for production simulations is T42, corresponding to $2.8^{\circ} \times 2.8^{\circ}$ in longitude and latitude. The highest vertical resolution available is 90 layers between surface and top of the atmosphere (L90).

One focus of atmospheric research at the DLR is to study the effects of transport emissions on climate. Aircrafts usually fly at levels between 200-350 hPa and all other anthropogenic emissions occur in the surface boundary layer. Therefore, we do not rely on a high vertical resolution at altitudes above 100 hPa and, hence, can benefit from the advantage associated with fewer vertical layers. We use a 41 layer setup (L41) instead of L90. L41 has a higher vertical resolution between surface and 100 hPa, and simulations are about 2.4 times faster and have less than half the storage requirements than L90. A speed-up by 2.4 is important as a typical 30-year simulation with L90 and comprehensive chemistry would require about 360 kCPUh (see Table 1).

On the downside, the vertical resolution of L41 decreases above 100 hPa. The atmosphere above 10 hPa is represented by 36 model layers in L90, but by just one layer in L41. A single layer is not sufficient to describe the chemical processes involving Br, N2O, CH4, and CFCs, in that altitude range correctly. Reaction rates are calculated for the 5 hPa centre level and assumed to be vertically constant throughout the whole uppermost layer in L41 which ranges from 10 to 0 hPa. However, real reaction rates depend on altitude non-linearly. Therefore, the 5 hPa rates are unlikely to represent the bulk rates for the 10 to 0 hPa layer. This affects 13 different multi-component gas phase reactions and 10 different photolytic reactions.

We developed a technique to account for the drawbacks associated with L41. Artificial reservoir species are introduced, which are not transported and exist in the uppermost layer only. The products from the degradation reactions of N2O, CFCs, CH_4 and Br- compounds go into their respective reservoirs, with rate coefficients modified in order to approximate the L90 rates in L41. Reservoir species are then transformed into reactive species with a turn-around time of five years, taking into account stratospheric transport this way.

Currently, we are basing the correction factors on the inter-comparison of two simulations; on the cumulative losses in the upper 36 layers of a L90 simulation and on the losses in the uppermost layer of a L41 simulation. Both simulations had two years of spin-up and there are three years for evaluation. This is relatively short for climatological averaging, but the noise is reasonably low due to slow stratospheric dynamics and fixed boundary conditions controlling photolysis reactions. It is desirable to start future simulations from a steady state with non-empty reservoirs. Also, strong alterations of the chemical configuration may represent a problem since the upper boundary parameterisation is currently based on only two simulations with a specific setup. Finally, a comparison with observations will be the ultimate test of our parameterisation.

3.2 Decoupling of Dynamics and Chemistry

Global chemistry-climate models account for the full non-linear coupling between atmospheric dynamics and chemistry. However, identifying the climate effect of short-lived anthropogenic substances, for instance of nitrogen oxide ($NO_x = NO + NO_2$), requires a comparison of at least two modelling setups: one without a particular NO_x source and one with. This results in long and costly integration times due to noise from atmospheric transport interfering with the NO_x signal [18]. The problem is particularly prominent in case the focus lies on the separate climate impacts by multiple simultaneous NO_x emission sources such as road traffic, shipping, or aviation.

A promising way to deal with such problems is to neglect any feedbacks between atmospheric chemistry and dynamics, i.e., to run the model in chemistry-transport mode. Leaving any coupling processes in the model is not sensible as model dynamics is affected by any slightest disturbance. In chemistry-transport mode, simulations with different NO_x sources display different chemical properties but have binary identical meteorological conditions and transport characteristics. As a result, the integration time needed to detect a given NO_x signal becomes much shorter and in some cases might even be possible solely with this approach. And finally, the approach allows for a quantification of signal feedbacks as it is possible to compare coupled against decoupled simulations.

We have made disengageable any chemistry-climate feedbacks and by far the major hurdle turned out to be feedbacks associated with polar stratospheric clouds (PSCs). PSCs represent an important feature of the global climate system and should not be neglected. Yet, occurrence and characteristics of PSCs depend on the stratospheric chemical composition, for instance on HNO_3 concentrations, and therefore on NO_x emissions. PCSs in turn affect atmospheric water through heterogeneous chemical reactions as well as sedimentation of water-containing particles. Water, finally, transfers the emission signal to atmospheric dynamics via both the hydrological cycle and radiative heating. The result is a chemistry-climate feedback which is not wanted for this purpose.

To resolve the PSC issue, we decided to impose predefined climatological tendencies for liquid water and ice. The tendencies are based on EMAC simulations in chemistry-climate mode. Imposing the tendencies into chemistry-transport simulations must be done water mass preserving. Our approach is to force artificial phase changes, i.e. to create missing ice contents from the model-inherent water vapor concentrations while preserving column-integrated total water. The latter is important because sedimentation redistributes water vertically.

The code has been implemented and tested on LRZ-ALTIX. Clearly, the water mass is preserved and the water tendencies change only little when taken from a chemistry-climate simulation and incorporated into a chemistry-transport simulation. Also, the procedure conserves the overall tendency patterns.

3.3 Atmosphere-Ocean Feedbacks

For studies involving climate feedbacks and climate response it is important to use a chemistry-climate model that includes an ocean module. For our studies EMAC is coupled to a slab ocean model with an oceanic mixed layer of 50 m depth (MLO) [16]. Any heat exchange with the deep ocean is represented by a prescribed climatological annual cycle. A thermodynamic sea-ice model is also integrated in this module [12], enabling albedo feedbacks via changes in ice coverage and ice thickness. An ECHAM5/MLO model configuration has already been used to investigate the importance of physical feedbacks, e.g., cloud feedback, for climate sensitivity. The role of chemical feedbacks, e.g., from ozone, has never been quantified by similar simulations so far.

The MLO module was implemented into the EMAC model system and test simulations were performed on LRZ/ALTIX. Equilibrium climate change simulations driven by external forcings involving both physical and chemical feedbacks have thus become possible [11]. While the slab ocean does not need substantially more CPU time than the pure atmospheric model, equilibrium climate change simulations require 20 spin-up years to adjust the perturbed system to the changes in forcings and feedbacks and at least another 20 simulation years are required in order to study the deviation between the equilibrium climates of the perturbed and the reference simulation. This means that a model system like this has considerable larger demands to computational economy than atmosphere/ocean models without chemistry or chemistry-climate simulations without the ocean component. A specific setup of the chemistry sub-model MECCA1 [17] has been prepared for the intended studies that offers an adequate compromise between chemical complexity and computational resources. We use EMAC with 41 vertical layers, L41DLR, in T42 horizontal resolution (see Section 1). Simplifications with regard to chemistry include the omission of reactions involving non-methane hydro carbons and bromine.

3.4 Lagrangian Modelling

Due to the increasing demand for interactive tracers in climate chemistry simulations it becomes necessary to use global models which meet the needs of a fast and exact tracer transport scheme. Commonly used methods to describe the large-scale transport in a general circulation model of the atmosphere follow the Eulerian method. However, Lagrangian transport schemes offer several advantages with respect to tracer transport: they show no numerical diffusion, they maintain steep gradients and they are efficient in the treatment of a large number of tracers. Therefore, the Lagrangian transport model ATTILA (Atmospheric Tracer Transport in a Lagrangian Model) was developed [13]. It runs in the framework of a general circulation model and advects the centroids of air parcels whose number depends on the selected vertical and horizontal resolution. The latest version was implemented in the EMAC climate model and completed by a Lagrangian convection scheme. This model system is run in T21 and T42 and with 31 or 41 vertical levels. The coarse resolution was used mainly for sensitivity studies related to the resolution of the Lagrangian space with a certain number of Lagrangian air parcels. The computer time increases nearly linearly with the number of Lagrangian air parcels in the T21L19 model resolution requiring about 4 hours/CPU per simulated year for 90000 Lagrangian parcels.

Further research is being done towards the development of a pure Lagrangian dynamical core in the EMAC climate model with the aim of calculating all transport with the same transport scheme. So far, energy and momentum are calculated by the dynamical spectral core of the EMAC model, but tracers are advected by a Eulerian approach to ensure mass conservation. This results in a certain degree of inconsistency, which we plan to overcome. The new Lagrangian dynamical core will be realized with the concept of the Finit-Mass Method [5]. The atmosphere is subdivided into small mass packets each of which is equipped with a finite number of internal degrees of freedom. In ATTILA the Lagrangian air parcel are just mass points without spatial extension. These mass packets move under the influence of internal and external forces and can change their shape to follow the motion. The total mass density results from the superposition of the individual mass packets. The implementation of the Finite Mass Method requires the development of a new submodel for the temporal integration of the deformation matrix. In a first step we will analyse the characteristics of the simulated parcel distribution from a set of climate simulations over several years.

4 Production-Oriented Simulations

The sample of scientific results presented here is based on some of our modelling activities on LRZ/ALTIX. Our findings refer to the environmental impact of anthropogenic volatile particles (aerosol), their precursor gases, and other gaseous substances. Section 4.1 demonstrates how particle aging changes aerosol characteristics and Section 4.2 is devoted to the impact on climate and public health by shipping emissions.

4.1 Atmospheric Ice Nuclei

Black carbon (BC) and mineral dust particles are among the most important atmospheric aerosol types forming ice crystals by heterogeneous nucleation (so called ice nuclei). Anthropogenic changes in the availability of such ice nuclei can have important climate effects. When emitted, most BC and dust particles are externally mixed with other aerosol compounds. Through coagulation with particles, condensation of gases, as well as cloud processing, externally mixed particles gain a liquid coating and are transferred to an internal mixture. These aging processes are essential for the effects of BC and dust particles on climate, since the coating changes the radiative and hygroscopic properties of the particles and, therefore, their cloud activation ability and lifetime. Moreover, laboratory studies have shown that a liquid coating influences the freezing properties of the particles and hence their behavior as ice nuclei [2, 10]. In the upper troposphere and lowermost stratosphere (UTLS) BC is emitted by aircraft engines. With the rapid increase in the amount of air traffic its contribution to the BC concentration is gaining more and more importance [6].

Due to large computational resources required, global climate models mostly parameterise the aging of BC and dust by using estimated turnover times, rather than simulating the aging processes explicitly. The LRZ/ALTIX offers extraordinarily high computational capacities. Hence, we took advantage of this opportunity and further developed a new aerosol model, MADEsoot, which allows for a representation of BC and dust particles and their different states of mixing as well as the relevant aging processes of externally mixed particles. MADEsoot is based on the

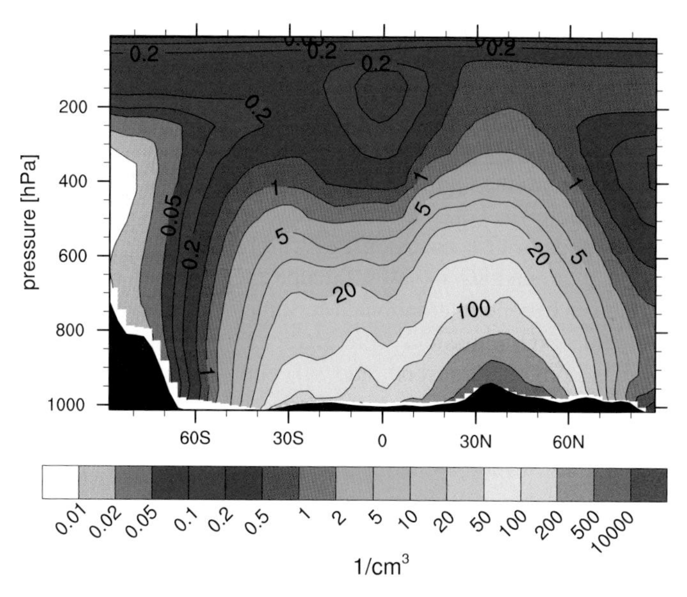

Fig. 1 Zonal annual mean number concentration of potential ice nuclei

standard EMAC aerosol module MADE [9]. We implemented MADEsoot in the global chemistry-climate model EMAC [7, 14] and performed a large number of simulations to assess the concentration, size distribution and mixing state of BC and dust particles in the global atmosphere. The mass and number concentration of internally and externally mixed BC and mineral dust particles were simulated separately for particles in two modes; in the Aitken mode, particles smaller than 0.1 μm, and in the accumulation mode, particles larger than 0.1 μm but smaller than 1 μm. Moreover MADEsoot uses two additional modes for BC and dust free particles and a coarse mode for particles larger than 1 μm.

Figure 1 shows the vertical distribution of the zonal mean of the number concentration of potential ice nuclei (BC and dust particles). The simulation has been performed on LRZ/ALTIX with EMAC-MADEsoot. While at surface the number concentration of potential ice nuclei reaches values between 500 and 1000 particles/cm^3, at the 300 hPa level, which is located at altitudes where cirrus clouds form, it never exceeds 10 particle/cm^3. The highest values are reached over South-West China, which is characterised by very high BC emissions (not shown).

The global aerosol-climate model EMAC-MADEsoot requires about 600 CPUh for each simulated year using a T42L19 grid. A climatologically significant simulation must cover at least 10 years. Furthermore, several different simulations are required to study separately the influence of each process or parameterisation on the model results. A set of 5 sensitivity experiments will consume about 30000 CPUh.

4.2 Impact of Ship Emissions on Atmospheric Composition and Climate

Emissions from international shipping contribute significantly to the total budget of anthropogenic emissions and have been recognised as a growing problem by both scientists and policymakers. Already in 2000, shipping contributed with around 2.7% to all anthropogenic CO_2 emissions, with around 15% to nitrogen oxide (NO_x) emissions and with around 8% to sulfur dioxide (SO_2) emissions [3]. If no control measures were taken beyond International Maritime Organisation (IMO) regulations that existed in 2005, NO_x emissions were predicted to further increase to values of today's emissions from road transport, and SO_2 emissions were predicted to double until 2050 [4]. We aim at quantifying the impact of gaseous and particulate ship emissions on the chemical composition of the atmosphere and on climate for present day conditions and for several scenarios of future development.

A variety of important results have been achieved. One of the major findings was that the potential of particle emissions or their precursors from shipping to modify the microphysical and optical properties of clouds (the so-called *indirect aerosol effect*) is significant [9]. The additional aerosol particles brighten the clouds above the oceans, which then are able to reflect more sunlight back into space (see Fig. 2). The model results indicate that the cooling due to altered clouds far outweighs the warming effects from greenhouse gases such as CO_2 or ozone from shipping, overall causing a negative radiative forcing today. The indirect aerosol effect of ships on climate is found to be far larger than previously estimated, contributing up to 39% to the total indirect effect of anthropogenic aerosols. This contribution is high because ship emissions are released in regions with frequent low marine clouds in an otherwise clean environment and the potential impact of particulate matter on the radiation budget is larger over the dark ocean surface than over polluted regions over land. The main reason for the high impact on clouds is the high average sulfur content in maritime fuels.

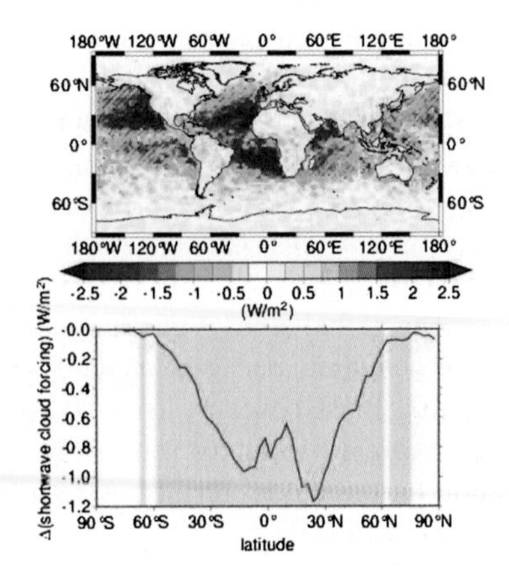

Fig. 2 Multi-year average of simulated changes in short-wave cloud forcing due to shipping at the top of the atmosphere in W/m^2. Upper panel shows the geographical distribution, lower panel zonal averages. Hatched areas (upper panel) and light-red shaded areas (lower panel) show differences which are significant at the 99% confidence level compared to the inter-annual variability (from [9])

We are planning to study how climate impacts of particulate and gaseous emissions from large diesel engines change when fuels from renewable sources are used. As a first step we have repeated the [9] simulations on LRZ/ALTIX and are now running simulations where heavy fuel oil is replaced with bio fuels in the international shipping fleet. Each experiment encompasses a period of 6 years (1999–2004), preceded by a spin-up time of one year, and uses the relaxation technique. The cost of each modelled year is around 400 CPUh, with a total output for each simulation of about 450 GB. Preliminary simulations suggest that replacing heavy fuel oil with bio fuels (e.g. palm oil) significantly reduces the near-surface sulfate (SO_4) concentration as well as the effect of shipping emissions in enhancing clould reflectivity.

In 2010, we plan to study the effect of parameterising the gas phase chemical processes in subgrid scale engine exhaust plumes of aviation and shipping. This study should allow a better representation of the impact of these emissions in particular on the tropospheric ozone burden.

5 Final Remarks

The high-performance computing facility LRZ/ALTIX represents an excellent platform for the numerical-simulation activities by the Institute of Atmospheric Physics of the German Aerospace Center. We use the model system EMAC to conduct simulations for a variety of scientific tasks that are related to questions concerning Earth's climate, atmospheric chemical composition, and aerosol characteristics. EMAC's major benefit is its ample flexibility to incorporate new code due to a modular architecture and the possibility of doing coupled chemistry-climate simulations.

In its current setups, the performance of EMAC on LRZ/ALTIX allows for multi-decadal simulations which is necessary to obtain climatologically significant results. EMAC is a model system that comprises various different numerical concepts, from spectral transformation and time integration to linear algebra and Lagrangian methods, with different types of distributed-memory parallelisation. Hence, the good performance demonstrates that LRZ/ALTIX represents a flexible computing platform.

Many of our modelling activities on LRZ/ALTIX are aiming at creating EMAC setups required for present and future scientific tasks. These setups comprise a new upper-boundary representation, a decoupling of chemistry and climate to allow for chemistry-transport simulations, the inclusion of a mixed-layer ocean, and full-Lagrangian transport and dynamics.

The future is going to bring even higher demands on computing performance. One reason is the need for higher model resolution in order to resolve more and more of the small-scale parameterised physics. Another reason is the need for climate simulations that do not only account for radiation-coupled atmospheric dynamics and chemistry but also for ocean feedback. Future versions of EMAC will have improved parallel scalability so that simulations on a greater number of cores will be feasible.

References

1. Damian, V., Sandu, A., Damian, M., Potra, F., Carmichael, G.: The kinetic preprocessor KPP - a software environment for solving chemical kinetics. Comput. Chem. Eng. **26**, 1567–1579 (2002)
2. DeMott, P., Chen, Y., Kreidenweis, S., Rogers, D., Sherman, D.: Ice formation by black carbon particles. Geophys. Res. Lett. **26**, 2429–2432 (1999)
3. Eyring, V., Köhler, H., van Aardenne, J., Lauer, A.: Emissions from international shipping: 1. the last 50 years. J. Geophys. Res. **110**, D17305 (2005). DOI 10.1029/2004JD005619
4. Eyring, V., Köhler, H., Lauer, A., Lemper, B.: Emissions from international shipping: 2. impact of future technologies on scenarios until 2050. J. Geophys. Res. **110**, D17306 (2005)
5. Gauger, C., Leinen, P., Yserantant, H.: The finite mass method. SIAM **37**, 1768–1799 (2000)
6. Hendricks, J., Kärcher, B., Lohmann, U., Ponater, M.: Do aircraft black carbon emissions affect cirrus clouds on the global scale? Geophys. Res. Lett. **32**, L12814 (2005)
7. Jöckel, P., Sander, R., Kerkweg, A., Tost, H., Lelieveld, J.: Technical note: the modular earth submodel system (messy) - a new approach towards earth system modelling. Atmos. Chem. Phys **5**, 433–444 (2004)
8. Jöckel, P., Tost, H., Pozzer, A., Brühl, C., Buchholz, J., Ganzeveld, L., Hoor, P., Kerkweg, A., Lawrence, M., Sander, R., Steil, B., Stiller, G., Tanarhte, M., Taraborrelli, D., van Aardenne, J., Lelieveld, J.: The atmospheric chemistry general circulation model echam5/messy1: consistent simulation of ozone from the surface to the mesosphere. Atmos. Chem. Phys. **6**(12), 5067–5104 (2006)
9. Lauer, A., Eyring, V., Hendricks, J., Jöckel, P., Lohmann, U.: Global model simulations of the impact of ocean-going ships on aerosols, clouds, and the radiation budget. Atmos. Chem. Phys. **7**, 5061–5079 (2007)
10. Möhler, O., Buettner, S., Linke, C., Schnaiter, M., Saathoff, H., Stetzer, O., Wagner, R., Kraemer, M., Mangold, A., Ebert, V., Schurath, U.: Effect of sulfuric acid coating on heterogeneous ice nucleation by soot aerosol particles. J. Geophys. Res. **110**, D11210 (2005)
11. Ponater, M.: Cloud radiative feedback and global warming: experiences from the echam5 global model (2009). (Poster presented at GlobCloud Workshop, 25-27 Mar, Berlin, Germany, available at: http://elib.dlr.de/63755/)
12. Ponater, M., Sausen, R., Feneberg, B., Roeckner, E.: Climate effect of ozone changes caused by present and future air traffic. Clim. Dyn. **15**, 631–642 (1999)
13. Reithmeier, C., Sausen, R.: Attila: Atmospheric tracer transport in a lagrangian model. Tellus **54B**, 278–299 (2002)
14. Roeckner, E., Bäuml, G., Bonaventura, L., Brokopf, R., Esch, M., Giorgetta, M., Hagemann, S., Kirchner, T., Kornblueh, L., Manzini, E., Rhodin, A., Schlese, U., Schulzweida, U., Tompkins, A.: The atmospheric general circulation model echam5: Part I. model description. Max-Plank-Institut für Meteorologie, Hamburg, Report No.349 (2003)
15. Roeckner, E., Brokopf, R., Esch, M., Giorgetta, M., Hagemann, S., Kornblueh, L., Manzini, E., Schlese, U., Schulzweida, U.: Sensitivity of simulated climate to horizontal and vertical resolution in the echam5 atmosphere model. J. Climate **19**, 3771–3791 (2006)
16. Roeckner, E., Siebert, T., Feichter, J.: Climate response to anthropogenic sulfate forcing simulated with a general circulation model. In: R. Charlson, J. Heintzenberg (eds.) Aerosol forcing of climate, pp. 349–362. Wiley and Sons, Chichester (1995)
17. Sander, R., Kerkweg, A., Jöckel, P., Lelieveld, J.: Technical note: the new comprehensive atmospheric chemistry module MECCA. Atmos. Chem. Phys. **5**, 445–450 (2005)
18. Unger, N., Shindell, D., Koch, D., Streets, D.: Air pollution radiative forcing from specific emissions sectors at 2030. J. Geophys. Res. **113**, D02306 (2008). DOI 10.1029/2007JD008683

Ab Initio Path Integral Simulations of Floppy Molecular Systems

Alexander Witt, Sergei D. Ivanov, and Dominik Marx

Abstract Protonated methane, CH_5^+, is one of the smallest representatives of the so–called floppy molecules whose treatment challenges both experiment and theory for decades. Recently, we succeeded in understanding the IR spectrum of the per–protonated parent system, i.e. isolated CH_5^+ itself. More recently, the IR spectra of all its H/D isotopologues, i.e. CD_5^+, CHD_4^+, $CH_2D_3^+$, $CH_3D_2^+$, CH_4D^+ and CH_5^+ have been measured in a *tour de force* experiment by our collaborators and now wait for interpretation. It has been shown both computationally and experimentally that nuclear quantum effects are crucial, which implies that they cannot be neglected when computing infrared spectra subject to H/D isotopic substitution. Thus, our investigations are carried out in the framework of *ab initio* path integral simulations together with the adiabatic centroid molecular dynamics extension which readily allow for nuclear quantum effects and yield access to the quasi–classical dynamics.

1 Introduction

Computing and understanding the infrared (IR) spectra of protonated methane, CH_5^+, and its isotopologues at finite temperatures is an extremely challenging and demanding task. Our particular interest in protonated methane is due to several reasons. CH_5^+ serves as the prototype for non–classical carbocations and three–center two–electron bonding [1], as well as being a key intermediate in superacid and hypercarbon chemistry [2, 3], and highly relevant in areas reaching from astrophysics to plasmachemistry [4–6]. Nonetheless, understanding this enigmatic molecule continues to present a challenge to both experiment and theory. This can be traced

Alexander Witt · Sergei D. Ivanov · Dominik Marx
Lehrstuhl für Theoretische Chemie Ruhr-Universität Bochum, D-44780 Bochum, Germany
e-mail: alexander.witt@theochem.rub.de
e-mail: sergei.ivanov@theochem.rub.de
e-mail: dominik.marx@theochem.rub.de

S. Wagner et al. (eds.), *High Performance Computing in Science and Engineering, Garching/Munich 2009*, DOI 10.1007/978-3-642-13872-0_56,

back to its unusually flat potential energy landscape thus enabling complex large–amplitude motion to take place. The resulting "fluxionality" even persists at very low temperatures due to the quantum–mechanical nature of the highly correlated hydrogen scrambling motion of the five protons [7]. Thus CH_5^+ certainly is a paradigmatic example of the wide class of floppy or fluxional molecules as reviewed in the Introduction of Ref. [8], see also Ref. [9]. Last but not least, IR spectra of CH_5^+ and all its isotopologues have been recently measured by the experimentalists group of Professor Schlemmer in Köln via Laser Induced Reaction (LIR) technique and these spectra are now waiting for theoretical interpretation; the analysis of the bare CH_5^+ was published in a joint effort [10] (see also Ref. [8] for a detailed account of the theory part).

From the simulations' point of view CH_5^+ is an extremely challenging system. Its interactions cannot be parameterised in terms of force fields. It has been shown both computationally and experimentally that nuclear quantum effects are crucial for these systems which implies that they cannot be neglected when computing infrared spectra. Finally, even temperature effects in the sense of thermal fluctuations of the nuclei on top of their quantum-mechanical excitations cannot be neglected. Thus, an "integrated" method that allows one to include, at least approximately, all these aspects is called for. Over the last two decades, "classical"[1] *ab initio* molecular dynamics (MD) methods [11, 12] (allowing for all aforementioned features except for the quantum nature of nuclei) have had success in simulating such complex molecular systems. One possible route to include quantum effects on nuclear motion is provided by the formalisms based on Feynman's path integral (PI) quantum statistical mechanics [13, 14], since they allow for quantum delocalisation of nuclei in an elegant and, in principle, exact fashion. Simulation methods based on PIs exploit the *quantum–classical isomorphism* [15], namely, the isomorphism between the quantum partition function, represented as an imaginary time path integral [13, 14] and the classical configurational integral over closed paths. Static properties of the quantum objects can thus be obtained by simulation of more complicated, but *classical* objects (independently of the way of treating electrons) using the full power of well established classical simulation techniques such as Monte Carlo or molecular dynamics. The aforementioned classical object turns out to consist of many replicas (often referred to as *beads*) of the system, subject to nearest–neighbour harmonic interactions. The convergence with respect to number of beads can be reached for any given system and temperature.

Nevertheless, methods that allow for computation of dynamic properties are called for in the framework of PIs. One promising approach is known as the "centroid (path integral) molecular dynamics" (CMD) method, see Refs. [16, 17] for reviews. Based on an earlier idea of Gillan [18], Cao and Voth [19] worked out the formulation of statistical mechanics in terms of the *centroid density* of Feynman paths. It turned out that the corresponding *dynamics of centroids* represents quasiclassical real-time evolution [16, 17, 19]. Most recently, another idea called "ring polymer molecular dynamics" (RPMD) has been proposed by Craig and Manolopou-

[1] Here and subsequently the term "classical" refers to classical nuclei, i.e. point particles, and not to interactions.

los [20] and tested on several model systems. According to this approach, path integral molecular dynamics is directly used to extract approximate quantum time–correlation functions thus circumventing the concept of centroids alltogether.

Marx, Tuckerman and Martyna developed an *ab initio* generalisation of adiabatic centroid path integral molecular dynamics [21] based on the underlying *ab initio* path integral method [22–24], as explained and reviewed in Refs. [11, 25]. Note, that the same idea can be naturally extended from *ab initio* CMD to *ab initio* RPMD as well; see [12] for more details about the state-of-the-art in this field.

Before applying CMD and RPMD methods to a "real" and therefore computationally demanding problem on top of the underlying *ab initio* technique, we investigated their applicability to IR spectroscopy on a set of test systems at chemically relevant temperatures and isotope substitutions [26]. It turned out that both methods have intrinsic problems in this framework. In particular, in RPMD artificial frequencies arising from the ring polymer itself crop up in the IR frequency range and their resonances with the physical frequencies led to in general unpredictable consequences. The major problem is that the number of the ring polymer's frequencies in the IR range increases with decreasing temperature and therefore it is highly probable that at least one of those frequencies would interfere with the physical IR peaks of CH_5^+.

In CMD, the discovered "curvature problem" leads to i) inapplicability of the method below certain temperature (the threshold temperature was estimated to be around 200 K for CH_5^+); ii) to artificial, but nearly linear in T, shifts of the stretching bands above the breakdown temperature.

These conclusions provide us with solid grounds to investigate CH_5^+ and its isotopologues by means of CMD at 200 K. It is stressed here that these are *extremely* demanding quantum dynamics simulations due to the underlying rather fine path integral discretisation, the required small timesteps in order to allow for adiabatic sampling of the centroid forces, and the need to average of many initial conditions in order to sample the canonical average.

2 Simulation Techniques and Technical Details

The standard protocol necessary for computing IR spectra as well as other dynamic properties consists of the following steps. Firstly, one has to perform the full *canonical*, i.e., NVT equilibration of the system using *ab initio* path integral sampling techniques [22–24]. Secondly, one has to generate such an equilibrium thermal distribution, namely to perform a NVT simulation starting from the previously equilibrated state. This alone allows one to compute static thermodynamic properties only. Thirdly, dynamical *ab initio* CMD simulations [21] are performed starting from uncorrelated initial conditions sampled from the thermal NVT distribution. Note, that only the dynamics of centroids should be microcanonical [21], which allows us to fully thermostat the non-centroid modes, thus providing efficient sampling. Along such a trajectory the time-autocorrelation function of the dipole centroid is computed, from which the IR spectrum *including quantum effects, anharmonicities, and mode coupling* is obtained via the Fourier transform. Finally, the latter step has to be repeated for a sufficient number of initial conditions in order to assure ergodicity and to converge the intensities of the resulting IR spectrum. The latter is a necessary but very demanding aspect of all such calculations.

The distribution functions and radii of gyration were computed for every single run and averaged afterwards for all cases except for PIMD simulations at 110 K and 200 K where a single NVT trajectory was computed. For all PIMD simulations a timestep of 14 a.u. was employed while the rest of the setup did not change.

An important technical aspect in the case of CMD is that one has to provide an *adiabaticity* decoupling of all non-centroid modes from the centroid mode as analysed theoretically in Ref. [21] with practical recipes given in Ref. [26]. This is done by an artificial scaling of the masses of the non-centroid modes. Ultimately, it results in the decrease of the time step, proportional to the square root of the adiabaticity parameter. Based on our experience with model potential systems, we chose the adiabaticity parameter to be 256. This, in turn, would reduce the time step by a factor of 16, i.e., down to 1 a.u. In order to have a comparable spectral resolution to the experimental one ($10\,cm^{-1}$), the individual run length was set to approximately 3 ps.

3 Results

Recently, the IR spectrum of bare CH_5^+ has been computed, measured and understood [8, 10]. Simulations there were performed by means of classical *ab initio* molecular dynamics at a higher (room) temperature than the experimental one (110 K) in order to somehow mimic quantum–mechanical fluctuation effects. The same approach worked out successfully for the other "pure" species, namely CD_5^+, as expected. In order to include nuclear quantum effects properly, *ab initio* PIMD simulations treating simultaneously both electrons and nuclei as quantum particles at finite temperature were carried out.

PIMD yields formally exact static properties, but it cannot be straightforwardly applied to simulations of real–time dynamics, which is needed for computing IR spectra. Therefore, an extension of the method, namely CMD was employed. In the following subsections we will compare different properties of CH_5^+ computed via PIMD, CMD and classical MD.

3.1 Static Properties

From previous investigations it is well known that it is crucial to include nuclear quantum effects such as zero–point fluctuations to properly describe the fluxionality of CH_5^+. One way to investigate the impact of these effects on the fluctional

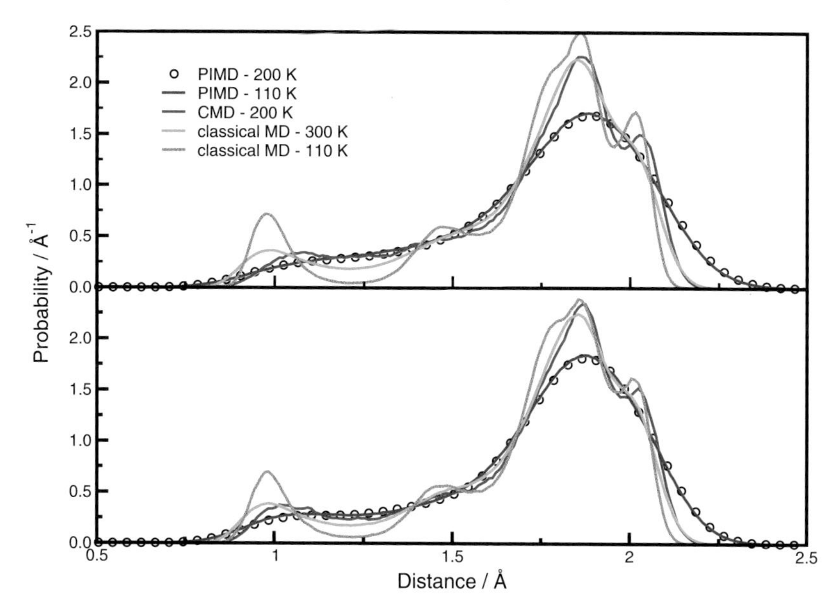

Fig. 1 H–H distribution functions (upper panel) and D–D distribution functions (lower panel) computed with classical MD at 110 K: brown lines, 300 K: green lines; CMD at 200 K: red lines and PIMD at 110 K: blue lines, 200 K: black circles. Note that the quantum distribution functions are shown for PIMD whereas the centroid distribution functions are shown for CMD here and in following figures

behaviour of CH_5^+ is to consider the structural properties such as H–H or C–H distances by means of their distribution functions. It is noted in passing that these functions are very similar for both CH_5^+ and CD_5^+.

The H–H distribution function obtained with classical MD at 110 K features a clean peak at about 1 Å which corresponds to the moiety distance, see Fig. 1. The remaining part of the distribution function is fairly structured as well indicating that CH_5^+ is not fluxional under these circumstances. Moreover the clean separation of the "moiety" peak shows that the scrambling motion is nearly completely hindered at such low temperatures. Increasing the temperature to 300 K results in a less structured distribution function which indicates an increase of fluxionality. A similar result is found when PIMD (at 110 and 200 K) is applied and the H–H distribution function is even less structured than in the classical high temperature case; note, that the distribution functions nearly coincide for the two different temperatures, The CMD result lies in between the PIMD and the classical results and is of comparable quality. This behaviour was anticipated as CMD is a quasi–classical approximation, which should therefore yield results which are "better" than the classical ones but not as good as the true quantum results.

The same trend is observed for the C–H distribution functions although the discrepancy between the classical and the CMD results on the one hand side and the PIMD result on the other hand side are larger. Although both the CMD and the clas-

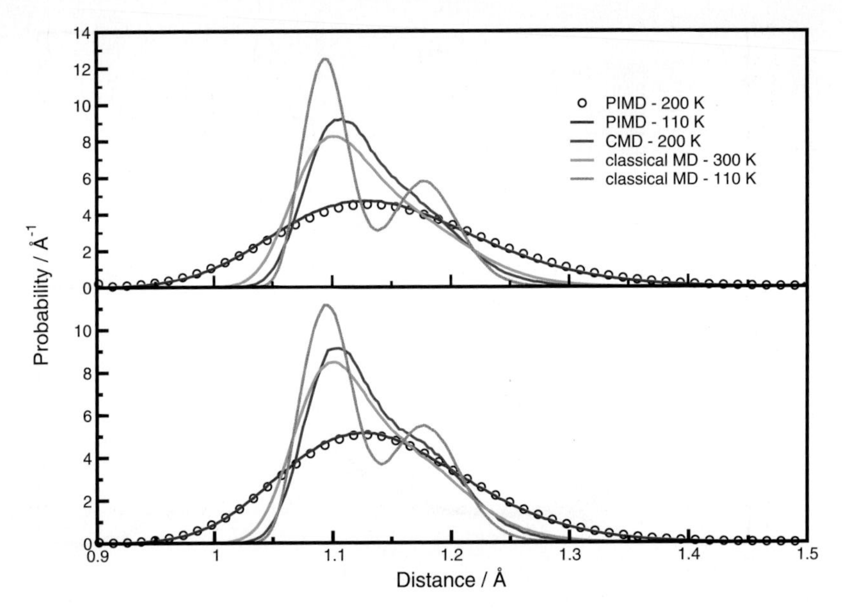

Fig. 2 C–H distribution functions (upper panel) and C–D distribution functions (lower panel) computed with classical MD at 110 K: brown lines, 300 K: green lines; CMD at 200 K: red lines and PIMD at 110 K: blue lines, 200 K: black circles

sical distribution functions are unimodal their PIMD counterpart is much broader. However, the correct trend that fluxionality increases upon taking nuclear quantum effects into account is observed for all cases studied.

3.2 Radii of Gyration

In order to quantify the "quantumness" or the particles' uncertainty their radii of gyration are computed. This property is nothing else but the spread of the ring polymer of a quantum particle. Therefore, the radius of gyration is a direct measure for the spacial extension of a quantum particle. Its definition reads

$$r_{gyr} = \sqrt{\frac{1}{2P^2} \sum_{i,j=1}^{P} (\mathbf{r}_i - \mathbf{r}_j)^2} \tag{1}$$

where P is the number of beads, $\mathbf{r}_i$ and $\mathbf{r}_i$ are the positions of beads with numbers i and j, respectively. This quantity can be evaluated analytically for the case of a free particle yielding

$$r_{gyr}^{free} = \frac{\hbar}{2\sqrt{mk_B T}} \tag{2}$$

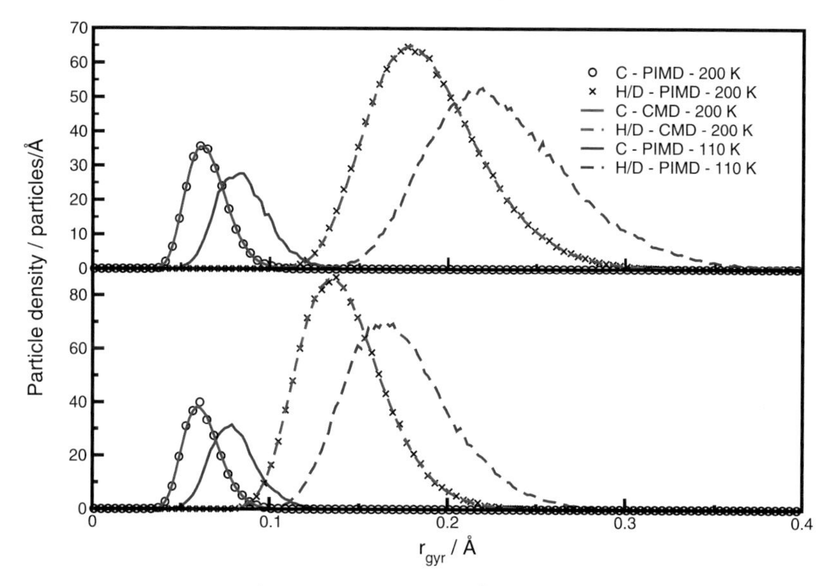

Fig. 3 Gyration, r_{gyr}, for CH_5^+ (upper panel) and CD_5^+ (lower panel) computed with PIMD at 110 K: blue lines, 200 K: black circles/crosses and CMD at 200 K: red lines

revealing its close connection to the de Broglie wavelength. The expectation values for a gyration radius of the free particle at 200 K with its mass corresponding to the hydrogen, deuterium and carbon atom's mass are 0.25, 0.17 and 0.07 Å respectively. As in CH_5^+ all atoms are confined by potentials originating from chemical bonds the radii of gyration should be smaller than for a free particle which indeed is the case as illustrated in Fig. 3. Moreover, the radii of gyration are identical for the PIMD and CMD simulations at the same temperature as expected. The values at 110 K are larger than at 200 K reflecting the increased quantumness of the nuclei. Although we have to perform our simulations at a target temperature of 200 K we still can describe the quantum behaviour of protons and deuterons and can preserve the important trends, namely that the lower the mass is the larger is the delocalisation of the quantum particle.

3.2.1 IR-Spectra

To access the dynamical performance of CMD, IR spectra have been simulated and we will compare those with experimental ones and with IR spectra computed using classical MD. In Fig. 4 preliminary results of the IR spectra of CH_5^+ and CD_5^+ computed with classical *ab initio* MD, with *ab initio* path integral centroid MD and the experimental result are shown.

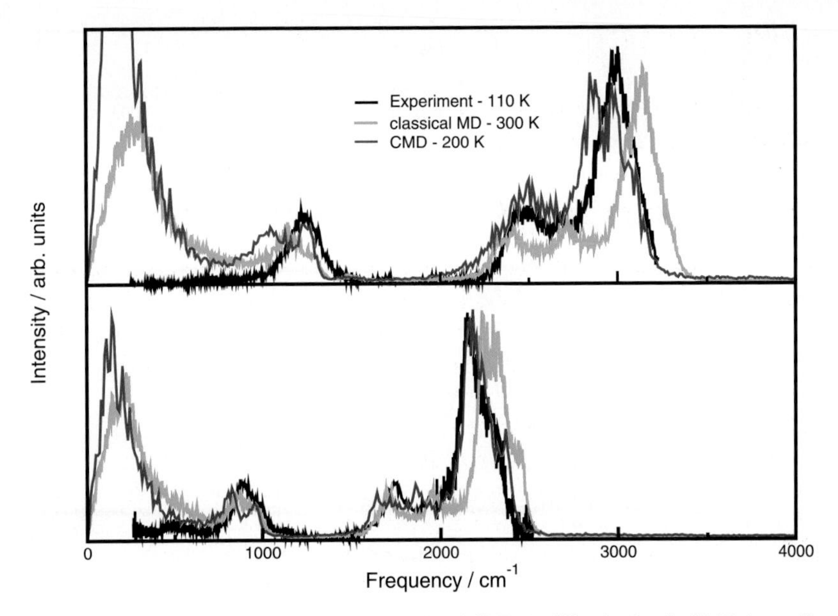

Fig. 4 Experimental (LIR) (black lines) and simulated CMD (red lines), classical MD (green lines) IR Spectra for CH_5^+ (upper panel) and CD_5^+(lower panel); note that LIR setup looses its sensitivity below about $500\,\mathrm{cm}^{-1}$

The first thing to mention is that in the experiment it is not possible to measure below $500\,\mathrm{cm}^{-1}$ due to loss of intensity in this region. One can see, that the shapes of CMD spectra match the experimental results extremely well "out of the box", i.e. without any *a posteriori* adjustment!

Three major bends can be identified in the IR spectra: The first bend which extends from 0 to about $500\,\mathrm{cm}^{-1}$ contains rearrangement motions like rotation and scrambling/pseudo–rotation. A second bend occurs at about $1000-1200\,\mathrm{cm}^{-1}$ and $800-1000\,\mathrm{cm}^{-1}$ in the case of CH_5^+ and CD_5^+, correspondingly. In a nutshell, bending motions like wagging, twisting, etc, contribute to this feature. The CH/CD stretching modes of the moiety and the tripod constitute the third part which extends from about $2000/1800\,\mathrm{cm}^{-1}$ to about $3500/2500\,\mathrm{cm}^{-1}$.

Both methods are able to reproduce the overall shape of the spectra impressively well. In the case of classical MD a blue shift of the high frequency peak is observed which is due to the DFT functional (and an underconverged cutoff) employed. The CMD spectrum closely follows the shape of the experimental one but due to a reason which is the result of a cancellation of errors: In Ref. [26] we have shown that the curvature problem which results in CMD leads to an artificial red shift of high frequency motions. In combination with the blue shift resulting from the functional this effect is nearly compensated which, in the end, leads to a very good agreement with experiment.

3.3 Analysis of CH_5^+

We have developed a new analysis technique [27] that allowed us to decompose bare IR spectra for each given isotopologue into the spectral contributions of all possible isotopomers. This technique has successfully been applied to IR spectra computed with classical *ab initio* MD. The spectra of the isotopomers can then be re-weighted using quantum probabilities obtained from fully *ab initio* PIMD simulations. The resulting quantum re-weighted spectra match very well the experimental results [28].

Having obtained the spectra of all isotopomers for each isotopologue, we performed the next analysis step, namely we assigned each and every spectral feature to molecular motions. This became possible using another analysis technique developed most recently in our group in a very different context, based on the quasi–rigidity assumption, i.e., the presence of a single well-defined reference structure. In other words we reduced the problem of analysing a highly fluxional CH_5^+ molecule to a set of problems each analysing a quasi–rigid molecule. The results of this analysis explains all features of the spectra of all isotopologues. Thus, the analysis performed is able to attribute spectral features to molecular motions in the simulated spectra and, hence, in the experimental spectra due to their close match [28].

3.4 Performance and Scaling

The `CPMD` package [29] is a density functional theory based molecular dynamics code which uses a plane–wave basis set and pseudopotentials as explained in much detail in Ref. [11]. There exist several levels of parallelisation which are illustrated in Fig. 5. First of all, any PI problem parallelises intrinsically with respect to the number of replicas (beads) of the system. Beads interact only via classical harmonic springs and this interaction requires negligible communication effort. Therefore, on the PI level the parallelisation is almost ideal, i.e. linear even on a Gigabit interconnect.

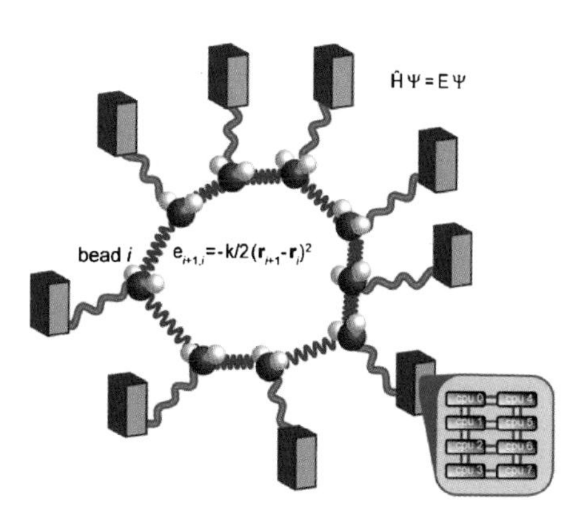

Fig. 5 Parallelisation scheme for Path Integral simulations

For each individual bead, the Kohn-Sham equations of DFT have to be either solved iteratively (Born-Oppenheimer MD) within a self consistent field (SCF) cycle or propagated in an extended Lagrangian scheme (Car-Parrinello MD), see Ref. [11]. This computationally demanding part of the calculation is assigned to processor groups (PGr), which require a high band width/low latency interconnect. Within a PGr parallelisation is realized via MPI for inter-node communication and either MPI or OpenMP/Vector processing for intra-node parallelisation. Thus, this hierarchical parallelisation strategy is highly adaptive and can be optimised for the individual platform.

Finally, we would like to mention that each individual bead is computationally equivalent to the copy of a system with the nuclei treated classically. `CPMD` [29] has proven its efficiency for simulations of this type on a wide variety of platforms during the last decade [11] including HLRB-II. The typical setup for a CMD run employed 128 processors, resulting in a nearly linear scaling regime. The total amount of cpu hours for a single run with a length of $\approx 3\,\text{ps}$ was around 7500.

4 Outlook and Planned Work

We presented CMD spectra for two out of six isotopologues of CH_5^+, namely CH_5^+, and CD_5^+, which match closely the experimental spectra. The quality of the obtained results encourage us to compute CMD spectra for the other cases and to converge one chosen case (CHD_4^+) in terms of intensities for better comparison with experiment (where all isotopologues' spectra are available to us) and for a possibility to apply further analysis techniques.

Another core issue which is of great interest is the microsolvation of CH_5^+ with hydrogen molecules. Whereas CH_5^+ itself is widely studied, there is only scarce information concerning its microsolvation available in the literature [30–35]. Although the influence of such additional solvating molecules is not yet fully clarified, there is an interesting body of evidence reported that solvation by as few as three hydrogen molecules slows down scrambling considerably, or even stops it completely [33–35]. However, in the computations underlying this interpretation quantum effects of nuclei were not taken into account. Moreover, classical simulations for bare CH_5^+ demonstrated that at 100 K the scrambling has nearly stopped without any additional hydrogen molecules attached [8]. Therefore, methods which allow for quantum behaviour of nuclei have to be considered and PIMD seems to be a perfectly suitable, although computationally demanding choice.

In order to prepare ourselves for running *ab initio* simulations on the microsolvated CH_5^+ species we investigated the performance of various DFT functionals and explored the plethora of energetic minima and saddle points as well as possible detachment channels for up to five hydrogen molecules attached to the carbonium core, which is published in Ref. [36]. Here, we show that no GGA functional is able to describe microsolvation appropriately (compared to quantum–chemical benchmark data generated by us), whereas a so–called meta–GGA functional, TPSS, is indeed

Table 1 Scaling tests for $CH_5^+(H_2)_3$ with 32 beads

NProc	NProc per bead	single SCF step (in seconds)	average MD time step (in seconds)
32	1	5.3	68.4
64	2	2.4	37.7
128	4	0.7	9.8
256	8	0.34	4.6
512	16	0.19	2.9

close to the MP2 reference data. Moreover, after publishing this paper we found out that the so–called Grimme dispersion correction [37] together with the standard GGA setup used by us for bare CH_5^+, yields results of MP2 quality for binding energies, scrambling barriers and geometries. Luckily, this dispersion–corrected GGA combination is nearly an order of magnitude cheaper than the setup with the TPSS functional. Thus, we are encouraged to used this more economical setup to perform computations employing classical *ab initio* MD at the experimental temperature of 110 K for the carbonium core solvated by one to five hydrogen molecules. Last but not least the increase of the physical system size gives us an access to larger processor numbers that can be employed in a single run, see Table 1. It is clear that the case with 3 hydrogen molecules attached scales reasonably up to 512 processors with an optimum at 256 processors.

Acknowledgements We gratefully acknowledge fruitful interactions with our experimental partner group, in particular Stephan Schlemmer and Oskar Asvany from Universität zu Köln. In addition, we are grateful for partial financial support of this work by DFG (Normalverfahren MA 1547/4) and by FCI (Doctoral Fellowship to A.W. as well as general support of D.M.).

References

1. G. A. Olah and G. Rasul, Acc. Chem. Res. **30**, 245 (1997)
2. G. A. Olah, G. K. S. Prakash, and J. Sommer, *Superacids* (Wiley, New York, 1995), see in particular Chapters 3 and 5
3. G. A. Olah, G. K. S. Prakash, R. E. Williams, L. D. Field, and K. Wade, *Hypercarbon Chemistry* (Wiley, New York, 1987), see in particular Chapters 1, 5, and 7
4. T. Oka, Phil. Trans. R. Soc. Lond. A **324**, 81 (1988)
5. E. Herbst, J. Phys. Chem. A **109**, 4017 (2005)
6. D. Gerlich, Phys. Chem. Chem. Phys. **7**, 1583 (2005)
7. D. Marx and M. Parrinello, Nature **375**, 216 (1995)
8. P. Kumar P and D. Marx, Phys. Chem. Chem. Phys. **8**, 573 (2006)
9. D. Marx and M. Parrinello, Science **284**, 59 (1999)
10. O. Asvany, P. Kumar P, B. Redlich, I. Hegemann, S. Schlemmer, and D. Marx, Science **309**, 1219 (2005)
11. D. Marx and J. Hutter, in *Modern Methods and Algorithms of Quantum Chemistry*, ed. by J. Grotendorst, John von Neumann Institute for Computing, Jülich, (2000), vol. 3 of *NIC Series*, pp. 329–477, 2nd ed.
12. D. Marx and J. Hutter, *Ab Initio Molecular Dynamics: Basic Theory and Advanced Methods* (Cambridge University Press, Cambridge, 2009)

13. R. P. Feynman and A. R. Hibbs, *Quantum Mechanics and Path Integrals* (McGraw-Hill, New-York, 1965)
14. R. P. Feynman, *Statistical Mechanics* (Addison-Wesley, Redwood City, 1972)
15. D. Chandler and P. G. Wolynes, J. Chem. Phys. **74**, 4078 (1981)
16. G. A. Voth, Adv. Chem. Phys. **93**, 135 (1996)
17. R. Ramírez and T. López-Ciudad, in *Quantum Simulations of Complex Many–Body Systems: From Theory to Algorithms*, ed. by J. Grotendorst, D. Marx, and A. Muramatsu, John von Neumann Institute for Computing, (2002), vol. 10 of *NIC Series*, pp. 325–375
18. M. J. Gillan, J. Phys. C Solid State **20**, 3621 (1987), see in particular Sec. 5 and 6
19. J. Cao and G. A. Voth, J. Chem. Phys. **99**, 10070 (1993)
20. I. R. Craig and D. E. Manolopoulos, J. Chem. Phys. **121**, 3368 (2004)
21. D. Marx, M. E. Tuckerman, and G. J. Martyna, Comput. Phys. Commun. **118**, 166 (1999)
22. D. Marx and M. Parrinello, Z. Phys. B (Rapid Note) **95**, 143 (1994)
23. D. Marx and M. Parrinello, J. Chem. Phys. **104**, 4077 (1996)
24. M. E. Tuckerman, D. Marx, M. L. Klein, and M. Parrinello, J. Chem. Phys. **104**, 5579 (1996)
25. D. Marx, in *Computer Simulations in Condensed Matter: From Materials to Chemical Biology*, ed. by M. Ferrario, K. Binder, and G. Ciccotti, no. 704 in Lect. Notes Phys. (Springer, Berlin, 2006), pp. 507–539
26. A. Witt, S. D. Ivanov, M. Shiga, H. Forbert, and D. Marx, J. Chem. Phys. **130**, 194510 (2009)
27. in preparation
28. in preparation
29. J. Hutter bet al., *CPMD*, IBM Corp 1990–2008, MPI für Festkörperforschung Stuttgart (1997–2001), `http://www.cpmd.org`
30. K. Hiraoka and P. Kebarle, J. Am. Chem. Soc. **97**, 4179 (1975)
31. K. Hiraoka and T. Mori, Chem. Phys. Lett. **161**, 111 (1989)
32. K. Hiraoka, I. Kudaka, and S. Yamabe, Chem. Phys. Lett. **184**, 271 (1991)
33. D. W. Boo and Y. T. Lee, Chem. Phys. Lett. **211**, 358 (1993)
34. D. W. Boo, Z. F. Liu, A. G. Suits, J. S. Tse, and Y. T. Lee, Science **269**, 57 (1995)
35. D. W. Boo and Y. T. Lee, J. Chem. Phys. **103**, 520 (1995)
36. A. Witt, S. D. Ivanov, H. Forbert, and D. Marx, J. Phys. Chem. A **112**, 12510 (2008)
37. S. Grimme, J. Comput. Chem. **27**, 1787 (2006)

Statistically Converged Properties of Water from *Ab Initio* Molecular Dynamics Simulations

Matthias Heyden and Martina Havenith

Abstract We present results from an *ab initio* molecular dynamics study of pure water. Using the resources available at the HLRB2 we were able to produce the first trajectories of sufficient length and number, that allow to study dynamical processes on the picosecond timescale with statistically reliable results. Additionally we computed a statistically converged infrared absorption spectrum from 0–4000 cm^{-1}, which is in good agreement with the experimental observation. In particular, at THz frequencies the spectra qualitatively reproduce important features, whereas, in contrast, force field based simulations have been shown to utterly fail. In order to compensate for the neglect of quantum effects for the nuclei in classical dynamics simulations on an *ab initio* potential energy surface and potential overbinding in conjunction with the used electron structure method, we applied an increased intrinsic temperature of 400K in order to obtain structural and dynamical properties corresponding to an experimental temperature of 300K.

1 Introduction

Water - the Liquid Water is the most important substance on earth. It exhibits numerous well known physical and chemical anomalies and many of those must be regarded as being essential to the existence of life as we know it. Liquid water can therefore be referred to as the 'matrix of life' [1].

The unique properties of water can be traced back to its large dipole moment for a molecule of its size and its extraordinary capacity of building networks of hydrogen bonds. Hydrogen bonds are strong enough to prevent the liquid from evaporating and to promote the formation of a very open, tetrahedral crystal structure

Matthias Heyden · Martina Havenith
Ruhr-University Bochum, Physical Chemistry II, Universitätsstr. 150, 44780 Bochum, Germany
e-mail: matthias.heyden@rub.de
e-mail: martina.havenith@rub.de

S. Wagner et al. (eds.), *High Performance Computing in Science and Engineering, Garching/Munich 2009*, DOI 10.1007/978-3-642-13872-0_57,

upon freezing. However those hydrogen bonds are also weak enough to be broken and reformed continuously under ambient conditions in the liquid phase, causing a viscosity of liquid water at 20°C which is only slightly higher than for other liquids with much less favorable molecule-molecule interactions. The properties of water that we experience as being ideal for the existence of life are therefore the result of a well balanced relation between rigidity and flexibility of the hydrogen bond network.

Furthermore water is not always the same. The term 'biological water' [20] refers to water which is solvating biomolecules, e.g. proteins and other solutes found in large amounts in living organisms. The properties of water are greatly influenced by the presence of such disturbances. Depending on the solute it may form strong hydrogen bonds with water molecules itself or it may not, creating a hole in the hydrogen bond network. In both cases the hydrogen bond network is influenced leading to altered properties of this type of water. The same is true for water close to any kind of surface. Although numerous experimental studies have been carried out in order to investigate water in these special environments [5, 6, 13, 14, 16, 20] one is usually limited to measure either an overall average of water properties, e.g. in a solution, or to introduce artificial probes, e.g. chromophores, which will have an effect on water properties on their own. The interpretation of the experimental findings therefore heavily relies on accompanying simulations.

Molecular Dynamics Simulations Molecular dynamics (MD) simulations have emerged as an invaluable tool in order to provide an understanding of experimentally observed processes on a molecular level. The special benefit of MD simulations, that separate it from other approaches such as Monte Carlo (MC), is that not only static properties can be studied, but also dynamical processes can be observed. This allows to follow diffusion processes, extract relaxation times and also to compute vibrational spectra. Therefore MD simulations can be related and compared to a large variety of experiments.

In simulations there is no restriction in accessing properties of separate parts of the system. In the case of 'biological water' it is therefore possible to separate properties of water molecules that are part of the solvation shell, of e.g. a protein, of those which are essentially uninfluenced by the presence of the solute.

The basis of any type of molecular simulation however is a proper description of the potential energy. This is commonly achieved with empirical force field potentials, involving simple analytic pairwise interaction potentials and increments for intramolecular energies. The first MD simulations using such an approach were reported already in the 1970's [22]. With the increasing computational power available today the simulation of large systems containing millions of atoms for timescales of microseconds is in reach. However the simplicity of these descriptions of the potential energy is also one of its major drawbacks, because many characteristics of intra- and intermolecular interactions can not be described correctly within the used analytical models. Polarization effects play an important role but their inclusion in force field models is non-trivial. In the past decade improved models have been proposed [7] that try to capture polarization effects. However with increasing model complexity also the number of parameters increases, while the number of experimental observables that can be used for the validation remains constant. Electronic structure calculations on different levels of theory are therefore also more and more utilized in order to provide data for the fitting process [7].

Parametrized force fields additionally have the disadvantage, that they are often adjusted to certain conditions, e.g. certain points in density and temperature space or chemical compositions (pure substances). They are not necessarily transferable to other conditions. Similarly high level electron structure calculations are usually limited to small numbers of molecules, e.g. clusters in vacuum. Force field models derived from them therefore do not necessarily perform well for bulk liquids. In addition to the limited transferability to different conditions, force field models are usually fitted to a specific set of experimental observables. Hence they tend to succeed reproducing these observables with reasonable accuracy. However their predictive power for observables which were not involved in the fitting process is limited.

***Ab Initio* MD Simulations** With the rise of parallel computers and high performance computing facilities an alternative approach towards molecular dynamics simulations became feasible within the last ten years. In *ab initio* MD simulations [18] the potential energy of the system is computed with an appropriate electron structure method on the fly at each step of the simulation. Due to an optimal trade off between computational costs and accuracy the Kohn-Sham formulation [15] of density functional theory (DFT) with gradient corrected exchange-correlation functionals such as the Perdew-Becke-Enzerhoff (PBE) [21] or Becke-Lee-Yang-Parr (BLYP) [3, 17] functionals are commonly used. Although the computational cost that is connected with such an approach is tremendous compared to the use of empirical force fields, it has many advantages. It is in principle free of fitted parameters. Because no parameters are adjusted to certain external conditions or chemical compositions, transferability is therefore a vital key feature of such an approach. Furthermore properties that are predicted by such a simulation are not the product of a preceding fitting process. Another main advantage is that effects which are difficult to model in empirical force fields, e.g. polarizability, are naturally included in this approach.

The main drawback of this approach however is, that due to the immense computational cost usually a trade off has to be made between sampling statistics and computing time. Especially the first *ab initio* MD studies were reported for very small systems and short trajectories, which lead to non-negligible finite size effects and insufficient statistics to predict many properties. The latest generation of high performance computing facilities such as HLRB2 in combination with efficient parallelization schemes however helped to overcome such difficulties.

Other problems have a more intrinsic character. The nuclei are, as well as in most force field based approaches, treated as classical particles. This is especially harmful for light atoms such as hydrogen. Because force fields are usually parametrized to experimental observables, which of course contain all influences due to the quantum character of light atoms, this is already accounted for in the fitting process. This is not the case in *ab initio* based approaches. Together with putative inaccuracies in the potential energy surface as described by the applied density functional, the neglect of the zero point energy (ZPE) and zero point motion leads to overstructuring and dynamical processes slowed down by one order of magnitude [10, 24]. A reasonable strategy in order to compensate for these effects without abandoning the classical treatment of all nuclei is to increase the kinetic energy in the system, hence raising the intrinsic temperature in order to speed up dynamical processes [8]. This can be done accordingly to produce dynamics that correspond to the actual target temperature.

IR Absorption at THz Frequencies An example which demonstrates the improvement due to the use of *ab initio* MD simulations is given by the THz spectrum of water. This frequency range of the IR absorption spectrum of water has drawn the attention of numerous experimental and theoretical studies recently, because it could be shown that the absorption of the intermolecular vibrational modes at this frequency are extremely sensitive for dynamical properties of water [5, 13]. THz-absorption studies could therefore be used to study 'biological water' solvating different kinds of biomolecules, ranging from amino acids towards carbohydrates [13, 14] and proteins [5, 6]. Especially in the case of proteins it could be shown that the solvating water molecules are strongly influenced by the solute. In addition it could be shown that dynamical properties of water solvating proteins are affected e.g. during protein folding and enzyme catalysis [16].

Force field based simulations however fail to reproduce the THz absorption spectrum of water. Especially an absorption band at roughly 200 cm^{-1} or 6 THz, assigned to vibrational modes of the hydrogen bond network, is absent in force field based simulations or appears only as a weak shoulder. This is still the case for very recent polarizable force field models. As a consequence the influence of changed water dynamics, e.g. due to the presence of a solute, on the THz absorption is also described incorrectly [14].

In contrast to this, it could already be shown in pioneering *ab initio* MD studies of water by Silvestrelli et al. [26] that the THz spectrum of water is described correctly in *ab initio* based MD simulations. However this study was limited to a small system size (32 molecules) and a short simulation time (12 ps). Their computed spectrum is therefore not statistically converged. More recently another study by Sharma et al. [25] presented similar results. They simulated a larger system (64 molecules) and produced a slightly longer trajectory (16 ps). However both studies suffer from unconverged results. The vibrational modes at THz frequencies have rather long periods, e.g. 1 THz = 33 cm^{-1} corresponds to a period of 1 ps. Thus simulations of 10-20 ps total length cover only very few vibrations in the THz region and, correspondingly, intensities and vibrational frequencies in this region are ill described. Furthermore, the vibrational modes in this frequency range are usually characterized as collective vibrations of the hydrogen bond network. The system sizes used by Silvestrelli et al. and Sharma et al. may therefore be insufficient to prevent a collective oscillation of the whole simulated system, which may lead to large finite size effects.

The perpetual increase in the availability of computational resources however allows today, to perform *ab initio* MD simulations with a larger system for longer timescales. This will enable us for the first time to compute e.g. THz absorption spectra and other properties of water, which are statistically converged. Besides offering the ability to study the nature of the vibrational modes at this frequency in great detail, such a result is needed as a reference for subsequent studies of aqueous solutions, where the influence of a solute on dynamical properties of water and the intermolecular vibrational modes will be studied. Only from such a comparison, the connection between the dynamical properties of water and the intermolecular vibrational modes can be revealed, which is observed in the THz studies of 'biological water'. Providing this reference is the main goal of the work presented here.

2 Methods

MD Protocol and Sampling Strategy A system of 128 water molecules was used in our Born-Oppenheimer *ab initio* MD simulation with the software package CP2K and the Quickstep [28] algorithm. We applied the PBE [3] exchange correlation functional with a TZV2P mixed Gaussian plane wave basis set and norm-conserving pseudopotentials by Goedecker, Teter and Hutter (GTH) [9, 11]. A plane wave energy cutoff of 400 Rydberg and a time step of 0.5 fs was used.

Using a massive Nosé chain thermostat with a resonance frequency of 2085 cm^{-1} we produced a long trajectory of 90 ps length in the canonical (NVT) ensemble. To compensate for the neglection of the ZPE and possible overbinding effects in the DFT description of the potential energy surface the simulations were performed at a raised temperature of 400 K. This is justified by a comparison of the oxygen-oxygen radial distribution function (RDF) from simulations at 300 K and 400 K with experimental results by Soper et al. [27] in Fig. 1, where a strongly increased agreement with the experiment is observed at 400K.

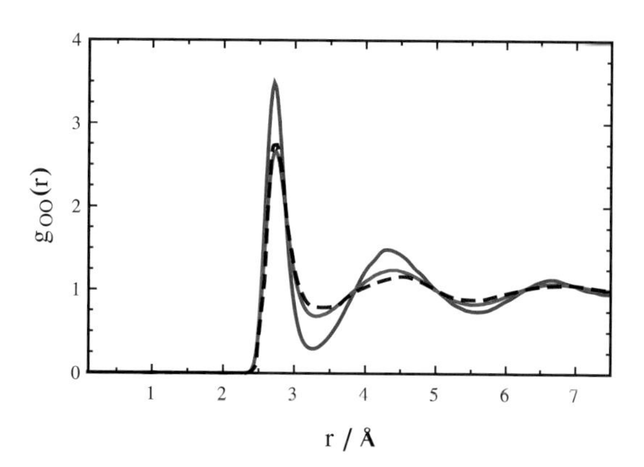

Fig. 1 Comparison of oxygen-oxygen radial distribution functions $g_{OO}(r)$ as obtained from *ab initio* MD simulations at 300K (blue line) and 400K (red line) to experimental results [27] (dashed black line)

In order to sample properties of water in the canonical (NVT) ensemble without introducing artifacts due to the applied thermostat, we selected 16 statistically independent snapshots separated by at least 4 ps from the NVT trajectory as starting points for simulations in the micro-canonical (NVE) ensemble. The NVE trajectories were simulated for 20 ps in farming jobs of several simultaneous runs. With this approach we could achieve an additional parallelization in order to use the capacities of HLRB2 with maximal efficiency. Average properties computed from all NVE trajectories correspond to the NVT ensemble at 400 K, because the thermal averaging of different energies exhibited by the system at 400K is done in the preceding NVT simulation that produced the starting points.

Using this approach we were able to submit 16 times 20 ps of *ab initio* MD simulation trajectories of a system containing 128 water molecules to our analysis. This is 20 times more than is usually reported in the literature so far. Furthermore the system is about twice as large as the commonly used ones, which yield 32-64 water molecules. This represents a significant advance in the use of *ab initio* MD simulations. For the first time we are thus able to extract data from our simulations that are statistically reliable.

Performance We thoroughly tested the parallel performance of the CP2K simulation code on HLRB2. The Message Passing Interface (MPI) was used for communication between parallel processes. Numerical libraries crucial to the overall performance of CP2K are the FFTW library for multidimensional fast Fourier transformations and the linear algebra libraries BLAS, LAPACK and SCALAPACK. The scalability of the computing performance increases with the size of the simulated system. The use of a large system like 128 water molecules therefore resembles an advantage provided that the necessary resources are available.

For 128 water molecules we observe a linear strong scaling for up to 64 processes, where 89% of the ideal scaling performance is obtained. For an increasing number of processes this decreases to 72% for 96 processes and 55% for 128 processes as shown in Fig. 2. Our simulations were therefore routinely carried out with 64 processes.

One iteration of the self consistent field (SCF) equations used to optimize the Kohn-Sham-wavefunction was completed in average within 3.8 seconds. The Quickstep algorithm implemented in CP2K utilizes an extrapolation method in order to produce an improved first guess for this wavefunction from the history of preceding optimized wavefunctions. Using a third order extrapolation energy convergence was reliably achieved within 8 SCF cycles within the simulation. One timestep of 0.5 fs (90 ps + 16 x 20 ps total simulation time) was therefore completed within 30.3 seconds.

In order to study electronic properties of the system the electron density was subjected on the fly to a localization procedure using maximally localized Wannier functions (MLWFs) and Wannier centers [18, 19] during the production simulations. The Wannier centers of the MLWFs can be regarded as centers of the charge density of the electron pairs in local orbitals. With this approach the individual dipole moments of the water molecules as well as the total dipole moment of the system

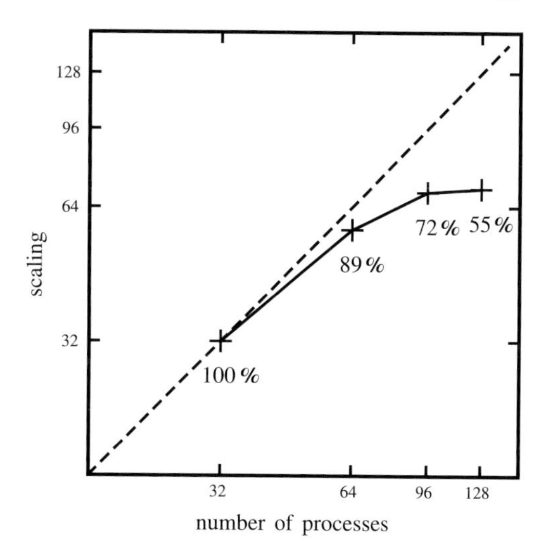

Fig. 2 Strong scaling of the computational performance for the simulation of 128 water molecules with an increasing number of processes (solid line and crosses) and the ideal scaling (dashed line)

can be accessed. This was done in every fourth NVE time step and was in average completed within 15.4 seconds.

3 Results

Diffusion As outlined before it is the goal of this study to provide statistically converged properties of water in *ab initio* MD simulations. This is a prerequisite for the study of water under special conditions, e.g. in the solvation shell of a biomolecule. High accuracy is needed in order to be able to distinguish between normal water properties and properties of water that are the result of an influence, e.g. the distortion of the hydrogen bond network of water due to the presence of a solute.

In Fig. 3 the average mean squared displacement (MSD) of a water molecule is shown as a function of time. It describes how fast water molecules translate through the system and therefore its slope in the linear regime is directly related to the self-diffusion coefficient via the Einstein relation

$$D = \frac{1}{2N} \lim_{t\to\infty} \frac{\mathrm{d}}{\mathrm{d}t} \left\langle |\mathbf{r}_i(t) - \mathbf{r}_i(0)|^2 \right\rangle , \tag{1}$$

where N is the number of dimensions and $\mathbf{r}_i(t)$ the position vector of particle i at time t. The $\langle\rangle$ brackets denote the ensemble average.

The averages of the MSD obtained from different numbers of trajectories are shown if Fig. 3. The deviations of averages obtained for one or two trajectories from the average of the full ensemble, strongly indicate smaller sized simulation approaches are not capable of producing reliable and statistically meaningful results.

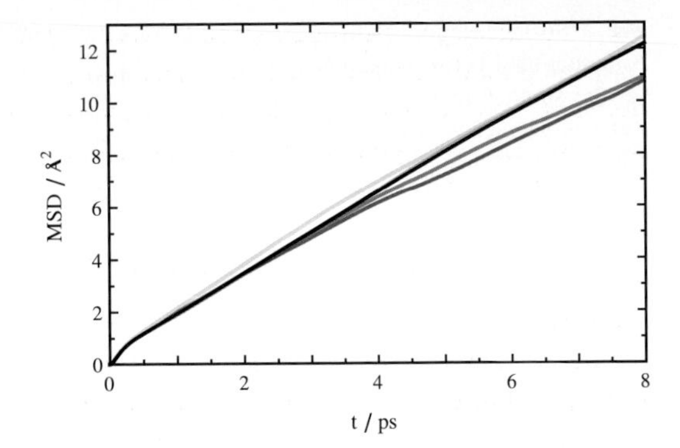

Fig. 3 Convergence of the mean square displacement, which describes diffusion processes; averages are obtained from increasing numbers of trajectories;
red line: 1 trajectory
orange line: 2 trajectories
yellow line: 4 trajectories
green line: 8 trajectories
black line: 16 trajectories

The average diffusion coefficient as obtained from our study is 2.22 $\text{nm}^2\,\text{ns}^{-1}$. The small deviation to the experimental value of 2.30 $\text{nm}^2\,\text{ns}^{-1}$ [12] suggests, that our approach to compensate for the lack of quantum effects and putative overbinding in the DFT potential energy by raising the simulation temperature was successful.

Hydrogen Bond Dynamics In order to characterize the dynamical properties of the hydrogen bond network we additionally studied hydrogen bond correlation functions. They describe the flexibility of hydrogen bonds and the dynamics of hydrogen bond rearrangements.

We applied the weak hydrogen bond correlation function

$$C_{HB}(t) = \langle h(0)h(t) \rangle \ , \tag{2}$$

where $h(t)$ is an operator that yields one if a hydrogen bond between two water molecules is intact at time t and zero otherwise [2]. The weak hydrogen bond correlation takes into account that initially broken hydrogen bonds often recover if the involved molecules do not become separated due to diffusional motions. The weak hydrogen bond correlation function therefore contains information about several dynamical properties, ranging from rotational motions, which are often involved in the initial breaking of hydrogen bonds, to translational diffusion. The results are

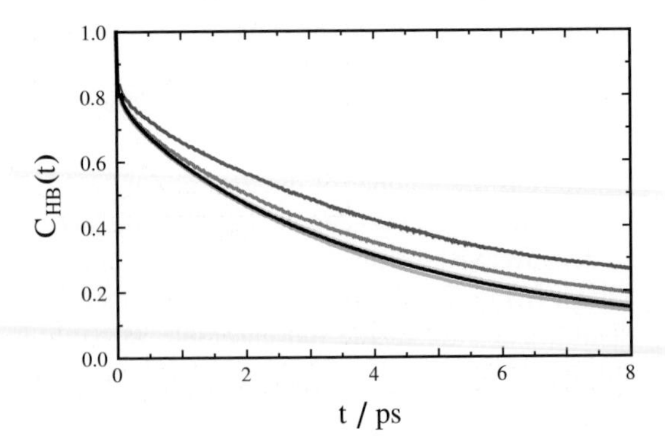

Fig. 4 Convergence of the hydrogen bond correlation function $C_{HB}(t)$; averages are obtained from increasing numbers of trajectories;
red line: 1 trajectory
orange line: 2 trajectories
yellow line: 4 trajectories
green line: 8 trajectories
black line: 16 trajectories

presented in Fig. 4. Hydrogen bonds were defined using a geometric criterion. A hydrogen bond between a pair of water molecules was considered as intact if the donor-acceptor distance was less than 0.35 nm and the O-H-O angle formed by the acceptor, the involved hydrogen and the donor was less than 150 degrees.

As illustrated in Fig. 4 also for the characterization of hydrogen bond dynamics single trajectories are insufficient. Taking into account all trajectories however, we can estimate the average decay time with 3.17 ps.

IR Spectrum Using Wannier centers to access molecular dipole moments and the dipole moment of the whole system, we furthermore computed the infrared absorption spectrum of water. The infrared spectrum contains information about all vibrational modes in the system that involve a change of the dipole moment. It is furthermore routinely accessible in experiments and provides an invaluable tool for comparison between simulation and experimental observation. It can be accessed via the Fourier transformed dipole auto-correlation function

$$\alpha(\omega) = \omega \frac{\hbar\omega}{n(\omega)k_{\mathrm{B}}T} \int_t e^{i\omega t} \langle \mathbf{M}(0)\mathbf{M}(t) \rangle \, \mathrm{d}t \,, \tag{3}$$

where $n(\omega)$ is the refractive index, k_{B} is Boltzmann's constant, T the temperature and $\mathbf{M}(t)$ the systems dipole moment at time t. The choice of the frequency dependent prefactor corresponds to the harmonic limit approximation [23].

The result is presented in Fig. 5. All spectra were computed with a frequency resolution of 20 cm^{-1}, which was achieved via a convolution with a Gaussian kernel function. Figure 5 illustrates how the computed spectrum converges, when more and more trajectories are taken into account. While the qualitative agreement with the experimental spectrum [4] can already be observed for single trajectories, statistically converged absorption spectra are only obtained by taking all available trajectories into account. As outlined before this accuracy is indispensable, when the

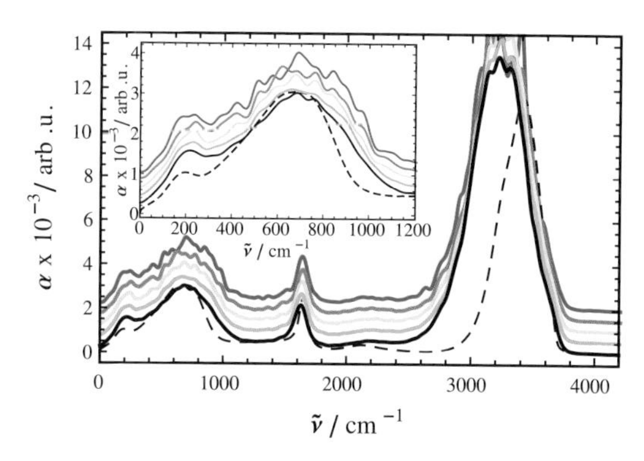

Fig. 5 Convergence of IR absorption spectra (inset: zoom into the far infrared region); averages are obtained from increasing numbers of trajectories; red line: 1 trajectory; orange line: 2 trajectories; yellow line: 4 trajectories; green line: 8 trajectories; black line: 16 trajectories; the computed spectra are shifted along the y-axis for improved clarity; black dashed line: experimental spectrum [4]

changes in IR absorbance due to the influence of solute molecules will be computed in the second part of the project in order to investigate how modified dynamical properties of the solvating water relate to the experimentally observed THz absorbance.

4 Discussion & Outlook

Using *ab initio* molecular dynamics simulations and the high performance computing capacities made available by HLRB2, we were able for the first time to produce pure water trajectories of sufficient length *and* in sufficient number in order to extract statistically reliable and converged dynamical properties on picosecond timescales. Exemplary we have shown in Sect. 3 how diffusion properties, hydrogen bond rearrangement dynamics and the IR absorption spectrum depend on the number of statistically independent trajectories taken into account using our sampling scheme outlined in Sect. 2. Taking into account 16 statistically independent trajectories in the microcanonical ensemble, whose starting points were sampled from an underlying simulation in the canonical ensemble at the target temperature, we could achieve statistical convergence for each of the introduced properties.

The achievement of statistical convergence is especially critical, when aiming at comparisons to aqueous solutions in subsequent studies, in order to investigate e.g. the underlying mechanisms of changed dynamical water properties in the solvation shell of biomolecules. Force field based simulations have been shown to fail in describing the change in THz absorption induced by the presence of a solute, in addition to a qualitatively incorrect description of the IR absorption in this frequency region [14]. In order to understand the connection between the experimentally observed increase of the THz absorption and the changed water dynamics it is necessary to have a model, which describes the IR absorption spectrum of water and aqueous solutions correctly, especially at THz frequencies. Such a model is given by *ab initio* MD simulations.

Besides their importance as a reference, the trajectories produced within this study offer a unique way to analyze the nature of the underlying vibrational modes at THz frequencies. Due to the implicit inclusion of the polarizability, the influence of molecular polarization and oscillating induced dipole moments can be separated from atomic motions. Comparing the results to force field models, which are restricted to describe the atomic motions, the sources of error leading to incorrect descriptions of IR absorption processes can be revealed in order to propose better models. In order to study vibrational modes and IR absorption mechanisms, specifically the collectivity of dipole oscillations and atomic vibrations, we have developed a spatially resolved cross-correlation analysis method which is able to relate collective oscillations and vibrations to the average structure in the environment of a water molecule, which is illustrated in Fig. 6. Although only pure water was simulated, this analysis revealed characteristics of vibrational modes at THz frequencies, that

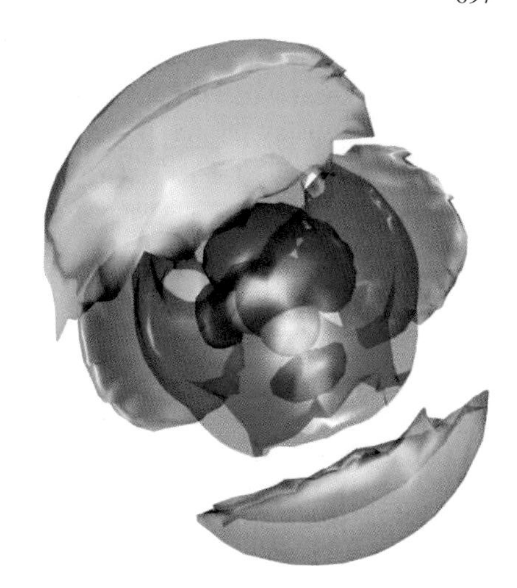

Fig. 6 3-dimensional structure of the average environment of a water molecule displayed in three isosurfaces of the relative oxygen atom number density; red: first solvation shell (1.75); blue: second solvation shell (1.26); green: third solvation shell (1.10); in parenthesis the isosurface value relative to the average number density is given

may yield an explanation for the strong sensitivity of the THz absorbance of water to the presence of solvents, distortions of the hydrogen bond network and changed dynamical properties. These results will be published shortly.

Acknowledgements M. Heyden acknowledges financial support by the German National Academic Foundation. The authors also thank Dr. Gerald Mathias and Prof. Dr. Dominik Marx for their close and fruitful cooperation within this project. Furthermore the authors gratefully acknowledge the Leibniz Computer Center (LRZ) in Garching for providing the necessary computational resources for this study within project h1242. We thank Dr. Helmut Satzger from the remote visualization department at LRZ for very helpful discussions regarding data evaluation and visualization. Dr. Jian Sun, Dr. Harald Forbert and Dr. Gerhard Schwaab are acknowledged for valuable discussions.

References

1. Ball P (2008) Water as an Active Constituent in Cell Biology. Chem Rev 108:74–108
2. Bagchi B (2005) Water Dynamics in the Hydration Layer around Proteins and Micelles. Chem Rev 105:3197–3219
3. Becke AD (1988) Density-functional exchange-energy approximation with correct asymptotic behavior. Phys Rev A 38:3098–3100
4. Bertie JE, Lan Z (1996) Infrared Intensities of Liquids XX: The Intensity of the OH Stretching Band of Liquid Water Revisited, and the Best Current Values of the Optical Constants of $H_2O(l)$ at 25°C between 15,000 and 1 cm^{-1}. Appl Spec 50:1047–1057
5. Ebbinghaus S, Kim SJ, Heyden M, Yu X, Heugen U, Gruebele M, Leitner DM, Havenith M (2007) An extended dynamical hydration shell around proteins. Proc Natl Acad Sci 104:20749–20752
6. Ebbinghaus S, Kim SJ, Heyden M, Yu X, Gruebele M, Leitner DM, Havenith M (2008) Protein Sequence- and pH-Dependent Hydration Probed by Terahertz Spectroscopy. J Am Chem Soc 130:2374–2375

7. Fanourgakis GS, Xantheas SS (2008) Development of transferable interaction potentials for water. V. Extension of the flexible, polarizable, Thole-type model potential (TTM3-F, v. 3.0) to describe the vibrational spectra of water clusters and liquid water. J Chem Phys 128:074506
8. Fernández-Serra MV, Artacho E (2004) Network equilibration and first-principles liquid water. J Cehm Phys 121:11136–11144
9. Goedecker S, Teter M, Hutter J (1996) Separable dual-space Gaussian pseudopotentials. Phys Rev B 54:1703–1710
10. Grossman JC, Schwegler E, Draeger EW, Gygi F, Galli G (2004) Towards an assessment of the accuracy of density functional theory for first principles simulations of water. J Chem Phys 120:300–311
11. Hartwigsen C, Goedecker S, Hutter J (1998) Relativistic separable dual-space Gaussian pseudopotentials from H to Rn. Phys Rev B 58:3641–3662
12. Hertz HG (1973) Nuclear magnetic relaxation spectroscopy. *In* Water: A Comprehensive Treatise, Vol. 3. F. Franks, editor. Plenum Press, New York 301–395.
13. Heugen U, Schwaab G, Bründermann E, Heyden M, Yu X, Leitner DM, Havenith M (2006) Solute-induced retardation of water dynamics probed directly by terahertz spectroscopy. Proc Natl Acad Sci 103:12301–12306
14. Heyden M, Niehues G, Heugen U, Leitner DM, Havenith M (2008) The long range influence of carbohydrates on the solvation dynamics of water - answers from THz spectroscopic measurements and molecular modelling simulations. J Am Chem Soc 130:5773–5779
15. Kohn W, Sham LJ (1965) Self-Consistent Equations Including Exchange and Correlation Effects. Phys Rev 140:A1133–A1138
16. Kim SJ, Born B, Havenith M, Gruebele M (2008) Real-Time Detection of Protein-Water Dynamics upon Protein Folding by Terahertz Absorption Spectroscopy. Angew Chem Int Ed 47:6486–6489
17. Lee C, Yang W, Parr RG (1988) Development of the Colle-Salvetti correlation-energy formula into a functional of the electron density. Phys Rev B 37:785–789
18. Marx D, Hutter J (2009) *Ab initio* Molecular Dynamics - Basic Theory and Advanced Methods. Cambridge University Press
19. Marzari N, Vanderbilt D (1997) Maximally localized generalized Wannier functions for composite energy bands. Phys Rev B 56:12847–12865
20. Pal SK, Peon J, Bagchi B, Zewail AH (2002) Biological Water: Femtosecond Dynamics of Macromolecular Hydration. J Phys Chem B 106:12376–12395
21. Perdew JP, Burke K, Ernzerhof M, (1996) Generalized Gradient Approximation Made Simple. Phys Rev Lett 108:74–108
22. Rahman A, Stillinger FH (1974) Propagation of sound in water. A molecular-dynamics study. Phys Rev A 10:368–378
23. Ramírez R, López-Ciudad T, Kumar P, Marx D (2004) Quantum corrections to classical time-correlation functions: Hydrogen bonding and anharmonic floppy modes. J Chem Phys 121:3973–3983
24. Schwegler E, Grossman JC, Gygi F, Galli G (2004) Towards an assessment of the accuracy of density functional theory for first principles simulations of water. II. J Chem Phys 121:5400–5409
25. Sharma M, Resta R, Car R (2005) Intermolecular Dynamical Charge Fluctuations in Water: A Signature of the H-Bond Network. Phys Rev Lett 95:187401
26. Silvestrelli PL, Bernasconi M, Parrinello M (1997) *Ab initio* infrared spectrum of liquid water. Chem Phys Lett 277:478–482
27. Soper AK (2000) The radial distribution functions of water and ice from 220 to 673 K and at pressures up to 400 MPa. Chem Phys 258:121–137
28. VandeVondele J, Krack M, Mohamed F, Parrinello M, Chassaing T, Hutter J (2005) Quickstep: Fast and accurate density functional calculations using a mixed Gaussian and plane waves approach. Comput Phys Commun 167:103–128

Ab Initio Molecular Dynamics Simulations of Aqueous Glycine Solutions: Solvation Structure and Vibrational Spectra

Jian Sun, Harald Forbert, David Bosquet, and Dominik Marx

Abstract Aqueous Glycine solutions are studied with ab initio molecular dynamics simulations to investigate the structural aspects of the solvation shell and the vibrational spectrascopy. The individual contributions to the total IR spectrum from the glycine and the water solvent are decomposed systematically using schemes in terms of maximally localized Wannier orbitals to define approximate molecular dipole moments in solution.

1 Introduction

The dynamical and thus entropic effects of solvation in aqueous solutions are of fundamental interest to physics, chemistry, and biology. They are particularly important for understanding the mechanisms of key processes, for instance ionic solubilities and protein folding. Amino acids not only are the building blocks of proteins, but also have a variety of roles in the metabolism. Therefore, amino acids are critical to life and particularly important in biochemistry.

There are several aspects that are very interesting concerning amino acids in solution, for instance the de– and re–protonation reactions involving the zwitterion, the solute-solvent interactions (more precisely, the hydrogen-bonding interactions with the environment), the IR spectroscopy of their solutions to name but a few. As the simplest amino acid, in which the amino group and the carboxyl group attach on the same carbon and the "side chain" is only one proton, glycine and its solvation behaviorhave already been investigated with different theoretical methods. [2, 6, 8, 10, 18, 19, 31, 33, 35] However, the effects of hydrogen-bonding and solvation on its IR vibrational spectroscopic properties remain largely in the dark.

Jian Sun · Harald Forbert · David Bosquet · Dominik Marx
Lehrstuhl für Theoretische Chemie, Ruhr–Universität Bochum, 44780 Bochum, Germany
e-mail: Jian.Sun@theochem.rub.de

S. Wagner et al. (eds.), *High Performance Computing in Science and Engineering, Garching/Munich 2009*, DOI 10.1007/978-3-642-13872-0_58,

Glycine occurs in the neutral form in the gas phase or the isolated state, while in solution or in the solid state it prefers the zwitterionic structure due to the interactions between the surrounding molecules through hydrogen bonds. [3, 15] Therefore, the stability of the zwitterionic state and the proton transfer path between the neutral state and zwitterionic form are interesting. It has been approved theoretically that at least two water molecules are necessary to stablize the zwitterionic form of glycine. [14] Recently, Leung and Rempe [19] studied the intramolecular proton transfer in glycine solution and found that the average hydration water surrounding the glycine is not constant along the reaction coordinate, but change from about five in the neutral molecule to about eight for the zwitterion.

Due to the demanding nature of electronic structure based calculations, most molecular dynamics simulations of glycine solutions are performed with classical force fields or with quantum chemical models which only contain a few water molecules. For instance, Ramaekers et al. [31] investigated the glycine/H_2O interaction in isolated conditions with DFT calculations and experimental matrix IR spectroscopies. Takayanagi et al. [33] studied the dynamics of glycine/water with semiempirical potentials whereas Leung and Rempe [19] used ab initio molecular dynamics.

In this paper, we investigate the dynamics of zwitterionic glycine in aqueous solution and the resulting IR vibrational spectroscopy using ab initio molecular dynamics [24]. In particular, the interactions between the solute and the environment are analyzed in terms of the hydration numbers to different atoms and individual contributions to the IR spectra stemming from solute and solvent are decomposed.

2 Methods

With the development of modern theoretical methodology and high performance computing, atomistic simulations have achieved a high level such that the microscopic details of chemical events can now be treated on a routine basis. One significant landmark in this area is the so-called ab initio molecular dynamics (AIMD) method, which originated from the efficient and elegant algorithm proposed by Car and Parrinello in 1985 [9]. AIMD combines finite temperature atomistic molecular dynamics with internuclear forces obtained from accurate electronic structure calculations performed "on the fly" as the MD simulation proceeds (see Ref. [22–24] for reviews).

These simulation algorithms together with high-speed computer hardware open an avenue to studying chemical systems and their dynamics that involve many coupled degrees of freedom, such as those taking place in liquids, in the "Virtual Lab" [21]. AIMD has been used to study a wide variety of chemically interesting systems, including but of course not limited to, calculations of the structure and IR spectroscopy of pure bulk water [32], biosystems such as bacteriorhodopsin [5, 27], complex molecular systems, i.e. large-amplitude motion in CH_5^+ [4, 17], and of spectral fingerprints of microsolvated HCl/water clusters [12, 26]. Many other such applications also attract the attention of the biology community, for instance, proton transport in aqueous acidic and basic environments [25], calculation of NMR properties of proteins [30], the structure of nucleic acids [13], and proton transport in hydrogen-bonded liquids [28].

Ab initio molecular dynamics simulations based on the Born-Oppenheimer approach are performed using the Gaussian and plane waves method (GPW) and its augmented extension (GAPW) using the Quickstep [34] module in the CP2K package [1]. A dual basis of atom-centred Gaussian orbitals and plane waves (regular grids) are applied in this approach. The former is used to represent the Kohn–Sham orbitals whereas the latter are used to represent the electronic density. Triple-ζ TZV2P–type Gaussian basis sets [34] and Goedecker-Teter-Hutter (GTH) pseudopotentials [11, 16] with the Perdew-Burke-Ernzerhof (PBE) [29] exchange-correlation functional are employed. A plane wave cutoff of 400 Ry for the finest grid level and a grid of 20 Ry to map the Gaussians are found to be sufficient for the system. A time step of 0.5 fs is used for the integration of the equations of motion. A cubic periodic supercell with a fixed box length of 9.85 Å hosting one glycine molecule and 30 water molecules is employed. The temperature is set to 400 K in order to correct for the known understimation of the temperature by density functionals such as PBE by roughly 20 to 30 % such as to achieve agreement with the room temperature experimental properties such as the oxygen–oxygen radial distribution function $g(r)$ and the IR spectrum of pure bulk water (data not shown). In order to be able to compute canonical time–correlation functions as required for IR spectroscopy a set of independent initial conditions for microcanonical NVE runs have been sampled from a long trajectory produced in the canonical ensemble NVT by coupling massive Nosé–Hoover chain [20] thermostats to the nuclei, while each independent NVE trajectory is 20 ps long.

The Maximally Localized Wannier Functions (MLWF) have been used instead of standard delocalized Bloch orbitals. Wannier centers give us a natural and intuitive sight of the electronic states of the system since there is a close connection between MLWFs and the shape and symmetry of chemical bonds. See the monograph [24] for a detailed presentation.

As shown in Fig. 1 by marking their centers, there are four doubly occupied MLWFs per water molecule which corresponds to a charge of $q_{\mathrm{MLWF}} = -2\mathrm{e}$. These centers are chemically meaningful in the sense that they correspond to the covalent bonds and the lone pairs, thus allowing a direct visualization of the tetrahedral hydrogen–bonding network. Wannier centers can be treated like quasi–particles, thus the molecular dipole can be naturally defined in a classical way by associating the charge and position of all nuclei and Wannier centers

$$\mu = \sum_{particle} q_{particle} \times \overrightarrow{r}_{particle} \tag{1}$$

that belong to a certain molecule, where the summation includes all nuclei and Wannier centers that belong to a given molecule, either glycine or water.

As it is well known, the spectral features of a system are just a frequency domain representation of the underlying time–evolution of the system. For IR absorption,

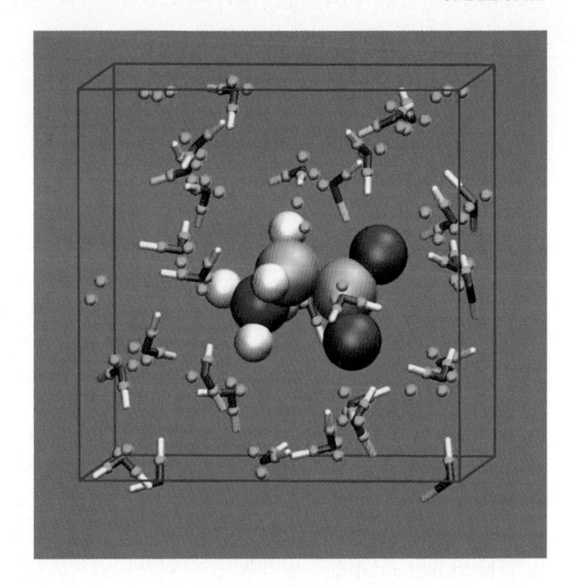

Fig. 1 A snapshot of the glycine/water system as used in our simulations. Water molecules are represented by a ball–and–stick representation, glycine by larger spheres, and the Wannier function centers are shown by small green spheres; note that some Wannier centers seem to be decoupled from the corresponding molecular skeleton which stems from folding back all sites into the central cubic box as a result of applying periodic boundary conditions

the spectrum encodes the time–dependent changes of the molecular dipole moment of the system, which in turn depends on the positions of electrons and nuclei. The lineshape function defining the IR spectrum is then given by the Fourier transformation of the auto correlation function of the dipole operator, i.e.

$$I(\omega) := \omega^2 \times \frac{1}{2\pi} \int_{-\infty}^{+\infty} \langle \mu_{tot}(t) \times \mu_{tot}(0) \rangle dt \ , \tag{2}$$

which is treated here in the classical approximation subject to a Quantum Correction Factor to restore detailed balance (see below); note that the absorption coefficient $\alpha(\omega)$ is proportional to $I(\omega)$ within a known wavelength–dependent prefactor. See the monograph [24] for a detailed presentation.

3 Results and Discussion

The molecular structure of amino acids themselves is very interesting since it depends qualitatively on the environment. In the gas phase or in isolation glycine exists in the neutral from while it prefers the zwitterionic form in solution or in the solid. This zwitterionic form is more stable compare to the neutral form due to the interaction with the surrounding molecules through the hydrogen bonding. Simulations in this work primarily focus on the dynamics in aqueous solution which implies that the zwitterionic form is analyzed.

Figure 2 depicts the radial distribution functions for the pairs of heavy atoms of glycine and the oxygen atoms of water molecules. As we can see, the glycine molecule is like a short stick, the average distances from the two ends (N and O

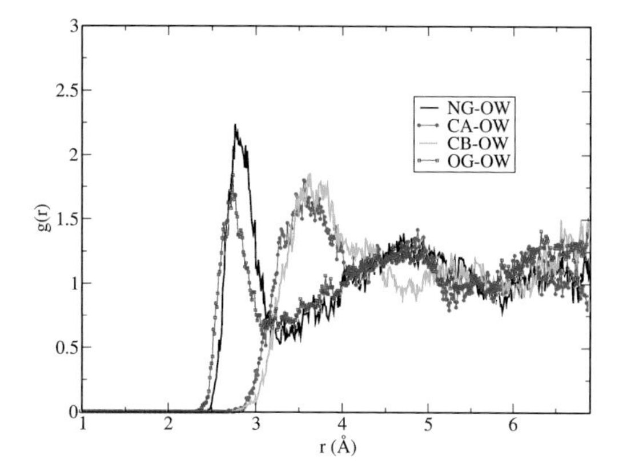

Fig. 2 Radial distribution functions of the heavy atoms in glycine (NG, CA, CB, OG) to the oxygen atoms of the water molecules where CA and CB represents the α–carbon and the carbon of the carboxylate group, respectively

atoms) to water are shorter than that from the carbon atoms in the middle, the shorter average distance is around 2.8 Å, while the larger ones are around 3.5 - 3.8 Å. This will help us to define the solvation shell of glycine properly and to disentangle the dipole autocorrelation function and thus the spectra into individual parts, i.e. the part that comes mainly from the solvent (water molecules) while others stem from the solute (i.e. glycine).

From the pair distribution functions of the N atom in glycine to the oxygen and hydrogen atoms in the water molecules, as we shown in Fig. 3, the extent of the first solvation shell can be determined and is found to amount to about 3.35 Å. This is almost the same as size of the first solvation shell in bulk water. From the integration of the the radial distribution function, there is a shoulder indicating that the number of the hydration water water surrounding the amino group is about three. The shell structure of the $g(r)$ for N to the H atoms of the water molecule (Fig. 3 right) is not as clear as NG-OW on the left. The peak of the first shell is further out, not so dominant, and there is no clear shoulder in the running integration. This

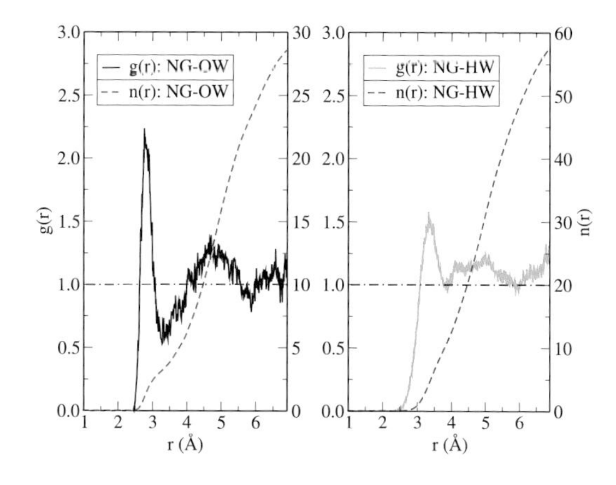

Fig. 3 Radial distribution functions of the N atom in glycine to the oxygen and hydrogen atoms of the water molecules; dashed lines are the corresponding running integration numbers here and in the following

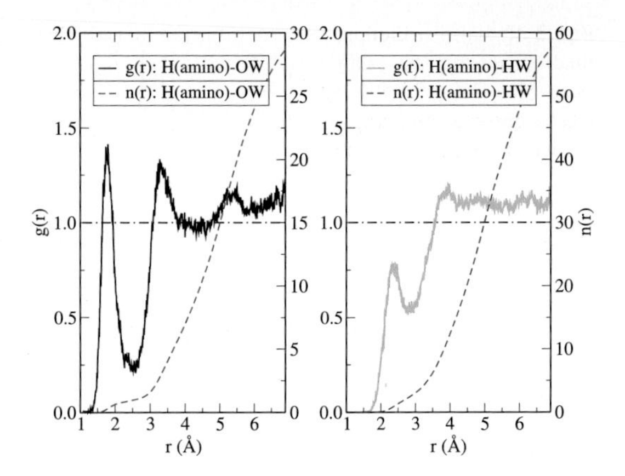

Fig. 4 Radial distribution functions of the H atoms of the amino group in glycine to the oxygen and hydrogen atoms of water molecules

suggests that the oxygen atoms preferrentially show towards the N site of glycine thus forming three hydrogens bonds, i.e. one per H atom of the amino group.

Since the hydrogen atoms in the amino group form hydrogen bonds with water molecules directly, the first solvation shell comes out clearly from $g(r)$: H(amino)-OW (Fig. 4 left). The peak of the first solvation shell is pronounced and the cutoff of it is extends to around 2.5 Å, which is exactly the hydrogen bond length. The integration clearly shows that there is only one water molecule hydrogen bonding to every hydrogen atom in the amino group. Compared to the first solvation shell of the amino group, the $g(r)$ after the first minimum becomes quite flat, indicating that the second shell is not so well–defined when compared to the first shell.

Now, we turn to the other end of the glycine molecule, which is the carboxyl(ate) group. In aqueous solutions, the proton of the carboxyl group transfer to the amino group and forms a –COO$^{\ominus}$ and a –NH3$^{\oplus}$ in the same molecule, the so-called zwitterionic from, in which the net charge of the whole molecule is still zero but a strong intramolecular charge separation and thus dipole moment results. The –COO$^{\ominus}$ and -NH3$^{\oplus}$ are of course hydrophilic, the former can accept hydrogen bonds and the latter can donate. As we can see in Fig. 4 and Fig. 5, the first solvation shell from these two ends to the water molecules are quite clear. Baesd on Fig. 3 and also Fig. 4, we know that there are three water molecules in the first solvation shell of the –NH3$^{\oplus}$ group. In addition, from the running integration of $g(r)$ (Fig. 5), one finds that every oxygen atom in the –COO$^{\ominus}$ group has two nearest water molecules. The average OG–OW distance is 2.7 Å and the extend of the first and second solvation shells of O in carboxylate group amount to around 3.3 and 5.5 Å, respectively. The average OG–HW distance is 1.8 and the first minimum of $g(r)$ is around 2.5 Å, again, the H in water is more disordered and therefore it is differecult to characterize the second solvation shell from the $g(r)$ of OG–HW pairs.

The average distance from the carbon atom in the carboxylate group to oxygen in water is around 3.7 Å, while the first minimum is around 4.7 Å. But the first peak is not as dominant and as we see from the $g(r)$ of OG–OW or NG–OW there is no shoulder in the running integration of $g(r)$: CB–OW.

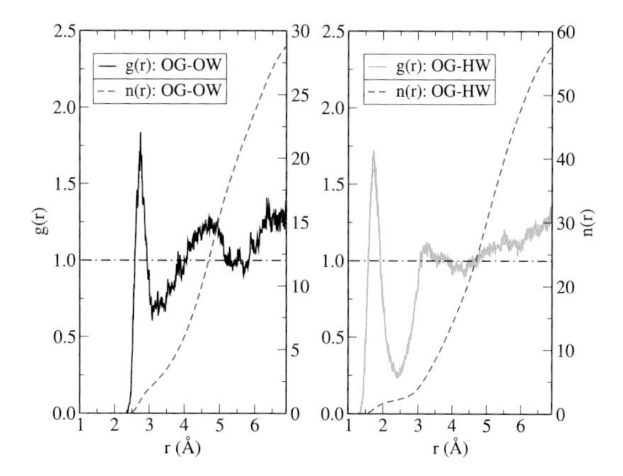

Fig. 5 Radial distribution functions of O atoms of the carboxylate group in glycine to the oxygen and hydrogen atoms of the water molecules

We now turn to the IR spectroscopy of this glycine in water solution. Fig. 7 shows the average IR spectra obtained from eight independent NVE trajectories with a length of 20 ps for each one. The black, red and green colors represent the IR spectra of the whole system, the glycine molecule in solution (but without any solvent around) and of all the solvent water molecules as obtained by partitioning the total dipole moment into approximate molecular dipole moments of gylcine and every individual water molecule using the Wannier decomposition technique. The dominant peak of the total spectrum at around 3400 cm^{-1} mainly comes from the stretching modes of the many water molecules. Some features coming from glycine are at located around 1350 and 1500 cm^{-1}, while the bending mode from water molecules is at around 1630 cm^{-1}. Note that the experimental data [7] for the molecular bending of liquid water is at around 1640 cm^{-1} which underscores the validity of our ab initio simulation approach; comparision the OH stretching modes is difficult since they are heavily affected by shifts due to quantum motion of the nuclei which is not capture by the classical dynamics underlying this investigation.

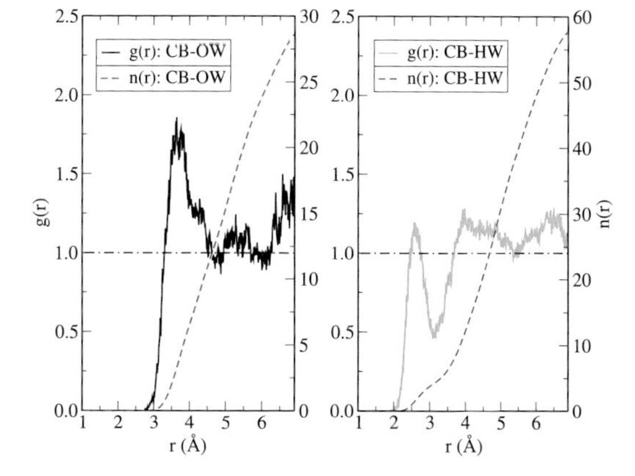

Fig. 6 Radial distribution functions of the C atom of the carboxylate group in glycine to the oxygen and hydrogen atoms of water molecules

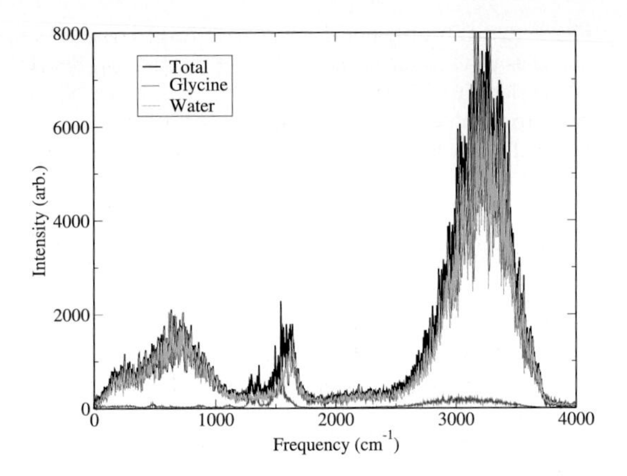

Fig. 7 Infrared spectra of the total system and decomposed into the solvated glycine molecule and the solvent water molecules

4 Conclusions and Outlook

Ab initio molecular dynamics relying on density functional theory is applied in order to investigate the dynamics and the solute/solvent interactions of glycine in aqueous solution both in terms of structure and vibrational properties. From our simulations the average number of hydration water molecules around the amino group is three, while the two oxygen atoms in the carbonxylate group of this zwitterion accept four hydrogen bonds. The α–carbon and the hydrogen attached to it, i.e. the so–called alkyl group, are hydrophobic and do not tend to form hydrogen–bonding contacts with surrounding water molecules. The IR spectra and the vibrational density of states demonstrate that with our methods, given sufficient statistics in terms of averaging over initial conditions, it will be possible to separate the individual contributions from the solute and from the solvent, in particular at low (THz) frequencies.

Acknowledgements This work benefitted from discussions with Marcel Baer. J. S. gratefully acknowledges the financial support of Alexander von Humboldt (AvH) foundation. The simulations were carried out at HLRB II (München) and at BOVILAB@RUB (Bochum).

References

1. CP2K code. `http://cp2k.berlios.de`
2. Aikens, C.M., Gordon, M.S.: Incremental solvation of nonionized and zwitterionic glycine. J. Am. Chem. Soc. **128**, 12835–12850 (2006)
3. Albrecht, G., Corey, R.B.: J. Am. Chem. Soc. **61**, 1087 (1939)
4. Asvany, O., Kumar, P., Redlich, B., Hegemann, I., Schlemmer, S., Marx, D.: Understanding the infrared spectrum of bare CH_5^+. Science **309**, 1219–1222 (2005)
5. Baer, M., Mathias, G., Kuo, I.F.W., Tobias, D.J., Mundy, C.J., Marx, D.: Spectral signatures of the pentagonal water cluster in bacteriorhodopsin. Chem. Phys. Chem. **9**, 2703–2707 (2008)
6. Balta, B., Aviyente, V.: Solvent effects on glycine II. Water-assisted tautomerization. Journal of Computational Chemistry **25**(5), 690–703 (2004)

7. Bertie, J.E., Lan, Z.: Infrared intensities of liquids XX: the intensity of the OH stretching band of liquid water revisited, and the best current values of the optical constants of H2O(l) at 25°C between 15,000 and 1 cm^{-1}. Appl. Spectrosec. **50**, 1047–1057 (1996)
8. Brancato, G., Barone, V., Rega, N.: Theoretical modeling of spectroscopic properties of molecules in solution: toward an effective dynamical discrete/continuum approach. Theoretical Chemistry Accounts **117**(5-6), 1001–1015 (2007)
9. Car, R., Parrinello, M.: Phys. Rev. Lett. **55**, 2471 (1985)
10. Fernández-Ramos, A., Smedarchina, Z., Siebrand, W., Zgierski, M.Z.: A direct-dynamics study of the zwitterion-to-neutral interconversion of glycine in aqueous solution. J. Chem. Phys. **113**, 9714–9721 (2000)
11. Goedecker, S., Teter, M., Hutter, J.: Separable dual-space Gaussian pseudopotentials. Phys. Rev. B **54**, 1703 (1996)
12. Gutberlet, A., Schwaab, G., Birer, O., Masia, M., Kaczmarek, A., Forbert, H., Havenith, M., Marx, D.: Aggregation-induced dissociation of HCl(H2O)4 below 1 K: The smallest droplet of acid. Science **324**, 1545–1548 (2009)
13. Hutter, J., Carloni, P., Parrinello, M.: Nonempirical calculations of a hydrated RNA duplex. J. Am. Chem. Soc. **36**, 8710–8712 (1996)
14. Jensen, J.H., Gordon, M.S.: On the number of water-molecules necessary to stabilize the glycine zwitterion. Journal of the American Chemical Society **117**(31), 8159–8170 (1995)
15. Jonsson, P.G., Kvick, A.: Acta Crystallogr., Sect. B **28**, 1827 (1972)
16. Krack, M.: Pseudopotentials for H to Kr optimized for gradient-corrected exchange-correlation functionals. Theor. Chem. Acc. **114**, 145–152 (2005)
17. Kumar, P., Marx, D.: Understanding hydrogen scrambling and infrared spectrum of bare CH_5^+ based on ab initio simulations. Phys. Chem. Chem. Phys. **8**, 573–586 (2006)
18. Kundrat, M.D., Autschbach, J.: Ab initio and density functional theory modeling of the chiroptical response of glycine and alanine in solution using explicit solvation and molecular dynamics. J. Chem. Theory and Comput. **4**, 1902–1914 (2008)
19. Leung, K., Rempe, S.B.: Ab initio molecular dynamics study of glycine intramolecular proton transfer in water. Journal of Chemical Physics **122**(18) (2005)
20. Martyna, G.J., Klein, M.L., Tuckermann, M.: J. Chem. Phys. **97**, 2635 (1992)
21. Marx, D.: Theoretical Chemistry in the 21st Century: The "Virtual Lab". In: W. Greiner, J. Reinhardt (eds.) Proceedings of the "Idea–Finding Symposium: Frankfurt Institute for Advanced Studies", pp. 139–153. EP Systema, Debrecen (2004)
22. Marx, D.: An Introduction to Ab Initio Molecular Dynamics Simulations. In: D.M.J. Grotendorst, S. Blugel (ed.) Computational Nanoscience: Do it yourself!, pp. 195–244. NIC, FZ Jülich (2006)
23. Marx, D., Hutter, J.: in Modern Methods and Algorithms of Quantum Chemistry, edited by J. Grotendorst. NIC, FZ Jülich (2000). For downloads see `http://www.theochem.rub.de/go/cprev.html`
24. Marx, D., Hutter, J.: Ab Initio Molecular Dynamics: basic theory and Advanced Methods. Cambridge University Press, New York (2009)
25. Marx, D., Tuckerman, M.E., Hutter, J., Parrinello, M.: The nature of the hydrated excess proton in water. Nature **397**, 601–604 (1999)
26. Masia, M., Forbert, H., Marx, D.: Connecting structure to infrared spectra of molecular and autodissociated HCl water aggregates. J. Phys. Chem. A **111**, 12181–12191 (2007)
27. Mathias, G., Marx, D.: Structures and spectral signatures of protonated water networks in bacteriorhodopsin. Proc. Natl. Acad. Sci. U.S.A. **104**(17), 6980–6985 (2007)
28. Morrone, J.A., Hasllinger, K.E., Tuckerman, M.E.: Ab initio molecular dynamics simulation of the structure and proton transport dynamics of methanol-water solutions. J. Phys. Chem. B **110**, 3712–3720 (2006)
29. Perdew, J.P., Burke, K., Ernzerhof, M.: Phys. Rev. Lett. **77**, 3865 (1996). Erratum: Phys. Rev. Lett., **78**, 1396 (1997)

30. Piana, S., Sebastiani, D., Carloni, P., Parrinello, M.: Ab initio molecular dynamics-based assignment of the protonation state of pepstatin A/HIV-1 protease cleavage site. J. Am. Chem. Soc. **123**, 8730–8737 (2001)
31. Ramaekers, R., Pajak, J., Lambie, B., Maes, G.: Neutral and zwitterionic glycine.H2O complexes: A theoretical and matrix-isolation fourier transform infrared study. J. Chem. Phys. **120**, 4182–4193 (2004)
32. Silvestrelli, P., Bernasconi, M., Parrinello, M.: Ab initio infrared spectrum of liquid water. Chem. Phys. Lett. **277**, 478–482 (1997)
33. Takayanagi, T., Yoshikawa, T., Kakizaki, A., Shiga, M., Tachikawa, M.: Molecular dynamics simulations of small glycine-(H2O)(n) (n=2-7) clusters on semiempirical PM6 potential energy surfaces. Journal of Molecular Structure-theochem **869**(1-3), 29–36 (2008)
34. VandeVondele, J., Krack, M., Mohamed, F., Parrinello, M., Chassaing, T., Hutter, J.: Quickstep: Fast and accurate density functional calculations using a mixed Gaussian and plane waves approach. Comput. Phys. Commun. **167**, 103–128 (2005)
35. Wood, G.P.F., Gordon, M.S., Radom, L., Smith, D.M.: Nature of glycine and its alpha-carbon radical in aqueous solution: A theoretical investigation. Journal of Chemical Theory and Computation **4**(10), 1788–1794 (2008)

Cyclodimerization of DNA and RNA Bases: Ab Initio Study of the Cyclodimerization of the Uracil Dimer Through a Butane-Like Conical Intersection

Vassil B. Delchev and Wolfgang Domcke

Abstract The mechanism of the photo-induced cyclodimerization of uracil through a butane-like conical intersection was studied with the CASPT2, CC2, and DFT ab initio electronic-structure methods. The linear-interpolation-in-internal-coordinates (LIIC) approach was applied to construct reaction paths and potential-energy profiles connecting reactants and products. The structures of the *cis-syn* cyclodimer, the S_0/S_1 conical intersection and the stacked dimer of uracil were optimized at the CASSCF level. Our calculations support a qualitative mechanistic picture of the relaxation mechanisms of the $^1\pi\pi^*$ state of the cyclodimer and the stacked dimer of uracil through a S_0/S_1 conical intersection.

1 Introduction

Photoinduced chemical reactions in DNA/RNA nucleobases after UV light absorption are important processes in living organisms. For example, a particularly high photostability of the nucleobases is essential for minimizing the photodamage of nucleic acids, making them optimal constituents for the storage of the genetic code [6]. On the other hand, the intrastrand dimerization of neighboring thymines is one of the most common processes leading to DNA damage under UV irradiation [2, 5]. The formation of mutagenic cyclic Pyr<>Pyr photoproducts (see Fig. 1) can disrupt the function of DNA and thereby trigger complex biological responses, including apoptosis, carcinogenesis and immune suppression [16, 18].

In order to recognize and repair these lesions in DNA, the organisms use different enzymatic pathways. DNA photolyases which are activated by concurrent or subsequent exposure to near-UV and visible light (300-500 nm), catalyse the cleav-

Vassil B. Delchev · Wolfgang Domcke
Lehrstuhl für Theoretische Chemie, Technische Universität München, Lichtenbergstr. 4, 85748 Garching bei München, Germany
e-mail: vdelchev@hotmail.com

S. Wagner et al. (eds.), *High Performance Computing in Science and Engineering, Garching/Munich 2009*, DOI 10.1007/978-3-642-13872-0_59,

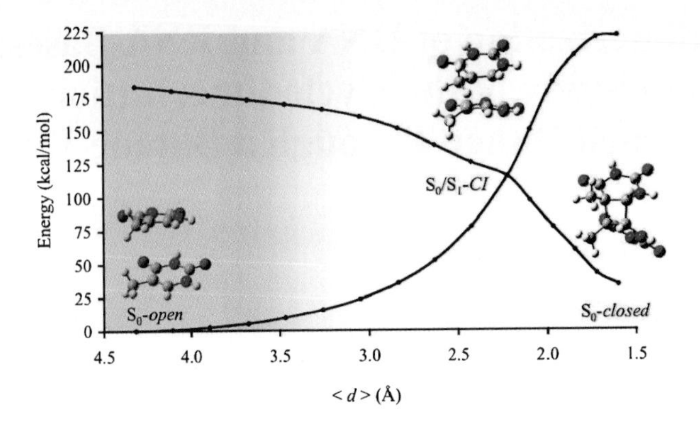

Fig. 1 Singlet excited-state potential-energy profile for the photochemical [2+2] cycloaddition of two stacked thymines. The shaded area indicates the region expected to be altered by the DNA environment. $< d >$ represents the average distance of the two active C-C bonds [4]

age of the cyclobutane bonds in Pyr<>Pyr and restore the initial stacked Pyr+Pyr dimers [14].

Light which is initially absorbed by nucleic acids leads to excited states of the same multiplicity, i.e., to singlet excited states. The excited-state reaction along the cyclodimerization path of pyrimidines occurs in competition with internal conversion processes to the electronic ground state [15] and with intersystem crossing to triplet states. Eisinger *et al.* [7] have shown that both triplet and singlet states are involved in the photodimerization of pyrimidines. Despite the fact that triplet formation has a low quantum yield, Merchan *et al.* [12] have postulated that the triplet state of cytosine is populated by an intersystem crossing mechanism. On the other hand, the time-resolved experiment of Marguet *et al.* [11] has shown that the photodimerization of thymine (to the T<>T dimer) in the oligonucleotide $(dT)_{20}$ leads to the cyclobutane dimer in less than 200 ns. Triplet states are seemingly not involved in the mechanism. The mechanisms of pyrimidine photodimer formation are thus controversial and poorly understood at the molecular level.

There are four possibilities of [2+2] cyclodimerization of uracil for the formation of a cyclobutane ring. They lead to the formation of *cis-syn*, *trans-syn*, *cis-anti* and *trans-anti* cyclodimers [21] (see Fig. 2). Uracil dimers have been isolated from TMV-RNA (tobacco mosaic virus ribonucleic acid) [13]. The experimental analysis (NMR, chromatography, etc.) has shown that the major form of the dimeric photoproduct isolated from irradiated frozen solutions and from native DNA is the *cis-syn* cyclodimer [1]. The *trans-syn* cyclodimer may be obtained upon irradiating denaturated DNA [3]. The spatial arrangement of the *cis-syn* cyclodimer U<>U is of particular interest in deducing what may happen to the local structure of DNA upon irradiation. The stacked face-to-face uracil dimer is a simplified model which is useful for the understanding of the photodimerization of thymine which is abundant in DNA.

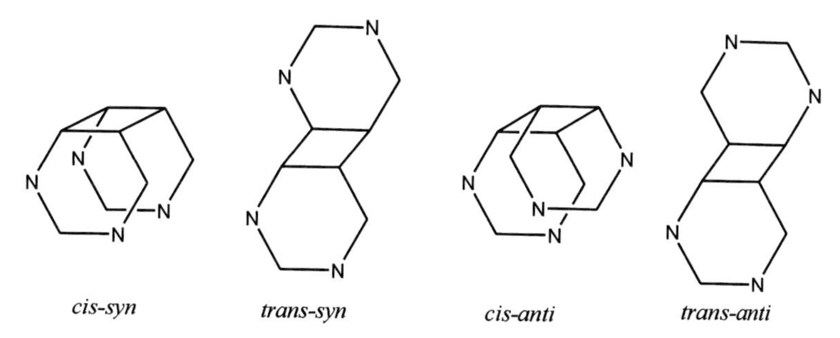

Fig. 2 Different types of cyclodimers of uracil [1]

Despite a considerable number of papers discussing the photoinduced cyclodimerization of pyrimidine bases, there is a lack of knowledge about the role of S_0/S_1 conical intersections in the process. Therefore, the main goal of the present research was the clarification of the mechanism of cyclodimerization of uracil through the excited singlet electronic states. We have performed a comparative study of the mechanism of cyclodimerization of uracil with the CASPT2, CC2 and TDDFT ab initio electronic-structure methods.

2 Computational Methods

The ground-state equilibrium geometries of the cyclo (U<>U) and stacked (U+U) dimers of uracil were optimized at the complete-active-space self-consistent-field (CASSCF) level, with an active space of six electrons in six orbitals. The internal coordinates of the U<>U cyclodimer were allowed to relax during the optimization, while one distance ($C3 \cdots C4$) of the stacked dimer was fixed at 4.0 Å. The two uracil molecules were orientated face-to-face in order to form the *cis-syn* cyclodimer. MP2 and B3LYP optimizations of the ground-state equilibrium geometries of the cyclodimer and the stacked dimer were also performed for the calculation of the vertical excitation energies with these methods. The vertical excitation energies of the cyclodimer and the stacked dimer were calculated at the optimized geometries of the two systems at the MP2, CASSCF and B3LYP levels.

The geometry of the lowest S_0/S_1 conical intersection was optimized at the CASSCF(6,6) level, allowing full relaxation of the internal coordinates during the optimization. The mutual orientation of the two uracil molecules is face-to-face. No symmetry restrictions were employed in the structure optimization. The gradient difference vector ($|\mathbf{GD}|$) and the derivative coupling vector ($|\mathbf{DC}|$) of the conical intersection were determined at the CASSCF(6,6) level.

Excited-state reaction paths and energy profiles were constructed with the linear-interpolation-in-internal-coordinates (LIIC) approach, using the CASPT2, CC2 and DFT (B3LYP) energies of the states. The study was performed with the CASSCF optimized structures in two steps: interpolation between the cyclodimer U<>U and the conical intersection; interpolation between the stacked dimer U+U and the conical intersection. The internal coordinates of the interpolated structures were obtained by the equation

$$q_i = q_i(R) + \varepsilon\{q_i(P) - q_i(R)\} \quad (1)$$

where $q_i(R)$ is an internal coordinate of the reactant, and $q_i(P)$ is the corresponding coordinate of the product (in this case, the S_0/S_1 conical intersection). The interpolation parameter ε varies in the interval from 0 (for the reactant) to 1 (for the product).

The active space for the CASPT2 calculations consists of five p-orbitals and one n-orbital. This active space is consistent with the one used for the geometry optimization of the conical intersection. The structures of the cyclodimer, the stacked dimer and the conical intersection, which were used for the LIIC path, have been optimized with the 6-31G* basis set. The same basis set was employed for the calculation of the vectors $|\mathbf{GD}|$ and $|\mathbf{DC}|$. The correlation-consistent cc-pVDZ basis set was applied for the calculation of the vertical excitation energies and the energy profiles of the excited-state and S_0 reaction paths.

The CASSCF and CASPT2 calculations were performed with the MOLPRO [20] ab initio program package. The program systems TURBOMOLE [9, 10, 17, 19] and GAUSSIAN 03 [8] were applied for the CC2 calculations and the CASSCF optimization of the S_0/S_1 conical intersection, respectively.

3 Results and Discussion

3.1 Geometry Optimization

The optimized ground-state equilibrium geometries of the cyclodimer and the stacked dimer at the CASSCF(6,6)/6-31G* level are shown in Fig. 3. The cyclobutane ring in the *cis-syn* cylodimer of uracil is twisted by an angle $\angle(C_3C_4C_9C_{10})$ of 20.8°. The conformation of the two pyrimidine rings is of the envelope type, which is the result of the repulsion of the oxygen and nitrogen atoms of the two rings. All C−C bonds of the cyclobutane ring are longer than 1.5 Å.

In the stacked dimer, the angle $\angle(C_3C_4C_9C_{10})$ is 16.2°. A weak $H_{21}\cdots O_{14}$ hydrogen bond supports the stacking of the rings, keeping them almost co-planar. The conical intersection shown in Fig. 3 was optimized with fully relaxed internal coordinates. Comparing the distances $C_3\cdots C_4$ of all structures given in Fig. 3, one can see that the geometry of CI_{34} is closer to that of the cyclodimer than to the structure of the stacked dimer. The $\angle(C_3C_4C_9C_{10})$ space angle of CI_{34} is 11.6°. Each of the two uracil rings slightly deviates from the planar structure. Only the hydrogen atoms that are directly bound to the carbon atoms which form the cyclobutane ring deviate considerably from the plane of the corresponding uracil

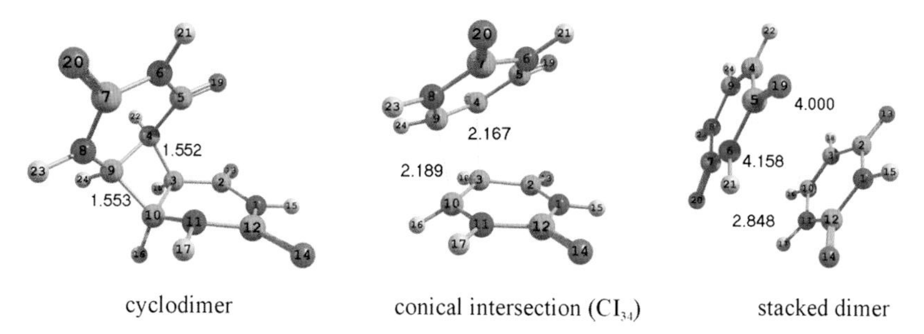

Fig. 3 Optimized structures of the cyclodimer, the S_0/S_1 conical intersection and the stacked dimer

ring: $\angle(H_{16}C_{10}C_3C_2) = -136.5°$, $\angle(H_{18}C_3C_{10}N_{11}) = 163.8°$, $\angle(H_{22}C_4C_9N_8) = -147.0°$ and $\angle(H_{24}C_9C_4C_5) = 157.5°$. No hydrogen bonds between the two rings can be detected.

3.2 Vertical Excitation Energies

The calculated vertical excitation energies of the cyclodimer and the stacked dimer are listed in Table 1. It can be seen that the excitation energies of the cyclodimer are higher than those of the stacked dimer. The CC2 and DFT calculations predict two dark $^1n\pi^*$ states which are very close in energy. The CASPT2(6,6) method pre-

Table 1 Vertical excitation energies (in eV), oscillator strengths and transition dipole moments.*CASPT2 calculations were performed with the (12,10) active space

method	state	E, eV	f	$\lvert\mu\rvert$, D
		cyclodimer		
CC2	$^1n\pi^*$	5.247	0.0007	0.2
	$^1n\pi^*$	5.316	0.0015	0.3
	$^1\pi\pi^*$	6.284	0.0063	0.5
CASPT2*	$^1n\pi^*$	5.478	0.0015	0.3
	$^1n\pi^*$	5.504	0.0005	0.2
	$^1\pi\pi^*$	7.077	0.4193	4.4
B3LYP-TD	$^1n\pi^*$	4.906	0.0002	0.1
	$^1n\pi^*$	5.007	0.0009	0.2
	$^1\pi\pi^*$	5.450	0.0036	0.4
		stacked dimer		
CC2	$^1n\pi^*$	4.825	0.0001	0.1
	$^1n\pi^*$	4.849	0.0002	0.1
	$^1\pi\pi^*$	5.479	0.0857	2.0
CASPT2*	$^1n\pi^*$	5.185	0.0007	0.2
	$^1n\pi^*$	5.386	0.0003	0.2
	$^1\pi\pi^*$	6.096	0.4107	4.2
B3LYP-TD	$^1n\pi^*$	4.581	0.0001	0.1
	$^1n\pi^*$	4.604	0.0001	0.1
	$^1\pi\pi^*$	4.800	0.0008	0.2

dicts only one $^1n\pi^*$ state, since the limited active space includes only one n-orbital. When the active space is enlarged (to twelve electrons in ten orbitals), the two $^1n\pi^*$ states are reproduced. Unfortunately, this active space leads to severe convergence problems in the construction of reaction energy profiles.

3.3 Reaction-Path Energy Profiles

The LIIC energy profiles obtained at the CASPT2, CC2 and DFT levels are shown in Fig. 4. The zero of the energy axis is defined by the energy of the ground-state equilibrium structure of the stacked dimer: -827.216271 a.u. (CASPT2(6,6)/cc-pVDZ); -827.450853 a.u. (CC2/cc-pVDZ); and -827.704030 a.u. (B3LYP/cc-pVDZ). As can be seen, the stacked dimer has a higher energy than the cyclodimer according to the CASPT2 calculations. The data at the CC2 and DFT levels give the opposite ordering. The vertical excitation energies obtained at the CASPT2 level are rather high (more than 7 eV for the bright $^1\pi\pi^*$ state). Two points (Fig. 4a – dashed lines) along the $^1\pi\pi^*$ and $^1n\pi^*$ curves had to be interpolated because of severe mixing of

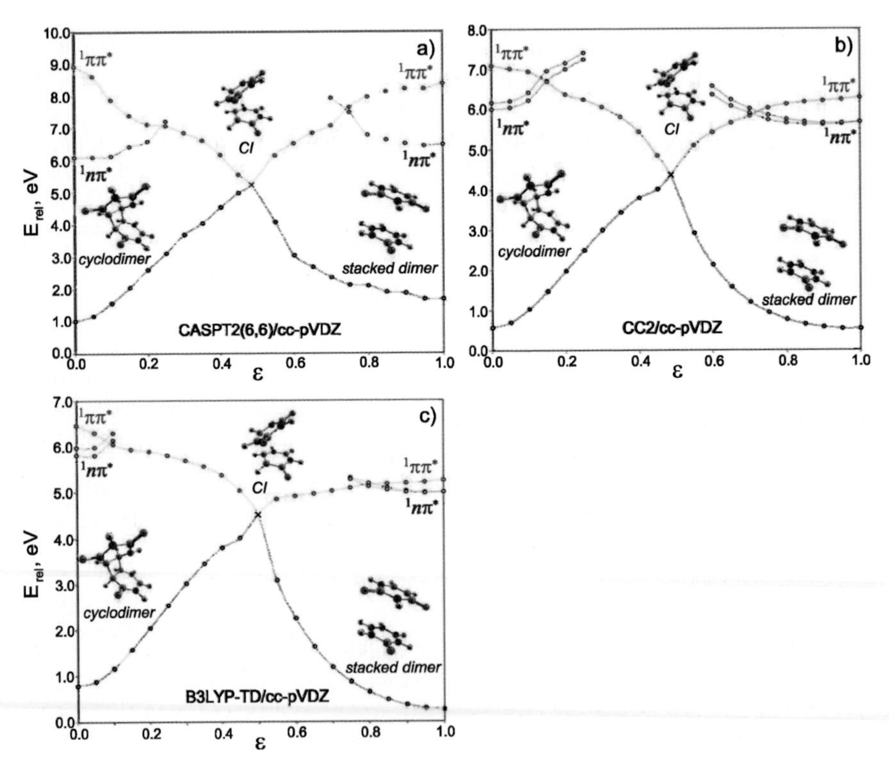

Fig. 4 Cyclodimer to stacked dimer reaction profiles obtained at the levels: a) CASPT2(6,6); b) CC2; and c) DFT, all with the cc-pVDZ basis set

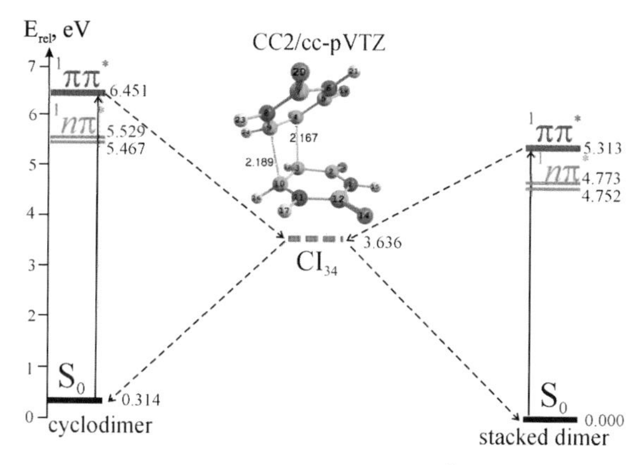

Fig. 5 Energy-level diagram of the relaxation pathways of the $^1\pi\pi^*$ excited state of the cyclodimer (U<>U) and the stacked dimer (U+U)

several excited states in this region (near the equilibrium geometry of the stacked dimer).

The potential curves given in Fig. 4 clearly show that the $^1\pi\pi^*$ excited state can relax in a barrierless manner to the ground state through the conical intersection CI_{34}. The calculations at all levels reveal that the excited cyclodimer system can be trapped in the $^1n\pi^*$ states via the conical intersections $CI_{(^1\pi\pi^*,^1n\pi^*)}$. These conical intersections are not discussed here for the sake of brevity. The CC2 and DFT methods predict two closely situated $^1n\pi^*$ states that cross the $^1\pi\pi^*$ curve. The same findings were established for the excited stacked dimer (right-hand sides in Fig. 4).

The curves in Fig. 4 indicate that the stacked dimer and the cyclodimer are not photostable: they can be transformed photochemically into the complementary form. The photoinduced transformation of the stacked dimer to the cyclodimer represents the dominant photodamage process in DNA. The reverse reaction is achieved by the CPD-photolyase upon the absorption of blue light.

The largest gradients of the $^1\pi\pi^*$ curves are obtained at the CASPT2 level, especially for the energy profile from the Franck-Condon region of the U+U stacked dimer to the conical intersection CI_{34} (Fig. 4a). At the DFT level, this curve is almost horizontal (Fig. 4c). The DFT calculations predict $^1n\pi^*$ states very close to the $^1\pi\pi^*$ state. The smoothest curves were obtained at the CC2 level (Fig. 4b). The potential-energy profiles calculated at the CASPT2 level strongly depend on the chosen active space.

Fig. 5 gives an overview of the energetics of the photocyclodimerization of the uracil dimer obtained with CC2/cc-pVTZ calculations. As can be seen, the excited states of the cyclodimer and the stacked dimer are connected by barrierless reaction paths via the conical intersection CI_{34} with the ground-state equilibrium geometries of the complementary forms.

4 Conclusions

The relaxation mechanisms of the excited states of the *cis-syn* cyclodimer and the stacked dimer of uracil have been investigated at three levels of ab initio electronic structure theory (CC2, CASPT2, and DFT). The energy profiles indicate a barrierless relaxation of the $^1\pi\pi^*$ excited state through a S_0/S_1 conical intersection to the equilibrium geometries of the cyclodimer or the stacked dimer. These reaction mechanisms are of fundamental importance for the photodamage as well as light-supported repair of DNA. The calculated gradient-difference and derivative-coupling vectors of the conical intersection do not show a preferred direction for the deactivation of the $^1\pi\pi^*$ excited state. The reaction is expected to proceed towards the cyclodimer or the stacked dimer with comparable probabilites.

5 Technical Support

In preliminary work (using the test account), the HLRB supercomputer was used for CC2 geometry optimizations of the $^1\pi\pi^*$ excited-state geometries along the reaction path from the stacked dimer to the conical intersection. The results obtained (see Fig. 6) encouraged us to apply for computing time on the HLRB II in order to use the parallel version of the TURBOMOLE program package. However, our attempts to study the minimum-energy reaction path from the cyclodimer to the stacked dimer through the conical intersection failed, since the optimizations of the $^1\pi\pi^*$ excited state geometries did not converge at the CC2 level on the path from the cyclodimer to the conical intersection. This is the reason why we did not exhaust

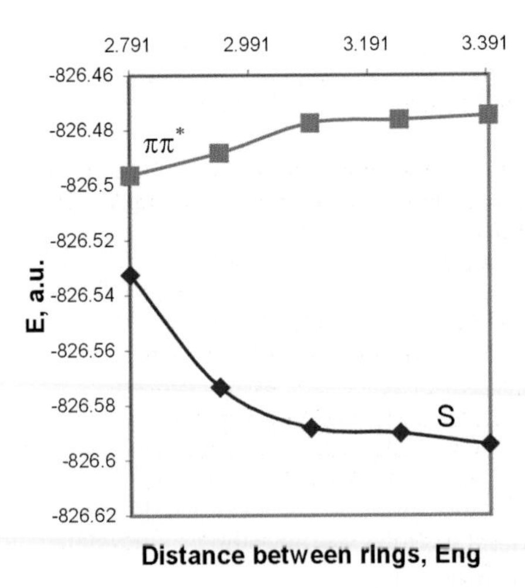

Fig. 6 Coordinate-driven potential-energy profiles of the S_0 and $^1\pi\pi^*$ states of the uracil dimer as a function of the distance of the two rings

Table 2 Geometry optimizations of the uracil dimer calculated on the HLRB with the test account

processors	wall-time (h)	CPU (real) time (h)	CPU-time (h)
1	576.06	202.49	202.49
4	265.70	135.95	543.80
8	89.30	69.40	555.20
10	89.30	70.12	701.20
16	81.98	69.40	1110.4

the CPU budget of the project (pr23yi). To overcome this problem, we decided to perform single point-calculations along the LIIC reaction path on the Linux-cluster. The investigation of the relaxation pathways was completed with serial computations using MOLPRO and GAUSSIAN as well as parallel computations (mostly with four processors) using TURBOMOLE on the Linux-cluster.

On the HRLB, the calculations were performed with eight processors, since this was estimated (using the test account) to be the optimum for our problem (Table 2). It can be seen that the use of more than eight processors does not reduce neither the CPU-time nor the wall-time. A good scaling of the CPU time is achieved when using up to eight processors.

Acknowledgements This work was financially supported by the Deutsche Forschungsgemeinschaft through Sonderforschungsbereich 749. The computations were performed at the Leibniz Rechenzentrum (LRZ) on the Linux-cluster and the HLRB II.

References

1. Adman, E., Jensen, L.H.: Acta Cryst. B **26**, 1326 (1970)
2. Anusiewicz, I., Berdys, J., Sobszyk, M., Shurski, P., Simons, J.: J. Phys. Chem. A **108**, 11381 (2004)
3. Ben-Hur, E., Beh-Ishai, R.: Biochim. Biophys. Acta **166**, 9 (1968)
4. Boggio-Pasqua, M., Groenhof, G., Schäfer, L.V., Grubmüller, H., Robb, M.A.: J. Am. Chem. Soc. **129**, 10996 (2007)
5. Cadet, J., Vigny, P.: Bioorganic photochemistry. Wiley, New York (1990)
6. Crespo-Hernández, C.E., Cohen, B., Hare, P.M., Kohler, B.: Chem. Rev. **104**, 1977 (2004)
7. Eisinger, J., Shulman, R.G.: Science **161**, 1311 (1968)
8. Frisch, M.J., et al.: Gaussian 03, revision e.01 (2004). Inc., Wallingford CT
9. Haase, F., Ahlrichs, R.: J. Comput. Chem. **14**, 907 (1993)
10. Hättig, C., Weigend, F.: J. Chem. Phys. **113**, 5154 (2000)
11. Marguet, S., Markovitsi, D.: J. Am. Chem. Soc. **127**, 5780 (2005)
12. Merchan, M., Serrano-Andres, L., Robb, M.A., Blancafort, L.: J. Am. Chem. Soc. **127**, 1820 (2005)
13. Merriam, V., Gordon, M.P.: Photochem. Photobiol. **6**, 309 (1967)
14. Sancar, A.: Biochem. **33**, 2 (1994)
15. Serrano-Andres, L., Merchan, M.: Radiation Induced Molecular Phenomena in Nucleic Acids. Springer (2008)
16. Taylor, J.S.: Acc. Chem. Res. **27**, 76 (1994)

17. TURBOMOLE V6.0 2009, a development of University of Karlsruhe and Forschungszentrum Karlsruhe GmbH, 1989-2007, TURBOMOLE GmbH, since 2007: Available from http://www.turbomole.com
18. Vink, A.A., Roza, L.: J. Photochem. Photobiol. B **65**, 101 (2001)
19. Weigend, F., Häser, M.: Theor. Chem. Acc. **97**, 331 (1997)
20. Werner, H.J., et al.: Molpro, version 2006.1, a package of ab initio programs. See http://www.molpro.net
21. Wulff, D.L., Fraenkel, G.: Biochim. Biophys. Acta **51**, 33 (1961)

Numerical Simulation of Electric Field Gradient Focusing and Separation of Analytes in Microchannels with Embedded Bipolar Electrode

Dzmitry Hlushkou, Richard M. Crooks, and Ulrich Tallarek

Abstract A new method for simultaneously concentrating and separating analytes in a microfluidic channel with embedded floating electrode is proposed. The complex interplay of electrophoretic, electroosmotic, bulk convective, and diffusive mass/charge transport in the microchannel is analyzed by numerical simulations. The thin floating electrode attached locally to the wall of the straight microchannel results in a redistribution of local field strength after the application of an external electric field. Together with bulk convection based on cathodic electroosmotic flow, an extended field gradient is formed in the anodic microchannel segment. It imparts a spatially dependent electrophoretic force on charged analytes and, in combination with the bulk convection, results in electric field gradient focusing at analyte-specific positions. Analyte molecules having different electrophoretic mobilities are focused at different locations within the channel.

1 Introduction

Multifunctional microchip devices for chemical separation and analysis provide significant advantages in performance, resulting in faster and cheaper analytical procedures by requiring small amounts of sample and reagents. However, the use of smaller geometries also means that the number of molecules to be analyzed is re-

D. Hlushkou · U. Tallarek
Fachbereich Chemie, Philipps-Universität Marburg, Hans-Meerwein-Strasse, 35032 Marburg, Germany
e-mail: dzimitry.hlushkou@staff.uni-marburg.de
e-mail: tallarek@staff.uni-marburg.de

Richard M. Crooks
Department of Chemistry and Biochemistry, Center of Electrochemistry, The University of Texas at Austin, 1 University Station, A5300, Austin, TX, 78712-0165, USA
e-mail: crooks@cm.utexas.edu

S. Wagner et al. (eds.), *High Performance Computing in Science and Engineering, Garching/Munich 2009*, DOI 10.1007/978-3-642-13872-0_60,

duced so that their detection can become a challenging task. Apart from the use of highly sensitive detectors, another option is to employ analyte preconcentration before (off-line) or after (on-line) sample injection. The integration of sample pretreatment into microfluidic devices is one of the remaining hurdles towards achieving true miniaturized total analysis systems.

In the literature, numerous sample preconcentration methods have been reported for miniaturized devices. They include approaches based on electrokinetic equilibrium techniques like field amplified sample stacking, temperature gradient focusing, isotachophoresis, isoelectric focusing, conductivity gradient focusing, and electric field gradient focusing. Other methods utilize specific interactions (e.g., electrostatic, hydrophilic, affinity) between analytes and the surface of the microchannel walls, or employ size- and electrostatic exclusion from, as well as concentration polarization at nanoporous membranes and discrete nanochannels.

Electrofocusing is an accompanying effect of electrophoretic equilibrium gradient methods, a subset of separation techniques in which a net restoring force acts against dispersive forces to simultaneously separate and concentrate. Because most electrofocusing techniques are not generally applicable, the only widely used technique in this family is isoelectric focusing (IEF). IEF is based on the fact that the net charge of molecules becomes zero if the pH of the surrounding solution is equal to the molecules isoelectric point (pI). Thus, if a pH gradient exists in the system, analytes can be immobilized and accumulated where pH = pI. However, the application of IEF-methods is limited. This technique can only be employed with an analyte having a well-defined pI. In addition, the solubility of proteins at their pI is low, that handicaps the application of IEF-methods to biochemical systems.

The scarcity of available electrofocusing-based methods has motivated recent interest in the development of new such techniques [3], especially those which can be applied to microelectromechanical platforms. [4] Recently, we reported a novel electric field gradient focusing technique for use in a straight microchannel device containing an embedded floating gold electrode which, at any time, is no part of the external circuitry. [2] This experimentally simple approach enables analyte enrichment up to 45 in 200 s within the 6 mm long, 100 μm wide, and ~20 μm high microchannel. To quantitatively interpret the experimental results we have developed a theoretical model of relevant physicochemical processes occurring in this device, including buffer and faradaic reactions, and carried out numerical simulations. [6]

2 Experimental Section

The hybrid poly(dimethylsiloxane)(PDMS)/glass microfluidic device (a layout is shown in Figure 1a) and bipolar electrode were fabricated using standard lithographic techniques. [9] The microfluidic channel was made by attaching a PDMS mold, containing a microchannel (6 mm long, 100 μm wide, and ~20 μm high) that connects two macroscale cylindrical reservoirs A and B (referred to as ResA and ResB) of ~2.5 mm diameter, to a microscope glass slide by O_2 plasma bonding

(60 W, model PDC-32G, Harrick Scientific, Ossining, NY). Before fabricating the microfluidic device, a floating electrode was prepared by depositing 100 nm of Au (no adhesion layer) onto the glass slide. Then, photolithographic and etching methods were used to pattern a single 500 μm × 1000 μm electrode.

Prior to each experiment, the microchannel was rinsed by filling ResA with 1 mM Tris-HCl buffer (pH 8.1) and applying a vacuum at ResB for 2 min. Following the rinsing process, the microchannel and reservoirs were filled with 5 μM BODIPY disulfonate in the same buffer. A 30 V bias was then applied between two platinum electrodes immersed in ResA (grounded; cathode) and ResB (at a positive potential; anode) to generate an electric field within the channel. The average electric field along the channel was estimated to be $\sim 5.0\,\mathrm{kV\,m^{-1}}$.

Due to contact with the Tris-HCl buffer, the surface of the glass and PDMS walls of the channel acquire negative electric charge. Therefore the generated electroosmotic flow (EOF) is cathodic, i.e., the motion of the liquid is directed to the cathode (to ResA in Fig. 1a). Since the electrophoretic mobility of the BODIPY disulfonate molecules is less than the bulk electroosmotic mobility, the tracer moves with the EOF from ResB to ResA in the absence of the floating electrode. [2] In contrast, the dye is concentrated to the right of the electrode, in the anodic segment of the channel, if the floating electrode is embedded into the microchannel (Fig. 1b). This

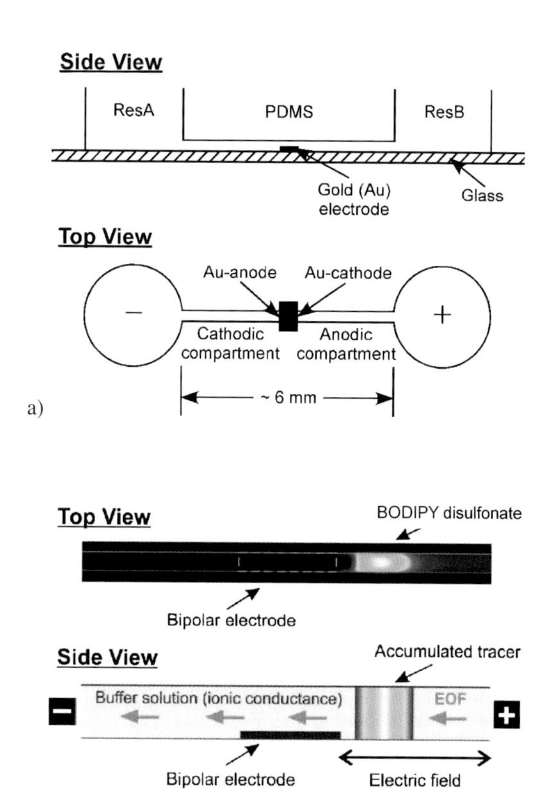

Fig. 1 (a) Layout of the hybrid PDMS/glass microfluidic device. The Au electrode (500 μm × 1000 μm) is at the center of the (6 mm long, 100 μm wide, and ∼ 20 μm high) microchannel connecting the two macroscale cylindrical reservoirs (∼ 2.5 mm in diameter), ResA and ResB.

(b) Optical fluorescence micrograph of the microchannel with the embedded Au electrode. The micrograph shows the concentration distribution of BODIPY disulfonate in 1 mM Tris-HCl buffer at pH 8.1 after applying a potential bias of 30 V for 240 s (top). Schematic illustration of the proposed mechanism of tracer accumulation in the microchannel with a bipolar electrode (bottom)

accumulation of the negatively charged analyte indicates that its transport by the EOF in the system with the floating electrode is locally offset by counter-directional electrophoretic motion.

When the initial concentration of BODIPY disulfonate (c_0) is 5 µM, its concentration in the enriched zone asymptotically tended with time to some specific value and did not change after approximately 425 s. In contrast, for the lower initial concentration, $c_0 = 0.1\,\mu\text{M}$, the dye concentration in the enriched zone continued to grow almost linearly with time.

3 Theoretical Background

Spatiotemporal variations in the concentrations of ionic species of the buffer solution and the tracer molecules in the studied system are governed by balance equations

$$\frac{\partial D_i}{\partial t} = \nabla(D_i \nabla c_i) - \frac{z_i F}{RT}\nabla(D_i \nabla \phi) - c_i \nabla \mathbf{v} + r_i\,, \qquad (1)$$

where c_i is the molar concentration of species i, D_i and z_i are its diffusion coefficient and valency, respectively, ϕ is the local electric potential; F, R and T represent the Faraday constant, molar gas constant and temperature, respectively, $\mathbf{v}$ is the flow velocity, and r_i is the homogeneous reaction term for species i.

The local concentrations of the charged species and the local electric potential are related by the Poisson equation

$$\nabla^2 \phi = -\frac{F}{\varepsilon_0 \varepsilon_r} \sum_i z_i c_i \qquad (2)$$

where ε_0 and ε_r are the vacuum permittivity and dielectric constant. Assuming that the liquid is incompressible, the generalized Navier-Stokes equation establishes the relation between the local flow velocity, pressure, and the electric body force

$$\rho\left(\frac{\nabla \mathbf{v}}{\nabla t} + \mathbf{v} \cdot \nabla \mathbf{v}\right) = -\nabla p + \mu \nabla^2 \mathbf{v} - (\nabla \phi) \sum_i z_i c_i \qquad (3)$$

where ρ and μ are the mass density and dynamic viscosity of the liquid, and p is hydrostatic pressure.

The numerical solution of coupled Eqs. 1–3 is a nontrivial task as it has to be accomplished at different spatiotemporal scales ranging from the order of the electric double layer thickness of typically 1–10 nm to the length of the channel of several millimeters. These widely disparate spatial scales require huge computational resources. One possibility for avoiding this limitation is to assume that the electric double layer thickness is negligibly small compared with the channel dimensions (thin-double-layer approximation). [8] The studied system meets the above requirement as the electric double layer thickness is ~ 30 nm in the presence of 1 mM Tris buffer. With the thin-double layer approximation the no-slip condition at the solid-liquid interface for the fluid flow is replaced by the velocity boundary condition formulated according to the Helmholtz-Smoluchowski equation. In this picture, the EOF slips past the solid-liquid interface. Finally, we assume that the geometry can be reduced to a two-dimensional configuration, where the width of the channel is infinite and all parameters vary only along the x- and y-axes.

The two-dimensional geometry of the simulated device is shown in Fig. 3. It is assumed that the microchannel is terminated by two relatively large reservoirs containing the electrodes — ResA with the cathode, ResB with the anode. Equations (1) and (2) were complemented by the corresponding boundary conditions: For the species balance equation (1), we specify zero flux across the channel walls and constant species concentrations at the left and right edges of the channel; for the Poisson equation (2), fixed potentials are assumed at the left and right edges of the channel, while at the solid-liquid interface the following boundary conditions were imposed

$$E_{liquid}^{\|} = E_{solid}^{\|} \quad \text{and} \quad \varepsilon_{liquid} E_{solid}^{\perp} = \varepsilon_{solid} E_{liquid}^{\perp} \tag{4}$$

where $\|$ and $\perp$ denote the tangential and normal components of the local electric field. In this study the dielectric constants of the aqueous electrolyte solution and the channel walls are assumed as 80 and 3, respectively.

In order to represent the metallic bipolar electrode we used the electron gas model. According to this approach, valence electrons of the atoms comprising a metal are able to leave them and move through the whole electrode accounting for electrostatic interactions. Thus, the metal electrode is assumed to be composed of two kinds of interacting charges: Mobile negatively charged (electron gas; however, it is not allowed to leave the electrode) and immobile positively charged species (ions of the lattice). In the absence of an external electric field, the whole electrode is electrically neutral because of the presence of the same amount of both kinds of charges. If an external electric field is applied to the metal-"environment" interface, electrons are redistributed to compensate this external distortion and form an electrostatic shield restraining their further electromigration within the electrode. In turn, this leads to the formation of two regions with local non-zero volume charge density at the two opposite sides of the electrode. The side towards ResA (to the external cathode) is characterized by a depletion of electrons. Therefore, it can be

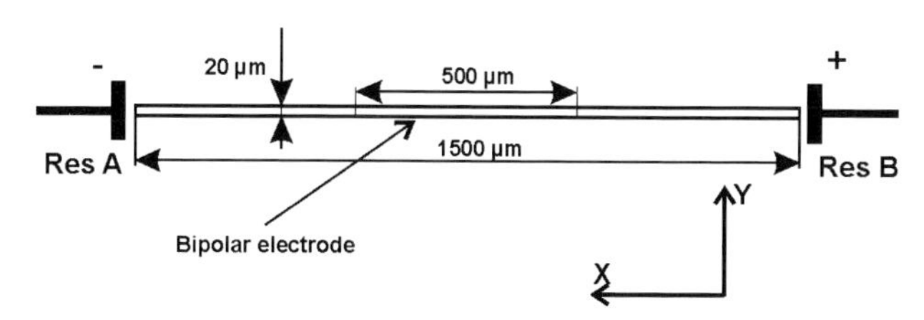

Fig. 2 Illustration of the 2D system used in the simulations for analyzing tracer concentration enrichment in a microfluidic device containing a flat bipolar electrode (cf. Fig. 1a). This electrode is attached locally to the bottom wall from $x = 500 - 1000\,\mu\text{m}$ and has a thickness (in the y-direction) of $3\,\mu\text{m}$

considered as electron sink for faradaic reactions or as an "induced anode". The opposite side of the bipolar electrode towards ResB (to the external anode) can be considered as an "induced cathode" because of an excess of electrons in this region.

Initially, the microchannel and reservoirs are assumed to be filled with 5 µM BODIPY disulfonate in 1 mM Tris-HCl buffer (pH 8.1). Thus, the species of interest include neutral Tris, its protonated form ($TrisH^+$), hydroxide and hydronium ions (OH^- and H_3O^+), and the tracer molecules (BODIPY disulfonate; twice negatively charged). If no external electric field is applied along the microchannel water protolysis

$$2H_2O \leftrightarrow H_3O^+ + OH^- \tag{5}$$

and the buffer reaction

$$\mathrm{Tris} + H_2O \leftrightarrow \mathrm{TrisH}^+ + OH^- \tag{6}$$

results in the following uniform (initial) species concentrations in the system:

$$\begin{aligned}
[H_3O^+] &= 10^{-\mathrm{pH}} = 7.943 \times 10^{-6}\,\mathrm{mM} \\
[OH^-] &= 10^{(\mathrm{pH}-14)} = 1.259 \times 10^{-3}\,\mathrm{mM} \\
[\mathrm{Tris}] &= 0.515\,\mathrm{mM} \\
[\mathrm{TrisH}^+] = [\mathrm{Cl}^-] &= 0.485\,\mathrm{mM}
\end{aligned}$$

The reaction terms in (1) for each species can be represented as

$$\begin{aligned}
r_{H_3O^+} &= \frac{\partial[H_3O^+]}{\partial t} = k_1[H_2O] - k_{-1}[H_3O^+][OH^-] \\
r_{OH^-} &= \frac{\partial[OH^-]}{\partial t} = k_1[H_2O] - k_{-1}[H_3O^+][OH^-] \\
&\quad + k_2[\mathrm{Tris}] - k_{-2}[\mathrm{TrisH}^+][OH^-] \\
r_{\mathrm{TrisH}^+} &= \frac{\partial[\mathrm{TrisH}^+]}{\partial t} = k_2[\mathrm{Tris}] - k_{-2}[\mathrm{TrisH}^+][OH^-] \\
r_{\mathrm{Tris}} &= \frac{\partial[\mathrm{Tris}]}{\partial t} = -k_2[\mathrm{Tris}] + k_{-2}[\mathrm{TrisH}^+][OH^-] \\
r_{\mathrm{Traser}^{2-}} &= \frac{\partial[\mathrm{Traser}^{2-}]}{\partial t} = 0,
\end{aligned}$$

where k_1 and k_{-1} denote the rate constants for the forward and backward reaction of water self-ionization (5), while k_2 and k_{-2} are the rate constants for the forward and backward buffer reaction (6).

Apart from the bulk reactions (Eqs. (5) and (6)), the local species concentrations can change due to faradaic reactions on the surface of the electrodes. Typical electrode reactions for an aqueous solution involve anodic oxygen and cathodic hydrogen evolution and the concomitant production of H_3O^+ and OH^- due to protolysis of water. The final balance for the reactions on the anode and on the cathode are respectively

$$6H_2O + 4e^- \leftrightarrow O_2 + 4H_3O^+ \quad \text{and} \quad 2H_2O + 2e^- \leftrightarrow H_2 + 2OH^- \tag{7}$$

We assume that the rates of the faradaic reactions are much higher than those of the chemical reactions (Eqs. (5) and (6)). It should be noted that faradaic reactions can also take place at the anodic and cathodic electrodes in ResA and ResB. However, most of the applied potential is dropped in the channel and the volume of the reservoirs is large compared to the channel, so the impact of electrolysis in the reservoirs should be small. Therefore, in our mathematical model, we assume that species concentrations in ResA and ResB are constant.

4 Numerical Methods

The mathematical description of the processes in the studied system was implemented as an iterative model based on discrete spatiotemporal schemes optimized for parallel computations. In particular, for the solution of the Navier-Stokes, Poisson and Nernst-Planck equations, respectively, the lattice-Boltzmann approach [5] and the numerical approaches proposed by Warren [11] and by Capuani et al. [1] were employed. In all numerical schemes, a time step of 2×10^{-5} s and a space step of 10^{-6} m were used. At each time step, the Poisson equation was solved 10 times with an under-relaxation factor of 0.25 in order to ensure numerical stability. A single simulation required about 16 hrs at 64 processors of an SGI Altix 4700 supercomputer to analyze the temporal behavior of the system for 500 s. Physiochemical parameters used in the simulations are given in Tables 1 and 2.

Table 1 Species properties considered in the simulations

Species	Diffusion coefficient	Charge
Tris	$0.785 \times 10^{-9}\ m^2\,s^{-1}$	0
$TrisH^+$	$0.785 \times 10^{-9}\ m^2\,s^{-1}$	+1
Cl^-	$2.033 \times 10^{-9}\ m^2\,s^{-1}$	−1
H_3O^+	$5.286 \times 10^{-9}\ m^2\,s^{-1}$	+1
OH^-	$5.286 \times 10^{-9}\ m^2\,s^{-1}$	−1
Tracer molecules	$0.427 \times 10^{-9}\ m^2\,s^{-1}$	−2

Table 2 Reaction rate constants for Eqs. (5) and (6)

k_1 (Eq. (5), forward)	$2 \times 10^{-5}\ s^{-1}$
k_{-1} (Eq. (5), backward)	$1.11 \times 10^{8}\ m^3\,mol^{-1}\,s^{-1}$
k_2 (Eq. (6), forward)	$2 \times 10^{-6}\ s^{-1}$
k_{-2} (Eq. (6), backward)	$1.11 \times 10^{7}\ m^3\,mol^{-1}\,s^{-1}$

5 Results and Discussion

To correspond as closely as possible with experimental data, the complete microchannel and both reservoirs in the simulated system were assumed to be filled uniformly with the tracer in 1 mM Tris-HCl buffer (pH 8.1) at an initial concentration of $c_0 = 5\,\mu$M. Then, a bias of 7.5 V ($E = 5\,\mathrm{kV\,m^{-1}}$) was applied to the channel and the evolution of two-dimensional distributions of electric potential and species concentrations was recorded every 2 s. We assume that the electrokinetic potential at the microchannel wall is −50 mV. This magnitude is typical for PDMS surfaces, taking further into account the actual pH and ionic strength of the electrolyte solution. [7]

The data simulated under these conditions demonstrate a close correlation between local species concentrations (Fig. 3) and local electric field strength along the microchannel (Fig. 4). Nonuniform species concentration distributions can be understood in terms of the nonuniformity of the local electric field and, as a result, differences in local transport rates of the individual species. In Fig. 3, bulk convection originating in the EOF is from the right to the left, also above the electrode. This has important consequences for the electrophoretic transport of charged species: They accelerate or decelerate depending on the local field strength. For cations, the electroosmotic and electrophoretic flux components are co-directional (from ResB to ResA), and hence their sum is proportional to the local field strength. Because field strengths in the anodic and cathodic microchannel segments and in the region above the bipolar electrode are related by $E_{anode} > E_{cathode} > E_{bipolar}$, steady-state concentrations of cations in these regions must correspond to $c_{bipolar} > c_{cathode} > c_{anode}$ in order to conserve charge flux. For anions, these fluxes due to electroosmotic flow and electrophoresis are counter-directional and their sum decreases as the local field strength increases and even becomes negative in the anodic segment (i.e., net transport is directed from ResA to ResB).

Specifically, the chloride ions in the anodic microchannel segment ($x < 500\,\mu$m, Fig. 3) move towards the right (to the anode), because their electrophoretic mobility is decidedly higher than the general EOF mobility in the system. In what follows it

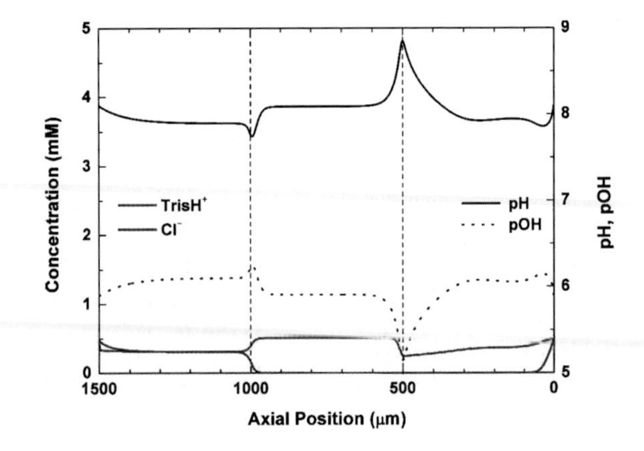

Fig. 3 Simulated profiles of the background ion concentrations (in mM), and pH and pOH for $t = 200$ s. Applied field strength: $5\,\mathrm{kV\,m^{-1}}$. Initial tracer concentration, $c_0 = 5\,\mu$M. Profiles represent the distribution along the geometrical axis of the channel ($y = 10\,\mu$m). The bipolar electrode is located at the wall from $x = 500 - 1000\,\mu$m

is most important that their local withdrawal cannot be supplied by other chloride ions which would be positioned at $x > 500\,\mu\text{m}$. In the neighboring microchannel compartment ($500 < x < 1000\,\mu\text{m}$), which contains the bipolar electrode along the bottom wall (cf. Fig. 2), the electric field strength becomes zero (Fig. 4). Thus, electrophoretic motion of chloride ions is not possible in this bipolar electrode segment where charge transport in the liquid phase is dominated by bulk convection through the microchannel from right to left. As a consequence, there are no chloride ions above the bipolar electrode that could compensate the local withdrawal of chloride ions in the anodic microchannel compartment which here ($x < 500\,\mu\text{m}$), due to their dominating electrophoretic motion and strong local field strengths up to 15–17 kV m^{-1} (with respect to the applied 5 kV m^{-1}, Fig. 4) move towards the external anode. This depletion region stops where (along this extended field gradient, towards $x = 0\,\mu\text{m}$) electrophoretic and electroosmotic contributions for the chloride ions compensate each other (for $x < 50\,\mu\text{m}$ in Figs. 3 and 4; local field strength here drops below 5 kV m^{-1}). For the TrisH^+ ions (Fig. 4) there is an increased concentration in the region $x \approx 500 - 1000\,\mu\text{m}$. This is caused by bulk convection, which dominates charge transport above the bipolar electrode, while in the cathodic compartment ($x = 1000 - 1500\,\mu\text{m}$) TrisH^+ ions experience (co-directional) flow and electrophoresis to the left electrode (cathode). The concentration profile of the TrisH^+ ions in the anodic compartment stems from their buffer reaction with the OH^- ions which are produced at the bipolar electrodes cathode ($x = 500\,\mu\text{m}$) and swept towards the external anode. Thus, the concentration of TrisH^+ ions in the anodic microchannel compartment is reduced, mostly at/close to the bipolar electrodes cathode ($x = 500\,\mu\text{m}$) and then recovers towards the anodic reservoir.

Since the local electric field strength is inversely proportional to the conductivity, the significantly reduced and nonuniform concentration of background electrolyte ions in the anodic segment leads to the formation of an extended electric field gradient with field strength increasing from the external anode (ResB) to the right edge of the bipolar electrode (Fig. 4). This is the key to understanding this device in view of analyte concentration enrichment: The extended field gradient forms due to the

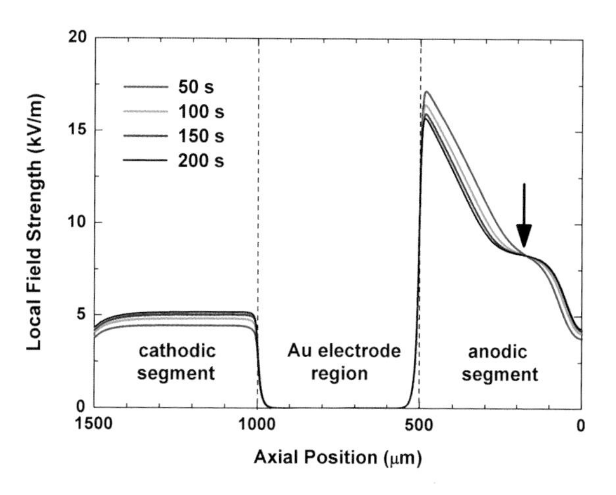

Fig. 4 Simulated profiles of the local axial electric field for $t = 50$, 100, 150, and 200 s. Applied field strength: 5 kV m^{-1}. Initial tracer concentration, $c_0 = 5\,\mu\text{M}$. Profiles represent the distribution of field strength along the geometrical axis of the channel ($y = 10\,\mu\text{m}$)

presence of the EOF and the bipolar electrode, which "divides" the microchannel into three segments with different transport characteristics. We have seen in complementary simulations that without the EOF an extended field gradient as in Fig. 4 (which covers the complete anodic microchannel segment) is not formed. The field gradient then is limited to the rapid drop in field strength close to the boundaries of the bipolar electrode segment (around $x = 500$ and $1000\,\mu m$) also seen in Fig. 4.

The implication of the nonuniform distribution of the local electric field along the microchannel (Fig. 4) is that tracer molecules, which come from the anodic reservoir due to the initially (more precisely, locally) dominant bulk convection based on the EOF, encounter an increasing electrophoretic force in the opposite direction (that is, back towards ResB) as they approach the bipolar electrode. This is illustrated by Fig. 5 which shows how the local electroosmotic (v_{eo}) and electrophoretic (v_{ep}) velocities of the tracer molecules change with axial position. The nonuniform distribution of the EOF velocity along the microchannel (shown in Fig. 5 for its geometrical axis, $y = 10\,\mu m$) is a direct result of local pressure gradients which develop as a consequence of a nonuniform electrokinetic body force. In particular, in the segment containing the Au electrode, where the electric field drops to zero, there will be no electroosmotic force and the flow will be pressure-driven. As a result, convex or concave velocity profiles form in the different segments of the microchannel, and local flow recirculation near the bipolar electrode (segment boundaries) results even in negative velocity components. Figure 5 also provides a means to compare the absolute values of $v_{ep}(x)$ with the EOF velocity for $t = 200$ s: At a certain axial position (here, at $x \approx 50\,\mu m$, but depending on the elapsed time and y-position), the two counter-directional motions (i.e., anodic electrophoresis vs. cathodic electroosmosis) are balanced; tracer molecules become quasi-stationary and locally enriched (Fig. 6).

The step-like character of the field strength profile in the anodic microchannel segment (see arrow in Fig. 3) can be explained by the concentration enrichment history of the negatively charged tracer shown in Fig. 6. The actual tracer concentration becomes comparable with the background ion concentration (1 mM Tris-HCl buffer

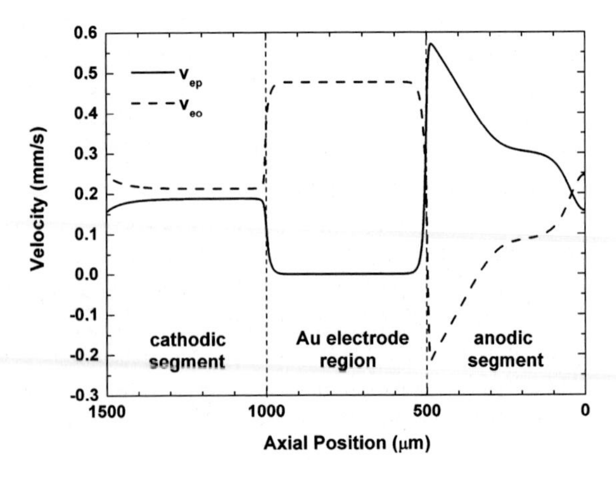

Fig. 5 Simulated profiles of local electrophoretic (vep) and electroosmotic (veo) velocities of the tracer for $t = 200$ s. Applied field strength: $5\,kV\,m^{-1}$. Initial tracer concentration, $c_0 = 5\,\mu M$. Profiles represent the distribution of velocities along the geometrical axis of the channel ($y = 10\,\mu m$)

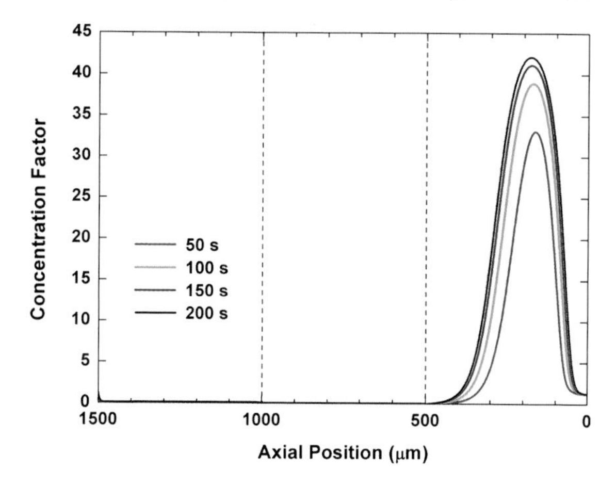

Fig. 6 Simulated tracer concentration profiles for $t = 50$, 100, 150, and 200 s. Applied field strength: $5\,\mathrm{kV\,m^{-1}}$. Initial tracer concentration, $c_0 = 5\,\mu\mathrm{M}$. Profiles represent the distribution of the concentration factor along the geometrical axis of the channel ($y = 10\,\mu\mathrm{m}$)

solution). As a consequence, the local field strength is a function of the tracer concentration as well as that of the buffer. Following its increase with time, the interplay of fluxes due to bulk convection, electrophoresis, and diffusion results in a broadening of the developing tracer zone. The axial position and the width of this zone at any time (Fig. 6) exactly correspond to those of the "steps" (plateaus) in the field strength profiles indicated by the arrow in Fig. 3.

6 Conclusion

We have used detailed numerical simulations to analyze an electric field gradient focusing mechanism in view of analyte concentration enrichment recently observed in a hybrid PDMS/glass straight microchannel containing a floating (bipolar) electrode. The unique properties of the embedded electrode due to its electronic conductance, which contrasts with the much lower ionic conductance of the surrounding electrolyte solution, result in a redistribution of the local field strength along the microchannel after the application of an external electric field. Together with bulk convection of liquid through the whole microchannel based on cathodic EOF, an extended field gradient is formed in the anodic microchannel compartment of the device between external anode and the bipolar electrodes cathode. It imparts a spatially dependent electrophoretic force on charged analytes and, in combination with the bulk convection, results in their focusing at analyte-specific positions. The accumulation of charged analytes in a specific region of the system is a result of zero net driving force considered as the sum of electrophoretic and convective components. The accompanying faradaic and buffering reactions, as well as water autodissociation, can further affect the local species concentrations. The device is useful for fast and scaleable concentration enrichment. We anticipate the demonstration of simultaneous concentration enrichment and separation of different analytes in the near future.

The simulations show that with a given buffer ionic strength the tracer concentration in the enriched zone tends towards the same upper value regardless of reservoir concentration. The latter only determines the temporal domain of enrichment. At the same time, our results demonstrate that the enrichment factor alone does not allow to characterize and compare rigorously the efficiency of the investigated device in view of concentration enrichment. More comprehensively, the absolute concentration of enriched analytes and/or the preconcentration rate should be provided as more informative (and fair) parameters. Because the underlying mechanisms resulting in concentration enrichment are of fundamental nature this concern also addresses preconcentration steps in related devices.

Acknowledgements We acknowledge the U.S. Department of Energy, Office of Basic Energy Sciences (Contract No. DE-FG02-01ER15247), and the Deutsche Forschungsgemeinschaft DFG (Bonn, Germany) for support of this work. Simulations were run at the "Leibniz-Rechenzentrum der Bayerischen Akademie der Wissenschaften" (Garching, Germany), supported by project HLRB pr26wo. We thank Prof. Henry S. White (University of Utah) for helpful discussions.

References

1. Capuani F., Pagonabarraga, I., Frenkel, D., J. Chem. Phys. **121**, 973–986 (2004)
2. Dhopeshwarkar, R., Hlushkou, D., Nguyen, M., Tallarek, U., Crooks, R.M., J. Am. Chem. Soc. **130**, 10480–10481 (2008)
3. Ivory, C.F., Sep. Sci. Technol. **35**, 1777–1793 (2000)
4. Ivory, C.F., Electrophoresis **28**, 15–25 (2007)
5. Higuera, F.J., Jiménez, V., Europhys. Lett. **9**, 663–668 (1989)
6. Hlushkou, D., Perdue, R.K., Dhopeshwarkar, R., Crooks, R.M., Tallarek, U., Lab Chip **9**, 1903–1913 (2009)
7. Kirby, B.J., Hasselbrink, Jr, E.F., Electrophoresis, **25**, 203–213 (2004)
8. Lyklema, J.: Fundamentals of Interface and Colloid Science, Vol. II: Solid-Liquid Interfaces, Academic Press, London, (1995)
9. McDonald, J.C., Duffy, D.C., Anderson, J.R., Chiu, D.T., Hongkai, W., Schueller, O.J.A., Whitesides, G.M., Electrophoresis, **21**, 27–40 (2000)
10. Videnova-Adrabinska, V., Coord. Chem. Rev., **251**, 1987–2016 (2007)
11. Warren, P.B., Int. J. Mod. Phys. C, **8**, 889–898 (1997)
12. Wei, W., Xue, G., Yeung, E.S., Anal. Chem., **74**, 934–940 (2002)

Part VIII
Bio Sciences

Annotation of Entirely Sequenced Genomes

Guy Yachdav, László Kaján, and Burkhard Rost

Abstract All the computational tasks initiated in this project began with running a battery of sequence analysis and protein structure and function prediction programs for entirely sequenced organisms. We started with a set of 45,000 proteins of particular biological interest; next, we will select a few representative eukaryotic and prokaryotic organisms (about 100) with about 500,000 proteins in total.

We pursue two important research objectives by this project. The first is related to the selection of targets for large-scale structural genomics. Structural genomics is an initiative that attempts to experimentally determine high-resolution structures of proteins for which we today have no experimental and little or no in silico evidence about structure.

The second scientific objective pertains to the study of proteins that have a particular structural feature, namely that they are natively unstructured, i.e. adopt regular three-dimensional structures only upon folding to substrates. Such proteins are particularly abundant in higher eukaryotes. In fact, the abundance of these proteins in higher eukaryotes is, aside from alternative splicing and the number of proteins, the most dramatic difference of the proteomes of higher eukaryotes and of simple bacteria.

All data generated will be made publicly available via - in addition to other channels - a new automatic tagging system named Reflect.

Guy Yachdav
Columbia University in the City of New York, Department of Biochemistry & Molecular Biophysics and Center for Computational Biology, 1130 St. Nicholas Ave, 8th. floor, Room 805 New York, N.Y. 10032, USA
e-mail: gyachdav@rostlab.org

László Kaján · Burkhard Rost
Technische Universität München, Arcisstrasse 21, 80333 München, Germany
e-mail: lkajan@rostlab.org
e-mail: rost@rostlab.org

S. Wagner et al. (eds.), *High Performance Computing in Science and Engineering, Garching/Munich 2009*, DOI 10.1007/978-3-642-13872-0_61,

1 Methods

In the proposed bioinformatics project we aim to provide automatic annotations to large and complete sets of proteins. The first task will be the annotation of a set of about 45,000 (45K) proteins of particular biological interest. Those annotations will be linked indirectly from the web version of the premier journal in molecular biology, namely CELL. The 45k proteins are selected from publicly available databases and were aligned against SWISS-PROT, TrEMBL [10] and PDB [9]. The following annotations are provided: secondary structure, transmembrane helices, coiled coils, regions of low complexity, signal peptides, PROSITE motifs, nuclear localization signals and classes of cellular function. Proteins that contain long regions without regular secondary structure are also identified.

All the proteins annotated in this project are (I) aligned against the SWISS-PROT, TrEMBL and PDB using pairwise BLAST [7], PSI-BLAST [8] (II) assigned secondary structure and other sequence based predictions by employing a set of neural network algorithms, and (III) assigned predicted subcellular localization according to LOCtree [19] using Bayesian networks. The structural and functional features we analyze include the method described in the following paragraphs.

1.1 Iterated Profile-Based Search (PSI-BLAST)

PSIblast is a fast, yet sensitive database search program [8]. We are running the iterated PSI-BLAST on a subset of the BIG database with SWISS-PROT + TrEMBL + PDB sequences. The number of iterations, the cut-off thresholds and the particular details of which sequences are used from BIG has been optimized in our group.

1.2 Functional Sequence Motifs (ProSite)

ProSite is a method of determining what is the function of uncharacterised proteins translated from genomic or cDNA sequences [14]. It consists of a database of biologically significant sites, patterns and profiles that help reliably identify to which known protein family (if any) a new sequence belongs.

1.3 Low-Complexity Regions (SEG)

SEG divides sequences into contrasting segments of low-complexity and high-complexity [33]. Low-complexity segments defined by the algorithm represent “simple sequences” or “compositionally-biased regions”. Locally-optimized low-complexity segments are produced at defined levels of stringency, based on formal definitions of local compositional complexity. The segment lengths and the number of segments per sequence are determined automatically by the algorithm.

1.4 Domain Assignment (ProDom)

The ProDom protein domain database consists of an automatic compilation of homologous domains detected in the SWISS-PROT database by the DOMAINER algorithm [32]. It has been devised to assist with the analysis of the domain arrangement of proteins. ProDom 'domains' are inferred on the basis of conserved subsequences as found in various proteins. Such a conservation corresponds frequently, though not always, to genuine structural domains: therefore domain boundaries should be treated with caution. For some domain families experts have been asked to correct domain boundaries on the basis of both sequence and structural information. This expertise will complement the automated process and improve the quality of ProDom domain families.

1.5 Secondary Structure (PHDsec)

Secondary structure is predicted by a system of neural networks rating at an expected average accuracy $> 72\%$ for the three states helix, strand and loop [4, 23–25]. Evaluated on the same data set, PHDsec is rated at ten percentage points higher three-state accuracy than methods using only single sequence information, and at more than six percentage points higher than, e.g., a method using alignment information based on statistics [17].

PHDsec predictions have three main features:

1. Improved accuracy through evolutionary information from multiple sequence alignments.
2. Improved beta-strand prediction through a balanced training procedure.
3. More accurate prediction of secondary structure segments by using a multi-level system.

1.6 Solvent Accessibility (PHDacc)

Solvent accessibility is predicted by a neural network method rating at a correlation coefficient (correlation between experimentally observed and predicted relative solvent accessibility) of 0.54 cross-validated on a set of 238 globular proteins [1, 26]. The output of the neural network codes for 10 states of relative accessibility. Expressed in units of the difference between prediction by homology modeling (best method) and prediction at random (worst method), PHDacc is some 26 percentage points superior to a comparable neural network using three output states (buried, intermediate, exposed) and using no information from multiple alignments.

1.7 Globularity of Proteins (GLOBE)

An additional result from the prediction of solvent accessibility is that of protein globularity. That method is not published, yet. For more information, you may have a look at the preliminary preprint.

1.8 Transmembrane Helices (PHDhtm)

Transmembrane helices in integral membrane proteins are predicted by a system of neural networks. The shortcoming of the network system is that often too long helices are predicted. These are cut by an empirical filter. The final prediction [2, 21] has an expected per-residue accuracy of about 95%. The number of false positives, i.e., transmembrane helices predicted in globular proteins, is about 2% [20].

The neural network prediction of transmembrane helices (PHDhtm) is refined by a dynamic programming-like algorithm. This method resulted in correct predictions of all transmembrane helices for 89% of the 131 proteins used in a cross-validation test; more than 98% of the transmembrane helices were correctly predicted. The output of this method is used to predict topology, i.e., the orientation of the N-term with respect to the membrane. The expected accuracy of the topology prediction is $>$ 86%. Prediction accuracy is higher than average for eukaryotic proteins and lower than average for prokaryotes. PHDtopology is more accurate than all other methods tested on identical data sets [3, 22].

1.9 Secondary Structure (PROFsec) and Solvent Accessibility (PROFacc)

Secondary structure (PROFsec) and solvent accessibility (PROFacc) are both predicted by a system of neural networks rating at an expected average accuracy $>$ 78% for the three states "helix", "strand" and "loop" (Rost, 2000, unpublished) predicted by PROFsec and for the two states "exposed" and "buried" (Rost, 2000, unpublished) predicted by PROFacc. Evaluated on the same data set PROFsec is rated at 6-8 percentage points higher three-state accuracy than PHDsec (see 1.5). PROFacc is rated at about five percentage points higher two-state accuracy than PHDacc (see 1.6).

1.10 Coiled-Coil Regions (COILS)

COILS is a program that compares a sequence to a database of known parallel two-stranded coiled-coils and derives a similarity score [18]. By comparing this score to the distribution of scores in globular and coiled-coil proteins, the program then calculates the probability that the sequence will adopt a coiled-coil conformation.

1.11 Disulphide Bridges (DISULFIND)

DISULFIND is a server for predicting the disulphide bonding state of cysteines and their disulphide connectivity starting from sequence alone. Disulphide bridges play a major role in the stabilization of the folding process for several proteins. Prediction of disulphide bridges from sequence alone is therefore useful for the study of structural and functional properties of specific proteins. In addition, knowledge about the disulphide bonding state of cysteines may help the experimental structure determination process and may be useful in other genomic annotation tasks [11].

1.12 Structural Switches (ASP)

ASP identifies amino acid subsequences that are the most likely to switch between different types of secondary structure. The program was developed by MM Young, K Kirshenbaum, KA Dill and S Highsmith. ASP was designed to identify the location of conformational switches in proteins with known switches. It is NOT designed to predict whether a given sequence does or does not contain a switch. For best results, ASP should be used on sequences of length >150 amino acids with >10 sequence homologues in the SWISS-PROT data bank. ASP has been validated against a set of globular proteins and may not be generally applicable. Please see [15, 34] for details and for how best to interpret this output. We consider ASP to be experimental at this time, and would appreciate any feedback from our users.

1.13 Localisation Classification (LOCtree)

LOCtree is a novel system of support vector machines (SVMs) that predict the subcellular localization of proteins, and DNA-binding propensity for nuclear proteins, by incorporating a hierarchical ontology of localization classes modeled onto biological processing pathways [19]. Biological similarities are incorporated from the description of cellular components provided by the gene ontology consortium (GO). GO definitions have been simplified and tailored to the problem of protein sorting. Technically the ontology has been implemented using a decision tree with SVMs as the nodes. LOCtree was extremely successful at learning evolutionary similarities among subcellular localization classes and was significantly more accurate than other traditional networks at predicting subcellular localization. Whenever available, LOCtree also reports predictions based on the following: 1) Nuclear localization signals found by PredictNLS, 2) Localization inferred using Prosite motifs and Pfam domains found in the protein, and 3) SWISS-PROT keywords associated with a protein. Localization is inferred in the last two cases using the entropy-based LOCkey algorithm.

1.14 Prediction of Nuclear Localisation Signal (PredictNLS)

PredictNLS is an automated tool for the analysis and in silico determination of Nuclear Localization Signals (NLS) [12]. In NLS discovery mode, PredictNLS searches a query protein for known and potential NLS's in PredictNLS NLSdb to determine if a protein is likely to be targeted to the nucleus. If the protein is determined to be nuclear, the program also reports if a known DNA binding motif is found. In Motif detection mode, the program can help you decide if a sequence motif is likely to act as a nuclear localization signal. The PredictNLS website also documents the largest collection of experimentally determined NLS's.

1.15 Keyword Based Prediction of Cellular Localization (LOCkey)

LOCkey is a database of subcellular localization of eukaryotic proteins inferred using SWISSPROT keywords. LOCkey was the first fully automated algorithm for inferring subcellular localization from database annotations. LOCkey outperformed semi-automated methods relying on expert annotators in benchmark tests. NLSdb is a database of nuclear localization signals (NLSs) and of nuclear proteins targeted to the nucleus by NLS motifs. NLSdb contains over 12,500 predicted nuclear proteins and over 1,500 DNA-binding proteins from six entirely sequenced eukaryotic proteomes (human, mouse, fly, worm, grass and yeast). ER/Golgi Localization: Analysis of experimentally characterized endoplasmic reticulum and Golgi apparatus retrieval motifs and estimates of their specificity to classify subcellular localization for the ER and Golgi. Further investigation of inferring ER and Golgi localization from homology-transfer sequence similarity of ER and Golgi localized proteins.

1.16 Prediction of Unstructured Loops (NORSnet)

NORSnet is a neural network based method that focuses on the identification of unstructured loops [28]. NORSnet was trained to distinguish between very long contiguous segments with non-regular secondary structure (NORS regions) and well-folded proteins. NORSnet was trained on predicted information rather than on experimental data. Therefore, it was optimized on a large data, which is not biased by today's experimental means of capturing disorder. Thus, NORSnet reached into regions in sequence space that are not covered by the specialized disorder predictors. One disadvantage of this approach is that it is not optimal for the identification of the "average" disordered region.

1.17 Prediction of Natively Unstructured Regions Through Contacts (Ucon)

Ucon is a method that combines protein-specific internal contacts with generic pairwise energy potentials to accurately predict long and functional unstructured regions [29]. One advantage of Ucon over statistical-potential based methods is that it incorporates the contribution of the specific order of the amino-acids rather than the amino acid composition alone.

1.18 Meta-Disorder Predictor (MD)

MD is a neural-network based meta-predictor that uses different sources of information predominantly obtained from orthogonal approaches [30]. MD significantly outperformed its constituents, and compared favourably to other top prediction methods. MD is capable of predicting disordered regions of all "flavors", and identifying new ones that are not captured by other predictors.

1.19 Prediction of Flexibility (PROFbval)

PROFbval is a neural-network method that aimed at predicting flexible and rigid residues in proteins from sequence alone (Fig. 1) [31]. PROFbval was trained on B-factor data from PDB- Xray structures and, to an extent, can capture disordered residues. Additionally, surface residues that are predicted to be rigid by PROFbval are correlated with the location of enzyme active sites.

1.20 Reflect

Reflect [6] and OnTheFly [5] (Fig. 2) create a system that enables users to tag genes, proteins, and small molecules in any web page, PDF, or Microsoft-Office document within a few seconds. Clicking on a tag opens a popup window with a concise

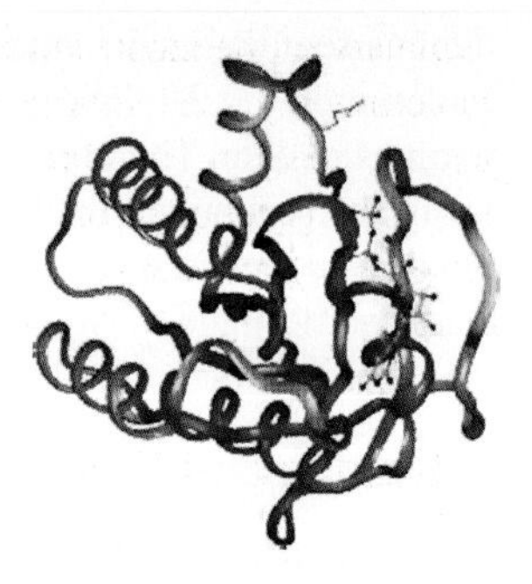

Fig. 1 PROFbval prediction. Explicitly predicting normalized B-values enables the implicit identification of flexible and rigid regions that relate to protein function. The crucial residues in the switch II region of the Ras family need to be very flexible for these proteins to function properly (more red=more flexible)

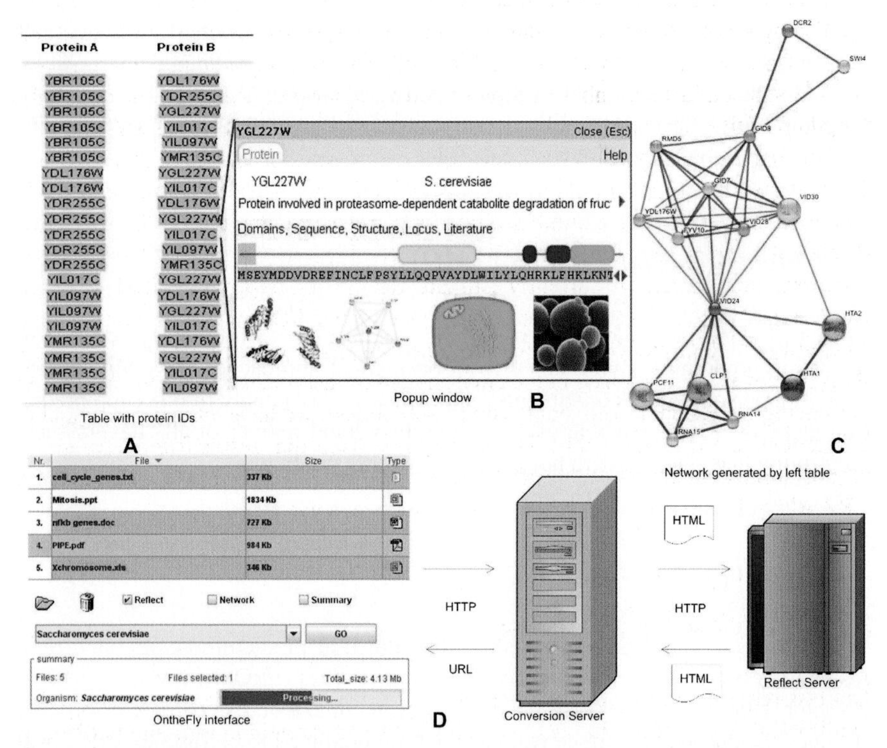

Fig. 2 OnTheFly interface. The figure shows an annotated table (A) of an PDF full text article, the generated popup window with information about the protein YGL227W (B), and an automatically generated protein-protein interaction network (C) of associated entities for the proteins shown in part (A). For demonstration purposes, we isolated the table from the pdf file and processed the table separately. Part (D) shows the architecture and functionality. A user can drag and drop files in the OnTheFly applet. The "GO" button sends the selected documents to the conversion server that converts the according file formats into HTML pages, which will then be sent to the tagging server. A URL pointing to the generated HTML document is returned. The organism selection drop-down list enables users to define a species protein dictionary to be used by default. The "Network" and "Summary" option will extract the STITCH [16] derived networks of associations of the recognized entities in the document(s) and produce a summary page listing the recognized entities

summary of the most significant information about each entity and with direct links to commonly used source data entries. The service has a strong focus on ease-of-use and ease-of-installation, and has shown to be useful to the general life scientist, not just to computational biologists. Figure 2 presents a visual description of the OnTheFly interface.

2 Results and Discussion

The planned duration of the first phase of this project should take about two weeks including waiting times for the processing resources to become available on the Altix cluster. We have setup a mechanism that moves the generated data to our publishing server soon after the data becomes available. Hence once the annotations are complete we expect to have all results ready for consumption. In the next phase, we will select a large number of representative organisms from all branches of the kingdom of life. The completion of that phase will carry us into fall 2009.

The major challenge is to run many relatively short jobs and to do this in a fashion that avoids killing the file system by making input and output the bottleneck. The first challenge is pure number crunching: create annotations for the exploding dataset of entirely sequenced organism. This challenge is met by just having many processors in parallel. We currently estimate that the first basics of our task will consume some 160K CPU-hours. The second challenge is I/O in a cluster environment. In the old-day approach of one machine/one disk, most of our jobs will spend less than 10% of the time on I/O. In the environment of fast nodes that we have been experiencing at Columbia University (up to 400 nodes), we have learned how quickly I/O problems can reverse this. There are some fixes that result from temporarily storing some data on nodes, but we hope that the advanced file systems used by the LRZ Altix cluster will boost significantly what we can get out.

2.1 Parallelization Techniques

Our task required the processing of around 45,000 individual protein sequences. Our application - PredictProtein [27] - is called with each protein in turn. PredictProtein itself is a collection of tools arranged into a pipeline where each tool is executed in a sequential manner.

A certain amount of parallelization would be possible within the component tools themselves. For example NCBI Blast [7, 8] - a tool used my many PredictProtein components - has a command line option that allows control over the number of threads that are used by the application. "hmmsearch" [13] - another search tool that uses Hidden Markov Models - also has multi threaded operating mode. But the scalability of these tools is not very well known to us. Using as many as 510 threads is possible on an Altix partition, however we have been cautioned by more experienced collaborators that Blast performance plateaus at as few as 6-8(!) threads. Another way to parallelize processing would be possible over the components of PredictProtein. Instead of calling components sequentially from a governor script, interdependencies of components could be expressed as make rules. An appropriately prepared make file could not only call the protein feature prediction tools in the correct order but could also automatically guide make to parallelize calls at a required level. However the number of PredictProtein components limits the degree of parallelization achievable in this way to around 10.

Both of the above mentioned techniques would require us to change PredictProtein code and would yield limited result as described above. Therefore we decided to parallelize over the individual PredictProtein runs: we arranged to run as many jobs as possible - each for a single protein - in parallel. On a traditional cluster using the Sun Grid Engine we would achieve our goal by simply submitting each of our proteins in a separate job and let the queuing system handle resource allocation and parallelization. On the Altix cluster in addition to queuing our jobs we had to manage the considerable CPU resources allocated to a single job. A simple and fast-to-implement technique that lent itself to the task of job allocation to hundreds of cores was to use make with the –jobs option. After reserving the desired number of cores with the appropriate qsub option we parametrized make to spawn the respective number of PredictProtein jobs. It is true that we had no previous knowledge about the efficiency make scales with to hundreds of jobs. However we have no reason to suspect a possible breakdown here. We took care to avoid encountering file system IO bottlenecks by storing frequently used large input files such as blast databases in memory with the help of an in-memory file system (tmpfs) mounted at /dev/shm. We also stored temporary files on this file system saving only a handful of final output files onto permanent storage.

In order to avoid having an excessive number of files in one directory we arranged our protein jobs into a tree structure. We took the first four letters of the hexadecimal hash signature of our jobs to designate four levels of the job directory tree. For example all files related to the protein of signature "44442f046c59110e5df85ba2 c998f134d2ad63d0" are stored under the path "4/4/4/4". This structure resulted in less than one job in one directory on average for the task at hand, therefore the job tree is sparse. With around 14 files per job the directories in our tree are quite empty, however we must remember that this is only a pilot project. Eventually we would like to analyze proteomes with hundreds of thousands of proteins altogether (c.f. the human alone now seems to have around 20,000 proteins).

2.2 Number of Processors Used/Degree of Parallelization

Our parallelization mechanism allowed the use of an entire Altix partition for one job, i.e. 510 cores. We noticed that our jobs got scheduled much earlier when asking for 256 or 128 cores instead of 510. Since no single job could process all of our data in the default processing limit of 48 wall clock hours (which we did not alter) we submitted our make job several times as we saw the expiry of earlier jobs. We also submitted non-overlapping parts of the job tree as separate jobs. Our jobs therefore used 128, 256 or 510 cores each achieving up to 510 fold parallelization on the level of one queue job. All queue jobs scheduled in parallel yield a parallelization value even higher. The upper limit for the degree of parallelization is 19 (number of partitions) times 512 (cores per partition) = 9,728 supposing all cores on all partitions are allocated to processing. While our pilot project did not come close to this degree of parallelization we see no reason why it should not be achievable.

2.3 *Job Run Times*

Using 39,665 CPU hours we were able to process 13,328 proteins on the Altix cluster using the parallelization techniques detailed above. This gives around 3 hours per protein processing time. This value is comparable to the performance of the Linux Cluster (LC) at the LRZ. Having run out of CPU time on the Altix we ran the remaining over 31,000 proteins on the Itanium processors of the Linux Cluster. This took 14,783 hours resulting in a 0.5 hour per job average. Why is the large difference? We believe the clock speeds of processors in both clusters are the same. One possible reason for the difference in processing times is that the Altix and the LC worked on different sets of proteins. While true, we have no reason to believe that this is a considerable factor. The partitioning of jobs for the Altix and LC was done practically at random as the ordering of jobs in the job tree is the result of the first four letters of hashes produced by the SHA1 hashing algorithm applied to protein sequences. This hashing algorithm is known to create hashes far apart for similar - or different - input sequences. Thus we believe that the size and composition of the job sets run on the Altix and LC is sufficient to gauge performance. The question of the origin of the discrepancy therefore remains open.

3 Conclusion

The goal of this pilot project was to support the new automatic tagging system Reflect [6] with an initial set of protein feature predictions generated by our bioinformatics toolset PredictProtein [27]. We decided to execute this task on the Altix at the LRZ in order to gauge cluster performance. A set of 45,000 proteins of particular biological interest was chosen. We parallelized our prediction runs using the parallel execution feature of "make". This allowed us to run up to 510 predictions in parallel per batch job. Before exhausting our CPU time we were able to obtain predictions for 13,328 proteins in 39,665 CPU hours - a roughly 3 hours per protein average. We submitted the remaining proteins one protein per job to the Itanium nodes of the Linux cluster at the LRZ. We obtained predictions for 29,217 proteins in 14,783

CPU hours giving a roughly half an hour per protein average. This is a factor of 6 improvement.

We might say our toolset PredictProtein is a typical bioinformatics application. Looking at these results one tends to conclude that - in general - we should not focus more than $1/7^{th}$ of our attention to the Altix since it is expected to deliver only around $1/7^{th}$ of all predictions submitted at the LRZ. Put in another way Itanium cores allocated to the Altix cluster serve only $1/6^{th}$ as much our purpose than those allocated to the Linux cluster.

We suspect - and conclude - that the Altix is specialized for applications that have fundamentally different requirements than ours. It also seems that - for our purposes in general - an extension of the Linux cluster would result in several times higher performance gain. In the future we plan to extend our prediction runs to computing resources similar to the Linux cluster, restricting our runs on the Altix to probing into techniques that may lead to improved performance.

References

1. Evaluation of PHDacc accuracy. URL http://www.predictprotein.org/Dtab/phd_acc.html
2. Evaluation of PHDhtm accuracy. URL http://www.predictprotein.org/Dtab/phd_htm.html
3. Evaluation of PHDhtmtop accuracy. URL http://www.predictprotein.org/Dtab/phd_htmtop.html
4. Evaluation of PHDsec accuracy. URL http://www.predictprotein.org/Dtab/phd_sec.html
5. OnTheFly. URL http://onthefly.embl.de/
6. Reflect. URL http://reflect.ws/
7. Altschul, S.F., Gish, W., Miller, W., Myers, E.W., Lipman, D.J.: Basic local alignment search tool. J Mol Biol **215**(3), 403–410 (1990)
8. Altschul, S.F., Koonin, E.V.: Iterated profile searches with psi-blast–a tool for discovery in protein databases. Trends Biochem Sci **23**(11), 444–447 (1998)
9. Berman, H.M., Westbrook, J., Feng, Z., Gilliland, G., Bhat, T.N., Weissig, H., Shindyalov, I.N., Bourne, P.E.: The protein data bank. Nucleic Acids Res **28**(1), 235–242 (2000)
10. Boutet, E., Lieberherr, D., Tognolli, M., Schneider, M., Bairoch, A.: UniProtKB/Swiss-Prot. Methods Mol Biol **406**, 89–112 (2007). URL http://eutils.ncbi.nlm.nih.gov/entrez/eutils/elink.fcgi?cmd=prlinks&dbfrom=pubmed&retmode=ref&id=18287689
11. Ceroni, A., Passerini, A., Vullo, A., Frasconi, P.: DISULFIND: a disulfide bonding state and cysteine connectivity prediction server. Nucleic Acids Res **34**(Web Server issue), W177–W181 (2006). URL http://eutils.ncbi.nlm.nih.gov/entrez/eutils/elink.fcgi?cmd=prlinks&dbfrom=pubmed&retmode=ref&id=16844986
12. Cokol, M., Nair, R., Rost, B.: Finding nuclear localization signals. EMBO Rep **1**(5), 411–415 (2000)
13. Eddy, S.R.: Profile hidden markov models. Bioinformatics **14**(9), 755–763 (1998)
14. Falquet, L., Pagni, M., Bucher, P., Hulo, N., Sigrist, C.J.A., Hofmann, K., Bairoch, A.: The prosite database, its status in 2002. Nucleic Acids Res **30**(1), 235–238 (2002)
15. Kirshenbaum, K., Young, M., Highsmith, S.: Predicting allosteric switches in myosins. Protein Sci **8**(9), 1806–1815 (1999)

16. Kuhn, M., von Mering, C., Campillos, M., Jensen, L.J., Bork, P.: Stitch: interaction networks of chemicals and proteins. Nucleic Acids Res **36**(Database issue), D684–D688 (2008)
17. Levin, J.M., Pascarella, S., Argos, P., Garnier, J.: Quantification of secondary structure prediction improvement using multiple alignments. Protein Eng **6**(8), 849–854 (1993)
18. Lupas, A.: Prediction and analysis of coiled-coil structures. Methods Enzymol **266**, 513–525 (1996)
19. Nair, R., Rost, B.: Mimicking cellular sorting improves prediction of subcellular localization. J Mol Biol **348**(1), 85–100 (2005)
20. Rost, B., Casadio, R., Fariselli, P.: Refining neural network predictions for helical transmembrane proteins by dynamic programming. Proc Int Conf Intell Syst Mol Biol **4**, 192–200 (1996)
21. Rost, B., Casadio, R., Fariselli, P., Sander, C.: Transmembrane helices predicted at 95% accuracy. Protein Sci **4**(3), 521–533 (1995)
22. Rost, B., Fariselli, P., Casadio, R.: Topology prediction for helical transmembrane proteins at 86% accuracy. Protein Sci **5**(8), 1704–1718 (1996)
23. Rost, B., Sander, C.: Improved prediction of protein secondary structure by use of sequence profiles and neural networks. Proc Natl Acad Sci U S A **90**(16), 7558–7562 (1993)
24. Rost, B., Sander, C.: Prediction of protein secondary structure at better than 70% accuracy. J Mol Biol **232**(2), 584–599 (1993). URL `http://eutils.ncbi.nlm.nih.gov/entrez/eutils/elink.fcgi?cmd=prlinks&dbfrom=pubmed&retmode=ref&id=8345525`
25. Rost, B., Sander, C.: Combining evolutionary information and neural networks to predict protein secondary structure. Proteins **19**(1), 55–72 (1994)
26. Rost, B., Sander, C.: Conservation and prediction of solvent accessibility in protein families. Proteins **20**(3), 216–226 (1994)
27. Rost, B., Yachdav, G., Liu, J.: The PredictProtein server. Nucleic Acids Res **32**(Web Server issue), W321–W326 (2004). URL `http://eutils.ncbi.nlm.nih.gov/entrez/eutils/elink.fcgi?cmd=prlinks&dbfrom=pubmed&retmode=ref&id=15215403`
28. Schlessinger, A., Liu, J., Rost, B.: Natively unstructured loops differ from other loops. PLoS Comput Biol **3**(7), e140 (2007)
29. Schlessinger, A., Punta, M., Rost, B.: Natively unstructured regions in proteins identified from contact predictions. Bioinformatics **23**(18), 2376–2384 (2007)
30. Schlessinger, A., Punta, M., Yachdav, G., Kajan, L., Rost, B.: Improved disorder prediction by combination of orthogonal approaches. PLoS ONE **4**(2), e4433 (2009). URL `http://eutils.ncbi.nlm.nih.gov/entrez/eutils/elink.fcgi?cmd=prlinks&dbfrom=pubmed&retmode=ref&id=19209228`
31. Schlessinger, A., Yachdav, G., Rost, B.: Profbval: predict flexible and rigid residues in proteins. Bioinformatics **22**(7), 891–893 (2006)
32. Sonnhammer, E.L., Kahn, D.: Modular arrangement of proteins as inferred from analysis of homology. Protein Sci **3**(3), 482–492 (1994)
33. Wootton, J.C., Federhen, S.: Analysis of compositionally biased regions in sequence databases. Methods Enzymol **266**, 554–571 (1996). URL `http://eutils.ncbi.nlm.nih.gov/entrez/eutils/elink.fcgi?cmd=prlinks&dbfrom=pubmed&retmode=ref&id=8743706`
34. Young, M., Kirshenbaum, K., Dill, K.A., Highsmith, S.: Predicting conformational switches in proteins. Protein Sci **8**(9), 1752–1764 (1999). URL `http://eutils.ncbi.nlm.nih.gov/entrez/eutils/elink.fcgi?cmd=prlinks&dbfrom=pubmed&retmode=ref&id=10493576`

Molecular Dynamics Simulation of the Nascent Peptide Chain in the Ribosomal Exit Tunnel

Lars Bock and Helmut Grubmüller

Abstract The ribosome is a large macromolecular complex which synthesizes all proteins in the cell according to the genetic code and thereby the central nanomachine of life. Nascent peptide chains are polymerized by the ribosome and exit through a tunnel in the large subunit. This tunnel is an important target for antibiotic action hindering the movement of the peptide chain. The recently solved structure of the ribosome led to a better understanding of the tunnel's geometry and the binding sites of the antibiotics. Yet the dynamics of the polypeptide's conformation and the interactions with the large subunit and the antibiotics remain unclear. We approach these questions with molecular dynamics simulations of the large subunit and the nascent chain in explicit water under physiological conditions. This allows us to investigate the dynamics the ribosome and the polypeptide at an atomistic level, which will lead to a huge improvement in the understanding of the translation process and the antibiotic action.

1 Introduction

The ribosome is the largest macromolecular complex, composed of rRNA and proteins. The ribosome, which is present in every cell, reads mRNA and polymerizes amino acids to synthesize proteins according to the sequence of nucleotide triplets. This essential role in translation also renders it an important target for antibiotics, which suppress gene expression by inhibiting the ribosome.

Prokaryotic ribosomes consist of a small subunit (30S) and a large subunit (50S). The 30S subunit decodes the information contained by the mRNA. The 50S subunit catalyzes the peptidyl transferase reaction connecting the amino acids and releases the protein. The nascent peptide chain exits the ribosome through a tunnel,

Lars Bock · Helmut Grubmüller
MPI for Biophysical Chemistry, Am Fassberg 11, 37077 Göttingen, German
e-mail: lbock@gwdg.de
e-mail: hgrubmu@gwdg.de

S. Wagner et al. (eds.), *High Performance Computing in Science and Engineering, Garching/Munich 2009*, DOI 10.1007/978-3-642-13872-0_62,

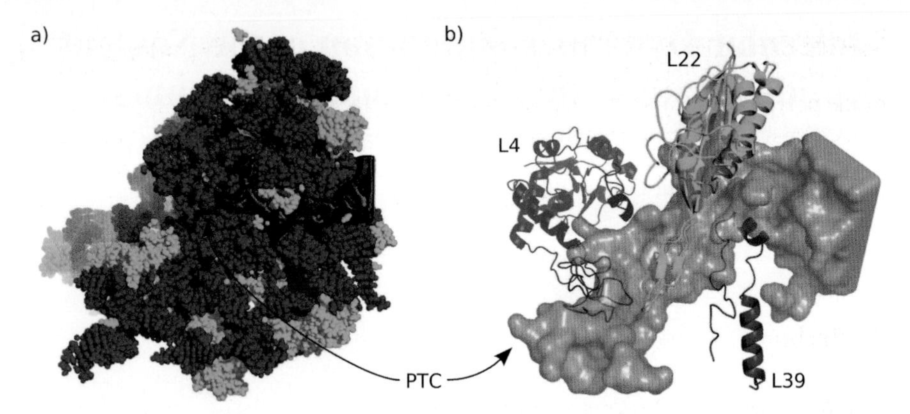

Fig. 1 (a) Cross-section of the 50S subunit, RNA drawn in red, proteins in green and the tunnel in black. (b) Surface plot of the tunnel with the proteins which form a part of the tunnel wall, L4, L22, and L39

which stretches through the large ribosomal subunit. The tunnel begins at the peptidyl transferase center (PTC), where the peptide bonds are formed and exits the ribosome at the opposite site of the subunit (figure 1).

Since the crystal structure of the large ribosomal subunit of *Haloarcula Marismortui* has been determined in 2000 [1] at atomic resolution, full-detail molecular dynamics (MD) simulations are now possible. In our work we focus on the investigation of the functionality of antibiotics, which interact with the tunnel and the nascent peptide chain. To examine the conformational flexibility and dynamics of nascent peptide chains we simulated peptide chains in the tunnel near the PTC using MD.

Certain sequence motifs of the nascent chains were experimentally shown to affect protein elongation and peptide termination [9, 11]. The polypeptide SecM induces a complete translation arrest. The arresting sequence motif was found to be FXXXXWIXXXXGIRAGP [10]. Further, mutants of the ribosome which allow completion of SecM have mutations of residues facing the inner wall of the exit tunnel. FRET measurements reveal a compaction of the c-terminus upon elongation from amino acid 164 to 166 [13]. Measurements with mutant ribosomes showed that the conformational change is induced by the ribosomal tunnel. SecM mutations that prevent nascent chain compaction (proline substitution 157–160) impair the translation arrest. Alanine substitution of residues which are sufficient for the arrest also lead to a compaction. This suggests that translation arrest requires a conformational change and key residue interaction.

The exact mechanism of this arrest remains unknown, it was proposed that the polypeptides strongly binds to the tunnel surface or that the conformation of the ribosome changes due to the polypeptide inside the tunnel and thus stops the translation [8].

Berisio et al. [3] observed that the loop of L22 which builds a part of the tunnel wall flips into the tunnel when the antibiotic troleandomycin is bound in the tunnel and they suggested that this flipping could also be induced by secM and thereby block nascent peptide passage.

Since there is no cyrstal structure of ribosomes with nascent peptides available, this study investigates if their conformations and dynamics depend on the sequence of the polypeptides with the help of molecular dynamics simulations (MD) [4]. We also compare different eleongation states of the arresting sequence of secM to study differences which may lead to the arrest. Due to the combination of high resolution (2.4 Å) and the low R_{free} factor (0.222) we chose to use the *Haloarcula Marismortui* crystal structure [6].

2 Methods

2.1 Principles of Molecular Dynamics

Molecular dynamics (MD) is a form of computer simulation, which describes a molecular system as a system of atoms and a potential acting upon them. Newton's equations are integrated over time to obtain information about the dynamics of the system. This approach is used in many scientific fields, especially for the description of the atomistic motion of biomolecules.

The time-dependent Schrödinger equation describes the exact motion of atoms, but analytic approaches even fail to solve the equation for the helium atom. Numerical methods are available, but computationally intense and therefore only applicable to systems comprising few atoms.

To reduce the computational effort, several approximations are required. The first is the Born-Oppenheimer approximation, which separates the electronic motion from the nucleic motion. This approximation rests on the fact that the mass of an electron is three magnitudes larger than the mass of a single nucleon. Therefore the dynamics of the electrons is much faster than the dynamics of the nuclei, because the relaxation of the electrons within the ground state is fast with respect to the nuclear motion. Accordingly, it is sufficient to only describe the movement of the nuclei.

The second approximation is the use of a potential $V(\mathbf{r}_1, \mathbf{r}_2, ..., \mathbf{r}_N)$ which describes the interatomic energies with simple functions [4]. This potential is called force field and the one used in this work has the following form:

$$V(\mathbf{r}_1,\mathbf{r}_2,...,\mathbf{r}_N) = \underbrace{\sum_{\text{bonds}} \frac{1}{2} K_b (b - b_0)^2}_{\text{bondstretching}} + \underbrace{\sum_{\text{angles}} \frac{1}{2} K_\theta (\theta - \theta_0)^2}_{\text{angle bending}}$$

$$+ \underbrace{\sum_{\text{improper dihedrals}} \frac{1}{2} K_\xi (\xi - \xi_0)}_{\text{improper dihedral angle}} + \underbrace{\sum_{\text{dihedrals}} \frac{1}{2} K_\phi 1 + \cos(n\phi - \delta)}_{\text{dihedral angle}}$$

$$+ \sum_{\text{pairs}(i,j)} (\underbrace{\frac{C_{12}(i,j)}{r_{ij}^{12}} - \frac{C_6(i,j)}{r_{ij}^{6}}}_{\text{van der Waals}} + \underbrace{\frac{q_i q_j}{4\pi\varepsilon_0\varepsilon_r r_{ij}}}_{\text{Coulomb}}),$$

where N is the number of atoms, $\mathbf{r}_i$ and q_i are the postion and the charge of atom i, respectively. In the force field, there are four forces which act upon the atoms which are covalently bonded. The bond stretching term describes the force which occurs when the bond length changes with a harmonic potential. The minimum energy bond length is b_0 and the force constant is K_b. The second term describes the bond angle bending interaction with a harmonic potential which depends on the angle between two bonds involving three atoms. The equilibrium bond angle is θ_0. There are two terms which specify interactions between four atoms. The first is the improper dihedral angle term which describes forces acting upon the atoms according to the angles between two planes, e.g., to keep aromatic rings planar. The second term is the dihedral angle term which describes the forces occurring due to the torsion around a bond.

The last term is a sum of non-bonded interactions over all pairs of atoms. The van der Waals interaction and the Pauli repulsion are described by the Lennard-Jones potential. The charges, which are smeared across molecules, are described as partial point charges q_i which are assigned to the atoms. The electron cloud is thereby simplified to a set of point charges, and the electrostatic forces are calculated by Coulomb's law.

There are two ways to obtain the parameters for the potential $V(\mathbf{r}_1, \mathbf{r}_2, ..., \mathbf{r}_N)$. One way is to fit them to results of ab-initio quantum mechanics calculations and another way is to fit the parameters to various experimental data like free energies of solvation, NMR or x-ray data. For most potentials, and also for the one used in this work, a mixture of both approaches is used.

The third approximation is to describe the motion of all atoms $i = 1, ..., N$ by Newton's equation of motion,

$$m_i \frac{d^2 \mathbf{r}_i(t)}{dt^2} = \nabla_i V(\mathbf{r}_1, ..., \mathbf{r}_N), \tag{1}$$

where m_i is the mass of atom i and V is the potential described above. This equation is integrated in discrete time steps with an integration step length of Δt. For all simulations described in this work we used an integration step length of $\Delta t = 2$ fs.

Velocities $\mathbf{v}_i$ and positions $\mathbf{r}_i$ of the atoms were updated with the Verlet algorithm [12],

$$\mathbf{v}_i(t + \frac{\Delta t}{2}) = \mathbf{v}_i(t - \frac{\Delta t}{2}) + \frac{\mathbf{F}_i(t)}{m_i} \cdot \Delta t$$

$$\mathbf{r}_i(t + \Delta t) = \mathbf{r}_i(t) + \mathbf{v}_i(t + \frac{\Delta t}{2}) \cdot \Delta t,$$

where $\mathbf{F}_i = -\nabla_i V(\mathbf{r}_1, \mathbf{r}_2, ..., \mathbf{r}_N)$ is the force acting on atom i.

2.2 Periodic Boundary Condition

The number of atoms in a simulation is limited due to limited computational resources. To minimize artefacts due to the resulting small system size and surface effects, periodic boundary conditions were applied in all simulations. Accordingly, the atoms are put into a space-filling simulation box, which is surrounded by translated images of itself. Atoms leaving the simulation box on one side are put back into the box on the opposite side. Similarly, for the calculation of the potential also atoms which are on the other side of the boundary are taken into account.

The simulation box size has been chosen sufficiently large to avoid that molecules interact with their images. The Debye-Hückel length gives a good estimate for the range of this interaction. The Debye-Hückel length for the ion concentration in our system (2 Mol/l) is 0.31 nm, such that the chosen distance of 1.5 nm between the ribosome and the walls of the simulation box guarantees that interactions with the images are small.

2.3 Temperature and Pressure Coupling

Under normal conditions cells have a close to constant temperature and pressure, which is described by an NpT-ensemble. To achieve this ensemble, we need to couple temperature and pressure to given reference values, because given a constant energy and a constant volume, the simulation would be in an microcanonical (NEV) ensemble. Consequently, we simulate with temperature and pressure coupling. To account for the time scales of energy and pressure fluctuations in the system, the coupling is not instantaneous, but a coupling time constant is introduced.

For temperature coupling we used the Berendsen temperature coupling scheme [2], where the velocity of every particle v is scaled to λv in every step with

$$\lambda = \sqrt{1 + \frac{\Delta t}{\tau_T}\left(\frac{T_0}{T} - 1\right)}, \tag{2}$$

where Δt is the integration step length, τ_T is the temperature coupling time constant, $T_0 = 300$ K is the reference temperature and T is the instantaneous temperature derived from the kinetic energy of all atoms.

Pressure coupling to the reference pressure $P_0 = 1$ atm was achieved by the Berendsen pressure coupling method [2], where the edges of the simulation box and the coordinates of the atoms are scaled with the factor μ.

$$\mu = 1 - \frac{\Delta t}{3\tau_P}\kappa(P_0 - P),$$

where τ_P is the pressure coupling time constant, κ is the isothermal compressibility of water and P is the instantaneous pressure derived from the velocities and forces of all atoms via the virial theorem.

2.4 Parallelisation Techniques

To achieve large timescales of many particle systems, efficient parallelisation of the simulation is crucial. We here present two important methods, domain decomposition and dynamic load balancing, used in the simulation toolkit GROMACS [5].

Domain Decomposition

The simulation system is divided into spatial domains and each domain is assigned to one processor. Due to the fact that most interactions in MD are local this decompositon reduces the requirement for communication between the processors. Calculation of forces between particles residing in two seperate domains is done with the eighth shell method [7].

Dynamic Load Balancing

Computation of the forces is the most time consuming step in MD simulations. The computational load may vary among the processors due to e.g. an inhomogeneous particle distribution or an inhomogeneous distribution of interaction calculation cost. When the computational load is not distributed equally among the processors, all processors have to wait for the one which takes the most time. With the dynamic load balancing method the volume of these domains is dynamically adjusted such that the load is balanced among the processors.

2.5 Preparation of Simulation System

To equilibrate the structure of the ribosome in solution we have started MD simulations with the large ribosomal subunit from *Haloarcula Marismortui* (coordinates from the Protein Data Bank entry 1S72) in explicit water with ions. We use the GROMACS simulation software [5]. Our simulation box contains $\sim 200,000$ protein and RNA atoms and $\sim 1,100,000$ water atoms and ions yielding altogether $\sim 1,300,000$ atoms.

We modeled three different polypeptides into the exit tunnel. The first polypeptide was a segment of the SecM sequence, amino acids 132–166, from now on referred to as SecM166. Pro166 is the amino acid where the translation arrest occurs and the length of the polypeptide was chosen such that the end of the peptide is outside the tunnel. The second polypeptide consists of the amino acids 132–164 of the SecM protein, from now on referred to as SecM164. This polypeptide describes the translation state two amino acids before the translation arrest occurs. As the third polypeptide we chose amino acids 2–26 of bovine pancreatic polypeptide (Bpp), a short peptide, which forms an α-helix in solution. Bpp was chosen, because its length is similar to the other polypeptides used, because we can observe whether it forms an α-helix inside the tunnel or not, and because it is a non-arresting sequence.

The modeled starting structures are equilibrated. From the trajectory snapshots were taken at 0.5 ns, 1 ns, and 1.5 ns, respectively. With these snapshots as starting structures, new simulations were started with new velocities.

3 Results

3.1 Achieved Performance on the HLRB system

Up to now we have used 125000 cpu · h for the simualtions of the 3 different polypeptide inside the ribosome. We ran 81 simulations, each on 32 cpus in parallel, with an average yielding an average speed of 188 ps/day.

Table 1 shows the possible scaling of ensemble MD simulations consisting of 4 simulations, which was calculated during test runs.

3.2 Conformations of Polypeptides Inside the Tunnel

Our simulations show that the whole system is reasonably stable. Figure 2 shows the rmsds of the polypeptides, SecM166, SecM164, and Bpp. As can be seen, the rmsd increases throughout the wole trajectories, showing that the polypeptides are not completely equilibrated yet.

To analyze whether the dynamics of the polypetides inside the tunnel depend on their sequence, we performed a PCA analysis on the trajectories of these twelve simulations. The backbone atoms of amino acids 1–26, counted from the PTC, were taken into account. To investigate the dynamics of the polypetides with respect to the tunnel, the trajectories were fitted to positions of Cα and P atoms of amino acids or nucleotides closer than 3 Å to the polypeptides. The covariance matrices of the backbone atoms were calculated using the combined trajectory of all simulations to obtain a set of eigenvectors where the eigenvectors with the largest eigenvalues describe the largest fluctuations of the backbone atoms of the three polypeptides.

Figure 3 shows the projections of the trajectories on the eigenvectors from the PCA of the motion with respect to the tunnel. The projections of trajectories of different polypetides are clearly separated. This could be due to the fact that the simulations were not converged or that the dynamics of the polypetides are different.

Table 1 Scaling of ensemble MD simulations

CPUs	$\frac{time}{step}$ [s]	scaling
128	1.62	100%
256	**0.84**	**97%**
512	0.48	84%

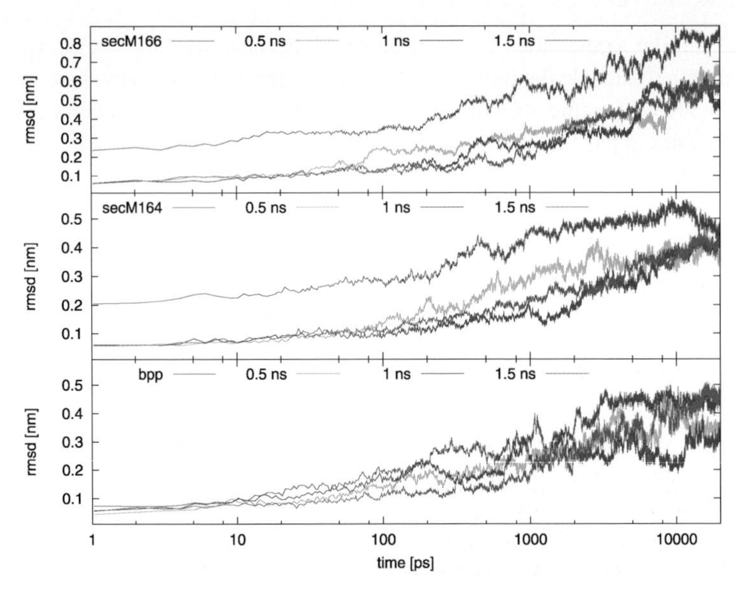

Fig. 2 Equilibration of polypetides inside the tunnel: Rmsd of the polypeptides, SecM166, SecM164, and Bpp, inside the tunnel for four different starting structures each

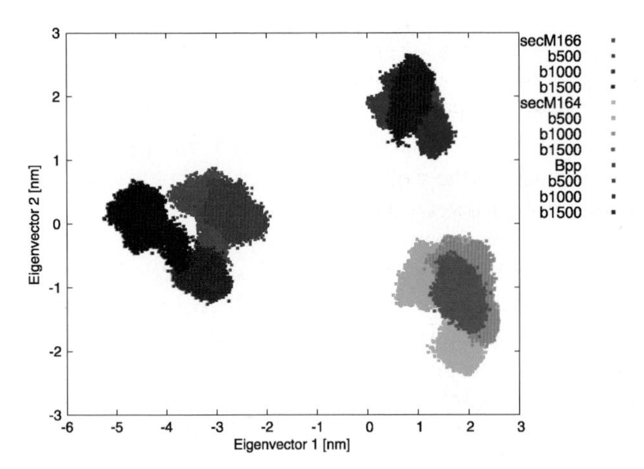

Fig. 3 Dynamics of the polypeptides in reference to the tunnel: Trajectories projected on the first two eigenvectors

The sampled regions in the first two eigenvectors of the individual trajectories of one polypeptide, represented by different shades, are close to each other and overlap, which shows that polypetides with the same sequence have the similar dynamics during our simulations.

The equilibrating polypeptides show different conformations and different secondary structure formation which suggests that the folding inside the tunnel is strongly depending on the sequence (figure 4).

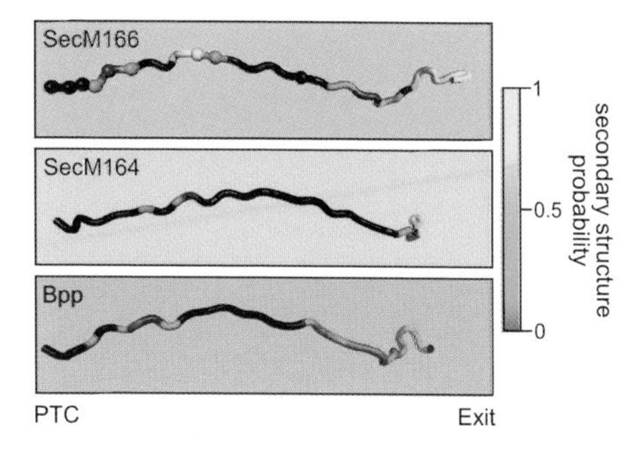

Fig. 4 Cartoon representation of the polypeptide conformations averaged over the last 5 ns of simulation. The color of the backbone denotes the probability of secondary structure formation. The Cα atoms of the residues necessary for the arrest sequence are drawn as spheres

References

1. Ban, N., Nissen, P., Hansen, J., Moore, P., Steitz, T.: The complete atomic structure of the large ribosomal subunit at 2.4 a resolution. Science **289**(5481), 905 (2000)
2. Berendsen, H., Postma, J., van Gunsteren, W., DiNola, A., Haak, J.: Molecular dynamics with coupling to an external bath. The Journal of Chemical Physics **81**, 3684 (1984)
3. Berisio, R., Schluenzen, F., Harms, J., Bashan, A., Auerbach, T., Baram, D., Yonath, A.: Structural insight into the role of the ribosomal tunnel in cellular regulation. Nature Structural Biology **10**(5), 366–370 (2003)
4. van Gunsteren, W., Berendsen, H., et al.: Computer simulation of molecular dynamics: Methodology, applications, and perspectives in chemistry. Angewandte Chemie International Edition in English **29**(9), 992–1023 (1990)
5. Hess, B., Kutzner, C., van der Spoel, D., Lindahl, E.: Gromacs 4: Algorithms for highly efficient, load-balanced, and scalable molecular simulation. J. Chem. Theory Comput **4**(3), 435–447 (2008)
6. Klein, D., Moore, P., Steitz, T.: The roles of ribosomal proteins in the structure assembly, and evolution of the large ribosomal subunit. Journal of Molecular Biology **340**(1), 141–177 (2004)
7. Liem, S., Brown, D., Clarke, J.: Molecular dynamics simulations on distributed memory machines. Computer Physics Communications **67**, 261–267 (1991)
8. Mitra, K., Schaffitzel, C., Fabiola, F., Chapman, M., Ban, N., Frank, J.: Elongation arrest by secm via a cascade of ribosomal RNA rearrangements. Molecular Cell **22**(4), 533–543 (2006)
9. Nakatogawa, H., Ito, K.: Secretion monitor, secm, undergoes self-translation arrest in the cytosol. Molecular Cell **7**(1), 185–192 (2001)
10. Nakatogawa, H., Ito, K.: The ribosomal exit tunnel functions as a discriminating gate. Cell **108**(5), 629–636 (2002)
11. Tenson, T., Ehrenberg, M.: Regulatory nascent peptides in the ribosomal tunnel. Cell **108**(5), 591–594 (2002)
12. Verlet, L.: Computer experiments on classical fluids. II. Equilibrium correlation functions. Phys. Rev **165**(1), 201–14 (1968)
13. Woolhead, C., Johnson, A., Bernstein, H.: Translation arrest requires two-way communication between a nascent polypeptide and the ribosome. Molecular Cell **22**(5), 587–598 (2006)

Preparing RAxML for the SPEC MPI Benchmark Suite

Michael Ott and Alexandros Stamatakis

Abstract Inferring phylogenetic trees from molecular sequence data is considered to be a *grand challenge* in Bioinformatics due to its immense computational requirements. Efficient parallel implementations of programs for phylogenetic inference and the utilization of high performance computing infrastructure are therefore of paramount importance for facilitating large-scale phylogenetic analyses. RAxML is a widely used tool for phylogenetic inference and its PThreads- and MPI-based parallelizations have already been shown to scale up to thousands of processors. The MPI version has recently been chosen to become a part of the upcoming SPEC MPI 2.0 benchmark suite.

1 Introduction

Phylogenetic trees are used to represent the evolutionary history of a set of n organisms. An alignment of DNA or protein sequences consisting of n sequences (taxa) and m alignment columns can be used as input for phylogenetic inference. The output is an unrooted binary tree; the n taxa are located at the leaves of the tree and the inner nodes represent common extinct ancestors. Evolutionary events such as mutations, are modeled to occur along the branches of the tree, between ancestor and descendant. The branch lengths essentially represent the relative time of evolution between nodes in the tree. Phylogenetic trees have many important applications in medical and biological research (see [2] for a summary) ranging from the analysis of emerging infectious diseases [10] to the tests of whether Caribbean frogs have

Michael Ott · Alexandros Stamatakis
Technische Universität München, Department of Computer Science, Boltzmannstr. 3, 85748 Garching b. München, Germany
e-mail: ottmi@cs.tum.edu
e-mail: stamatak@cs.tum.edu

S. Wagner et al. (eds.), *High Performance Computing in Science and Engineering, Garching/Munich 2009*, DOI 10.1007/978-3-642-13872-0_63,

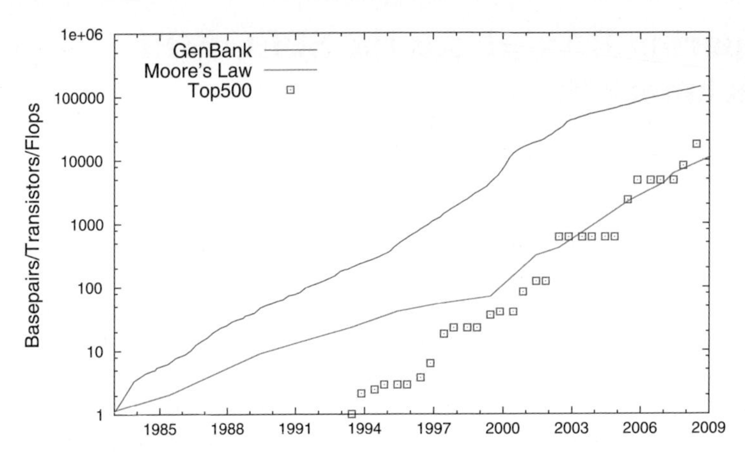

Fig. 1 Bio-Gap: Accumulation of new sequence data in GenBank (number of basepairs), performance-gain according to Moore's Law (number of transistors), and performance of the #1 system in the Top500 supercomputer list (FLOPs). All values have been scaled accordingly

a single common origin or have diversified via multiple independent invasions of islands [9].

Due to the continuously accelerating accumulation of sequence data, which is driven by novel wet-lab techniques such as, e.g., pyrosequencing [14] or advances in Expressed Sequence Tag techniques (ESTs [1]), there is an increasing demand to compute large trees which often comprise thousands of organisms, each represented by data from several genes, thousands of SNPs (Single Nucleotide Polymorphisms), or whole genomes. Since alignment matrices continuously grow in both dimensions (number of organisms *and* alignment length), efficient parallel phylogeny programs are required to handle memory demands (which are primarily a function of alignment length) and growing inference times (which are primarily a function of the number of taxa). Moreover, there is an increasing gap, which we call the "Bio-Gap" between the speed at which molecular data accumulates and the speed at which computer architectures become faster according to Moore's law (see Figure 1). Hence, we urgently need to devise appropriate parallelization schemes to deploy high performance computers for keeping pace with the biological data flood.

RAxML [15] (Randomized Axelerated Maximum Likelihood) is an open-source code for large-scale Maximum Likelihood (ML) based inference of evolutionary trees [7] using multiple alignments of binary, DNA, RNA secondary structure, multi-state morphological, or AA (Amino Acid/Protein) sequences as well as combinations thereof. It is becoming increasingly popular in real-world biological studies and forms an integral component of the CIPRES (CyberInfrastructure for Phylogenetic RESearch, www.phylo.org) project and the greengenes workbench [5] (greengenes.lbl.gov).

The Standard Performance Evaluation Corporation (SPEC) offers numerous benchmarks suites that aim at providing standardized mechanisms for assessing the performance of state-of-the-art computer systems. The benchmarks suites comprise the source code of several real-world applications. The well-known SPECint and SPECfp benchmarks allow for assessing single-core integer and floating point arithmetic performance. In addition, the SPEC High Performance Group (HPG) provides two benchmark suites for multiprocessor systems: SPEC OMP for shared memory systems (based on OpenMP) and SPEC MPI for distributed memory systems (based on the Message Passing Interface MPI). Version 2.0 of the MPI suite will be released at the Supercomputing conference in November 2009 and will include RAxML as a new component which is denoted as 125.raxml.

RAxML exhibits properties that make it a well-suited candidate for a benchmark suite: *Firstly,* the communication to computation ratio can easily be controlled and adapted by using input alignments of variable length; *secondly* it exhibits a relatively fine-grain sequence of global synchronization and reduction operations; *thirdly* the parallelization strategies adopted in RAxML are generally applicable to *all* Likelihood-based programs for phylogenetic inference, including Bayesian methods, i.e., they are not program-specific but can be used for a general purpose phylogenetic likelihood kernel library.

2 The Phylogenetic Likelihood Kernel

The central component of every Maximum Likelihood-based program for phylogenetic inference is the phylogenetic likelihood kernel (PLK) It allows for evaluating the likelihood of a given tree topology and reduces its complexity to a single numerical value (the likelihood score) that represents a statistical measure for the fit of the tree to the data. In order to compute the likelihood score on a fixed topology several components are required:

- The input sequence alignment, represented by an $n \times m$ matrix.
- A statistical model that describes the process of evolution from one sequence into another sequence over time. It consists of the instantaneous nucleotide substitution matrix Q which contains the transition probabilities for time dt between nucleotide (4x4 matrix) or Amino Acid (20x20 matrix) characters, the prior probabilities of observing the nucleotides, e.g., $\pi_A, \pi_C, \pi_G, \pi_T$ for DNA data, which can for instance be determined empirically from the alignment, and finally the α shape parameter that forms part of the Γ model [19] of rate heterogeneity which accounts for the biological fact that different columns in the alignment evolve at different speeds.
- The lengths of the tree's $2n-3$ branches.

Under the assumption that the tree topology, the branch lengths, and all model parameters are fixed, one initially needs to compute the entries for all internal (ancestral) likelihood vectors that are located at the inner nodes, in order to compute the likelihood of an *unrooted* binary tree. The vectors contain the probabilities $P(A), P(C), P(G), P(T)$ of observing an `A`, `C`, `G`, or `T` (for DNA data; for AA data there are 20 probability values) at each site m of the input alignment at a specific ancestral node. They are computed bottom-up from the tips towards a virtual root that can be placed into any branch of the tree. Under certain standard model restrictions (time-reversibility of the model) the final likelihood score will be the same regardless of the placement of the virtual root. Once the partial likelihood arrays to the left and right of the virtual root have been computed, the log likelihood score can then be calculated by combining and summing over the entries in the two likelihood arrays (please refer to [7] for more details).

In order to compute the *Maximum* Likelihood value for a fixed tree topology, all individual branch lengths, as well as the parameters of the Q matrix and the α shape parameter, must be optimized via an ML estimate. For the Q matrix and the α shape parameter the most common approach in "classic" ML implementations consists in using Brent's algorithm [3]. In order to evaluate changes in Q or α, the entire tree needs to be re-traversed, i.e., a full tree traversal needs to be conducted in order to correctly re-compute the likelihood. For the optimization of branch lengths the Newton-Raphson method is commonly used. In order to optimize the branches of a tree, the branches are repeatedly visited and optimized one-by-one until the achieved likelihood improvement is smaller than some pre-defined ε. Note that, if only one branch length is changed, the tree does not need to be completely re-traversed, since the likelihood score is invariant to the placement of the virtual root.

Current tree search algorithms typically alternate between tree search phases and model optimization phases. During the tree search phase, the model parameters are fixed and the ML score is improved by changing the topology and re-estimating only the branch lengths that are immediately affected by the topological rearrangements, i.e., in the neighborhood of topological changes. During the model optimization phase, the parameters of the evolutionary model are re-estimated, and typically the branch lengths of the whole tree are optimized as well.

3 Sources of Parallelism in Phylogenetic Analyses and Related Work

Any real-world phylogenetic analysis (see [8] for an example) exhibits several sources of parallelism that can be exploited:

- **Embarrassing Parallelism** The Maximum Likelihood (ML) phylogeny problem is NP-hard [4]. This means that the phylogenetic search space for a dataset grows exponentially with the number of species in that dataset. Consequently, all programs for phylogenetic analyses implement heuristics to reduce the search space. The downside of this approach is, that, none of these heuristic search algorithms is guaranteed to find *the* ML tree, and will hence only yield a *best-known* ML tree. Therefore, one usually needs to perform multiple tree searches from distinct starting points (starting trees) in order to explore the search space more thoroughly and to avoid local maxima.

Moreover, real-world phylogenetic analyses also require the execution of several (100–1,000) bootstrap (BS) tree searches. They are required to assign confidence values to the inner nodes of the best-known/best-found ML tree on the original alignment. This allows to determine how well-supported certain parts of the tree are and is important for deriving biological conclusions. It is analogous to conducting multiple tree searches on the original alignment; the only difference is that inferences are conducted on a randomly re-sampled alignment (a BS replicate), i.e., a certain number of alignment columns are re-weighted for every bootstrap replicate.
All those individual tree searches, be it bootstrap or multiple inferences, are completely independent from each other can thus be performed concurrently.
- **Inference Parallelism** The individual tree search strategies, as implemented in some state-of-the-art tree inference programs may allow for exploiting some degree of parallelism (see below for some examples).
- **Loop Level Parallelism** The functions of the PLK typically account for over 95% of total execution time in most programs for phylogenetic tree inference. The main bulk of these functions consists of `for`-loops over the likelihood vectors. An important property of the PLK is the assumption, that sites evolve independently, hence all entries `i` of the likelihood arrays, where `i=1...m` can be computed independently. This inherent parallelism of the likelihood function can be exploited at two levels of granularity: at a relatively coarse-grain thread level and at a fine-grain CPU level via SIMD instructions.

Most parallel implementations of ML programs [6, 11, 16, 18, 20] have mainly focused on inference/search algorithm parallelism. These implementations are mainly based upon highly algorithm-specific and mostly MPI-based parallelizations of various hill-climbing, genetic, as well as divide-and-conquer search algorithms. Typically, such parallelizations yield a lower parallel efficiency compared to the embarrassing and loop-level types due to partially hard-to-resolve dependencies in the respective search algorithms. Moreover, these parallelizations are program-specific and therefore not generally applicable. Minh *et al.* [12] implemented a hybrid OpenMP/MPI version of IQPNNI which exploits loop-level *and* inference parallelism.

4 Efficient Parallelization of the Phylogenetic Likelihood Kernel

Exploiting the embarrassing parallelism of real-world phylogenetic analyses is straight-forward. In the simplest case, several instances of sequential programs can be executed in parallel with different starting trees or bootstrapped input datasets. We will therefore not elaborate on embarrassing parallelism here. However, we described a hybrid approach for conveniently exploiting embarrassing parallelism and loop level parallelism simultaneously in [13].

As mentioned in the previous Section, the exploitation of inference parallelism can be challenging and usually does not allow for scaling up to hundreds or thousands of processors. Since the implementations of the PLK are very similar in most ML-based programs and because almost the entire program runtime is spent in the PLK, exploiting the inherent loop level parallelism allows for a more generic and highly scalable parallelization. Hence we will focus on exploiting the loop level parallelism in the PLK for which we propose a master/worker scheme. It can be implemented with MPI and PThreads and, provided appropriate software engineering, even allows for easily deriving one implementation from the other, i.e., the PThreads version can be transformed into an MPI version and vice versa in a straight-forward way. Besides that, the master/worker scheme also allows for a more efficient parallelization than with OpenMP because a significant amount of synchronization events can be avoided (see below).

4.1 Master/Worker Scheme

The likelihood vectors are distributed equally to the workers: each of the p workers holds a fraction of m'/p vector entries (where m' is the number of unique columns in the sequence alignment). The data–structures required for the likelihood vectors typically account for $\approx 90\%$ of the overall memory footprint. The memory space required for a phylogenetic analysis is therefore equally distributed among the workers. As the memory footprint of large phylogenetic analyses can easily exceed 100GB, this allows for overcoming memory limitations for such analyses. However, the likelihood arrays are not distributed in consecutive chunks to the workers, but in a cyclic round-robin fashion: worker #1 holds the columns $\{0, p, 2p, ...\}$, worker #2 $\{1, p+1, 2p+1, ...\}$, and so on (see Figure 2). This allows for better and easier load distribution, especially for partitioned analyses of multi–gene datasets. By using a striped allocation every worker will have an approximately balanced portion of columns for each partition. More importantly, this also applies to partitioned analyses of DNA and protein data, since the computation of the likelihood score for a single site (alignment column) for protein data is significantly more compute-intensive than for nucleotide data.

The master orchestrates the distribution of the data structures at start-up and steers the search algorithm as well as the iterative branch length and model parameter optimization. The actual computations on the likelihood vectors are performed solely by the workers. Thus, the master simply has to broadcast commands such as e.g.: optimize the length of branch b_4 between nodes 4 and 5 as depicted in Figure 2.

In general, there are three frequent operations which require interaction between the master and the workers:

1. **Computation of Likelihood Vectors:** This operation computes the entries of the likelihood vector located at an inner node by combining the values of the likelihood vectors and branch lengths of its two descendants.
2. **Computation of Log Likelihood Score:** This function just combines the values of two likelihood vectors at the nodes located at either end of the branch where the virtual root has been placed and computes the log likelihood score.

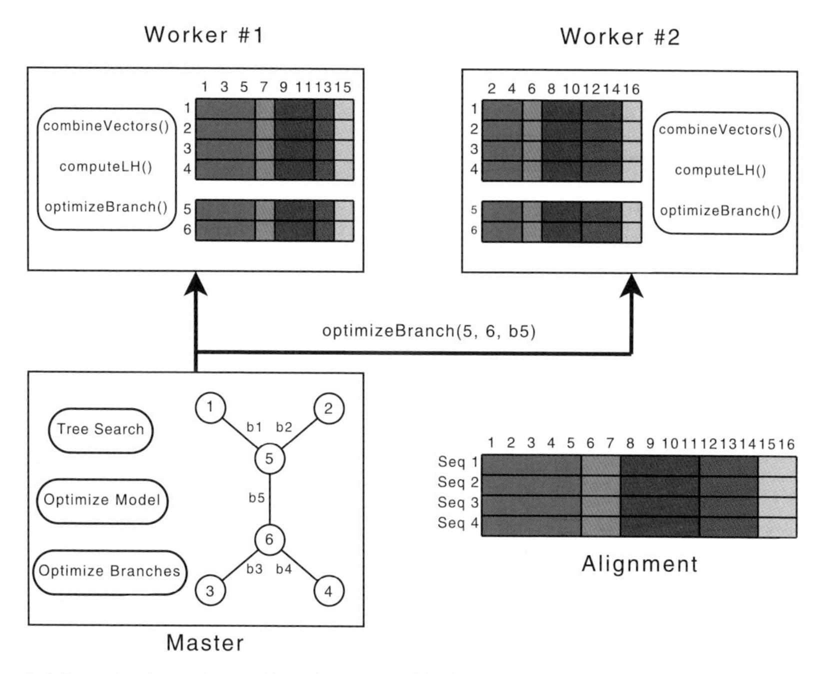

Fig. 2 Master/worker scheme: The alignment with 4 sequences, 16 columns, and 5 partitions is distributed round-robin to two workers. The master holds the tree topology and steers the tree search

3. **Branch Length Optimization:** This operation optimizes a specific branch between two nodes of the tree via a Newton-Raphson procedure.

During the tree search phase, the master and workers continuously execute the following operations: After a change in tree topology the master generates a partial tree traversal list that contains references to the inner likelihood vector arrays that need to be recomputed and then transfers it to the workers which execute operation #1 on the respective nodes. After the likelihood vectors have been updated, the master will usually request the workers to compute the tree's likelihood score (operation #2) which requires a reduction operation. The master uses the updated likelihood score to decide whether the change in tree topology was expedient or whether it should be reversed. Note that, the master/worker scheme allows for a higher parallel efficiency than a pure OpenMP parallelization of the loops over the likelihood vectors in operation #1. This is due to the fact that the computation of several likelihood vectors can be aggregated into a single job which results in a better computation to communication ratio.

A change in the tree topology also requires the optimization of the local branch lengths (operation #3) that are most affected by the change. As the branch lengths are optimized via an iterative Newton-Raphson procedure, this requires a synchronizing reduction operation after each individual iteration at each branch.

During the model optimization phase the master will re–estimate the parameters of the model of sequence evolution. Each parameter is optimized via Brent's algorithm which also represents an iterative method. After each iteration, the master needs to transmit the current model parameters to the workers, which then recompute *all* likelihood vectors using the updated model parameters and calculate the new likelihood score of the tree. Additionally, the master will initiate a global optimization of all branch lengths in the tree during the model optimization phase. This step is analogous to the branch length optimization during the search phase with the only difference that *all* branches are being optimized.

Special care needs to be taken for multi-gene phylogenomic datasets. For such datasets, the alignment and the likelihood model can be divided into distinct partitions that correspond, for instance, to individual genes. For every partition, the Q matrix, α shape parameter, and the branch lengths will need to be optimized separately. Evidently, the number of iterations required by the optimizations procedures to converge will be different for each partition. However, synchronization is needed after each iteration at each partition. Hence, in order to keep the total number of synchronization events as low as possible, it is important to perform the computations required by individual iterations concurrently on all partitions for as long as possible. That is, all workers will start by performing the first iteration on all partitions. After the first iteration over all partitions has been completed, the workers synchronize with the master to determine whether the iterative optimization of a specific parameter for a partition has converged. Thus, the master needs to keep track of the convergence state of partitions and send an additional boolean vector to the workers after each iteration. The workers can then ignore the partitions that are marked as converged in the vector and only process the partitions that have not converged so far. As we have shown in [17], this procedure is essential for obtaining a high degree of parallel efficiency when performing partitioned phylogenetic analyses.

5 Experimental Setup and Results

In order to asses the performance of our parallelization approach, we used the same datasets as in previous experiments (see [13] for a detailed description): two datasets with 50 sequences, one with 50,000 base-pairs and one with 500,000 base-pairs (d50_50000 and d50_500000), and one dataset with 250 sequences and 500,0000 base-pairs (d250_500000). The experiments have been carried out on the following systems:

- The InfiniBand Cluster at Technische Universität München: A custom Linux cluster consisting of 32 4-way AMD 2.4 GHz Opteron 850 processors, Infiniband interconnect (referred to as TUM IC in the following).
- The BlueGene/L system at Iowa State University: A one–rack machine with 1,024 nodes (2,048 CPUs) and a peak performance of 5.734 Teraflops (BlueGene/L).

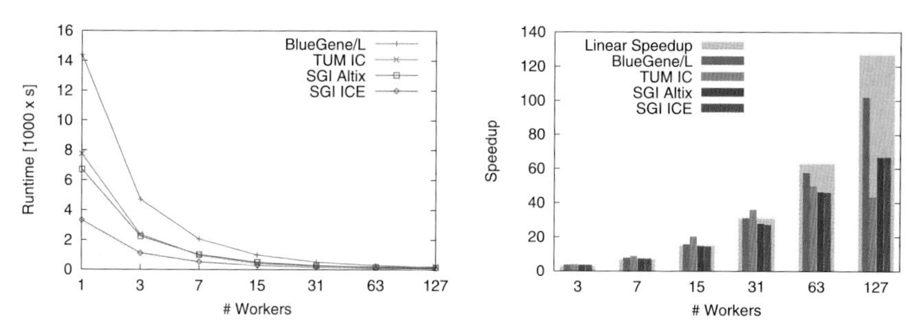

Fig. 3 Runtimes and Speedups on the d50_50000 dataset

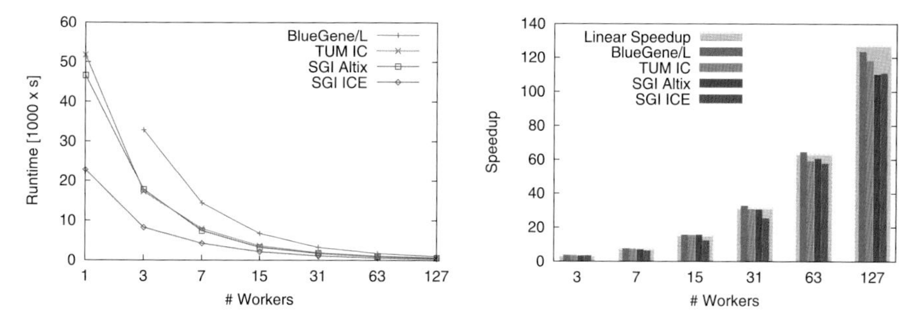

Fig. 4 Runtimes and Speedups on the d50_500000

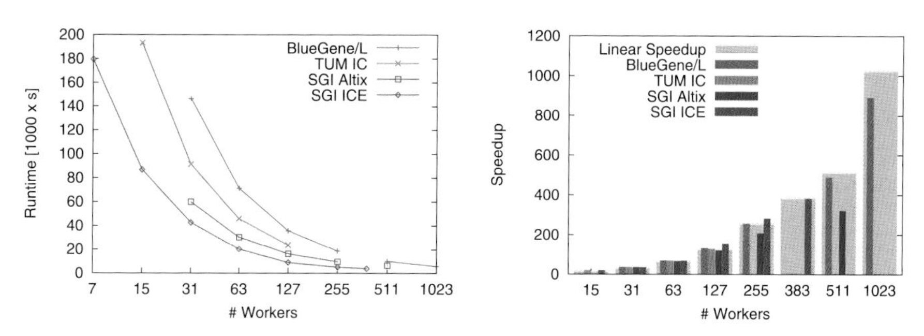

Fig. 5 Runtimes and Speedups on the d250_500000

- The HLRB2 at Leibniz Rechenzentrum: An SGI Altix 4700 system with a total of 9,728 Intel Itanium2 Montecito cores, an aggregated peak performance of 62.3 Teraflops, and 39 Terabyte of main memory (SGI Altix).
- The PRACE prototype system at Leibniz Rechenzentrum: An SGI ICE system consisting of 48 nodes, each node equipped with two Intel 2.53 GHz Xeon E5540 quad-core processors (Nehalem), Infiniband interconnect (SGI ICE).

Figures 3 and 4 depict runtime and speedup graphs for the 50 sequences datasets. The absolute runtimes on the SGI ICE are approx. 75% lower than on the BlueGene/L and more than 50% lower than on the SGI Altix and the Opteron Cluster. Since RAxML is memory-bandwidth bound, these results were expected, given the high aggregated memory bandwidth of the Intel Nehalem. All four systems show good scalability, with the BlueGene/L performing slightly better because of the more favorable communication to computation ratio which is due to the very low latency network in combination with moderate per-CPU computing power.

Due to runtime and/or memory limitations, the experiments on the 250 sequences dataset could not be conducted with a small number of workers. Therefore, a few numbers are missing in the plots and the scalability evaluations for this dataset are relative to the run with the least feasible amount of workers. Compared to the BlueGene/L and TUM IC, the runtimes of the SGI ICE are respectively 70% and 55% lower for this dataset which is consistent with the results on the smaller 50 sequences datasets. However, compared to the SGI Altix, the performance advantage of the ICE system decreases to 29%. This is probably due to the fact that the larger working set of this dataset benefits from the large L3 caches of the Itanium-based Altix system. Scalability-wise, the ICE prototype clearly outperforms the SGI Altix system and scales linearly up to 383 workers. Though the numbers are not necessarily comparable, the ICE system also appears to outperform the BlueGene/L which was not the case for the smaller datasets.

In general, one can conclude that the scalability directly depends on the length of the alignment, or rather the number of distinct patterns, as the communication to computation ratio increases with the length of the likelihood vectors. Increasing the number of sequences in the input alignment increases the pressure on the communication subsystem. This is due to the fact that an increased number of internal nodes and branches in the tree results in more time being spent in the model optimization phase which is communication-wise more expensive than the tree search phase.

6 Conclusion

We have presented the most recent developments of the parallel fine-grain version of RAxML, which is a widely used tool for phylogenetic inference. We have demonstrated how to resolve load-balance problems on typical partitioned phylogenomic datasets and assessed the scalability of our approach on a large variety of supercomputer systems.

RAxML is an optimal benchmark for assessing the performance of single-core, multi-core, and large distributed memory computers: the computations it performs are very floating point intensive and therefore allow for assessing the computational performance of a computer's microprocessor. It's irregular data access patterns are well suited to test the performance of the computer's memory subsystem. The MPI version of RAxML makes heavy use of collective communication routines; the messages that are sent are very short. It therefore allows for evaluating the network's topology and it's latency in large distributed memory systems. By modifying the dimensions of the input sequence alignment, the load can be moved between different components of the computer: long sequences stress the floating point unit and the memory subsystem, increasing the number of sequences put more pressure on the interconnect.

Current work focuses on large-scale collaborative analyses with Biologists to resolve, e.g., the phylogeny of seed plants. Future work will focus on the design of novel parallelization strategies for datasets with many taxa and comparatively few genes as well as the design of a scalable, transparent, and portable BLAS-like library for the PLK as well as for statistical alignment methods. We also intend to devise novel algorithms for dynamic addition of new sequences (as they become available in GenBank) to existing organismal reference trees.

Acknowledgements This work is funded by the Bavarian Competence Network for Technical and Scientific High Performance Computing (KONWIHR) and under the auspices of the Emmy-Noether program by the German Science Foundation (DFG).

References

1. Adams, M., Kelley, J., Gocayne, J., Dubnick, M., Polymeropoulos, M., Xiao, H., Merril, C., Wu, A., Olde, B., Moreno, R., et al.: Complementary DNA sequencing: expressed sequence tags and human genome project. Science **252**(5013), 1651–1656 (1991)
2. Bader, D.A., Moret, B.M.E., Vawter, L.: Industrial applications of high-performance computing for phylogeny reconstruction. In: Proceedings of SPIE ITCom, vol. 4528, pp. 159–168 (2001)
3. Brent, R.: Algorithms for Minimization without Derivatives. Prentice Hall (1973)
4. Chor, B., Tuller, T.: Maximum likelihood of evolutionary trees: hardness and approximation. Bioinformatics **21**(1), 97–106 (2005)
5. DeSantis, T., Hugenholtz, P., Larsen, N., Rojas, M., Brodie, E., Keller, K., Huber, T., Dalevi, D., Hu, P., Andersen, G.: Greengenes, a Chimera-Checked 16S rRNA Gene Database and Workbench Compatible with ARB. Appl. Environ. Microbiol. **72**(7), 5069–5072 (2006)
6. Du, Z., Lin, F., Roshan, U.: Reconstruction of large phylogenetic trees: a parallel approach. Computational Biology and Chemistry **29**(4), 273–280 (2005)
7. Felsenstein, J.: Evolutionary trees from DNA sequences: a maximum likelihood approach. J. Mol. Evol. **17**, 368–376 (1981)
8. Grimm, G.W., Renner, S.S., Stamatakis, A., Hemleben, V.: A Nuclear Ribosomal DNA Phylogeny of Acer Inferred with Maximum Likelihood, Splits Graphs, and Motif Analyses of 606 Sequences. Evolutionary Bioinformatics Online **2**, 279–294 (2006)
9. Heinicke, M.P., Duellman, W.E., Hedges, S.B.: From the Cover: Major Caribbean and Central American frog faunas originated by ancient oceanic dispersal. Proceedings of the National Academy of Sciences **104**(24), 10092 (2007)
10. Janies, D., Hill, A.W., Guralnick, R., Habib, F., Waltari, E., Wheeler, W.C.: Genomic Analysis and Geographic Visualization of the Spread of Avian Influenza. Systematic Biology **56**(2), 321–329 (2007)
11. Minh, B., Vinh, L., Haeseler, A., Schmidt, H.: pIQPNNI: parallel reconstruction of large maximum likelihood phylogenies. Bioinformatics **21**(19), 3794–3796 (2005)
12. Minh, B., Vinh, L., Schmidt, H., Haeseler, A.: Large Maximum Likelihood Trees. In: Proc. of the NIC Symposium 2006, pp. 357–365 (2006)
13. Ott, M., Zola, J., Aluru, S., Stamatakis, A.: Large-scale Maximum Likelihood-based Phylogenetic Analysis on the IBM BlueGene/L. In: Proc. of IEEE/ACM Supercomputing Conference 2007 (SC2007) (2007)

14. Ronaghi, M.: Pyrosequencing Sheds Light on DNA Sequencing. Genome Research **11**(1), 3–11 (2001)
15. Stamatakis, A.: RAxML-VI-HPC: maximum likelihood-based phylogenetic analyses with thousands of taxa and mixed models. Bioinformatics **22**(21), 2688–2690 (2006)
16. Stamatakis, A., Ludwig, T., Meier, H.: Parallel Inference of a 10.000-taxon Phylogeny with Maximum Likelihood. In: Proc. of Euro-Par 2004, pp. 997–1004 (2004)
17. Stamatakis, A., Ott, M.: Load Balance in the Phylogenetic Likelihood Kernel. In: Proceedings of the 38th International Conference on Parallel Processing (ICPP 2009) (2009)
18. Stewart, C., Hart, D., Berry, D., Olsen, G., Wernert, E., Fischer, W.: Parallel Implementation and Performance of FastDNAml – A Program for Maximum Likelihood Phylogenetic Inference. In: Proc. of SC2001. Denver, CO (2001)
19. Yang, Z.: Maximum likelihood phylogenetic estimation from DNA sequences with variable rates over sites. J. Mol. Evol. **39**, 306–314 (1994)
20. Zwickl, D.: Genetic Algorithm Approaches for the Phylogenetic Analysis of Large Biological Sequence Datasets under the Maximum Likelihood Criterion. Ph.D. thesis, University of Texas at Austin (2006)

Parallel Computing with the R Language in a Supercomputing Environment

Markus Schmidberger and Ulrich Mansmann

Abstract R is an open-source programming language and software environment for statistical computing and graphics. During the last decade a great deal of research has been conducted on parallel computing techniques with the R language. Two packages (**snow** and **Rmpi**) stand out as particularly useful for general use on computer clusters and the **multicore** package for the use on multi-core machines.

This article describes the operation of the R language at the supercomputer HLRB2 hosted at the Leibniz-Rechenzentrum in Munich, Germany. Additional, a small benchmark is provided and the article explains and discusses two parallel biostatistical applications calculated at the HLRB2. The indirect comparison of interaction graph example outlines the requirements for more than 10.000 processors.

1 Introduction

R is an open-source programming language and software environment for statistical computing and graphics [2, `http://www.R-project.org`]. The core R installation provides the language interpreter and many statistical and modeling functions. R was originally created by R. Ihaka and R. Gentleman in 1993 and is now being developed by the R Development Core Team. It is highly extensible through the use of packages. These are libraries for specific functions or specific areas of study, frequently created by R users and distributed under suitable open-source licenses. A large number of packages is available at the Comprehensive R Archive Network (CRAN) at `http://CRAN.R-project.org` and packages for the analysis and comprehension of genomic data are stored in the Bioconductor repository [7] at

Markus Schmidberger · Ulrich Mansmann
Division of Biometrics and Bioinformatics, IBE, Ludwig-Maximilians-Universität München, 81377 München, Germany
e-mail: `Markus.Schmidberger@ibe.med.uni-muenchen.de`
e-mail: `Mansmann@ibe.med.uni-muenchen.de`

S. Wagner et al. (eds.), *High Performance Computing in Science and Engineering, Garching/Munich 2009*, DOI 10.1007/978-3-642-13872-0_64,

Table 1 Number of packages in the mostly used R repositories, duplicates removed (October 12, 2009)

	Number of packages
http://cran.rakanu.com	1779
http://www.omegahat.org/R	67
http://www.bioconductor.org/packages/2.4/bioc	305
http://R-Forge.R-project.org	448

`http://www.bioconductor.org`. Table 1 shows the latest numbers of R packages in the commonly used R repositories. Duplicate packages, e.g. between CRAN and the development repository R-Forge, are removed.

The R language was developed to provide a powerful and extensible environment for statistical and graphical techniques. However, the development of the R language was not aimed at providing a software for parallel computing. Nonetheless, during the last decade a great deal of research has been conducted on parallel computing techniques with R. There are several reasons for the increased focus on high-performance computing [5]: larger data sets like genomic data, increased computational requirements stemming from more sophisticated methodologies like Bootstrapping, and latest developments in computer chip production like quad-core processors. Especially in microprocessor technology there is currently performing a paradigm change from large single processor machines to many small cores working together in one machine.

This article describes the operation of the R language on the supercomputer HLRB2 hosted in the Leibniz-Rechenzentrum (LRZ) in Munich, Germany. In Section 2 more general details about parallel computing with the R language are examined and the use of the R language at the HLRB2 is explained. Additional, it contains a short benchmark example using the classical bootstrap. Section 3 explains two parallel biostatistical applications calculated at the HLRB2. The article ends with an outlook for further research and the requirements for more than 10.000 processors in a hybrid parallelized resampling example.

2 R and Parallel Computing

Due to the mentioned reasons, during the last decade more and more research has focused on using parallel computing techniques with R. The paper [16] reviews the developments in the last years and presents an overview of techniques for parallel computing with R on computer clusters, on multi-core systems, and in grid computing environments. It lists sixteen different packages, comparing them on their state of development, the parallel technology used, as well as on usability, acceptance, and performance. Two packages stand out as particularly useful for general use on computer clusters: **Rmpi** [23], which is a direct interface to different implementations of the MPI (Message Passing Interface) standard; **snow** [12, Simple Network of Workstations], which provides an abstraction layer by hiding the different supported communication types. Both packages have acceptable usability, support a spectrum of functionality for parallel computing with R, and deliver good performance. Other packages try to improve usability, but so far usability gains have usually been achieved at the expense of lower functionality. All packages for parallel computing at computer clusters use the manager-worker, often called master-slave [18], concept. Packages for grid computing are still in development, with only one package currently available to the end user. For multi-core systems four different packages exist, but a number of issues pose challenges to early adopters. In January 2009 a new and promising package for the use of multi-core systems with R was released: **multicore** [20], which provides a way of running parallel computations in R on machines with multiple cores using the `fork` system call.

For a detailed description and installation advises see the corresponding package specific documentations and publications. A hands-on tutorial is provided in [3].

2.1 R at the HLRB2

At all computing resources hosted in the LRZ the latest R version is provided or will be installed by the hlrb-admin team. Default R and Bioconductor libraries are installed, too. Due to the use of the environment module approach to manage the user environment for different software, library or compiler versions, first of all the corresponding R module has to be loaded.

```
module load R/2.9.2_mpi
```

Modules with different R versions and with different compiler settings are provided. For parallel computing the "R/X.X.X_mpi" version build with the GNU compiler `gcc` has to be used.

After several compiling modifications the LRZ team now provides a working **Rmpi** package version, which cooperates with the Altix MPI implementation and supports parallel computing with all 9728 processors. In detail, the **Rmpi** package has to be build with the `mpicc` compiler. The following example code outlines a very simple PBS Batch job file, which can be submitted with the command `qsub`.

```
#!/bin/bash
#PBS -l select=51:ncpus=1
. /etc/profile.d/modules.sh
module load R/2.9.2_mpi
export R_PROFILE=$R_BASE/lib/R/library/Rmpi/Rprofile
echo "
    library(Rmpi)
    # my R code ...
" | mpiexec -n 51 omplace R --no-save -q > mycode.out
```

After loading the R module a new R profile environment has to be exported to assure the correct start of the worker R sessions. The different R sessions are started with the mpiexec command and omplace controls the placement of the processes to the processors. This example script starts one manager and 50 workers with the no-save and quite (-q) options. The R code lines are passed into the R session and the output is written to the mycode.out file. In the R code the **Rmpi** package is loaded and no more specific declaration or start–stop command for the cluster is required. The mpi.xxx() commands can be used directly.

Using the **Rmpi** package for communication, the **snow** package works, too. Alternatively, the **snow** package works with raw socket connections as communication method. But at the HLRB2 only 150 connections per processor are allowed, therefore the number of possible processors is limited.

```
#!/bin/bash
#PBS -l select=51:ncpus=1
. /etc/profile.d/modules.sh
module load R/2.9.2_mpi
export R_PROFILE=$R_BASE/lib/R/library/RMPISNOWprofile
echo "
    library(snow)
    cl <- getMPIcluster()
    # my R code ...
" | mpiexec -n 51 omplace R --no-save -q > mycode.out
```

The use of the **snow** package is very similar to the **Rmpi** version. After loading the R module again a new R profile environment has to be exported. In the R code the **snow** library is loaded and the cluster object cl is initialized with the command getMPIcluster(). Note, this cluster object is required for all further **snow** commands.

For parallel computing at the HLRB2 the **multicore** package works on one partition with up to 512 processors, but there are some problems with the correct assignment of processes, generated with the fork system call, to the reserved processors. In detail, in some cases more than one R session runs on one processor and therefore limits the performance.

```
#!/bin/bash
#PBS -l select=1:ncpus=51
. /etc/profile.d/modules.sh
module load R/2.9.2_mpi
export OMP_NUM_THREADS=1
echo "
    library(multicore)
    options(cores = 50)
    # my R code ...
" | omplace R --no-save -q > mycode.out
```

This code reserves 51 processors, loads the R module and starts an R session with the no-save and quite option. The omplace command and the export variable OMP_NUM_THREADS=1 should control the placement of processes. The correct assignment of processes to processors is ongoing work. In the R code the **multicore** package is loaded and the `cores` parameter for the corresponding number of processors is set. Without `cores` declaration (default) the **multicore** package detects 512 processors and not the number of reserved processors.

The combination of the **snow** and **multicore** packages provides a very simple and efficient interface for hybrid parallel computing.

2.2 Benchmark

To review the speed and performance of the HLRB2 cluster environment with the R language a simple bootstrap-benchmark is implemented. Bootstrapping is a classic example of a time-consuming but simple to parallelize computation [5]. Bootstrap replicates of a generalized linear model fit for data on the cost of constructing nuclear power plants are generated. Available code from the **boot** [1] package and the proposed parallelization in the article [12] is used. For the parallelization the latest versions of the packages **Rmpi**, **snow** with raw sockets and **Rmpi** for communication, and **multicore** and R version 2.9.2 are applied. 5000 bootstrap replicates are calculated on different numbers of processors and the calculation is replicated 10 times to adjust for external influences (network traffic, other computations).

First of all, the serial computation time of one bootstrap sample is compared to R versions with different compiler settings and to another cluster environment at the Institute for Medical Informationtechnology, Biometry and Epidemiology (IBE, University of Munich, Germany). For high-performance computing with the R language it is very important to use an R version build with for performance optimized compiler settings, e.g. profiling support [19] should be disabled. In this case, calculations in R are up to 10 times faster (0.25 sec and 2.3 sec with profiling support). Differences in the calculation time (HLRB2: 0.25 sec; IBE: 0.17 sec) between the two different cluster environments can be explained by the different hardware and chip technology (HLRB2: Intel Itanium2 Montecito Dual Core, 1.6 GHz; IBE: Intel Xeon E5150, 2.66GHz). Differences due to other compilation parameters or different version numbers of the R language or packages were not analysed.

Due to the short computation time for one bootstrap sample, using more than 200 processors the complete calculation time for this benchmark example gets very small (< 10 sec) and communication time gets bigger than calculation time. The execution time for starting and stopping R sessions, loading data sets and libraries is not included in this measurement. Only the time for the parallel execution (`apply()` command) is measured. These additional operations require only a short time, for example starting an R session takes less than one second. In Figure 1 the relative speedup curves are visualized up to 500 processors.

The best performance is achieved with the **snow** package and raw sockets as communication mechanism. But this is limited up to 150 processors, due to the socket connection limitations. The computation time of the packages using the MPI standard is very similar and up to 150 processors there is an almost linear speedup. The performance of the **multicore** package is slower, because the starting process

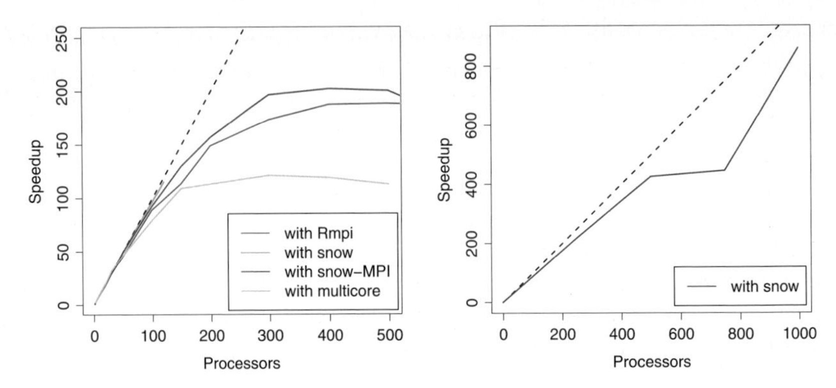

Fig. 1 Relative speedup curves for bootstrap benchmark example (left) and indirect comparison of interaction graphs data (right) at the computer environment HLRB2. Dotted lines visualize theoretical linear speedup

(`fork`) of the workers is included in the `mcapply()` command and therefore, included in the time measurement. But up to 150 processors there are only low differences in the speedup. For more than 200 processors all speedup curves in the bootstrap-benchmark example drop due to the mentioned small computation times and increased network traffic.

3 Applications

In (bio-) statistics, especially in all sampling approaches, computation in each sample often is independent of the others and parallelization is quite easy. Problems like that are called "embarrassingly parallel", as no changes on the base algorithms are needed to take the programs to multiple processors. As well, there are several problems in bioinformatics and (bio-) statistics which require more complex parallelization strategies to achieve good performance [5, 13].

3.1 Indirect Comparison of Interaction Graphs

In many applications from medicine and biology the detection of changes in the conditional independence structure (CIS) between elements under different conditions is of interest. For example, the elements could be the genes in a living organism which are annotated to a certain pathway. The conditions may be defined by two different diseases and two data sets containing the corresponding gene expression information measured in tissues from the respective patients. The CIS between the genes of the pathway may be explicitly estimated by an appropriate method, e.g. PC-Algorithm [9]. The detection of nodes which show differences in the way they connect to other nodes is straightforward by visual inspection of both graphs. But, it is difficult to decide which of the detected nodes show a differential CIS between both conditions caused by systematic differences and which are statistical artifacts caused by the algorithm or random fluctuation in the data.

In Ref. [10] a strategy is presented to detect a set of nodes with differential CIS under a controlled error rate. The proposed strategy to detect the set of nodes is called `indirect` because an explicit estimation of the CIS between nodes is avoided. The use of an indirect strategy is motivated by the potential bias inherent to all available graph estimation algorithms, which is caused by fixing specific regularization parameters to estimate the graph. An R package with the corresponding code is in preparation and for more theoretical details see [10].

3.1.1 Parallelization

For the explicit and implicit comparison of the CIS a permutation test is required. A total of 1000 samples are created to perform the comparison of the CIS between both conditions. The calculation is very computer intensive and sample calculations are independent from each other. This is a typical scenario for an MPI parallelization and the **snow** package with the **Rmpi** package for communication is used. To guarantee a different random number stream in every R session, an additional package **rlecuyer** [17] is applied.

In the proposed indirect approach it is necessary to fit a L1-penalized logistic or linear regression model in each sampling step, for both conditions, and each node of the graph. The working horse of this Lasso Regression is a computationally efficient gradient-ascent algorithm as proposed and implemented in [8]. Due to the analysed data set in every sample more than 100 independent models have to be estimated. Therefore, using a second level of parallelization (hybrid) the algorithm can be further improved. For the parallelization in the samples or workers the **multicore** package is used.

The calculation time for one sample in the permutation test without hybrid parallelization is about 1600 seconds and for the comparison of the fitted models between both conditions about 100 seconds are required. Therefore, the complete serial computation time would be about 20 days. Using 1000 processors the calculation time can be reduced to about 30 minutes. The relative speedup curve is visualized in Figure 1 and is nearly optimal. For 500 and 750 processors there is a similar computation time. All samples have a similar computation time and due to the equal sample distribution there is a bad load-balancing.

Due to the described problems with the **multicore** package the acceleration of the hybrid parallelization provides only little benefit. In theory the computation time can be approximated with $T(n,m) = \frac{1000(100+1600/m)}{n}$, where n is the number of workers and m the number of processors per worker. With 25 workers and 5 processors per worker a computation time of 5 hours is achieved, which is an improvement of factor 3.76 (computation time with 25 workers and 1 processor per worker: 18.8 hours).

3.1.2 Example with ICF Data

This example studies a multivariate binary situation and is taken from the field of rehabilitation science. Functioning and disability are universal human experiences. Over the life span people may experience limitations in functioning in relation to health conditions including an acute disease or injury, a chronic condition, or aging. A standard language for the analysis of functioning is provided by the International Classification of Functioning, Disability and Health [22, ICF]. A secondary analysis of observational cross sectional data of patients from five early post-acute rehabilitation units is performed. The ICF is used to measure functioning and contextual factors. The presence of an impairment or limitation is binary coded for each of the 122 categories (elements) considered. 616 patients (mean age 63 years, 46% male) are included.

Besides the comparison of functional profiles it is important to understand stability and distinctiveness of functioning across health conditions. This can be achieved by revealing patterns of associations between distinct aspects of functioning. Based on the proposal of Wainwright et al. [21], CIS graphs can be estimated for each condition and visually compared as done in Figure 2. Besides a simple visual inspection for differential CIS we apply the combination between our indirect test and a hierarchical testing procedure [11]. The hierarchy is defined by the ICF itself [22]. Nodes with significant differential CIS on a 5% significance level are marked with arrows.

The indirect strategy presented offers an explorative tool to detect nodes in a graph with the potential of a relevant impact on the regulatory process between interacting units in a complex process.

3.2 Parallel Computing in Microarray Data

Studies of gene expression using high-density oligonucleotide microarrays have become standard in a variety of biological and clinical fields. They enable scientists to investigate the functional relationship between the cellular and physiological processes of organisms by studying transcription at genome-wide system levels. Affymetrix GeneChip® arrays are a very common variant of high-density oligonucleotide expression microarrays. The data recorded by means of the microarray technique are characterized by the typical levels of noise induced by the preparation, hybridization and measurement processes as well as a specific structure. Removing the sources of bias needs a specific preprocessing of the raw data as far as the steps background correction, normalization, and summarization are concerned. For more details and a brief introduction see e.g. [6].

Processing a large number of microarrays is generally limited by the available computer hardware. The main memory limits the number of arrays analysed. Furthermore, most of the existing preprocessing methods are very time consuming.

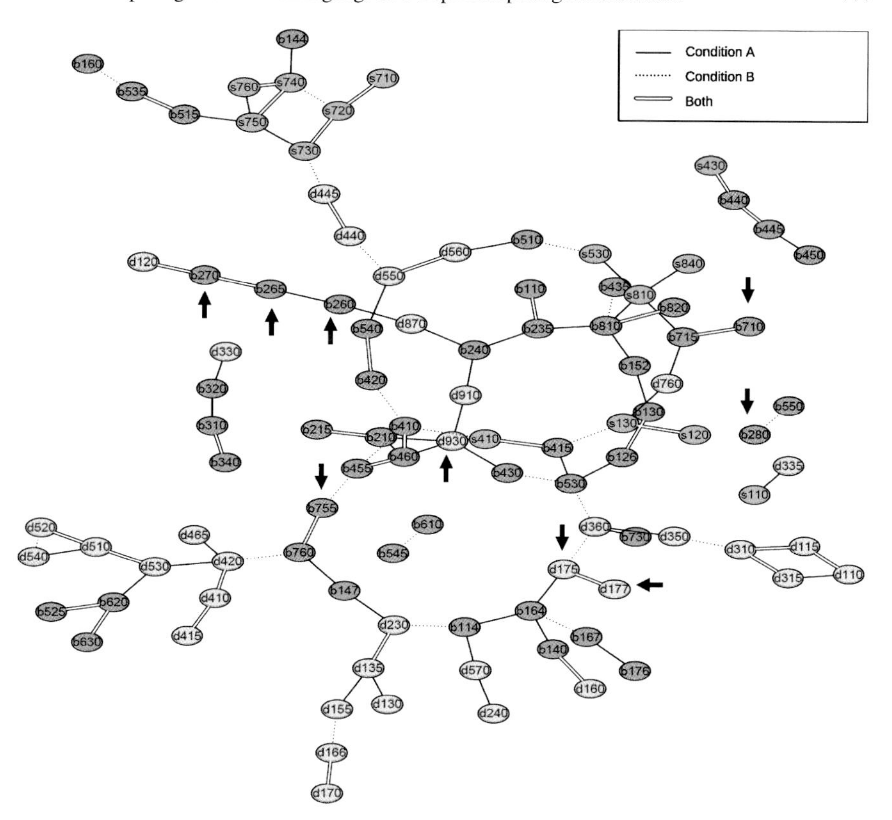

Fig. 2 Estimated CIS graphs and visual presentation of common edges as well as different edges within both graphs. The arrows point to the nodes with significant differential CIS on a 5% significance level. The three ICF components are presented in different colors: orange: Body Functions, green: Body Structures, blue: Activities & Participation

A further challenge is the fact that microarray experiments are becoming increasingly popular and large in number of arrays [14].

Distributed computing is one of the most promising solution to solve these problems, because it accelerates the methods and it solves the main memory problems with distributing data to different machines [13, 15]. Therefore, a new Bioconductor package called **affyPara** [14, 15] implements strategies using parallel computing for preprocessing of microarray data. Parallel computing for microarray data distributes arrays to different processors, performs demanding computations on smaller sets of microarrays and communicates the results between the processors efficiently to achieve the needed overall result. Microarray data are stored in a matrix structure. Using the block cyclic distribution the arrays are distributed equally to all nodes. Therefore, the amount of required main memory per processor gets smaller. Basic Bioconductor packages can be used on single processors since their data structure is array oriented. For all parallelized methods in the **affyPara** package using 5 to 10 processors (or more) and more than 150 microarrays there is an obvious acceleration compared to using the **affy** package [4]. The user-interface is simple and extends the functionality of the **affy** package. Using the **snow** package the package works on computer clusters as well as on multi-core architectures. It supports standard preprocessing steps and provides tools for quality assessment of huge numbers of microarray data. The **affyPara** package is an open source package – under GPL license – and available form the Bioconductor project at `www.bioconductor.org`. A user guide and examples are provided with the package.

A large cancer study with more than 5000 microarrays is ongoing work and the data are preprocessed with the **affyPara** package at the HLRB2. For this, 50 processors with 4 GB main memory each are used. In high-level analysis the gene-gene interaction graphs for different pathways are estimated and compared. The described direct comparison approach with the PC-algorithm is used, the newly developed indirect approach in ongoing work.

4 Conclusion

R is an open-source programming language and software environment for statistical computing and graphics and provides several useful packages for parallel computing. The R base installation and several packages are installed at the supercomputer HLRB2. The packages **snow** (with **Rmpi** for communication) and **multicore** are very user-friendly and support the fast development of parallel code in the R language. After several adaptions of the R software the performance at the HLRB2 is good.

Both presented examples demonstrate the usefulness for parallel computing in biostatistical applications. The resampling approach for the indirect comparison of interaction graphs uses hybrid parallelization and demonstrates the requirements for more than 100.000 processors (1000 samples * 100 model estimations). In this example communication costs and additional costs (e.g. loading libraries) have only low influence to the performance of parallel computing with R. Especially the extension to microarray data will increase the number of elements and the computation time per element. The microarray example, especially the **affyPara** package, solves the main memory problems in preprocessing of microarray data and demonstrates a parallel implementation handling huge amount of data. In these microarray examples the number of required processors is low (smaller 500), but more than 2 GB main memory per processor is required.

Acknowledgements The work is supported by the LMUinnovative collaborative centre 'Analysis and Modelling of Complex Systems in Biology and Medicine'. Computing ressources are provided from the LRZ project 'pr32di'. Special thanks goes to Klaus Rüschstroer (system administrator at the IBE) and Ferdinand Jamitzky (system administrator at the HLRB2) for the installation and administration of the R versions.

References

1. Canty, A., Ripley, B.: **boot**: Bootstrap R (S-Plus) functions (2009). URL `http://nws-r.sourceforge.net/`. R package version 1.2-39
2. R Development Core Team: R: A Language and Environment for Statistical Computing. R Foundation for Statistical Computing, Vienna, Austria (2009). URL `http://www.R-project.org`. ISBN 3-900051-07-0
3. Eugster, M.J.A., Knaus, J., Porzelius, C., Schmidberger, M., Vicedo, E.: Hands-on tutorial for parallel computing with R. Computational Statistics - submitted (2009)
4. Gautier, L., Cope, L., Bolstad, B.M., Irizarry, R.A.: **affy** – analysis of affymetrix genechip data at the probe level. Bioinformatics **20**(3), 307–315 (2004). 10.1093/bioinformatics/btg405. URL `http://dx.doi.org/10.1093/bioinformatics/btg405`
5. Gentle, J.E., Härdle, W.: Handbook of Computational Statistics. Springer (2004)
6. Gentleman, R.C., Carey, V., Huber, W., Irizarry, R., Dudoit, S.: Bioinformatics and Computational Biology Solutions Using R and Bioconductor, 1 edn. Springer (2005)
7. Gentleman, R.C., Carey, V.J., Bates, D.M., Bolstad, B., Dettling, M., Dudoit, S., Ellis, B., Gautier, L., Ge, Y., Gentry, J., Hornik, K., Hothorn, T., Huber, W., Iacus, S., Irizarry, R., Leisch, F., Li, C., Maechler, M., Rossini, A.J., Sawitzki, G., Smith, C., Smyth, G., Tierney, L., Yang, J.Y., Zhang, J.: Bioconductor: Open software development for computational biology and bioinformatics. Genome Biology **5** (2004). URL `http://genomebiology.com/2004/5/10/R80`
8. Goeman, J.: L1 penalized estimation in the cox proportional hazards model. Biometrical Journal **51** (2009)
9. Kalisch, M., Bühlmann, P.: Estimating high-dimensional directed acyclic graphs with the PC-Algorithm. Journal of Machine Learning Research **8**, 613–636 (2007)
10. Mansmann, U., Schmidberger, M., Strobl, R., Jurinovic, V.: Statistical Modelling and Regression Structures – Festschrift in Honour of Ludwig Fahrmeir, chap. Indirect Comparison of Interaction Graphs. Springer (2009)
11. Meinshausen, N.: Hierarchical testing of variable importance. Biometrika **95**, 265–276 (2008)
12. Rossini, A.J., Tierney, L., Li, N.M.: Simple parallel statistical computing in R. Journal of Computational and Graphical Statistics **16**(2), 399–420 (2007)
13. Schmidberger, M.: Parallel computing for biological data. Ph.D. thesis, University of Munich, [Epub ahead of print] (2009). URL `http://edoc.ub.uni-muenchen.de`
14. Schmidberger, M., Mansmann, U.: Parallelized preprocessing algorithms for high-density oligonucleotide arrays. In: Proc. IEEE International Symposium on Parallel and Distributed Processing IPDPS 2008, pp. 1–7 (2008)
15. Schmidberger, M., Mansmann, U.: **affyPara** - a Bioconductor package for parallelized preprocessing algorithms of affymetrix microarray data. Bioinformatics and Biology Insights **3** (2009)
16. Schmidberger, M., Morgan, M., Eddelbuettel, D., Yu, H., Tierney, L., Mansmann, U.: State of the art in parallel computing with R. Journal of Statistical Software **31**(1) (2009). URL `http://www.jstatsoft.org/v31/i01/`
17. Sevcikova, H., Rossini, T.: **rlecuyer**: R interface to RNG with multiple streams (2009). URL `http://www.iro.umontreal.ca/~lecuyer/myftp/papers/streams00.pdf`. R package version 0.2
18. Sloan, J.: High Performance Linux Clusters with OSCAR, Rocks, OpenMosix, and MPI (Nutshell Handbooks). O'Reilly Media, Inc. (2004)
19. Tierney, L.: **proftools**: Profile Output Processing Tools for R. R package (2007). URL `http://CRAN.R-project.org/package=proftools`. R package version 0.0-2
20. Urbanek, S.: **multicore**: Parallel processing of R code on machines with multiple cores or CPUs (2009). URL `http://www.rforge.net/multicore/`. R package version 0.1-3

21. Wainwright, M.J., Ravikumar, P., Lafferty, J.D.: High dimensional graphical model selection using L1-regularized logistic regression. Proceedings of Advances in neural information processing systems **9**, 1465–1472 (2006)
22. WHO: International classification of functioning, disability and health (ICF). Nature Genetics (2001)
23. Yu, H.: **Rmpi**: Parallel Statistical Computing in R. R News **2**(2), 10–14 (2002). URL `http://CRAN.R-project.org/doc/Rnews/`

Printing: Ten Brink, Meppel, The Netherlands
Binding: Stürtz, Würzburg, Germany